Communications and Control Engineering

Communications and Control Engineering is a high-level academic monograph series publishing research in control and systems theory, control engineering and communications. It has worldwide distribution to engineers, researchers, educators (several of the titles in this series find use as advanced textbooks although that is not their primary purpose), and libraries.

The series reflects the major technological and mathematical advances that have a great impact in the fields of communication and control. The range of areas to which control and systems theory is applied is broadening rapidly with particular growth being noticeable in the fields of finance and biologically-inspired control. Books in this series generally pull together many related research threads in more mature areas of the subject than the highly-specialised volumes of *Lecture Notes in Control and Information Sciences*. This series's mathematical and control-theoretic emphasis is complemented by *Advances in Industrial Control* which provides a much more applied, engineering-oriented outlook.

Publishing Ethics: Researchers should conduct their research from research proposal to publication in line with best practices and codes of conduct of relevant professional bodies and/or national and international regulatory bodies. For more details on individual ethics matters please see:
https://www.springer.com/gp/authors-editors/journal-author/journal-author-helpdesk/publishing-ethics/14214

More information about this series at http://www.springer.com/series/61

Torsten Söderström

Errors-in-Variables Methods in System Identification

Springer

Torsten Söderström
Division of Systems and Control,
Department of Information Technology
Uppsala University
Uppsala
Sweden

ISSN 0178-5354 ISSN 2197-7119 (electronic)
Communications and Control Engineering
ISBN 978-3-030-09125-5 ISBN 978-3-319-75001-9 (eBook)
https://doi.org/10.1007/978-3-319-75001-9

Mathematics Subject Classification (2010): 62F10, 93E12

Printed on acid-free paper

This Springer imprint is published by the registered company Springer International Publishing AG part of Springer Nature
The registered company address is: Gewerbestrasse 11, 6330 Cham, Switzerland

To Andreas, Christer, David, Elisabet, Frida, Gunnar, Gustav, Hjalmar, Johanna, Klara, Marianne and Olof

Preface

This book is intended to give a comprehensive overview of errors-in-variables (EIV) problems in system identification. This problem is about modeling of dynamic systems when all measured variables and signals are noise-corrupted. A number of different approaches are described and analyzed. The area has been a central one in my own research for a long time, and this experience has influence my own way of thinking of how to describe and categorize the many proposed methods available in the literature. The area continues to be active today, and there is a steady inflow of articles on EIV for dynamic systems, to leading conferences as well as to journals.

As a proper background the reader is expected to have at least elementary knowledge of system identification. The textbooks (1999) and Söderström and Stoica (1989) can be recommended. They cover much more than what is required in this context.

This book starts with giving a background for the errors-in-variables (EIV) problem. First static systems are treated in some detail. The dominating part of this book copes with dynamic systems. The EIV problem as such is carefully analyzed, and it is demonstrated that some additional assumption(s) must be imposed if a unique solution is to be found. Several approaches and EIV methods are presented and analyzed. This book ends with a chapter on users' perspectives for applying EIV methods in practice. See also Sect. 1.2 for a more detailed description of this book.

It is a pleasure to express my gratitude to many colleagues with whom I over the years have discussed, learned from, and published work together with on errors-in-variables problems. These colleagues include Juan Carlos Agüero, Brian D. O. Anderson, Theodore Anderson, Keith Burnham, Mats Cedervall, Han-Fu Chen, Bart De Moor, Manfred Deistler, Roberto Diversi, Mats Ekman, Hugues Garnier, Marion Gilson, Tryphon Georgiou, Graham Goodwin, Roberto Guidorzi, Christiaan Heij, Håkan Hjalmarsson, Mei Hong Bjerstedt, Alireza Karimi, Erlendur Karlsson, David Kreiberg, Alexander Kukush, Tomas Larkowski, Erik K. Larsson, Jens Linden, Kaushik Mahata, Ivan Markovsky, Magnus Mossberg, Rik Pintelon, Agnes Rensfelt, Cristian Rojas, Wolfgang Scherrer, Johan Schoukens, Virginija Šimonytė, Joachim

Sorelius, Umberto Soverini, Petre Stoica, Stephane Thil, Klaske van Heusden, Sabine Van Huffel, Kiyoshi Wada, Fan Yang Wallentin, and Wei Xing Zheng.

I am truly indebted to many colleagues who have read earlier versions of the book, in whole or in part, and pointed out a large number of errors and unclear points, and also generously given me proposals for additional ideas. For example, the last chapter of the book was written and included based on such feedback. The reviewers who provided this indispensable help are Juan Carlos Agüero, Roberto Diversi, Håkan Hjalmarsson, David Kreiberg, Ivan Markovsky, Magnus Mossberg, Giorgio Picci, Rik Pintelon, Johan Schoukens, and Umberto Soverini.

I would also like to thank the personnel at Springer (Oliver Jackson, Meertinus Faber, Geethajayalaxmi Govindarjan, Komala Jaishankar, Ravikrishnan Karunanandam, and Balaganesh Sukumar) for a smooth cooperation in producing the book.

It happens sometimes that I have found manuscripts with 'error-in-variables', rather than 'errors-in-variables' in the title (yes, it has happened also in my own draft papers!). As is explained in the book, for EIV systems it is indeed a key aspect that there are errors on both input and output measurements, and therefore one must use plural! During the work with the book manuscript I have corrected quite a number of errors, and it is my sincere hope that not too many remain, even though there may still be more than a single one! During my scientific career I have mainly been active in the control community, and therefore I believe in feedback. In particular, I would welcome the readers' comments on the text and possibly pointing out any remaining error. I can be reached on the e-mail address: `torsten.soderstrom@it.uu.se`.

Some years ago I told my family that I was planning to write another book. Some suggested that this time I should write a thriller. A plot was laid out about a murder that was detected at the opening ceremony of a major control conference. I quickly turned down this idea, that was not so serious anyway. It would demonstrate my inability to write something exciting from a fiction point of view. Furthermore, the theme does not match my general impressions from almost half a century with the control community. I have mainly found it to be characterized by friendly and helpful people. I have dedicated this book to my extended family (grandchildren are included): **A**ndreas,..., **O**lof. You are the A and O to me!

Vattholma and Uppsala, Sweden Torsten Söderström
December 2017

Contents

1 Introduction 1
1.1 Four Motivating Examples 2
1.2 Outline of the Book 8
1.3 Some Important Concepts in System Identification 10
1.4 Some Notations 11
1.5 Extensions and Bibliographical Notes 12

2 The Static Case 15
2.1 Line Fitting 15
2.1.1 Some System Theoretic Considerations of Identifiability 20
2.2 Confirmatory Factor Analysis 21
2.2.1 The Modeling Part 22
2.2.2 Estimation Part 24
2.3 The Frisch Scheme 30
2.4 Extensions and Bibliographical Notes 30
2.A Further Details 35
2.A.1 Further Results for Line Fitting 35
2.A.2 Consistency of the CFA Estimate 46

3 The Errors-in-Variables Problem for Dynamic Systems 49
3.1 The EIV Problem 49
3.2 About Numerical Examples 55
3.3 Two Special Cases 57
3.4 Some Naïve Approaches 59
3.4.1 Neglecting the Input Noise 59
3.4.2 Estimating the Noise-Free Input Signal 62
3.4.3 Rewriting the Model into Standard Form 66
3.5 Extensions and Bibliographical Notes 68

4 Identifiability Aspects 71
4.1 Some General Aspects 71
4.2 Identifiability Analysis for Parametric Models 78
4.3 Identifiability When Using Multiple Experiments 82
4.4 Closed-Loop Operation 84
4.5 Extensions and Bibliographical Notes 87

5 Modeling Aspects 89
5.1 Problem Statement and Notations 89
5.2 Using Models with an Arbitrary Delay 92
5.3 Continuous-Time EIV Models and Conversion to Discrete-Time 92
5.4 Modeling the Noise Properties 95
5.5 Frequency Domain Models 96
5.6 Modeling the Total System 97
5.7 Models for Multivariable Systems 99
5.8 Classification of Estimators Based on Data Compression 102
5.9 Model Order Determination 103
5.9.1 Introduction 103
5.9.2 Some Approaches 104
5.9.3 About the Rank Tests 107
5.9.4 Discussion 108
5.10 Extensions and Bibliographical Notes 110
5.A Further Details 112
5.A.1 Discrete-Time Model Approximation 112
5.A.2 Analyzing Effects of Small Singular Values 118

6 Elementary Methods 121
6.1 The Least Squares Method 121
6.2 The Instrumental Variable Method 123
6.2.1 Description 123
6.2.2 Consistency Analysis 124
6.2.3 User Choices. Examples of Instrumental Vectors 125
6.2.4 Instrumental Variable Methods Exploiting Higher-Order Statistics 128
6.2.5 Other Instrumental Variable Techniques 130
6.3 Extensions and Bibliographical Notes 133

7 Methods Based on Bias-Compensation 135
7.1 The Basic Idea of Bias-Compensation 135
7.2 The Bias-Eliminating Least Squares Method 138
7.2.1 Introduction 138
7.2.2 White Output Noise 139
7.2.3 Correlated Output Noise 142

7.3 The Frisch Scheme . . . 145
7.3.1 General Aspects . . . 145
7.3.2 White Output Noise . . . 145
7.3.3 Correlated Output Noise . . . 150
7.3.4 Using an Alternating Projection Algorithm . . . 150
7.4 The Generalized Instrumental Variable Method . . . 153
7.4.1 General Framework . . . 153
7.4.2 Various Examples . . . 157
7.4.3 GIVE Identification of MIMO Models . . . 161
7.5 Extensions and Bibliographical Notes . . . 164
7.5.1 BELS . . . 164
7.5.2 The Frisch Scheme . . . 166
7.5.3 GIVE . . . 167
7.A Further Details . . . 168
7.A.1 Proof of Lemma 7.1 . . . 168
7.A.2 Algorithm for the Canonical Form . . . 168

8 Covariance Matching . . . 171
8.1 The Basic Idea of Covariance Matching . . . 171
8.2 The Covariance Matching Method . . . 173
8.3 Expressions for the Covariance Elements . . . 177
8.4 User Choices in the Algorithm . . . 181
8.4.1 General Aspects . . . 181
8.4.2 Compatibility and Identifiability Conditions . . . 182
8.5 Applying Confirmatory Factor Analysis Modeling for EIV Identification . . . 185
8.6 Extensions and Bibliographical Notes . . . 187
8.A Further Details . . . 188
8.A.1 The Rank Condition . . . 188
8.A.2 An Alternative Parameterization . . . 192

9 Prediction Error and Maximum Likelihood Methods . . . 197
9.1 The Basic Ideas . . . 197
9.2 Time Domain Formulation . . . 199
9.3 Frequency Domain Formulation . . . 204
9.4 A Frequency Domain Maximum Likelihood Method . . . 207
9.4.1 The Frequency Domain ML Estimator . . . 209
9.4.2 Maximization with Respect to λ_u . . . 209
9.4.3 Minimization with Respect to $\mathbf{U}_0$ . . . 210
9.4.4 Minimization with Respect to τ . . . 210
9.4.5 Minimization with Respect to θ . . . 211
9.4.6 The ML Algorithm . . . 213
9.5 An Extended Frequency Domain ML Method . . . 214
9.6 Extensions and Bibliographical Notes . . . 218

10 Frequency Domain Methods . . . 221
10.1 Introduction . . . 221
10.2 Nonparametric Methods . . . 222
10.2.1 Estimating the Spectrum of the Observed Signals . . . 223
10.2.2 Estimating the Transfer Function $G(e^{i\omega})$ . . . 225
10.2.3 Estimating the Noise Variances . . . 225
10.2.4 An Estimate of θ . . . 227
10.3 A Frisch Scheme-Based Method in the Frequency Domain . . . 228
10.4 A Parametric Method . . . 228
10.5 A Frequency Domain GIVE Method . . . 231
10.5.1 Description . . . 231
10.5.2 Some Comments . . . 233
10.5.3 Some Analysis . . . 234
10.6 Extensions and Bibliographical Notes . . . 236

11 Total Least Squares . . . 237
11.1 The Total Least Squares Problem . . . 237
11.2 Computation of the TLS Estimate . . . 238
11.3 Using the TLS Estimate for System Identification . . . 239
11.4 The Structured Total Least Squares (STLS) Estimate . . . 240
11.5 Analysis of the TLS Estimate . . . 242
11.5.1 General Aspects . . . 242
11.5.2 Analysis of the TLS Estimate in an EIV Setting . . . 242
11.5.3 Analysis of the STLS Estimate in an EIV Setting . . . 244
11.6 Extensions and Bibliographical Notes . . . 248
11.A Further Details . . . 251
11.A.1 The Eckart–Young–Mirsky Lemma . . . 251
11.A.2 Characterization of the TLS Solution . . . 252
11.A.3 Proof of Lemma 11.2 . . . 253

12 Methods for Periodic Data . . . 255
12.1 Introduction . . . 255
12.2 Using Instrumental Variable Estimation . . . 257
12.2.1 Introduction . . . 257
12.2.2 Consistency Analysis . . . 258
12.3 The Sample Maximum Likelihood Method . . . 261
12.3.1 A Frequency Domain ML Method . . . 261
12.3.2 The SML Method . . . 264
12.4 Extensions and Bibliographical Notes . . . 266

13 Algorithmic Properties 269
13.1 Introduction 269
13.2 Algorithmic User Choices 271
13.3 Some General Concepts 271
13.4 Variable Projection Algorithms 274
13.5 Handling Overdetermined Systems of Equations 275
13.6 Recursive Algorithms 276
13.6.1 General Aspects 276
13.6.2 Recursive Version of the GIVE Estimate 279
13.6.3 Recursive Version of the Covariance Matching Estimate 280
13.6.4 Recursive Version of the Maximum Likelihood Estimate 282
13.7 Extensions and Bibliographical Notes 283
13.A Algorithmic Aspects of the GIVE Estimate 284
13.A.1 General Aspects 284
13.A.2 Use of a Variable Projection Algorithm for MIMO Systems 289
13.B Handling Overdetermined Systems of Equations 291
13.C Algorithmic Aspects of CFA-Based Estimators 296

14 Asymptotic Distributions 299
14.1 Background and General Considerations 299
14.2 CFA-Based Parameter Estimates 302
14.2.1 The ML Criterion $V_1(\vartheta)$ 303
14.2.2 The Criterion $V_2(\vartheta)$ 305
14.2.3 The Criterion $V_3(\vartheta)$ 305
14.2.4 Comparisons 306
14.2.5 The Matrix $\mathbf{C}_r$ 307
14.2.6 A Lower Bound on the Parameter Covariance Matrix 308
14.3 Instrumental Variable Methods 309
14.3.1 The Basic IV Estimator 309
14.3.2 Extensions 310
14.3.3 Optimal IV 311
14.4 Generalized Instrumental Variable Methods 314
14.4.1 The SISO Case 314
14.4.2 Evaluation of the Matrix $\mathbf{C}$ 316
14.4.3 The MIMO Case 318
14.5 Covariance Matching Methods 320
14.5.1 Asymptotic Distribution 320
14.5.2 The Matrix $\mathbf{R}$ 324

14.6 The Maximum Likelihood Method 330
14.6.1 The Prediction Error Method 330
14.6.2 The Maximum Likelihood Method in the Frequency Domain 333
14.6.3 The Extended Maximum Likelihood in the Frequency Domain 335
14.7 Methods for Periodic Data 336
14.7.1 The General Case 336
14.7.2 Instrumental Variable 339
14.7.3 The FML and SML Estimates 341
14.8 The Cramér–Rao Lower Bound For Maximum Likelihood Problems 342
14.8.1 Introduction 342
14.8.2 Algorithm for Computing the CRB for Arbitrary State Space Models 343
14.8.3 The Cramér–Rao Lower Bound for the Frequency Domain Maximum Likelihood Problem 346
14.8.4 Numerical Illustration of the Cramér–Rao Lower Bounds 352
14.9 Extensions and Bibliographical Notes 353
14.A Asymptotic Distribution of CFA Estimates 354
14.A.1 Proof of Lemma 14.1 354
14.A.2 Evaluation of $\mathbf{R}_{\varphi}(\tau)$ 355
14.B Asymptotic Distribution for IV Estimates 358
14.B.1 Proof of Lemma 14.2 358
14.B.2 Proof of Lemma 14.6 359
14.C Asymptotic Distribution for GIVE 361
14.C.1 The Sensitivity Matrix **S** for the SISO Case 361
14.C.2 Computation of the Matrix **C** 364
14.C.3 Non-Gaussian Distributed Data. Proof of Lemma 14.7 366
14.C.4 Proof of Lemma 14.8 367
14.D Asymptotic Accuracy for Models Obtained under Linear Constraints 368
14.E Asymptotic Distribution for the Covariance Matching Method 370
14.E.1 Covariance Matrix of the Extended Parameter Vector 370
14.E.2 Proof of Theorem 14.2 372
14.E.3 Proof of Lemma 14.9 378
14.F Asymptotic Distribution for PEM and ML Estimates 378
14.F.1 Asymptotic Covariance Matrix of the Parameter Estimates 378

14.F.2 Asymptotic Distribution for Frequency Domain ML Estimates ... 380
14.F.3 Asymptotic Distribution for the Extended ML Approach ... 384
14.G Asymptotic Distribution Results for Periodic Data ... 391
14.G.1 Proof of Lemma 14.12 ... 391
14.G.2 Proof of Lemma 14.13 ... 393
14.G.3 Proof of Corollary 14.3 ... 394
14.G.4 Proof of Lemma 14.14 ... 395
14.H The Cramér–Rao Lower Bound for the Frequency Domain ML Problem ... 397

15 Errors-in-Variables Problems in Practice ... 403
15.1 Comparing Performance of Some Estimators ... 403
15.2 User Choices in the Algorithms ... 406
15.3 The Role of Assumptions ... 410
15.4 Some General Guidelines ... 411
15.5 Guidelines Related to the Experimental Setup ... 412
15.6 Guidelines Related to the Measurement Noise ... 414
15.7 Guidelines Related to the Noise-Free Input Signal ... 418

Appendix A: General Background Results ... 421

References ... 447

Index of Cited Authors ... 475

Subject Index ... 483

Abbreviations

AIC	Akaike's information criterion
AR	Autoregressive
AR(n)	AR of order n
arg	Argument
ARMA	Autoregressive moving average
ARMA(n, m)	ARMA where AR and MA parts have orders n and m, respectively
ARMAX	Autoregressive moving average with exogenous input
BELS	Bias-eliminating least squares
BIC	Bayesian information criterion
CFA	Confirmatory factor analysis
CM	Covariance matching
cov	Covariance matrix
CRB	Cramér-Rao lower bound
deg	Degree
DFT	Discrete Fourier transform
DLS	Data least squares
EIV	Errors-in-variables
EIV	Extended instrumental variable
ETFE	Empirical transfer function estimate
FD	Frequency domain
FIR	Finite impulse response
FML	Frequency domain maximum likelihood
FOH	First-order hold
FPE	Final prediction error
GIVE	Generalized instrumental variable estimate
iid	Independent and identically distributed
IIR	Infinite impulse response
Imag	Imaginary part
IV	Instrumental variable
LHS	Left-hand side

LS	Least squares
MA	Moving average
MA(n)	MA of order n
MC	Monte Carlo
MIMO	Multi-input, multi-output
ML	Maximum likelihood
pdf	Probability density function
PEM	Prediction error method
PLR	Pseudo-linear regression
Real	Real part
RHS	Right-hand side
RMS	Root-mean-square
SEM	Structural equation modeling
SISO	Single-input, single-output
SML	Sample maximum likelihood
SNR	Signal-to-noise ratio
STLS	Structured total least squares
SVD	Singular value decomposition
TD	Time domain
TLS	Total least squares
TML	Time domain maximum likelihood
tr	Trace (of a matrix)
vec	vec($\mathbf{A}$) is the long column vector obtained by stacking the columns of $\mathbf{A}$ on top of each other
YW	Yule Walker
ZOH	Zero-order hold

Notation

$\mathscr{CN}$	Complex Gaussian distribution
E	Expectation operator
$\mathsf{E}[\cdot \vert \cdot]$	Conditional expectation
e	Basis of natural logarithm
$e(t)$	White noise (a sequence of independent random variables)
$\mathbf{e}_i$	Unit vector (ith element equal to 1)
$\mathscr{GP}(\mu, K)$	Gaussian process with mean value function μ and covariance function K
$G(q^{-1})$	Transfer function operator
$\mathbf{g}_N$	Vector of dimension N with all elements equal to 1
$H(q^{-1})$	Transfer function operator, noise shaping filter
h	Sampling interval
h_k	Weighting function coefficient, $H(q^{-1}) = \sum_{k=0}^{\infty} h_k q^{-k}$
$\mathbf{I}$	Identity matrix
$\mathbf{I}_n$	$n \times n$ identity matrix
i	Imaginary unit
$\mathbf{J}$	Information matrix
L	Log likelihood function
$\mathscr{L}$	Laplace transform
$m \times n$	Matrix has dimension m by n
N	Number of data points
$\mathscr{N}(\mathbf{m},\mathbf{P})$	Normal (Gaussian) distribution with mean value $\mathbf{m}$ and covariance matrix $\mathbf{P}$
$\mathscr{N}$	Null space
$\mathbf{P}$	Covariance matrix of state or state prediction error
Pr	Probability
q	Shift operator, $qx(t) = x(t+1)$
q^{-1}	Backward shift operator, $q^{-1}x(t) = x(t-1)$
$\mathscr{R}$	Range space

$\mathscr{R}^n$	Euclidean n-dimensional space
$\mathscr{R}^{n\times m}$	Linear space of $n \times m$-dimensional matrices
$\mathbf{R}$	Covariance matrix
$\hat{\mathbf{R}}$	Sample covariance matrix (estimate of $\mathbf{R}$)
$\hat{\mathbf{R}}_N$	Sample covariance matrix based on N data points
$\mathbf{r}$	Covariance vector
r	Noise variance ratio
$\mathscr{S}(A, B)$	Sylvester matrix associated with polynomials A and B
t	Time variable (integer valued for discrete-time models)
tr	Trace (of a matrix)
$u(t)$	Input signal (possibly vector-valued)
$u_0(t)$	Noise-free input signal
$\tilde{u}(t)$	Input measurement noise
V	Loss function, performance index
V_N	Loss function based on N data points
V_∞	Asymptotic loss function, $V_\infty = \lim_{N\to\infty} V_N$
$v(t)$	White noise (a sequence of independent random variables)
$\mathbf{W}$	Weighting matrix
$\mathbf{x}(t)$	State vector
Y^t	All available output measurements at time t, $Y^t = \{y(t), y(t-1)\ldots\}$
$y(t)$	Output signal (possibly vector-valued)
$y_0(t)$	Noise-free output signal
$\hat{y}(t\|t-1)$	Optimal one-step predictor of $y(t)$ given Y^{t-1}
$\tilde{y}(t)$	Output measurement noise
$\mathscr{Z}$	z transform
$\mathbf{z}(t)$	Instrumental variable vector
$\gamma(\mathbf{x}; \mathbf{m}, \mathbf{P})$	Pdf of normal (Gaussian) distribution with mean value $\mathbf{m}$ and covariance matrix $\mathbf{P}$
Δ	Difference
$\delta_{t,s}$	Kronecker delta (=1 if $s = t$, else $= 0$)
$\varepsilon(t)$	Prediction error
Λ	Covariance matrix of innovations
$\lambda_{\min}(\mathbf{A})$	Smallest eigenvalue of the symmetric matrix $\mathbf{A}$
λ	Variance of white noise
λ_u	Variance of input noise
λ_y	Variance of output noise
$\phi(z)$	Spectrum
$\phi(\omega)$	Spectral density
$\phi_u(\omega)$	Spectral density of the signal $u(t)$
$\phi_{yu}(\omega)$	Cross-spectral density between the signals $y(t)$ and $u(t)$
$\Phi(z)$	Spectrum for a multivariate signal
$\boldsymbol{\rho}$	Noise parameter vector
$\boldsymbol{\tau}$	Vector of parameters describing effects of initial and final values
τ	Time lag (in covariance function)

ω	Angular frequency
$\varphi(t)$	Regressor vector
θ	Parameter vector
θ_0	True value of θ
$\hat{\theta}$	Estimate of θ
$\hat{\theta}_N$	Estimate $\hat{\theta}$ based on N data points
ϑ	Total parameter vector $\vartheta = (\theta^T \; \boldsymbol{\rho}^T)^T$
Θ	Extended parameter vector ($\Theta = (1 \; \theta^T)^T$)
$\boldsymbol{\zeta}(t)$	Input–output data, $\boldsymbol{\zeta}(t) = (y(t) \; u(t) \;)^T$
$\oint$	Integration counterclockwise around the unit circle
$\mathbf{0}$	Zero matrix
$\mathbf{0}_{m\times n}$	Zero matrix of dimension $m \times n$
$O(x)$	Big ordo, $\lvert O(x)\rvert / \lvert x\rvert$ bounded when $x \to 0$
$o(x)$	Small ordo, $\lvert o(x)\rvert / \lvert x\rvert \; \to 0$ when $x \to 0$

Notational Conventions

$H^{-1}(q)$	$[H(q)]^{-1}$
$\mathbf{x}^T(t)$	$[\mathbf{x}(t)]^T$
$\mathbf{A}^{-T}$	$[\mathbf{A}^{-1}]^T$
$\mathbf{A} \geq \mathbf{B}$	The difference matrix $\mathbf{A} - \mathbf{B}$ is nonnegative definite
$\mathbf{A} > \mathbf{B}$	The difference matrix $\mathbf{A} - \mathbf{B}$ is positive definite
$\stackrel{\Delta}{=}$	Defined as
$\sim$	Distributed as
$\overline{w}$	Complex conjugate of w
$\mathbf{w}^T$	Transpose of $\mathbf{w}$
$\mathbf{w}^*$	Conjugate transpose of $\mathbf{w}$, $\mathbf{w}^* = \overline{\mathbf{w}}^T$
$\mathbf{w}^H$	Conjugate transpose of $\mathbf{w}$, $\mathbf{w}^H = \overline{\mathbf{w}}^T$
w^{-*}	$(w^{-1})^* = (w^*)^{-1}$
$\mathbf{A}^\dagger$	Pseudo-inverse of the matrix $\mathbf{A}$, $\mathbf{A}^\dagger = (\mathbf{A}^T\mathbf{A})^{-1}\mathbf{A}^T$
V'	Gradient of V
V''	Hessian of V
$\|\mathbf{A}\|_F$	Frobenius norm of matrix $\mathbf{A}$
$\|\mathbf{x}\|_\mathbf{W}$	Weighted norm, $\|\mathbf{x}\|_\mathbf{W}^2 = \mathbf{x}^T\mathbf{W}\mathbf{x}$

Summary of Assumptions

The different general assumptions introduced in the book are summarized here for the reader's convenience. References are also provided to where the definitions are introduced.

Assumptions on the System

AS1. The system is linear and asymptotically stable [Chap. 3].
AS2. The system is causal, so $y_0(t)$ depends on $u_0(s)$ for $s \leq t$, but not on future values of $u_0(\cdot)$ [Chap. 3].
AS3. The noise-free input and output signals are linked by

$$A(q^{-1})y_0(t) = B(q^{-1})u_0(t).$$

All system modes are observable and controllable; *i.e.*, $A(z)$ and $B(z)$ have no common factor. The polynomial degrees n_a and n_b are known [Chap. 3].
AS4. The system transfer function $G(z)$ has no pair of zeros reflected in the unit circle; that is, if $G(z_1) = 0$, then $G(z_1^{-1}) \neq 0$ [Chap. 4].
AS5. If the system is non-causal, then $G(z)$ has no pair of poles reflected in the unit circle; that is, p_1 and p_1^{-1} cannot both be poles of $G(z)$ [Chap. 4].
AS6. The order of the transfer functions fulfills

$$\text{order}(GH) = \text{order}(G) + \text{order}(H)$$

[Chap. 5].

Assumptions on the Noise

AN1. The noise sequences $\tilde{u}(t)$, $\tilde{y}(t)$ are stationary random processes, with zero mean values and spectra $\phi_{\tilde{u}}(\omega)$ and $\phi_{\tilde{y}}(\omega)$, respectively. Further, $\tilde{u}(t)$ and $\tilde{y}(t)$ are mutually uncorrelated [Chap. 3].
AN2. The measurement noises are Gaussian distributed [Chap. 4].

Naturally, at most one of the next three noise assumptions applies at a given situation.

AN3a. Both $\tilde{y}(t)$ and $\tilde{u}(t)$ are ARMA processes, as in (4.25) and (4.26) [Chap. 4].

AN3b. The output noise $\tilde{y}(t)$ is an ARMA process, while the input noise $\tilde{u}(t)$ is white. This means that $n_k = n_m = 0$ in (4.26) [Chap. 4].

AN3c. Both $\tilde{y}(t)$ and $\tilde{u}(t)$ are white noise sequences. This means that $n_f = n_h = 0$ in (4.25) and $n_k = n_m = 0$ in (4.26) [Chap. 4].

AN4. Both $\tilde{y}(t)$ and $\tilde{u}$ are white noise sequences. The ratio of their variances, $r = \lambda_y/\lambda_u$, is known [Chap. 4].

Assumptions on the Noise-free Input

AI1. The true input $u_0(t)$ is a stationary process of zero mean, with spectral density $\phi_{u_0}(\omega)$. The input $u_0(t)$ is assumed to be persistently exciting of a suitable order, which means that $\phi_{u_0}(\omega) > 0$ for a sufficient number of frequencies [Chap. 3].

AI2. The input $u_0(t)$ is uncorrelated with the measurement noise sources $\tilde{u}(t)$ and $\tilde{y}(t)$ [Chap. 4].

AI3. The true input $u_0(t)$ is Gaussian distributed [Chap. 4].

AI4. The true input $u_0(t)$ is an ARMA process; that is, it can be modeled as

$$D(q^{-1})u_0(t) = C(q^{-1})e(t) ,$$

where $e(t)$ is a white noise signal [Chap. 4].

AI5. The noise-free signal $u_0(t)$ is a periodic function. The length of the period is denoted N. It is assumed that M periods of the data $u(t), y(t)$ are available. Hence the total data length is NM. In each period $u_0(t)$ is a stationary process [Chap. 12].

AI6. The measurement noise signals $\tilde{u}(t)$ and $\tilde{y}(t)$ are uncorrelated with the noise-free input $u_0(s)$ for all t and s. Further, the measurement noise signals within different periods are uncorrelated [Chap. 12].

Assumptions on the Experimental Conditions

AE1. The data comes from one (single) experiment [Chap. 3].

AE2a. There is more than one experiment. The spectrum of the noise-free input is different in the different experiments [Chap. 4].

AE2b. There is more than one experiment. The measurement noises $\tilde{u}(t)$, $\tilde{y}(t)$ are uncorrelated between different experiments. The true noise-free input $u_0(t)$ is correlated between the experiments [Chap. 4].

AE3. The noise-free signal $u_0(t)$ is a periodic function. It is assumed that M periods of the data $u(t), y(t)$ are available. In each period $u_0(t)$ is a stationary process [Chap. 12].

Applicability

Many assumptions are assumed to be generally valid throughout the text, while a few are valid only locally, when explicitly stated so. The table below summarizes the status of the assumptions.

Validity **Topic**	General	Default	Locally
System	**AS1**, **AS2**	**AS3**	**AS4**, **AS5**, **AS6**
Noise	**AN1**	**AN3c**	**AN2**, **AN3a**, **AN3b**, **AN4**
Noise-free input	**AI1**, **AI2**		**AI3**, **AI4**, **AI5**
Experiment		**AE1**	**AE2a**, **AE2b**, **AE3**

Chapter 1
Introduction

Many different solutions have been presented for system identification of linear dynamic systems from noise-corrupted output measurements; see, for example, Ljung (1999) or Söderström and Stoica (1989). On the other hand, estimation of the parameters for linear dynamic systems when also the input is affected by noise is recognized as a more difficult problem. Representations where errors or measurement noises are present on both inputs and outputs are usually called 'errors-in-variables' (EIV) models. They play an important role when the purpose is the determination of the physical laws that describe the process, rather than the prediction of its future behavior.

The class of scientific disciplines which make use of such representations is very broad, as proved by the several applications collected in Van Huffel (1997), Van Huffel and Lemmerling (2002), such as time series modeling, array signal processing for direction-of-arrival estimation, blind channel equalization, multivariate calibration in analytical chemistry, image processing, astronomical data reduction, etc. In case of static systems, errors-in-variables representations are closely related to other well-known topics such as *latent variables models* and *factor models*, Fuller (1987), van Schuppen (1989).

Errors-in-variables models can be motivated in several situations.

- One such case is the modeling of the dynamics between the noise-free input and noise-free output. The reason can be to have a better understanding of the underlying relations, rather than to make a good prediction from noisy data. This is the 'classical' motivation used in econometrics and some other areas.
- Another situation is when a high-dimensional data vector is to be approximated by a small number of factors, which is the standard motivation for factor analysis.
- A third situation is when the user lacks enough information to classify the available signals into inputs and outputs and prefer to use a 'symmetric' system model. This is closely connected to the behavioral approach to modeling, Willems (1986), Heij

T. Söderström, *Errors-in-Variables Methods in System Identification*, Communications and Control Engineering,
https://doi.org/10.1007/978-3-319-75001-9_1

et al. (1997), Polderman and Willems (1998). A treatment of errors-in-variables problem for exact and approximate modeling within a behavioral framework appears in Markovsky et al. (2005b), Markovsky et al. (2006c).

EIV problems typically occur when the purpose of the modeling is to get physical insight into a process, rather than to have a model as a basis for prediction or control design. The identification experiment may not be under full 'control' of the modeler. In some applications, perhaps typically so in non-technical areas such as nature, biology, economics, it may be useful to regard the identification experiment as designed by somebody else, as the modeler has to work with given recorded input and output data. The issue on when errors-in-variables problem occur will be further discussed in Chap. 3.

With reference to these systems, the assumptions (*prejudices*) which lie behind the identification procedure have been thoroughly analyzed in Kalman (1982a), Kalman (1982b) with particular attention to the Frisch scheme, Frisch (1934). This scheme, in particular, assumes that each variable is affected by an unknown amount of additive noise and each noise component is independent of every other noise component and of every variable. As a consequence of this fact, in this case the solution is constituted by a whole family of models compatible with the set of noisy data, unlike other traditional approaches (e.g., least squares, total least squares, maximum likelihood) where the solution is characterized by a single model.

1.1 Four Motivating Examples

Example 1.1 This example is described in much more detail in Zhang et al. (2013). Consider the problem of identifying transmissibility functions in a mechanical mass-spring system. For example, the mass-spring system in Zhang et al. (2013) consists of a network with 6 masses and 11 springs and dampers. The total system can be modeled as

$$\mathbf{M}\frac{\mathrm{d}^2\mathbf{y}(t)}{\mathrm{d}t^2} + \mathbf{C}\frac{\mathrm{d}\mathbf{y}(t)}{\mathrm{d}t} + \mathbf{K}\mathbf{y}(t) = \mathbf{f}(t) \ , \tag{1.1}$$

where $\mathbf{y}(t)$ is a vector of nodal displacements, and $\mathbf{f}(t)$ a vector of external forces. Further, $\mathbf{M}$ is the mass matrix, $\mathbf{C}$ is the damping matrix, while $\mathbf{K}$ contains the spring constants. This model corresponds to Euler–Lagrange equations of motion. See also Meirovitch (1997) for models of the type (1.1).

The transmissibility function is a ratio of two noise-free outputs, e.g., the jth output and kth output, with a single force applied at the kth element. This leads to

$$G_{jk}(\omega) = \frac{Y_j(\mathrm{e}^{-\mathrm{i}\omega})}{Y_k(\mathrm{e}^{-\mathrm{i}\omega})} \ ,$$

where $Y_j(e^{-i\omega})$ is the discrete Fourier transform of the signal $y_j(t)$. The transmissibility function takes a rational form and can be seen as an underlying model between two output observations. When one takes into account that the signals cannot be observed exactly but are subject to measurement noise, there is clearly an identification setting where both input and output measurements are noise corrupted.

The above application is an example where the purpose of the modeling is to gain insight and knowledge about the internal properties of the system in question, rather than to get a basis for feedback design. A more general study of EIV problems in dynamic networks and the possibilities to identify certain subsystems is the topic in Dankers et al. (2015). See also Sect. 4.5 for more references on identification of dynamic networks. ■

A characteristic of the above example and many other EIV problems is that the system under study is not excited by directly manipulating the input. Rather, the excitation enters at some other point, perhaps not even arranged by the user or the experimentalist. The input–output data are just to be recorded and can all be subject to measurement errors (i.e., noise).

Example 1.2 A model used for forecasting in macroeconomics is based on a so-called approximate dynamic factor model, due to Stock and Watson (2002). The idea is that there are many variables (time series) that are observed with some noise, but they depend all on just a few essential factors. The variable to be forecasted contains some dynamics and depends also on the factors. To apply forecasting it is then of interest to also estimate or determine the unknown factors. The underlying model may be written in the form

$$\mathbf{y}(t) = \boldsymbol{\Lambda}\mathbf{f}(t) + \tilde{\mathbf{y}}(t) , \tag{1.2}$$

where $\mathbf{y}(t)$ is the vector-valued observations at time t, $\boldsymbol{\Lambda}$ is a matrix of so-called loadings, and $\mathbf{f}(t)$ is the factor vector. The term $\tilde{\mathbf{y}}(t)$ denotes some measurement uncertainty, or noise. Hence, in (1.2) $\mathbf{y}(t)$ is measured, while $\mathbf{f}(t)$ and $\tilde{\mathbf{y}}(t)$ are unknown.

Typically in the macroeconomic models, the dimension of $\mathbf{y}(t)$ is much larger than that of $\mathbf{f}(t)$. The variable to be forecasted, say $x(t+1)$, is assumed to follow a dynamic model of the form

$$x(t+1) = \boldsymbol{\beta}^T\mathbf{f}(t) + \gamma(q^{-1})x(t) + v(t) . \tag{1.3}$$

Here, $\boldsymbol{\beta}^T$ is a row vector of coefficients, $\gamma(q^{-1})$ a polynomial in the backward shift operator q^{-1}, and $v(t)$ a disturbance term.

To exemplify, in Chen and Lin (2011) this type of model is used for forecasting industrial production growth rate, unemployment rate, and inflation rate in Taiwan's economy. These quantities then correspond to $x(t+1)$. The recorded data $\mathbf{y}(t)$ consisted of 95 quarterly time series for Taiwan, 53 quarterly time series for the USA and 36 for Japan (USA and Japan are the main trade partners of Taiwan), all measured over the period 1981–2006. The data series represent quantities such as tax, labor market, stock market, exchange rate, interest rate, price index, and wages. The maximum number of factors considered was 12, so indeed in this case $\mathbf{f}(t)$ has much lower dimension than $\mathbf{y}(t)$. ∎

Example 1.3 This example is about direct data-driven controller design. The basic idea is the following. Consider a system with a transfer function G. Assume that a compensator K is to be designed so that the closed-loop system from reference to output is M. All these transfer functions are assumed to reside in discrete-time and can be seen as functions of either discrete-time frequency or the backward shift operator q^{-1}. The design problem is thus to achieve K such that

$$M = \frac{KG}{1+KG} . \tag{1.4}$$

If both the desired M and the system G are known, a direct solution of equation (1.4) leads to

$$K^* = \frac{M}{G(1-M)} . \tag{1.5}$$

The star in (1.5) denotes the *ideal controller.* It may not be feasible as G can be allowed to have very high order, and there is no guarantee that the closed-loop system will be internally stable.

The idea behind data-driven controller design is trying to design an appropriate compensator K by somehow approximating (1.4) and not to exploit any knowledge about G. Instead the compensator is parameterized using a parameter vector $\boldsymbol{\rho}$, and the search for $K(\boldsymbol{\rho})$ is formulated as an identification problem. The description to follow below summarizes a longer analysis carried out in van Heusden et al. (2011b). For some further treatments, see, for example, Karimi et al. (2007), van Heusden et al. (2011a), Cerone et al. (2017b). Needless to say, this approach does not have an 'exact' solution, but some assumption or simplification is needed in order to achieve a treatable problem.

The sensitivity function for the ideal controller (1.5) becomes

$$\frac{1}{1+K^*G} = 1 - M , \tag{1.6}$$

and it is assumed that also the designed controller satisfies this relation:

$$\frac{1}{1+K(\boldsymbol{\rho})G} = 1 - M . \tag{1.7}$$

The deviation of M from the obtained closed-loop system $K(\boldsymbol{\rho})G/(1+K(\boldsymbol{\rho})G)$ can then expressed with the criterion

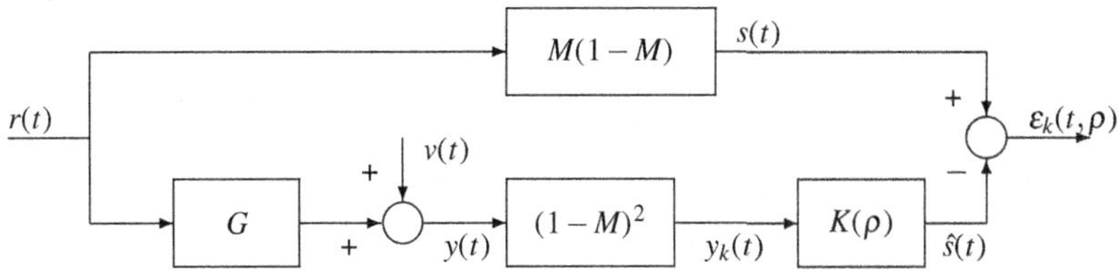

Fig. 1.1 Generalized tuning scheme for a model reference problem using only one experiment

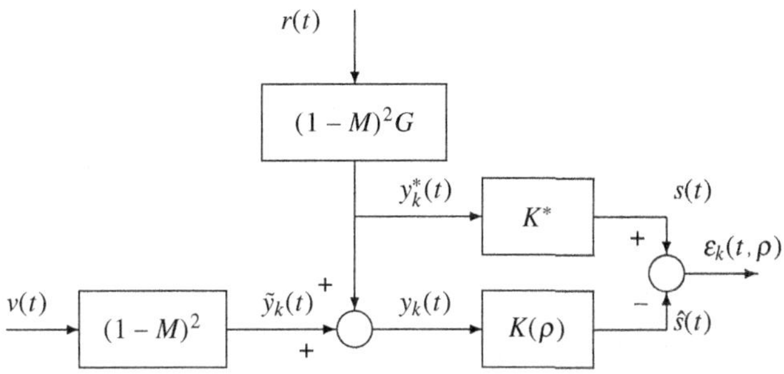

Fig. 1.2 Alternative representation of data-driven controller tuning scheme

$$
\begin{aligned}
\| M - \frac{K(\rho)G}{1+K(\rho)G} \|_2^2 &= \| \frac{K^*G}{1+K^*G} - \frac{K(\rho)G}{1+K(\rho)G} \|_2^2 \\
&= \| \frac{K^*G - K(\rho)G}{(1+K^*G)(1+K(\rho)G)} \|_2^2 \\
&\approx \| \frac{K^*G - K(\rho)G}{(1+K^*G)^2} \|_2^2 \\
&= \| (1-M)\left[M - K(\rho)(1-M)G\right] \|_2^2 \stackrel{\Delta}{=} J(\rho) \,. \qquad (1.8)
\end{aligned}
$$

The controller design problem is then to find the parameter vector $\boldsymbol{\rho}$ that minimizes the criterion $J(\boldsymbol{\rho})$ in (1.8).

The tuning problem is illustrated in Fig. 1.1, where the criterion $J(\boldsymbol{\rho})$ in (1.8) corresponds to $\mathsf{E}\left\{\varepsilon_k^2(t, \boldsymbol{\rho})\right\}$.

To formulate this as an identification problem, consider a linear system of the form

$$y(t) = G(q^{-1})r(t) + v(t) \,, \qquad (1.9)$$

where $r(t)$ denotes the input signal and $v(t)$ is the total effect of process disturbances and measurement noise.

The minimization problem in Fig. 1.1 can be redrawn as in Fig. 1.2. Here the unknown signals are $v(t)$, $y_k^*(t)$ and the noise signal $\tilde{y}_k(t)$ given by

$$\tilde{y}_k(t) = (1 - M)^2 v(t) \,. \tag{1.10}$$

The known signals are $r(t)$, $y_k(t) = (1 - M)^2 y(t)$ and $s(t)$ given by

$$s(t) = (1 - M)^2 G K^* r(t) = (1 - M) M r(t) \,. \tag{1.11}$$

From an identification point of view, the non-standard aspect is the presence of the *input noise* $\tilde{y}_k(t)$. The criterion to be minimized using measured data would be

$$\sum_{t=1}^{N} \varepsilon_k^2(t, \rho) \,,$$

where

$$\varepsilon_k(t, \rho) = K^* y_k^*(t) - K(\rho) y_k^*(t) - K(\rho) \tilde{y}_k(t) \,. \tag{1.12}$$

For identifying the controller it is important to take the effect of the noise model $K(\rho)\tilde{y}_k(t)$ into account.

Now recall that the known signals are $s(t)$ and $y_k^*(t)$, while $\tilde{y}_k(t)$ is unknown. The transfer function $K(\rho)$ is to be estimated by minimizing

$$\sum_{t=1}^{N} \varepsilon_k^2(t, \rho) \,.$$

This is just a standard output error identification, interpreting $s(t)$ as output and $y_k(t)$ as input. What is special in this case is the absence of any disturbance on the output side. On the other hand, there is a non-measurable noise contribution to the input side, namely $\tilde{y}_k(t)$. This is a problem that is very closely related to the errors-in-variables problem, but does not really belong to the EIV class. This problem is further discussed in the end of Sect. 3.1. ■

Example 1.4 The standardization of fifth generation (5G) wireless networks is expected to lead to a dramatic increase of networked feedback control applications over packet-switched combined fiber and wireless interfaces. The tactile Internet where force feedback is employed to enable remote sensing of objects is one of the most promising applications, IEEE (2017), while wireless closed-loop feedback control of robotic machinery in manufacturing plants constitutes another application, Delgado et al. (2017). Two characteristics of such feedback control applications are that they are subject to network delay in the downlink and uplink, and to quantization of control and feedback signal information; see Fig. 1.3. The network delay is partially unknown and varying which is why a call for a renewed focus on delay in a broad sense was issued in Samad (2016).

To support the need for models for controller design, networked identification is expected to become increasingly important. In view of the above discussion,

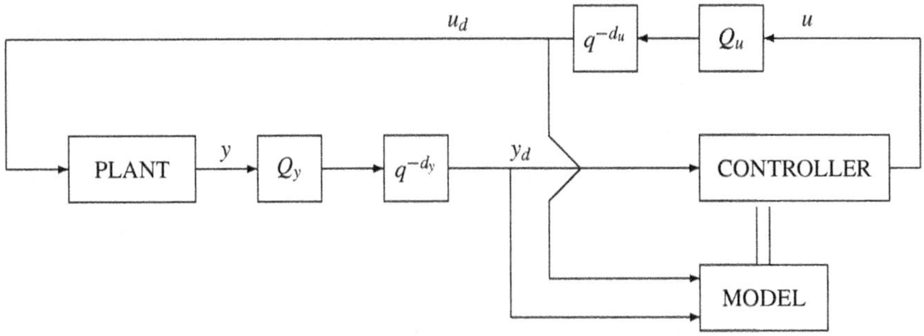

Fig. 1.3 A typical 5G wireless architecture for networked control, where identification is applied. The input to the plant is denoted by u, its quantized and delayed value by u_d, the output from the plant by y, and its quantized and delayed value by y_d

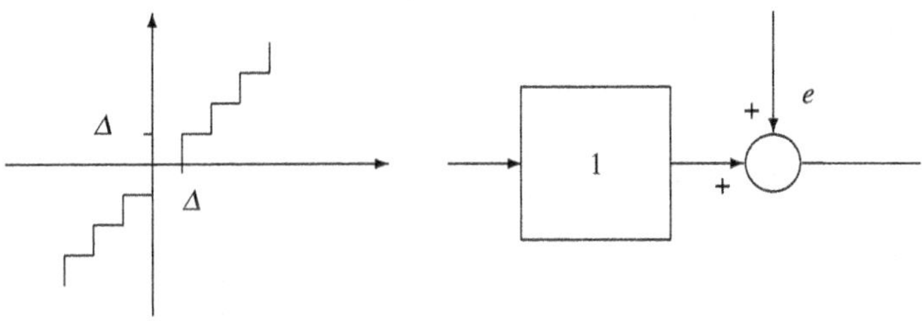

Fig. 1.4 Approximation of a quantizer as a unit gain with an additive noise, using the Sheppard correction

networked identification algorithms need to be able to model both delay and quantization. Figure 1.3 depicts the architecture of such an identification algorithm intended to identify a linear model subject to control and feedback signal quantization by means of the quantization functions Q_u and Q_y. As can be seen the control signal quantization is performed in the controller node, while the feedback signal quantization of the plant node is represented also in the controller node.

The problem to identify the plant (using the block labeled MODEL in Fig. 1.3) may be considered as an EIV problem. Then the quantizers are approximated. A quantizer with step Δ may be approximated as a unit gain with an additive white noise, uniformly distributed over $(-\Delta/2, \Delta/2)$; see Fig. 1.4. This is called the Sheppard correction; see, for example, Söderström (2002). It implies that the noise in Fig. 1.4 has variance $\mathsf{E}\{e^2(t)\} = \Delta^2/12$. The approximation can be expected to be quite accurate if Δ is small, and the signals are typically varying more than Δ between the sampling instants.

For some examples treating networked control problems, see Wigren (1998), Wigren (2017), Yasini and Wigren (2018). It can be noted that EIV was in fact applied to an identification problem with input and output quantization as early as 1995 in Krishnamurthy (1995). ■

Examples of other applications where errors-in-variables problems are solved include:

- Electromagnetic mineral exploration, Lau et al. (2007), Lau et al. (2008), Lau et al. (2009)
- Estimation of time-varying fading channels in mobile communication systems, Jamoos et al. (2007).
- Roll dynamics of a ship is considered in Linder (2017).
- Applications of oil well testing are described in Mansoori et al. (2014).
- Wave propagation in pavements are considered in Hostettler et al. (2014). The estimated model describing the mechanical properties from non-destructive testing can be used for wear prediction and maintenance. Other possible aspects include traffic monitoring from seismic activity recorded in sensors and caused by passing vehicles.
- Modeling and tracking autonomous underwater vehicles (AUVs) is treated in Yan et al. (2014).
- The paper Yu et al. (2013) deals with blind system identification using quantized observations, and the quantization errors may be regarded as causing an EIV problem.
- Errors-in-variables regression models are used for gene detection in RNA sequencing and micro-array techniques, as described in Xu et al. (2013).
- Structural health monitoring can be used to check the status of (large) mechanical constructions and buildings. An general aim can be to find early signs for damages and to monitor changes in the dynamics. Often this is done by fitting dynamic models to observed (noisy) data, and then comparing these models with reference ones. Recent examples of this techniques are given in Guidorzi and Diversi (2013), Guidorzi et al. (2015). Some earlier work on this type of monitoring includes Rougée et al. (1987), Basseville et al. (1987).
- A more recent application concerns thermal models for many-core systems-on-chip. Computers based on many-cores have better performances in terms of compromising between high clock frequency (fast computation) and low energy consumption (low heat dissipation) than traditional single-core computers. To achieve good compromises accurate models are required, and due to measurement noise, this turns out to be an EIV problem. It is treated, for example, in Diversi et al. (2013a, b), Diversi et al. (2014).
- Linear switching systems are treated in Nazari et al. (2015). An EIV formulation is used to detect and isolate abrupt changes.

1.2 Outline of the Book

The book is organized in the following way.

- The first group of chapters set the scene. Static problems are coped with in Chap. 2, including a detailed analysis of straight line fitting. The results can to a large extent be generalized to dynamic systems, but the analysis here is considerably simpler. Chapter 3 presents the EIV problem for dynamic systems and explains why some naïve approaches to solve it will lead to biased estimates.

- Next follow two chapters with a general analysis of the problem as such. Identifiability, that is, the possibility to recover the true system dynamics from the noisy measurements, is the topic of Chap. 4. Among other things it is demonstrated that some specific assumptions must be made in order to reach identifiability. These assumptions though can be made in some different ways, and the choice of way leads to different classes of estimation methods treated in later chapters. Chapter 5 treats various modeling aspects, including characterization of noise properties, extension to multivariable systems and to continuous-time models, etc. The chapter also presents some options for how to estimate the model order in an EIV problem.
- Several chapters deal with various identification methods, a few of which are tied to specific assumptions. Most of these chapters provide both derivations of the estimators and analysis of the consistency properties. Chapter 6 handles least squares and instrumental variable methods. Methods based on bias-compensation is the topic of Chap. 7, while Chap. 8 treats so-called covariance matching methods. Chapter 9 describes prediction error and maximum likelihood methods for EIV problems. Several methods can be formulated using frequency domain data, and a number of such examples are given in Chap. 10. Methods based on total least squares is the topic of Chap. 11. It happens in some applications that periodic data are available, or more generally, multiple experiments can be performed. Such a situation is often quite favorable from an identifiability point of view. Identification methods based on availability of periodic data are treated in Chap. 12.
- Two chapters give various types of analysis for the many identification methods presented so far. Chapter 13 copes with some algorithmic properties, including how to enhance convergence, solving overdetermined systems of equations and to derive time-recursive updating of the parameter estimates. Chapter 14 presents the asymptotic distribution of many of the estimates previously derived. The emphasis is to find the asymptotic covariance matrix of the parameter estimates for the case of large data sets.
- The last chapter, Chap. 15, has another character. It takes the user's perspective and describes how the previous analysis and the many different results on various methods can be useful when dealing with practical applications. The chapter contains a number of user guidelines.
- Many chapters include an appendix which contains proofs and lengthy derivations of results.
- The chapters end with a section where possible extensions are briefly discussed and many bibliographical notes are provided.
- Appendix A summarizes various general background materials such as some matrix results, discrete Fourier transforms, and useful results from probability and statistics.
- There are several numerical examples in the book, illustrating many of the different introduced identification methods. To facilitate comparisons the same data sets are used in most of these studies. The data sets are introduced and described in Sect. 3.2.

1.3 Some Important Concepts in System Identification

It is assumed that the reader is familiar with the basics of system identification. There are many textbooks on the area that may be consulted if needed, Ljung (1999) and Söderström and Stoica (1989) being two examples.

Some topics of identification that are assumed familiar to the reader are the ones listed below. The very brief descriptions here are merely aimed as a refresher, while many details and more precise treatments can be found in the general textbooks mentioned earlier.

- *Consistency*. An estimated parameter vector based on N data points, say $\hat{\boldsymbol{\theta}}$, is said to be consistent if it converges to its 'true' value $\boldsymbol{\theta}_0$ as the number of measured data grows without any limit,

$$\hat{\boldsymbol{\theta}}_N \rightarrow \boldsymbol{\theta}_0, \quad N \rightarrow \infty\,. \tag{1.13}$$

  Consistency is, of course, a desired property of an estimation method.
- *Ergodic process*. A process (or signal) is said to be ergodic if its sample covariances converge to their expected values as the number of data points tends to infinity. Let the signal be denoted by $x(t)$. Then a sample covariance element based on N data points can be taken as

$$\hat{r}_x(\tau) = \frac{1}{N-\tau} \sum_{t=1}^{N-\tau} x(t+\tau)x(t) \tag{1.14}$$

  for an arbitrary positive integer τ. For an ergodic process it holds

$$\hat{r}_x(\tau) \rightarrow \mathsf{E}\{x(t+\tau)x(t)\}, \quad N \rightarrow \infty\,. \tag{1.15}$$

- *Identifiability*. Identifiability concerns the principal possibility to retrieve (identify) the unknown dynamics from the measured data. Normally, the asymptotic case of an infinite amount of data is considered. Then the identifiability question is whether or not the unknown dynamics can be uniquely found from the spectrum of the measured data. Identifiability for EIV problems is specifically studied in Sect. 2.1.1 and in Chap. 4.
- *Persistent excitation*. This concept concerns the properties of the signal exciting the system, which normally is the input signal. The concept describes to what degree the signal varies in a sufficiently irregular way to allow identification of the system. Consider an ergodic signal $x(t)$. It is said to be persistently exciting (p.e.) of order n if its spectrum is strictly positive for n distinct frequencies. As illustration, it can be mentioned that a signal being constant in time is p.e. of order 1, a noise-free sine wave is p.e. of order 2, while an ARMA process is p.e. of infinite order.

- *Spectral factorization*. Consider a stationary process with spectral density $\phi(e^{i\omega})$, and assume that ϕ is a rational function of $e^{i\omega}$. Then there is a unique factorization of ϕ of the form

$$\phi(e^{i\omega}) = \lambda^2 H(e^{i\omega}) H(e^{-i\omega}) = \lambda^2 |H(e^{i\omega})|^2 \tag{1.16}$$

where H and H^{-1} are asymptotically stable, finite order filters, and $H(0) = 1$. This is called spectral factorization. The important consequence is that in the time domain the process can be represented as

$$y(t) = H(q^{-1}) e(t) \tag{1.17}$$

where $H(q^{-1})$ is a discrete-time transfer function using the backward shift operator q^{-1}, and $e(t)$ is white noise of zero mean and variance λ^2. It is not necessarily so that (1.17) describes how $y(t)$ is physically generated. Indeed, $y(t)$ may be the collected effect of several noise sources. The important aspect is that the model (1.17) gives a perfect description of all second-order moments of $y(t)$, and hence in particular its covariance function and spectrum.

1.4 Some Notations

Some general comments on the notations used in the book follow.

Vectors and matrices are generally set in bold face. A vector or a matrix with all elements equal to zero will be written as $\mathbf{0}$ or $\mathbf{0}_{m\times n}$, where in the latter case it is emphasized that the matrix has m rows and n columns. Similarly, the identity matrix is denoted $\mathbf{I}$ or $\mathbf{I}_n$. A unit vector, with a one in position k, will be denoted $\mathbf{e}_k$.

For a stationary random process $x(t)$, define its covariance function $r_x(\tau)$ as:

$$r_x(\tau) = \mathsf{E}\{x(t)x(t-\tau)\} , \tag{1.18}$$

where E denotes the expectation operator.

Further, the cross-covariance matrix between two random vectors $\mathbf{x}(t)$ and $\mathbf{y}(t)$ and the cross-covariance vector between a random vector $\mathbf{x}(t)$ and a random variable $z(t)$ are denoted as

$$\mathbf{R}_{\mathbf{xy}} = \mathsf{E}\{\mathbf{x}(t)\mathbf{y}^T(t)\}, \quad \mathbf{r}_{\mathbf{x}z} = \mathsf{E}\{\mathbf{x}(t)z(t)\} . \tag{1.19}$$

The estimates of these covariances from measured data are denoted as

$$\hat{\mathbf{R}}_{\mathbf{xy}} = \frac{1}{N}\sum_{t=1}^{N} \mathbf{x}(t)\mathbf{y}^T(t), \quad \hat{\mathbf{r}}_{\mathbf{x}z} = \frac{1}{N}\sum_{t=1}^{N} \mathbf{x}(t)z(t) . \tag{1.20}$$

For stationary (and quasistationary) signals, ergodicity applies; see Söderström and Stoica (1989), Ljung (1999). This means that

$$\lim_{N\to\infty} \hat{\mathbf{R}}_{\mathbf{xy}} = \mathbf{R}_{\mathbf{xy}}, \quad \lim_{N\to\infty} \hat{\mathbf{r}}_{\mathbf{x}z} = \mathbf{r}_{\mathbf{x}z} \ . \tag{1.21}$$

Occasionally, an index N will be added in the notation to emphasize the dependence on the number of data points. Then

$$\mathbf{R}_N = \frac{1}{N}\sum_{t=1}^{N} \mathbf{x}(t)\mathbf{y}^T(t) \ . \tag{1.22}$$

Many identification methods are tied to minimization of a loss function that depends quadratically on the equation errors or the prediction errors $\varepsilon(t, \boldsymbol{\theta})$:

$$V(\boldsymbol{\theta}) = V_N(\boldsymbol{\theta}) = \frac{1}{N}\sum_{t=1}^{N} \varepsilon^2(t, \boldsymbol{\theta}) \ . \tag{1.23}$$

The notation V_N with the subscript N is used in circumstances when the dependence on N, the number of data points, is to be emphasized. For ergodic processes, the loss function converges (at least pointwise) to a limit,

$$V_\infty(\boldsymbol{\theta}) = \lim_{N\to\infty} V_N(\boldsymbol{\theta}) \ . \tag{1.24}$$

Several further notational conventions are listed after the preface of the book and in Sect. 5.1.

1.5 Extensions and Bibliographical Notes

When searching on the Internet for 'errors-in-variables' Google gives more than 80 million hits. Various publication databases such as Science Citation Index or Elsevier Science Direct give a few thousand different references each on the subject. The area is therefore quite extensive and is also quickly growing with time. The vast majority of papers are written from an application perspective and can deal with biomedicine, chemistry, chemical engineering, earth sciences, econometrics, managements, mechanical engineering, finance, ecology, geosciences, image systems, time series analysis, etc. Most of the papers published in automatic control journals and conference proceedings focus on methodology, as do some of the many papers in various statistical journals.

One of the first solutions for the identification of dynamic EIV models was given in Levin (1964) and Aoki and Yue (1970). After these pioneering works, this topic has been re-proposed by several authors with different approaches. An overview of

EIV identification methods can be found in Söderström (2007a). See also Söderström et al. (2002), Söderström (2012b) and Guidorzi et al. (2008).

Other papers deal with a related problem called 'errors-in-variables filtering.' This refers to a standard filtering problem, but where both input and output measurements are contaminated by noise. This topic is treated, for example, in Goodwin et al. (2002), Guidorzi et al. (2003), Diversi et al. (2003b), Diversi et al. (2003a), Diversi et al. (2005b), Markovsky et al. (2002), Markovsky and De Moor (2005), Diversi et al. (2005a), Diversi (2012), Sumisławska et al. (2011) but falls outside the scope of this book.

Chapter 2
The Static Case

The theme of this book is errors-in-variables modeling of dynamic systems. As a prelude a simple static case is first studied in Sect. 2.1, namely line fitting. From that case there are some lessons to be learned that apply also to more complicated situations. Sections 2.2 and 2.3 deal with some more general aspects for static errors-in-variables problems.

2.1 Line Fitting

Assume that a set of points in the x, y plane that correspond to measurements $(x_1, y_1), \ldots, (x_N, y_N)$ is given. To make the problem really simple, assume that a straight line should be fitted to the data and that the line has to pass the origin. In practice, all the measurements x_i and y_i are assumed to be noisy. Therefore model the situation as follows. Let the line be given by

$$y = ax \ , \tag{2.1}$$

where a is the unknown slope parameter to be estimated. Further assume that the data can be described by one exact part with noise added. Mathematically, this means that the data are described by the model

$$\begin{aligned} y_i &= y_{0i} + \tilde{y}_i \ , \\ x_i &= x_{0i} + \tilde{x}_i \ , \qquad i = 1, \ldots, N \ . \\ y_{0i} &= a_0 x_{0i} \ , \end{aligned} \tag{2.2}$$

The problem is to estimate the slope a_0 as accurately as possible (one would like the estimate of a to be close to the true value a_0). The measurement errors, or noise, $\{\tilde{y}_i\}$ and $\{\tilde{x}_i\}$ are assumed to be sequences of independent random variables of zero mean and variances λ_y and λ_x, respectively. The situation is depicted in Fig. 2.1, where

T. Söderström, *Errors-in-Variables Methods in System Identification*, Communications and Control Engineering,
https://doi.org/10.1007/978-3-319-75001-9_2

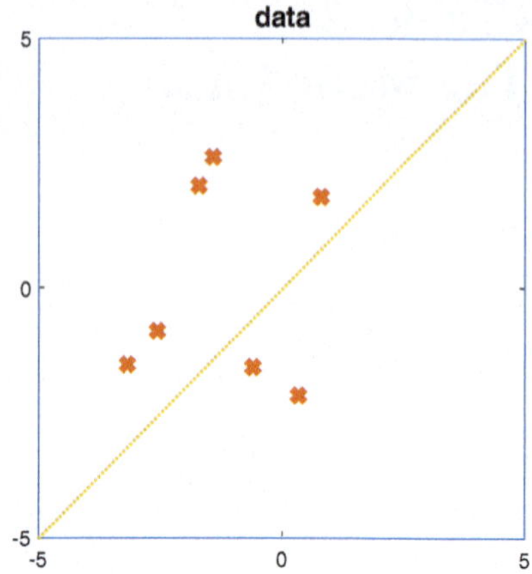

Fig. 2.1 A set of noisy data

some noisy observations are shown. In particular one would like the estimate $\hat{a}$ of a_0 to be consistent, which means

$$\hat{a} \to a_0, \; N \to \infty \,. \tag{2.3}$$

This means that the estimate $\hat{a}$ will be arbitrarily close to the true value a_0 when the number of data points is large enough.

There are several approaches for how to estimate the parameter a. A standard way would be to use least squares. Then one neglects the errors in the x variable, and determine a so that the sum of the squared errors in the y direction is minimized:

$$\hat{a}_{\mathrm{LS}} = \arg \min_a \sum_{i=1}^{N} (y_i - a x_i)^2 \;, \tag{2.4}$$

which leads to the estimate

$$\hat{a}_{\mathrm{LS}} = \frac{\sum_i y_i x_i}{\sum_i x_i^2} \;. \tag{2.5}$$

Another possibility would be to minimize the sum of the squared errors in the x direction. This is sometimes called data least squares. One would then have

$$\hat{a}_{\mathrm{DLS}} = 1/\hat{c} \;, \tag{2.6}$$

$$\hat{c} = \arg \min_c \sum_{i=1}^{N} (x_i - c y_i)^2 \;, \tag{2.7}$$

which leads to the estimate

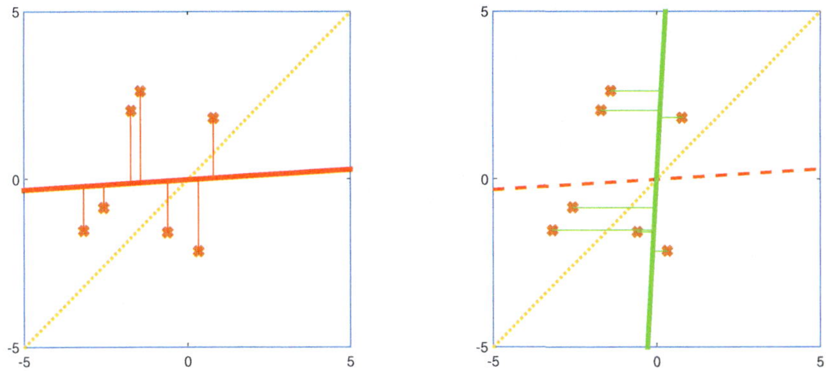

Fig. 2.2 The fitted lines. Left: Least squares estimate (solid), true line (dotted). Right: Data least squares (solid), least squares (dashed), true line (dotted)

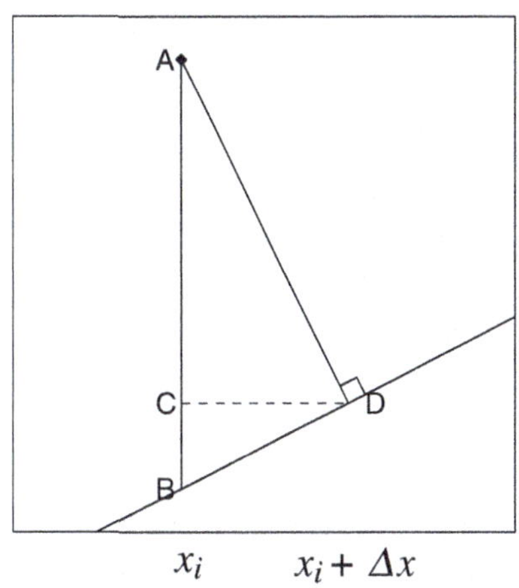

Fig. 2.3 The distance to be expressed in orthogonal regression is |AD|

$$\hat{a}_{\text{DLS}} = \frac{\sum_i y_i^2}{\sum_i x_i y_i} . \tag{2.8}$$

The two estimated lines for the data set introduced in Fig. 2.1 are displayed in Fig. 2.2. One can immediately see that the two estimates differ considerably. Furthermore, it apparently holds in this case that

$$\hat{a}_{\text{LS}} < \hat{a}_{\text{DLS}} . \tag{2.9}$$

A third option is to use orthogonal regression. The idea is then to minimize neither the vertical, nor the horizontal distances to the line, but the orthogonal ones. To formulate the appropriate criterion, consider Fig. 2.3. The different distances are as follows:

$$
\begin{aligned}
|\mathrm{AB}| &= s \stackrel{\Delta}{=} |y_i - ax_i| \ , \\
|\mathrm{CD}| &= \Delta x \ , \\
|\mathrm{BC}| &= |a|\Delta x \ , \\
|\mathrm{AC}| &= \frac{1}{|a|}\Delta x \ .
\end{aligned}
$$

Applying the Pythagorean theorem on the triangle ABD gives

$$
\begin{aligned}
|\mathrm{AB}|^2 &= |\mathrm{BD}|^2 + |\mathrm{AD}|^2 \\
&= 2|\mathrm{CD}|^2 + |\mathrm{AC}|^2 + |\mathrm{BC}|^2
\end{aligned}
\tag{2.10}
$$

leading to

$$
s^2 = 2(\Delta x)^2 + a^2(\Delta x)^2 + \frac{1}{a^2}(\Delta x)^2 \ , \tag{2.11}
$$

and

$$
(\Delta x)^2 = \frac{s^2}{\left(|a| + \frac{1}{|a|}\right)^2} \ . \tag{2.12}
$$

Finally, the sought distance as $|AD|$ is obtained from

$$
\begin{aligned}
|\mathrm{AD}|^2 &= |\mathrm{CD}|^2 + |\mathrm{AC}|^2 = (\Delta x)^2 \left(1 + \frac{1}{a^2}\right) \\
&= \frac{s^2 a^2}{(a^2+1)^2} \frac{a^2+1}{a^2} = \frac{s^2}{a^2+1} \ .
\end{aligned}
\tag{2.13}
$$

This means that the parameter a is determined as

$$
\hat{a} = \arg \min_a \frac{\sum_{i=1}^n (y_i - ax_i)^2}{1 + a^2} \ . \tag{2.14}
$$

The result for the given example is displayed in Fig. 2.4. The estimate (2.14) is normally called the total least squares (TLS) estimate, and it will be described in some more detail in Chap. 11. The minimization in (2.14) can be carried out explicitly in this simple example. The stationary points of the criterion do in fact satisfy the second-order equation

$$
a^2 \sum_i y_i x_i + a\left(\sum_i x_i^2 - \sum_i y_i^2\right) - \sum_i x_i y_i = 0 \ . \tag{2.15}
$$

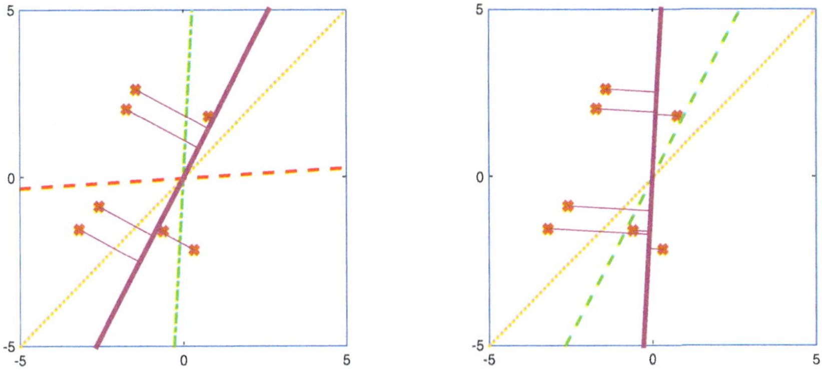

Fig. 2.4 Lines fitted with the total least squares. Left: TLS estimate (solid), LS estimate (dashed), DLS estimate (dash-dotted), true line (dotted). Right: Rescaled TLS estimate (solid), original TLS estimate (dashed), true line (dotted). The y-data were rescaled with a factor of 10

In a more general setting, one should look for the singular value decomposition of the data matrix

$$\mathbf{A} = \begin{pmatrix} y_1 & -x_1 \\ \vdots & \vdots \\ y_N & -x_N \end{pmatrix} . \tag{2.16}$$

Then, the right singular vector associated with the smallest singular value of $\mathbf{A}$ can, after a normalization, be written as $\left(1\ \hat{a}_{\mathrm{TLS}}\right)^T$. This singular vector is also an eigenvector of the matrix

$$\mathbf{A}^T\mathbf{A} = \begin{pmatrix} \sum_i y_i^2 & -\sum_i y_i x_i \\ -\sum_i y_i x_i & \sum_i x_i^2 \end{pmatrix} .$$

In the above example, it turns out that the total least squares estimate $\hat{a}_{\mathrm{TLS}}$ is much better than $\hat{a}_{\mathrm{LS}}$ and $\hat{a}_{\mathrm{DLS}}$. Is $\hat{a}_{\mathrm{TLS}}$ always to be preferred? Not necessarily. For the TLS estimate to work well it is necessary that the errors in the x and the y variables are of the same order. Expressed differently, the TLS estimate is not invariant to scaling, Wald (1940). As an illustration, assume that the observations y_i correspond to a distance given in meters. If the problem is scaled so that y is given in decimeters, and the total least squares estimate is applied, the situation changes completely; see Fig. 2.4 for illustration. It is clear that rescaling of one of the variables can significantly change the estimate.

So far, it is clear from this simple example that from second-order statistics with no further information, one cannot successfully determine the slope of the line in a reliable way.

The problem can be extended in several ways. One extension is to let x and a be vector-valued. Of further relevance to system identification is to include dynamics, so while y would correspond to the current value of the system output, the different x components would correspond to delayed values of both the system input and the system output. As will be seen in the coming chapters, the dynamic case introduces some couplings between the variables that may in fact facilitate the possibilities for the estimation. Still it turns out that also in the dynamic case, the problem will become much simplified when the variance ratio λ_y/λ_x is known.

Some further results for the static case are described in Appendix 2.A. The conclusions of that analysis are as follows.

- For the model (2.1) the parameter a is identifiable from first- and second-order moments only if $\lim_{N\to\infty} \frac{1}{N}\sum_{i=1}^{N} x_{0i} \neq 0$.
- If higher-order moments are included as well to identify the parameter a in the model (2.1), it can be determined up to the sign; that is, only a^2 is identifiable.
- If the model is extended to an affine relation between inputs and outputs in the sense

$$y = ax + b \,, \tag{2.17}$$

  then the parameters are not identifiable from first- and second-order moments. This also applies in the extended case when x, y, and b are vectors, and a is a matrix.
- If the ratio λ_y/λ_x of the noise variances is known, then the parameters a and b in the model (2.17) are identifiable.
- If the ratio λ_y/λ_x is known, and the noise is Gaussian, the maximum likelihood estimate of λ_x is *not consistent*. Rather, it holds, see Lindley (1947)

$$\lim_{N\to\infty} \hat{\lambda}_x = \lambda_x/2 \,. \tag{2.18}$$

It will be seen in the later chapters how many of the above properties carry over to the more complex situation where dynamic models are fitted to noisy input–output data.

2.1.1 Some System Theoretic Considerations of Identifiability

As was seen from the previous analysis identifiability does not hold in general, unless some additional assumption is applied. Considerations of such additional assumption are discussed in Chap. 15 for the dynamic case. In particular it is illustrated what can happen when such an additional assumption is made in order to get a well-defined estimate, but that the real data do *not* fulfill the assumption.

The following example give some characterization of the possible estimates for the simple line-fitting case.

Example 2.1 Let $r_y,\ r_x,\ r_{yx}$ be the second-order moments of the (noisy) data. Then any estimate should fulfill

$$r_y = a^2\sigma + \lambda_y \ , \tag{2.19}$$

$$r_{yx} = a\sigma \ , \tag{2.20}$$

$$r_x = \sigma + \lambda_x \ , \tag{2.21}$$

with side conditions

$$0 \le \lambda_x,\ \ 0 \le \lambda_y,\ \ 0 \le \sigma \ . \tag{2.22}$$

One gets from (2.21) and (2.22):

$$\sigma = r_x - \lambda_x \le r_x \ , \tag{2.23}$$

and from (2.19), (2.20) and (2.22):

$$r_y = r_{yx}^2/\sigma + \lambda_y \ \Rightarrow \sigma \ge r_{yx}^2/r_y \ . \tag{2.24}$$

Next one can use (2.20) to transform (2.23) and (2.24) into inequalities in the estimate a. Assume, for convenience only, that $r_{yx} > 0$ (this corresponds precisely to $a > 0$).

It apparently holds

$$a \ge r_{yx}/\max(\sigma) = r_{yx}/r_x = \hat{a}_{\text{LS}} \ , \tag{2.25}$$

$$a \le r_{yx}/\min(\sigma) = r_y/r_{yx} = \hat{a}_{\text{DLS}} \ . \tag{2.26}$$

This calculation shows that the extreme values of the possible estimates that are compatible with the data are given precisely by the estimates $\hat{a}_{\text{LS}}$ (where all uncertainties are assumed to be in the y measurements) and $\hat{a}_{\text{DLS}}$ (where all uncertainties are assumed to be in the x measurements). Compare with Fig. 2.2 and (2.9). ■

2.2 Confirmatory Factor Analysis

In multivariate statistics, confirmatory factor analysis (CFA) is a well-established technique for handling estimation problems involving latent variables, i.e., models where key variables are not directly measurable. For descriptions and some treatment of such modeling, see Jöreskog (1970) and Bartholomew et al. (2011). The basics of the CFA approach is summarized below.

2.2.1 The Modeling Part

Consider the model

$$\mathbf{y} = \mathbf{A}(\boldsymbol{\vartheta})\boldsymbol{\eta} + \boldsymbol{\varepsilon} \ , \tag{2.27}$$

$$\text{cov}(\boldsymbol{\eta}) = \mathbf{R}_{\eta}(\boldsymbol{\vartheta}), \quad \text{cov}(\boldsymbol{\varepsilon}) = \mathbf{R}_{\varepsilon}(\boldsymbol{\vartheta}) \ . \tag{2.28}$$

This model is called a *confirmatory factor analysis* (CFA) *model.*

Here $\boldsymbol{\eta}$ is a vector of latent (unobservable) variables, while $\mathbf{y}$ is an observed vector. The noise vector $\boldsymbol{\varepsilon}$ is assumed to have zero mean (an assumption that can be relaxed). The matrix $\mathbf{A}(\boldsymbol{\vartheta})$ can be fully or partly unknown. This matrix as well as the covariance matrices $\mathbf{R}_{\eta}(\boldsymbol{\vartheta})$ and $\mathbf{R}_{\varepsilon}(\boldsymbol{\vartheta})$ is parameterized with a parameter vector $\boldsymbol{\vartheta}$. As soon as $\boldsymbol{\vartheta}$ is determined, these matrices are fully known. The elements of the coefficient matrix $\mathbf{A}(\boldsymbol{\vartheta})$ are called loadings, and the elements of $\boldsymbol{\eta}$ are called factors. Typically $\mathbf{y}$ has (much) higher dimension than $\boldsymbol{\eta}$.

The idea is hence to describe (a large set of) observations as due to a (modest) number of independent factors. Further, to avoid ambiguities in the formulation, it is often assumed that the covariance matrix of $\boldsymbol{\eta}$ is identity, that is

$$\mathbf{R}_{\eta} = \mathsf{E}\left\{\boldsymbol{\eta}\boldsymbol{\eta}^T\right\} = \mathbf{I} \ . \tag{2.29}$$

In addition it may be assumed that the covariance matrix of the errors $\mathbf{R}_{\varepsilon} = \mathsf{E}\left\{\boldsymbol{\varepsilon}\boldsymbol{\varepsilon}^T\right\}$ is diagonal. The aim is to find and characterize the matrix $\mathbf{A}(\boldsymbol{\eta})$ of loadings. Using second-order statistics, it follows that

$$\mathbf{R}_{\mathbf{y}} = \mathbf{A}(\boldsymbol{\vartheta})\mathbf{A}^T(\boldsymbol{\vartheta}) + \mathbf{R}_{\varepsilon} \ . \tag{2.30}$$

The left-hand side of (2.30) can be estimated from repeated observations of $\mathbf{y}$. Then to estimate $\mathbf{A}$ becomes a problem of (approximate) factorization, i.e., to find $\mathbf{A}$ and the diagonal $\mathbf{R}_{\varepsilon}$ satisfying (2.30). For a given positive definite matrix $\mathbf{R}_{\mathbf{y}}$, the decomposition implied in (2.30) into a low-rank part and a diagonal part is non-trivial. For example, if $\mathbf{A}$ is a column vector, no solution exists in general. If $\mathbf{A}$ is a square matrix, many solutions exist. In fact, the decomposition problem in (2.30) is not well posed; see Bottegal and Picci (2015).

Remark 2.1 A special case of the model (2.27) appears in array processing. This field concerns the ability of using and combining data from different sensors (often called antennas). Both temporal and spatial processing of recorded signals can take place. Applications appear in radar, sonar, seismic exploration, and wireless communications. The noise sources (the components of $\boldsymbol{\varepsilon}$) are assumed to be white noise of equal variance. Further, the matrix $\mathbf{A}(\boldsymbol{\vartheta})$ typically has the structure

$$\mathbf{A}(\boldsymbol{\vartheta}) = \left(\mathbf{a}(\boldsymbol{\vartheta}_1)\ \mathbf{a}(\boldsymbol{\vartheta}_2)\ \ldots\ \mathbf{a}(\boldsymbol{\vartheta}_n)\right) \ , \tag{2.31}$$

where the column $\mathbf{a}(\boldsymbol{\vartheta})$ is called the steering vector, and the scalars $\vartheta_1, \ldots, \vartheta_n$ are the direction of arrivals of an impinging wavefront. Observing the outputs $\mathbf{y}$ at many time points, one can estimate the covariance matrix

$$\mathbf{R}_y = \mathbf{A}(\vartheta_0)\mathbf{R}_x\mathbf{A}^*(\vartheta_0) + \sigma^2\mathbf{I} . \tag{2.32}$$

As data are often treated in complex-valued form, the factor $\mathbf{A}^*$ rather than $\mathbf{A}^T$ appears in (2.32). ■

Remark 2.2 Structural equation modeling (SEM) is a more complex modeling technique that is sometimes used in multivariate statistics, in particular for social science applications. It is described by the model

$$\eta = \mathbf{B}\eta + \boldsymbol{\Gamma}\xi + \zeta , \tag{2.33}$$
$$\mathbf{y} = \boldsymbol{\Lambda}_y\eta + \boldsymbol{\varepsilon} , \tag{2.34}$$
$$\mathbf{x} = \boldsymbol{\Lambda}_x\xi + \boldsymbol{\delta} . \tag{2.35}$$

Here η and ξ are latent (unobserved) variables, while $\mathbf{y}$ and $\mathbf{x}$ are variables that are observed. All these quantities are vector-valued. In the system, the variable ξ is exogenous (i.e., independent), while η is endogenous (i.e., dependent). The model is such that $\mathbf{I} - \mathbf{B}$ is nonsingular, so (2.33) can be used to uniquely describe how η depends on ξ and ζ. The dimensions of the matrices containing the model parameters are as follows:

$$\dim(\mathbf{B}) = m \times m, \ \dim(\boldsymbol{\Gamma}) = m \times p, \ \dim(\boldsymbol{\Lambda}_y) = n \times m, \ \dim(\boldsymbol{\Lambda}_x) = q \times n . \tag{2.36}$$

and all the vectors are column vectors with compatible dimensions.

The noise, or disturbance terms, is assumed here for simplicity to have zero means (an assumption that can be relaxed) and covariance matrices

$$\mathsf{E}\left\{\zeta\zeta^T\right\} = \boldsymbol{\Psi} , \tag{2.37}$$
$$\mathsf{E}\left\{\xi\xi^T\right\} = \boldsymbol{\Phi} , \tag{2.38}$$
$$\mathsf{E}\left\{\boldsymbol{\varepsilon}\boldsymbol{\varepsilon}^T\right\} = \boldsymbol{\Theta}_\epsilon , \tag{2.39}$$
$$\mathsf{E}\left\{\boldsymbol{\delta}\boldsymbol{\delta}^T\right\} = \boldsymbol{\Theta}_\delta . \tag{2.40}$$

It is assumed, again for simplicity, that the random vectors ζ, $\boldsymbol{\varepsilon}$, $\boldsymbol{\delta}$ are independent.

The matrices $\mathbf{B}$, $\boldsymbol{\Gamma}$, $\boldsymbol{\Lambda}_y$, $\boldsymbol{\Lambda}_x$, $\boldsymbol{\Psi}$, $\boldsymbol{\Phi}$, $\boldsymbol{\Theta}_\epsilon$, $\boldsymbol{\Theta}_\delta$ are fully or partly unknown. They are parameterized with a parameter vector $\boldsymbol{\vartheta}$. As soon as $\boldsymbol{\vartheta}$ is determined, these matrices are fully known.

In fact, the full SEM model (2.33)–(2.35) can be transformed into the CFA form (2.27) by redefining variables. To see this, write (2.34) and (2.35) as

$$\begin{pmatrix} \mathbf{y} \\ \mathbf{x} \end{pmatrix} = \begin{pmatrix} \Lambda_y & \mathbf{0} \\ \mathbf{0} & \Lambda_x \end{pmatrix} \begin{pmatrix} \eta \\ \xi \end{pmatrix} + \begin{pmatrix} \varepsilon \\ \delta \end{pmatrix} .$$

Next one can write (2.33) as

$$\begin{pmatrix} \mathbf{I} - \mathbf{B} & -\Gamma \\ \mathbf{0} & \mathbf{I} \end{pmatrix} \begin{pmatrix} \eta \\ \xi \end{pmatrix} = \begin{pmatrix} \zeta \\ \xi \end{pmatrix} ,$$

and therefore

$$\begin{pmatrix} \mathbf{y} \\ \mathbf{x} \end{pmatrix} = \underbrace{\begin{pmatrix} \Lambda_y & \mathbf{0} \\ \mathbf{0} & \Lambda_x \end{pmatrix} \begin{pmatrix} \mathbf{I} - \mathbf{B} & -\Gamma \\ \mathbf{0} & \mathbf{I} \end{pmatrix}^{-1}}_{\mathbf{A}} \begin{pmatrix} \zeta \\ \xi \end{pmatrix} + \begin{pmatrix} \varepsilon \\ \delta \end{pmatrix} .$$

which takes the form of a CFA model. Which form to use, the full SEM model or the simplified CFA model, is a matter of which parameterization of the matrices that is convenient. ■

2.2.2 *Estimation Part*

Assume that there is a number of observations $\mathbf{y}_i, i = 1, \ldots, N$, available of $\mathbf{y}$, and one would like to estimate the parameter vector ϑ. From the observations, one computes the sample covariance matrix

$$\hat{\mathbf{R}} = \frac{1}{N} \sum_{i=1}^{N} \mathbf{y}_i \mathbf{y}_i^T . \tag{2.41}$$

From the expression above, the model covariance matrix implied by the model is found to be

$$\mathbf{R}(\vartheta) = \mathsf{E}\left\{\mathbf{y}\mathbf{y}^T\right\} \tag{2.42}$$

$$= \mathbf{A}(\vartheta)\mathbf{R}_\eta(\vartheta)\mathbf{A}^T(\vartheta) + \mathbf{R}_\varepsilon(\vartheta) \tag{2.43}$$

The estimation problem is then to find a vector ϑ that is 'compatible' with the observations in the sense that

$$\hat{\mathbf{R}} \approx \mathbf{R}(\vartheta) . \tag{2.44}$$

The relation (2.44) can be interpreted as a system of equations. One has to take into account that both sides of (2.44) are symmetric matrices, so the 'effective' number of equations is (at most)

$$n_e = \frac{n(n+1)}{2} \, . \tag{2.45}$$

where $n = \dim(\mathbf{y})$.

The matrix $\mathbf{R}(\boldsymbol{\vartheta})$ may have a special structure. For example, $\mathbf{R}(\boldsymbol{\theta})$ is called Toeplitz if its elements are equal along the diagonals. This means that it takes the form

$$\mathbf{R}(\boldsymbol{\vartheta}) = \begin{pmatrix} r_{11}(\boldsymbol{\vartheta}) & r_{12}(\boldsymbol{\vartheta}) & \dots & r_{1n}(\boldsymbol{\vartheta}) \\ r_{21}(\boldsymbol{\vartheta}) & r_{11}(\boldsymbol{\vartheta}) & \dots & r_{1,n-1}(\boldsymbol{\vartheta}) \\ \vdots & & \ddots & \\ r_{n1}(\boldsymbol{\vartheta}) & \dots & \dots & r_{11}(\boldsymbol{\vartheta}) \end{pmatrix} . \tag{2.46}$$

If the model is such that the matrix $\mathbf{R}(\boldsymbol{\vartheta})$ is fully or partly Toeplitz, then the effective number of equations is indeed smaller than given by (2.45). For the estimation problem to be feasible one must require the number of elements in $\boldsymbol{\vartheta}$ to be at most as large as the quantity in (2.45), or rather the number of different elements in $\mathbf{R}(\boldsymbol{\vartheta})$.

Assuming the estimation problem is feasible, one can define an estimate of $\boldsymbol{\vartheta}$ as the minimizing element of a measure expressing how much $\mathbf{R}(\boldsymbol{\vartheta})$ differs from $\hat{\mathbf{R}}$.

Generally, write such a loss function (often called 'fit function' in the CFA literature) as

$$V(\boldsymbol{\vartheta}) = f(\hat{\mathbf{R}}, \mathbf{R}(\boldsymbol{\vartheta})) \, . \tag{2.47}$$

There are several ways for how such loss functions may be constructed:

- One possible measure is

$$V_1(\boldsymbol{\vartheta}) = \operatorname{tr}\left(\hat{\mathbf{R}}\mathbf{R}^{-1}(\boldsymbol{\vartheta})\right) + \log\left(\det(\mathbf{R}(\boldsymbol{\vartheta}))\right) \, . \tag{2.48}$$

  If the data are assumed to be Gaussian and *independently distributed*, this criterion is equivalent to a maximum likelihood estimation (based on the observation/information in $\hat{\mathbf{R}}$). However, the criterion may be used also for other distributions of the data, even if it is then no longer an ML estimate. In the SEM literature, this criterion is called ML; see, for example, Jöreskog (1970), Jöreskog and Goldberger (1975). The same approach, but applied for the somewhat simpler factor model, is also described in Jöreskog (1967).
  Some caution has to be exercised when using this criterion. During the numerical optimization, it is necessary to constrain the elements of $\boldsymbol{\vartheta}$ such that the matrix $\mathbf{R}(\boldsymbol{\vartheta})$ remains positive definite. If this constraint is not considered, the numerical search may easily diverge and not lead to an appropriate estimate. See Appendix 2.A.2 for more details.
- A second possibility is to consider the criterion

$$V_2(\boldsymbol{\vartheta}) = \operatorname{tr}\left(\left(\hat{\mathbf{R}} - \mathbf{R}(\boldsymbol{\vartheta})\right)\mathbf{W}_1\left(\hat{\mathbf{R}} - \mathbf{R}(\boldsymbol{\vartheta})\right)\mathbf{W}_2\right) , \tag{2.49}$$

where $\mathbf{W}_1$ and $\mathbf{W}_2$ are some positive definite weighting matrices. These matrices can be viewed as user choices.
When $\mathbf{W}_1 = \mathbf{W}_2 = \mathbf{I}$ the approach is known as unweighted least squares (ULS), and the choice $\mathbf{W}_1 = \mathbf{W}_2 = \hat{\mathbf{R}}^{-1}$ is known as the generalized least squares (GLS), Jöreskog (1972), Jöreskog et al. (1981), Bollen (1989). In the CFA literature the choice $\mathbf{W}_1 = \mathbf{W}_2 = \mathbf{R}^{-1}(\boldsymbol{\vartheta})$ is referred to as an ML-type estimation procedure.

- A third alternative would be to vectorize the difference between the covariance matrices in (2.44). Set

$$\tilde{\mathbf{r}}(\boldsymbol{\vartheta}) = \operatorname{vec}\big(\hat{\mathbf{R}} - \mathbf{R}(\boldsymbol{\vartheta})\big) , \tag{2.50}$$

and then form the criterion

$$V_3(\boldsymbol{\vartheta}) = \tilde{\mathbf{r}}^T(\boldsymbol{\vartheta})\mathbf{W}\tilde{\mathbf{r}}(\boldsymbol{\vartheta}) , \tag{2.51}$$

where $\mathbf{W}$ is a positive definite weighting matrix to be chosen by the user. This criterion is known as weighted least squares (WLS), Bollen (1989), even though this term normally is reserved for the special case (2.56) below.
This approach can be modified, taking various constraints into account:

  - One such natural possibility is to note that the matrix $\hat{\mathbf{R}} - \mathbf{R}$ is symmetric, and thus let the vec-operation apply only to the lower triangular part of the matrix.
  - A still further constraint can be imposed for cases where the matrix $\hat{\mathbf{R}} - \mathbf{R}$ is known to be block Toeplitz. Then it is fully characterized by still fewer elements.

To take such constraints into account one can write

$$\tilde{\mathbf{r}}(\boldsymbol{\vartheta}) = \mathbf{F}\mathbf{x}(\boldsymbol{\vartheta}) , \tag{2.52}$$

where $\mathbf{F}$ is a tall matrix of full column rank and $\mathbf{x}(\boldsymbol{\vartheta})$ is a random vector with positive definite covariance matrix.

Example 2.2 For illustration, assume that $\mathbf{R}$ is a 2×2 matrix, and let the constraint be that $\hat{\mathbf{R}}$ (and $\mathbf{R}(\boldsymbol{\vartheta})$) is symmetric. Then (2.52) can be written as

$$\tilde{\mathbf{r}}(\boldsymbol{\vartheta}) = \begin{pmatrix} \tilde{r}_{11} \\ \tilde{r}_{21} \\ \tilde{r}_{12} \\ \tilde{r}_{22} \end{pmatrix} = \underbrace{\begin{pmatrix} 1\,0\,0 \\ 0\,1\,0 \\ 0\,1\,0 \\ 0\,0\,1 \end{pmatrix}}_{\mathbf{F}} \underbrace{\begin{pmatrix} x_1 \\ x_2 \\ x_3 \end{pmatrix}}_{\mathbf{x}} . \tag{2.53}$$

If it also holds that $\hat{\mathbf{R}}$ is Toeplitz, then $r_{22} = r_{11}$, and

Table 2.1 User choices in the SEM criteria

Criterion	User choices
V_1	none
V_2	$\mathbf{W}_1,\ \mathbf{W}_2$
V_3	$\mathbf{W}$
V_{3c}	$\mathbf{W}_c$

$$\tilde{\mathbf{r}}(\boldsymbol{\vartheta}) = \begin{pmatrix} \tilde{r}_{11} \\ \tilde{r}_{21} \\ \tilde{r}_{12} \\ \tilde{r}_{22} \end{pmatrix} = \underbrace{\begin{pmatrix} 1 & 0 \\ 0 & 1 \\ 0 & 1 \\ 1 & 0 \end{pmatrix}}_{\mathbf{F}} \underbrace{\begin{pmatrix} x_1 \\ x_2 \end{pmatrix}}_{\mathbf{x}} . \tag{2.54}$$

■

It is clear that in more general cases, it is possible to express constraints in the form (2.52) where $\mathbf{F}$ is a known matrix (typically with elements being either 0 or 1), and $\mathbf{x}(\boldsymbol{\vartheta})$ is an unknown random vector.

Imposing the constraint (2.52) to the criterion (2.51) leads to the modified criterion

$$V_{3c}(\boldsymbol{\vartheta}) = \mathbf{x}^T(\boldsymbol{\vartheta})\mathbf{F}^T\mathbf{W}\mathbf{F}\mathbf{x}(\boldsymbol{\vartheta}) , \tag{2.55}$$

or

$$V_{3c}(\boldsymbol{\vartheta}) \stackrel{\Delta}{=} \mathbf{x}^T(\boldsymbol{\vartheta})\mathbf{W}_c\mathbf{x}(\boldsymbol{\vartheta}) . \tag{2.56}$$

What are the user choices for the different criteria? In all cases, it is necessary to specify the model parameterization, which shows up in how the matrix $\mathbf{R}(\boldsymbol{\vartheta})$ depends on the parameter vector $\boldsymbol{\vartheta}$. The remaining user choices are as given in Table 2.1.

To emphasize the influence of the user choices, one may write the various criteria as

$$V_1(\boldsymbol{\vartheta}),\ V_2(\boldsymbol{\vartheta}; \mathbf{W}_1, \mathbf{W}_2),\ V_3(\boldsymbol{\vartheta}; \mathbf{W}),\ V_{3c}(\boldsymbol{\vartheta}; \mathbf{W}_c) .$$

By inspection and from the above reasoning, the following relations can be established (recall the definitions (2.49) and (2.51)):

-
$$V_2(\boldsymbol{\vartheta}; \mathbf{W}_1, \mathbf{W}_2) = V_2(\boldsymbol{\vartheta}; \mathbf{W}_2, \mathbf{W}_1) . \tag{2.57}$$

-
$$V_2(\boldsymbol{\vartheta}; \mathbf{W}_1, \mathbf{W}_2) = V_3(\boldsymbol{\vartheta}; \mathbf{W}_1 \otimes \mathbf{W}_2) , \tag{2.58}$$

where $\otimes$ denotes Kronecker product; see Sect. A.1.2.

- Assuming that the constraint (2.52) applies, it also holds

$$V_3(\boldsymbol{\vartheta}, \mathbf{W}) = V_{3c}(\boldsymbol{\vartheta}, \mathbf{F}^T\mathbf{W}\mathbf{F}) . \tag{2.59}$$

The following general results are reported in Bartholomew et al. (2011):

- The estimates are consistent for all the criteria listed above.
- ULS and GLS can be seen as special cases of WLS, obtained by particular choices of the weighting matrix $\mathbf{W}$.
- Only WLS and ML are asymptotically efficient (under weak assumptions). For the WLS, the weighting matrix has then to be chosen as

$$\mathbf{W} = \left[\text{cov}(\tilde{\mathbf{r}})\right]^{-1} . \tag{2.60}$$

When the constraints are imposed, the relation will be modified to

$$\mathbf{W}_c = [\text{cov}(\mathbf{x})]^{-1} . \tag{2.61}$$

These weak assumptions include that the observations are mutually independent.

Based on the earlier considerations, one finds that the criteria V_2 and V_3 are in fact the same, provided that the weightings are related as

$$\mathbf{W} = \mathbf{W}_2 \otimes \mathbf{W}_1 . \tag{2.62}$$

It is trivial to see that

- If $\mathbf{W}_1, \mathbf{W}_2$ are given, then $\mathbf{W}$ is uniquely found from (2.62).
- If $\mathbf{W}$ is given, then there is in general no pair of matrices $\mathbf{W}_1, \mathbf{W}_2$ satisfying (2.62). Note that, in the right-hand side of (2.62), the matrix blocks are all proportional to $\mathbf{W}_1$, while this is not the case for the left-hand side in general.
- If $\mathbf{W}$ is given, then the criterion $V_{3c}(\boldsymbol{\vartheta})$ in (2.56) is easily found using

$$\mathbf{W}_{3c} = \mathbf{F}^T\mathbf{W}\mathbf{F} . \tag{2.63}$$

Conversely, however, for a given $\mathbf{W}_{3c}$ there is no unique matrix $\mathbf{W}$ fulfilling (2.63), but many.

An alternative estimation method, popular in array signal processing applications, is weighted subspace fitting. It cannot be formulated as in (2.47). The idea is as follows. Let the matrix $\mathbf{R}_y$, as given by (2.32), have the eigendecomposition

$$\mathbf{R}_y = \sum_{i=1}^{m} \lambda_i \mathbf{e}_i \mathbf{e}_i^* = \mathbf{E}_s \boldsymbol{\Lambda}_s \mathbf{E}_s^* + \mathbf{E}_n \boldsymbol{\Lambda}_n \mathbf{E}_n^* , \tag{2.64}$$

where the eigenvalues are

$$\lambda_1 > \lambda_2 > \cdots > \lambda_{d+1} = \cdots = \lambda_n = \sigma^2 .$$

Further,

$$\boldsymbol{\Lambda}_{\mathrm{s}} = \mathrm{diag}\left(\lambda_1 \cdots \lambda_d\right), \quad \boldsymbol{\Lambda}_{\mathrm{n}} = \sigma^2 \mathbf{I} . \tag{2.65}$$

Comparing (2.32) and (2.64) one can note that the columns of $\mathbf{A}(\boldsymbol{\vartheta}_0)$ and $\mathbf{E}_{\mathrm{s}}$ must span the same d-dimensional subspace. A natural criterion is then to fit an estimate of $\boldsymbol{\vartheta}$ and a transformation $\mathbf{T}$ as

$$\hat{\boldsymbol{\vartheta}}, \hat{\mathbf{T}} = \arg\min_{\boldsymbol{\vartheta}, \mathbf{T}} \| \hat{\mathbf{E}}_{\mathrm{s}} \mathbf{W}^{1/2} - \mathbf{A}(\boldsymbol{\vartheta})\mathbf{T} \|_{\mathrm{F}}^2 , \tag{2.66}$$

where $\| \quad \|_{\mathrm{F}}$ denotes the Frobenius norm, and $\mathbf{W}^{1/2}$ is a symmetric square root of a user chosen weighting matrix $\mathbf{W}$. Further, $\hat{\mathbf{E}}_{\mathrm{s}}$ is formed as in (2.64), but using instead the estimated covariance matrix $\hat{\mathbf{R}}_{\mathrm{y}}$.

The criterion in (2.66) is quadratic in $\mathbf{T}$. The minimization with respect to the nuisance variable $\mathbf{T}$ can therefore be carried out directly and substituted back to give a criterion in $\boldsymbol{\vartheta}$ only. This leads to the estimate

$$\hat{\boldsymbol{\vartheta}} = \arg\min_{\boldsymbol{\vartheta}} V(\boldsymbol{\vartheta}) , \tag{2.67}$$

$$V(\boldsymbol{\vartheta}) = \mathrm{tr}\left[\mathbf{P}_{\mathbf{A}}^{\perp}(\boldsymbol{\vartheta})\hat{\mathbf{E}}_{\mathrm{s}}\mathbf{W}\hat{\mathbf{E}}_{\mathrm{s}}^*\right] , \tag{2.68}$$

$$\mathbf{P}_{\mathbf{A}}^{\perp}(\boldsymbol{\vartheta}) = \mathbf{I} - \mathbf{A}(\boldsymbol{\vartheta})\left(\mathbf{A}^*(\boldsymbol{\vartheta})\mathbf{A}(\boldsymbol{\vartheta})\right)^{-1}\mathbf{A}^*(\boldsymbol{\vartheta}) . \tag{2.69}$$

Note that $\mathbf{P}_{\mathbf{A}}^{\perp}$ is the orthogonal projection on $\mathscr{R}(\mathbf{A})^{\perp} = \mathscr{N}(\mathbf{A}^*)$, that is, the null space of $\mathbf{A}^*$.

It has been shown in the literature that there is a best choice (a choice giving the smallest covariance matrix of $\hat{\boldsymbol{\theta}}$ in the asymptotic case) of the weighting matrix $\mathbf{W}$, leading also to asymptotically efficiency for Gaussian distributed waveforms $\boldsymbol{\eta}(t)$. This choice can be written as

$$\mathbf{W} = \left(\hat{\boldsymbol{\Lambda}}_{\mathrm{s}} - \hat{\sigma}^2\mathbf{I}\right)^2 \hat{\boldsymbol{\Lambda}}_{\mathrm{s}}^{-1} , \tag{2.70}$$

where $\hat{\boldsymbol{\Lambda}}_{\mathrm{s}}$ is obtained from the eigendecomposition of $\hat{\mathbf{R}}_y$, and $\hat{\sigma}^2$ is any consistent estimate of the noise variance σ^2.

2.3 The Frisch Scheme

An important formulation of the static EIV problem is the so-called Frisch problem, Frisch (1934). It has been analyzed (for the static case) in Guidorzi (1991) and Beghelli and Soverini (1992). A similar problem for the dynamic case will be described in Sect. 7.3. Assume that a vector $\varphi(t)$ is observed and that it has one deterministic part, $\varphi_0(t)$, and one part due to measurement noise, $\tilde{\varphi}(t)$. From the observed data,

$$\varphi(t) = \varphi_0(t) + \tilde{\varphi}(t), \qquad t = 1, \ldots, N , \tag{2.71}$$

one can estimate the covariance matrix $\mathbf{R} = \mathsf{E}\left\{\varphi(t)\varphi^T(t)\right\}$. Due to (2.71) this matrix be decomposed as

$$\mathbf{R} = \mathbf{R}_0 + \tilde{\mathbf{R}} . \tag{2.72}$$

It is assumed that the noise-free part satisfies one or more linear relations, so

$$\varphi_0^T(t)\mathbf{A} = \mathbf{0} , \tag{2.73}$$

where the number of columns of the matrix $\mathbf{A}$ is equal to the number of linearly independent relations. It is also assumed that the different components of the noise vector $\tilde{\varphi}(t)$ are independent, and hence $\tilde{\mathbf{R}}$ is known to be a diagonal matrix.

The problem under study is to assume $\mathbf{R}$ to be given, and to characterize the possible decompositions (2.72) such that (2.73) holds (where $\mathbf{A}$ is unknown), $\mathbf{R}_0$ is positive semidefinite and $\tilde{\mathbf{R}}$ is diagonal. The problem is treated with a geometrical analysis and expressed in the parameter space (represented by the matrix $\mathbf{A}$) and in the noise space (represented by the covariance matrix $\tilde{\mathbf{R}}$). The concept of corank plays an essential role. The maximal corank of $\mathbf{R}$ is

$$\max_{\tilde{\mathbf{R}}} \left(n - \operatorname{rank}(\mathbf{R} - \tilde{\mathbf{R}})\right) ,$$

where $n = \dim(\mathbf{R})$, and the previous constraints on $\mathbf{R}_0$ and $\tilde{\mathbf{R}}$ hold. The problem is much harder for $\operatorname{corank}(\mathbf{R}) \geq 2$ than for $\operatorname{corank}(\mathbf{R}) = 1$.

2.4 Extensions and Bibliographical Notes

General Aspects

There is a rich literature dealing with the errors-in-variables problem for the static case.

The papers Adcock (1877, 1878) are recognized to be the first ones to talk about measurement error models. Adcock's least squares method is now known as orthogonal regression (in statistics) and as total least squares (in engineering and numerical analysis; see Sect. 11.1) in its simplest form where the variance is the same in both variables.

Some classical works in the area include Frisch (1934), Koopmans (1937), Reiersøl (1950), Hannan (1963), Madansky (1959), and Moran (1971).

Extensive analysis of the general EIV problem is given in the papers Anderson (1984, 1989), Anderson et al. (1996). The topic is also well treated in books that cover many different aspects on identifiability and estimation aspects; see Cheng and Van Ness (1999), Fuller (1987). For overviews, see also Gillard (2010) and Chen et al. (2015).

Various aspects on the EIV problem for static systems are considered in Klepper and Leamer (1984) and for the multivariable case in Croux et al. (2010). Robust algorithms for the static EIV case are considered in Watson (2007). See also Nyquist (1987). Further general papers on the EIV problem for static systems include Song et al. (2015), Li et al. (2014), Schennach (2014).

A heteroscedastic model, where the total error covariance structure is known up to a scalar factor (but the variance of the measurement noise is allowed to vary with time), is treated in Kukush and Van Huffel (2004).

It has been noted early, see Wald (1940), that TLS is not invariant to scaling. Further, Wald (1940) provides a sufficient condition for consistency. This condition is based on 'grouping' the observations into two groups that are 'sufficiently different' in a precise mathematical sense. Unfortunately, the condition is restrictive in practice. It is, for example, not fulfilled for a stationary process.

Use of higher-order statistics for EIV problems may first have been proposed in Reiersøl (1950).

Confidence regions for the parameters in a partially linear EIV model are considered in Li and Xue (2008). The static case, with a quite general formulation, is treated in Stoica et al. (1996). The thesis by Musekiwa (2005) treats the static case with line fitting.

The paper by Gleser (1981) provides a statistical analysis, dealing with consistency. However, the assumptions are quite restrictive and do not apply for the case of identifying dynamic systems.

The result in (2.18) that the estimate of the noise level for the line-fitting problem is off by 50 percent is analyzed in Lindley (1947).

Instrumental Variable and Nonlinear EIV Models

An early use of instrumental variable estimation, although that name was not used, appears in Wald (1940). Devereux (2007) considers static EIV models and shows that a standard estimator is exactly equivalent to the jackknife instrumental variable estimator. The maximum likelihood method for treating a nonlinear static EIV model is treated in Amemiya and Fuller (1988). See also Fedorov (1974).

Identification of nonlinear EIV models is also treated in Fan and Truong (1993), Schennach (2007), De Nadai and Lewbei (2016), D'Haultfoeuille and Fevier (2015), Chen and Gao (2015), Zinde-Walsh (2014), Schennach (2013), Schennach and Hu (2013).

The Frisch Scheme

The Frisch scheme has its roots in a classical algebraic estimation problem; see Frisch (1934). There are several papers dealing with the Frisch scheme for static problems, see Beghelli and Soverini (1992) and Guidorzi (1991) which provide geometric analysis and interpretations. Other treatments of the static case include Soverini and Beghelli (2001) and Guidorzi (1995).

An interesting paper giving a deep analysis of the Frisch scheme and its relation to EIV problems and to factor analysis is Ning et al. (2015). A starting point in the treatment is that $\mathbf{R}_0$ in (2.72) is made as rank-deficient as possible; that is, $\mathbf{A}$ in (2.73) has a maximum number of linearly independent columns. This is called the Frisch-Kalman dictum; see also Kalman (1982b). This rank minimization problem has been approximated by several approaches, often including convex relaxations.

Another way of approaching the problem, not only for the static case, is to determine the maximal corank of a covariance matrix of noisy data in the context of the Frisch scheme. The locus of solutions (in the parameter space) simply 'explodes' when leaving the condition max corank $= 1$. The paper Guidorzi (1995) works in the noise space instead than in the parameter space, and this proved to be fruitful because no discontinuities appear in the transition from corank 1 to higher values. It is proved in the paper that points corresponding to corank k are accumulation points for those corresponding to corank $(k - 1)$. The paper also gives geometrical/analytical conditions for the determination of the maximal number of linear relations under the Frisch scheme. The paper Woodgate (1995) gives an upper bound to this number. See also Guidorzi and Pierantoni (1995), Castaldi et al. (1996), and De Moor and Vandewalle (1986).

Factor Models

Factor models are treated in depth in Skrondal and Rabe-Hesketh (2004) and Child (2006). See also Shapiro (1985).

Generalized factor analysis (GFA) models concern the case when dim $(\mathbf{y})$ in (2.27) grows to infinity. There are applications in econometrics, where there may be a huge number of 'measurements' of economic developments (possibly also spatially distributed). In the GFA model the matrices $\mathbf{A}(\eta)$ and $\mathbf{R}_\varepsilon(\eta)$ in (2.30) get in the limit an infinite number of rows. For the decomposition (2.30) to be tractable, one has to require that $\mathbf{R}_\varepsilon$ corresponds to a bounded operator; see Bottegal and Picci (2015).

The paper Lee and Jennrich (1979) relates to factor analysis and structural equation modeling and compares different criteria, as in the examples of Sect. 2.2.2, and different numerical algorithms.

Generalized Dynamic Factor Models

A dynamic factor model is an extension of the factor model (2.27). In this case the vector $\mathbf{f}$ of factors consists of delayed (i.e., lagged) values of a variable:

$$\eta = \begin{pmatrix} \eta(t) \\ \vdots \\ \eta(t-q) \end{pmatrix}. \tag{2.74}$$

See Forni and Lippi (2001). Typically one lets the dimension of $\mathbf{y}$ tend to infinity. Another formulations is based on writing the model as

$$\mathbf{y}(t) = \mathbf{A}\boldsymbol{\eta}(t) + \boldsymbol{\varepsilon}(t), \quad t = 1, 2, \ldots \tag{2.75}$$

and letting $\boldsymbol{\eta}(t)$ be the output of a finite-order linear model of a stationary stochastic process.

(Generalized) dynamic factor models are treated in Gewecke (1977), Picci and Pinzoni (1986a), Picci (1989), Forni et al. (2000), Forni and Lippi (2001), Forni et al. (2004), Deistler and Zinner (2007), Anderson and Deistler (2008a), Deistler et al. (2010b), Bottegal and Picci (2011), Barhoumi et al. (2014), Bottegal and Picci (2015).

Generalized factor analysis applied to dynamic EIV models is treated in Picci and Pinzoni (1986b).

Other works study the generalized dynamic factor models and show their relation to autoregressive models with singular spectra; see Deistler et al. (2010a), Ninness and Wills (2010), Andersson et al. (2012), Andersson and Deistler (2009), Anderson et al. (2009), Anderson and Deistler (2008b). In particular it is shown that when the number of outputs exceeds the number of inputs, the spectrum of the noise-free data becomes singular. In such cases it holds generically (i.e., for almost all parameter values) that the transfer function from inputs to outputs, which is represented by a tall transfer function matrix, becomes zero-free (following the usual definition of zeros for a linear multivariable system). Further, the corresponding spectral factor of the noise-free spectrum can be represented by a pure autoregressive model. This type of models is sometimes used in econometrics. In the above papers there is also some theory for the asymptotic case when the number of outputs (i.e., observations) tend to infinity. The papers Andersson et al. (2017) and Koelbl et al. (2016) deal also with multivariate AR processes with singular spectra. In addition, the case of mixed frequency data is considered, where different output variables can be measured at a few different rates.

Related work is carried out in Zorzi and Ferrante (2012), Zorzi and Chiuso (2017) and treats estimation of structured covariance matrices.

Sensor Array Processing and Weighted Subspace Fitting

For background and more material, see, e.g., Haykin (1985), Haykin (1991b), Viberg and Ottersten (1991), Viberg et al. (1991), Ottersten et al. (1992), Krim and Viberg (1996), Jansson et al. (1998), Bengtsson and Ottersten (2001).

Structural Equation Modeling (SEM)

For general texts on structural equation modeling, see Bollen (1989), Bartholomew et al. (2011), Bagozzi and Yi (2012), Westland (2015).

Classical papers on SEM include Jöreskog (1967), Jöreskog (1970), Jöreskog (1972), Jöreskog and Goldberger (1975), Jöreskog et al. (1981).

How to use SEM for identifying dynamic systems is discussed in Kreiberg et al. (2013), Kreiberg et al. (2016). See also Sect. 8.5 for this topic.

Nonlinear Static Models

This book treats linear models only. The case of nonlinear models adds further difficulties. See Example 2.12 and Vajk and Hetthéssy (2003).

A simple case extending the static linear models to cover nonlinear relations is presented in Kukush et al. (2004). Some nonlinear static EIV problems related to curve fitting are dealt with in Delaigle and Hall (2008).

Nonlinear EIV identification for static models is treated in Wolter and Fuller (1982), Hsiao (1989), Hsiao et al. (1997), Schennach (2004), Matei and Meer (2006), Schennach (2007), Hu (2008), Hu and Schennach (2008), Jukić and Marković (2010).

Static polynomial type of EIV models are considered in Zavala et al. (2007). The papers Song (2007), Song (2009) concern testing EIV regression models with validation data. The nonlinear static case is considered.

Maximum Likelihood Methods

The paper Wiesel et al. (2008) deals with the static case, where there is some Gaussian uncertainties in the model matrix. The relationships between ML estimates and regularized LS estimators are explored.

The ML approach to factor analysis, for the case of white 'measurement noise' of constant variance, is treated in Stoica and Jansson (2009). Some weaknesses with classical papers treating this problem are pointed out.

The paper Solari (1969) demonstrates that there is indeed no ML solution to the static problem in the general case! This case can be compared to Stoica and Li (2005) who demonstrate that the ML estimate does not exist for a certain blind multichannel identification problem. It corresponds in EIV terms for dynamic models to estimating the time series of the noise-free input.

Latent Variables

Latent variables are sometimes used as a concept in statistics, as opposed to observable variables. When latent variables correspond to physical concepts, they may also be called hidden variables. The term 'latent variables' is more commonly used in social sciences, though. In economics, variables, such as quality of life, business confidence, happiness, and conservatism, can be viewed as examples of latent variables.

A certain class of latent or hidden variables constitutes hidden Markov models; see Anderson (1999), Lawrence (1989), Cappé et al. (2005), and Elliott et al. (1995) for treatment of such models.

Appendix

2.A Further Details

2.A.1 Further Results for Line Fitting

This appendix presents a closer analysis of the line-fitting example introduced in this chapter. Some various situations are considered as examples.

Example 2.3 Consider the case of a scalar x and a scalar y, and assume that

$$y_0 = a_0 x_0 \tag{2.76}$$

holds.

Now examine the identifiability properties from first- and second-order moments of the observed data. Let $\tilde{x}_i$ and $\tilde{y}_i$ have zero means. Let $\{x_0\}_i$ all have mean m and variance σ.

The first-order moments of the observations will be

$$m_y = \mathsf{E}\{y_i\} = a_0 m \ , \tag{2.77}$$

$$m_x = \mathsf{E}\{x_i\} = m \ , \tag{2.78}$$

while the second-order moments are

$$r_y = \text{var}(y_i) = a_0^2 \sigma + \lambda_y \ , \tag{2.79}$$

$$r_x = \text{var}(x_i) = \sigma + \lambda_x \ , \tag{2.80}$$

$$r_{yx} = \text{cov}(y_i, x_i) = a_0 \sigma \ . \tag{2.81}$$

The left-hand sides of equations (2.77)–(2.81) can be estimated from the data in a straightforward manner. Therefore, to analyze identifiability, regard the left-hand sides of (2.77)–(2.81) as known, and these equations as the information available to determine the sought parameter a_0. Apparently there are 5 equations for determining the 5 unknowns $a_0,\ m,\ \sigma,\ \lambda_y$, and λ_u. A somewhat closer look reveals that if $m \neq 0,\ a_0 \neq 0$, these equations have a unique solution, and hence, the system is then identifiable. On the other hand, if $m = 0$, then (2.77) no longer carries any information about a_0, and the system becomes unidentifiable.

How would the estimates be determined in the case of $m \neq 0$? The slope a is estimated as

$$\hat{a} = \frac{\frac{1}{N}\sum_i y_i}{\frac{1}{N}\sum_i x_i} \ . \tag{2.82}$$

Based on the data model one then finds that

$$\hat{a} - a_0 = \frac{\frac{1}{N}\sum_i (y_i - a_0 x_i)}{\frac{1}{N}\sum_i x_i} = \frac{\frac{1}{N}\sum_i (\tilde{y}_i - a_0 \tilde{x}_i)}{\frac{1}{N}\sum_i (x_{0i} + \tilde{x}_i)} . \tag{2.83}$$

If the number of points, N, becomes large, then noise terms such as $(1/N)\sum_i \tilde{y}_i$ and $(1/N)\sum_i \tilde{x}_i$ will converge to zero. Assuming that $(1/N)\sum_i x_{0i}$ has a nonzero limit will therefore lead to a consistent estimate of the parameter a. ■

If the data are not Gaussian, use of higher-order moments can be used to improve the identifiability properties. This is illustrated with an example.

Example 2.4 Reconsider the setup of Example 2.3 and assume that $m = 0$. Then the system is not identifiable from (first- and) second-order moments. Assume that $\tilde{x}_i, \tilde{y}_i$ are Gaussian, with zero mean, so

$$\mathsf{E}\left\{\tilde{x}_i^2\right\} = \lambda_x, \qquad \mathsf{E}\left\{\tilde{x}_i^4\right\} = 3\lambda_x^2, \qquad \mathsf{E}\left\{\tilde{y}_i^2\right\} = \lambda_y, \qquad \mathsf{E}\left\{\tilde{y}_i^4\right\} = 3\lambda_y^2 . \tag{2.84}$$

Assume further that x_{oi} has a symmetric distribution around $x = 0$ and that it is not Gaussian. Then

$$\mathsf{E}\left\{x_{0i}\right\} = 0, \qquad \mathsf{E}\left\{x_{0i}^2\right\} = \sigma, \qquad \mathsf{E}\left\{x_{0i}^4\right\} = 3\sigma^2 + \gamma , \tag{2.85}$$

where $\gamma \neq 0$. Now express the second- and fourth-order moments of the data:

$$r_y = \mathsf{E}\left\{y_i^2\right\} = a_0^2\sigma + \lambda_y , \tag{2.86}$$

$$r_x = \mathsf{E}\left\{x_i^2\right\} = \sigma + \lambda_x , \tag{2.87}$$

$$r_{yx} = \mathsf{E}\left\{y_i x_i\right\} = a_0\sigma , \tag{2.88}$$

$$\begin{aligned} v_y = \mathsf{E}\left\{y_i^4\right\} &= \mathsf{E}\left\{\left[a_0 x_{0i} + \tilde{y}_i\right]^4\right\} \\ &= a_0^4(3\sigma^2 + \gamma) + 6a_0^2\sigma\lambda_y + 3\lambda_y^2 \\ &= a_0^4\gamma + 3(a_0^2\sigma + \lambda_y)^2 , \end{aligned} \tag{2.89}$$

$$\begin{aligned} v_x = \mathsf{E}\left\{x_i^4\right\} &= \mathsf{E}\left\{\left[x_{0i} + \tilde{x}_i\right]^4\right\} \\ &= 3\sigma^2 + \gamma + 6\sigma\lambda_x + 3\lambda_x^2 \\ &= \gamma + 3(\sigma + \lambda_x)^2 , \end{aligned} \tag{2.90}$$

$$\begin{aligned} v_{yx} = \mathsf{E}\left\{y_i^2 x_i^2\right\} &= \mathsf{E}\left\{\left[a_0 x_{0i} + \tilde{y}_i\right]^2 \left[x_{0i} + \tilde{x}_i\right]^2\right\} \\ &= a_0^2(3\sigma^2 + \gamma) + \lambda_y\sigma + \lambda_x a_0^2\sigma + \lambda_x\lambda_y . \end{aligned} \tag{2.91}$$

The left-hand sides of (2.86)–(2.91) can be determined from data with arbitrarily good accuracy when $N \to \infty$.

Keep a_0 as an unknown for the moment, and use (2.88) to solve for σ, (2.86) to solve for λ_y, (2.87) to solve for λ_x, and (2.90) to solve for γ. This gives

$$\begin{aligned}
\sigma &= r_{yx}/a_0 \,, \\
\lambda_y &= r_y - a_0 r_{yx} \,, \\
\lambda_x &= r_x - r_{yx}/a_0 \,, \\
\gamma &= v_x - 3r_x^2 \,.
\end{aligned} \tag{2.92}$$

Now use (2.92) into (2.89) and (2.91) to determine the remaining unknown, namely a_0, which after some simplification gives

$$v_y = a_0^4(v_x - 3r_x^2) + 3r_y^2 \,, \tag{2.93}$$

$$v_{yx} = a_0^2(v_x - 3r_x^2) + (2r_{yx}^2 + r_x r_y) \,. \tag{2.94}$$

As (2.93) and (2.94) are functions of even powers of a_0, it is found that a_0^2 can be determined uniquely, but not the sign of a_0. ■

Example 2.5 For the problem treated in Example 2.3, it holds

$$|\hat{a}_{\text{LS}}| \leq |\hat{a}_{\text{TLS}}| \leq |\hat{a}_{\text{DLS}}| \,. \tag{2.95}$$

See also the numerical example of Sect. 2.1. Introduce the notations

$$\hat{r}_x = \frac{1}{N}\sum_{i=1}^{N} x_i^2, \;\; \hat{r}_y = \frac{1}{N}\sum_{i=1}^{N} y_i^2, \;\; \hat{r}_{yx} = \frac{1}{N}\sum_{i=1}^{N} y_i x_i \,. \tag{2.96}$$

Now prove (2.95) for the case $\hat{r}_{yx} > 0$. Then all estimates in (2.96) take positive values. It holds

$$\hat{a}_{\text{LS}} = \frac{\hat{r}_{yx}}{\hat{r}_x}, \;\; \hat{a}_{\text{DLS}} = \frac{\hat{r}_y}{\hat{r}_{yx}} \,. \tag{2.97}$$

From the Cauchy–Schwarz inequality it holds

$$\hat{r}_{yx}^2 \leq \hat{r}_x \hat{r}_y \,, \tag{2.98}$$

and the relation

$$\hat{a}_{\text{LS}} \leq \hat{a}_{\text{DLS}} \tag{2.99}$$

follows directly.

The estimate $\hat{a}_{\text{TLS}}$ follows from (2.15). To avoid dealing with messy expressions including square roots, it is more convenient to utilize (2.14). One needs to prove that the criterion

$$V(a) = \frac{\hat{r}_y + a^2\hat{r}_x - 2a\hat{r}_{yx}}{1 + a^2} \tag{2.100}$$

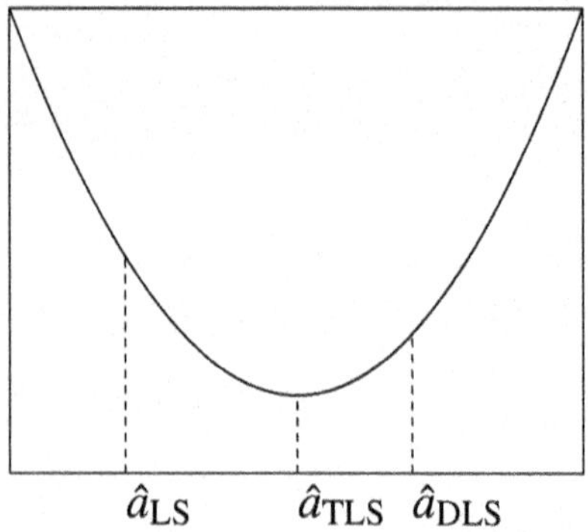

Fig. 2.5 The criterion $V(a)$ versus a

which has a minimum for $a = \hat{a}_{TLS}$, behaves as in Fig. 2.5.

To verify (2.95), it then is enough to show that

$$V'(\hat{a}_{LS}) \leq 0, \quad V'(\hat{a}_{DLS}) \geq 0 . \tag{2.101}$$

However, straightforward differentiation gives

$$\begin{aligned} V'(a) &= \frac{(2a\hat{r}_x - 2\hat{r}_{yx})(1+a^2) - 2a(\hat{r}_y + a^2\hat{r}_x - 2a\hat{r}_{yx})}{(1+a^2)^2} \\ &= \frac{2}{(1+a^2)^2}\left(a^2\hat{r}_{yx} + a(\hat{r}_x - \hat{r}_y) - \hat{r}_{yx}\right) , \end{aligned} \tag{2.102}$$

$$\begin{aligned} V'(\hat{a}_{LS}) &= \frac{2}{(1+\hat{a}_{LS}^2)^2}\left[\frac{\hat{r}_{yx}^3}{\hat{r}_x^2} + \frac{\hat{r}_{yx}}{\hat{r}_x}(\hat{r}_x - \hat{r}_y) - \hat{r}_{yx}\right] \\ &= \frac{2}{(1+\hat{a}_{LS}^2)^2}\frac{\hat{r}_{yx}}{\hat{r}_x^2}\left(\hat{r}_{yx}^2 - \hat{r}_x\hat{r}_y\right) \leq 0 , \end{aligned} \tag{2.103}$$

$$\begin{aligned} V'(\hat{a}_{DLS}) &= \frac{2}{(1+\hat{a}_{DLS}^2)^2}\left[\frac{\hat{r}_y^2}{\hat{r}_{yx}} + \frac{\hat{r}_y}{\hat{r}_{yx}}(\hat{r}_x - \hat{r}_y) - \hat{r}_{yx}\right] \\ &= \frac{2}{(1+\hat{a}_{DLS}^2)^2}\frac{\hat{r}_x\hat{r}_y - \hat{r}_{yx}^2}{\hat{r}_{yx}} \geq 0 , \end{aligned} \tag{2.104}$$

which completes the proof of (2.101). See also Sect. 2.1.1. ■

Next evaluate the asymptotic expressions for the estimates $\hat{a}_{LS}$, $\hat{a}_{DLS}$, and $\hat{a}_{TLS}$ when N, the number of data points, tends to infinity.

Example 2.6 Assume

$$\mathsf{E}\left\{x_{0i}^2\right\} = r . \tag{2.105}$$

Then in the limit as $N \to \infty$

$$\begin{aligned} r_x &= r + \lambda_x \ , \\ r_y &= a_0^2 r + \lambda_y \ , \\ r_{yx} &= a_0 r \ . \end{aligned} \tag{2.106}$$

Hence,

$$\hat{a}_{\text{LS}} = \frac{a_0 r}{r + \lambda_x} = a_0 + \frac{-a_0 \lambda_x}{r + \lambda_x} \ , \tag{2.107}$$

$$\hat{a}_{\text{DLS}} = \frac{a_0^2 r + \lambda_y}{a_0 r} = a_0 + \frac{\lambda_y}{a_0 r} \ . \tag{2.108}$$

Apparently, both $\hat{a}_{\text{LS}}$ and $\hat{a}_{\text{DLS}}$ differ from a_0.

The estimate $\hat{a}_{\text{TLS}}$ is the solution to, cf. (2.15),

$$a^2 a_0 r + a \left(r + \lambda_x - a_0^2 r - \lambda_y \right) - a_0 r = 0 \ ,$$

which can be rearranged as

$$(a - a_0)(1 + a a_0) r + a \left(\lambda_x - \lambda_y \right) = 0 \ . \tag{2.109}$$

In particular one finds that $\hat{a}_{\text{TLS}} = a_0$ *only if* $\lambda_x = \lambda_y$.

In summary, all the three estimates under consideration are biased and not consistent. ■

It is important to note that the line-fitting examples above, even if very simple, still are a bit special. A specific property is that the line is constrained to go through the origin. Another way to express this constraint is to add the origin as a further given data point with no measurement error. Now generalize to an arbitrary straight line.

Example 2.7 Consider the same situation as in Example 2.3, but allow the line to pass outside the origin; that is, the model is changed to

$$y = ax + b \ , \tag{2.110}$$

and the data is assumed to fulfill

$$\begin{aligned} y_i &= y_{0i} + \tilde{y}_i \ , \\ x_i &= x_{0i} + \tilde{x}_i \ , \qquad i = 1, \ldots, N \ . \\ y_{0i} &= a_0 x_{0i} + b_0 \ , \end{aligned} \tag{2.111}$$

To examine the identifiability properties from first- and second-order moments, let x_{0i} have mean m and variance σ. Let the noise terms $\tilde{y}_i$ and $\tilde{x}_i$ be independent of x_{0j} (for all i and j) and have zero means and variances λ_y and λ_x, respectively. Consider the equations

$$m_y = \mathsf{E}\{y_i\} = b_0 + a_0 m \ , \tag{2.112}$$
$$m_x = \mathsf{E}\{x_i\} = m \ , \tag{2.113}$$

while the second-order moments are

$$r_y = \text{var}(y_i) = a_0^2\sigma + \lambda_y \ , \tag{2.114}$$
$$r_x = \text{var}(x_i) = \sigma + \lambda_x \ , \tag{2.115}$$
$$r_{yx} = \text{cov}(y_i, x_i) = a_0\sigma \ . \tag{2.116}$$

Now there are 6 unknowns (a_0, b_0, m, σ, λ_x, and λ_y) and still only 5 equations. *Hence identifiability is lost.*

To express an attempt for estimation, consider the least squares estimate of $\boldsymbol{\theta} = \begin{pmatrix} a_0 & b_0 \end{pmatrix}^T$:

$$\hat{\boldsymbol{\theta}}_{\text{LS}} = \arg\min_{\theta} \frac{1}{N}\sum_{i=1}^{N}(y_i - ax_i - b)^2 \ , \tag{2.117}$$

which leads to the normal equations

$$\frac{1}{N}\begin{pmatrix} \sum_i x_i^2 & \sum_i x_i \\ \sum_i x_i & \sum_i 1 \end{pmatrix}\begin{pmatrix} a \\ b \end{pmatrix} = \frac{1}{N}\begin{pmatrix} \sum_i x_i y_i \\ \sum_i y_i \end{pmatrix} . \tag{2.118}$$

In the limit when $N \to \infty$, the solution to (2.118) becomes

$$\begin{aligned} \begin{pmatrix} \hat{a} \\ \hat{b} \end{pmatrix} &= \begin{pmatrix} m^2 + \sigma + \lambda_x & m \\ m & 1 \end{pmatrix}^{-1} \begin{pmatrix} a_0(m^2 + \sigma) + b_0 m \\ a_0 m + b_0 \end{pmatrix} \\ &= \begin{pmatrix} a_0 \\ b_0 \end{pmatrix} + \frac{a_0\lambda_x}{\sigma + \lambda_x}\begin{pmatrix} -1 \\ m \end{pmatrix} . \end{aligned} \tag{2.119}$$

It is apparent that the estimate is biased, also in the asymptotic case. This is an illustration of what is already known: *the system is not identifiable.* ■

Can the situation improve in the multivariable case? Unfortunately, the answer is negative, as shown in the following example.

Example 2.8 Consider the general affine case, where $\mathbf{x}$ is a vector of dimension n_x and $\mathbf{y}$ is a vector of dimension n_y. Postulate the model

$$\begin{aligned} \mathbf{y}_i &= \mathbf{y}_{0i} + \tilde{\mathbf{y}}_i \ , \\ \mathbf{x}_i &= \mathbf{x}_{0i} + \tilde{\mathbf{x}}_i \ , \qquad i = 1, \ldots, N \ , \\ \mathbf{y}_{0i} &= \mathbf{A}_0\mathbf{x}_{0i} + \mathbf{b}_0 \ , \end{aligned} \tag{2.120}$$

where $\mathbf{b}_0$ is an n_y-dimensional vector, and $\mathbf{A}_0$ is an $n_y \times n_x$-dimensional matrix. Let $\mathbf{x}_{0i}$ have mean $\mathbf{m}$ and a positive definite covariance matrix $\mathbf{R}_x$, and let $\tilde{\mathbf{x}}_i$ have zero mean and covariance matrix $\boldsymbol{\Lambda}_\mathbf{x} > 0$. Similarly, let $\tilde{\mathbf{y}}_i$ have zero mean and covariance matrix $\boldsymbol{\Lambda}_\mathbf{y} > 0$. For this case the first-order moments become, comparing (2.112) and (2.113),

$$\mathsf{E}\{\mathbf{y}_i\} = \mathbf{b}_0 + \mathbf{A}_0\mathbf{m} \, , \tag{2.121}$$

$$\mathsf{E}\{\mathbf{x}_i\} = \mathbf{m} \, , \tag{2.122}$$

while the second-order moments become

$$\text{cov}(\mathbf{y}_i) = \mathbf{A}_0\mathbf{R}_\mathbf{x}\mathbf{A}_0^T + \boldsymbol{\Lambda}_\mathbf{y} \, , \tag{2.123}$$

$$\text{cov}(\mathbf{x}_i) = \mathbf{R}_\mathbf{x} + \boldsymbol{\Lambda}_\mathbf{x} \, , \tag{2.124}$$

$$\text{cov}(\mathbf{y}_i, \mathbf{x}_i) = \mathbf{A}_0\mathbf{R}_\mathbf{x} \, . \tag{2.125}$$

The number of equations in (2.121)–(2.125) is altogether (taking the symmetry of (2.123) and (2.124) into account)

$$\begin{aligned} \#\text{equations} &= n_y + n_x + \frac{n_y(n_y+1)}{2} + \frac{n_x(n_x+1)}{2} + n_y n_x \\ &= \frac{n_x^2}{2} + \frac{3n_x}{2} + \frac{n_y^2}{2} + \frac{3n_y}{2} + n_y n_x \, . \end{aligned} \tag{2.126}$$

As both $\mathbf{R}_x$ and $\boldsymbol{\Lambda}_x$ have $n_x(n_x+1)/2$ unknowns each, the total number of unknowns (for $\mathbf{A}_0, \ \mathbf{b}_0, \ \mathbf{m}, \ \boldsymbol{\Lambda}_y, \ \mathbf{R}_x, \ \boldsymbol{\Lambda}_x$) becomes

$$\begin{aligned} \#\text{unknowns} &= n_y n_x + n_y + n_x + \frac{n_y(n_y+1)}{2} + 2 \times \frac{n_x(n_x+1)}{2} \\ &= n_x^2 + 2n_x + \frac{n_y^2}{2} + \frac{3n_y}{2} + n_y n_x \, , \end{aligned} \tag{2.127}$$

which apparently always exceeds the number of equations given in (2.126). More precisely, the number of degrees of freedom in the solution will be

$$\#\text{unknowns} \; - \#\text{equations} \; = \frac{n_x(n_x+1)}{2} \, . \tag{2.128}$$

Therefore, *the system is not identifiable*. The degrees of freedom in the solution will grow with the dimension of $\mathbf{x}$. ■

In the above examples, the parameters of the model (the parameters a and b) were considered to be estimated from second-order moments of the data (the time series $\{x_i\}$ and $\{y_i\}$. The unknown x coordinates were characterized by their mean (m) and variance (σ). It is also possible, however, to change the setting and formulate another identification problem. This is done next.

Example 2.9 Assume that the measurement noise $\tilde{x}_i$, $\tilde{y}_i$ are independent and Gaussian, and consider the maximum likelihood (ML) estimation of the unknowns. The unknowns are a and b in the model (2.110). In addition, choose to treat the noise-free x-values $\{x_{0i}\}_{i=1}^N$ as auxiliary unknowns.

The likelihood function L turns out to satisfy

$$\log(L) = -\frac{1}{2\lambda_x}\sum_{i=1}^N (x_i - x_{0i})^2 - \frac{1}{2\lambda_y}\sum_{i=1}^N (y_i - ax_{0i} - b)^2$$
$$-\frac{N}{2}\log\lambda_x - \frac{N}{2}\log\lambda_y - \frac{N}{2}\log(2\pi)\ . \tag{2.129}$$

The ML estimate is obtained as the parameters that maximize the likelihood function.

However, in this case, if L is reconsidered for $x_{0i} = x_i$, $i = 1, \ldots N$, then apparently $L \to \infty$, when $\lambda_x \to 0$. Hence the optimization problem is not well posed. In fact, the ML estimate does not exist here, and the parameters are not identifiable. ■

Return to the situation treated in Example 2.9. One might believe that the root of the identifiability problem is that all the N unknowns $\{x_{oi}\}$ need to be determined as auxiliary unknowns. This is not so, as shown next.

Example 2.10 Reconsider the situation treated in Example 2.9, but assume that the noise ratio

$$\lambda_y/\lambda_x = r \tag{2.130}$$

is known.

Then *the parameters a and b are identifiable*. This can be realized in three different ways.

Geometrically, the assumption (2.130) means that one can scale the measurements so that the uncertainties in the x and y directions are the same. Then it will be feasible to apply orthogonal regression, given by the total least squares estimate.

Treating the problem instead *algebraically*, one has now 6 equations (namely (2.112)–(2.116), (2.130)) and still 6 unknowns. As the equations are nonlinear, one needs a more detailed examination to conclude whether or not there is a unique solution with respect to a_0 and b_0. Eliminating first b, λ_y and m, one arrives at three equations for determining a, σ, and λ_x:

$$\begin{cases} r_y = a^2\sigma + r\lambda_x\ , \\ r_x = \sigma + \lambda_x\ , \\ r_{yx} = a\sigma\ . \end{cases} \tag{2.131}$$

One can next solve for σ and λ_x, and finally arrive at a second-order equation for a:

$$0 = a^2 r_{yx} + a(rr_x - r_y) - rr_{yx}\ . \tag{2.132}$$

This equation has two roots: one equal to the true value $a = a_0$, and one false, $a = -r/a_0$. The false one can though be neglected as it leads to a *negative* estimate of the variable σ, which must be constrained to be positive as it stands for a variance.

For a *statistical* treatment of the problem, make use of the (2.130) in the logarithm of the likelihood function, (2.129) which now becomes (removing a trivial constant term)

$$\begin{aligned}\log(L) &= -\frac{1}{2\lambda_x}\sum_{i=1}^{N}(x_i - x_{0i})^2 - \frac{1}{2r\lambda_x}\sum_{i=1}^{N}(y_i - ax_{0i} - b)^2 \\ &\quad - \frac{N}{2}\log\lambda_x - \frac{N}{2}\log(r\lambda_x) - \frac{N}{2}\log(2\pi) \\ &\triangleq -\frac{N}{2\lambda_x}V(a, b, x_{0i}) - \frac{N}{2}\left[2\log(\lambda_x) + \log(r)\right] - \frac{N}{2}\log(2\pi)\,.\end{aligned} \tag{2.133}$$

where

$$V(a, b, x_{0i}) = \frac{1}{N}\sum_{i=1}^{N}\left[(x_i - x_{0i})^2 + \frac{1}{r}(y_i - ax_{0i} - b)^2\right]\,. \tag{2.134}$$

One can then find the ML estimates of a, b and x_{0i}, $i = 1, \ldots N$ by minimizing $V(a, b, x_{0i})$. First minimize with respect to x_{0i}:

$$0 = \frac{\partial V}{\partial x_{0i}} = -2(x_i - x_{0i}) - \frac{2a}{r}(y_i - ax_{0i} - b)\,, \tag{2.135}$$

leading directly to

$$x_{0i} = \frac{r}{a^2 + r}x_i + \frac{a}{a^2 + r}(y_i - b)\,, \tag{2.136}$$

$$x_i - x_{0i} = \frac{a}{a^2 + r}(ax_i - y_i + b)\,, \tag{2.137}$$

$$y_i - ax_{0i} - b = \frac{r}{a^2 + r}(y_i - ax_i - b)\,. \tag{2.138}$$

Inserting this into (2.134) leads to a criterion depending on just the primary parameters a and b,

$$
\begin{aligned}
\bar{V}_N(a,b) &\overset{\Delta}{=} \min_{x_{0i}} V(a,b,x_{0i}) \\
&= \frac{1}{N}\sum_{i=1}^{N} \frac{1}{(a^2+r)^2}\left[a^2\left(ax_i - y_i + b\right)^2 + \frac{1}{r}r^2\left(y_i - ax_i - b\right)^2\right] \\
&= \frac{1}{(a^2+r)}\frac{1}{N}\sum_{i=1}^{N}\left(y_i - ax_i - b\right)^2 . \qquad (2.139)
\end{aligned}
$$

The minimizing arguments of $\bar{V}_N(a,b)$ is precisely the ML estimate of a and b. Compared to (2.14) one finds that using $\bar{V}_N(a,b)$ is nothing but the orthogonal regression (or total least squares) criterion.

Letting the true values of the parameters be denoted a_0 and b_0, one gets in the asymptotic case when the number of observation data tends to infinity

$$
\begin{aligned}
\bar{V}_\infty(a,b) &= \lim_{N\to\infty} \bar{V}_N(a,b) \\
&= \frac{1}{a^2+r}\lim_{N\to\infty}\frac{1}{N}\sum_{i=1}^{N}\left[a_0 x_{0i} + b_0 + \tilde{y}_i - a(x_{i0} + \tilde{x}_i) - b\right]^2 \\
&= \frac{1}{a^2+r}\left[((a_0 - a)m\sigma + (b_0 - b))^2 + (a_0 - a)^2\sigma + (r + a^2)\lambda_x\right] , \qquad (2.140)
\end{aligned}
$$

which apparently has minimum for the true values,

$$a = a_0, \quad b = b_0 .$$

This analysis illustrates the consistency of the parameter estimates. ■

The estimate of the noise variance λ_x is treated next.

Example 2.11 It is of interest to continue Example 2.10 to find also the estimate of the noise variance λ_x. The estimate is the minimization of (2.133)

$$\hat{\lambda}_x = \arg\min_{\lambda_x} \log(L) . \qquad (2.141)$$

Direct differentiation of (2.133) leads to

$$\frac{N}{2\hat{\lambda}_x^2} V(\hat{a},\hat{b},\hat{x}_{0i}) - \frac{N}{\hat{\lambda}_x} = 0 ,$$

and hence

$$\hat{\lambda}_x = \frac{1}{2} V(\hat{a},\hat{b},\hat{x}_{0i}) = \frac{1}{2}\bar{V}_N(\hat{a},\hat{b}) .$$

In the asymptotic case therefore

$$\lim_{N\to\infty} \hat{\lambda}_x = \frac{1}{2}\bar{V}_\infty(a_0, b_0) = \frac{1}{2}\lambda_x . \tag{2.142}$$

It is striking that the factor $1/2$ appears in (2.142) and that the noise variance estimate is *not consistent*. This is a classical result; see, for example, Lindley (1947). An intuitive explanation may be that both the number of observed data and the number of free variables increases linearly with N. This makes the ML problem differ from the 'standard' case, where the number of independent parameters does not depend on N. ■

So far, linear models have been considered. When even a simple nonlinearity is introduced, the complexity is increased. This is illustrated in the next example.

Example 2.12 Consider the following generalization of Example 2.7.

$$\begin{aligned} y_i &= y_{0i} + \tilde{y}_i , \\ x_i &= x_{0i} + \tilde{x}_i , \qquad\qquad i = 1, \dots, N . \\ y_{0i} &= a x_{0i}^2 + b x_{0i} + c , \end{aligned} \tag{2.143}$$

Assume that $\tilde{y}_i$ and $\tilde{x}_i$ are zero mean Gaussian noise with variance λ_y and λ_x, respectively. Further, let x_{0i} have zero mean and a symmetric distribution. Set

$$r_k = \mathsf{E}\left\{x_{0i}^k\right\} . \tag{2.144}$$

Obviously, $r_k = 0$ for k odd.

The first-order moments of the data (they can of course be estimated from time records) are given by

$$\mathsf{E}\{y_i\} = \mathsf{E}\{y_{0i}\} = a r_2 + c , \tag{2.145}$$

$$\mathsf{E}\{u_i\} = 0 . \tag{2.146}$$

Of course, equation (2.146) does not bring any information at all. From first-order moments, there is hence 1 equation (namely (2.145)) and 4 unknowns (namely a, b, c, r_2).

Considering second-order moments of the data leads to the additional equations

$$\mathsf{E}\left\{y_i^2\right\} = \mathsf{E}\left\{y_{0i}^2\right\} + \lambda_y = a^2 r_4 + b^2 r_2 + c^2 + 2acr_2 + \lambda_y , \tag{2.147}$$

$$\mathsf{E}\{y_i x_i\} = \mathsf{E}\{y_{0i} x_{0i}\} = b r_2 , \tag{2.148}$$

$$\mathsf{E}\left\{x_i^2\right\} = r_2 + \lambda_x . \tag{2.149}$$

Considering now equations (2.145)–(2.149), there are 4 useful equations but 7 unknowns (a, b, c, r_2, r_4, λ_y, λ_u). Hence there is still three degrees of freedom in the solution (leaving aside whether or not there is any further ambiguity when solving for the unknowns.)

By adding some further moments one gets also

$$\mathsf{E}\left\{y_i x_i^2\right\} = \mathsf{E}\left\{y_{0i}\left(x_{0i}^2 + \tilde{x}_i^2\right)\right\} = \lambda_x(ar_2 + c) + ar_4 + cr_2 \;, \tag{2.150}$$
$$\mathsf{E}\left\{x_i^4\right\} = \mathsf{E}\left\{x_{0i}^4 + 6x_{0i}^2\tilde{x}_i^2 + \tilde{x}_i^4\right\} = r_4 + 6r_2\lambda_x + 3\lambda_x^2 \;, \tag{2.151}$$
$$\mathsf{E}\left\{y_i^2 x_i\right\} = \mathsf{E}\left\{y_{0i}^2 x_{0i}\right\} = 2abr_4 + 2bcr_2 \;. \tag{2.152}$$

Then the number of equations becomes equal to the number of unknowns (and equal to 7).

As the purpose of the example is to illustrate that the complexity of the equations grows rapidly when nonlinear models are treated, the issue whether or not the above 7 equations in this example has a unique solution is not considered here. ■

Needless to say, a parabola as treated in Example 2.12 is an extremely simple case of a nonlinear static model. To base an estimate for a nonlinear EIV model in general on higher-order moments of the data will soon become very complex when more advanced parameterizations are considered, if at all possible.

2.A.2 *Consistency of the CFA Estimate*

Consider here the general CFA-based estimator introduced in Sect. 2.2. In order to emphasize the influence of the sample length N, introduce the notations V_N for the loss function used and $\hat{\mathbf{R}}_N$ for the sample covariance matrix of the data.

Given the model, assume the model to be identifiable. This means that

$$\mathbf{R}(\boldsymbol{\theta}) = \mathbf{R}(\boldsymbol{\theta}_0) \;\Rightarrow\; \boldsymbol{\theta} = \boldsymbol{\theta}_0 \;. \tag{2.153}$$

According to the general assumptions, the data is assumed ergodic, and then

$$\lim_{N\to\infty} \hat{\mathbf{R}}_N \to \mathbf{R}_\infty = \mathbf{R}(\boldsymbol{\vartheta}_0) \;. \tag{2.154}$$

Consistency is hence essentially proved if the asymptotic loss function $V_\infty(\boldsymbol{\vartheta})$ has a global minimum for $\boldsymbol{\vartheta} = \boldsymbol{\vartheta}_0$. This means that the following inequality holds true

$$V_\infty(\boldsymbol{\vartheta}) \geq V_\infty(\boldsymbol{\vartheta}_0) \;. \tag{2.155}$$

- For the criteria $V_2(\boldsymbol{\vartheta})$, (2.49), and $V_3(\boldsymbol{\vartheta})$, (2.51), it is trivial to see that (2.155) applies.
- For the criterion $V_1(\boldsymbol{\vartheta})$, (2.48), the inequality (2.155) turns out to be equivalent to

$$\begin{aligned} \operatorname{tr}\left(\mathbf{R}(\boldsymbol{\vartheta}_0)\mathbf{R}^{-1}(\boldsymbol{\vartheta})\right) + \log(\det \mathbf{R}(\boldsymbol{\vartheta})) &\geq n + \log(\det \mathbf{R}(\boldsymbol{\vartheta}_0)) \Leftrightarrow \\ \operatorname{tr}\left(\mathbf{R}(\boldsymbol{\vartheta}_0)\mathbf{R}^{-1}(\boldsymbol{\vartheta})\right) + \log(\det \mathbf{R}(\boldsymbol{\vartheta})\mathbf{R}^{-1}(\boldsymbol{\vartheta}_0)) &\geq n \;. \end{aligned} \tag{2.156}$$

Now set

$$\mathbf{S} = \mathbf{R}^{1/2}(\boldsymbol{\vartheta}_0)\mathbf{R}^{-1}(\boldsymbol{\vartheta})\mathbf{R}^{1/2}(\boldsymbol{\vartheta}_0) \, , \tag{2.157}$$

which is positive definite by construction. Let its eigenvalues be denoted $\lambda_1, \ldots, \lambda_n$. The relation (2.156) is equivalent to

$$\begin{aligned} &\mathrm{tr}(\mathbf{S}) - \log(\det \mathbf{S}) \geq n \Leftrightarrow \\ &\sum_{i=1}^{n} \lambda_i - \log\left(\prod_{i=1}^{n} \lambda_i\right) \geq n \Leftrightarrow \\ &\sum_{i=1}^{n} \left[\lambda_i - \log(\lambda_i) - 1\right] \geq 0 \, , \end{aligned} \tag{2.158}$$

which holds true, as

$$\lambda > 0 \Rightarrow \log(\lambda) \leq \lambda - 1 \, . \tag{2.159}$$

Note that in this analysis it is crucial that λ is positive. This illustrates also that in the numerical search for minimizing $V(\boldsymbol{\theta})$ only values of $\boldsymbol{\vartheta}$ that keeps $\mathbf{R}(\boldsymbol{\vartheta})$ positive definite must be considered. In case values of $\boldsymbol{\vartheta}$ making the matrix $\mathbf{R}(\boldsymbol{\vartheta})$ indefinite were allowed, lower values than the right-hand side of (2.155) could be obtained.

Chapter 3
The Errors-in-Variables Problem for Dynamic Systems

This chapter starts in Sect. 3.1 with a formal description of the EIV problem for dynamic systems. In Sect. 3.2 it is discussed how numerical examples based on Monte Carlo simulations are used in the book. Section 3.3 is devoted to discussion of two special cases, while some simple attempts for solution are presented in Sect. 3.4. Further comments about extensions and bibliographical notes are given in Sect. 3.5.

3.1 The EIV Problem

As a typical model example, consider the linear dynamic system depicted in Fig. 3.1 with noise-corrupted input and output measurements.

The noise-free input is denoted by $u_0(t)$ and the undisturbed output by $y_0(t)$. Assume that the observations are corrupted by additive measurement noises $\tilde{u}(t)$ and $\tilde{y}(t)$. The available signals appear in discrete-time and are of the form

$$\begin{aligned} u(t) &= u_0(t) + \tilde{u}(t) \; , \\ y(t) &= y_0(t) + \tilde{y}(t) \; . \end{aligned} \tag{3.1}$$

The noise-free signals are related through a transfer function operator $G(q^{-1})$, where $G(q^{-1})$ is a function of the backward shift operator q^{-1}, and

$$\begin{aligned} y_0(t) &= G(q^{-1})u_0(t) \\ &= \left[\sum_{k=1}^{\infty} g_k q^{-k} \right] u_0(t) \\ &= \sum_{k=1}^{\infty} g_k u_0(t-k) \; . \end{aligned} \tag{3.2}$$

T. Söderström, *Errors-in-Variables Methods in System Identification*, Communications and Control Engineering,
https://doi.org/10.1007/978-3-319-75001-9_3

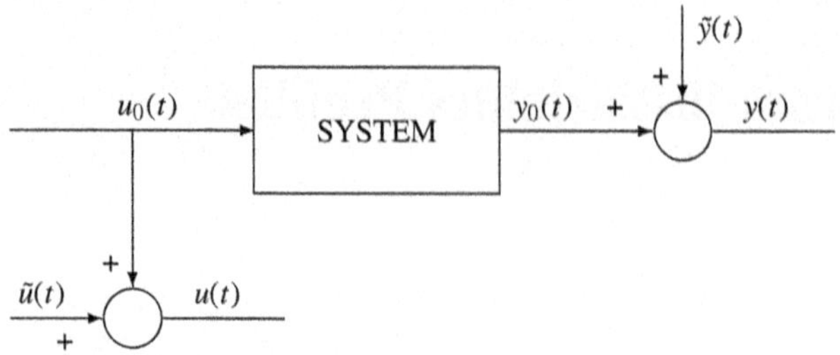

Fig. 3.1 The basic setup for a dynamic errors-in-variables problem

The general problem is to determine the system characteristics, such as the system transfer function $G(q^{-1})$ from measurements of the noisy signals $y(1), u(1), \dots, y(N), u(N)$.

In many parts of the book, a finite-order model will be postulated, so that $G(q^{-1})$ can be expressed as a ratio of two polynomials

$$G(q^{-1}) = \frac{B(q^{-1})}{A(q^{-1})} = \frac{b_1 q^{-1} + \dots + b_{n_b} q^{-n_b}}{1 + a_1 q^{-1} + \dots + a_{n_a} q^{-n_a}} . \tag{3.3}$$

In such a case, the estimation of $G(q^{-1})$ is equivalent to determination of the parameter vector

$$\boldsymbol{\theta} = \begin{pmatrix} a_1 & \dots & a_{n_a} & b_1 & \dots & b_{n_b} \end{pmatrix}^T . \tag{3.4}$$

Remark 3.1 The description (3.3) may be modified to include a direct term, that is a term b_0 in the numerator. Then there is no delay at all in the transfer function. Such an extension is analyzed in Sect. 5.2. As (3.3) is more standard in other system identification contexts, it is kept here. ■

In order to proceed, some further assumptions must be introduced. To some degree they can be made more or less restrictive. In the following a number of assumptions are introduced, that are partly alternative ones. It will then be discussed what assumptions that are necessary for different results, and what the consequences will be if the assumptions are relaxed.

Assumptions will be labeled as follows

AS1, **AS2**, ... assumptions related to the system
AN1, **AN2**, ... assumptions related to the measurement noises $\tilde{u}(t)$, $\tilde{y}(t)$
AI1, **AI2**, ... assumptions related to the noise-free input $u_0(t)$
AE1, **AE2**, ... assumptions related to the experimental condition

For the reader's convenience, a summary of all the general definitions introduced in this text is also listed in the beginning of the book.

Concerning the system, impose the following assumption.

AS1. The system is linear and asymptotically stable.

In many cases the following assumption is also used (even though it may be relaxed).

AS2. The system is strictly causal, so $y_0(t)$ depends on $u_0(s)$ for $s \leq t-1$, but not on future values of $u_0(\cdot)$.

In most cases t will denote time, and then it is most natural to require the system to be causal as in **AS2**. However, the estimation techniques can also be applied if t has the meaning, say, of a spatial variable, and the model describes for example some cross-directional property of a material. For such cases, non-causal models can make perfect sense. Then (3.2) has to be adjusted so that the summation starts at a lower value of k.

A third assumption on the system is

AS3. The noise-free input and output signals are linked by

$$A(q^{-1})y_0(t) = B(q^{-1})u_0(t) \; . \tag{3.5}$$

All system modes are observable and controllable, i.e., $A(z)$ and $B(z)$ have no common factor. The polynomial degrees n_a and n_b are known.

AS4. The system transfer function $G(z)$ has no pair of zeros reflected in the unit circle, that is, if $G(z_1) = 0$, then $G(z_1^{-1}) \neq 0$.

AS5. If the system is non-causal, then $G(z)$ has no pair of poles reflected in the unit circle, that is, p_1 and p_1^{-1} cannot both be poles of $G(z)$.

AS6. Let the noise-free input be written as a filtered white noise

$$u_0(t) = H(q^{-1})v(t) \tag{3.6}$$

where $v(t)$ is white noise. Then the order of the transfer functions fulfills

$$\begin{aligned} \text{order}(GH) &\stackrel{\Delta}{=} n_T = n_G + n_H \\ &= \text{order}\,(G) + \text{order}\,(H) \; . \end{aligned} \tag{3.7}$$

Here 'order' denotes the order (the MacMillan degree) of a system represented by its transfer function. Further, G is the transfer function (3.3) and H is the transfer

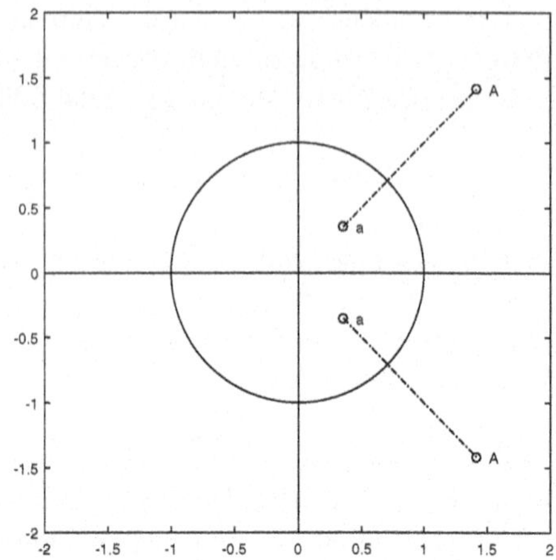

Fig. 3.2 Illustrating reflection of points in the unit circle. The reflections of the points marked 'a' are the points marked 'A' and vice versa

function in (3.6). The condition (3.7) means precisely that there is no pole–zero cancellation in the total transfer function GH.

Remark 3.2 The assumption **AS3** of known polynomial degrees n_a and n_b is crucial in the analysis of many identification methods. It is discussed in Sect. 5.9 how these integers may be determined from experimental data in an EIV context. ■

Example 3.1 The meaning of reflected zeros (**AS4**) and reflected poles (**AS5**) is illustrated in Fig. 3.2. ■

Next introduce assumptions on the noise and the noise-free input.

AN1. The noise sequences $\tilde{u}(t)$, $\tilde{y}(t)$ are stationary random processes, with zero mean values and spectra $\phi_{\tilde{u}}(\omega)$ and $\phi_{\tilde{y}}(\omega)$, respectively. Further, $\tilde{u}(t)$ and $\tilde{y}(t)$ are mutually uncorrelated. [In a few cases to be specified later, the noise sequences are allowed to be cross-correlated].

AN2. The measurement noises are Gaussian distributed.

Naturally, at most one of the following noise assumptions applies at any given situation.

AN3a. Both $\tilde{y}(t)$ and $\tilde{u}(t)$ are ARMA processes, see (4.25) and (4.26).

AN3b. The output noise $\tilde{y}(t)$ is an ARMA process, while the input noise $\tilde{u}$ is white. This means that $n_k = n_m = 0$ in (4.26).

AN3c. Both $\tilde{y}(t)$ and $\tilde{u}(t)$ are white noise sequences. This means that $n_f = n_h = 0$ in (4.25) and $n_k = n_m = 0$ in (4.26).

To motivate the options **AN3b** and **AN3c**, note that measurement noise often is white (uncorrelated in time and of a high bandwidth compared to the sampling frequency).

Occasionally (see, e.g., Sects. 9.4 and 11.5) a further assumption on the noise is introduced.

AN4. Both $\tilde{y}(t)$ and $\tilde{u}(t)$ are white noise sequences. The ratio of their variances, $r = \lambda_y/\lambda_u$ is known.

Next follows some assumptions on the noise-free input signal.

AI1. The true input $u_0(t)$ is a stationary process of zero mean, with spectral density $\phi_{u_0}(\omega)$. The input $u_0(t)$ is assumed to be persistently exciting of a sufficient order, which means that $\phi_{u_0}(\omega) > 0$ for a sufficient number of frequencies.

AI2. The input $u_0(t)$ is uncorrelated with the measurement noise sources $\tilde{u}(t)$ and $\tilde{y}(t)$.

AI3. The true input $u_0(t)$ is Gaussian distributed.

AI4. The true input $u_0(t)$ is an ARMA process, that is, it can be modeled as

$$D(q^{-1})u_0(t) = C(q^{-1})e(t) \ , \tag{3.8}$$

where $e(t)$ is a white noise signal.

The assumption of zero mean is a weak one and rather made for convenience. A situation where $u_0(t)$ and thereby also $y_0(t)$ has nonzero mean is in fact easier from an identifiability point of view, as the static gain of the system can then be determined separately. Should the basic setup be modified so that an unknown constant level is added to the output, the mean values of $u_0(t)$ and $y_0(t)$ determine the working point of the system, while the second-order statistics will describe the dynamics of the deviations from the operating point.

Next introduce assumptions about the experimental condition. Some alternative assumptions include:

AE1. The data comes from one (single) experiment.

AE2a. There is more than one experiment. The spectrum of the noise-free input is different in the different experiments.

AE2b. There is more than one experiment. The measurement noises $\tilde{u}(t)$, $\tilde{y}(t)$ are uncorrelated between different experiments. The true noise-free input $u_0(t)$ is correlated between the experiments.

Table 3.1 Characters of the different assumptions

Topic	Validity		
	General	Default	Locally
System	**AS1, AS2**	**AS3**	**AS4, AS5**
Noise	**AN1**	**AN3c**	**AN2, AN3a, AN3b**
Noise-free input	**AI1, AI2**		**AI3, AI4**
Experiment		**AE1**	**AE2a, AE2b**

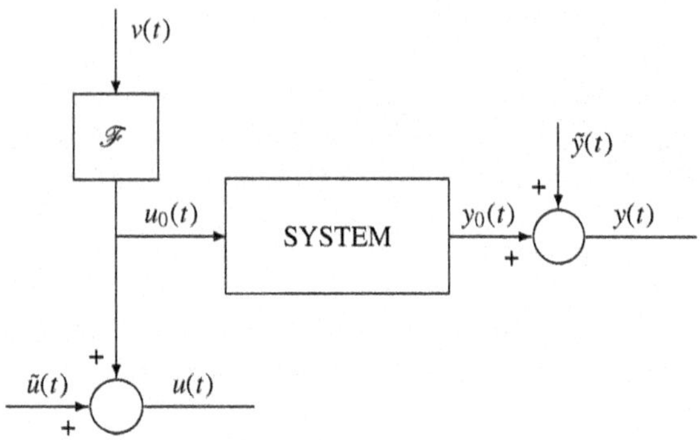

Fig. 3.3 The basic setup, including input generation

Many assumptions are assumed to be generally valid throughout the text, while a few are valid only locally, when explicitly stated so. Table 3.1 summarizes the status of the assumptions.

When do errors-in-variables problems occur? Consider the following extension of Fig. 3.3. One can distinguish between some different situations or options.

1. One option is to assume one experiment only, and that $u_0(t)$ cannot be affected by the user. The experiment is 'arranged' by nature, or the considered system is just a part of a larger system and excited at some other point. In this case both $v(t)$ and $\mathscr{F}$ are unknown. Possibly the true input $u_0(t)$ can be modeled as a stationary stochastic process with rational spectrum. This means that in Fig. 3.3, $\mathscr{F}$ is a finite order, unknown linear filter, and $v(t)$ is an (unmeasurable) white noise sequence. Hence for this option assumption **AE1** applies.
2. Another option is to assume that the signal $v(t)$ is fully accessible to the user, but that the filter $\mathscr{F}$ is an unknown and possibly nonlinear dynamic system, so that $u_0(t)$ can neither be chosen freely, nor computed. Nevertheless, in such scenarios it is possible to make repeated experiments with the same $v(t)$, and hence with the same $u_0(t)$. In such cases the assumption **AE2b**, cf. Sect. 4.3, applies.
3. There is a third option, where $v(t)$ can be considered as an accessible control variable, while $\mathscr{F}$ describes some actuator with potentially complicated dynamics. For example, the signal $v(t)$ may be the command signal to a valve, while $u_0(t)$ is the flow through the valve, and the 'system' is a part of a process plant. Assume that the purpose of the identification is to get a model as basis for feedback design,

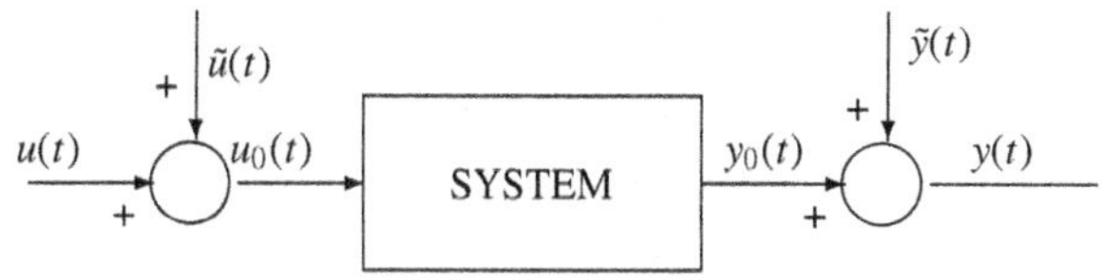

Fig. 3.4 A system affected by input noise

and the feedback is to be applied using the command signal $v(t)$. In such a case it is more meaningful to estimate a model with $v(t)$ as input, and then the problem is no longer an errors-in-variables problem.

Finally, consider a related problem, as depicted in Fig. 3.4. that, however, is *not* an EIV situation. Recall the following differences between the situations in Figs. 3.1 and 3.4.

- In Fig. 3.4, the system is affected by the input noise $\tilde{u}(t)$, while this is not the case in Fig. 3.1.
- The situation in Fig. 3.4 can be redrawn with disturbances acting *only* at the output side, in this case the total output disturbance being $\tilde{y}(t)+G(q^{-1})\tilde{u}(t)$, and standard system identification methods may be applied.
- In Fig. 3.4 the measured $u(t)$ and the actual input $u_0(t)$ do effect the system in identical ways, and this is not the case in Fig. 3.1.

It is worth mentioning that alternative EIV problems exist. For example, in Cerone et al. (2017a) the system is modeled by its weighting function, and a nonparametric identification is considered. Further, the measurement noises are assumed to be bounded, with known upper bounds.

3.2 About Numerical Examples

There will be several numerical examples in the book, aimed to illustrate the properties of various methods to be introduced. To facilitate comparisons, the same data sets will be used throughout the examples. The second-order system given by the equation below will be considered

$$y_0(t) - 1.5y_0(t-1) + 0.7y_0(t-2) = 2.0u_0(t-1) + 1.0u_0(t-2) \,. \tag{3.9}$$

The noise-free input $u_0(t)$ is taken as a second-order ARMA process:

$$u_0(t) - 1.0u_0(t-1) + 0.2u_0(t-2) = v(t) + 0.7v(t-1) \,, \tag{3.10}$$

where $v(t)$ is Gaussian white noise of variance $\lambda_v = 1$. Further, there is white observation noise on both the input side and the output side:

$$y(t) = y_0(t) + \tilde{y}(t), \quad \mathsf{E}\{\tilde{y}(t)\} = 0, \quad \mathsf{E}\{\tilde{y}^2(t)\} = \lambda_y, \tag{3.11}$$

$$u(t) = u_0(t) + \tilde{u}(t), \quad \mathsf{E}\{\tilde{u}(t)\} = 0, \quad \mathsf{E}\{\tilde{u}^2(t)\} = \lambda_u, \tag{3.12}$$

The numerical values are chosen as

$$\lambda_y = 10, \quad \lambda_u = 1.81 \ . \tag{3.13}$$

This choice of parameters leads to the signal-to-noise ratios

$$\mathrm{SNR}_u = \frac{\mathsf{E}\{u_0^2(t)\}}{\mathsf{E}\{\tilde{u}^2(t)\}} = \frac{9.06}{1.81} = 5 = 7.0 \text{ dB} \ , \tag{3.14}$$

$$\mathrm{SNR}_y = \frac{\mathsf{E}\{y_0^2(t)\}}{\mathsf{E}\{\tilde{y}^2(t)\}} = \frac{2436}{10} = 243.6 = 23.9 \text{ dB} \ . \tag{3.15}$$

By purpose the signal-to-noise ratio is lowest on the input side, in order to emphasize the effects of the input noise on various estimates.

In the Monte Carlo simulations the resulting performance is assessed as follows. The system was simulated using $N = 1000$ data points. Further, $M = 100$ independent realizations were used, so as to be able to evaluate the statistical accuracy in terms of, say, standard deviations of the parameter estimates. The parameter vectors obtained in realization i are denoted by $\hat{\boldsymbol{\theta}}^{(i)}, i = 1, \ldots, M$. Arithmetic mean values and standard deviations over these estimates are presented in tables. Further, as a scalar measure, the root-mean-square value

$$\mathrm{RMS} = \sqrt{\frac{1}{M} \sum_{i=1}^{M} \| \hat{\boldsymbol{\theta}}^{(i)} - \boldsymbol{\theta}_0 \|^2} \ , \tag{3.16}$$

is also considered.

How will the data be used in the examples? Typically an estimator will be applied to the full data sets. Then the statistics in terms of empirical means and standard deviations will be computed from the different realizations. For consistent estimates, the mean values should be close to the true parameter values, in the sense that the difference should be small compared to the standard deviations. For biased estimators on the other hand, the mean values are expected to deviate considerably (measured in terms of number of standard deviations) from the true values.

In most of the coming chapters where different methods are introduced and presented the behavior is illustrated using Monte Carlo simulations as described above. For some of the methods there are design parameters to be set by the user. In a few examples it is studied what the influence of these design parameters will be on the performance.

Later, in Chap. 14, theoretical results are presented for what standard deviations to expect. These theoretical accuracy results are then compared to what is achieved from the Monte Carlo simulations.

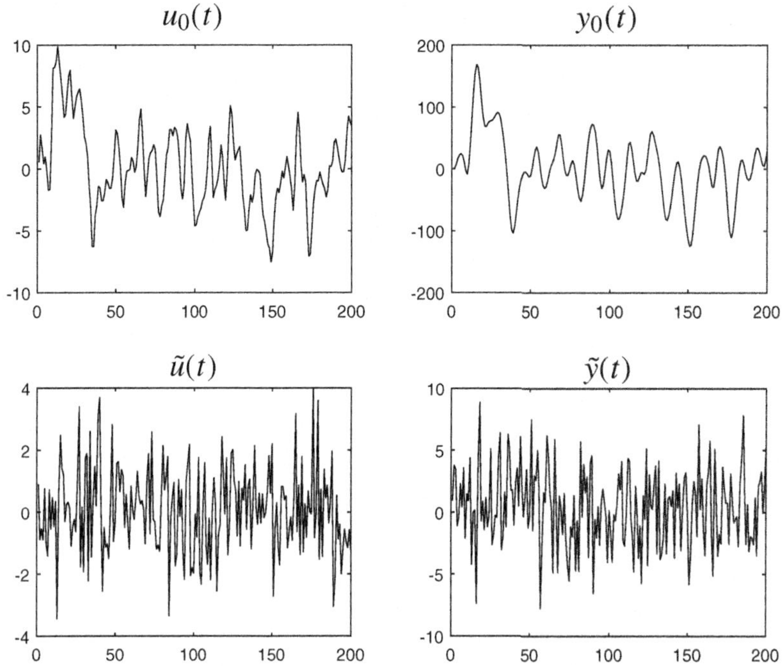

Fig. 3.5 Part of the signals for one realization: $u_0(t)$ upper left, $y_0(t)$ upper right, $\tilde{u}(t)$ lower left, $\tilde{y}(t)$ lower right

As a basic illustration of the system under consideration, the behavior of the signals for part of one realization is given in Fig. 3.5.

3.3 Two Special Cases

It was found in Chap. 2 that the EIV problem does not have a unique solution in the static case unless some additional assumption is imposed. As will become clear in Chap. 4, this is also true when dynamic systems are treated. One special case that occasionally is treated is that of the measurement noises $\tilde{u}(t)$ and $\tilde{y}(t)$ both being white (**AN3c** applies) and with a *known* variance ratio $r = \lambda_y/\lambda_u$.

This assumption may be restrictive from a practical point of view. Still it improves considerably the identifiability properties and simplifies the derivation of estimation algorithms. In the book, this special case is treated in Sect. 9.4 and Chap. 11.

Another special case, or rather a related and simpler problem, is that of noisy measurements of one signal only. In such a case the measurements are modeled as

$$y(t) = y_0(t) + \tilde{y}(t) , \tag{3.17}$$

and the measurement noise is assumed to be white. The signal $y_0(t)$ is modeled as an AR or possibly an ARMA process.

Some restrictions on the ARMA model for $y_0(t)$, say

$$D(q^{-1})y_0(t) = C(q^{-1})e(t) \tag{3.18}$$

apply. Let

$$\begin{aligned} C(q^{-1}) &= 1 + c_1 q^{-1} + \cdots + c_{n_c} q^{-n_c} , \\ D(q^{-1}) &= 1 + d_1 q^{-1} + \cdots + d_{n_d} q^{-n_d} . \end{aligned} \tag{3.19}$$

Noting that $e(t)$ and $\tilde{y}(t)$ are white noises of variance λ_e and λ_y, respectively, it is found that the spectrum of the observations $y(t)$ can be written as

$$\phi_y = \frac{CC^*}{DD^*}\lambda_e + \lambda_y = \frac{MM^*}{DD^*}\lambda_\varepsilon , \tag{3.20}$$

where the monic polynomial

$$M = M(q^{-1}) = 1 + m_1 q^{-1} + \cdots + m_n q^{-n} \tag{3.21}$$

with $n = \max(n_c, n_d)$ is obtained as the unique stable factor of the spectral factorization

$$CC^*\lambda_e + DD^*\lambda_y = MM^*\lambda_\varepsilon . \tag{3.22}$$

A consequence of (3.20) is that, as far as the second-order properties of the observations are concerned, the measurements can be equivalently regarded as an ARMA process,

$$D(q^{-1})y(t) = M(q^{-1})\varepsilon(t) . \tag{3.23}$$

In the original model parameterization (3.17), (3.18), the unknown parameters are $c_1, \ldots, c_{n_c}, d_1, \ldots, d_{n_d}, \lambda_e, \lambda_y$, that is in total

$$n_\theta = n_c + n_d + 2 \tag{3.24}$$

parameters. From the second-order properties of $y(t)$ one can (at best) determine the parameters of (3.23), which is

$$n^* = n_d + n_m + 1 \tag{3.25}$$

parameters. For identifiability, that is to be able to uniquely recover the original parameters, an obvious necessary condition is therefore

$$
\begin{aligned}
n_\theta \le n^* &\Leftrightarrow n_c + 1 \le n_m \\
&\Leftrightarrow n_c + 1 \le \max(n_c, n_d) \\
&\Leftrightarrow n_c < n_d \, .
\end{aligned} \tag{3.26}
$$

The problem of estimating the parameters of (3.17), (3.18) lies outside the scope of this book. It is a topic treated in depth in the signal processing literature, see Sect. 3.5 for some examples. Several of the identification methods to be presented in this book can be simplified to handle this special case as well.

3.4 Some Naïve Approaches

This section will analyze some simple and straightforward attempts to deal with the errors-in-variables problem for dynamic systems. It will be shown that they lead to difficulties and that they are not feasible solutions when accurate models are desired.

3.4.1 Neglecting the Input Noise

A first simple approach is to just neglect the presence of input noise. It was illustrated for the line-fitting example in Chap. 2 that doing so leads to a non-consistent estimate when a least squares method is applied, cf (2.107).

This happens to hold for a much more general setting. To treat a fairly general case, consider the case when a linear model structure

$$
y(t) = G(q^{-1})u(t) + H(q^{-1})\varepsilon(t) \tag{3.27}
$$

is considered. Here, $G(q^{-1}) = G(q^{-1}; \boldsymbol{\theta})$ and $H(q^{-1}) = H(q^{-1}; \boldsymbol{\theta})$ are filters in the backward shift operator q^{-1}, and they are parameterized with a parameter vector $\boldsymbol{\theta}$. One possible case is that (3.27) stands for an ARMAX model, but also more general parameterizations are possible. Further, $H(q^{-1})\varepsilon(t)$ is assumed to be an innovations model for the disturbances, so it is assumed that both $H(q^{-1})$ and $H^{-1}(q^{-1})$ are asymptotically stable, and that the innovations sequence $\varepsilon(t), t = 1, 2, \ldots$ is white noise.

Assume further that a prediction error approach is applied to determine the estimate of $\boldsymbol{\theta}$, that is

$$
\hat{\boldsymbol{\theta}} = \arg\min_{\boldsymbol{\theta}} V_N(\boldsymbol{\theta}) \, , \tag{3.28}
$$

$$
V_N(\boldsymbol{\theta}) = \frac{1}{N} \sum_{t=1}^{N} \left[H^{-1}(q^{-1}) \left(y(t) - G(q^{-1})u(t) \right) \right]^2 \, . \tag{3.29}
$$

Recall that the optimal one-step prediction error of $y(t)$ is

$$y(t) - \hat{y}(t|t-1) = \frac{1}{H(q^{-1})} \left[y(t) - G(q^{-1})u(t) \right] , \tag{3.30}$$

see Söderström (2002).

The underlying data are assumed to be described by

$$y_0(t) = G_0(q^{-1})u_0(t) , \tag{3.31}$$
$$y(t) = y_0(t) + \tilde{y}(t) , \tag{3.32}$$
$$u(t) = u_0(t) + \tilde{u}(t) . \tag{3.33}$$

Here, $y_0(t)$ and $u_0(t)$ refer to the noise-free (unmeasurable) output and input, respectively. It is assumed that the signals $u_0(t)$, $\tilde{y}(t)$ and $\tilde{u}(t)$ are independent. Further assume that the parameterization of the system transfer function is pertinent in the sense that there is a 'tentatively true' parameter vector $\boldsymbol{\theta}_0$ such that

$$G(q^{-1}, \boldsymbol{\theta}_0) \equiv G_0(q^{-1}) . \tag{3.34}$$

To examine identifiability, consider the asymptotic case when $N \to \infty$. The loss function (3.29) becomes under the given assumptions

$$\begin{aligned} V_\infty(\boldsymbol{\theta}) &= \mathsf{E}\left\{ \left[H^{-1}(q^{-1}) \left(y(t) - G(q^{-1})u(t) \right) \right]^2 \right\} \\ &= \mathsf{E}\left\{ \left[H^{-1}(q^{-1}) \left(G_0(q^{-1})u_0(t) + \tilde{y}(t) - G(q^{-1}) \left\{ u_0(t) + \tilde{u}(t) \right\} \right) \right]^2 \right\} \\ &= \mathsf{E}\left\{ \left[H^{-1}(q^{-1}) \left(G_0(q^{-1}) - G(q^{-1}) \right) u_0(t) \right]^2 \right\} \\ &\quad + \mathsf{E}\left\{ \left[H^{-1}(q^{-1})\tilde{y}(t) \right]^2 \right\} + \mathsf{E}\left\{ \left[H^{-1}(q^{-1})G(q^{-1})\tilde{u}(t) \right]^2 \right\} . \end{aligned} \tag{3.35}$$

Consider first the 'classical' case of no input noise, and assume that the model parameterization is flexible enough. Then the last term in (3.35) vanishes. The sum of the other two is minimized if the parameters are chosen so that (3.34) holds, and the innovation form for $\tilde{y}(t)$ happens to be $\tilde{y}(t) = H(q^{-1})e(t)$.

However, in the errors-in-variables case, where input noise is present, the third term in (3.35) cannot be neglected. As a consequence, at the minimum point of the criterion, (3.34) will no longer hold, and there will be a bias of $\hat{\boldsymbol{\theta}}$, also in the limiting case when $N \to \infty$.

One can also show that the gradient of the criterion *does not vanish* for the true parameter vector, that is

$$V'_\infty(\boldsymbol{\theta}_0) \neq 0 \tag{3.36}$$

unless the input is measured exactly (so that $\lambda_u = 0$). Let

$$G_{\boldsymbol{\theta}}(q^{-1}) = \frac{\partial}{\partial \boldsymbol{\theta}} G(q^{-1}; \boldsymbol{\theta}) \bigg|_{\boldsymbol{\theta}=\boldsymbol{\theta}_0},$$
$$H_{\boldsymbol{\theta}}(q^{-1}) = \frac{\partial}{\partial \boldsymbol{\theta}} H(q^{-1}; \boldsymbol{\theta}) \bigg|_{\boldsymbol{\theta}=\boldsymbol{\theta}_0}.$$

Then the gradient of the asymptotic loss function (3.35) is given by

$$\begin{aligned}
V'_{\infty}(\boldsymbol{\theta}_0) &= 2\mathsf{E}\left\{\left[\frac{1}{H(q^{-1};\boldsymbol{\theta})}\tilde{y}(t)\right]\left[\frac{-H_{\boldsymbol{\theta}}(q^{-1})}{H^2(q^{-1};\boldsymbol{\theta})}\tilde{y}(t)\right]\right\} \\
&\quad +2\mathsf{E}\left\{\left[\frac{G(q^{-1};\boldsymbol{\theta})}{H(q^{-1};\boldsymbol{\theta})}\tilde{u}(t)\right]\left[\frac{-H_{\boldsymbol{\theta}}(q^{-1})G(q^{-1};\boldsymbol{\theta})}{H^2(q^{-1};\boldsymbol{\theta})}\tilde{u}(t) + \frac{G_{\boldsymbol{\theta}}(q^{-1})}{H(q^{-1};\boldsymbol{\theta})}\tilde{u}(t)\right]\right\} \\
&= -2\int \frac{H_{\boldsymbol{\theta}}(e^{i\omega})}{H(e^{i\omega})}\frac{1}{|H(e^{i\omega})|^2}\phi_{\tilde{y}}(\omega)d\omega \\
&\quad -2\int \left[\frac{H_{\boldsymbol{\theta}}(e^{i\omega})|G(e^{i\omega})|^2}{H(e^{i\omega})|H(e^{i\omega})|^2} + \frac{G_{\boldsymbol{\theta}}(e^{i\omega})G(e^{-i\omega})}{|H(e^{i\omega})|^2}\right]\phi_{\tilde{u}}(\omega)d\omega \\
&= -2\int \frac{H_{\boldsymbol{\theta}}(e^{i\omega})}{H(e^{i\omega})}\left[\lambda_{\varepsilon}^2 + \frac{G(e^{-i\omega})}{H(e^{-i\omega})}\phi_{\tilde{u}}(\omega)\right]d\omega \\
&\quad -2\int \left[\frac{G_{\boldsymbol{\theta}}(e^{i\omega})G(e^{-i\omega})}{|H(e^{i\omega})|^2}\right]\phi_{\tilde{u}}(\omega)d\omega \\
&= -2\int \frac{G_{\boldsymbol{\theta}}(e^{i\omega})G(e^{-i\omega})}{|H(e^{i\omega})|^2}\phi_{\tilde{u}}(\omega)d\omega\,.
\end{aligned} \tag{3.37}$$

Here, (3.50) below was utilized. The last equality follows since $H_{\boldsymbol{\theta}}(z)/H(z)$ has all poles inside the unit circle. In conclusion, $V'_{\infty}(\boldsymbol{\theta}_0) \neq 0$ in general, unless $\phi_{\tilde{u}}(\omega) \equiv 0$. That is, $\boldsymbol{\theta}_0$ is not a minimum point, unless there is no noise on the input measurements.

It is another matter, whether or not this lack of consistency is serious or not. If the EIV modeling is used as a step in a control design, an uncertain open-loop system may not be so harmful. A controller is often designed so that the closed-loop system is insensitive to uncertainties in the open-loop dynamics, and this is a major advantage of feedback control.

In other applications though, the aim is to extract detailed information about the system itself, and then it can be important to avoid systematic errors, as those caused by neglecting the presence of input disturbances.

Example 3.2 In this example the least squares method is applied to the standard data sets used in the book. The numerical results obtained are presented in Table 3.2.

There is apparently a clear bias in all the parameter estimates. The deviation of the mean value from the corresponding true value is always much larger than the standard deviation. This is just as expected from theory. ■

Table 3.2 Least squares estimates for the standard data sets

Parameter	True value	Mean	Standard dev.
a_1	−1.5	−1.414	0.015
a_2	0.7	0.619	0.014
b_1	2.0	1.599	0.103
b_2	1.0	1.361	0.129

3.4.2 Estimating the Noise-Free Input Signal

Another simple approach to solve the EIV problem is to estimate the noise-free input signal $u_0(t)$ as additional, auxiliary, variables.

It was shown in Chap. 2 and Appendix 2.A, that such an approach is not feasible for the line-fitting problem, and for other static cases, unless one knows the ratio of the noise variances. See Examples 2.9 and 2.10. Similar results hold in much more general settings for dynamic systems, as is now shown.

Consider the data situation given by (3.31)–(3.33), and assume that the noise sources $\tilde{y}(t)$ and $\tilde{u}(t)$ are white, with unknown variances λ_y^0 and λ_u^0, respectively. When the noise is Gaussian, and neglecting transient effects, the negative log-likelihood function is given by

$$L_N = \frac{N}{2}\log(\lambda_y) + \frac{N}{2}\log(\lambda_u) + \frac{1}{2\lambda_y}\sum_{t=1}^{N}\left(y(t) - G(q^{-1})u_0(t)\right)^2$$
$$+\frac{1}{2\lambda_u}\sum_{t=1}^{N}(u(t) - u_0(t))^2 \ . \tag{3.38}$$

To proceed, use Parseval's relation, and describe the sums using the discrete Fourier transforms as

$$L_N = \frac{N}{2}\log(\lambda_y) + \frac{N}{2}\log(\lambda_u) + \frac{1}{2\lambda_y}\sum_{k=0}^{N-1}\left[\ \left|Y(\omega_k) - G(\mathrm{e}^{\mathrm{i}\omega_k})U_0(\omega_k)\right|^2\right.$$
$$\left.+\frac{1}{2\lambda_u}|U(\omega_k) - U_0(\omega_k)|^2\right] , \tag{3.39}$$

where, for $k = 0, \ldots, N-1$,

$$\omega_k = 2\pi\frac{k}{N} \tag{3.40}$$

and

$$U(\omega_k) = \frac{1}{\sqrt{N}}\sum_{j=0}^{N-1} u(j)\mathrm{e}^{-\mathrm{i}2\pi\frac{jk}{N}} \ . \tag{3.41}$$

The advantage of using (3.39) instead of (3.38) is that the dynamics is present as an algebraic multiplication rather than as a weighted sum.

In the formulation (3.39) the transient influence of initial values was neglected. This is reasonable, as the asymptotic case is considered when the number of data points grows without limit, $N \to \infty$.

Remark 3.3 For a more complete analysis taking the initial conditions into account, see Sects. 5.5, 9.4 and 10.5. ∎

To simplify the notations, use the symbols

$$\begin{aligned}&Y_k = Y(\omega_k), \quad U_k = U(\omega_k), \quad U_{0k} = U_0(\omega_k), \quad G_k = G(\mathrm{e}^{\mathrm{i}\omega_k}) \,, \\ &k = 0, \dots, N-1 \,.\end{aligned} \tag{3.42}$$

In the future analysis five different cases will be treated:

1. Both λ_y and λ_u are known.
2. Both λ_y and λ_u are unknown.
3. The variance λ_y is known, and λ_u is unknown.
4. The variance λ_y is unknown, and λ_u is known.
5. Both variances are unknown, but the ratio $r = \lambda_y/\lambda_u$ is known.

Next proceed to minimize L with respect to the sequence $U_{0k}, k = 0 \dots, N-1$. As this dependence is quadratic, the procedure is straightforward:

$$L_N = \frac{N}{2} \log(\lambda_y \lambda_u) + V_N \,, \tag{3.43}$$

$$\begin{aligned} V_N &= \frac{1}{2\lambda_y} \sum_{k=0}^{N-1} (Y_k - G_k U_{0k})^* (Y_k - G_k U_{0k}) + \frac{1}{2\lambda_u} \sum_{k=0}^{N-1} (U_k - U_{0k})^* (U_k - U_{0k}) \\ &= \frac{1}{2} \sum_{k=0}^{N-1} \left[U_{0k}^* U_{0k} \left(\frac{1}{\lambda_y} G_k^* G_k + \frac{1}{\lambda_u} \right) + U_{0k}^* \left(-\frac{1}{\lambda_y} G_k^* Y_k - \frac{1}{\lambda_u} U_k \right) \right. \\ &\quad \left. + U_{0k} \left(-\frac{1}{\lambda_y} G_k Y_k^* - \frac{1}{\lambda_u} U_k^* \right) + \frac{1}{\lambda_y} Y_k^* Y_k + \frac{1}{\lambda_u} U_k^* U_k \right] . \end{aligned} \tag{3.44}$$

Hence L_N and V_N are minimized with respect to $U_{0k}, k = 0, 1, \dots, N-1$, for

$$U_{0k} = \frac{\frac{1}{\lambda_y} G_k^* Y_k + \frac{1}{\lambda_u} U_k}{\frac{1}{\lambda_y} G_k^* G_k + \frac{1}{\lambda_u}} = \frac{\lambda_u G_k^* Y_k + \lambda_y U_k}{\lambda_u G_k^* G_k + \lambda_y} \,. \tag{3.45}$$

This leads to

$$Y_k - G_k U_{0k} = \frac{\lambda_y (Y_k - G_k U_k)}{\lambda_u G_k^* G_k + \lambda_y} , \tag{3.46}$$

$$U_k - U_{0k} = \frac{\lambda_u G_k^* (Y_k - G_k U_k)}{\lambda_u G_k^* G_k + \lambda_y} , \tag{3.47}$$

$$\begin{aligned} \overline{V}_N \triangleq \min_{\{U_{0k}\}} V_N &= \frac{1}{2\lambda_y} \sum_{k=0}^{N-1} \frac{\lambda_y^2 |Y_k - G_k U_k|^2}{(\lambda_u |G_k|^2 + \lambda_y)^2} + \frac{1}{2\lambda_u} \sum_{k=0}^{N-1} \frac{\lambda_u^2 |G_k|^2 |Y_k - G_k U_k|^2}{(\lambda_u |G_k|^2 + \lambda_y)^2} \\ &= \frac{1}{2} \sum_{k=0}^{N-1} \frac{|Y_k - G_k U_k|^2}{(\lambda_u |G_k|^2 + \lambda_y)} . \end{aligned} \tag{3.48}$$

To proceed, express the asymptotic criterion $\overline{V}_\infty$ in the time domain. For this purpose introduce the filter $H(q^{-1})$ and the positive scalar λ through the spectral factorization

$$\lambda H(z) H(z^{-1}) \equiv \lambda_u G(z) G(z^{-1}) + \lambda_y . \tag{3.49}$$

This can also be written as

$$|H(e^{i\omega})|^2 \lambda \equiv \phi_{\tilde{y}}(\omega) + |G(e^{i\omega})|^2 \phi_{\tilde{u}}(\omega) , \tag{3.50}$$

where $H(q^{-1})$ and $H^{-1}(q^{-1})$ are asymptotically stable filters. The innovation $\varepsilon(t)$, or the prediction error, can then be introduced as

$$\varepsilon(t) = \frac{1}{H(q^{-1})} \left[y(t) - G(q^{-1}) u(t) \right] = \frac{1}{H(q^{-1})} \left[\tilde{y}(t) - G(q^{-1}) \tilde{u}(t) \right] . \tag{3.51}$$

In the last equality it is assumed that the filter $G(q^{-1})$ is evaluated for the true parameters, that is, (3.34) applies.

It follows from (3.48) that the asymptotic loss function can be written as

$$\overline{V}_\infty = \frac{1}{2\lambda} \mathsf{E} \left\{ \left[H^{-1}(q^{-1}) \left(y(t) - G(q^{-1}) u(t) \right) \right]^2 \right\} . \tag{3.52}$$

Consider first the case when both noise variances are known, so that $\lambda_y = \lambda_y^0$, $\lambda_u = \lambda_u^0$ holds. Then

$$\begin{aligned} \overline{V}_\infty &= \frac{1}{2\lambda} \mathsf{E} \left\{ \left[H^{-1}(q^{-1}) \left(G_0(q^{-1}) u_0(t) + \tilde{y}(t) - G(q^{-1}) \{ u_0(t) + \tilde{u}(t) \} \right) \right]^2 \right\} \\ &= \frac{1}{2\lambda} \mathsf{E} \left\{ \left[H^{-1}(q^{-1}) \left(G_0(q^{-1}) - G(q^{-1}) \right) u_0(t) \right]^2 \right\} \\ &\quad + \frac{1}{2\lambda} \mathsf{E} \left\{ \left[H^{-1}(q^{-1}) \left(\tilde{y}(t) - G(q^{-1}) \tilde{u}(t) \right) \right]^2 \right\} \\ &= \frac{1}{2\lambda} \left[\mathsf{E} \left\{ \left[H^{-1}(q^{-1}) \left(G_0(q^{-1}) - G(q^{-1}) \right) u_0(t) \right]^2 \right\} + \lambda \right] . \end{aligned} \tag{3.53}$$

In this case the criterion $\overline{V}_\infty$ is minimized with respect to $\boldsymbol{\theta}$ when (3.34) holds, i.e., $G(q^{-1}) = G_0(q^{-1})$.

Next consider the case when the noise variance λ_u is unknown. Examine the derivative of the normalized criterion

$$L_\infty = \lim_{N\to\infty} \frac{1}{N} L = \frac{1}{2}\log(\lambda_y\lambda_u) + \overline{V}_\infty(\boldsymbol{\theta}, \lambda_y, \lambda_u) \tag{3.54}$$

with respect to λ_u and evaluate it at the true parameters. In case the derivative is nonzero, this shows that the true parameter values are not retrieved, and that the system is not identifiable in such a case. Under the given assumption

$$L_\infty = \frac{1}{2}\log(\lambda_y\lambda_u) + \frac{1}{2\lambda}\mathsf{E}\left\{\left[H^{-1}(q^{-1})\left(y(t) - G(q^{-1})u(t)\right)\right]^2\right\}, \tag{3.55}$$

$$\frac{\partial L_\infty}{\partial \lambda_u} = \frac{1}{2\lambda_u} - \frac{1}{2\lambda^2}\frac{\partial \lambda}{\partial \lambda_u}\lambda = \frac{1}{2\lambda_u\lambda}\left(\lambda - \lambda_u\frac{\partial\lambda}{\partial\lambda_u}\right) \neq 0\,. \tag{3.56}$$

To realize (3.56), consider how $\lambda = \lambda(\lambda_u)$ is a function defined implicitly by (3.49). If $\lambda - \lambda_u \frac{\partial\lambda}{\partial\lambda_u} = 0$ it must hold

$$\lambda(\lambda_u) = C\lambda_u \tag{3.57}$$

for some constant C. However, (3.57) is not compatible with (3.49), and (3.56) follows.

This calculation shows that the system is not identifiable if λ_u is unknown. It can be straightforward extended to cover all the cases 2, 3, and 4 above (the system is not identifiable in any of these cases).

The problem of minimizing L in (3.43) is in fact not well posed. To realize this, consider specifically the arguments $\boldsymbol{\theta}$ arbitrary, λ_y arbitrary, $\lambda_u \to 0$, $U_{0k} = U_k, k = 0, \ldots, N-1$. Then $L \to \infty$. This means that *no* maximizing element of L can be found, and that the maximum likelihood estimate does not exist. This fact was first pointed out in Anderson and Rubin (1956). Further details of this type of example can be found in Solari (1969), Stoica and Li (2005), and Söderström et al. (2002), Eising et al. (1983).

Finally, consider Case 5, where the noise variance ratio is known. Describe this as

$$\lambda_y = \alpha\lambda_y^0, \ \lambda_u = \alpha\lambda_u^0\,, \tag{3.58}$$

where the parameter α is unknown. In this case the asymptotic normalized log-likelihood function fulfills

$$L_\infty = \frac{1}{2}\log(\alpha^2\lambda_y^0\lambda_u^0) + \frac{1}{2\alpha\lambda}\mathsf{E}\left\{\left[H^{-1}(q^{-1})\left(y(t) - G(q^{-1})u(t)\right)\right]^2\right\}. \tag{3.59}$$

The minimization with respect to $\boldsymbol{\theta}$ is straightforward, and it is seen that also in this case, the minimum is achieved for the true parameters, so then (3.34) holds, and the system is identifiable. Minimization with respect to α gives though

$$\frac{\partial L_\infty}{\partial \alpha} = \frac{1}{\alpha} - \frac{1}{2\alpha^2 \lambda}\lambda \Rightarrow \alpha = 0.5 \ . \tag{3.60}$$

This means that the noise variance level is underestimated with a factor of 0.5. This is indeed the same result as for the line-fitting case, see Example 2.11.

The results of the analysis can now be summarized as follows:

(a) The estimate $\hat{\theta}$ will not be consistent in the cases (corresponds to cases 2, 3, and 4 above)

- λ_u and λ_y are both unknown,
- One of the variances λ_u and λ_y is known, and the other is unknown.

This statement follows from the above reasoning.

(b) The estimate $\hat{\theta}$ will be consistent, if either (corresponds to cases 1 and 5 above)

- both the noise variances λ_y and λ_u are known.
- their ratio λ_y/λ_u is known.

Note that in both the cases here, it is in fact possible to consistently identify the system, even if there are $2N$ data values, and more than N unknowns. It is hence not the fact that the number of unknowns grows as fast as the number of data points that causes the consistency problem.

It is worth pointing out that the example refers to a situation of nonparametric treatment of the undisturbed input. It does not matter if the estimation is carried out in the time domain or in the frequency domain. However, in case $u_0(t)$ is described by a parametric model (say with a fixed finite number of parameters), then the situation is different. See Sect. 4.2 for details.

3.4.3 Rewriting the Model into Standard Form

The spectral factorization (3.50) may be used to rewrite the underlying dynamics and lumping all disturbances to the output in some sense. Introducing the innovations sequence $\varepsilon(t)$ as in (3.51), the noisy input–output data are found to satisfy

$$y(t) = G(q^{-1})u(t) + H(q^{-1})\varepsilon(t) \ . \tag{3.61}$$

This is a standard input–output model with the disturbances given in innovations form. It is then tempting to apply a prediction error method, Ljung (1999), Söderström and Stoica (1989) in a standard way. It is known though from Sect. 3.4.1 that this will lead to a bias. Here, the reasons for this approach to fail are analyzed from a slightly different perspective. More precisely, examine the cross-covariance function between the input $u(t)$ and the prediction error $\varepsilon(t)$:

$$r_{\varepsilon u}(\tau) = \mathsf{E}\left\{\varepsilon(t+\tau)u(t)\right\} \ . \tag{3.62}$$

If a system is operating in open loop, then for an identifiable parameterization the cross-covariance function will be identically zero. The system may be identifiable also under causal feedback, in which case it is required

$$r_{\varepsilon u}(\tau) = 0 \ \text{ for } \tau > 0\,. \tag{3.63}$$

The cross-covariance function of the system under current investigation is to be evaluated for the true parameter vector $\boldsymbol{\theta}_0$. In the particular case under study, the cross-covariance function becomes

$$\begin{aligned} r_{\varepsilon u}(\tau) &= \mathsf{E}\left\{\varepsilon(t+\tau)u(t)\right\} \\ &= \mathsf{E}\left\{\left[\frac{1}{H(q^{-1})}\left[\tilde{y}(t+\tau) - G(q^{-1})\tilde{u}(t+\tau)\right]\right]\left[u_0(t) + \tilde{u}(t)\right]\right\} \\ &= -\mathsf{E}\left\{\frac{G(q^{-1})}{H(q^{-1})}\tilde{u}(t+\tau)\tilde{u}(t)\right\} \neq 0\,. \end{aligned} \tag{3.64}$$

The expected value (3.64) becomes nonzero, as $\lambda_u > 0$. The noise term $[G(q^{-1})/H(q^{-1})]\tilde{u}(t+\tau)$ will be correlated with all past values of $\tilde{u}(t)$. Note that the relation (3.64) means that the process (3.61) has a *non-causal* feedback, which is the reason why identifiability is lost.

A related analysis is provided in Castaldi and Soverini (1998) for the single-input single-output case, with extension to the multivariable case in Castaldi et al. (1999). The analysis can be interpreted as follows. Consider a general linear model

$$y(t) = \hat{G}(q^{-1};\boldsymbol{\theta})u(t) + \hat{H}(q^{-1};\boldsymbol{\theta})\varepsilon(t)\,, \tag{3.65}$$

where $\hat{G}(q^{-1};\boldsymbol{\theta})$ may be non-causal, and $\hat{H}(q^{-1};\boldsymbol{\theta})$ as well as $\hat{H}^{-1}(q^{-1};\boldsymbol{\theta})$ are restricted to be asymptotically stable. Hence the model (3.65) can be said to be in innovations form, if $\varepsilon(t)$ is a white noise process.

Consider the optimization problem

$$\min_{\hat{G},\,\hat{H}} \mathsf{E}\left\{\varepsilon^2(t)\right\} \tag{3.66}$$

under the above restrictions, and require that the spectrum ϕ_ε is constant and that the cross-spectrum $\phi_{\varepsilon u}$ vanishes. The solution is found to be

$$\hat{G}(\mathrm{e}^{\mathrm{i}\omega}) = G(\mathrm{e}^{\mathrm{i}\omega})\frac{W_0(\mathrm{e}^{\mathrm{i}\omega})W_0(\mathrm{e}^{-\mathrm{i}\omega})}{W(\mathrm{e}^{\mathrm{i}\omega})W(\mathrm{e}^{-\mathrm{i}\omega})}\,, \tag{3.67}$$

where

$$\begin{aligned} W_0(\mathrm{e}^{\mathrm{i}\omega})W_0(\mathrm{e}^{-\mathrm{i}\omega}) &= \phi_{u_0}(\omega)\,, \\ W(\mathrm{e}^{\mathrm{i}\omega})W(\mathrm{e}^{-\mathrm{i}\omega}) &= \phi_u(\omega) = \phi_{u_0}(\omega) + \lambda_u\,, \end{aligned} \tag{3.68}$$

Table 3.3 Prediction error estimates for the standard data sets

Parameter	True value	Mean	Standard dev.
a_1	−1.5	−1.526	0.010
a_2	0.7	0.714	0.008
b_1	2.0	1.725	0.085
b_2	1.0	0.872	0.091
c_1	–	−0.559	0.032
c_2	–	0.222	0.032

and $\hat{H}(e^{i\omega})$ is given by the spectral factorization

$$\hat{H}(e^{i\omega})\hat{H}(e^{-i\omega}) = G(e^{i\omega})G(e^{-i\omega})\frac{W_0(e^{i\omega})W_0(e^{-i\omega})}{W(e^{i\omega})W(e^{-i\omega})}\lambda_u + \lambda_y \,. \tag{3.69}$$

One can interpret the obtained model (3.65) as the best non-causal linear open-loop model given in innovations form. The model is non-causal as $\hat{G}(q^{-1};\boldsymbol{\theta})$ has poles in the instability region. It is in the innovations form as $\hat{H}(q^{-1};\boldsymbol{\theta})$ and $\hat{H}^{-1}(q^{-1};\boldsymbol{\theta})$ are asymptotically stable and the spectrum of $\varepsilon(t)$ can be shown to be constant for all frequencies ω.

Example 3.3 The prediction error method used in a standard fashion using an ARMAX(2, 2, 2) model is applied to the standard data sets. The numerical results obtained are presented in Table 3.3.

There is some bias in the estimates, although much smaller than the bias in Example 3.2 where the least squares method is used. The bias now is significant in the estimates of a_1 (about 2.5 standard deviations) and b_1 (about 3 standard deviations). ■

3.5 Extensions and Bibliographical Notes

General Aspects

General treatments of the EIV problem for dynamic systems are given in, for example, Söderström (2007a). See also Söderström et al. (2002), Söderström (2012b), and Guidorzi et al. (2008).

Stoica and Li (2005) demonstrate that the ML estimate does not exist for a certain blind multi-channel identification problem. It corresponds in EIV terms to estimating the time series of the noise-free input.

The Case of Known Noise Variance Ratio

There are many papers on this topic. For some general treatments see, for example, Stoica et al. (1995b) and Söderström (2013).

Some classical papers on identification of dynamic systems with known noise variance ratio include Koopmans (1937) and Levin (1964).

Autoregressive Process Observed With Measurement Noise

This topic has been treated in many papers. Some handles also the extension to having both the signal and the measurement noise being ARMA processes. In the case of AR process with white measurement noise, identifiability always applies, but in the extended case elaborated conditions on various polynomial degrees apply for identifiability, see Söderström (1980).

For papers treating various approaches to identification of AR processes with additional observation noise, see Kay (1979), Sakai and Arase (1979), Kay (1980), Kay (1988), Nehorai and Stoica (1988), Davila (1998), Zheng (1998b), Zheng (1999b), Zheng (2000a), Zheng (2005), Zheng (2006a), Zheng (2006b), Labarre et al. (2006), Mahmoudi and Karimi (2010), Xia and Zheng (2015), Soverini and Söderström (2015b).

The are also many papers dealing with identification of ARMA models with additional measurement noise. One example is Chen (2008). Needless to say, in this case some restriction on the polynomial orders has to be imposed in order to achieve identifiability. Compare the analysis leading to (3.26) and the forthcoming Example 4.2. A recent treatment can be found in Diversi et al. (2017).

It follows from the developments in the chapter that an MA process observed in noise is not identifiable. An approach to identify such a process by approximating it with a (very) high-order AR model is considered in Youcef et al. (2015).

Chapter 4
Identifiability Aspects

In this chapter identifiability aspects for the errors-in-variables problem are covered in some different settings. Some comparisons with the findings from the simple line-fitting problem treated in Chap. 2 will be pointed out.

General aspects are given in Sect. 4.1. More specific analysis is provided for parametric modeling in Sect. 4.2 and for multiple experiments in Sect. 4.3. Section 4.4 discusses closed-loop operation for EIV problems. Extensions and bibliographical notes appear in Sect. 4.5.

4.1 Some General Aspects

Identifiability concerns the possibilities to correctly identify the unknown system from measured data. The identifiability problem is normally formulated for the asymptotic case when the number of measurements tends to infinity. Assume therefore that the spectrum (the spectral density) of the measured signals is known and available. The identifiability question is then if the unknown dynamics can be uniquely retrieved from the spectrum.

Identifiability depends on two factors:

- *The parameterization of the underlying model.* This specifies what are the unknown quantities to be estimated. The model may be parametric (say, a finite-order model) or nonparametric (say, a transfer function without specifying any particular structure).
- *The excitation of the unknown system.* In the EIV context this refers to the noise-free input signal $u_0(t)$. If this signal is absent, or constant, or persistently exciting of a quite low order, this will prevent identifiability.

T. Söderström, *Errors-in-Variables Methods in System Identification*, Communications and Control Engineering,
https://doi.org/10.1007/978-3-319-75001-9_4

Note that so far identifiability is not tied to the use of any particular identification method.

To treat the identifiability issue first in a general setting, introduce the following assumptions.

AN2. The measurement noises are Gaussian distributed.

AI2. The noise-free input is Gaussian distributed.

Under the assumptions **AN2** and **AI2** only first- and second-order moments carry information about the distribution, and higher-order moments do not bring further information. One may alternatively say that the study is limited for the time being to infer information from second-order statistics.

It turns out that in such cases, without introducing more explicit assumptions, it is not possible to uniquely identify the system. In fact, when only second-order statistics are exploited the identification of errors-in-variables models cannot, in general, admit a single solution, Anderson and Deistler (1984), Anderson and Deistler (1987).

As an illustration of the identifiability difficulties, consider the following example.

Example 4.1 Let the measurement noises be autocorrelated, with spectral densities $\phi_{\tilde{u}}(\omega)$, $\phi_{\tilde{y}}(\omega)$, respectively, but mutually uncorrelated. Denote the observations as

$$\boldsymbol{\zeta}(t) = \begin{pmatrix} y(t) \\ u(t) \end{pmatrix} . \tag{4.1}$$

Their joint spectrum is

$$\boldsymbol{\Phi}_{\zeta}(\omega) = \begin{pmatrix} GG^* & G \\ G^* & 1 \end{pmatrix} \phi_{u_0}(\omega) + \begin{pmatrix} \phi_{\tilde{y}}(\omega) & 0 \\ 0 & \phi_{\tilde{u}}(\omega) \end{pmatrix} . \tag{4.2}$$

The variables in (4.2) are all functions of the angular frequency ω. For briefness use the symbols

$$G = G(\mathrm{e}^{\mathrm{i}\omega}), \qquad G^* = [G(\mathrm{e}^{\mathrm{i}\omega})]^* = G^T(\mathrm{e}^{-\mathrm{i}\omega}) \tag{4.3}$$

in (4.2) and in what follows. Further drop, for the same reason, the argument ω in the spectra.

From the measured data $\boldsymbol{\zeta}(1), \ldots, \boldsymbol{\zeta}(N)$, one can (at least in principle) determine the spectrum $\boldsymbol{\Phi}_{\zeta}(\omega)$. With $\boldsymbol{\Phi}_{\zeta}$ given and G, ϕ_{u_0}, $\phi_{\tilde{u}}$, $\phi_{\tilde{y}}$ as unknowns, there is no unique solution to (4.2). Recall that as $\boldsymbol{\Phi}_{\zeta}$ is Hermitian, and hence the non-diagonal elements of $\boldsymbol{\Phi}_{\zeta}$ bring identical information.

To be more explicit, let the estimates of the aforementioned variables be denoted by $\hat{G}$, $\hat{\phi}_{u_0}$, $\hat{\phi}_{\tilde{u}}$, $\hat{\phi}_{\tilde{y}}$. The equations determining the estimates are

$$\boldsymbol{\Phi}_{\zeta} = \begin{pmatrix} \hat{G}\hat{G}^* & \hat{G} \\ \hat{G}^* & 1 \end{pmatrix} \hat{\phi}_{u_0} + \begin{pmatrix} \hat{\phi}_{\tilde{y}} & 0 \\ 0 & \hat{\phi}_{\tilde{u}} \end{pmatrix} . \tag{4.4}$$

Note that this is in fact, for each frequency, 3 equations with 4 unknowns. [In case complex-valued quantities are substituted with their real and imaginary parts, there are 4 equations and 5 unknowns.] There is hence one degree of freedom (for each frequency) in the solution. Recall from Chap. 2 that the line-fitting problem is also under-determined, with precisely one degree of freedom, cf. (2.128).

Equating (4.2) and (4.4), one can with straightforward calculations derive

$$\hat{G} = G\frac{\phi_{u_0}}{\hat{\phi}_{u_0}} , \tag{4.5}$$

$$\hat{\phi}_{\tilde{u}} = \phi_{\tilde{u}} + \phi_{u_0} - \hat{\phi}_{u_0} , \tag{4.6}$$

$$\hat{\phi}_{\tilde{y}} = \phi_{\tilde{y}} + GG^* \left(\phi_{u_0} - \frac{\phi_{u_0}^2}{\hat{\phi}_{u_0}} \right) . \tag{4.7}$$

Requiring that the estimated noise spectra are positive,

$$\hat{\phi}_{\tilde{u}}(\omega) \geq 0, \qquad \hat{\phi}_{\tilde{y}}(\omega) \geq 0 \tag{4.8}$$

will give bounds on the possible values of $\hat{\phi}_{u_0}$. Note that (4.8) is to hold for all values of ω. Straightforward calculations give

$$\hat{\phi}_{\tilde{u}} \geq 0 \Leftrightarrow \hat{\phi}_{u_0} - \phi_{u_0} \leq \phi_{\tilde{u}} , \tag{4.9}$$

$$\begin{aligned}
\hat{\phi}_{\tilde{y}} \geq 0 &\Leftrightarrow \phi_{\tilde{y}} + GG^* \frac{\phi_{u_0}}{\hat{\phi}_{u_0}} \left(\hat{\phi}_{u_0} - \phi_{u_0} \right) \geq 0 \\
&\Rightarrow \phi_{\tilde{y}} + GG^* \phi_{u_0} \geq \frac{GG^* \phi_{u_0}^2}{\hat{\phi}_{u_0}} \\
&\Rightarrow \hat{\phi}_{u_0} \geq \frac{GG^* \phi_{u_0}^2}{\phi_{\tilde{y}} + GG^* \phi_{u_0}} \\
&\Rightarrow \hat{\phi}_{u_0} - \phi_{u_0} \geq \frac{-\phi_{\tilde{y}} \phi_{u_0}}{\phi_{\tilde{y}} + GG^* \phi_{u_0}} .
\end{aligned} \tag{4.10}$$

The solution (4.5)–(4.8) is parameterized using $\hat{\phi}_{u_0}$. Combining (4.9) and (4.10) gives bounds for how the variable $\hat{\phi}_{u_0}$ can be chosen. Note that when the input noise is small ($\phi_{\tilde{u}}$ small), it cannot be much larger than the true noise-free input spectrum ϕ_{u_0}. Similarly, when the output noise is small, it cannot be much smaller than ϕ_{u_0}. ■

It is possible though to make a stricter analysis of the possible solutions to (4.4). Here it has so far only been exploited that spectra must be positive, but one can also use the fact that spectra are symmetric functions of frequency.

Note that in (4.5) the factor $\phi_{u_0}/\hat{\phi}_{u_0}$ will generally vary with frequency and will not correspond to a causal system.

Under the assumptions of **AS2** and **AS3**, or **AS3** and **AS4**, the solution in (4.5) must be such that the ratio $\phi_{u_0}/\hat{\phi}_{u_0}$ is a constant and does not vary with frequency.

This means that the general solution to (4.4) can be written with a scalar, frequency-independent degree of freedom as

$$\hat{G}(e^{i\omega}) = G(e^{i\omega})\frac{1}{\alpha} , \tag{4.11}$$

$$\hat{\phi}_{\tilde{u}}(\omega) = \phi_{\tilde{u}}(\omega) + \phi_{u_0}(\omega)(1-\alpha) , \tag{4.12}$$

$$\hat{\phi}_{\tilde{y}}(\omega) = \phi_{\tilde{y}}(\omega) + |G(e^{i\omega})|^2\phi_{u_0}(\omega)\left(1-\frac{1}{\alpha}\right) . \tag{4.13}$$

It remains to find out which values the parameter α can attain. Using again the conditions $\hat{\phi}_{\tilde{y}}(\omega) \geq 0$, $\hat{\phi}_{\tilde{u}}(\omega) \geq 0$ for all ω, one gets

$$\begin{aligned}
\hat{\phi}_{\tilde{u}} \geq 0 \quad \forall\omega &\Rightarrow \alpha\phi_{u_0}(\omega) \leq \phi_{u_0}(\omega) + \phi_{\tilde{u}}(\omega) \quad \forall\omega \\
&\Rightarrow \alpha \leq \frac{\phi_{u_0}(\omega) + \phi_{\tilde{u}}(\omega)}{\phi_{u_0}(\omega)} \quad \forall\omega \\
&\Rightarrow \alpha \leq \alpha^* \stackrel{\Delta}{=} \inf_{\omega} \frac{\phi_{u_0}(\omega) + \phi_{\tilde{u}}(\omega)}{\phi_{u_0}(\omega)} , \qquad (4.14) \\
\hat{\phi}_{\tilde{y}} \geq 0 \quad \forall\omega &\Rightarrow \frac{1}{\alpha}|G|^2\phi_{u_0} \leq \phi_{\tilde{y}} + |G|^2\phi_{u_0} \quad \forall\omega \\
&\Rightarrow \alpha \geq \frac{|G|^2\phi_{u_0}}{\phi_{\tilde{y}} + |G|^2\phi_{u_0}} \quad \forall\omega \\
&\Rightarrow \alpha \geq \alpha_* \stackrel{\Delta}{=} \sup_{\omega} \frac{|G(e^{i\omega})|^2\phi_{u_0}(\omega)}{\phi_{\tilde{y}}(\omega) + |G(e^{i\omega})|^2\phi_{u_0}(\omega)} . \qquad (4.15)
\end{aligned}$$

Remark 4.1 The result so far found is that, under assumption **AS3**, all possible solutions to (4.4) satisfy

$$\hat{G}(e^{i\omega}) = G(e^{i\omega})/\alpha , \tag{4.16}$$

where the constant α satisfies

$$\alpha_* \leq \alpha \leq \alpha^* . \tag{4.17}$$

Needless to say, the value $\alpha = 1$, which corresponds to a perfect model, is always a possible solution. This is easily seen from (4.17), as by construction $\alpha_* < 1$, $\alpha^* > 1$. ■

Remark 4.2 The solution (4.5) can be written as

$$\hat{G}(e^{i\omega}) = G(e^{i\omega})\frac{1}{\alpha(\omega)} , \tag{4.18}$$

where the function $\alpha(\omega)$ satisfies

$$\frac{|G(e^{i\omega})|^2\phi_{u_0}(\omega)}{\phi_{\tilde{y}}(\omega) + |G(e^{i\omega})|^2\phi_{u_0}(\omega)} \leq \alpha(\omega) \leq \frac{\phi_{u_0}(\omega) + \phi_{\tilde{u}}(\omega)}{\phi_{u_0}(\omega)} . \tag{4.19}$$

The bounds (4.17) are much sharper than those in (4.19). In particular for high frequencies, typically both $\phi_{u_0}(\omega)$ and $|G(e^{i\omega})|$ become small, and the differences between the bounds (4.17) and (4.19) become very pronounced. ■

Remark 4.3 The one degree of freedom solution (4.11)–(4.13) with the bounds (4.19) can also be derived using an explicit spectral factorization of the spectra, see Agüero et al. (2005), Agüero (2005), Agüero and Goodwin (2006) for details. ■

To summarize so far, it has been established that *there is a fundamental lack of identifiability in the errors-in-variables problem, as long as only quite general assumptions are imposed.* This holds no matter what identification method that would be applied.

There are different possibilities for how to reflect about this fact.

1. One possibility is to 'accept' the situation and not make any further assumptions. Instead of looking for a unique estimate, one has to deal with the whole set of estimates (4.11) where the constant α is bounded as in (4.17). This choice is in the spirit of set membership estimation, Milanese and Vicino (1991), where one characterizes the set of possible models rather than a unique estimate. In Agüero et al. (2005), the specific choice

$$\hat{\alpha} = \frac{1}{2}\left(\alpha_* + \alpha^*\right) \tag{4.20}$$

is presented. It minimizes the H^∞ norm of the estimation error $\hat{G}(e^{i\omega}) - G(e^{i\omega})$. Example of set membership identification for EIV problems is given in Cerone et al. (2012) and Cerone et al. (2018).

Recall in this context the treatment of identifiability for the line-fitting example in Sect. 2.1.1. There it is shown that any model

$$y_{0i} = a_0 x_{0i} \tag{4.21}$$

where a_0 lies in the interval spanned by the extreme estimates $\hat{a}_{\rm LS}$ (corresponding to all uncertainties attributed to the y measurements) and $\hat{a}_{\rm DLS}$ (all uncertainties being in the x measurements). See (2.25) and (2.26).
2. A second option is to introduce some additional assumptions, as exemplified a little later.

One may think of such additional assumptions in different ways:

- One possibility is to let them reflect some additional a priori information that the user has about the system.
- A second possibility is to let them correspond to what is just 'believed' to hold true. In such cases one may speak about prejudices.

- A third possibility is to introduce some convenient assumption just to achieve identifiability, and hope for the best.

In Chap. 15 there are several examples illustrating what may happen if such an additional assumption is imposed on the identification problem and for the estimator, but not valid for the data.

It is important to realize that from the experimental data to be used for identification it is in most cases impossible to assess whether or not these additional assumptions are fulfilled.

Examples of an additional assumption introduced to achieve identifiability are:

- One option is to modify at least one of the assumptions **AN2**, **AI3** on Gaussian distributed data. When the data are not Gaussian distributed, higher-order statistics can be employed to gain additional information about the system. This option is used in Tugnait (1992), Tugnait and Ye (1995).
  Some aspects on using higher-order statistics are presented in Sect. 6.2.4.
- Another option is to impose more detailed models for the measurement noises and for the noise-free input $u_0(t)$. Typically, $\tilde{u}(t)$, $\tilde{y}(t)$ and $u_0(t)$ are then modeled as ARMA processes of specified orders. Then the decomposition of the spectrum $\boldsymbol{\Phi}_{\zeta}(\omega)$ in (4.4) into one part depending on the noise-free input and a second part due to the measurement noise may have a unique solution in some cases. More details about this option are presented in Sect. 4.2.
- A third possibility is to consider the more restricted problem where the noise-free input signal $u_0(t)$ is periodic. This case turns out to be similar to the situation where there is more than one experiment with the noises in the different experiments being mutually uncorrelated, while the input signal is mutually correlated. Such a situation is analyzed in Sect. 4.3.
- A fourth option is to consider another restricted problem, namely that the noise variance ratio $r = \lambda_y/\lambda_u$ is known. It is known already from Chap. 2 that such an assumption leads to identifiability. Most of the methods developed and presented in the book can be adjusted (i.e., simplified) to handle the case with known noise variance ratio.

The above classification of different situations from an identifiability point of view has a direct impact on the organization of the coming chapters of the book:

- The second option with parametric models of the noises is pursued in Chaps. 6, 7, 8 and 9. In Chap. 11 a model for the noise-free input is used as well. The option of parametric models is also applied in Chap. 10.
- The fourth option dealing with known variance ratio is used in Chap. 11 and also in Sect. 9.4.
- The third option of periodic data is treated in Chap. 12, where some different estimators exploiting this option are presented.

What do the results on identifiability tell for a practical case? It is important to realize that the identifiability problem refers to a situation where there is an infinite number of data ($N \rightarrow \infty$). In practice N is finite. This means that the true spectrum is not exactly known, but an estimate is available, that is, an approximation. The minimum phase factor can in principle be determined from the spectrum. However, should the spectrum be uncertain for some frequency region, it can mean that the poles and zeros of the spectral factor that have the dominating effect on the spectrum in that region, can be quite uncertain. In such situations, it is not obvious if it is possible to successfully allocate poles and zeros to the different spectral terms in (4.2).

This section is finished by giving some comments on the general assumptions. The assumptions can sometimes be relaxed. For a few identification methods additional assumptions need to be imposed.

- The measurement noises may be somewhat correlated. Some estimation methods can be extended to handle that $\tilde{u}(t)$ is a moving average process, $\tilde{y}(t)$ is an ARMA process, or that $\tilde{u}(t)$ and $\tilde{y}(t)$ are correlated.
- Most of the methods can be extended to multivariable systems.
- The problem will be significantly simplified if the noise variances λ_u and λ_y are known, or even only the ratio $r = \lambda_y/\lambda_u$.
- For a maximum likelihood or prediction error approach, there is a need to postulate also a model parameterization of the unperturbed input signal $u_0(t)$, see Sect. 9.2. Note that Assumption **AI3** implies that $u_0(t)$ can be modeled as an ARMA process.

Some comparisons between the different main approaches treated in the coming chapters are summarized in Table 4.1. (One should bear in mind that the performance judgments in the table are crude and possibly subjective.) Nevertheless, the

Table 4.1 A summarizing comparison of some approaches for solving the EIV identification problem

Method	Noise condition	Experimental condition	Computational complexity	Statistical accuracy
Basic IV (Chap. 6)	$\tilde{u}(t)$ MA	–	Very low	Low
GIVE (Chap. 7)	$\tilde{u}(t)$, $\tilde{y}(t)$ White	–	Low	Medium
CM (Chap. 8)	$\tilde{u}(t)$, $\tilde{y}(t)$ White	–	Medium	Medium-high
PEM, ML (Chap. 9)	–	–	High	High
Freq. domain (Chap. 12)	–	Repeated exp.	Medium	Medium-high
TLS (Chap. 11)	$\tilde{u}(t)$, $\tilde{y}(t)$ White		Medium	Medium
	$r = \lambda_y/\lambda_u$ known	Repeated exp.		

table shows some important relations between the different approaches. The methods GIVE and CM can both be extended to handle correlated output noise.

4.2 Identifiability Analysis for Parametric Models

It was found in Sect. 4.1 that there is a fundamental lack of identifiability in the errors-in-variables problem, at least as long as nonparametric models are used. This section considers the identifiability problem when some classes of parametric models are used.

It is important to realize that the errors $\tilde{u}(t)$ and $\tilde{y}(t)$ can have several causes. One possible cause is measurement errors. It seems often realistic to assume such errors to be uncorrelated in time, and therefore relevant to model as white noise processes. However, the output error $\tilde{y}(t)$ must also accommodate effects of process disturbances and modeling errors. Both these types of contributions are typically autocorrelated in time. Therefore, it is natural to model the output error as an ARMA process in a more general case.

As an ARMA process is a general model for describing a stationary, or quasistationary, process with a rational spectra, such a model may also be postulated for describing the noise-free input $u_0(t)$. See Ljung (1999) for discussions on quasistationary processes.

One hence arrives at a *parametric* problem. Model the system as a finite-order one. Assume that the noise-free input $u_0(t)$ as well as the input noise $\tilde{u}(t)$ and output noise $\tilde{y}(t)$ are ARMA processes. The total model is described in Fig. 4.1.

More specifically, let the system transfer function be described as, cf (3.3)

$$G(q^{-1}) = \frac{B(q^{-1})}{A(q^{-1})} , \tag{4.22}$$

$$\begin{aligned} A(q^{-1}) &= 1 + a_1 q^{-1} + \cdots + a_{n_a} q^{-n_a} , \\ B(q^{-1}) &= b_1 q^{-1} + \cdots + b_{n_b} q^{-n_b} . \end{aligned} \tag{4.23}$$

Further, in Fig. 4.1, the noise-free input signal is the ARMA process

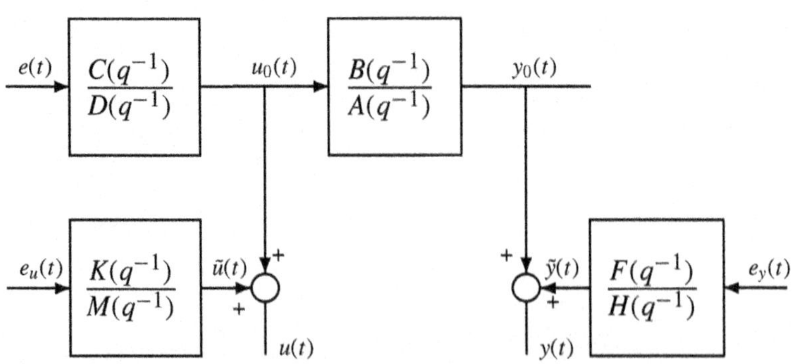

Fig. 4.1 Modeling a finite-order system, with $u_0(t)$, $\tilde{u}(t)$ and $\tilde{y}(t)$ as ARMA processes

$$\begin{aligned} D(q^{-1})u_0(t) &= C(q^{-1})e(t) , \\ C(q^{-1}) &= 1 + c_1 q^{-1} + \cdots + c_{n_c} q^{-n_c} , \\ D(q^{-1}) &= 1 + d_1 q^{-1} + \cdots + d_{n_d} q^{-n_d} , \\ \mathsf{E}\{e(t)e(s)\} &= \lambda_e \delta_{t,s} . \end{aligned} \tag{4.24}$$

The output noise model is

$$\begin{aligned} H(q^{-1})\tilde{y}(t) &= F(q^{-1})e_y(t) , \\ F(q^{-1}) &= 1 + f_1 q^{-1} + \cdots + f_{n_f} q^{-n_f} , \\ H(q^{-1}) &= 1 + h_1 q^{-1} + \cdots + h_{n_h} q^{-n_h} , \\ \mathsf{E}\left\{e_y(t)e_y(s)\right\} &= \lambda_y \delta_{t,s} , \end{aligned} \tag{4.25}$$

and the input noise model is

$$\begin{aligned} M(q^{-1})\tilde{u}(t) &= K(q^{-1})e_u(t) , \\ K(q^{-1}) &= 1 + k_1 q^{-1} + \cdots + k_{n_k} q^{-n_k} , \\ M(q^{-1}) &= 1 + m_1 q^{-1} + \cdots + m_{n_m} q^{-n_m} , \\ \mathsf{E}\{e_u(t)e_u(s)\} &= \lambda_u \delta_{t,s} . \end{aligned} \tag{4.26}$$

The total parameter vector to be estimated is

$$\begin{aligned} \boldsymbol{\vartheta} = \big(&a_1 \ldots a_{n_a}\ b_1 \ldots b_{n_b}\ c_1 \ldots c_{n_c}\ d_1 \ldots d_{n_d}\ f_1 \ldots f_{n_f}, \\ &h_1 \ldots h_{n_h}\ k_1 \ldots k_{n_k}\ m_1 \ldots m_{n_m}\ \lambda_e\ \lambda_y\ \lambda_u\big)^T . \end{aligned} \tag{4.27}$$

The identifiability problem will then concern whether or not the parameter vector $\boldsymbol{\vartheta}$ can be uniquely recovered from the spectrum $\boldsymbol{\Phi}_\zeta(\omega)$, (4.2), of the measured input–output data. Assuming for simplicity that the polynomial degrees are known, and using ˆ to denote the estimated (or model) quantities, the equations determining the identifiability properties will be

$$\frac{\hat{B}\hat{B}^*}{\hat{A}\hat{A}^*}\frac{\hat{C}\hat{C}^*}{\hat{D}\hat{D}^*}\hat{\lambda}_e + \frac{\hat{F}\hat{F}^*}{\hat{H}\hat{H}^*}\hat{\lambda}_y \equiv \frac{BB^*}{AA^*}\frac{CC^*}{DD^*}\lambda_e + \frac{FF^*}{HH^*}\lambda_y , \tag{4.28}$$

$$\frac{\hat{B}}{\hat{A}}\frac{\hat{C}\hat{C}^*}{\hat{D}\hat{D}^*}\hat{\lambda}_e \equiv \frac{B}{A}\frac{CC^*}{DD^*}\lambda_e , \tag{4.29}$$

$$\frac{\hat{C}\hat{C}^*}{\hat{D}\hat{D}^*}\hat{\lambda}_e + \frac{\hat{K}\hat{K}^*}{\hat{M}\hat{M}^*}\hat{\lambda}_u \equiv \frac{CC^*}{DD^*}\lambda_e + \frac{KK^*}{MM^*}\lambda_u . \tag{4.30}$$

The identities in (4.28)–(4.30) are to hold for all frequencies. The right-hand sides represent the components of the true spectra. The left-hand sides are rational functions, parameterized with the vector $\boldsymbol{\vartheta}$, (4.27). Equations (4.28)–(4.30) are rewritten by comparing how the both sides are written in factorized form, a system of nonlinear equations with $\boldsymbol{\vartheta}$ as unknown is obtained. Some simple examples illustrating the principle are given below. Trivially, the true value of the parameter vector will satisfy the identities. The system is identifiable if there is no other value of the parameter vector ϑ that satisfies the identities.

Now consider some different scenarios, describing different special cases.

AN3a. Both $\tilde{y}(t)$ and $\tilde{u}(t)$ are ARMA processes, as in (4.25) and (4.26).

AN3b. The output noise $\tilde{y}(t)$ is an ARMA process, while the input noise $\tilde{u}$ is white. This means that $n_k = n_m = 0$ in (4.26).

AN3c. Both $\tilde{y}(t)$ and $\tilde{u}(t)$ are white noise sequences. This means that $n_f = n_h = 0$ in (4.25) and $n_k = n_m = 0$ in (4.26).

In order to give some insight, some simple cases are presented first, before going to a more general result.

Example 4.2 Assume that Assumption **AN3b** applies and further that

$$n_d > n_c \tag{4.31}$$

holds. Cf. also (3.26). In this case, it is not difficult to find out that the system is identifiable. Note that here $\phi_{\tilde{u}} = \lambda_u$. As the polynomials C, D, $\hat{C}$ and $\hat{D}$ have all zeros inside the unit circle, it follows by considering the denominators in (4.30) that $\hat{D} = D$. Further, (4.30) then implies

$$\hat{C}\hat{C}^*\hat{\lambda}_e + DD^*\hat{\lambda}_u \equiv CC^*\lambda_e + DD^*\lambda_u \ .$$

The terms here consist of sums over $\mathrm{e}^{ik\omega}$, with k ranging from $-n_d$ to n_d. Examining the specific terms with $k = n_d$ and invoking (4.31) it follows that $\hat{\lambda}_u = \lambda_u$. Then spectral factorization, see, e.g., Söderström (2002), gives $\hat{C} = C$, $\hat{\lambda}_e = \lambda_e$. So far, the conclusion is that the spectrum of the measured input can be uniquely decomposed in the effect of the input noise and the spectra of the noise-free input. It then follows from (4.29) that $\hat{A} = A$ and $\hat{B} = B$, while (4.28) finally gives $\hat{\phi}_{\tilde{y}} = \phi_{\tilde{y}}$, leading to $\hat{F} = F$, $\hat{H} = H$, $\hat{\lambda}_y = \lambda_y$. ■

In the above example, the degree condition (4.31) is crucial. When this is not fulfilled the analysis becomes a bit more complicated, and one cannot use the identifiability equations (4.28)–(4.30) just one by one as in Example 4.2.

Example 4.3 Assume still that Assumption **AN3b** applies and consider the specific case when the noise-free input signal is white noise. This means that

$$n_c = 0, \ n_d = 0 \ .$$

In this case the identifiability equations (4.28)–(4.30) become

$$\frac{\hat{B}\hat{B}^*}{\hat{A}\hat{A}^*}\hat{\lambda}_e + \hat{\phi}_{\tilde{y}} \equiv \frac{BB^*}{AA^*}\lambda_e + \phi_{\tilde{y}} \ , \tag{4.32}$$

$$\frac{\hat{B}}{\hat{A}}\hat{\lambda}_e \equiv \frac{B}{A}\lambda_e \ , \tag{4.33}$$

$$\hat{\lambda}_e + \hat{\lambda}_u \equiv \lambda_e + \lambda_u \ . \tag{4.34}$$

Here, (4.33) implies

$$\hat{A} = A, \ \hat{B} = \alpha B, \ \hat{\lambda}_e = \lambda_e/\alpha \ , \tag{4.35}$$

where α so far is an arbitrary constant. Equation (4.34) does not give any useful additional information, but implies

$$\hat{\lambda}_u = \lambda_u + \left(1 - \frac{1}{\alpha}\right)\lambda_e \ . \tag{4.36}$$

If the system is identifiable, then the constant α must be determined from (4.32) which in rewritten form is

$$\alpha \frac{BB^*}{AA^*}\lambda_e + \hat{\phi}_{\tilde{y}} \equiv \frac{BB^*}{AA^*}\lambda_e + \phi_{\tilde{y}} \ . \tag{4.37}$$

In case $\tilde{y}(t)$ is white noise, and $n_b \leq n_a$, then it follows from (4.37) as in Example 5.1 that $\alpha = 1$ and $\hat{\phi}_{\tilde{y}} = \phi_{\tilde{y}}$.

Another possibility to construct sufficient conditions for identifiability is to exploit the ARMA model (4.25) of $\tilde{y}(t)$. Equation (4.37) can then be rewritten as

$$\frac{\hat{F}\hat{F}^*}{\hat{H}\hat{H}^*}\hat{\lambda}_y \equiv \frac{FF^*}{HH^*}\lambda_y + (1-\alpha)\frac{BB^*}{AA^*}\lambda_e \ . \tag{4.38}$$

Assume that A has at least one zero that is not a zero of H. To have compatible denominator degrees on both sides of (4.38) it then follows that $\alpha = 1$, which quickly leads to $\hat{F} = F, \ \hat{H} = H, \ \hat{\lambda}_y = \lambda_y$. ■

Finally treat the case when the noise-free input is a general ARMA process, (4.24). Instead of conditions on the degrees of different polynomials, a certain condition on the system zeros is introduced.

Example 4.4 Let $\tilde{u}(t)$ be white noise (**AN3b** applies), and let $u_0(t)$ be a general ARMA process, (4.24), where no specific conditions between n_c and n_d is imposed, except $n_c + n_d > 0$ (meaning that $u_0(t)$ is not white noise). Assume **AS3** applies, so that the system has no pair of zeros reflected in the unit circle, that is

$$B(z_1) = 0 \Rightarrow B(z_1^{-1}) \neq 0 \ . \tag{4.39}$$

As in Example 5.1 one finds that $\hat{D} = D, \ \hat{A} = A$. Then (4.29) can be rewritten as

$$\hat{B}\hat{C}\hat{C}^*\hat{\lambda}_e \equiv BCC^*\lambda_e \ . \tag{4.40}$$

Using the assumption (4.39), it then follows that B cannot have any common factor with $\hat{C}\hat{C}^*$, and one can conclude that $\hat{C} = C$. Equation (4.30) leads to

$$CC^*\left(\hat{\lambda}_e - \lambda_e\right) \equiv DD^*\left(\lambda_u - \hat{\lambda}_u\right) \tag{4.41}$$

and due to the assumption $n_c + n_d > 0$, C and D being coprime, one can conclude that $\hat{\lambda}_e = \lambda_e$, $\hat{\lambda}_u = \lambda_u$ and finally (4.40) implies $\hat{B} = B$. ■

To summarize so far, sufficient conditions for identifiability have been established by requiring that the system zeros do not appear in pairs reflected in the unit circle. Alternatively, one has to impose some constraints on the polynomial degrees. It is also an underlying assumption that the correct polynomial degrees are known.

It is possible to extend the identifiability analysis in Examples 4.2–4.4. The most general result known is due to Agüero (2005), Agüero and Goodwin (2006), and runs as follows.

Theorem 4.1 *Let the noise-free input be an ARMA process as in (4.24), the output noise an ARMA process as in (4.25) and the input noise an ARMA process as (4.26). (Hence Assumptions* **AN3a** *and* **AI4** *apply). Assume that*

- *$B(z)$ has no zero that is mirrored in the unit circle (i.e., it is not allowed to be also a zero of $B(z^{-1})$),*
- *$B(z)$ has no zero that is also a zero of $D(z^{-1})$,*
- *$A(z)$ has no zero that is also a zero of $D(z)$.*

Then the system is identifiable, if any of the following additional assumptions holds:

1. *There exists at least one zero of $D(z)$ that is not a zero of $M(z)$.*
2. *There exists at least one zero of $A(z)$ that is not a zero of $H(z)$.*
3. *There exists at least one zero of $D(z)$ that is not a zero of $H(z)$.*
4. *The polynomial degrees satisfy*

$$n_m - n_k > n_d - n_c \ . \tag{4.42}$$

5. *The polynomial degrees satisfy*

$$n_h - n_f > (n_d - n_c) + (n_a - n_b) \ . \tag{4.43}$$

Note that the expressions in the inequalities (4.42) and (4.43) are expressed in terms of the pole excess of various filters. ■

4.3 Identifiability When Using Multiple Experiments

As seen in the previous sections, there is in general a fundamental lack of identifiability for error-in-variables system, unless some additional assumption or condition is added. In this section two cases are considered, both discussed in the literature, where data are available from two or more experiments, and these experiments have some valuable features to exploit.

Example 4.5 One possibility is to consider the case where there are two experiments, for which the noise-free input has *different* characters, while the measurement noise properties remain the same. This means that Assumption **AE2a** applies. Such a situation is treated in Markovsky et al. (2006a), although it is there described as that $u_0(t)$ changes character at some point of time in a single experiment.

Let $\phi_1(\omega)$ and $\phi_2(\omega)$ denote the spectrum $\phi_{u_0}(\omega)$ in experiment 1 and 2, respectively. In this case the equations governing the identifiability properties can be written, cf (4.4):

$$\begin{pmatrix} G \\ 1 \end{pmatrix} \begin{pmatrix} G^* & 1 \end{pmatrix} \phi_k + \begin{pmatrix} \phi_{\tilde{y}} & 0 \\ 0 & \phi_{\tilde{u}} \end{pmatrix} \equiv \begin{pmatrix} \hat{G} \\ 1 \end{pmatrix} \begin{pmatrix} \hat{G}^* & 1 \end{pmatrix} \hat{\phi}_k + \begin{pmatrix} \hat{\phi}_{\tilde{y}} & 0 \\ 0 & \hat{\phi}_{\tilde{u}} \end{pmatrix}, \quad k = 1, 2 \, . \tag{4.44}$$

Evaluating the 12-block for $k = 1, 2$ gives

$$\hat{G} = G \frac{\phi_1}{\hat{\phi}_1} = G \frac{\phi_2}{\hat{\phi}_2} \, , \tag{4.45}$$

which directly leads to

$$\hat{\phi}_1 = \phi_1 \frac{\hat{\phi}_2}{\phi_2} \, . \tag{4.46}$$

Using this when evaluating the 22-block for $k = 1, 2$ gives

$$\begin{aligned} \phi_1 + \phi_{\tilde{u}} &= \phi_1 \frac{\hat{\phi}_2}{\phi_2} + \hat{\phi}_{\tilde{u}} \, , \\ \phi_2 + \phi_{\tilde{u}} &= \hat{\phi}_2 + \hat{\phi}_{\tilde{u}} \, . \end{aligned}$$

Subtracting the two equations above leads to

$$\phi_1 - \phi_2 = \frac{\hat{\phi}_2}{\phi_2} (\phi_1 - \phi_2) \, .$$

As it was assumed that $\phi_1 - \phi_2 \neq 0$, one can conclude that

$$\frac{\hat{\phi}_2}{\phi_2} = 1 \Rightarrow \hat{\phi}_2 = \phi_2 \Rightarrow \hat{\phi}_1 = \phi_1 \Rightarrow \hat{G} = G \, . \tag{4.47}$$

Further (4.47) implies that $\hat{\phi}_{\tilde{y}} = \phi_{\tilde{y}}$ while $\hat{\phi}_{\tilde{u}} = \phi_{\tilde{u}}$ follows from the above calculations. ■

Example 4.6 Another scenario with more than one experiment is based on Assumption **AE2b**. Then the measurement noises $\tilde{u}(t)$, $\tilde{y}(t)$ are uncorrelated between different experiments. The true noise-free input $u_0(t)$ is correlated between the experiments.

Such a situation may occur when making repeated experiments for determining some system properties as explained in Schoukens et al. (1997), Pintelon and Schoukens (2001). In contrast to the analysis done in Example 4.5, one here exploits the correlation of data between the experiments to show identifiability. Let u_k, u_{0k}, y_k, y_{0k} denote u, u_0, y, y_0 in experiment k. It is easy to see that the cross-spectra satisfy for example

$$\phi_{y_{01}u_{02}} = G\phi_{u_{01}u_{02}} . \tag{4.48}$$

Note that $\phi_{u_{01}u_{02}}$ is nonzero (for almost all frequencies) due to the assumption of correlation between the experiments. As the noise contribution is independent between the experiments, it also holds that

$$\phi_{u_1u_2} = \phi_{u_{01}u_{02}}, \qquad \phi_{y_1u_2} = \phi_{y_{01}u_{02}} . \tag{4.49}$$

and therefore

$$\phi_{y_1u_2} = G\phi_{u_1u_2} , \tag{4.50}$$

from which the system transfer function G can be determined. ■

4.4 Closed-Loop Operation

It may seem as a paradox to consider identification of an EIV system from closed-loop operations, as EIV systems are typically characterized by the true input not being accessible to the experimentalist, while a feedback system (with a known feedback) would provide precisely a known input.

Two different cases will be discussed in this section, and it will be shown that based on the particular situation, identifiability may or may not apply.

In both cases, identifiability from second-order statistics is studied, as done earlier in this chapter. What matters is therefore the spectrum $\boldsymbol{\Phi}_\zeta(\omega)$ of the measured data.

The simplest closed-loop case considered here is the one depicted in Fig. 4.2. The feedback will in this case be given by

$$u_0(t) = r(t) - Fy(t) . \tag{4.51}$$

Here, it is reasonable to assume that both the transfer function F and $r(t)$ are unknown, as otherwise $u_0(t)$ would be known (which is a contradiction). Hence, this case is characterized by

- **Known**: Measurements of $u(t),\ y(t)$.
- **Unknowns**: $G,\ \phi_{u_0},\ \lambda_u,\ \lambda_y,\ r(t),\ F$.

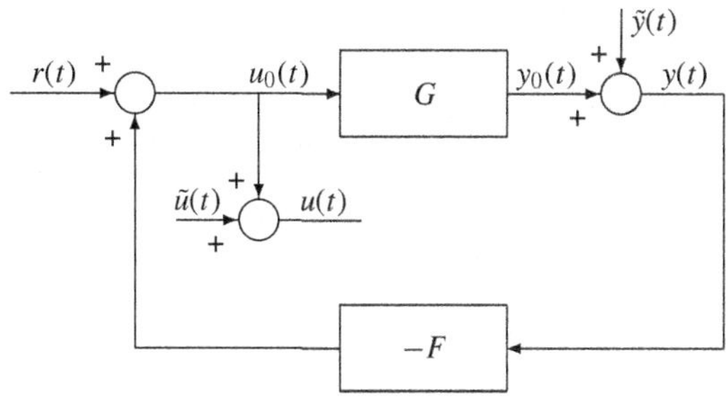

Fig. 4.2 An errors-in-variables system under closed-loop operation. Simplest case

Based on block diagram-related algebra, it is straightforward to show that the recorded input–output data is described by

$$y(t) = \frac{G}{1+GF} r(t) + \frac{1}{1+GF} \tilde{y}(t) , \tag{4.52}$$

$$u(t) = \frac{1}{1+GF} r(t) + \frac{-F}{1+GF} \tilde{y}(t) + \tilde{u}(t) . \tag{4.53}$$

Then the spectrum of $\zeta(t)$, based on this relation, is

$$\begin{aligned} \boldsymbol{\Phi}_\zeta = &\begin{pmatrix} G \\ 1 \end{pmatrix} \begin{pmatrix} G^* & 1 \end{pmatrix} \frac{1}{|1+GF|^2} \phi_r \\ &+ \begin{pmatrix} 1 \\ -F \end{pmatrix} \begin{pmatrix} 1 & -F^* \end{pmatrix} \frac{1}{|1+GF|^2} \phi_{\tilde{y}} + \begin{pmatrix} 0 & 0 \\ 0 & \lambda_u \end{pmatrix} . \end{aligned} \tag{4.54}$$

Note that the open-loop situation would correspond to the special case $F = 0$. Due to the increase of the number of unknowns, the identifiability problem is much worse in the closed-loop case compared to the open-loop situation. The same set of known variables apply in both cases, but for closed-loop operation, in addition the feedback transfer function is also unknown. Identification is therefore *not feasible* in this setup.

A somewhat more complicated case is shown in Fig. 4.3. There are physical interpretations for this case. The block H can be seen as a controller followed by an actuator, with unknown and even complicated dynamics. The block F may represent the known or calibrated sensor characteristics and the signal $r(t)$ can be the output of a waveform generator used as the reference signal to the closed-loop control system. Possible applications include

- Determining an impedance from noisy measurements of current and voltage.
- Determining the mechanical correspondence by exciting a shaker in a mechanical circuit and using noisy measurements of force and acceleration.

See Söderström et al. (2013) for details of these two applications.

Consider the setup in Fig. 4.3, and assume H to be unknown, but $r(t)$ and F to be known and under control by the experimentalist. Thus this case can be characterized as follows.

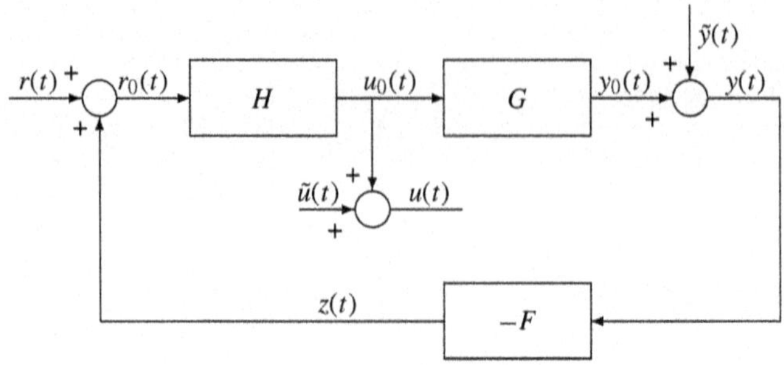

Fig. 4.3 An errors-in-variables system under closed-loop operation. Cascaded case

- **Known**: Measurements of $u(t),\ y(t),\ F,\ r(t),\ r_0(t)$ (or equivalently measurements of $u(t),\ y(t),\ z(t)$ and $r(t)$).
- **Unknowns**: $G,\ \phi_{u_0},\ \lambda_u,\ \lambda_y,\ H$.

Based on the block diagram in Fig. 4.3, it is found that in this case

$$y(t) = \frac{GH}{1+GHF} r(t) + \frac{1}{1+GHF} \tilde{y}(t) \ , \tag{4.55}$$

$$u(t) = \frac{H}{1+GHF} r(t) + \frac{-FH}{1+GHF} \tilde{y}(t) + \tilde{u}(t) \ . \tag{4.56}$$

Use this to write the spectrum of $\boldsymbol{\zeta}(t)$ as

$$\begin{aligned} \boldsymbol{\Phi}_\zeta &= \begin{pmatrix} G \\ 1 \end{pmatrix} \begin{pmatrix} G^* & 1 \end{pmatrix} \frac{|H|^2}{|1+GHF|^2} \phi_r \\ &+ \begin{pmatrix} 1 \\ -FH \end{pmatrix} \begin{pmatrix} 1 & -F^*H^* \end{pmatrix} \frac{1}{|1+GHF|^2} \phi_{\tilde{y}} + \begin{pmatrix} 0 & 0 \\ 0 & \lambda_u \end{pmatrix} \ . \end{aligned} \tag{4.57}$$

The fact that $r(t)$ is known or can be designed is very important. It leads to two different particular consequences.

- First, one can use the measured signal values of $r(t)$ and estimate the transfer function directly as a ratio between two cross-spectra

$$G = \frac{\phi_{yr}}{\phi_{ur}} \ , \tag{4.58}$$

  which also shows that the system is identifiable.
- A second meaning is that the experimentalist can design the external input $r(t)$ to be used for the experiments. In particular, one can let $r(t)$ be a periodic signal, which corresponds to a series of experiments where $r(t)$ is the same, but the noise sequences can be assumed to be uncorrelated between the experiments. This fact can be exploited to give quite general estimators, such as the sample maximum

likelihood (SML), Schoukens et al. (1997) and Sect. 12.3.2, and the IV-related approach described in Söderström and Hong (2005) and Sect. 12.2.

It follows from the analysis of the approaches in Schoukens et al. (1997) and Söderström and Hong (2005) that these approaches can even be extended and applied if the link H is nonlinear. What matters is not a linear dependence of $y(t)$ and $u(t)$ on $r(t)$, but the fact that all influence of $r(t)$ on $u(t)$ and $y(t)$ lead to periodic functions. Further, in these estimators the possible knowledge of the feedback block F and the signal $r(t)$ is *not exploited*. Indeed, it is worth stressing that knowing that the reference is periodic is less demanding than assuming that it is completely known.

4.5 Extensions and Bibliographical Notes

The paper Anderson and Hsiao (1981) presents an early analysis of identifiability and the use of the maximum likelihood method for a simple EIV case. In the characterization of the general solutions in this chapter, Assumptions **AS2** and **AS3** were imposed. A more detailed identifiability analysis is carried out in Anderson (1985), Deistler (1986), Anderson et al. (1987), Deistler and Anderson (1989). In general, there are $n_z + 1$ degrees of freedom for characterizing the class of systems that match a given spectrum of the measured signals. Here n_z is the number of non-minimum phase zeros. Identifiability of multivariable systems is treated in Green and Anderson (1986). An identifiability analysis covering also non-causal systems can be found in Anderson and Deistler (1984). The paper Bloch (1989) treats identifiability of dynamic errors-in-variables problems using the factor analysis approach.

A general system theoretic description of all observationally equivalent systems is found in Scherrer and Deistler (1998). Starting from the fact that the knowledge of the second-order moments is not sufficient for uniquely identifying the underlying system, a class of observationally equivalent systems is derived and the mapping relating this class to the spectral densities of the observations is investigated.

The parametric identifiability problem has been dealt with by several authors. An extensive analysis with various previous results as special cases are given in Agüero (2005), Agüero and Goodwin (2006), which extends results in Agüero et al. (2005). Further identifiability conditions for EIV problems are derived in Agüero and Goodwin (2008). In the frequency domain an essential assumption for identifiability is that the noisy input–output signals $u(t)$, $y(t)$ have rational spectra, Castaldi and Soverini (1996). In this case the identifiability of the EIV system is ensured even if the orders of the processes are not *a priori* known, provided that no zero/pole cancellation occurs between the transfer function $G(q^{-1})$ and the ARMA model of the noise-free input $u_0(t)$, and all the ARMA processes involved in the EIV representation do not share common poles. Identifiability of systems with non-stationary data is examined in Agüero et al. (2008). In particular, it is shown how identifiability holds under weak assumptions if the non-stationarity is due to some shifts in the system structure.

An analysis of the identifiability of EIV problems, with an emphasis of checkable conditions (from measured data, or rather the joint input–output spectrum), is given in Bottegal et al. (2011). For an early examination of the non-uniqueness decompositions such as (4.4), see Picci et al. (1993).

Other results are more specific for various special cases. For example, identifiability under the noise assumption **AN3c** is investigated in Castaldi and Soverini (1996), Söderström (2003), Stoica and Nehorai (1987). The situation of the output noise being an ARMA process and the input noise being white, Assumption **AN3b**, is treated in Söderström (1980) and Solo (1986). The more general case **AN3a** where both $\tilde{u}(t)$ and $\tilde{y}(t)$ are ARMA processes is coped with in Nowak (1985), Nowak (1993), Castaldi and Soverini (1996), Agüero (2005), Agüero and Goodwin (2006). Generalization to the multivariate case is considered in Nowak (1992). Analysis of a set membership approach, where models consistent with given equations and constraints on measurements and error bounds, is considered in Casini et al. (2014).

Identification of continuous-time EIV models was considered in Mahata and Garnier (2006). The essence of this approach is to use BELS techniques (see Chap. 7) to handle the noise effects and state variable filters to handle the discrete-time data combined with a continuous-time model.

Use of higher-order statistics is proposed in Deistler (1986).

Somewhat related to the EIV problem is to consider identification of cascade systems, from noisy output measurements. See Wahlberg et al. (2008), Wahlberg et al. (2009a), Wahlberg et al. (2009b).

Closed-loop Operation

The analysis in Section 4.4 is developed in more detail in Söderström et al. (2013). More recent works treat more elaborated feedback structures. In particular, EIV problems in networks where certain blocks are to be identified from noisy measurements is the topic of Dankers et al. (2015). See also Dankers et al. (2014), Mansoori et al. (2014), Dankers et al. (2016), Risuleo et al. (2017d), Weerts et al. (2017), Galrinho et al. (2017), Linder and Enqvist (2017a), Gevers et al. (2017), Yue et al. (2017), Dankers et al. (2017) and Weerts et al. (2018).

An early consideration of how to treat identification of a system operating in closed loop in an EIV setting is given in Pintelon et al. (1992). Estimation of time-variant frequency responses of certain multivariate time-variant systems operating in feedback is considered in Pintelon et al. (2013b), Pintelon et al. (2017).

The paper Geng et al. (2011) relates to the approach of using a model set rather than one model as the outcome and treats closed-loop operation. A similar analysis of MIMO models with bounded noise appears in Geng et al. (2012).

The paper Wang et al. (2007) treats identification of EIV problems where finite impulse response (FIR) models of $G(q^{-1})$ are sought. The methodology is based on the assumption that the noise-free input $u_0(t)$ is cyclo-stationary.

Identification of EIV systems in closed loop with a known controller and using continuous-time models is studied in Victor et al. (2017).

Chapter 5
Modeling Aspects

In this chapter modeling and further parameterization aspects are discussed in some more detail. Notations are also introduced, in particular for linear regression models that will be used throughout the book.

The chapter is organized as follows. In the next section the EIV problem as introduced in Chap. 3 is restated, and the linear regression modeling and pertinent notations are introduced. Section 5.2 will discuss how more general structures including an arbitrary delay can easily be handled. Using continuous-time models and (approximate) conversions to discrete-time models is the topic of Sect. 5.3. In several cases there is a need also to model the properties of the noise, which is treated in Sect. 5.4. Section 5.5 is devoted to modeling the dynamics in the frequency domain. Sometimes also the noise-free input is modeled. Then the total system can be described using state space techniques, as explained in Sect. 5.6. Extension of the linear regression model to multivariable systems is presented in Sect. 5.7. Section 5.8 shows a way to classify different possible estimation methods for the EIV problem based on how the data are used. Section 5.9 describes some approaches for how to determine the model order from empirical data. (In the rest of the book it is assumed that the model order is known.) Finally, Sect. 5.10 contains some bibliographical notes and comments on extensions.

5.1 Problem Statement and Notations

Now briefly repeat the EIV problem as stated in Chap. 3.

Consider the system depicted in Fig. 5.1 with noise-corrupted input and output measurements. The noise-free input is denoted by $u_0(t)$ and the undisturbed output

T. Söderström, *Errors-in-Variables Methods in System Identification*, Communications and Control Engineering,
https://doi.org/10.1007/978-3-319-75001-9_5

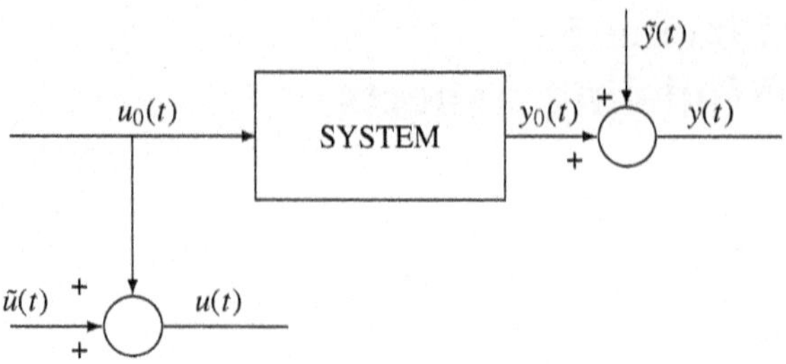

Fig. 5.1 The basic setup for an errors-in-variables problem

by $y_0(t)$. They are linked through the linear difference equation; see Assumption **AS3**,

$$A(q^{-1})\, y_0(t) = B(q^{-1})\, u_0(t)\ , \tag{5.1}$$

where $A(q^{-1})$ and $B(q^{-1})$ are polynomials in the backward shift operator q^{-1}, i.e., $q^{-1}x(t) = x(t-1)$. More precisely,

$$\begin{aligned} A(q^{-1}) &= 1 + a_1 q^{-1} + \cdots + a_{n_a} q^{-n_a}\ , \\ B(q^{-1}) &= b_1 q^{-1} + \cdots + b_{n_b} q^{-n_b}\ . \end{aligned} \tag{5.2}$$

Assume that the observations are corrupted by additive measurement noises $\tilde{u}(t)$ and $\tilde{y}(t)$ of zero mean. The available signals are thus of the form

$$\begin{aligned} u(t) &= u_0(t) + \tilde{u}(t)\ , \\ y(t) &= y_0(t) + \tilde{y}(t)\ . \end{aligned} \tag{5.3}$$

The estimation problem is as follows. Given the noisy input–output data $u(1)$, $y(1), \ldots u(N)$, $y(N)$, determine an estimate of the system parameter vector

$$\boldsymbol{\theta} = \begin{pmatrix} a_1 & \ldots & a_{n_a} & b_1 & \ldots & b_{n_b} \end{pmatrix}^T . \tag{5.4}$$

In the model it is assumed that the integers fulfill $n_a \geq 0,\ n_b \geq 1$.

The following notations will be convenient. Introduce the regressor vector

$$\boldsymbol{\varphi}(t) = \begin{pmatrix} -y(t-1) & \ldots & -y(t-n_a) & u(t-1) & \ldots & u(t-n_b) \end{pmatrix}^T , \tag{5.5}$$

which is compatible with the parameter vector (5.4). Then one may write the system as a *linear regression*

$$y(t) = \boldsymbol{\varphi}^T(t)\boldsymbol{\theta} + \varepsilon(t)\ , \tag{5.6}$$

where the noise effects are collected in the term

$$\varepsilon(t) = A(q^{-1})\tilde{y}(t) - B(q^{-1})\tilde{u}(t) \,. \tag{5.7}$$

Further, use the following conventions:

- $\boldsymbol{\theta}_0$ denotes the true parameter vector, and $\hat{\boldsymbol{\theta}}$ its estimate.
- Similarly, let $A_0(q^{-1}),\ B_0(q^{-1})$ denote the true values of $A(q^{-1})$, $B(q^{-1})$, respectively.
- $\boldsymbol{\varphi}_0(t)$ denotes the noise-free part of the regressor vector:

$$\boldsymbol{\varphi}_0(t) = \big(-y_0(t-1) \ldots -y_0(t-n_a)\ u_0(t-1) \ldots u_0(t-n_b)\big)^T \,. \tag{5.8}$$

- $\tilde{\boldsymbol{\varphi}}(t)$ denotes the noise-contribution to the regressor vector. This means that

$$\tilde{\boldsymbol{\varphi}}(t) = \big(-\tilde{y}(t-1) \ldots -\tilde{y}(t-n_a)\ \tilde{u}(t-1) \ldots \tilde{u}(t-n_b)\big)^T \,, \tag{5.9}$$

and it holds that

$$\boldsymbol{\varphi}(t) = \boldsymbol{\varphi}_0(t) + \tilde{\boldsymbol{\varphi}}(t) \,. \tag{5.10}$$

Further introduce the following partitions of the parameter vector

$$\boldsymbol{\theta} = \begin{pmatrix} \mathbf{a} \\ \mathbf{b} \end{pmatrix}, \quad \mathbf{a} = \begin{pmatrix} a_1 \\ \vdots \\ a_{n_a} \end{pmatrix}, \quad \mathbf{b} = \begin{pmatrix} b_1 \\ \vdots \\ b_{n_b} \end{pmatrix}, \quad \bar{\mathbf{a}} = \begin{pmatrix} 1 \\ \mathbf{a} \end{pmatrix}. \tag{5.11}$$

For future use, introduce also the extended parameter vector

$$\overline{\boldsymbol{\theta}} = \begin{pmatrix} 1 \\ \boldsymbol{\theta} \end{pmatrix} \tag{5.12}$$

and the extended regressor vector

$$\begin{aligned} \overline{\boldsymbol{\varphi}}(t) &= \Big(-y(t) \ldots \ -y(t-n_a)\ u(t-1) \ldots \ u(t-n_b)\Big)^T \\ &= \Big(-y(t)\ \ \boldsymbol{\varphi}^T(t)\Big)^T \,. \end{aligned} \tag{5.13}$$

Using the above notations and the system description (5.1), it follows that

$$\overline{\boldsymbol{\varphi}}_0^T(t)\overline{\boldsymbol{\theta}}_0 = \Big(-y_0(t)\ \ \boldsymbol{\varphi}_0^T(t)\Big)\begin{pmatrix} 1 \\ \boldsymbol{\theta}_0 \end{pmatrix} = -A(q^{-1})y_0(t) + B(q^{-1})u_0(t) = 0 \,. \tag{5.14}$$

5.2 Using Models with an Arbitrary Delay

Sometimes one would like to use a more arbitrary model than (5.1), of the form

$$A(q^{-1})y_0(t) = q^{-k}B(q^{-1})u_0(t) \, , \tag{5.15}$$

where the extra delay k is an integer that can take positive or negative values. For example, by setting $k = -1$ one arrives in (5.15) to a model with no delay. As will be seen in Sect. 5.3 there are in fact good reasons to apply such models with $k = -1$ in practice.

Next it is shown that by a simple transformation one can easily convert the more general model of the form (5.15) to one with $k = 0$, that is (5.1). This will be an argument why the perhaps more standard form (5.1) has been adopted in the future treatment of the EIV problem.

Given the signals $y(t)$ and $u(t)$, introduce transformed signals as

$$\begin{aligned} \overline{y}(t) &= y(t) \, , \\ \overline{u}(t) &= q^{-k}u(t) = u(t-k) \, . \end{aligned} \tag{5.16}$$

The transformed signals consist of a noise-free part and a noisy part, just as the untransformed signals. Hence

$$\begin{aligned} \overline{y}(t) &= y_0(t) + \tilde{y}(t) \, , \\ \overline{u}(t) &= u_0(t-k) + \tilde{u}(t-k) \, . \end{aligned} \tag{5.17}$$

Apparently it holds

$$A(q^{-1})\overline{y}_0(t) = B(q^{-1})\overline{u}_0(t) \, , \tag{5.18}$$

$$A(q^{-1})\left(\overline{y}(t) - \tilde{y}(t)\right) = B(q^{-1})\left(\overline{u}(t) - \tilde{u}(t-k)\right) \, . \tag{5.19}$$

As a delay in the input noise $\tilde{u}$ does not change its properties, and it is never directly observed, the model (5.19) has precisely the same statistical properties as the simpler 'standard' model (5.1). Phrased differently, by just applying a shift to the input data as in (5.16) one can effectively apply the more general model (5.15) by using the standard model (5.1) on the shifted data (5.16).

5.3 Continuous-Time EIV Models and Conversion to Discrete-Time

System identification is mostly applied to sampled, discrete-time data. However, the underlying dynamics is often continuous in time. If the input signal is under control of the experimenter, its behavior during the sampling intervals is often known. It is, for example, common that it is kept constant between the sampling instants. Then a continuous-time (linear) model can be transformed into a discrete-time one without any approximation. In other situations, such as for errors-in-variables problems, the input behavior during the sampling intervals is typically unknown and mostly not under full control by the experimenter.

When the input behavior during the sampling intervals is not known, it becomes trickier how to carry out an identification exercise. Further, for a given continuous-time system it is not obvious what the (best) associated discrete-time model should be. The discrete-time (or sampled) model should aim at describing the behavior between the input and the output variables at the sampling instants.

- As the noise-free input $u_0(t)$ cannot be measured, it may be appropriate to model it in continuous-time, while the measurement mechanism generates discrete-time data $\{y(t)\}$, $\{u(t)\}$. Various approaches for such continuous-time modeling of an EIV problem from discrete-time data are given in Söderström (2009). Some specific algorithms based on Shinbrot functionals have appeared earlier in Pearson and Lee (1985), Pearson and Shen (1993).
- Continuous-time EIV modeling is also considered in Mahata and Garnier (2005). State variable filtering of the observations is applied, as a means to avoid explicit differentiation. Two different state variables filters are used with the purpose of determining the noise variances λ_u and λ_y, in addition to the system dynamics.

Various aspects on approximating continuous-time models with discrete-time ones are given in Sect. 5.A.1 of Appendix 5.A.

When the inter-sample variation of the input is not known, one has to work with approximations. If the input is changing relatively slowly, it is shown that most approximation methods produce rather similar results for low frequencies.

One might therefore believe from this observation that a reasonable way of finding a continuous-time model from discrete-time data would be to first fit a discrete-time model and then convert it to continuous-time using the inverse operation of zero-order-hold (ZOH) sampling. *This is though not a good idea when the inter-sample behavior of the input is not known.* As an illustration a simple example is given.

Example 5.1 Consider a first-order continuous-time system given by

$$\dot{y}(t) + ay(t) = bu(t), \quad a = b = 1 \ . \tag{5.20}$$

Let it be excited by a (continuous-time) sine wave,

$$u(t) = \sin(\omega t) \ , \tag{5.21}$$

and assume the input u and the output y are observed in discrete-time (sampling interval being h), without any measurement noise. Let a discrete-time first-order model

$$y(t+h) + \alpha y(t) = \beta u(t) \tag{5.22}$$

be fitted using the least squares method, and consider the asymptotic case $N \to \infty$. The estimates are hence given by

$$\begin{pmatrix} \hat{\alpha} \\ \hat{\beta} \end{pmatrix} = \left(\lim_{N\to\infty} \frac{1}{N} \sum_{t=1}^{N} \begin{pmatrix} -y(t) \\ u(t) \end{pmatrix} \begin{pmatrix} -y(t) & u(t) \end{pmatrix} \right)^{-1} \times \left(\lim_{N\to\infty} \frac{1}{N} \sum_{t=1}^{N} y(t+h) \begin{pmatrix} -y(t) \\ u(t) \end{pmatrix} \right) \tag{5.23}$$

The model (5.22) is the converted into the continuous-time model

$$\dot{y} + \hat{a}y = \hat{b}u \tag{5.24}$$

using zero-order hold sampling, that is the relations

$$\hat{\alpha} = -\exp(-\hat{a}h) \tag{5.25}$$

$$\frac{\hat{\beta}}{1+\hat{\alpha}} = \frac{\hat{b}}{\hat{a}} \tag{5.26}$$

are exploited. In Fig. 5.2 it is illustrated how the estimates $\hat{\alpha}$, $\hat{\beta}$, $\hat{a}$, $\hat{b}$ vary with the angular frequency ω. One can note that there is a considerable bias (deviation from the true values) also for small values of ω. ■

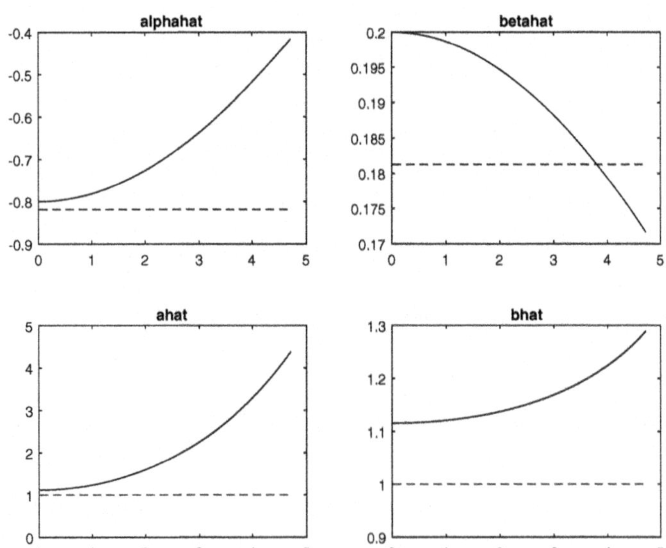

Fig. 5.2 Estimates $\hat{\alpha}$, $\hat{\beta}$ in discrete-time (upper part) and $\hat{a}$, $\hat{b}$ in continuous-time (lower part) as functions of the angular frequency ω of the sine wave. The sampling interval is $h = 0.2$. The dashed lines give the true values

In cases where important information is to be extracted from a continuous-time model and the inter-sample behavior of the input is not known, it is advisable to fit a continuous-time model directly to data.

In a sense it is natural that when the inter-sample behavior of the input signal is unknown, an accurate discrete-time model of the system should also model the input behavior, cf. Appendix 5.A.1.4.

5.4 Modeling the Noise Properties

In case $\tilde{y}(t)$ is correlated, see Assumption **AN3b**, there is often a need to estimate part of its covariance elements. For this reason introduce the noise parameter vector ρ as

$$\rho = \begin{pmatrix} r_{\tilde{y}}(0) \ \dots \ r_{\tilde{y}}(m-2) \ \lambda_u \end{pmatrix}^T . \tag{5.27}$$

which is a vector of dimension m. For the special case when the output measurement noise is white, that is Assumption **AN3c** holds, $m = 2$ applies, and one can use the notation

$$\rho = \begin{pmatrix} \lambda_y \ \lambda_u \end{pmatrix}^T . \tag{5.28}$$

The appropriate value of m in (5.27) when $\tilde{y}(t)$ is correlated will be dealt with later.

A special case of correlated output noise is so called *shock error models*. These terms appear in some econometric literature, and the linear regression model (5.1) is augmented with a noise term $v(t)$, called shock error,

$$A(q^{-1})y_0(t) = B(q^{-1})u_0(t) + v(t) . \tag{5.29}$$

This implies in particular that the output measurements

$$\begin{aligned} y(t) &= y_0(t) + \tilde{y}(t) \\ &= \frac{B(q^{-1})}{A(q^{-1})}u_0(t) + \frac{1}{A(q^{-1})}v(t) + \tilde{y}(t) \end{aligned} \tag{5.30}$$

has two noise components. Assuming $v(t)$ to be white noise, or in the generalized case an ARMA process, and regarding only the second-order properties of the data, it is found that the total output noise gets the form of a *structured ARMA process*. This means that $(1/A)v(t) + \tilde{y}(t)$ has the same covariance function as an ARMA process, but can be parameterized with fewer independent parameters. This is the same reasoning as for $y(t)$ in Sect. 3.3.

It will also be convenient to introduce the total parameter vector as

$$\boldsymbol{\vartheta} = \begin{pmatrix} \boldsymbol{\theta} \\ \boldsymbol{\rho} \end{pmatrix}. \tag{5.31}$$

In the following $\boldsymbol{\vartheta}$, or $\boldsymbol{\theta}$ and $\boldsymbol{\rho}$, are invariably used as arguments, as will be convenient.

5.5 Frequency Domain Models

Let the noise-free input–output data be $u_0(t), y_0(t), t = 0, \ldots, N-1$. Then the discrete Fourier transforms of these signals are

$$U_k \triangleq U(\omega_k) = \frac{1}{\sqrt{N}} \sum_{j=0}^{N-1} u(j) \mathrm{e}^{-\mathrm{i}2\pi\frac{jk}{N}}, \; k = 0, \ldots, N-1 , \tag{5.32}$$

and similarly for Y_k. Here the k:th value of the angular frequency is $\omega_k = 2\pi k/N$. Let

$$A_k = A(\mathrm{e}^{-\mathrm{i}\omega_k}), \;\; B_k = B(\mathrm{e}^{-\mathrm{i}\omega_k}), \; k = 0, \ldots, N-1 . \tag{5.33}$$

Then the exact input–output relation

$$A(q^{-1})y(t) = B(q^{-1})u(t) \tag{5.34}$$

with some initial conditions will in the frequency domain become

$$A_k Y_k = B_k U_k + T_k , \tag{5.35}$$

where

$$T_k = = T(\mathrm{e}^{-\mathrm{i}\omega_k}) , \tag{5.36}$$

$$T(q^{-1}) = \tau_0 + \tau_1 q^{-1} + \cdots + \tau_{n-1} q^{-n+1} \tag{5.37}$$

for some coefficients $\tau_0, \ldots, \tau_{n-1}$. The term T_k in (5.35) represents the collected effects of the initial and final conditions.

The result (5.35) has been obtained in Pintelon et al. (1997). It follows also by applying Example A.2 to the time domain model (5.34).

The relation (5.35) will be called the frequency domain model of the system. It can be used, for example, when one would like to explicitly include the effects of the transients in the modeling and estimation. Another example where it is used is an maximum likelihood approach in the frequency domain; see Sect. 9.4.

Remark 5.1 The term T_k in (5.35) is sometimes called the system leakage term. ■

Remark 5.2 Equation (5.35) describes the precise effect of transients in the data when using the frequency domain. For alternative descriptions of transient effects in the time domain, see Söderström and Stoica (1989), Ljung (1999), Galrinho et al. (2015). ■

Remark 5.3 Models in the frequency domain can be easily modified to handle continuous-time data by just replacing $e^{i\omega}$ by $i\omega$. ■

5.6 Modeling the Total System

Sometimes not only the system dynamics and the noise statistics are modeled, but also the statistics of the noise-free input. This is, for example, the case when a maximum likelihood or a prediction error method (PEM) is applied; see Chap. 9.

Assume that the noise-free input is an ARMA process, see Assumption **AI4**, so

$$D(q^{-1})u_0(t) = C(q^{-1})e(t) . \tag{5.38}$$

The total system can then be described through

$$\zeta(t) = \begin{pmatrix} y(t) \\ u(t) \end{pmatrix} = \begin{pmatrix} \frac{B(q^{-1})}{A(q^{-1})}\frac{C(q^{-1})}{D(q^{-1})} & 1 & 0 \\ \frac{C(q^{-1})}{D(q^{-1})} & 0 & 1 \end{pmatrix} \begin{pmatrix} e(t) \\ \tilde{y}(t) \\ \tilde{u}(t) \end{pmatrix} . \tag{5.39}$$

To apply a PEM, it is needed to rewrite the system (5.39) into innovation form

$$\zeta(t) = \mathbf{S}(q^{-1})\mathbf{w}(t), \quad \mathsf{E}\left\{\mathbf{w}(t)\mathbf{w}^T(s)\right\} = \mathbf{Q}\delta_{t,s} , \tag{5.40}$$

where the filters $\mathbf{S}(q^{-1})$ and $\mathbf{S}^{-1}(q^{-1})$ are asymptotically stable and $\mathbf{w}(t)$, $t = 1, 2, \ldots,$ (the innovations) form a white noise sequence. This can be done using polynomial techniques, Kučera (1979), but using a state space formalism is probably a more common route to choose.

Example 5.2 Assume that the input and output noise $\tilde{u}(t)$, $\tilde{y}(t)$ are white and independent, that is Assumption **AN3c** applies. To simplify notations, set

$$n \stackrel{\Delta}{=} \max(n_a, n_b) , \tag{5.41}$$
$$m \stackrel{\Delta}{=} \max(n_c + 1, n_d) , \tag{5.42}$$
$$a_i = 0 \text{ if } n_a < i \leq n , \tag{5.43}$$
$$b_i = 0 \text{ if } n_b < i \leq n , \tag{5.44}$$
$$c_i = 0 \text{ if } n_c + 1 < i \leq m , \tag{5.45}$$
$$d_i = 0 \text{ if } n_d < i \leq m . \tag{5.46}$$

Then write the total system (5.39) in state space form as

$$\mathbf{x}(t+1) = \begin{pmatrix} -a_1 & 1 & \mathbf{0} & b_1 & \mathbf{0} & & \\ \vdots & & \ddots & \vdots & & & \\ & & 1 & & & & \\ -a_n & & \mathbf{0} & b_n & & & \\ \mathbf{0} & & & -d_1 & 1 & & \mathbf{0} \\ & & & \vdots & & \ddots & \\ & & & & & & 1 \\ & & & -d_m & \mathbf{0} & & \end{pmatrix} \mathbf{x}(t) + \begin{pmatrix} 0 \\ \vdots \\ \vdots \\ 0 \\ 1 \\ \vdots \\ \vdots \\ c_m \end{pmatrix} e(t) , \tag{5.47}$$

$$\boldsymbol{\zeta}(t) = \begin{pmatrix} 1\ 0 \ldots 0 & \mathbf{0} \\ \mathbf{0} & 1\ 0 \ldots 0 \end{pmatrix} \mathbf{x}(t) + \begin{pmatrix} 1 & 0 \\ 0 & 1 \end{pmatrix} \begin{pmatrix} \tilde{y}(t) \\ \tilde{u}(t) \end{pmatrix} . \tag{5.48}$$

Write (5.47) and (5.48) for short as

$$\mathbf{x}(t+1) = \mathbf{F}\mathbf{x}(t) + \mathbf{G}e(t) , \tag{5.49}$$

$$\boldsymbol{\zeta}(t) = \mathbf{H}\mathbf{x}(t) + \tilde{\boldsymbol{\zeta}}(t) , \tag{5.50}$$

$$\mathsf{E}\{e^2(t)\} = \lambda_e, \quad \operatorname{cov}(\tilde{\boldsymbol{\zeta}}(t)) = \operatorname{cov}\left(\begin{pmatrix} \tilde{y}(t) \\ \tilde{u}(t) \end{pmatrix}\right) = \boldsymbol{\Lambda}_\zeta = \begin{pmatrix} \lambda_y & 0 \\ 0 & \lambda_u \end{pmatrix} . \tag{5.51}$$

The innovations form (5.40) can be obtained by first finding the positive definite solution to the algebraic Riccati equation; see, e.g., Söderström (2002)

$$\mathbf{P} = \mathbf{F}\mathbf{P}\mathbf{F}^T + \mathbf{G}\mathbf{G}^T\lambda_e - \mathbf{F}\mathbf{P}\mathbf{H}^T \left(\mathbf{H}\mathbf{P}\mathbf{H}^T + \boldsymbol{\Lambda}_\zeta\right)^{-1} \mathbf{H}\mathbf{P}\mathbf{F}^T , \tag{5.52}$$

leading to

$$\mathbf{K} = \mathbf{F}\mathbf{P}\mathbf{H}^T \left(\mathbf{H}\mathbf{P}\mathbf{H}^T + \boldsymbol{\Lambda}_\zeta\right)^{-1} , \tag{5.53}$$

$$\mathbf{Q} = \mathbf{H}\mathbf{P}\mathbf{H}^T + \boldsymbol{\Lambda}_\zeta . \tag{5.54}$$

The innovations form can then be phrased in state space form as

$$\hat{\mathbf{x}}(t+1) = \mathbf{F}\hat{\mathbf{x}}(t) + \mathbf{K}\boldsymbol{\varepsilon}(t) , \tag{5.55}$$

$$\boldsymbol{\zeta}(t) = \mathbf{H}\mathbf{x}(t) + \boldsymbol{\varepsilon}(t) , \tag{5.56}$$

$$\mathsf{E}\{\boldsymbol{\varepsilon}(t)\boldsymbol{\varepsilon}^T(t)\} = \mathbf{Q} . \tag{5.57}$$

The filter $\mathbf{S}(q^{-1})$ in (5.40) is easily found to be linked to the above state space model, (5.55) and (5.56) as

$$\mathbf{S}(q^{-1}) = \mathbf{I} + \mathbf{H}\,(q\mathbf{I} - \mathbf{F})^{-1}\,\mathbf{K} . \tag{5.58}$$

If any of the noise sources $\tilde{u}(t)$, $\tilde{y}(t)$ is correlated, say an ARMA process, the same idea can be used but further state variables have to be introduced. ■

5.7 Models for Multivariable Systems

In this section it is investigated how the previous linear regression model (5.1) may be generalized to multivariable systems.

For multi-input multi-output system assume that the noise-free data are related as, cf. (5.1),

$$\mathbf{A}(q^{-1})\mathbf{y}_0(t) = \mathbf{B}(q^{-1})\mathbf{u}_0(t) , \tag{5.59}$$

$$\mathbf{A}(q^{-1}) = \mathbf{I}_{n_y} + \mathbf{A}_1 q^{-1} + \cdots + \mathbf{A}_{n_a} q^{-n_a} , \tag{5.60}$$

$$\mathbf{B}(q^{-1}) = \mathbf{B}_1 q^{-1} + \cdots + \mathbf{B}_{n_b} q^{-n_b} . \tag{5.61}$$

Note that here $\mathbf{u}_0(t)$ and $\mathbf{y}_0(t)$ are both column vectors. This description may not be parsimonious, as the number of delayed signal values is assumed to be the same in all the channels. Nevertheless, it is useful and has computational advantages. It is called the 'full polynomial form' in Söderström and Stoica (1983).

In this case introduce the system parameter matrix (rather than a parameter vector) $\boldsymbol{\Theta}$ as

$$\boldsymbol{\Theta} = \begin{pmatrix} \mathbf{A}_1 & \dots & \mathbf{A}_{n_a} & \mathbf{B}_1 & \dots & \mathbf{B}_{n_b} \end{pmatrix}^T . \tag{5.62}$$

Keep the definitions of $\boldsymbol{\varphi}(t)$, repeated here for convenience,

$$\boldsymbol{\varphi}(t) = \begin{pmatrix} -\mathbf{y}(t-1) \\ \vdots \\ -\mathbf{y}(t-n_a) \\ \mathbf{u}(t-1) \\ \vdots \\ \mathbf{u}(t-n_b) \end{pmatrix} , \tag{5.63}$$

cf. (5.5). Then write the system equation (5.59) as

$$\mathbf{y}_0(t) = \boldsymbol{\Theta}^T \boldsymbol{\varphi}_0(t) . \tag{5.64}$$

The residual is

$$\boldsymbol{\varepsilon}(t, \boldsymbol{\theta}) = \mathbf{y}(t) - \boldsymbol{\Theta}^T \boldsymbol{\varphi}(t) . \tag{5.65}$$

It will be convenient to introduce also a parameter vector describing the noise properties. Assume, for simplicity, that the noise assumption **AN4** applies. Set

$$\boldsymbol{\rho} = \begin{pmatrix} \text{vec}(\boldsymbol{\Lambda}_y) \\ \text{vec}(\boldsymbol{\Lambda}_u) \end{pmatrix} , \tag{5.66}$$

where $\boldsymbol{\Lambda}_y = \mathsf{E}\left\{\tilde{\mathbf{y}}(t)\tilde{\mathbf{y}}^T(t)\right\}$, $\boldsymbol{\Lambda}_u = \mathsf{E}\left\{\tilde{\mathbf{u}}(t)\tilde{\mathbf{u}}^T(t)\right\}$. Here vec denotes the vectorization operator, that is the columns of the matrix are stacked in one long column vector. This implies in particular that one neglects the constraints that the noise covariance matrices are symmetric and positive definite.

In case the output noise is correlated, introduce a noise parameter vector instead as

$$\boldsymbol{\rho} = \begin{pmatrix} \text{vec}(\mathbf{R}_{\tilde{\mathbf{y}}}(0)) \\ \text{vec}(\mathbf{R}_{\tilde{\mathbf{y}}}(1)) \\ \vdots \\ \text{vec}(\mathbf{R}_{\tilde{\mathbf{y}}}(n_a)) \\ \text{vec}(\mathbf{R}_{\tilde{\mathbf{u}}}(0)) \end{pmatrix} . \tag{5.67}$$

Note that (5.67) can be viewed as an extension of (5.66) and that $\mathbf{R}_{\tilde{\mathbf{y}}}(0) = \boldsymbol{\Lambda}_y$ as well as $\mathbf{R}_{\tilde{\mathbf{u}}}(0) = \boldsymbol{\Lambda}_u$ are symmetric.

It will also be convenient to have a total parameter vector. Therefore introduce it as $\boldsymbol{\vartheta}$, which will in this case be written as

$$\boldsymbol{\vartheta} = \begin{pmatrix} \text{vec}(\boldsymbol{\Theta}^T) \\ \boldsymbol{\rho} \end{pmatrix} . \tag{5.68}$$

The model (5.65) may alternatively be rewritten as

$$\mathbf{y}(t) = \boldsymbol{\Phi}^T(t)\boldsymbol{\theta} + \boldsymbol{\varepsilon}(t) , \tag{5.69}$$

where now $\mathbf{y}(t)$ is a column vector and $\boldsymbol{\Phi}(t)$ is a regressor matrix of the form

$$\boldsymbol{\Phi}^T(t) = \begin{pmatrix} \boldsymbol{\varphi}^T(t) & & \mathbf{0} \\ & \ddots & \\ \mathbf{0} & & \boldsymbol{\varphi}^T(t) \end{pmatrix} = \mathbf{I}_{n_y} \otimes \boldsymbol{\varphi}^T(t) , \tag{5.70}$$

$$\boldsymbol{\varphi}^T(t) = \begin{pmatrix} -\mathbf{y}^T(t-1) \ \dots \ -\mathbf{y}^T(t-n_a) \ \mathbf{u}^T(t-1) \ \dots \ \mathbf{u}^T(t-n_b) \end{pmatrix} , \tag{5.71}$$

cf. (5.63) and $\otimes$ denotes Kronecker product, cf. Sect. A.1.2. Note that in (5.69) $\boldsymbol{\theta}$ is a large vector, while $\boldsymbol{\Theta}$ is a matrix in (5.62). They are related as

$$\boldsymbol{\Theta}^T = \begin{pmatrix} \boldsymbol{\theta}^{(1)\,T} \\ \vdots \\ \boldsymbol{\theta}^{(n_y)\,T} \end{pmatrix} , \quad \boldsymbol{\theta} = \begin{pmatrix} \boldsymbol{\theta}^{(1)} \\ \vdots \\ \boldsymbol{\theta}^{(n_y)} \end{pmatrix} . \tag{5.72}$$

Further, note that the principal treatment of state space models, see Sect. 5.6, can be generalized in a straightforward way to cover also multivariable systems.

Remark 5.4 The noise parameters in $\boldsymbol{\rho}$, (5.66) or (5.67), are treated as nuisance parameters. It is possible, and can be preferable, to impose conditions on $\boldsymbol{\rho}$ so that the covariance matrices $\boldsymbol{\Lambda}_u$, $\boldsymbol{\Lambda}_y$ are nonnegative definite, as well as the covariance sequence $\left\{\mathbf{R}_{\tilde{\mathbf{y}}}(k)\right\}_{k=0}^{n_a}$. This can be done by letting $\boldsymbol{\rho}$ be a function of a vector $\bar{\boldsymbol{\rho}}$ of lower dimension.

As illustration, consider (5.66) with the additional constraint that the components of $\tilde{\mathbf{u}}(t)$ and $\tilde{\mathbf{y}}(t)$ are all mutually independent. Then the elements of $\bar{\boldsymbol{\rho}}$ can be chosen as the standard deviations of these noise signals. For example, if $n_u = 3,\ n_y = 2$, one would then have

$$\begin{aligned}\boldsymbol{\rho} &= \left((\boldsymbol{\Lambda}_y)_{11}\ 0 : 0\ (\boldsymbol{\Lambda}_y)_{22} : (\boldsymbol{\Lambda}_u)_{11}\ 0\ 0 : 0\ (\boldsymbol{\Lambda}_u)_{22}\ 0 : 0\ 0\ (\boldsymbol{\Lambda}_u)_{33} \right)^T \\ &= \left(\bar{\rho}_1^2\ 0 : 0\ \bar{\rho}_2^2 : \bar{\rho}_3^2\ 0\ 0 : 0\ \bar{\rho}_4^2\ 0 : 0\ 0\ \bar{\rho}_5^2 \right)^T . \end{aligned} \tag{5.73}$$

If the parameterization (5.67) applies, one can use the Wiggins–Whittle–Robinson algorithm (an extension of the Levinson–Durbin algorithm for SISO systems); see Whittle (1963), Wiggins and Robinson (1966), Stoica and Söderström (1985), Söderström and Stoica (1989).

To impose the constraints on positive definiteness would no doubt make the obtained estimates more physically meaningful. The price to be paid is that the estimation algorithm will be computationally more complex. Under the listed assumptions the condition on positive definiteness is fulfilled for the true parameters. In particular, for asymptotic accuracy of consistent estimates the case $N \to \infty$ is considered, and therefore in such an analysis one can disregard these constraints. ■

The 'full polynomial form' is easy and computationally convenient to apply. However, it has some considerable restrictions, as pointed out, for example in Gevers (1986).

For the model (5.59) to be tractable, one needs to assume $\mathbf{A}(q^{-1})$ and $\mathbf{B}(q^{-1})$ to be left coprime. This implies that the model order (the degree of the pole polynomial) will be $n_y n_a$. Starting with a state space model for a multivariable system, the model order can be an arbitrary integer, not just a multiple of n_y.

As pointed out in Guidorzi (1975, 1981), Gevers (1986) it is possible to use a canonical parameterization of the system, using a minimal number of parameters. This requires that the left Kronecker indices of the system transfer function are known. The canonical form has a structure that, compared to (5.59)–(5.61), differs in the following way

- the identity matrix $\mathbf{I}_{n_y}$ in (5.60) is replaced by a lower triangular matrix with ones in the diagonal,
- some elements of the polynomial coefficients $\mathbf{A}_j$, $\mathbf{B}_j$ are constrained to be zero.

The specific degrees of the elements of $\mathbf{A}(q^{-1})_{i,j}$ and $\mathbf{B}(q^{-1})_{i,j}$ are determined via the Kronecker indices; see Guidorzi (1975, 1981), Gevers (1986) for details.

How may the model (5.69) be modified to treat the Guidorzi canonical form, or an alternative one?

First (due to the fact that $\mathbf{I}_{n_y}$ in (5.60) is to be replaced by a triangular matrix) introduce

$$\overline{\boldsymbol{\varphi}}(t) = \left(-\mathbf{y}^T(t)\ \boldsymbol{\varphi}^T(t)\right)^T , \tag{5.74}$$

$$\overline{\boldsymbol{\Phi}}^T(t) = \mathbf{I}_{n_y} \otimes \overline{\boldsymbol{\varphi}}^T(t) , \tag{5.75}$$

$$\overline{\boldsymbol{\theta}} = \operatorname{vec}\left(\mathbf{A}_0 \ldots \mathbf{A}_{n_a}\ \mathbf{B}_1 \ldots \mathbf{B}_{n_b}\right) . \tag{5.76}$$

Then one may write the canonical form as

$$\mathbf{y}(t) = \overline{\boldsymbol{\Phi}}^T(t)\overline{\boldsymbol{\theta}} + \boldsymbol{\varepsilon}(t) , \tag{5.77}$$

$$\mathbf{S}\overline{\boldsymbol{\theta}} = \mathbf{0} . \tag{5.78}$$

Here $\mathbf{S}$ is a matrix used to describe the parameters in $\overline{\boldsymbol{\theta}}$ that are set to zero. Typically, each row of $\mathbf{S}$ will have one nonzero element, which further is equal to 1. The matrix $\mathbf{S}$ will have full row rank. An advantage and tractable aspect with the form (5.77) and (5.78) is that the linear regression format is kept, which is combined with the linear constraints (5.78) on the parameter vector.

5.8 Classification of Estimators Based on Data Compression

Estimation methods can be classified and organized in different ways. Here, a classification based on an initial data compression step is presented. After a first 'preprocessing' of the data, some reduced information is set up and used for the final computation of the parameter estimates. In case the condensed information is really a sufficient statistic, cf. Convention A.8 in Appendix A, one would even be able to achieve statistical efficiency in the final step. Expressed in other words, the condensed information contains in such cases the same information about the unknown parameters as the full data set of measured data. Also when this does not apply, such an estimation scheme can still be useful, for example, due to low computational complexity.

The two steps of the estimators, with an initial data compression, are illustrated in the Fig. 5.3.

The different groups of methods that will be discussed in the chapters to follow differ in the way the data compression is carried out. The following cases will be treated.

(a) Using a covariance matrix. This case includes instrumental variable (Chap. 6), bias-eliminating least squares (Sect. 7.2), the Frisch scheme (Sect. 7.3), total least squares (Chap. 11), and other schemes such as covariance matching (Chap. 8). In

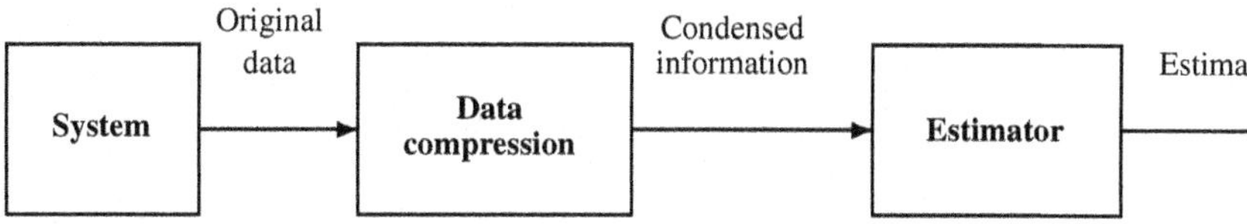

Fig. 5.3 Classification of the data compression prior to the estimation step

all these cases the condensed information is a small set of estimated covariance elements $\{\hat{r}_u(\tau)\}$, $\{\hat{r}_{yu}(\tau)\}$ and $\{\hat{r}_y(\tau)\}$, where τ typically takes small values only.

(b) Using the input-output spectrum and frequency domain data (Chap. 10). As compared to case a), one basically has a large set of estimated covariance elements here.

(c) Using the original time-series data. This case includes the use of prediction error and maximum likelihood techniques (Chap. 9). No data compression takes place.

5.9 Model Order Determination

5.9.1 Introduction

So far it has been assumed that the model order is known a priori. Can it be estimated from the measured data? This question is treated here. It is assumed that the measurement noise is white (Assumption **AN3c**) and that the noise-free input is an ARMA model (Assumption **AI4**), now written as

$$u_0(t) = H(q^{-1})v(t) = \frac{C(q^{-1})}{D(q^{-1})}e(t) , \tag{5.79}$$

with

$$\begin{aligned} C(q^{-1}) &= 1 + c_1 q^{-1} + \cdots + c_{n_c} q^{-n_c} , \\ D(q^{-1}) &= 1 + d_1 q^{-1} + \cdots + d_{n_d} q^{-n_d} , \end{aligned} \tag{5.80}$$

where $e(t)$ is white noise with variance λ_e and the monic polynomials $C(q^{-1})$, $D(q^{-1})$ are relatively prime and asymptotically stable. The order of $H(q^{-1})$ is

$$n_H = \max(n_c, n_d) . \tag{5.81}$$

In addition assume

AS6. The order of the transfer functions fulfills

$$\text{order}(GH) \triangleq n_T = n_G + n_H . \tag{5.82}$$

This is a technical assumption, that holds generically. It is applied for convenience. A case when it *does not* hold is if G is a FIR system, and H corresponds to an AR model. More generally, pole–zero cancellations in GH are not allowed.

5.9.2 Some Approaches

5.9.2.1 Direct Approach

A rather straightforward and simple approach is to note that according to the basic setup, the measured signals $u(t)$ and $y(t)$ can be interpreted as scalar ARMA processes.

The measured input signal $u(t)$ is in fact an ARMA(n_d, n_H) process (and also an ARMA(n_H, n_H) process). The reason is that the spectrum of $u(t)$ can be written in the form

$$\phi_u(z) = \frac{C(z)C(z^{-1})}{D(z)D(z^{-1})}\lambda_e + \lambda_u \stackrel{\Delta}{=} \frac{M(z)M(z^{-1})}{D(z)D(z^{-1})}\lambda_v \ , \tag{5.83}$$

where the polynomial $M(z)$ is generically of degree n_H, is monic and has all zeros inside the unit circle, and λ_v is a positive scalar. This follows from spectral factorization, cf. Sect. 3.3. Hence, the *measured* noisy input can be viewed as another ARMA process, namely

$$u(t) = \frac{M(q^{-1})}{D(q^{-1})}v(t) \tag{5.84}$$

as far as all second-order moments are concerned.

Any standard method for order determination of a scalar time series model can be used to estimate n_H from the measurements of $u(t)$. Similarly, the recorded output signal $y(t)$ is an ARMA(n_T, n_T) process. Again, several standard methods for scalar time series can be used to estimate the model order n_T. Examples of such standard methods are AIC and BIC; see Ljung (1999), Söderström and Stoica (1989), for example.

5.9.2.2 Correlation Approach

The basic idea of the Correlation approach is to compute the cross-covariance matrix relating delayed inputs and the regressor vector. If the model order is chosen too high, this matrix will be singular. More specifically, consider

$$\mathbf{R} = \mathsf{E}\left\{\boldsymbol{\psi}_1(t)\boldsymbol{\psi}_2^T(t)\right\} \ , \tag{5.85}$$

$$\boldsymbol{\psi}_1(t) = \left(u(t-\hat{n}-1) \ldots u(t-3\hat{n})\right)^T \ , \tag{5.86}$$

$$\boldsymbol{\psi}_2^T(t) = \left(\boldsymbol{\zeta}^T(t-1) \ldots \boldsymbol{\zeta}^T(t-\hat{n})\right) \ . \tag{5.87}$$

Here $\hat{n}$ is a user chosen integer that should be larger than the guessed model order. Hence,

$$p \stackrel{\Delta}{=} \hat{n} - n > 0 . \tag{5.88}$$

Apparently, the matrix $\mathbf{R}$ is square and of dimension $2\hat{n}$.

Remark 5.5 It is possible to extend the approach by including more delayed elements in $\boldsymbol{\psi}_1(t)$ and thus let $\mathbf{R}$ be a tall matrix. ■

It turns out that under the given general assumptions, $\mathbf{R}$ is singular with (at least) p singular values being zero. This can be seen as follows. As the measurement noise $\tilde{u}(t)$ is white, it follows that $\mathbf{R}$ is the cross-covariance matrix of the noise-free data,

$$\mathbf{R} = \mathsf{E}\left\{\boldsymbol{\psi}_{10}(t)\boldsymbol{\psi}_{20}^T(t)\right\} . \tag{5.89}$$

Next, introduce the reordered parameter vector

$$\overline{\boldsymbol{\theta}} = \begin{pmatrix} 1 \; 0 \; a_1 \; -b_1 \; \dots \; a_n \; -b_n \end{pmatrix}^T , \tag{5.90}$$

and form the $2\hat{n} \times p$-dimensional matrix

$$\mathbf{S} = \begin{pmatrix} \overline{\boldsymbol{\theta}} & \mathbf{0}_{2\times 1} & \dots \; \mathbf{0}_{(2p-2)\times 1} \\ \vdots & \overline{\boldsymbol{\theta}} & \vdots \\ \mathbf{0}_{(2p-2)\times 1} & \mathbf{0}_{(2p-4)\times 1} & \overline{\boldsymbol{\theta}} \end{pmatrix} . \tag{5.91}$$

As

$$\boldsymbol{\psi}_{20}^T(t)\mathbf{S} = \mathbf{0}_{1\times p} , \tag{5.92}$$

it follows that

$$\mathbf{RS} = \mathbf{0} . \tag{5.93}$$

By construction, $\mathbf{S}$ has full column rank (equal to p), and therefore $\mathbf{R}$ will have p singular values equal to zero.

To use this result for model order testing, first note that from data it is straightforward to estimate $\mathbf{R}$ in (5.89) as

$$\hat{\mathbf{R}} = \frac{1}{N}\sum_{t=1}^{N} \boldsymbol{\psi}_1(t)\boldsymbol{\psi}_2^T(t) . \tag{5.94}$$

One can then compute its singular values. Due to the presence of noise all singular values will be strictly positive, but p of them should be 'small' if the data set is large enough. What 'small' means and how a suitable test quantity can be constructed is described in Sect. 5.A.2 of Appendix 5.A. Once p is determined (a natural algorithm is to try $p = 1, 2, \dots, p_{\max}$ until no further 'small enough' singular values are detected), n can be determined from (5.88). There is though a limitation with the approach. From (5.93) it can be seen that the range space of $\mathbf{S}$ is contained in the null space of $\mathbf{R}$. Ideally, these two subspaces should be the same. However, the null space $\mathscr{N}(\mathbf{R})$ is indeed larger than $\mathscr{R}(\mathbf{S})$ if

$$n_H \leq n_G \,, \tag{5.95}$$

which is not an unrealistic situation. When (5.95) holds, the method can still work after some modifications. One has then to involve also the associated right singular vectors and use them to distinguish whether the cause of rank deficiency is due to the model order n being found, or the input dynamics has low order so that (5.95) does not hold, or both. Note that $n_H > n_G$ is not a sufficient condition for $\mathscr{N}(\mathbf{R}) = \mathscr{R}(\mathbf{S})$ to hold, but must be complemented with some condition on the properties of $u_0(t)$.

5.9.2.3 Realization Approach

A still further alternative way of finding the model order from the two-dimensional ARMA model would be to apply ideas from realization theory for stochastic systems.

Introduce

$$\mathbf{R}(\tau) = \mathsf{E}\left\{\zeta(t)\zeta^T(t-\tau)\right\} \tag{5.96}$$

and the matrix

$$\mathbf{R}_q = \begin{pmatrix} \mathbf{R}(1) & \mathbf{R}(2) & \dots & \mathbf{R}(q) \\ \mathbf{R}(2) & \mathbf{R}(3) & \dots & \\ \vdots & & \ddots & \\ \mathbf{R}(q) & & & \mathbf{R}(2q-1) \end{pmatrix} . \tag{5.97}$$

(Note that the subscript here relates to the number of blocks and not to the number of data points.) The rank of this block Hankel matrix is closely connected to stochastic realization algorithms (and to subspace identification methods). In particular its rank properties will carry information about the model order n_T for describing $\zeta(t)$. Provided $2q \geq n_T$, it holds

$$\operatorname{rank}(\mathbf{R}_q) = n_T \,. \tag{5.98}$$

Remark 5.6 Similar to (5.85) it is possible to extend the approach and to use different numbers of rows and columns in $\mathbf{R}_q$, although the square case allows neater notations in the analysis. ■

When constructing a test from data, the matrix $\mathbf{R}_q$ is replaced by the estimate

$$\hat{\mathbf{R}}_q = \frac{1}{N}\sum_{t=1}^{N} \varphi_1(t)\varphi_2^T(t)\,, \tag{5.99}$$

$$\varphi_1(t) = \left(\zeta^T(t)\ \ldots\ \zeta^T(t+q-1)\right)^T\,, \tag{5.100}$$

$$\varphi_2(t) = \left(\zeta^T(t-1)\ \ldots\ \zeta^T(t-q)\right)^T\,. \tag{5.101}$$

Similarly to the Correlation approach, the model order determination is based on computing $\hat{\mathbf{R}}_q$, making a singular value decomposition, and then examining the (number of) 'small' singular values.

5.9.2.4 Two-Dimensional ARMA Approach

Again, exploit Assumption **AI4** ($u_0(t)$ is an ARMA process). For simplicity it is still convenient to let the measurement noises be white, although extensions are possible.

For this situation consider the recorded data as a two-dimensional ARMA process, cf. Sect. 5.6. Its ARMA parameters do in fact have some internal structure. Some details of this ARMA model are presented in Söderström et al. (2003). Criteria that can be used to determine the model order of a multivariable ARMA process include FPE, AIC and BIC, Söderström and Stoica (1989). As the BIC criterion is known to give consistent order estimates under mild conditions, Schwarz (1978), Akaike (1977), it may be preferred.

When applying the BIC criterion for model order determination one penalizes the use of more than the minimally required number of parameters. The BIC principle can be written as

$$\hat{\mathscr{M}} = \arg\min_{\mathscr{M}} \left(-L + \dim(\boldsymbol{\vartheta})\frac{\log(N)}{2}\right)\,, \tag{5.102}$$

where L denotes the value of the likelihood function maximized over the parameter values, and $\dim(\boldsymbol{\vartheta})$ denotes the number of parameters. Further, $\hat{\mathscr{M}}$ denotes the estimated model structure (which includes the model order).

To compute the maximum value of L, an optimization problem has to be solved numerically; see Chap. 9 for details.

5.9.3 About the Rank Tests

There are large algebraic similarities between the Correlation and the Realization approaches. Indeed, (5.94) and (5.99) have the same algebraic form. As long as the meaning of the vectors $\varphi_j(t)$ are not explicitly specified, one can therefore analyze the two approaches jointly.

To test if the matrix $\hat{\mathbf{R}}$ has r 'small' singular values, one possibility is to compute the test quantity

$$T = \sum_{i=1}^{r} \sigma_i^2(\hat{\mathbf{R}}) , \tag{5.103}$$

where the summation is over the smallest singular values. Ideally, T should be 'negligible' for large data sets. However, one needs to set an appropriate test limit to decide whether or not the computed value of T should be regarded as small.

The preferred way to go would be to find the distribution of T and then use an appropriate statistical test. This is often not possible, as the distribution depends, among other things, on unknown quantities such as the distribution of the noise-free input $u_0(t)$. A somewhat simpler approach is to find the expected value of T and then construct the ratio

$$T_0 = T/\mathsf{E}\{T\} \tag{5.104}$$

or an estimate thereof. If T_0 is smaller than a chosen modest number, then it is reasonable to regard the r smallest singular values as sufficiently close to zero.

In case all elements of $\hat{\mathbf{R}} - \mathbf{R}$ would be independent and equally distributed, it turns out that a scaled version of T is χ^2 distributed; see Fujikoshi (1974), Rensfelt and Söderström (2010). The assumption of independence is not valid here. Still one might use this distribution result as a good approximation; see Camba-Mendez and Kapeanios (2004). See also Yuan and Bentler (2010) for a further discussion. Let the matrix $\mathbf{R}$ have dimension $q_0 \times q_0$, and let r eigenvalues be 'small'. Set

$$T \sim c\chi^2((q_0 - r)^2) \tag{5.105}$$

with c being a scaling constant. Then

$$\mathsf{E}\{T\} = c(q_0 - r)^2 , \tag{5.106}$$

and therefore

$$T \sim \frac{\mathsf{E}\{T\}}{(q_0 - r)^2} \chi^2((q_0 - r)^2) , \tag{5.107}$$

from which one can design appropriate test limit as soon as the expectation $\mathsf{E}\{T\}$ is known. Expressions for $\mathsf{E}\{T\}$ for the Correlation and Realization approaches, valid for large N, are given in Appendix 5.A.2.

5.9.4 *Discussion*

The four introduced approaches can be described in different ways. One way in which they differ is that:

- The Direct approach and the 2DARMA approach are both based on the BIC or a similar criterion. The ML estimates of all the model parameters have first to be determined for a number of candidate model orders, which requires some more or less complex optimization problems to be carried out.
- The Correlation and the Realization approaches are both based on finding the number of 'small' singular values of a covariance matrix. To assess what is meant by 'small' will require a test limit, which should be found based on some statistical grounds.

Concerning the user choices, for all the approaches, the user has to provide an upper limit on possible model orders. Needless to say, the higher this limit is set, the more different types of dynamics can be found, but the computational burden will also increase notably. The risk of choosing a wrong model order also increases.

For the 2DARMA approach, in its described form it delivers an estimate of the order n_T of the total system. It is possible to complement the algorithm with an estimate of the model order n_H for the input dynamics, and thereby allow also an estimate of the system order n_G. For the Correlation approach to work one has to respect the condition (5.95).

As far as the computational load is concerned, the Correlation and the Realization approaches, which both are based on a singular value decomposition, are those with the lowest amount of computations. Next comes the Direct approach, where a number of scalar ARMA models are to be estimated. The approach with the by far most heavy computational load is the 2DARMA approach which is based on estimating a two-dimensional ARMA model with quite a number of parameters, for a number of parameterizations.

The role of some of the assumptions is worth to comment upon.

Assumption **AI4** (the noise-free input being an ARMA process) is essential for all the methods. The model order n_H of that ARMA process influences the methods. For the Correlation approach it is less explicitly so in the algorithm, but the inequality (5.95) giving a bound for n_H must hold for this approach to work reliably in its simplest form. A more elaborate version involving also the right singular vectors is though possible when (5.95) is not satisfied.

Assumption **AN3c** (the output measurement noise being white) is somewhat restrictive, as in practice this noise would also capture process noise, and therefore in most cases be autocorrelated. The methods can be suitably modified to handle such a case as well. For the Direct approach, n_H would then be increased with the model order for the correlation of $\tilde{y}(t)$, cf. the spectral factorization reasoning in Sect. 5.6. Similarly, the underlying total model for the Realization and the ARMA approaches would need to be expanded with the correlation dynamics of $\tilde{y}(t)$. For the Correlation approach though, the changes are more subtle. The number of 'small' singular values, and how to compute the test quantity T, (5.103), would remain the same. The expression for its expectation would need to be changed, as this expression is partly based on the assumption of white output noise.

Some of the methods are closely linked to parameter estimation of G and H. This is not the case for the Direct approach, where the model (5.84) rather than (5.79) is identified. For the Correlation approach the right singular vectors of $\hat{\mathbf{R}}$ span the range space of $\mathbf{S}$, (5.91). This can be exploited to derive Yule–Walker (or instrumental variable) estimation of $\boldsymbol{\theta}$. The Realization approach is closely linked to subspace identification, Van Overschee and De Moor (1996), Katayama (2005) which though would lead to a parameter estimation of a 'full' model (5.49)–(5.51) without taking any internal structure into account. Finally, for the 2DARMA approach, the total parameter vector $\boldsymbol{\vartheta}$ is estimated as an integral part of (5.102).

5.10 Extensions and Bibliographical Notes

Continuous-Time Models (Sect. 5.3*)*

Sampling, or more generally time discretization, can be arranged in many different ways. Many general aspects on sampling and connections between continuous-time and discrete-time models are given in Feuer and Goodwin (1996).

A recent and comprehensive treatment of sampling is given in Yuz and Goodwin (2014). The text describes also many misconceptions about the topic. Numerical examples illustrating the effect of sampling appear in Agüero et al. (2009).

Some different sampling strategies and their impact for identifying a dynamic system is an EIV setting are considered and analyzed in Mossberg (2015, 2016).

Identification of continuous-time EIV models was considered in Mahata and Garnier (2006). The approach is, simplified, to used BELS-techniques (cf. Chap. 7) to handle the noise effects, and state variable filters to handle the discrete-time data combined with a continuous-time model. The Ph.D. thesis by Thil (2007) contains aspects of EIV on continuous-time models (and also on the use of higher-order statistics).

Noise Models (Sect. 5.4*)*

Treating correlated output noise can be done in many ways. One approach is to postulate an ARMAX structure to model the effect of process noise and then to add white measurement noise. Such a modeling parameterization is analyzed in Diversi et al. (2009).

Frequency Domain Models (Sect. 5.5*)*

The relations between time domain and frequency domain models have been investigated in detail by Pintelon and Schoukens (1997) and Agüero et al. (2010).

Models for Multivariable Systems (Sect. 5.7*)*

It is worth pointing out that there are alternative ways for estimating canonical MIMO models in an errors-in-variables setting; see Diversi and Guidorzi (2009), Guidorzi and Diversi (2009).

The special case of (many) more outputs than inputs is also treated by generalized dynamic factor models, as mentioned in Sect. 2.4.

Model Order Determination (Sect. 5.9)

Further developments of the analysis in Sect. 5.9 can be found in Söderström and Wang (2012) and Söderström (2012a). Related work on model validation is presented in Söderström and Yuz (2013).

The paper Stoica and Babu (2012) deals with the proper form of the criterion BIC. Standard expressions need to be modified if the sample length is not large enough.

Two further approaches for model order determination in an EIV context are proposed in Liu et al. (2017). They are based on modifications of the final prediction error and the output error, respectively. The idea is to take away the influence of the input noise by correlating these errors with the output signal.

Nonlinear Models

The papers Larkowski et al. (2008a), Mu and Chen (2013, 2014, 2015) deal with nonlinear models and recursive algorithms. Identification of nonlinear dynamic polynomial models for EIV problems is treated in Larkowski et al. (2009b), Jun and Bernstein (2007). The paper Song (2013) treats identification of linear EIV models where the output is observed through a static nonlinearity. The paper Sumisławska et al. (2012) deals with Hammerstein–Wiener models, where the output linearity is monotonic. Wahlberg et al. (2014) show how identification of a Wiener model with process noise can be interpreted as an EIV problem. The algorithm in Wang et al. (2017) for identification of nonlinear Hammerstein–Wiener systems from noisy data is based on combining ideas from bias-compensation and singular value decomposition.

EIV identification, where the measurement noise is given by a finite-state hidden Markov model, is treated in Krishnamurthy and Logothetis (1996).

The papers Chetoui et al. (2012, 2013b, a), Ivanov (2013), deal with fractional models, which may be regarded as a special case of nonlinear models.

Shock Error Models

Shock error models correspond to a special case of Assumption **AN3b**. More precisely, the output noise is modeled to be a sum of two contributions. First there is process noise, which together with the input–output dynamics is modeled as an ARX (autoregressive with an exogenous input) process, and there is in addition white measurement noise on the output side. This means that the output noise is an ARMA process, where the poles of the ARMA process coincide with those of the system. The zeros of the ARMA process depend in a somewhat complicated way on the poles and the ratio between the process noise variance and the measurement noise variance. See the discussion following (5.29).

Shock error models are treated, for example, in Hsiao (1977), Hsiao and Robinson (1978), and Maravall (1979). Another structured model is treated in Diversi et al. (2011) and concerns an ARMAX model with additional measurement noise.

Appendix

5.A Further Details

5.A.1 Discrete-Time Model Approximation

As approximation by sampling is quite general, it will not be explicitly treated for an EIV system. Rather, the input will in this appendix be denoted by $u(t)$ and the output by $y(t)$. More details of the analysis here can be found in Söderström (2009).

Let the continuous-time system transfer function to be discretized be given by

$$G_c(s) = \frac{B_c(s)}{A_c(s)} , \tag{5.108}$$

where

$$A_c(s) = s^n + a_1 s^{n-1} + \cdots + a_n , \tag{5.109}$$
$$B_c(s) = b_1 s^{n-1} + \cdots + b_n , \tag{5.110}$$

and allow the leading coefficients of $B_c(s)$ to be zero. The order of the system is apparently n. Let the poles of the system be denoted by $p_i^c, i = 1, \ldots n$, so

$$A_c(p_i^c) = 0 . \tag{5.111}$$

Further, let the sampling interval be denoted h. For a discrete-time model one should look for a linear dynamic relationship between the discrete-time signals $u(kh)$ and $y(kh)$.

As the inter-sample behavior of $u(t)$ is not known, *a unique discrete-time (sampled) model of* $G_c(s)$ *does not exist*. Rather, one has to work with approximations. In the coming subsections some elementary approximations are reviewed. Some more elaborated approaches are presented where the inter-sample behavior of the input is modeled. Such a modeling is shown to lead to an increased model order of the discrete-time model approximation of $G_c(s)$.

5.A.1.1 Model Approximations

There are several ways to get a discrete-time model from $G_c(s)$ by approximation. One may consider, for example,

1. Forward difference approximation by setting

$$s = \frac{z-1}{h} . \tag{5.112}$$

2. Backward difference approximation by setting

$$s = \frac{1 - z^{-1}}{h} = \frac{z - 1}{zh} . \tag{5.113}$$

3. Bilinear approximation by setting

$$s = \frac{2}{h} \frac{z - 1}{z + 1} . \tag{5.114}$$

Out of the three alternatives above, the bilinear transformation has the advantage of preserving the stability boundary (so that the stability boundary for the sampled model will be the inner part of the unit circle); see Åström and Wittenmark (1997).

For all the cases the transformed model will in discrete-time be of the form

$$G_d(z) = \frac{B_d(z)}{A_d(z)} . \tag{5.115}$$

The degree of the denominator polynomial $A_d(z)$ will always be n. For the forward difference approximation the degree of $B_d(z)$ will be $\deg(B_c)$, which is smaller than n. For the other two alternatives, the degree of $B_d(z)$ will be equal to n, and then $G_d(z)$ will automatically have a direct term, that is $b_0 \neq 0$, cf. Sect. 5.2.

When the input behavior between the sampling instants is not known, also the standard zero-order-hold (ZOH) procedure can be regarded as a model approximation. (ZOH implies that the input is constant during the sampling intervals.) Write the effect of this sampling strategy as

$$G_d(z) = \mathscr{S}_{\text{ZOH}} G_c(s) . \tag{5.116}$$

If $G_c(s)$ is represented in state space form as

$$\begin{aligned} \dot{x} &= \mathbf{A}x + \mathbf{B}u , \\ y &= \mathbf{C}x , \end{aligned} \tag{5.117}$$

then the ZOH sampled transfer function can be written as

$$G_d(z) = \mathbf{C}(z\mathbf{I} - \mathbf{F})^{-1}\mathbf{G} , \tag{5.118}$$

where

$$\mathbf{F} = \mathrm{e}^{\mathbf{A}h}, \quad \mathbf{G} = \int_0^h \mathrm{e}^{\mathbf{A}s}\mathbf{B}\mathrm{d}s . \tag{5.119}$$

One may alternatively write (5.116) as

$$G_d(z) = \frac{z - 1}{z} \mathscr{Z} \mathscr{L}^{-1} \frac{G_c(s)}{s} . \tag{5.120}$$

This is to be interpreted as follows. Let $g(t)$ be a continuous-time function, whose Laplace transform is

$$\mathscr{L}g(t) = \frac{G_c(s)}{s} \,. \tag{5.121}$$

Then the z transform of $g(t)$, evaluated at the sampling instants, $t = 0, h, 2h, \ldots$ is

$$\mathscr{Z}g(t) = \sum_{k=0}^{\infty} z^{-k} g(kh) \,. \tag{5.122}$$

First-order-hold (FOH) sampling occurs when the input signal is interpolated linearly between the sampling points. This can be seen as a ZOH sampling of an integrator followed by the system $G_c(s)$. Due to the ZOH block, the output of the integrator will vary linearly with time. In order to get the correct dynamics, the presence of the integrator is compensated for after the ZOH sampling. Mathematically, the discrete-time system can in this case be computed according to

$$G_d(z) = \frac{z-1}{h} \mathscr{S}_{\text{ZOH}} \left(\frac{1}{s} G_c(s) \right) \,. \tag{5.123}$$

The transfer function in (5.123) has normally *equal* polynomial degrees in numerator and denominator. In contrast, $G_d(z)$ in (5.118) has lower degree in the numerator than in the denominator.

A further possibility to perform time discretization is to assume that the input signal is band-limited. Details are developed in Pintelon and Schoukens (2001).

5.A.1.2 Modeling the Input as a Filtered Pulse-Train

Another possible approach for discretization is to assume that the continuous-time input is obtained by filtering a sequence of the filtered pulses, Schoukens et al. (1994), Pintelon and Schoukens (2001). The situation is displayed in Fig. 5.4. The pulses $\delta(t)$ are assumed to be constant over a sampling interval.

The signal $z(t)$ is constant over the sampling intervals, and the filter transfer function $G_u(s)$ is taken as

$$G_u(s) = \frac{B_u(s)}{A_u(s)} \,, \tag{5.124}$$

where

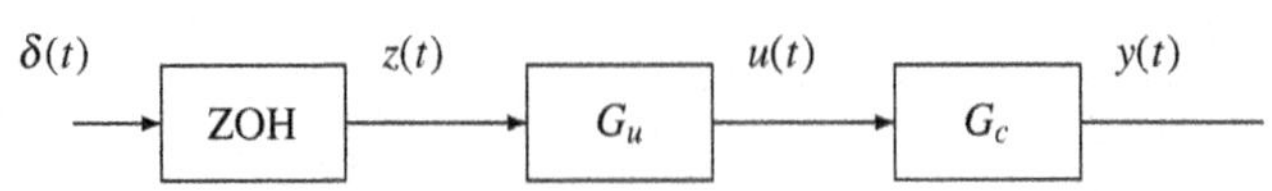

Fig. 5.4 Modeling the input as a filtered pulse-train

$$A_u(s) = s^m + a_1^u s^{m-1} + \cdots + a_m^u \,, \tag{5.125}$$
$$B_u(s) = b_1^u s^{m-1} + \cdots + b_m^u \,. \tag{5.126}$$

The coefficients of A_u and B_u can be regarded as user parameters or as hyperparameters. Again, allow the leading coefficients of $B_u(s)$ to possibly be zero. With the current setup, it is possible to find precisely how the discrete-time signals $u(kh)$ and $y(kh)$ behave, as one can apply standard ZOH sampling from the pulse-sequence $\{z(kh)\}$. Sampling in this way from z to y gives the discrete-time transfer function

$$G_2^d(z) = \frac{\tilde{B}(z)}{\tilde{A}(z)} \,. \tag{5.127}$$

Note that the degree of $\tilde{A}(z)$ will be $n + m$. Further, n of the discrete-time poles are given as $e^{p_i^c h}$, while the remaining m discrete-time poles are $e^{p_j^u h}$ with $p_j^u, j = 1, \ldots, m$ being zeros of the polynomial $A_u(s)$. The degree of $\tilde{B}(z)$ is generically $n+m-1$ (and never higher). This holds also if $\deg B_c(s) < n-1$ or $\deg B_c^u(s) < m-1$.

It holds, exactly, at the sampling points

$$y(kh) = \frac{\tilde{B}(q)}{\tilde{A}(q)} z(kh) \,, \tag{5.128}$$

where q denotes the forward shift operator. One can similarly sample the dynamics from z to u and arrives then at

$$u(kh) = \frac{B_u^d(q)}{A_u^d(q)} z(kh) \,. \tag{5.129}$$

Here, $A_u^d(z)$ is the uniquely defined polynomial of degree m with zeros in $e^{p_j^u h}$, $j = 1, \ldots m$. The polynomial $B_u^d(z)$ has generically degree $m - 1$.

Now combining (5.128) and (5.129) leads to

$$y(kh) = \frac{\tilde{B}(q)}{\tilde{A}(q)} z(kh) = \frac{\tilde{B}(q)}{\tilde{A}(q)} \frac{A_u^d(q)}{B_u^d(q)} u(kh) \,, \tag{5.130}$$

which is the sought discrete-time system transfer function. It is clear that $A_u^d(z)$ is factor of $\tilde{A}(z)$. In fact,

$$\tilde{A}(z) = A_u^d(z) A_y^d(z) \,. \tag{5.131}$$

Hence the discrete-time transfer function has the form

$$G_d(z) = \frac{\tilde{B}(q)}{A_y^d(q) B_u^d(q)} \,. \tag{5.132}$$

One can now make a number of observations.

1. The poles corresponding to the polynomial $A_y^d(z)$ in (5.132) are precisely $e^{p_j^c h}, j = 1, \dots n$.
2. The transfer function $G_d(z)$ has order $n+m-1$ in contrast to the continuous-time system $G_c(s)$, which is of order n.
3. Generically $\deg(A_y^d B_u^d) = \deg(\tilde{B}) = n + m - 1$. This implies that the transfer function $G_d(z)$ has a direct term, which is in contrast to the continuous-time representation (5.108).

5.A.1.3 Modeling the Input as a Stationary Stochastic Process

A further possible approach is to assume that the continuous-time input is a stationary stochastic process with rational spectrum. This idea has been used in an errors-in-variables context; see, for example, Söderström (1981, 2007a).

Set

$$\boldsymbol{\zeta}(t) = \begin{pmatrix} y(t) \\ u(t) \end{pmatrix} , \tag{5.133}$$

and let the input signal $u(t)$ in continuous-time have a spectrum that in factorized form is given as

$$\phi_u(\omega) = G_u(i\omega)G_u(-i\omega) . \tag{5.134}$$

Here, the filter $G_u(s)$ is of the form given by (5.124), (5.125), and (5.126). Further, it is constrained to have all poles and zeros strictly inside the left half plan. This just means that the innovations representation of $u(t)$ can be written as

$$u(t) = G_u(p)e_c(t) = \frac{B_u(p)}{A_u(p)} e_c(t) \tag{5.135}$$

with $e_c(t)$ being a continuous-time white noise process. It is worth to underline that the process $e_c(t)$ does not exist as a physical quantity (which would not be reasonable as continuous-time white noise has infinite variance). Rather, (5.135) is to be regarded as the time domain representation of a signal $u(t)$ whose spectrum is given as in (5.134). Note that it is essential that $\deg(B_u) < \deg(A_u)$ to ensure that the input $u(t)$ has finite variance (and the spectrum $\phi_u(\omega)$ to decay for large frequencies ω).

To describe the input–output relations for the sampled stochastic processes, one has to consider the input and the output jointly, that is the vector $\boldsymbol{\zeta}(t)$. Due to sampling, the discrete-time spectrum of $\boldsymbol{\zeta}(t)$ will be

$$\boldsymbol{\Phi}_\zeta^d(\omega) = \sum_{j=-\infty}^{\infty} h\boldsymbol{\Phi}_\zeta(\omega + j\frac{2\pi}{h}) . \tag{5.136}$$

It holds of course (in continuous-time) that

$$\boldsymbol{\Phi}_{\zeta}(\omega) = \begin{pmatrix} G_c(\mathrm{i}\omega) \\ 1 \end{pmatrix} \phi_u(\omega) \begin{pmatrix} G_c(-\mathrm{i}\omega) & 1 \end{pmatrix} . \tag{5.137}$$

Each term in (5.136) will be a 2×2 matrix of rank one. However, the different terms have different range space, so the sum will always be of full rank. As $\boldsymbol{\Phi}_{\zeta}^{d}(\mathrm{e}^{\mathrm{i}\omega h})$ has rank 2, *one cannot write* the sampled spectrum in the form

$$\boldsymbol{\Phi}_{\zeta}^{d}(\omega) = \begin{pmatrix} G_d(\mathrm{e}^{\mathrm{i}\omega h}) \\ 1 \end{pmatrix} \boldsymbol{\Phi}_{u}^{d}(\omega) \begin{pmatrix} G_d(\mathrm{e}^{-\mathrm{i}\omega h}) & 1 \end{pmatrix} \tag{5.138}$$

for any discrete-time filter G_d. Some ways for how to approximate $\boldsymbol{\Phi}_{\zeta}^{d}(\omega)$ in (5.136) with the right-hand side of (5.138) are presented in Söderström (2009).

5.A.1.4 Discussion and Illustration

Several remarks on the developments in Sects. 5.A.1.2 and 5.A.1.3 are in order.

1. It holds
$$\lim_{h \to 0} G_d(\mathrm{e}^{\mathrm{i}\omega h}) = G_c(\mathrm{i}\omega) \tag{5.139}$$
for any fixed ω. The relation (5.139) does not hold uniformly over the whole frequency axis of interest, that is for all ω in the interval $0 \le \omega \le \pi/h$.
2. In loose terms (5.139) means that any sensible approximation of $G_c(i\omega)$ will be almost perfect, if only the sampling interval h is chosen small enough. For small values of h the input signal does hardly change from one sampling instant to the next one, so any 'sensible' way of approximating or modeling the input behavior should make sense.
3. It is striking that the properties of G_d do depend on the input signal, through the dependence on $G_u(s)$. This raises also the question of how $G_u(s)$ is to be selected as a user choice. How sensitive is the choice of $G_u(s)$ for the discrete-time model $G_d(z)$? Are there any rules of thumb for how $G_u(s)$ may be selected when it is not a priori known, but rather a user design variable?
4. A quite heuristic explanation of why the described sampling approach leads to models G_d of higher degree than G_c is as follows. Instead of stating explicitly how the input signal $u(t)$ varies during the sampling interval, this behavior is indirectly specified in form of the input model
$$u(t) = G_u z(t) . \tag{5.140}$$
To get a full and a unique description of how the discrete-time samples of the process input and output values are related, one has to include also the input model (5.140).

A further analysis and examples are given in Söderström (2009).

5.A.2 Analyzing Effects of Small Singular Values

There are large algebraic similarities between the Correlation and the Realization approaches for model order estimation. Indeed, (5.94) and (5.99) have the same form. As long as one does not explicitly use the meaning of the vectors involved, one can therefore analyze the approaches jointly.

To test if the matrix $\hat{\mathbf{R}} = \hat{\mathbf{R}}_{12}$ has r 'small' singular values, one can compute the test quantity

$$T = \sum_{i=1}^{r} \sigma_i^2(\hat{\mathbf{R}}) , \tag{5.141}$$

where the summation is over the smallest singular values. Ideally, T should be 'negligible' for large data sets. However, one needs to find an appropriate test limit to decide whether or not the computed value of T should be regarded as small.

To proceed use a perturbation result for singular values, see Rensfelt and Söderström (2010), cf. also Sun (1998). The analysis leads to the following result.

Assume that the data vectors $\mathbf{x}_1(t), \mathbf{x}_2(t),\ t = 1, 2, \ldots$ are all jointly Gaussian distributed. Set

$$\mathbf{R}_{jk}(\tau) = \mathsf{E}\left\{\mathbf{x}_j(t+\tau)\mathbf{x}_k^T(t)\right\}, \quad j, k = 1, 2 . \tag{5.142}$$

The following asymptotic result applies

Lemma 5.1 *It holds*

$$\begin{aligned} T_1 \stackrel{\Delta}{=} \lim_{N\to\infty} NE(T) = \sum_{\tau=-\infty}^{\infty} \Big[& \operatorname{tr}\{\mathbf{R}_{22}(-\tau)\mathbf{P}_1\} \operatorname{tr}\{\mathbf{R}_{11}(\tau)\mathbf{P}_2\} \\ & + \operatorname{tr}\{\mathbf{R}_{12}(\tau)\mathbf{P}_1\mathbf{R}_{21}(\tau)\mathbf{P}_2\}\Big] , \end{aligned} \tag{5.143}$$

where

$$\mathbf{P}_2 = \mathbf{I} - \mathbf{U}_1\mathbf{U}_1^T , \tag{5.144}$$

and similarly,

$$\mathbf{P}_1 = \mathbf{I} - \mathbf{V}_1\mathbf{V}_1^T , \tag{5.145}$$

and $\mathbf{U}_1$ *and* $\mathbf{V}_1$ *are determined from the singular value decomposition*

$$\mathbf{R} = \begin{pmatrix} \mathbf{U}_1 & \mathbf{U}_2 \end{pmatrix} \begin{pmatrix} \mathbf{\Sigma}_1 & 0 \\ 0 & \mathbf{\Sigma}_2 \end{pmatrix} \begin{pmatrix} \mathbf{V}_1^T \\ \mathbf{V}_2^T \end{pmatrix} . \tag{5.146}$$

Here, $\mathbf{\Sigma}_2$ *corresponds to the 'small' singular values, and has dimension* $r \times r$*, and the other submatrices have compatible dimensions.*

Proof See Söderström (2012a). ∎

The expression in (5.143) can, at least approximately, be computed from measured data. From the singular value decomposition of $\hat{\mathbf{R}}$, the projection matrices $\mathbf{P}_1$ and $\mathbf{P}_2$ are easily found. Further, any element of the covariance functions $\mathbf{R}_{ij}(\tau),\ i, j = 1, 2$ is straightforward to estimate. In general an infinite sum is needed in (5.143). It is a non-trivial issue how it should be truncated. However, as will be shown below for the two treated approaches, most of the terms are indeed zero, and T_1 will be exactly equal to a *finite* sum.

However, it turns out that one can utilize specific properties of the matrices $\mathbf{R}$ and $\hat{\mathbf{R}}$ to simplify the expression (5.143). When doing so one has to distinguish between on one hand the Correlation approach and on the other hand the Realization approach.

Consider the Correlation approach. Assume that the null space of $\mathbf{R}_{12}$ is indeed given by the range space of $\mathbf{S}$, (5.91) and (5.93). (Recall that this assumption requires that $n_H > n_G$ holds; see (5.95)). In this case (5.143) can be substituted by a *finite* sum.

Lemma 5.2 *Under the given assumptions it holds for the Correlation approach*

$$T_1 = \sum_{\tau=-\hat{n}+1}^{\hat{n}-1} \left[\operatorname{tr}\left\{ \mathbf{R}_{22}(-\tau)\mathbf{P}_1 \right\} \operatorname{tr}\left\{ \mathbf{R}_{11}(\tau)\mathbf{P}_2 \right\} \right] , \tag{5.147}$$

where $\mathbf{P}_1$ *and* $\mathbf{P}_2$ *are formed as in (5.144) and (5.145).*

Proof See Söderström (2012a). ■

Next, consider the Realization approach for the EIV problem. Also in this case (5.143) can be substituted by a *finite* sum.

Lemma 5.3 *Under the given assumptions it holds for the Realization approach*

$$T_1 = \sum_{\tau=-\hat{n}+1}^{\hat{n}-1} \left[\operatorname{tr}\left\{ \mathbf{R}_{22}(-\tau)\mathbf{P}_1 \right\} \operatorname{tr}\left\{ \mathbf{R}_{11}(\tau)\mathbf{P}_2 \right\} \right] . \tag{5.148}$$

Proof See Söderström (2012a). ■

Chapter 6
Elementary Methods

This chapter describes and analyzes two types of elementary methods. The least squares method is treated in Sect. 5.6, while different forms of the instrumental variable method are covered in Sect. 6.2. Finally, some extensions and bibliographical notes are given in Sect. 6.3.

6.1 The Least Squares Method

The least squares (LS) method is applicable to linear regression models, see (5.6),

$$A(q^{-1})y(t) = B(q^{-1})u(t) + \varepsilon(t) , \tag{6.1}$$

with

$$\begin{aligned} A(q^{-1}) &= 1 + a_1 q^{-1} + \ldots + a_{n_a} q^{-n_a} , \\ B(q^{-1}) &= b_1 q^{-1} + \ldots + b_{n_b} q^{-n_b} . \end{aligned} \tag{6.2}$$

Here $\varepsilon(t)$ denotes an equation error, which can describe disturbances as well as unmodeled dynamics. The model (6.1) can be equivalently expressed as

$$y(t) = \boldsymbol{\varphi}^T(t)\boldsymbol{\theta} + \varepsilon(t) , \tag{6.3}$$

where the regressor vector $\boldsymbol{\varphi}(t)$ and the parameter vector $\boldsymbol{\theta}$ are given by, cf. (5.4), (5.5)

$$\boldsymbol{\varphi}(t) = (-y(t-1) \ \ldots \ -y(t-n_a) u(t-1) \ \ldots \ u(t-n_b))^T , \tag{6.4}$$

$$\boldsymbol{\theta} = (a_1 \ \ldots \ a_{n_a} \quad b_1 \ \ldots \ b_{n_b})^T . \tag{6.5}$$

T. Söderström, *Errors-in-Variables Methods in System Identification*, Communications and Control Engineering,
https://doi.org/10.1007/978-3-319-75001-9_6

The least squares estimate $\hat{\boldsymbol{\theta}}_{\text{LS}}$ of the parameter vector $\boldsymbol{\theta}$ from the data $u(1)$, $y(1),\ldots,u(N)$, $y(N)$ is defined as the minimizing argument of the sum of squared equation errors,

$$V_N(\boldsymbol{\theta}) = \frac{1}{N}\sum_{t=1}^{N}\varepsilon^2(t)\ . \tag{6.6}$$

Setting the gradient of $V_N(\boldsymbol{\theta})$ to zero leads to the so-called normal equations

$$\left[\frac{1}{N}\sum_{t=1}^{N}\boldsymbol{\varphi}(t)\boldsymbol{\varphi}^T(t)\right]\hat{\boldsymbol{\theta}}_{\text{LS}} = \left[\frac{1}{N}\sum_{t=1}^{N}\boldsymbol{\varphi}(t)y(t)\right]\ . \tag{6.7}$$

Now examine the LS estimate for consistency. Consider a linear system of an arbitrary order and write it as

$$A_0(q^{-1})y(t) = B_0(q^{-1})u(t) + v(t)\ , \tag{6.8}$$

or, equivalently

$$y(t) = \boldsymbol{\varphi}^T(t)\boldsymbol{\theta}_0 + v(t)\ . \tag{6.9}$$

Particularly for an EIV situation the disturbance term $v(t)$ can be evaluated as

$$\begin{aligned} v(t) &= A_0(q^{-1})y(t) - B_0(q^{-1})u(t) \\ &= A_0(q^{-1})[y_0(t) + \tilde{y}(t)] - B_0(q^{-1})[u_0(t) + \tilde{u}(t)] \\ &= A_0(q^{-1})\tilde{y}(t) - B_0(q^{-1})\tilde{u}(t)\ . \end{aligned} \tag{6.10}$$

The estimation error becomes

$$\begin{aligned} \hat{\boldsymbol{\theta}}_{\text{LS}} - \boldsymbol{\theta}_0 &= \left[\frac{1}{N}\sum_{t=1}^{N}\boldsymbol{\varphi}(t)\boldsymbol{\varphi}^T(t)\right]^{-1}\left[\frac{1}{N}\sum_{t=1}^{N}\boldsymbol{\varphi}(t)y(t) - \left\{\frac{1}{N}\sum_{t=1}^{N}\boldsymbol{\varphi}(t)\boldsymbol{\varphi}^T(t)\right\}\boldsymbol{\theta}_0\right] \\ &= \left[\frac{1}{N}\sum_{t=1}^{N}\boldsymbol{\varphi}(t)\boldsymbol{\varphi}^T(t)\right]^{-1}\left[\frac{1}{N}\sum_{t=1}^{N}\boldsymbol{\varphi}(t)v(t)\right]\ . \end{aligned} \tag{6.11}$$

Under weak conditions, the sums in (6.11) tend to the corresponding expected values as the number of data points, N, tends to infinity, Ljung (1999), Söderström and Stoica (1989). Hence $\hat{\boldsymbol{\theta}}_{\text{LS}}$ is consistent if

$$\mathsf{E}\left\{\boldsymbol{\varphi}(t)\boldsymbol{\varphi}^T(t)\right\} \text{ is non-singular,} \tag{6.12}$$

$$\mathsf{E}\left\{\boldsymbol{\varphi}(t)v(t)\right\} = \mathbf{0}\ . \tag{6.13}$$

The condition (6.12) is satisfied in most cases. There are a few exceptions; see Söderström and Stoica (1989).

The condition (6.13) is much more restrictive than (6.12). In case the disturbance $v(t)$ is white noise it will be independent of *all the past data values* and (6.13) will be satisfied. However, as $v(t)$ in the EIV case is given by (6.10), it is correlated with the delayed input and output values present in $\varphi(t)$, and (6.13) will be violated. In an EIV setting, the least squares estimate will therefore *not be consistent.*

The least squares method is a simple method for system identification and has some attractive properties. The estimate is easy to compute and has good robustness properties. The restrictive consistency property is the essential drawback and can be seen as the main reason for considering more advanced methods, including the instrumental variable method. Recall that the lack of consistency when LS is used in an EIV setting is illustrated numerically in Example 3.2.

6.2 The Instrumental Variable Method

6.2.1 Description

Instrumental variable methods can be seen as generalizations of the least squares estimates. One can say that the main idea is to modify the estimate so that it is consistent for an arbitrary disturbance. Consider ARX models (6.1), (6.2) and modify the normal equations of (6.7) into

$$\left[\frac{1}{N}\sum_{t=1}^{N}\mathbf{z}(t)\varphi^T(t)\right]\hat{\boldsymbol{\theta}}_{\text{IV}} = \left[\frac{1}{N}\sum_{t=1}^{N}\mathbf{z}(t)y(t)\right], \tag{6.14}$$

where $\mathbf{z}(t)$ is a vector of *instrumental variable*. This vector can be chosen in different ways (as will be exemplified below). It should fulfil certain conditions that guarantee the consistency of the estimate $\hat{\boldsymbol{\theta}}_{\text{IV}}$ defined as the solution to (6.14). These conditions will be specified later. Evidently the IV estimate is a generalization of the LS estimate: For the specific choice $\mathbf{z}(t) = \varphi(t)$, (6.14) reduces to (6.7).

The basic IV method can be generalized in different ways.

The *extended* IV estimates of $\boldsymbol{\theta}_0$ are obtained by generalizing (6.14) in two directions. Such IV estimation methods allow for an augmented $\mathbf{z}(t)$ vector (i.e., one can have $\dim \mathbf{z}(t) > \dim \varphi(t)$), as well as a prefiltering of the data. The extended IV estimate is given by

$$\hat{\boldsymbol{\theta}}_{\text{IV}} = \arg\min_{\boldsymbol{\theta}} \left|\left| \left[\sum_{t=1}^{N}\mathbf{z}(t)F(q^{-1})\varphi^T(t)\right]\boldsymbol{\theta} - \left[\sum_{t=1}^{N}\mathbf{z}(t)F(q^{-1})y(t)\right]\right|\right|_{\mathbf{W}}^2. \tag{6.15}$$

Here $\mathbf{z}(t)$ is the IV vector of dimension $n_{\mathbf{z}} \geq \dim \boldsymbol{\theta}$, $F(q^{-1})$ is an asymptotically stable (pre-)filter, and $||\mathbf{x}||_{\mathbf{W}}^2 = \mathbf{x}^T\mathbf{W}\mathbf{x}$, where $\mathbf{W}$ is a positive definite weighting matrix. When $F(q^{-1}) \equiv 1$ and $n_{\mathbf{z}} = n_{\theta}$, ($\mathbf{W} = \mathbf{I}$), the basic IV estimate (6.14) is obtained. Note that the estimate (6.15) is the weighted least squares solution of an overdetermined linear system of equations. The solution can readily be found to be

$$\hat{\boldsymbol{\theta}}_{\mathrm{IV}} = (\mathbf{R}_N^T \mathbf{W} \mathbf{R}_N)^{-1} \mathbf{R}_N^T \mathbf{W} \mathbf{r}_N \,, \tag{6.16}$$

where

$$\mathbf{R}_N = \frac{1}{N} \sum_{t=1}^{N} \mathbf{z}(t) F(q^{-1}) \boldsymbol{\varphi}^T(t), \quad \mathbf{r}_N = \frac{1}{N} \sum_{t=1}^{N} \mathbf{z}(t) F(q^{-1}) y(t) \,, \tag{6.17}$$

even if this is not the *numerically best way* to implement it.

Another generalization is to model multiple-output systems. Assume that the model (5.69) is used. For this form of the multivariable model, the extended IV estimate is quite similar to (6.15). It now holds

$$\hat{\boldsymbol{\theta}}_{\mathrm{IV}} = \arg\min_{\boldsymbol{\theta}} \left\| \left[\sum_{t=1}^{N} \mathbf{z}(t) \mathbf{F}(q^{-1}) \boldsymbol{\Phi}^T(t) \right] \boldsymbol{\theta} - \left[\sum_{t=1}^{N} \mathbf{z}(t) \mathbf{F}(q^{-1}) \mathbf{y}(t) \right] \right\|_{\mathbf{W}}^2 . \tag{6.18}$$

6.2.2 Consistency Analysis

For a general consistency analysis of the extended IV estimate (6.15) applied to a system of the form (6.9), assume that the data are stationary (follows from Assumptions **AI1** and **AS1**), and that $\mathbf{z}(t)$ is uncorrelated with the disturbances $v(s)$ for all t and s. Then it holds, similarly to (6.11), that when $N \to \infty$,

$$\begin{aligned}
\hat{\boldsymbol{\theta}}_{\mathrm{IV}} - \boldsymbol{\theta}_0 &= (\mathbf{R}_N^T \mathbf{W} \mathbf{R}_N)^{-1} (\mathbf{R}_N^T \mathbf{W} \mathbf{r}_N - \mathbf{R}_N^T \mathbf{W} \mathbf{R}_N \boldsymbol{\theta}_0) \\
&= (\mathbf{R}_N^T \mathbf{W} \mathbf{R})^{-1} \mathbf{R}_N^T \mathbf{W} \frac{1}{N} \sum_{t=1}^{N} \mathbf{z}(t) F(q^{-1}) \{ y(t) - \boldsymbol{\varphi}^T(t) \boldsymbol{\theta}_0 \} \\
&\to (\mathbf{R}^T \mathbf{W} \mathbf{R})^{-1} \mathbf{R}^T \mathbf{W} \, \mathsf{E} \left\{ \mathbf{z}(t) F(q^{-1}) v(t) \right\} = \mathbf{0} \,,
\end{aligned} \tag{6.19}$$

where

$$\mathbf{R} \triangleq \mathsf{E} \left\{ \mathbf{z}(t) F(q^{-1}) \boldsymbol{\varphi}^T(t) \right\} \tag{6.20}$$

is assumed to have full column rank. Hence the consistency conditions (6.12), (6.13) in the general case become here

$$\mathbf{R} \text{ has full column rank,} \tag{6.21}$$

$$\mathsf{E} \left\{ \mathbf{z}(t) F(q^{-1}) v(t) \right\} = \mathbf{0} \,. \tag{6.22}$$

For any reasonable choice of instrumental vector $\mathbf{z}(t)$ the rank condition (6.21) is satisfied for almost any (but not *all*) systems. One can hence say that the estimate $\hat{\boldsymbol{\theta}}_{\mathrm{IV}}$ is *generically consistent*. There are though counter-examples (based on certain combinations of input spectrum and system parameter $\boldsymbol{\theta}$ that leads to a singular matrix $\mathbf{R}$. Details are given in Söderström and Stoica (1983), Söderström and Stoica (1989).

6.2.3 User Choices. Examples of Instrumental Vectors

The IV estimator contains some user parameters, in addition to the choice of model order. These user parameters are the following:

1. A first choice is the instrumental variable vector $\mathbf{z}(t)$. This choice concerns both its dimension, and how the elements are formed from the measured data. See Söderström and Stoica (1983) for a number of examples. Due to the consistency conditions (6.21), (6.22), it can be said that the IV vector $\mathbf{z}(t)$ should be well correlated with the regressor $\boldsymbol{\varphi}(t)$, and uncorrelated with the disturbances $v(t)$. For the errors-in-variables case, consider specifically the case when both the input and the output noise are white (Assumption **AN3c**). Then due to the disturbance property (6.10) the instrumental vector $\mathbf{z}(t)$ should be constructed from the measured input and output data, so as not to contain the elements

$$u(t-1), \ldots, u(t-n_b), \;\; y(t), \ldots, y(t-n_a) \; .$$

   One way (out of many possible ways) to achieve this is to choose

$$\mathbf{z}(t) = \boldsymbol{\varphi}(t-n) + \boldsymbol{\varphi}(t+n), \quad n = \max(n_a, n_b) \; . \tag{6.23}$$

2. A second user choice is the prefilter $F(q^{-1})$.
3. The third user choice applies when the system of IV equations is overdetermined, that is when the IV vector has higher dimension than the regressor vector. In that case the weighting matrix $\mathbf{W}$ of the equations, see (6.16), needs to be chosen.

In the EIV case *for open-loop operation*, vectors like $\mathbf{z}(t)$ in (6.23) above, and more generally those whose elements are obtained by filtering and delaying the input signal will be independent of the disturbance $v(t)$ and mostly also satisfy the rank condition on $\mathbf{R}$. For *closed-loop operations* some modifications can be done to achieve consistent estimates provided an external signal, independent of the disturbances, can be measured.

Example 6.1 The simplest possible case is to let the instruments consist of delayed inputs only, so

$$\mathbf{z}(t) = \begin{pmatrix} u(t-n_b-1) \\ \vdots \\ u(t-n_b-p) \end{pmatrix} \tag{6.24}$$

with $p \geq n_a + n_b$. Then the second consistency condition (6.21) is satisfied also for correlated output noise. However, this choice implies some restrictions for the excitation signal $u_0(t)$. Assuming that no prefiltering takes place, $F(q^{-1}) = 1$, it holds

$$\begin{aligned}
\mathbf{R} &= \mathsf{E}\left\{\mathbf{z}(t)\boldsymbol{\varphi}^T(t)\right\} \\
&= \mathsf{E}\left\{\left(\begin{pmatrix} u_0(t-n_b-1) \\ \vdots \\ u_0(t-n_b-p) \end{pmatrix} + \begin{pmatrix} \tilde{u}(t-n_b-1) \\ \vdots \\ \tilde{u}(t-n_b-p) \end{pmatrix}\right)\boldsymbol{\varphi}^T(t)\right\} \\
&= \mathsf{E}\left\{\begin{pmatrix} u_0(t-n_b-1) \\ \vdots \\ u_0(t-n_b-p) \end{pmatrix}\boldsymbol{\varphi}_0^T(t)\right\}, \qquad (6.25) \\
\boldsymbol{\varphi}_0^T(t) &= \left(-y_0(t-1) \ \ldots \ -y_0(t-n_a) \ u_0(t-1) \ \ldots \ u_0(t-n_b)\right). \qquad (6.26)
\end{aligned}$$

Consider now the case when $u_0(t)$ is an ARMA process of order m,

$$\left(1 + d_1 q^{-1} + \cdots + d_m q^{-m}\right) u_0(t) = \left(1 + c_1 q^{-1} + \cdots + c_m q^{-m}\right) v(t). \tag{6.27}$$

If $n_b > m$, it follows that

$$\begin{aligned}
&\mathbf{R}\begin{pmatrix} \mathbf{0}_{n_a\times 1} \\ 1 \\ d_1 \\ \vdots \\ d_m \\ \mathbf{0}_{(n_b-m-1)\times 1} \end{pmatrix} \\
&= \mathsf{E}\left\{\begin{pmatrix} u_0(t-n_b-1) \\ \vdots \\ u_0(t-n_b-p) \end{pmatrix}\left(u_0(t-1) + d_1 u_0(t-2) + \cdots + d_m u_0(t-m-1)\right)\right\} \\
&= \mathsf{E}\left\{\begin{pmatrix} u_0(t-n_b-1) \\ \vdots \\ u_0(t-n_b-p) \end{pmatrix}\left(v(t-1) + c_1 v(t-2) + \cdots + c_m v(t-m-1)\right)\right\} \\
&= \mathbf{0}_{p\times 1}. \qquad (6.28)
\end{aligned}$$

The procedure may be repeated with the nonzero elements in the vector following $\mathbf{R}$ in (6.28) shifted downwards. These considerations show that the null space of $\mathbf{R}$ has at least dimension $n_b - m$, and therefore that

$$\text{rank } \mathbf{R} \leq n_a + \max(m, n_b) \ . \tag{6.29}$$

The conclusion is that $\mathbf{R}$ by necessity, for the input (6.27) is rank deficient as soon as

$$m < n_b \ . \tag{6.30}$$

Consistency therefore cannot be proven when (6.30) holds, as it violates the key condition (6.21).

Note that essentially the same analysis was carried out for the correlation approach to model order determination in Sect. 5.9.2. Then the condition (6.30) above corresponds to (5.95). The two problems are similar, related, but not identical. ■

Example 6.2 In this example a simple version of instrumental variable is applied to the standard data sets used in the book. The IV vector

$$\mathbf{z}(t) = \big(u(t-3) \ \dots \ u(t-6)\big)^T$$

is used, which in a sense is the simplest possible case. The numerical results obtained are presented in Table 6.1. For comparison, the results for using the LS method are repeated as well in the table.

The IV estimates seem to be consistent, as the deviations of the means from the true values are small compared to the standard deviations. However, the standard deviations are huge, in particular for the parameters b_1 and b_2. A closer look reveals that the main contribution for these statistics comes from a few of the 100 realizations. The numerical results when three realizations are removed are summarized in Table 6.2. The results are then much improved, in particular for the estimates of the B parameters.

One can summarize the findings as that the elementary IV estimate is easy to compute, avoids the bias problem of LS, but gives very low accuracy of the parameter estimates. It is thus of clear interest to find out if improvements can be achieved by applying somewhat more complex IV estimators. ■

Table 6.1 Instrumental variable estimates for the standard data sets

Parameter	True value	IV		LS	
		Mean	Standard dev.	Mean	Standard dev.
a_1	−1.5	−1.440	0.534	−1.414	0.015
a_2	0.7	0.670	0.302	0.619	0.014
b_1	2.0	−0.510	27.9	1.599	0.103
b_2	1.0	3.599	26.2	1.361	0.129

Table 6.2 Instrumental variable estimates for the standard data sets. IV* refers to statistics from 97 realizations, while IV refers to all 100 realizations

Parameter	True value	IV *		IV	
		Mean	Standard dev.	Mean	Standard dev.
a_1	−1.5	−1.485	0.177	−1.440	0.534
a_2	0.7	0.689	0.115	0.670	0.302
b_1	2.0	1.794	10.4	−0.510	27.9
b_2	1.0	1.308	9.1	3.599	26.2

It was mentioned in Sect. 6.2.1 that it is possible to use more than the minimal number of equations when defining the IV estimates, that is to let $\mathbf{z}(t)$ have higher dimension than $\boldsymbol{\theta}$. Then the system of equations will be overdetermined. Sometimes, but not always, it may be beneficial to use more than the minimal number of equations. This idea is examined in Example 6.3.

Example 6.3 The instrumental variable vector tested in this example is

$$\mathbf{z}(t) = \left(u(t-3) \ldots u(t-6-p)\right)^T , \tag{6.31}$$

where the user parameter p varies from $p = 0$ (which is indeed the situation treated in Example 4.5) to $p = 10$. The results are presented in Table 6.3.

One can see that when p is increased from $p = 0$ there are certainly some improvements in the accuracy of the estimates. After $p = 4$, however, the changes are at most minor.

The root-mean-square (RMS) values are displayed versus p in Fig. 6.1. It is clear that the RMS value in this example decreases when p is increased. After $p = 4$ the improvement is rather marginal. ■

6.2.4 *Instrumental Variable Methods Exploiting Higher-Order Statistics*

Sometimes it is an option to use higher-order statistics for identifying EIV systems. For this to be relevant, atleast one of the signals, say the noise-free input $u_0(t)$, has to be non-Gaussian, so that the higher-order moments carry some additional information. The possibility to make use of higher-order statistics for the modeling of dynamic systems was mentioned early in Deistler (1986). Various algorithms exploiting this possibility have appeared, Tugnait (1992), Tugnait and Ye (1995).

Example 6.4 Assume that the noise-free input $u_0(t)$ is independent of the noise sequences $\tilde{u}(t)$, $\tilde{y}(t)$, and that these noises are white and Gaussian; see Assumptions

Table 6.3 Instrumental variable estimates for increasing length of the instrumental variable vector using the standard data sets. The arithmetic mean values are given with standard deviations within parentheses

Parameter	a_1	a_2	b_1	b_2
p True	−1.5	0.7	2.0	1.0
0	−1.440 (0.534)	0.670 (0.302)	−0.510 (27.9)	3.599 (26.2)
1	−1.493 (0.042)	0.697 (0.029)	1.510 (4.56)	1.469 (3.77)
2	−1.495 (0.028)	0.698 (0.021)	1.734 (2.89)	1.271 (2.33)
3	−1.499 (0.020)	0.700 (0.016)	1.726 (2.18)	1.236 (1.75)
4	−1.500 (0.017)	0.700 (0.014)	1.862 (1.63)	1.122 (1.32)
5	−1.500 (0.017)	0.701 (0.014)	1.835 (1.54)	1.138 (1.23)
6	−1.500 (0.017)	0.701 (0.014)	1.895 (1.49)	1.086 (1.19)
7	−1.500 (0.017)	0.700 (0.013)	1.773 (1.13)	1.184 (0.91)
8	−1.500 (0.017)	0.700 (0.013)	1.761 (1.06)	1.188 (0.88)
9	−1.501 (0.017)	0.701 (0.013)	1.751 (1.01)	1.191 (0.83)
10	−1.501 (0.017)	0.701 (0.013)	1.845 (0.97)	1.115 (0.81)

ANI2, AN2, AN3c. Assume further that the noise-free input $u_0(t)$ has zero mean and nonzero third-order moments.

$$\mathsf{E}\left\{u_0^3(t)\right\} = \mu_u \neq 0 \, . \tag{6.32}$$

Then a possible instrumental vector can be constructed as

$$\mathbf{z}(t) = \left(u^2(t-1) \, \dots \, u^2(t-n_a-n_b)\right)^T \, . \tag{6.33}$$

For this choice of IV vector to work one needs to show that $\mathsf{E}\{\mathbf{z}(t)v(t)\} = \mathbf{0}$. However, for an arbitrary component it is indeed found that

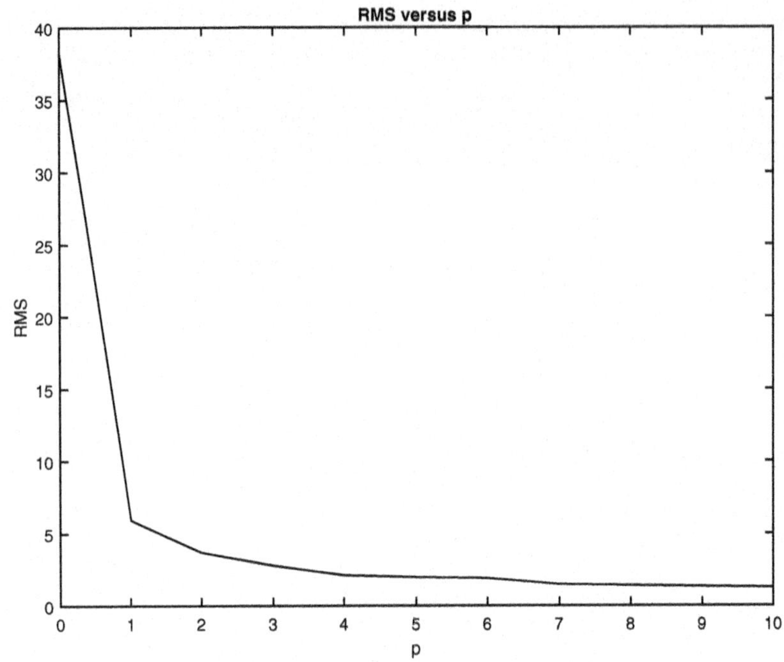

Fig. 6.1 RMS values versus p for the IV estimates

$$
\begin{aligned}
[\mathsf{E}\{\mathbf{z}(t)v(t)\}]_j &= \mathsf{E}\{u^2(t-j)v(t)\} \\
&= \mathsf{E}\{[u_0^2(t-j) + 2u_0(t-j)\tilde{u}(t-j) + \tilde{u}^2(t-j)] \\
&\quad \times [A_0(q^{-1})\tilde{y}(t) - B_0(q^{-1})\tilde{u}(t)]\} \\
&= 0 \,.
\end{aligned} \tag{6.34}
$$

A more detailed analysis can be found, for example, in Thil et al. (2008c). ■

The example can easily be modified to handle the case of $u_0(t)$ having zero third-order moments, by substituting $u^2(t-j)$ by $u^3(t-j)$, $j = 1, \ldots, n_a + n_b$ in (6.33).

6.2.5 *Other Instrumental Variable Techniques*

Previously in this chapter several estimation methods that are all based on a small set of covariance elements of the noisy input and the noisy output were presented. There are several other methods that also as a first step condense the information into a modest number of estimated covariance elements. A brief discussion is given here of two examples based on instrumental variable, while other constructions are described in Chap. 7.

Example 6.5 An IV algorithm based on prespecified filtering techniques is proposed and analyzed in Mahata and Söderström (2002). The input–output data $u(t),\ y(t)$ are prefiltered using two FIR filters $G_1(q^{-1})$, $G_2(q^{-1})$ of order $m > n \stackrel{\Delta}{=} \max(n_a, n_b)$. Set

$$\boldsymbol{\varphi}_i(t) = G_i(q^{-1})\boldsymbol{\varphi}(t), \quad y_i(t) = G_i(q^{-1})y(t), \quad i = 1, 2 . \tag{6.35}$$

Assume that the noise condition **AN3c** (both $\tilde{u}(t)$, $\tilde{y}(t)$ are white), and that the prefilters are chosen so that

$$\mathbf{R}_{\tilde{\varphi}_1, \tilde{\varphi}_2} = \mathbf{0} . \tag{6.36}$$

Then two possible IV estimators are

$$\hat{\boldsymbol{\theta}} = \hat{\mathbf{R}}^{\dagger}_{\varphi_1, \varphi_2} \hat{\mathbf{r}}_{\varphi_1, y_2} , \tag{6.37}$$

$$\hat{\boldsymbol{\theta}} = \hat{\mathbf{R}}^{\dagger}_{\varphi_2 \varphi_1} \hat{\mathbf{r}}_{\varphi_2, y_1} , \tag{6.38}$$

where $\mathbf{R}^{\dagger} = (\mathbf{R}^T\mathbf{R})^{-1}\mathbf{R}^T$ is the pseudo-inverse of $\mathbf{R}$. One way to design the prefilters so that the condition (6.36) is satisfied under the noise assumption **AN3c** is the following. Let the polynomial

$$H(q) \stackrel{\Delta}{=} q^m G_1(q^{-1}) G_2(q) = \sum_{j=0}^{2m} h_j q^j \tag{6.39}$$

satisfy

$$h_j = 0, \ m - n \leq j \leq m + n . \tag{6.40}$$

Except for proving consistency of the estimates (6.37) and (6.38), the asymptotic covariance matrix of these parameter estimates is derived in Mahata and Söderström (2002). It is shown that the estimates are considerably more accurate than straightforward overdetermined IV estimates using only delayed inputs as instrumental variables. ■

Example 6.6 It is possible to extend and improve the IV estimate by combining it with a weighted subspace fitting approach; see Stoica et al. (1995a). The noise assumption **AN3b** applies (and can be somewhat weakened: One can let $\tilde{u}(t)$ be an MA process, that is also finitely correlated with $\tilde{y}(t)$.) This combined instrumental variable weighted subspace fitting, or IV-WSF, can be viewed as an extension to the approach discussed above. Here the regressor vector $\boldsymbol{\varphi}(t)$ corresponds to a model order p such that $p \geq n \stackrel{\Delta}{=} \max(n_a, n_b)$. Let the null vector of dimension $p - n$ be denoted by $\boldsymbol{\rho}$ and $\boldsymbol{\theta}_p = \begin{pmatrix} 1 & \mathbf{a}^T & \boldsymbol{\rho}^T & \mathbf{b}^T & \boldsymbol{\rho}^T \end{pmatrix}^T$. Denote the first column of the $p-n+1$-dimensional identity matrix by $\boldsymbol{\mu}$. Consider the Toeplitz matrix $\mathbf{G}(\boldsymbol{\theta})$ having $\boldsymbol{\theta}_p$ as its first column and $\boldsymbol{\mu}^T$ as its first row. Then it can be shown that

$$\mathbf{R}_{z\overline{\varphi}}\mathbf{G}(\boldsymbol{\theta}) = \mathbf{0} \; . \tag{6.41}$$

An approach to exploit (6.41) is to fit $\mathbf{G}(\boldsymbol{\theta})$ to span the null space of $\mathbf{R}_{z\overline{\varphi}}$. Let the columns of $\hat{\mathbf{G}}$ be the right singular vectors corresponding to the least $p - n + 1$ singular values of $\hat{\mathbf{R}}_{z\overline{\varphi}}$. Then one way to determine $\hat{\boldsymbol{\theta}}$ is to minimize the distance between $\mathbf{G}(\hat{\boldsymbol{\theta}})$ and the span of $\hat{\mathbf{G}}$, i.e.,

$$\hat{\boldsymbol{\theta}} = \arg\min_{\boldsymbol{\theta}} \left(\min_{\mathbf{C}} \frac{1}{2} \left[\text{vec}(\mathbf{G}(\boldsymbol{\theta}) - \hat{\mathbf{G}}\mathbf{C}) \right]^T \mathbf{W} \left[\text{vec}(\mathbf{G}(\boldsymbol{\theta}) - \hat{\mathbf{G}}\mathbf{C}) \right] \right) \; , \tag{6.42}$$

where $\mathbf{C}$ is a non-singular $(p - n + 1) \times (p - n + 1)$ matrix, $\mathbf{W}$ is a positive definite weighting matrix and may be data-dependent, and vec$(\cdot)$ denotes the vectorization operation. This can be shown to be equivalent to (see Stoica et al. (1995a) for details)

$$\hat{\boldsymbol{\theta}} = \arg\min_{\boldsymbol{\theta}} \frac{1}{2} \left[\text{vec}(\mathbf{S}^T \mathbf{G}(\boldsymbol{\theta})) \right]^T \widetilde{\mathbf{W}} \left[\text{vec}(\mathbf{S}^T \mathbf{G}(\boldsymbol{\theta})) \right] \; , \tag{6.43}$$

where $\widetilde{\mathbf{W}}$ is a positive definite weighting matrix and may be data-dependent, and the columns of $\mathbf{S}$ are the right singular vectors corresponding to the largest $p+n$ singular values of $\hat{\mathbf{R}}_{z\overline{\varphi}}$. For optimal selection of the weighting matrix $\widetilde{\mathbf{W}}$ and the corresponding implementational aspects, see Stoica et al. (1995a). The optimal weight determination is computationally expensive.

A computationally economical suboptimal IV-WSF is presented in Cedervall and Stoica (1996). Instead of working with (6.41), one makes a singular value decomposition (SVD) of

$$\mathbf{R} = \mathbf{R}_{\mathbf{z}}^{-1/2} \mathbf{R}_{\mathbf{z}\overline{\varphi}} \mathbf{R}_{\overline{\varphi}}^{-1/2} \; , \tag{6.44}$$

where

$$\mathbf{z}(t) = \left(u(t - p - L) \; \ldots \; u(t - p - L - m + 1) \right)^T \; , \tag{6.45}$$

with $p > n$, and $m \geq p + n$. The SVD is written as

$$\mathbf{R} = \hat{\mathbf{U}}\hat{\boldsymbol{\Sigma}}\hat{\mathbf{S}}^T \; , \tag{6.46}$$

where $\hat{\mathbf{S}}$ is an $m \times (n + p)$-dimensional matrix.

In Cedervall and Stoica (1996) it is proposed to minimize the following criterion

$$V(\boldsymbol{\theta}) = \left\| \left(\hat{\mathbf{G}}^T \hat{\mathbf{R}}_{\varphi} \hat{\mathbf{G}} \right)^{-1/2} \hat{\mathbf{G}}^T \hat{\mathbf{R}}_{\varphi}^{1/2} \hat{\mathbf{S}} \hat{\boldsymbol{\Sigma}} \right\|_{\mathrm{F}}^2 \; . \tag{6.47}$$

This criterion depends on $\boldsymbol{\theta}$ in a highly nonlinear way. A multi-step but non-iterative procedure is described which is computationally much simpler, but have the same asymptotic properties as the minimizer of (6.47). ■

6.3 Extensions and Bibliographical Notes

General Classical References

General treatments and analysis of instrumental variable methods can be found in many books, for example, Söderström and Stoica (1983), Söderström and Stoica (1989), Ljung (1999).

A very early reference about the use of least squares is Gauss (1809). The papers Adcock (1877), Adcock (1878) are among the first to account for errors in both independent and dependent variables.

Instrumental Variable Methods and Correlation Techniques

Simple correlation techniques are described and illustrated in Söderström (1981), Van Huffel and Vandewalle (1989), Soverini and Söderström (2000), Söderström et al. (2002), Thil et al. (2008b).

Some early papers using IV techniques in system identification include Reiersøl (1941), Young (1965), Mayne (1967), Wong and Polak (1967), Young (1970), Rowe (1970), Finigan and Rowe (1974).

IV-based subspace identification algorithms employing the state space models are proposed in Chou and Verhaegen (1997). The method is based on the noise assumption **AN3b** ($\tilde{u}(t)$ white, $\tilde{y}(t)$ ARMA), but allows the noise terms to be correlated. These algorithms can be applied to MIMO systems operating in open or closed loop, where one has to account for the process noise also.

Subspace system identification methods may be regarded as advanced variants of instrumental variable. Some applications of subspace identification using also an EM algorithm for implementation are considered in ALMutawa (2008), ALMutawa (2009). The relation between subspace identification and dynamic principal component analysis (PCA) is treated in Li and Qin (2001). The paper Stanković et al. (2015) deals with decentralized identification of large-scale systems in an EIV setting. Modified Yule–Walker equations are used, and the methodology hence fits into the IV framework.

Higher-order Statistics

For a general treatment of identification methods using higher-order statistics, see the tutorial paper Nikias and Mendel (1993).

Examples where HOS methods are applied for identifying EIV models can be found in Mendel (1991), Tugnait (1992), Tugnait and Ye (1995). Another approach where cumulant equations are used for estimating EIV models appears in Erickson et al. (2014).

The paper Thil et al. (2007) deals with using HOS methods for continuous-time EIV models. Further, Thil et al. (2008a), Thil et al. (2009) consider several EIV estimators based on higher-order statistics.

Identification of an AR Process Observed in Noise

For autoregressive processes observed in noise, see also Mendel (1991), Tugnait (1992), Tugnait and Ye (1995). See also Sect. 3.5.

Chapter 7
Methods Based on Bias-Compensation

The chapter is organized as follows. The basic idea of bias-compensated least squares is presented in Sect. 7.1. After that, different ways to derive and interpret bias-compensation are introduced in the following sections. The bias-eliminating least squares method is presented in Sect. 7.2, and the Frisch scheme approach appears in Sect. 7.3. In Sect. 7.4 it is shown how both these approaches can be embedded into a generalized instrumental variable (GIVE) method. Depending on the choice of some user parameters in GIVE, various special cases appear, originally constructed and presented as specific methods. The chapter ends in Sect. 7.5 with extensions and bibliographical notes.

7.1 The Basic Idea of Bias-Compensation

The EIV system described in Chaps. 3 and 6 can be expressed as a linear regressor model

$$y(t) = \boldsymbol{\varphi}^T(t)\boldsymbol{\theta} + \varepsilon(t)\ , \tag{7.1}$$

where the equation error is $\varepsilon(t) = \tilde{y}(t) - \tilde{\boldsymbol{\varphi}}^T(t)\boldsymbol{\theta}$. Recall from Chap. 3 that $y(t)$ denotes the noisy output, $u(t)$ the noisy input, $\boldsymbol{\theta}$ the parameter vector, and the elements of the regressor vector $\boldsymbol{\varphi}(t)$ are composed of delayed values of $y(t)$ and $u(t)$.

For the standard least squares (LS) method, the estimate $\hat{\boldsymbol{\theta}}_{\text{LS}}$ is given by the solution to the normal equation

$$\hat{\mathbf{R}}_{\varphi}\hat{\boldsymbol{\theta}}_{\text{LS}} = \hat{\mathbf{r}}_{\varphi y}\ . \tag{7.2}$$

Assuming that the data are ergodic (this is the case for stationary processes), it is convenient to study the asymptotic case when $N \to \infty$. Then it holds

T. Söderström, *Errors-in-Variables Methods in System Identification*, Communications and Control Engineering,
https://doi.org/10.1007/978-3-319-75001-9_7

$$\begin{aligned}
\lim_{N\to\infty} \hat{\mathbf{R}}_{\varphi} &= \lim_{N\to\infty} \frac{1}{N}\sum_{t=1}^{N} \boldsymbol{\varphi}(t)\boldsymbol{\varphi}^T(t) \\
&= \mathsf{E}\left\{\boldsymbol{\varphi}(t)\boldsymbol{\varphi}^T(t)\right\} = \mathsf{E}\left\{\boldsymbol{\varphi}_0(t)\boldsymbol{\varphi}_0^T(t) + \tilde{\boldsymbol{\varphi}}(t)\tilde{\boldsymbol{\varphi}}^T(t)\right\} \\
&= \mathbf{R}_{\varphi_0} + \mathbf{R}_{\tilde{\varphi}} \,, \qquad (7.3)
\end{aligned}$$

$$\begin{aligned}
\lim_{N\to\infty} \hat{\mathbf{r}}_{\varphi y} &= \lim_{N\to\infty} \frac{1}{N}\sum_{t=1}^{N} \boldsymbol{\varphi}(t)y(t) \\
&= \mathsf{E}\left\{\boldsymbol{\varphi}(t)y(t)\right\} = \mathsf{E}\left\{\boldsymbol{\varphi}_0(t)y_0(t) + \tilde{\boldsymbol{\varphi}}(t)\tilde{y}(t)\right\} \\
&= \mathbf{r}_{\varphi_0 y_0} + \mathbf{r}_{\tilde{\varphi}\tilde{y}} \,. \qquad (7.4)
\end{aligned}$$

In this way the contribution to the covariance elements by the noise-free input u_0 and the measurement noise, respectively, are separated into two parts.

Note that

$$y_0(t) = \boldsymbol{\varphi}_0^T(t)\boldsymbol{\theta}_0 \,, \qquad (7.5)$$

and hence

$$\mathbf{r}_{\varphi_0 y_0} = \mathbf{R}_{\varphi_0}\boldsymbol{\theta}_0 \,. \qquad (7.6)$$

Recall from Sect. 5.1 that $\boldsymbol{\varphi}(t)$ denotes the noisy regressor, $\boldsymbol{\varphi}_0(t)$ its part due to the noise-free input, and $\tilde{\boldsymbol{\varphi}}(t)$ its part due to the measurement noise.

Considering the asymptotic case ($N \to \infty$) the deviation of the LS estimate $\hat{\boldsymbol{\theta}}_{\mathrm{LS}}$ from the true parameter vector $\boldsymbol{\theta}_0$ can be found from

$$\begin{aligned}
\mathbf{R}_{\varphi}\left(\hat{\boldsymbol{\theta}}_{\mathrm{LS}} - \boldsymbol{\theta}_0\right) &= \mathbf{r}_{\varphi y} - \mathbf{R}_{\varphi}\boldsymbol{\theta}_0 \\
&= \mathsf{E}\left\{\boldsymbol{\varphi}(t)\left[y(t) - \boldsymbol{\varphi}^T(t)\boldsymbol{\theta}_0\right]\right\} \\
&= \mathsf{E}\left\{\boldsymbol{\varphi}_0(t)\left[y_0(t) - \boldsymbol{\varphi}_0^T(t)\boldsymbol{\theta}_0\right]\right\} + \mathsf{E}\left\{\tilde{\boldsymbol{\varphi}}(t)\left[\tilde{y}(t) - \tilde{\boldsymbol{\varphi}}^T(t)\boldsymbol{\theta}_0\right]\right\} \\
&= \mathbf{r}_{\tilde{\varphi}\tilde{y}} - \mathbf{R}_{\tilde{\varphi}}\boldsymbol{\theta}_0 \,. \qquad (7.7)
\end{aligned}$$

To proceed with the key idea of bias-compensation, consider now specifically the case of white input and output noise (i.e., Assumption **AN3c** applies). The extension to handle correlated output noise is treated later in the coming sections. Under the white noise assumption it holds

$$\mathbf{r}_{\tilde{\varphi}\tilde{y}} = \mathsf{E}\left\{\begin{pmatrix} -\tilde{y}(t-1) \\ \vdots \\ -\tilde{y}(t-n_a) \\ \tilde{u}(t-1) \\ \vdots \\ \tilde{u}(t-n_b) \end{pmatrix} \tilde{y}(t)\right\} = \mathbf{0} \,, \qquad (7.8)$$

$$\mathbf{R}_{\tilde{\varphi}} = \begin{pmatrix} \lambda_y \mathbf{I}_{n_a} & \mathbf{0} \\ \mathbf{0} & \lambda_u \mathbf{I}_{n_b} \end{pmatrix} , \tag{7.9}$$

and (7.2) implies

$$\left(\mathbf{R}_{\varphi_0} + \mathbf{R}_{\tilde{\varphi}}\right) \left(\hat{\boldsymbol{\theta}}_{\text{LS}} - \boldsymbol{\theta}_0\right) \neq \mathbf{0} . \tag{7.10}$$

Therefore the estimate $\hat{\boldsymbol{\theta}}_{\text{LS}}$ is not consistent, as already known from Chap. 6.

To get consistent estimates, a bias-compensated least squares (BCLS) scheme, Stoica and Söderström (1982), can be introduced as, see (7.6)

$$\hat{\boldsymbol{\theta}}_{\text{BCLS}} = \left(\mathbf{R}_{\varphi} - \mathbf{R}_{\tilde{\varphi}}\right)^{-1} \left(\mathbf{r}_{\varphi y} - \mathbf{r}_{\tilde{\varphi}\tilde{y}}\right) . \tag{7.11}$$

Equation (7.11) is a common part of the equations used to define the estimates for all the methods to be considered in this chapter. It can be considered as an adjusted system of normal equations. Some further relations are needed in order to determine the noise parameter vector $\boldsymbol{\rho}$, (5.27), and thereby $\mathbf{R}_{\tilde{\varphi}}$ and $\mathbf{r}_{\tilde{\varphi}\tilde{y}}$. There are different ways to do so.

Recall that in the white noise case the vector $\boldsymbol{\rho}$ of noise parameters is

$$\boldsymbol{\rho} = \begin{pmatrix} \lambda_y \\ \lambda_u \end{pmatrix} . \tag{7.12}$$

There are different ways to derive bias-compensated estimates as in (7.11). In the white noise case, at least two more equations are needed, in addition to the adjusted normal equation, as both $\boldsymbol{\theta}$ and $\boldsymbol{\rho}$ in (7.12) are regarded as unknowns. If the output noise is correlated the number of additional equations is higher.

In the following sections some different derivations are presented.

- Section 7.2 presents the bias-eliminating approach. An expression for the minimal value of the LS criterion is used to derive one equation. The other equation is obtained by considering an over-parameterization. Assume for ease of illustration that A is extended in the model as

$$A(q^{-1}) = 1 + a_1 q^{-1} + \cdots + a_{n_a+1} q^{-n_a-1} . \tag{7.13}$$

  When constructing a bias-compensated estimate using A as in (7.13), it is postulated that $a_{n_a+1} = 0$. This idea gives the second needed equation.
- Section 7.3 presents the approach of using the Frisch scheme, as applied to the linear regression (7.1). Using the extended regressor, cf. (5.13),

$$\overline{\varphi}(t) = \left(-y(t) \ \ldots \ -y(t-n_a) \ u(t-1) \ \ldots \ u(t-n_b)\right)^T$$

  the equation

$$\left(\mathbf{R}_{\overline{\varphi}} - \mathbf{R}_{\tilde{\overline{\varphi}}}\right) \overline{\boldsymbol{\theta}} = \mathbf{0} \tag{7.14}$$

is set up. Requiring that the matrix in (7.14) is singular leads to a relation between λ_y and λ_u. Another relation between these two variables is derived by reconsidering the problem for an over-parameterized case, say be taking A as in (7.13).

- Finally, Sect. 7.4 is based on a generalized IV (GIVE) approach. For standard IV, an essential aspect is to construct the vector of instruments so that it becomes uncorrelated with the equation noise, see (6.13). Here this condition is relaxed, and a bias-compensation similar in style to (7.11) is considered.

At this stage it should be emphasized that the three approaches sketched above lead to similar results. Indeed, the GIVE formulation is shown in Sect. 7.4 to include the other two, and several explicit examples are presented. Still, the underlying ideas of each approach have their pedagogic merits. Of this reason, the three different approaches are all presented in some detail in the next three sections.

7.2 The Bias-Eliminating Least Squares Method

7.2.1 *Introduction*

The bias-eliminating least squares (BELS) method is one of the simpler approaches based on (7.11).

Consider first, as a background, a simplified case without input noise.

Remark 7.1 Consider first the case when there is no input noise, $\tilde{u}(t) \equiv 0$, and the output noise $\tilde{y}(t)$ is white. This is a special case of Assumption **AN3c**. Then $\mathbf{r}_{\tilde{\varphi}\tilde{y}} = 0$ and

$$\mathbf{R}_{\tilde{\varphi}} = \begin{pmatrix} \lambda_y \mathbf{I}_{n_a} & \mathbf{0} \\ \mathbf{0} & \mathbf{0} \end{pmatrix} .$$

To estimate $\mathbf{R}_{\tilde{\varphi}}$ means to estimate λ_y. One way to do so is to consider the matrix

$$\mathbf{R}_{\overline{\varphi}} - \begin{pmatrix} \hat{\lambda}_y \mathbf{I}_{n_a+1} & \mathbf{0} \\ \mathbf{0} & \mathbf{0} \end{pmatrix} , \tag{7.15}$$

where $\overline{\varphi}(t)$ is the extended regressor vector with a leading element equal to $-y(t)$, see (5.13).

When $\hat{\lambda}_y < \lambda_y$, the matrix in (7.15) is positive definite. When $\hat{\lambda}_y = \lambda_y$, it becomes singular, and when $\hat{\lambda}_y > \lambda_y$, it will be indefinite. Straightforward procedures exploiting this fact for determining λ_y and thereby designing a bias-compensation have been presented in Sagara and Wada (1977), Stoica and Söderström (1982), Zheng and Feng (1990, 1995), Feng and Zheng (1991), Zheng (2002b). To find $\hat{\lambda}_y$ so that the matrix in (7.15) is singular (and positive semidefinite) is a generalized eigenvalue problem.

For the truly errors-in-variables problem, with both input and output noise present, the problem of determining $\mathbf{R}_{\tilde{\varphi}}$ is in general a bit more complicated. In this case the matrix $\mathbf{R}_{\tilde{\varphi}}$ becomes

$$\mathbf{R}_{\tilde{\varphi}} = \begin{pmatrix} \lambda_y \mathbf{I}_{n_a} & \mathbf{0} \\ \mathbf{0} & \lambda_u \mathbf{I}_{n_b} \end{pmatrix}, \tag{7.16}$$

and it has two different parameters (λ_y and λ_u) to determine. ■

Remark 7.2 An exception to the treatment in the above remark is the case when both the input and the output noise are assumed to be white, and *the ratio of their variances is known* (Assumption **AN4** applies). For such a case one single scaling parameter of $\mathbf{R}_{\tilde{\varphi}}$ has to be determined. Identification algorithms for this case of EIV problems have been proposed by Koopmans (1937), Levin (1964), Aoki and Yue (1970), Guidorzi (1975, 1981), Eising et al. (1983), Fernando and Nicholson (1985). In some of these references, it is assumed that the input and output noise variances are equal, $\lambda_y = \lambda_u$. This is in principle equivalent to assuming that the variance ratio λ_y/λ_u is known, as the signals then can be scaled to have the same amount of noise.

A further analysis of such 'one degree of freedom' bias-compensating schemes has shown that they can be interpreted as a form of weighted instrumental variable methods, see Stoica et al. (1995b), Gilson and Van den Hof (2001), Söderström et al. (1999), Garnier et al. (2000). Compare also the treatment of the line-fitting problem in Chap. 2, where it was shown that the estimation problem simplifies drastically when the noise variance ratio is known. ■

In the following section the two cases of white output noise (Assumption **AN3c**) and correlated output noise (Assumption **AN3b**), respectively, will be treated.

7.2.2 *White Output Noise*

The BELS methods for the EIV problem were first proposed by Zheng in Zheng and Feng (1989), Zheng (1998a) for identifying EIV systems with white input and output measurement noise sequences.

Now apply Assumption **AN3c**, that is, when $\tilde{y}(t)$ and $\tilde{u}(t)$ are both white but with unknown variances. The modified normal equations (7.11) must be complemented with (at least) two more equations to determine also the two unknown noise variances λ_y and λ_u. This can be done in several ways. Here, the approach in Zheng (1998a) is presented.

One such additional relation can be derived from the minimal value of the least squares criterion:

$$\begin{aligned} V_{\text{LS}} &= \min_{\theta} \mathsf{E}\left\{\left[y(t) - \boldsymbol{\varphi}^T(t)\boldsymbol{\theta}\right]^2\right\} \\ &= \mathsf{E}\left\{[y(t) - \boldsymbol{\varphi}^T(t)\hat{\boldsymbol{\theta}}_{\text{LS}}]^2\right\} \end{aligned}$$

$$
\begin{aligned}
&= \lambda_y + \mathsf{E}\left\{[\boldsymbol{\varphi}_0^T(t)\boldsymbol{\theta}_0 - \boldsymbol{\varphi}^T(t)\hat{\boldsymbol{\theta}}_{\text{LS}}]^2\right\} \\
&= \lambda_y + \boldsymbol{\theta}_0^T \mathbf{R}_{\varphi_0}\boldsymbol{\theta}_0 + \hat{\boldsymbol{\theta}}_{\text{LS}}^T \mathbf{R}_{\varphi}\hat{\boldsymbol{\theta}}_{\text{LS}} - 2\hat{\boldsymbol{\theta}}_{\text{LS}}^T \mathbf{R}_{\varphi_0}\boldsymbol{\theta}_0 \, .
\end{aligned}
$$

From (7.2) it follows that

$$
\begin{aligned}
\mathbf{R}_{\varphi_0}\boldsymbol{\theta}_0 &= \mathbf{r}_{\varphi_0 y_0} \\
&= \mathbf{r}_{\varphi y} = \mathbf{R}_{\varphi}\hat{\boldsymbol{\theta}}_{\text{LS}}
\end{aligned}
\tag{7.17}
$$

and hence

$$
\begin{aligned}
V_{\text{LS}} &= \lambda_y + \boldsymbol{\theta}_0^T \mathbf{R}_{\varphi}\hat{\boldsymbol{\theta}}_{\text{LS}} + \boldsymbol{\theta}_0^T \mathbf{R}_{\varphi_0}\hat{\boldsymbol{\theta}}_{\text{LS}}^T - 2\hat{\boldsymbol{\theta}}_{\text{LS}}^T \mathbf{R}_{\varphi_0}\boldsymbol{\theta}_0 \\
&= \lambda_y + \boldsymbol{\theta}_0^T \mathbf{R}_{\tilde{\varphi}}\hat{\boldsymbol{\theta}}_{\text{LS}} \, .
\end{aligned}
\tag{7.18}
$$

Note that (7.18) can be seen as a *linear equation* in λ_y and λ_u, assuming $\boldsymbol{\theta}_0$ is known, cf. (7.16).

To get also a second relation for λ_y and λ_u, an extended model structure is considered. This is similar to the so-called Frisch scheme, Beghelli et al. (1990), see Sect. 7.3. For this purpose introduce extended versions of $\boldsymbol{\varphi}(t)$, $\boldsymbol{\theta}$ and $\boldsymbol{\theta}_0$ as

$$
\overline{\boldsymbol{\varphi}}(t) = \begin{pmatrix} \boldsymbol{\varphi}(t) \\ \underline{\boldsymbol{\varphi}}(t) \end{pmatrix}, \quad \overline{\boldsymbol{\theta}} = \begin{pmatrix} \boldsymbol{\theta} \\ \underline{\boldsymbol{\theta}} \end{pmatrix}, \quad \overline{\boldsymbol{\theta}}_0 = \begin{pmatrix} \boldsymbol{\theta}_0 \\ \mathbf{0} \end{pmatrix}
\tag{7.19}
$$

which are used within this section, rather than the previous conventions (5.12) and (5.13).

The model extension can, for example, mean that an additional A parameter is appended. In that case,

$$
\underline{\boldsymbol{\varphi}}(t) = -y(t - n_a - 1) \, , \qquad \underline{\boldsymbol{\theta}} = a_{n_a+1} \, .
\tag{7.20}
$$

Another possibility is to append an additional B parameter, leading to

$$
\underline{\boldsymbol{\varphi}}(t) = u(t - n_b - 1) \, , \qquad \underline{\boldsymbol{\theta}} = b_{n_b+1} \, .
\tag{7.21}
$$

The general case here covers (7.20) and (7.21) as special cases. Note in passing that although for simplicity $\underline{\boldsymbol{\varphi}}(t)$ is assumed to be a scalar, it is possible to generalize the analysis to $\underline{\boldsymbol{\varphi}}(t)$ being a vector. The number of new relations derived will be equal to the dimension of $\underline{\boldsymbol{\varphi}}(t)$.

Next consider least squares estimation in the extended linear regression model

$$
y(t) = \overline{\boldsymbol{\varphi}}^T(t)\overline{\boldsymbol{\theta}} + \varepsilon(t) \, ,
\tag{7.22}
$$

which leads to

$$\mathbf{R}_{\overline{\varphi}}\hat{\overline{\boldsymbol{\theta}}}_{\mathrm{LS}} = \mathbf{r}_{\underline{\varphi} y} \ . \tag{7.23}$$

Recall that $y(t) = y_0(t) + \tilde{y}(t)$, and $\overline{\varphi}(t) = \overline{\varphi}_0(t) + \tilde{\overline{\varphi}}(t)$. Hence

$$\begin{aligned} \mathbf{R}_{\overline{\varphi}}\hat{\overline{\boldsymbol{\theta}}}_{\mathrm{LS}} &= \mathbf{r}_{\overline{\varphi}_0 y_0} + \mathbf{r}_{\tilde{\overline{\varphi}}\tilde{y}} \\ &= \mathbf{R}_{\overline{\varphi}_0}\overline{\boldsymbol{\theta}}_0 \\ &= (\mathbf{R}_{\overline{\varphi}} - \mathbf{R}_{\tilde{\overline{\varphi}}})\overline{\boldsymbol{\theta}}_0 \ . \end{aligned} \tag{7.24}$$

Note that in the right-hand side $\mathbf{R}_{\underline{\tilde{\varphi}}}$ varies linearly with λ_y and λ_u, cf. (7.16). Set

$$\mathbf{H} = (\mathbf{0} \quad \mathbf{I}), \quad \mathbf{J}_0 = \begin{pmatrix} \mathbf{I}_{n_a+n_b} \\ \mathbf{0} \end{pmatrix}, \quad \overline{\boldsymbol{\theta}}_0 = \mathbf{J}_0\boldsymbol{\theta}_0 \ . \tag{7.25}$$

Recalling that $\mathbf{H}\mathbf{J}_0 = \mathbf{0}$ and hence $\mathbf{H}\overline{\boldsymbol{\theta}}_0 = \mathbf{0}$, it is found that (7.24) implies

$$\begin{aligned} \mathbf{H}\hat{\overline{\boldsymbol{\theta}}}_{\mathrm{LS}} &= \mathbf{H}\mathbf{R}_{\overline{\varphi}}^{-1}(\mathbf{R}_{\overline{\varphi}} - \mathbf{R}_{\tilde{\overline{\varphi}}})\overline{\boldsymbol{\theta}}_0 \\ &= -\mathbf{H}\mathbf{R}_{\overline{\varphi}}^{-1}\mathbf{R}_{\tilde{\overline{\varphi}}}\mathbf{J}_0\boldsymbol{\theta}_0 \ . \end{aligned} \tag{7.26}$$

Recall that the noise parameter vector is

$$\boldsymbol{\rho} = \begin{pmatrix} \lambda_y & \lambda_u \end{pmatrix}^T \ . \tag{7.27}$$

To summarize, the following set of equations

$$\mathbf{R}_{\varphi}\hat{\boldsymbol{\theta}}_{\mathrm{LS}} = [\mathbf{R}_{\varphi} - \mathbf{R}_{\tilde{\varphi}}(\boldsymbol{\rho})]\boldsymbol{\theta} \ , \tag{7.28}$$

$$V_{\mathrm{LS}} = \lambda_y + \hat{\boldsymbol{\theta}}_{\mathrm{LS}}^T\mathbf{R}_{\tilde{\varphi}}(\boldsymbol{\rho})\boldsymbol{\theta} \ , \tag{7.29}$$

$$\mathbf{H}\hat{\overline{\boldsymbol{\theta}}}_{\mathrm{LS}} = -\mathbf{H}\mathbf{R}_{\overline{\varphi}}^{-1}\mathbf{R}_{\tilde{\overline{\varphi}}}(\boldsymbol{\rho})\mathbf{J}_0\boldsymbol{\theta} \ . \tag{7.30}$$

with $\boldsymbol{\theta}$ and $\boldsymbol{\rho}$ as unknowns define the estimator.

Equations (7.28)–(7.30) turn out to be *bilinear in the unknowns* $\boldsymbol{\theta}$ and $\boldsymbol{\rho}$: They are linear in $\boldsymbol{\theta}$ and linear in $\boldsymbol{\rho}$. There are different ways to solve the equations. Two different algorithms for solving the equations are given in Appendix 13.A. An analysis of convergence of these algorithms is also provided. A few remarks are in order.

Remark 7.3 There are alternative BELS methods in the literature, see Sect. 7.5. ■

Remark 7.4 In case $\underline{\boldsymbol{\varphi}}(t)$ is a vector and not a scalar, the system of Eqs. (7.28)–(7.30) will be overdetermined. There are then a number of possibilities. One option is to solve the total set of equations in a least squares sense. Another possibility is to require some of the equations to hold exactly, while the remaining ones are to be solved in a least squares or weighted least squares sense. Various aspects on such choices as well as other algorithmic issues will appear later in Chap. 13, see Sect. 13.2.

At this stage though, let the estimation method be defined by the set of equations as such. Note that the properties of the estimates (the solution to the equations) do not depend on the numerical way the equations are solved. If the equations are overdetermined and weighted, the weighting will though influence the estimates.

The above type of comments will apply also for other estimators described in this chapter. ■

7.2.3 *Correlated Output Noise*

This BELS algorithm to be presented here is proposed in Zheng (1999a), see also Zheng (2002a).

Assume that **AN3b** applies, so that $\tilde{u}(t)$ is white while $\tilde{y}(t)$ is a stationary but arbitrarily correlated stochastic process.

One need then to estimate the system parameter vector $\boldsymbol{\theta}$ and in addition also the noise parameter vector, (5.27),

$$\boldsymbol{\rho} = \left(r_{\tilde{y}}(0)\ r_{\tilde{y}}(1) \ldots r_{\tilde{y}}(n_a)\ \lambda_u \right)^T . \tag{7.31}$$

The LS normal equations give

$$\begin{aligned} \mathbf{R}_{\varphi}\hat{\boldsymbol{\theta}}_{\mathrm{LS}} &= \mathbf{r}_{\varphi y} = E\left\{\boldsymbol{\varphi}(t)[y_0(t) + \tilde{y}(t)]\right\} \\ &= \mathbf{R}_{\varphi_0}\boldsymbol{\theta}_0 + \mathbf{r}_{\tilde{\varphi}\tilde{y}} \\ &= (\mathbf{R}_{\varphi} - \mathbf{R}_{\tilde{\varphi}})\boldsymbol{\theta}_0 + \mathbf{r}_{\tilde{\varphi}\tilde{y}} , \end{aligned} \tag{7.32}$$

and hence it holds

$$\begin{aligned} \mathbf{R}_{\varphi}(\hat{\boldsymbol{\theta}}_{\mathrm{LS}} - \boldsymbol{\theta}_0) &= -\mathsf{E}\left\{\tilde{\boldsymbol{\varphi}}(t)\tilde{\boldsymbol{\varphi}}^T(t)\right\}\boldsymbol{\theta}_0 + \mathsf{E}\left\{\tilde{\boldsymbol{\varphi}}(t)\tilde{y}(t)\right\} \\ &= -\begin{pmatrix} r_{\tilde{y}}(0) & \ldots & r_{\tilde{y}}(n_a-1) & \\ \vdots & \ddots & \vdots & \mathbf{0} \\ r_{\tilde{y}}(n_a-1) & \ldots & r_{\tilde{y}}(0) & \\ & \mathbf{0} & & \lambda_u \mathbf{I}_{n_b} \end{pmatrix} \begin{pmatrix} \mathbf{a} \\ \mathbf{b} \end{pmatrix} \\ &\quad + \begin{pmatrix} -r_{\tilde{y}}(1) \\ \vdots \\ -r_{\tilde{y}}(n_a) \\ \mathbf{0} \end{pmatrix} . \end{aligned} \tag{7.33}$$

The right-hand side of (7.33) can be expressed in the noise parameter vector $\boldsymbol{\rho}$. Let $\mathbf{e}_j$ denote the jth unit vector, and set

$$\mathbf{T}_j = \begin{pmatrix} \mathbf{0} & \mathbf{0} \\ \mathbf{I}_{n_a-j} & \mathbf{0} \end{pmatrix} .$$

Then

$$\begin{pmatrix} r_{\tilde{y}}(0) & \dots r_{\tilde{y}}(n_a-1) \\ \vdots & \ddots \\ r_{\tilde{y}}(n_a-1) & r_{\tilde{y}}(0) \end{pmatrix} \mathbf{a} = r_{\tilde{y}}(0)\mathbf{a} + \sum_{j=1}^{n_a-1} (\mathbf{T}_j + \mathbf{T}_j^T) r_{\tilde{y}}(j)\mathbf{a}$$

$$= \left(\mathbf{a}, (\mathbf{T}_1 + \mathbf{T}_1^T)\mathbf{a}, (\mathbf{T}_2 + \mathbf{T}_2^T)\mathbf{a}, \dots, (\mathbf{T}_{n_a-1} + \mathbf{T}_{n_a-1}^T)\mathbf{a}\right) \begin{pmatrix} r_{\tilde{y}}(0) \\ \vdots \\ r_{\tilde{y}}(n_a-1) \end{pmatrix} . \tag{7.34}$$

This leads to

$$\begin{aligned} \mathbf{R}_{\tilde{\varphi}}\boldsymbol{\theta}_0 - \mathbf{r}_{\tilde{\varphi}\tilde{y}} &= \left(\begin{matrix} \left[\mathbf{a}\mathbf{e}_1^T + \sum_{j=1}^{n_a-1} (\mathbf{T}_j + \mathbf{T}_j^T)\mathbf{a}\mathbf{e}_{j+1}^T\right] + \left[\mathbf{0}_{n_a\times 1}, \mathbf{I}_{n_a}, \mathbf{0}_{n_a\times 1}\right] \\ \mathbf{b}\mathbf{e}_{n_a+2}^T \end{matrix} \right) \boldsymbol{\rho} \\ &\stackrel{\Delta}{=} \mathbf{M}(\boldsymbol{\theta}_0)\boldsymbol{\rho} . \end{aligned} \tag{7.35}$$

Hence the normal equations (7.32) can be written as

$$\mathbf{R}_{\varphi}(\hat{\boldsymbol{\theta}}_{\mathrm{LS}} - \boldsymbol{\theta}_0) = -\mathbf{M}(\boldsymbol{\theta}_0)\boldsymbol{\rho} . \tag{7.36}$$

Next express the minimal value of the LS loss function. Using (7.32) and (7.35),

$$\begin{aligned} V_{\mathrm{LS}} &= \min_{\boldsymbol{\theta}} \mathsf{E}\left\{[y(t) - \boldsymbol{\varphi}^T(t)\boldsymbol{\theta}]^2\right\} \\ &= r_y(0) - \hat{\boldsymbol{\theta}}_{\mathrm{LS}}^T \mathbf{R}_{\varphi} \hat{\boldsymbol{\theta}}_{\mathrm{LS}} \\ &= r_{y_0}(0) + r_{\tilde{y}}(0) - \hat{\boldsymbol{\theta}}_{\mathrm{LS}}^T \left[\mathbf{R}_{\varphi}\boldsymbol{\theta}_0 - \mathbf{R}_{\tilde{\varphi}}\boldsymbol{\theta}_0 + \mathbf{r}_{\tilde{\varphi}\tilde{y}}\right] \\ &= r_{\tilde{y}}(0) + \hat{\boldsymbol{\theta}}_{\mathrm{LS}}^T \left[\mathbf{R}_{\tilde{\varphi}}\boldsymbol{\theta}_0 - \mathbf{r}_{\tilde{\varphi}\tilde{y}}\right] + r_{y_0}(0) - \mathbf{r}_{y\varphi}\boldsymbol{\theta}_0 \\ &= r_{\tilde{y}}(0) + \hat{\boldsymbol{\theta}}_{\mathrm{LS}}^T \mathbf{M}(\boldsymbol{\theta}_0)\boldsymbol{\rho} - \mathbf{r}_{\tilde{y}\tilde{\varphi}}\boldsymbol{\theta}_0 \\ &= r_{\tilde{y}}(0) + \hat{\boldsymbol{\theta}}_{\mathrm{LS}}^T \mathbf{M}(\boldsymbol{\theta}_0)\boldsymbol{\rho} - \left[0 \;\; \mathbf{a}_0^T \;\; 0\right]\boldsymbol{\rho} . \end{aligned} \tag{7.37}$$

Note that the right-hand side of (7.37) is a linear combination of the elements in $\boldsymbol{\rho}$. Symbolically write this as

$$V_{\mathrm{LS}} = \mathbf{q}^T(\hat{\boldsymbol{\theta}}_{\mathrm{LS}}, \boldsymbol{\theta}_0)\boldsymbol{\rho} . \tag{7.38}$$

Some further relations are needed to determine $\boldsymbol{\rho}$. For this purpose introduce an extended parameter vector, cf. (7.19)

$$\overline{\boldsymbol{\theta}} = \begin{pmatrix} \boldsymbol{\theta} \\ \mathbf{0}_{p\times 1} \end{pmatrix} , \tag{7.39}$$

which contains p zero elements, and a corresponding extended regressor

$$\begin{aligned} \overline{\boldsymbol{\varphi}}(t) &= \left(\boldsymbol{\varphi}^T(t) u(t-n_b-1) \ldots u(t-n_b-p) \right)^T \\ &\stackrel{\Delta}{=} \left(\boldsymbol{\varphi}^T(t) \underline{\boldsymbol{\varphi}}^T(t) \right)^T . \end{aligned} \tag{7.40}$$

Obviously,

$$\boldsymbol{\varphi}^T(t)\boldsymbol{\theta} = \overline{\boldsymbol{\varphi}}^T(t)\overline{\boldsymbol{\theta}} .$$

as $\mathbf{J}_0\overline{\boldsymbol{\theta}} = \boldsymbol{\theta}$ still applies, cf. (7.25). The LS estimate of the extended parameter vector becomes

$$\hat{\overline{\boldsymbol{\theta}}}_{\text{LS}} = \mathbf{R}_{\overline{\varphi}}^{-1} \mathbf{r}_{\overline{\varphi}y} , \tag{7.41}$$

and it satisfies, cf. (7.36),

$$\mathbf{R}_{\overline{\varphi}}(\hat{\overline{\boldsymbol{\theta}}}_{\text{LS}} - \overline{\boldsymbol{\theta}}_0) = -\overline{\mathbf{M}}(\overline{\boldsymbol{\theta}}_0)\boldsymbol{\rho} , \tag{7.42}$$

where

$$\overline{\mathbf{M}}(\overline{\boldsymbol{\theta}}) = \begin{pmatrix} \mathbf{M}(\boldsymbol{\theta}) \\ \mathbf{0}_{p\times(n_a+2)} \end{pmatrix} , \tag{7.43}$$

To get the additional equations, use (7.42) and exploit that the last p elements of $\overline{\boldsymbol{\theta}}_0$ are known to be zero. Using (7.35) and (7.43) gives

$$\mathbf{r}_{\overline{\varphi}y} - \mathbf{R}_{\overline{\varphi}} \begin{pmatrix} \boldsymbol{\theta}_0 \\ \mathbf{0} \end{pmatrix} = - \begin{pmatrix} \mathbf{M}(\boldsymbol{\theta})\boldsymbol{\rho} \\ \mathbf{0} \end{pmatrix} . \tag{7.44}$$

Partitionating this equation leads to

$$\begin{cases} \mathbf{r}_{\varphi y} = \mathbf{R}_{\varphi}\boldsymbol{\theta}_0 - \mathbf{M}(\boldsymbol{\theta})\boldsymbol{\rho} , \\ \mathbf{r}_{\underline{\varphi} y} = \mathbf{R}_{\underline{\varphi}\varphi}\boldsymbol{\theta}_0 . \end{cases} \tag{7.45}$$

Eliminating $\boldsymbol{\theta}_0$ from the first part of (7.45) leads finally to

$$\mathbf{r}_{\underline{\varphi} y} = \mathbf{R}_{\underline{\varphi}\varphi} \left[\hat{\boldsymbol{\theta}}_{\text{LS}} + \mathbf{R}_{\varphi}^{-1}\mathbf{M}(\boldsymbol{\theta})\boldsymbol{\rho} \right] . \tag{7.46}$$

The equations derived so far for determining $\boldsymbol{\theta}$ and $\boldsymbol{\rho}$ are

- (7.32) $n_a + n_b$ equations,
- (7.38) 1 equation,
- (7.46) p equations.

There are now $n_a + n_b$ unknowns in $\boldsymbol{\theta}$ and $n_a + 2$ further unknowns in $\boldsymbol{\rho}$, making the total number of unknowns equal to $2n_a + n_b + 2$. To have compatibility, one must require

$$n_a + n_b + 1 + p \geq 2n_a + n_b + 2 ,$$

or

$$p \geq n_a + 1 . \tag{7.47}$$

7.3 The Frisch Scheme

7.3.1 General Aspects

The Frisch scheme with roots in a classical algebraic estimation problem and in econometrics, Frisch (1934), was introduced in Chap. 2, see Sect. 2.3.

An extension of the Frisch scheme to identification of dynamic models appeared in Beghelli et al. (1990) and has later been refined and modified, for example, in Castaldi et al. (1995), Guidorzi (1996), Soverini and Söderström (2000), Söderström et al. (2002), Diversi et al. (2003b).

The Frisch scheme for dynamic EIV problems will be presented and discussed in the following subsections. As for BELS, it is convenient to separate between the cases of white and autocorrelated output noise, respectively.

7.3.2 White Output Noise

First note that, cf. (5.12), (5.13),

$$\overline{\boldsymbol{\varphi}}_0^T(t)\overline{\boldsymbol{\theta}}_0 = -A_0(q^{-1})y_0(t) + B_0(q^{-1})u_0(t) = 0 . \tag{7.48}$$

Further it holds that

$$\mathbf{R}_{\varphi} = \mathbf{R}_{\varphi_0} + \mathbf{R}_{\tilde{\varphi}}, \qquad \mathbf{R}_{\overline{\varphi}} = \mathbf{R}_{\overline{\varphi}_0} + \mathbf{R}_{\tilde{\overline{\varphi}}} . \tag{7.49}$$

It follows from (7.48) that

$$\mathbf{R}_{\overline{\varphi}_0}\overline{\boldsymbol{\theta}}_0 = \mathsf{E}\left\{\overline{\boldsymbol{\varphi}}_0(t)\overline{\boldsymbol{\varphi}}_0^T(t)\overline{\boldsymbol{\theta}}_0\right\} = \mathbf{0}\,. \tag{7.50}$$

Hence the matrix $\mathbf{R}_{\overline{\varphi}_0}$ is singular (positive semidefinite), with at least one eigenvalue equal to zero. The corresponding eigenvector is $\overline{\boldsymbol{\theta}}_0$. One can show that under the general assumptions **AS3** and **AI1**, the matrix $\mathbf{R}_{\overline{\varphi}_0}$ will in fact have only one eigenvalue in the origin.

The noise covariance matrix has a simple structure, as

$$\mathbf{R}_{\tilde{\overline{\varphi}}} = \begin{pmatrix} \lambda_y \mathbf{I}_{n_a+1} & \mathbf{0} \\ \mathbf{0} & \lambda_u \mathbf{I}_{n_b} \end{pmatrix}. \tag{7.51}$$

The relation (7.50) is the basis for the Frisch method. The idea is to have appropriate estimates of the noise variances and then determine the parameter vector $\boldsymbol{\theta}$ from

$$\left(\hat{\mathbf{R}}_{\overline{\varphi}} - \hat{\mathbf{R}}_{\tilde{\overline{\varphi}}}\right)\hat{\overline{\boldsymbol{\theta}}} = \mathbf{0}\,. \tag{7.52}$$

Remark 7.5 Note that (7.52) holds also for correlated output noise, even if the matrix $\hat{\mathbf{R}}_{\tilde{\overline{\varphi}}}$ then must be modified, cf. the calculations in (7.33).

Further, dropping the first equation in (7.52) does in fact give precisely the compensated normal equations (7.11). ■

7.3.2.1 Determining $\hat{\lambda}_y$ and $\hat{\boldsymbol{\theta}}$

Assume for a moment that an estimate $\hat{\lambda}_u$ of the input noise variance is available. Then the output noise variance λ_y is determined so that the matrix appearing in (7.52) is singular. More specifically, the following result holds true.

Lemma 7.1 *Write*

$$\overline{\boldsymbol{\varphi}}(t) = \left(\overline{\boldsymbol{\varphi}}_y^T(t)\ \boldsymbol{\varphi}_u^T(t)\right)^T\,. \tag{7.53}$$

where $\overline{\varphi}_y(t)$ and $\overline{\varphi}_u(t)$ contain delayed values of $y(t)$ and $u(t)$, respectively. Let the estimate $\hat{\lambda}_u$ satisfy

$$0 \le \hat{\lambda}_u \le \lambda_{\min}\left(\hat{\mathbf{R}}_{\varphi_u} - \hat{\mathbf{R}}_{\varphi_u\overline{\varphi}_y}\hat{\mathbf{R}}_{\overline{\varphi}_y}^{-1}\hat{\mathbf{R}}_{\overline{\varphi}_y\varphi_u}\right)\,, \tag{7.54}$$

where $\lambda_{\min}(\mathbf{C})$ denotes the minimal eigenvalue of the symmetric matrix $\mathbf{C}$. *Define*

$$\hat{\lambda}_y = \lambda_{\min}\left(\hat{\mathbf{R}}_{\overline{\varphi}_y} - \hat{\mathbf{R}}_{\overline{\varphi}_y\varphi_u}\left(\hat{\mathbf{R}}_{\varphi_u} - \hat{\lambda}_u\mathbf{I}_{n_b}\right)^{-1}\hat{\mathbf{R}}_{\varphi_u\overline{\varphi}_y}\right)\,. \tag{7.55}$$

Then the matrix

$$\hat{\mathbf{R}}_{\bar{\tilde{\varphi}}} = \begin{pmatrix} \hat{\mathbf{R}}_{\overline{\varphi}_y} & \hat{\mathbf{R}}_{\overline{\varphi}_y \varphi_u} \\ \hat{\mathbf{R}}_{\varphi_u \overline{\varphi}_y} & \hat{\mathbf{R}}_{\varphi_u} \end{pmatrix} - \begin{pmatrix} \hat{\lambda}_y \mathbf{I}_{n_a+1} & \mathbf{0} \\ \mathbf{0} & \hat{\lambda}_u \mathbf{I}_{n_b} \end{pmatrix} \tag{7.56}$$

is positive semidefinite with one eigenvalue in the origin.

Proof See Appendix 7.A.1. ■

An essential part of the Frisch algorithm is based on Lemma 7.1. Assume that an estimate $\hat{\lambda}_u$ of the input noise variance is available (how this estimate is to be found will be described below). The estimate $\hat{\lambda}_y$ is then found from (7.55). The estimate of the parameter vector $\boldsymbol{\theta}$ is next determined by solving equations $2, \ldots, n_a + n_b + 1$ of

$$\hat{\mathbf{R}}_{\bar{\tilde{\varphi}}} \hat{\bar{\boldsymbol{\theta}}} = \mathbf{0} \,, \tag{7.57}$$

where the matrix $\hat{\mathbf{R}}_{\bar{\tilde{\varphi}}}$ is given by (7.56). As $\hat{\mathbf{R}}_{\bar{\tilde{\varphi}}}$ by construction is singular, this means that $\hat{\boldsymbol{\theta}}$ is the solution to

$$\left(\hat{\mathbf{R}}_{\overline{\varphi}} - \begin{pmatrix} \hat{\lambda}_y \mathbf{I}_{n_a} & \mathbf{0} \\ \mathbf{0} & \hat{\lambda}_u \mathbf{I}_{n_b} \end{pmatrix} \right) \hat{\boldsymbol{\theta}} = \hat{\mathbf{r}}_{\varphi y} \,. \tag{7.58}$$

which is nothing but the bias-compensated normal equations (7.11).

To determine $\hat{\lambda}_u$, several alternatives have been proposed to get the needed second relation. In what follows three different possibilities are presented as examples.

Example 7.1 Similar to the BELS methods, one choice is to evaluate the Frisch equation (7.52) for an extended model with an additional regressor $\underline{\varphi}(t)$, cf. (7.19). This method was proposed in Beghelli et al. (1990). For this alternative the extended Frisch equation will be (7.52). Therefore the function $\lambda_y(\lambda_u)$ is evaluated both for the nominal model and for the extended model by utilizing the relation (7.55). The two functions correspond to two curves in the (λ_u, λ_y) plane. The curves will ideally have one unique common point, which defines the estimates. In case $\hat{\mathbf{R}}_{\overline{\varphi}}$ would be replaced by its true value $\mathbf{R}_{\overline{\varphi}}$ one would then get a situation as displayed in Fig. 7.1. In Fig. 7.1 curve A corresponds to the use of the true model order, while curve B applies for the increased model order. The coordinates of the common point P give precisely the true noise variances $\lambda_u, \ \lambda_y$.

For a finite data set ($N < \infty$) the situation is less ideal, and there is not a distinct point P where the curves A and B share a common point. It is referred to Beghelli et al. (1990), Soverini and Söderström (2000) for more detailed aspects on how the Frisch scheme can be implemented. ■

Example 7.2 The second alternative is to use overdetermined Yule–Walker equations, Diversi et al. (2006).

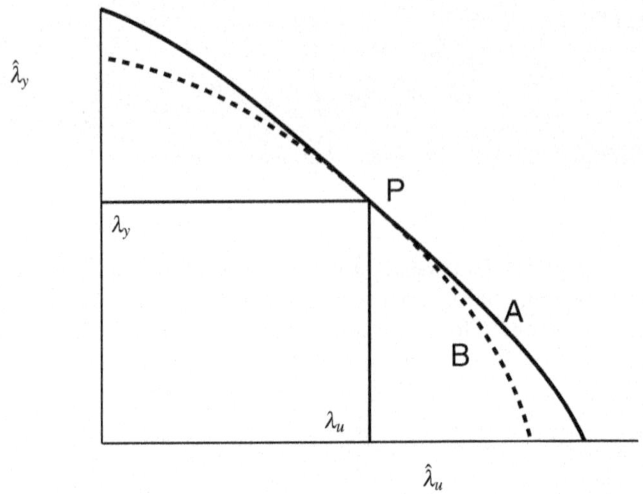

Fig. 7.1 Illustration of the principle for Frisch estimation

The additional equations are obtained by using the cross-correlation of residuals and past input,

$$\begin{aligned}
&\mathsf{E}\left\{\begin{pmatrix} u(t-n_b-1) \\ \vdots \\ u(t-n_b-p) \end{pmatrix} \big(A(q^{-1})y(t) - B(q^{-1})u(t)\big)\right\} = \mathbf{0} , \\
\Leftrightarrow\ &\mathsf{E}\left\{\underline{\varphi}(t)\varepsilon(t)\right\} = \mathbf{0} , \\
\Leftrightarrow\ &\mathbf{R}_{\underline{\varphi}\varphi}\boldsymbol{\theta} = \mathbf{r}_{\underline{\varphi}y} .
\end{aligned} \tag{7.59}$$

■

Example 7.3 A third alternative is to compute residuals and compare their statistical properties with what can be predicted from the model. This alternative was proposed in Diversi et al. (2003b). Define the residuals

$$\varepsilon(t, \hat{\boldsymbol{\theta}}) = \hat{A}(q^{-1})y(t) - \hat{B}(q^{-1})u(t) , \tag{7.60}$$

and compute sample covariance elements

$$\hat{r}_\varepsilon(k) = \frac{1}{N}\sum_{t=1}^{N} \varepsilon(t, \hat{\boldsymbol{\theta}})\varepsilon(t+k, \hat{\boldsymbol{\theta}}) . \tag{7.61}$$

Compute also theoretical covariance elements $\hat{r}_{\varepsilon_0}(k)$ based on the model

$$\varepsilon_0(t) = \hat{A}(q^{-1})\hat{\tilde{y}}(t) - \hat{B}(q^{-1})\hat{\tilde{u}}(t) , \tag{7.62}$$

where $\hat{\tilde{y}}(t)$, $\hat{\tilde{u}}(t)$ are white noise sequences with

$$\mathsf{E}\left\{\hat{\tilde{y}}^2(t)\right\} = \hat{\lambda}_y, \qquad \mathsf{E}\left\{\hat{\tilde{u}}^2(t)\right\} = \hat{\lambda}_u . \tag{7.63}$$

Next, define a criterion for comparing $\{\hat{r}_\varepsilon(k)\}$ and $\{\hat{r}_{\varepsilon_0}(k)\}$. A fairly general way to do this is to take

$$V_N(\hat{\lambda}_u) = \boldsymbol{\delta}^T \mathbf{W} \boldsymbol{\delta} , \tag{7.64}$$

where $\mathbf{W}$ is a user chosen, positive definite weighting matrix and the vector $\boldsymbol{\delta}$ is

$$\boldsymbol{\delta} = \begin{pmatrix} \hat{r}_\varepsilon(1) - \hat{r}_{\varepsilon_0}(1) \\ \vdots \\ \hat{r}_\varepsilon(n_\varepsilon) - \hat{r}_{\varepsilon_0}(n_\varepsilon) \end{pmatrix} . \tag{7.65}$$

The maximum lag n_ε used in (7.65) is also to be chosen by the user. The estimate $\hat{\lambda}_u$ is determined as the minimizing element of the criterion

$$\hat{\lambda}_u = \arg\min_{\lambda_u} V_N(\lambda_u) . \tag{7.66}$$

Hence it holds that

$$\left.\frac{d}{d\lambda_u} V_N(\lambda_u)\right|_{\lambda_u = \hat{\lambda}_u} = 0 . \tag{7.67}$$

In the evaluation of the loss function $V_N(\hat{\lambda}_u)$, also (7.55) and (7.58) are used to get $\hat{\lambda}_y$ and $\hat{\boldsymbol{\theta}}$, respectively. ■

Remark 7.6 It holds that

$$\hat{r}_\varepsilon(0) = \hat{r}_{\varepsilon_0}(0) . \tag{7.68}$$

There is hence no point to include the lag argument 0 in (7.65). To verify (7.68), straightforward calculations give

$$\hat{r}_\varepsilon(0) = \frac{1}{N}\sum_{t=1}^{N} \varepsilon^2(t, \hat{\boldsymbol{\theta}}) = \frac{1}{N}\sum_{t=1}^{N} \left[\overline{\boldsymbol{\varphi}}^T \hat{\overline{\boldsymbol{\theta}}}\right]^2 = \hat{\overline{\boldsymbol{\theta}}}^T \hat{\mathbf{R}}_{\overline{\varphi}} \hat{\overline{\boldsymbol{\theta}}} ,$$

and

$$\hat{r}_{\varepsilon_0}(0) = \hat{\lambda}_y \hat{\overline{\mathbf{a}}}^T \hat{\overline{\mathbf{a}}} + \hat{\lambda}_u \hat{\mathbf{b}}^T \hat{\mathbf{b}} = \hat{\overline{\boldsymbol{\theta}}}^T \begin{pmatrix} \hat{\lambda}_y \mathbf{I}_{na+1} & \mathbf{0} \\ \mathbf{0} & \hat{\lambda}_u \mathbf{I}_{nb} \end{pmatrix} \hat{\overline{\boldsymbol{\theta}}} .$$

It now follows from (7.57) that

$$\hat{r}_{\varepsilon}(0) - \hat{r}_{\varepsilon_0}(0) = \hat{\bar{\boldsymbol{\theta}}}^T \left(\hat{\mathbf{R}}_{\overline{\varphi}} - \begin{pmatrix} \hat{\lambda}_y \mathbf{I}_{na+1} & \mathbf{0} \\ \mathbf{0} & \hat{\lambda}_u \mathbf{I}_{nb} \end{pmatrix} \right) \hat{\bar{\boldsymbol{\theta}}} = 0 \, .$$

■

7.3.3 Correlated Output Noise

The Frisch scheme was extended for identifying EIV systems with correlated output noise in Söderström (2007b, 2008). The additional equations were obtained by using the cross-correlation of the residuals and the past input, cf. (7.59),

$$\begin{aligned} & \mathsf{E} \left\{ \begin{pmatrix} u(t-n_b-1) \\ \vdots \\ u(t-n_b-p_u) \end{pmatrix} (A(q^{-1})y(t) - B(q^{-1})u(t)) \right\} = \mathbf{0} \, , \\ \Leftrightarrow \; & \mathsf{E} \left\{ \underline{\boldsymbol{\varphi}}(t)\varepsilon(t) \right\} = \mathbf{0} \, , \\ \Leftrightarrow \; & \mathbf{R}_{\underline{\varphi}\varphi}\boldsymbol{\theta} = \mathbf{r}_{\underline{\varphi}y} \, . \end{aligned} \tag{7.69}$$

For this case, the effect of the output noise is no longer as shown in (7.55). Therefore, in this case one can instead express λ_u as a function of $r_{\tilde{y}}(0), \ldots r_{\tilde{y}}(n_a)$, with $m = n_a + 2$. The result is that (7.55) is to be substituted by

$$\lambda_u = \lambda_{\min} \left(\mathbf{R}_{\varphi_u} - \mathbf{R}_{\varphi_u \overline{\varphi}_y} \left(\mathbf{R}_{\overline{\varphi}_y} - \mathbf{R}_{\overline{\varphi}_{\tilde{y}}} \right)^{-1} \mathbf{R}_{\overline{\varphi}_y \varphi_u} \right) \, , \tag{7.70}$$

where

$$\mathbf{R}_{\tilde{\overline{\varphi}}_y} = \mathbf{R}_{\tilde{\overline{\varphi}}_y}(\rho) = \begin{pmatrix} r_{\tilde{y}}(0) & \ldots & r_{\tilde{y}}(n_a) \\ \vdots & \ddots & \\ r_{\tilde{y}}(n_a) & \ldots & r_{\tilde{y}}(0) \end{pmatrix} \, . \tag{7.71}$$

The constitutive equations for the Frisch methods are then (7.59), (7.70), and (7.32).

7.3.4 Using an Alternating Projection Algorithm

A somewhat different approach using the Frisch scheme than the one presented above is proposed in Sorelius and Stoica (1998). The starting point is still (7.49).

The identification algorithm consists of two steps.

Step 1. Decomposition

In the first step consider (7.49):

$$\mathbf{R}_{\overline{\varphi}} = \mathbf{R}_{\overline{\varphi}_0} + \mathbf{R}_{\tilde{\overline{\varphi}}} , \tag{7.72}$$

where $\mathbf{R}_{\overline{\varphi}}$ is given (or easily estimated from available data), $\mathbf{R}_{\overline{\varphi}_0}$ is rank deficient, and $\mathbf{R}_{\tilde{\overline{\varphi}}}$ is diagonal. Next formulate this in a somewhat more general form.

For a given matrix $\hat{\mathbf{R}}$, find matrices $\mathbf{R}_0$ and $\tilde{\mathbf{R}}$ such as

$$\hat{\mathbf{R}} = \mathbf{R}_0 + \tilde{\mathbf{R}} , \tag{7.73}$$

$$\tilde{\mathbf{R}} \quad \text{is positive definite and diagonal} , \tag{7.74}$$

$$\mathbf{R}_0 \quad \text{of dimension } m \times m, \text{ of rank } r < m . \tag{7.75}$$

Note that the squared Frobenius norm of the difference between a given matrix $\mathbf{R}$ and its closest rank r approximation is (see Lemma 11.3 in Appendix 11.A.1).

$$\sum_{r+1}^{m} \lambda_i^2(\mathbf{R}) , \tag{7.76}$$

where $\lambda_i, i = 1, \ldots, m$ are the eigenvalues in decreasing order of $\mathbf{R}$.

A reasonable formulation of the decomposition problem, used in Sorelius and Stoica (1998), is to let $\mathbf{x}$ denote a vector consisting of independent elements of the diagonal matrix $\tilde{\mathbf{R}}$. In this particular case, it is natural to take $\mathbf{x} = \boldsymbol{\rho} = \begin{pmatrix} \lambda_y & \lambda_u \end{pmatrix}^T$. The estimate of $\mathbf{x}$ is then selected as the solution to the following minimization problem

$$\hat{\mathbf{x}} = \arg \min_{\mathbf{x}} \sum_{i=r+1}^{m} \lambda_i(\hat{\mathbf{R}} - \tilde{\mathbf{R}}(\mathbf{x})) , \tag{7.77}$$

and the optimization is carried out over the set

$$C = \left\{ \mathbf{x} | \tilde{\mathbf{R}}(\mathbf{x}) \geq \mathbf{0}; \mathbf{R} - \tilde{\mathbf{R}}(\mathbf{x}) \geq \mathbf{0} \right\} . \tag{7.78}$$

It can be shown that the set C is convex. The paper Sorelius and Stoica (1998) suggests also an alternating projection algorithm that converges under weak conditions, even if convergence sometimes is slow.

Step 2. Estimation of θ

Note that the extended regressor vector $\overline{\varphi}_0(t)$ can be written as

$$\overline{\varphi}_0(t) = \begin{pmatrix} y_0(t) \\ \vdots \\ y_0(t-n_a) \\ u_0(t-1) \\ \vdots \\ u_0(t-n_b) \end{pmatrix} = \mathbf{S}\boldsymbol{\psi}_0(t) , \tag{7.79}$$

where

$$\mathbf{S}(\boldsymbol{\theta}) = \mathscr{S}(B, A) = \begin{pmatrix} b_1 & \dots & b_{n_b} & & \mathbf{0} \\ & \ddots & & \ddots & \\ \mathbf{0} & & b_1 & \dots & b_{n_b} \\ 1 & a_1 & \dots & a_{n_a} & \mathbf{0} \\ & \ddots & & \ddots & \\ \mathbf{0} & & 1 & \dots & a_{n_a} \end{pmatrix} , \tag{7.80}$$

$$\boldsymbol{\psi}_0(t) = \frac{1}{A(q^{-1})} \begin{pmatrix} u_0(t-1) \\ \vdots \\ u_0(t-n_a-n_b) \end{pmatrix} . \tag{7.81}$$

The Sylvester matrix $\mathscr{S}(B, A)$ is of rank $n_a + n_b$ as soon as the polynomials A and B are coprime, cf. Sect. A.1.5. Set

$$\mathbf{P} = \mathsf{E}\left\{\boldsymbol{\psi}_0(t)\boldsymbol{\psi}_0^T(t)\right\} . \tag{7.82}$$

Thus

$$\mathbf{R}_{\overline{\varphi}_0} = \mathbf{S}(\boldsymbol{\theta})\mathbf{P}\mathbf{S}^T(\boldsymbol{\theta}) . \tag{7.83}$$

From Step 1 of the algorithm an estimate of the rank-deficient matrix $\mathbf{R}_{\overline{\varphi}_0}$ is available, and hence also of its range space, which here is represented by the column vectors of a matrix $\hat{\mathbf{S}}$ of dimension $(n_a + n_b + 1) \times (n_a + n_b)$. Note that $\mathbf{S}$ is a function of $\boldsymbol{\theta}$. The parameter vector $\boldsymbol{\theta}$ is determined by solving the optimization problem (using the Frobenius norm)

$$\hat{\boldsymbol{\theta}} = \arg \min_{\boldsymbol{\theta}, \mathbf{M}} \| \hat{\mathbf{S}}\mathbf{M} - \mathbf{S}(\boldsymbol{\theta}) \|_F^2 . \tag{7.84}$$

This is a separable problem and

$$\mathbf{M} = \left(\hat{\mathbf{S}}^T\hat{\mathbf{S}}\right)^{-1} \hat{\mathbf{S}}^T\mathbf{S}(\boldsymbol{\theta}) , \tag{7.85}$$

and

$$\hat{\boldsymbol{\theta}} = \arg \min_{\boldsymbol{\theta}} \operatorname{tr} \left[\mathbf{S}^T(\boldsymbol{\theta}) \mathbf{Q} \mathbf{S}(\boldsymbol{\theta}) \right] , \tag{7.86}$$

$$\mathbf{Q} = \mathbf{I} - \hat{\mathbf{S}} \left(\hat{\mathbf{S}}^T \hat{\mathbf{S}} \right)^{-1} \hat{\mathbf{S}}^T . \tag{7.87}$$

Note that $\mathbf{S}$ is in fact an affine function of $\boldsymbol{\theta}$ and can be written as

$$\mathbf{S}(\boldsymbol{\theta}) = \mathbf{S}_0 + \sum_{i=1}^{n_a+n_b} \mathbf{S}_i \boldsymbol{\theta}_i . \tag{7.88}$$

Applying (7.88) in (7.86) leads to a criterion that contains quadratic, linear, and constant terms of $\boldsymbol{\theta}$. It is hence straightforward to solve. To be specific, it holds

$$\operatorname{tr} \left[\mathbf{S}^T(\boldsymbol{\theta}) \mathbf{Q} \mathbf{S}(\boldsymbol{\theta}) \right] = \boldsymbol{\theta}^T \mathbf{W}_1 \boldsymbol{\theta} + 2\boldsymbol{\theta}^T \mathbf{w}_2 + w_3 , \tag{7.89}$$

where

$$(\mathbf{W}_1)_{i,j} = \operatorname{tr} \mathbf{S}_i^T \mathbf{Q} \mathbf{S}_j \quad i, j = 1, \ldots, n_a + n_b , \tag{7.90}$$

$$(\mathbf{w}_2)_i = \operatorname{tr} \mathbf{S}_i^T \mathbf{Q} \mathbf{S}_0 \quad i = 1, \ldots, n_a + n_b , \tag{7.91}$$

$$w_3 = \operatorname{tr} \mathbf{S}_0^T \mathbf{Q} \mathbf{S}_0 . \tag{7.92}$$

The resulting parameter estimate that minimizes (7.89) is readily found to be

$$\hat{\boldsymbol{\theta}} = -\mathbf{W}_1^{-1} \mathbf{w}_2 . \tag{7.93}$$

7.4 The Generalized Instrumental Variable Method

7.4.1 General Framework

The generalized instrumental variable estimator (GIVE) has been introduced in Söderström (2011) as a class of estimators based on the bias-eliminating principle (7.11) containing many previously known methods as special cases.

Recall that the total parameter vector $\boldsymbol{\vartheta}$ is

$$\boldsymbol{\vartheta} = \begin{pmatrix} \boldsymbol{\theta} \\ \boldsymbol{\rho} \end{pmatrix} . \tag{7.94}$$

Introduce a generalized instrumental vector (IV) $\mathbf{z}(t)$, composed of delayed values of $y(t)$ and $u(t)$ and of dimension $n_\mathbf{z}$, where

$$n_\mathbf{z} \geq \dim(\boldsymbol{\vartheta}) = n_a + n_b + 2 \,. \tag{7.95}$$

Correlating $\mathbf{z}(t)$ with the equation error $\varepsilon(t)$ in (7.1), it is possible to set up the following overdetermined set of equations

$$\left(\mathbf{R}_{\mathbf{z}\varphi} - \mathbf{R}_{\tilde{\mathbf{z}}\tilde{\varphi}}(\boldsymbol{\rho})\right)\boldsymbol{\theta} = \left(\mathbf{r}_{\mathbf{z}y} - \mathbf{r}_{\tilde{\mathbf{z}}\tilde{y}}(\boldsymbol{\rho})\right) , \tag{7.96}$$

where the choice of the instrumental variable $\mathbf{z}(t)$ determines the structure of $\mathbf{R}_{\tilde{\mathbf{z}}\tilde{\varphi}}(\boldsymbol{\rho})$ and $\mathbf{r}_{\tilde{\mathbf{z}}\tilde{y}}(\boldsymbol{\rho})$, cf. (7.11). In order to determine the parameter vectors $\boldsymbol{\theta}$ and $\boldsymbol{\rho}$, some of the entries in $\mathbf{z}(t)$ must be correlated with $\varepsilon(t)$.

When $n_\mathbf{z}$ is chosen so that inequality applies in (7.95), the system of equations in (7.96) is overdetermined. A very common choice for the vector $\mathbf{z}(t)$ is

$$\mathbf{z}(t) = \left(y(t) \ldots y(t-n_a-p_y)\ u(t) \ldots u(t-n_b-p_u)\right)^T , \tag{7.97}$$

where p_u and p_y are user chosen variables, with $p_y \geq 0$, $p_u \geq 0$.

In the general case, for the GIVE method the parameter estimate $\hat{\boldsymbol{\vartheta}}_{GIVE}$ is defined as the solution to an optimization problem. The GIVE estimate of $\boldsymbol{\vartheta}$ is

$$\hat{\boldsymbol{\vartheta}}_{\text{GIVE}} = \arg\min_{\boldsymbol{\vartheta}} V_{\text{GIVE}}(\boldsymbol{\vartheta}) \,, \tag{7.98}$$

$$\begin{aligned} V_{\text{GIVE}}(\boldsymbol{\vartheta}) &= \|\hat{\mathbf{r}}_{\mathbf{z}y} - \mathbf{r}_{\tilde{\mathbf{z}}\tilde{y}}(\boldsymbol{\rho}) - \left(\hat{\mathbf{R}}_{\mathbf{z}\varphi} - \mathbf{R}_{\tilde{\mathbf{z}}\tilde{\varphi}}(\boldsymbol{\rho})\right)\boldsymbol{\theta}\|^2_{\mathbf{W}(\boldsymbol{\theta})} \\ &= \|\hat{\mathbf{r}}_{\mathbf{z}\varepsilon} - \mathbf{r}_{\tilde{\mathbf{z}}\varepsilon}(\boldsymbol{\theta}, \boldsymbol{\rho})\|^2_{\mathbf{W}(\boldsymbol{\theta})} \\ &\stackrel{\Delta}{=} \|\bar{\mathbf{r}}_{\mathbf{z}\varepsilon}(\boldsymbol{\theta}, \boldsymbol{\rho})\|^2_{\mathbf{W}(\theta)} \,. \end{aligned} \tag{7.99}$$

In its most general form, one uses a $\boldsymbol{\theta}$-dependent weighting matrix $\mathbf{W}(\boldsymbol{\theta})$, but often $\mathbf{W}(\boldsymbol{\theta})$ is chosen as a constant matrix.

When the weighting matrix $\mathbf{W}$ does not depend on $\boldsymbol{\theta}$, the minimization in (7.98) with respect to $\boldsymbol{\theta}$ is easy, as the criterion V_{GIVE} is quadratic in $\boldsymbol{\theta}$. The problem is then indeed a separable nonlinear least squares problem. This means that the estimate (7.98) can be obtained as the solution of an associated problem of *lower dimension*, cf. Sect. A.1.4:

$$\hat{\boldsymbol{\theta}}_{\text{GIVE}} = \left[\overline{\mathbf{R}}^T_{\mathbf{z}\varphi}\mathbf{W}\overline{\mathbf{R}}_{\mathbf{z}\varphi}\right]^{-1}\overline{\mathbf{R}}^T_{\mathbf{z}\varphi}\mathbf{W}\bar{\mathbf{r}}_{\mathbf{z}y}|_{\boldsymbol{\rho}=\hat{\boldsymbol{\rho}}_{\text{GIVE}}} , \tag{7.100}$$

$$\overline{\mathbf{R}}_{\mathbf{z}\varphi}(\boldsymbol{\rho}) = \hat{\mathbf{R}}_{\mathbf{z}\varphi} - \mathbf{R}_{\tilde{\mathbf{z}}\tilde{\varphi}}(\boldsymbol{\rho}) \stackrel{\Delta}{=} \overline{\mathbf{R}}(\boldsymbol{\rho}) \,, \tag{7.101}$$

$$\bar{\mathbf{r}}_{\mathbf{z}\varphi}(\boldsymbol{\rho}) = \hat{\mathbf{r}}_{\mathbf{z}y} - \mathbf{r}_{\mathbf{z}\tilde{y}}(\boldsymbol{\rho}) \stackrel{\Delta}{=} \bar{\mathbf{r}}(\boldsymbol{\rho}) \,, \tag{7.102}$$

$$\hat{\boldsymbol{\rho}}_{\text{GIVE}} = \arg\min_{\boldsymbol{\rho}} \bar{V}_{\text{GIVE}}(\boldsymbol{\rho}) \,, \tag{7.103}$$

$$\bar{V}_{\text{GIVE}}(\boldsymbol{\rho}) = \bar{\mathbf{r}}^T(\boldsymbol{\rho})\mathbf{W}\bar{\mathbf{r}}(\boldsymbol{\rho}) - \bar{\mathbf{r}}^T(\boldsymbol{\rho})\mathbf{W}\overline{\mathbf{R}}(\boldsymbol{\rho})\left[\overline{\mathbf{R}}^T(\boldsymbol{\rho})\mathbf{W}\overline{\mathbf{R}}(\boldsymbol{\rho})\right]^{-1}\overline{\mathbf{R}}^T(\boldsymbol{\rho})\mathbf{W}\bar{\mathbf{r}}(\boldsymbol{\rho}) \,. \tag{7.104}$$

The function $\bar{V}_{\text{GIVE}}(\boldsymbol{\rho})$ is called a concentrated loss function.

There is one special situation that is worth discussing for the case when the system of Eqs. (7.96) is overdetermined. It is not uncommon for such cases that one chooses to require some of the equations to hold exactly, while for the others the difference between the left and the right-hand sides is minimized in a weighted least squares sense.

To formulate such a case, split the vector $\mathbf{z}(t)$ as

$$\mathbf{z}(t) = \begin{pmatrix} \mathbf{z}_1(t) \\ \mathbf{z}_2(t) \end{pmatrix} \begin{matrix} \}n_1 \text{ elements} \\ \}n_2 \text{ elements} \end{matrix} \tag{7.105}$$

where it is required that $\bar{\mathbf{r}}_{\mathbf{z}_1\varepsilon} = \mathbf{0}$ holds exactly. For this to be meaningful, one must have

$$n_1 < \dim(\boldsymbol{\eta}) = n_a + n_b + 2 \,, \tag{7.106}$$

$$n_2 = n_{\mathbf{z}} - n_1 \,. \tag{7.107}$$

The optimization problem for finding $\hat{\boldsymbol{\vartheta}}_{\text{GIVE}}$ is then

$$\hat{\boldsymbol{\vartheta}}_{\text{GIVE}} = \begin{matrix} \arg\min_{\boldsymbol{\rho},\boldsymbol{\theta}} \|\bar{\mathbf{r}}_{\mathbf{z}_2\varepsilon}(\boldsymbol{\rho},\boldsymbol{\theta})\|^2_{\mathbf{W}_2} \\ \text{such that } \bar{\mathbf{r}}_{\mathbf{z}_1\varepsilon}(\boldsymbol{\rho},\boldsymbol{\theta}) = \mathbf{0} \end{matrix} \,. \tag{7.108}$$

This estimate can be seen as an extreme case of the general formulation (7.100)–(7.104), by choosing

$$\mathbf{W} = \begin{pmatrix} \alpha\mathbf{I}_{n_1} & 0 \\ 0 & \mathbf{W}_2 \end{pmatrix} \tag{7.109}$$

and letting α tend to infinity.

It was found in Hong et al. (2007b), Hong and Söderström (2009), Söderström (2011) that many bias-compensating schemes can be formulated as the general estimator (7.100)–(7.104), by appropriate choices of $\mathbf{z}_1(t)$, $\mathbf{z}_2(t)$ and $\mathbf{W}$. This means that *these formally different estimators are equivalent*. Several examples are presented in Sect. 7.4.2.

What does equivalence mean in this context?

1. First, one has to distinguish between *the equations* (7.100)–(7.104) defining the estimates, on the one hand, and *the choice of numerical algorithm* employed to solve them on the other hand. The properties of the estimates (the solution to the equations), such as statistical properties of the estimation error $\hat{\boldsymbol{\vartheta}} - \boldsymbol{\vartheta}_0$, do not depend on the way the equations are solved. That is, which particular algorithm that is used for finding the solution has no importance on the properties of the solution itself.
   The choice of the algorithm may still be important from a practical point of view. The amount of computations needed, the robustness to rounding errors and to the initial guesses for the nonlinear optimization in (7.98) or (7.103), can differ considerably between different algorithms. The use of the concentrated loss function formulation (7.100)–(7.104) is a preferable and robust way to solve the optimization problem.
2. The parameter estimates for different estimators defined by solving the same set of equations can differ somewhat for various reasons:
   - One aspect is that different weights $\mathbf{W}(\boldsymbol{\theta})$ may have been chosen.
   - A more subtle difference that should be of minor importance is the precise way the covariance estimates (7.96) are formed from the data. For example, are the time points where all elements of $u(t)$ and $y(t)$ are not available completely discarded, or are unavailable data replaced by zeros?
   - Another reason for minor differences is that all estimator algorithms include some iterative computations and that different stopping rules may be applied.

The estimator can be extended to some more general cases:

- One may consider the case when $\tilde{y}(t)$ is arbitrarily autocorrelated, but uncorrelated with the white input noise $\tilde{u}(t)$. For such a case, one might add covariance elements $r_{\tilde{y}}(\tau)$ for a number of τ−values in the $\boldsymbol{\rho}$ vector. However, it then turns out to be infeasible to use any delayed values of the output $y(t)$ in the vector $\mathbf{z}(t)$. For any new such vector element added, one has also to include a further unknown $r_{\tilde{y}}(\tau)$. Therefore, it only makes sense in such scenarios to use delayed input variables in the $\mathbf{z}(t)$ vector, so for example

$$\mathbf{z}(t) = \big(u(t) \ \dots \ u(t - n_b - p_u)\big)^T \tag{7.110}$$

  where, according to (7.95), $p_u \geq n_a$, as $\boldsymbol{\rho} = \lambda_u$ in this case.
- For the situation above, it is also possible to use only further delayed inputs in the $\mathbf{z}(t)$ vector

$$\mathbf{z}(t) = \big(u(t - n_b - 1) \ \dots \ u(t - n_b - p_u)\big)^T \tag{7.111}$$

  with $p_u \geq n_a + n_b$. In this case there is indeed no noise parameter vector needed, as $\mathbf{r}_{\tilde{\mathbf{z}}\tilde{y}}$ and $\mathbf{R}_{\tilde{\mathbf{z}}\tilde{\varphi}}$ both become zero. The estimate is the instrumental variable estimate described, e.g., in Söderström (1981). It can also be interpreted as a Yule–Walker estimate.

- A more general case allowing the white noise sequences $\tilde{y}(t)$ and $\tilde{u}(t)$ to be cross-correlated (the noise parameter vector $\boldsymbol{\rho}$ is to be augmented with the expectation $\mathsf{E}\{\tilde{y}(t)\tilde{u}(t)\} = \lambda_{yu}$) is treated in Söderström et al. (2014b).

7.4.2 Various Examples

It will now be illustrated how several methods earlier proposed in the literature fit into the general GIVE framework. For each method, it is specified how $\mathbf{z}_1(t)$, $\mathbf{z}_2(t)$, and $\mathbf{W}_2$ are selected.

Remark 7.7 Note that the signs of the elements in the vector $\mathbf{z}(t)$ are not significant, and these signs do not change the equations defining the estimators. For example, one may change a negative delayed output to its positive value, that is, to replace $-y(t-i)$ by $y(t-i)$, without affecting the estimator. This property will sometimes be used below, when deemed convenient. ■

Example 7.4 The bias-eliminating least squares method was introduced in Zheng (1998a, 2002a) and proposed in a number of variants. See also Sect. 7.2.1.

The algorithm BELS-1 of Zheng (1998a) corresponds to the choice $\mathbf{z}_1(t) = \left(y(t)\ \boldsymbol{\varphi}^T(t)\right)^T$, $\mathbf{z}_2(t) = y(t - n_a - 1)$. As $\mathbf{z}_2(t)$ is a scalar, there is no need for any weighting matrix $\mathbf{W}_2$.

Similarly, the algorithm BELS-2 of Zheng (1998a) corresponds to the choice $\mathbf{z}_1(t) = \left(y(t)\ \boldsymbol{\varphi}^T(t)\right)^T$, $\mathbf{z}_2(t) = u(t - n_b - 1)$. As $\mathbf{z}_2(t)$ also in this case is a scalar, there is no need for any weighting matrix $\mathbf{W}_2$. ■

Example 7.5 Another variant of the bias-eliminating least squares method is the algorithm BELS-II of Zheng (2002a), which copes with the case of arbitrarily correlated output noise. See also Sect. 7.2.3. The equations obtained by correlating past outputs, say $y(t-j)$, with the equation errors $\varepsilon(t) = y(t) - \boldsymbol{\varphi}^T(t)\boldsymbol{\theta}$ are then not 'useful,' in the sense that for each further equations used, the number of unknowns also increases by one. After eliminating all the equations involving the unknown correlation function $r_{\tilde{y}}(\tau)$ of the output disturbances, the algorithm leads to the use of

$$\mathbf{z}_1(t) = \begin{pmatrix} u(t) \\ \vdots \\ u(t-n_b) \end{pmatrix}, \quad \mathbf{z}_2(t) = \begin{pmatrix} u(t-n_b-1) \\ \vdots \\ u(t-n_b-n_a-1) \end{pmatrix}$$

for estimating the unknowns $\boldsymbol{\theta}$ and λ_u. No weighting is needed. ■

Example 7.6 The Frisch scheme for EIV identification has been proposed in several forms. See also Sect. 7.3.2. One of the first appeared in Beghelli et al. (1993). See Example 7.1. A common aspect for all these methods is that the adjusted normal equations are used. This means precisely that

$$\mathbf{z}_1(t) = \big(y(t) \ldots y(t-n_a)\; u(t) \ldots u(t-n_b)\big)^T . \tag{7.112}$$

The shifted relation criterion described in Diversi et al. (2004) is based on the following equation

$$\left(\mathbf{R}_{\overline{\varphi}} - \mathbf{R}_{\tilde{\overline{\varphi}}}(\rho_0)\right)\left(\mathbf{v}_1\; \mathbf{v}_2\right) = \mathbf{0} , \tag{7.113}$$

where

$$\overline{\varphi}(t) = \big(-y(t) \ldots -y(t-n_a-1)\; u(t) \ldots u(t-n_b-1)\big)^T , \tag{7.114}$$
$$\tilde{\overline{\varphi}}(t) = \big(-\tilde{y}(t) \ldots -\tilde{y}(t-n_a-1)\; \tilde{u}(t) \ldots \tilde{u}(t-n_b-1)\big)^T , \tag{7.115}$$

and

$$\mathbf{v}_1 = \big(1\; a_1 \ldots a_{n_a}\; 0\; b_1 \ldots b_{n_b}\; 0\big)^T , \tag{7.116}$$
$$\mathbf{v}_2 = \big(0\; 1\; a_1 \ldots a_{n_a}\; 0\; b_1 \ldots b_{n_b}\big)^T , \tag{7.117}$$

so that four additional relations are used besides the standard Frisch equations. The use of $\mathbf{v}_1$ leads to

$$\mathbf{z}_2(t) = \big(y(t-n_a-1)\; u(t-n_b-1)\big)^T ,$$

while the use of $\mathbf{v}_2$ leads to

$$\mathbf{z}_2(t) = \big(y(t+1)\; u(t+1)\big)^T .$$

Therefore, the use of both $\mathbf{v}_1$ and $\mathbf{v}_2$ corresponds to

$$\mathbf{z}_2(t) = \big(y(t-n_a-1)\; u(t-n_b-1)\; y(t+1)\; u(t+1)\big)^T . \tag{7.118}$$

Finally, in Beghelli et al. (1993) the choice $\mathbf{W}_2 = \mathbf{I}$ is made. ■

Example 7.7 Another variant of the Frisch scheme is to use additional Yule–Walker equations, Diversi et al. (2006). See Example 7.2.

This corresponds to

$$\mathbf{z}_1(t) = \big(y(t) \ldots y(t-n_a) u(t) \ldots u(t-n_b)\big)^T ,$$
$$\mathbf{z}_2(t) = \big(u(t-n_b-1) \ldots u(t-n_b-m)\big)^T .$$

Equal weighting, $\mathbf{W}_2 = \mathbf{I}$, is proposed in Diversi et al. (2006). The size m of the vector $\mathbf{z}_2(t)$ is normally chosen so that an overdetermined system is obtained.

Adaption of this approach to the general case with cross-correlated noise, $\lambda_{yu} \neq 0$, is treated in Diversi et al. (2012). ■

Example 7.8 A third variant of the Frisch scheme is based on comparing the correlation functions of the equations errors, using the model on the one hand and using the measured data on the other. See Example 7.3. It is shown in Söderström (2011) that it corresponds to

$$
\begin{aligned}
\mathbf{z}_1(t) &= \big(y(t)\dots y(t-n_a)u(t-1)\dots u(t-n_b)\big)^T\ ,\\
\mathbf{z}_2(t) &= \begin{pmatrix} \varepsilon(t-1,\boldsymbol{\theta}) \\ \vdots \\ \varepsilon(t-k,\boldsymbol{\theta}) \end{pmatrix}\\
&= \begin{pmatrix} 1 & a_1 & \dots & a_{n_a} & \mathbf{0} & b_1 & \dots & b_{n_b} & \mathbf{0} \\ & & \ddots & & \ddots & & \ddots & & \\ \mathbf{0} & 1 & a_1 & \dots & a_{n_a} & \mathbf{0} & b_1 & \dots & b_{n_b} \end{pmatrix}\\
&\quad \times \begin{pmatrix} y(t) \\ \vdots \\ y(t-n_a-k) \\ u(t-1) \\ \vdots \\ u(t-n_b-k) \end{pmatrix}\\
&\stackrel{\Delta}{=} \mathbf{M}(\boldsymbol{\theta}) \begin{pmatrix} \mathbf{z}_1(t) \\ \bar{\mathbf{z}}(t) \end{pmatrix} = \mathbf{M}_1(\boldsymbol{\theta})\mathbf{z}_1(t) + \mathbf{M}_2(\boldsymbol{\theta})\bar{\mathbf{z}}(t)\ , \qquad (7.119)\\
\bar{\mathbf{z}}(t) &= \big(y(t-n_a-1)\dots y(t-n_a-k)\\
&\qquad u(t-n_b-1)\dots u(t-n_b-k)\big)^T\ . \qquad (7.120)
\end{aligned}
$$

Further, as $\bar{\mathbf{r}}_{\mathbf{z}_1\varepsilon} = \mathbf{0}$, the criterion to be minimized can also be written as

$$
\begin{aligned}
\| \bar{\mathbf{r}}_{\mathbf{z}_2\varepsilon}(\boldsymbol{\theta}) \|^2 &= \| \mathbf{M}_1(\boldsymbol{\theta})\bar{\mathbf{r}}_{\mathbf{z}_1\varepsilon}(\boldsymbol{\theta}) + \mathbf{M}_2(\boldsymbol{\theta})\bar{\mathbf{r}}_{\bar{\mathbf{z}}\varepsilon}(\boldsymbol{\theta}) \|^2\\
&= \| \mathbf{M}_2(\boldsymbol{\theta})\bar{\mathbf{r}}_{\bar{\mathbf{z}}\varepsilon}(\boldsymbol{\theta}) \|^2\ . \qquad (7.121)
\end{aligned}
$$

One may therefore also identify the vector $\mathbf{z}_2(t)$ in the general algorithm with $\bar{\mathbf{z}}(t)$ in (7.120), and let the weighting matrix depend on the parameter vector $\boldsymbol{\theta}$ as

$$\mathbf{W}_2(\theta) = \mathbf{M}_2^T(\theta)\mathbf{M}_2(\theta)\ . \qquad (7.122)$$

An extension of this algorithm to handle the general case with correlated noise, $\lambda_{yu} \neq 0$, is presented in Diversi (2013). ■

Example 7.9 The extended compensated least squares (ECLS) method was proposed in Ekman (2005) and analyzed in Ekman et al. (2006). It corresponds to $\mathbf{z}_1(t)$ being absent, $\mathbf{z}_2(t)$ as in (7.97), with the weighting matrix $\mathbf{W}_2 = I$. ■

Table 7.1 Generalized instrumental variable estimates (GIVE) for the standard data sets

Parameter	True value	GIVE		IV	
		Mean	Standard dev.	Mean	Standard dev.
a_1	−1.5	−1.528	0.085	−1.440	0.534
a_2	0.7	0.709	0.040	0.671	0.302
b_1	2.0	3.57	2.82	−0.510	27.9
b_2	1.0	−0.69	3.30	3.599	26.2
λ_y	10.0	8.37	4.51	–	–
λ_u	1.81	2.00	0.64	–	–

Example 7.10 In this numerical example the GIVE vector is taken as

$$\mathbf{z}(t) = \big(y(t) \ \dots \ y(t-2) \ u(t-1) \ \dots \ u(t-3) \big)^T . \tag{7.123}$$

There will hence be 6 equations ($\mathbf{z}(t)$ has 6 elements) and 6 unknowns ($\boldsymbol{\theta}, \lambda_y, \lambda_u$) to estimate. The results using Monte Carlo simulations with the standard data sets are summarized in Table 7.1. The outcome of using IV is included as a reference.

One can see that there are large improvements when going from IV to GIVE. Further, for both methods the A parameters are estimated much more accurately than the B parameters.

Still a closer look show that the dominating influence on the standard deviations comes from some 15 of the 100 realizations. This means that for this example, GIVE is not so robust but can be sensitive to the data. In the next example it will be shown that when an overdetermined system of equations in used (corresponding to more than the lowest possible dimension of the vector $\mathbf{z}(t)$) the results are significantly more robust. ■

Example 7.11 Next consider GIVE with an extended vector $\mathbf{z}(t)$ that has more than the minimal number of elements. Specifically, the vector was given as

$$\mathbf{z}(t) = \big(y(t) \ \dots \ y(t-2-p) \ u(t-1) \ \dots \ u(t-3-p) \big)^T , \tag{7.124}$$

where the user design parameter p was considered ranging from $p = 0$ to $p = 10$. The results are summarized in Table 7.2.

Further, the RMS values are computed for the estimate of $\boldsymbol{\theta}$. They are displayed versus p in Fig. 7.2. It is seen that in this case $p = 2$ gives the smallest RMS value. It hence pays to have p larger than minimally possible, but nothing is gained by choosing p too large. ■

Table 7.2 Generalized instrumental variable estimates (GIVE) for increasing length of the instrumental variable vector $\mathbf{z}(t)$ using the standard data sets. The arithmetic mean values are given with standard deviations within parentheses

Parameter	a_1	a_2	b_1	b_2	λ_y	λ_u
p True	−1.5	0.7	2.0	1.0	10.0	1.81
0	−1.528	0.709	3.57	−0.69	8.37	2.00
	(0.085)	(0.040)	(2.82)	(3.30)	(4.51)	(0.64)
1	−1.500	0.696	2.359	0.652	10.56	1.97
	(0.018)	(0.011)	(1.01)	(1.08)	(1.77)	(0.23)
2	−1.498	0.698	2.213	0.807	10.80	1.95
	(0.013)	(0.010)	(0.33)	(0.37)	(1.17)	(0.22)
3	−1.499	0.698	2.225	0.793	10.80	1.95
	(0.014)	(0.010)	(0.40)	(0.44)	(1.16)	(0.26)
4	−1.499	0.698	2.208	0.811	10.82	1.94
	(0.015)	(0.010)	(0.46)	(0.51)	(1.20)	(0.28)
5	−1.499	0.698	2.197	0.819	10.82	1.94
	(0.016)	(0.010)	(0.56)	(0.62)	(1.20)	(0.30)
6	−1.499	0.698	2.189	0.826	10.84	1.93
	(0.017)	(0.010)	(0.69)	(0.73)	(1.26)	(0.36)
7	−1.499	0.698	2.187	0.823	10.84	1.93
	(0.018)	(0.011)	(0.84)	(0.88)	(1.33)	(0.38)
8	−1.500	0.698	2.173	0.833	10.87	1.95
	(0.018)	(0.011)	(1.01)	(1.02)	(1.46)	(0.42)
9	−1.500	0.699	2.165	0.835	10.89	1.91
	(0.019)	(0.012)	(1.16)	(1.16)	(1.62)	(0.47)
10	−1.500	0.699	2.153	0.843	10.91	1.89
	(0.019)	(0.012)	(1.25)	(1.23)	(1.76)	(0.51)

7.4.3 GIVE Identification of MIMO Models

Here it is discussed how the previous results may be extended to multivariable systems. As before the governing equations are the generalized IV equations, which now are written as

$$\mathbf{0} \approx \mathbf{f}(\boldsymbol{\vartheta}) \stackrel{\Delta}{=} \hat{\mathbf{r}}_{\mathbf{z}\boldsymbol{\varepsilon}}(\boldsymbol{\theta}) - \mathbf{r}_{\mathbf{z}\boldsymbol{\varepsilon}}(\boldsymbol{\theta}, \boldsymbol{\Lambda}_y, \boldsymbol{\Lambda}_u) , \tag{7.125}$$

where $\boldsymbol{\varepsilon}(t)$ is an n_y-dimensional vector given by (5.65):

$$\boldsymbol{\varepsilon}(t) = \mathbf{y}(t) - \boldsymbol{\Theta}^T \boldsymbol{\varphi}(t) , \tag{7.126}$$

and $\boldsymbol{\Theta}$ is given by (5.62).

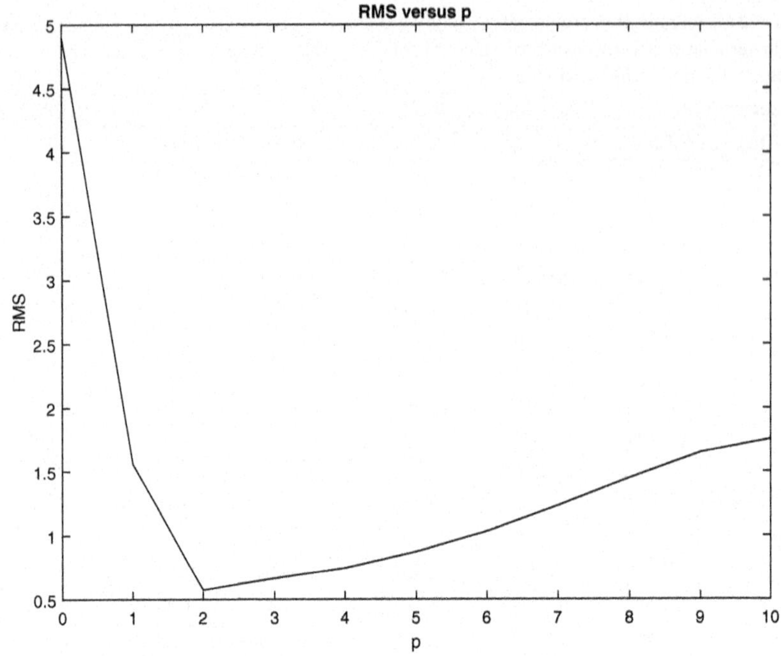

Fig. 7.2 RMS values versus p for the GIVE estimates

Note that here $\mathbf{f}(\boldsymbol{\vartheta})$ is an $m \times n_y$ matrix, where now

$$m = (n_a + 1 + p_y)n_y + (n_b + p_u)n_u \ . \tag{7.127}$$

It is desired to have $\mathbf{f}(\boldsymbol{\vartheta}) = \mathbf{0}$. To get compatibility one must require that the number of equations is at least as large as the number of unknowns. This gives the constraint (for the case of white output noise)

$$mn_y \geq n_a n_y^2 + n_b n_y n_u + n_y^2 + n_u^2 \ ,$$

leading to

$$p_y n_y^2 + p_u n_u n_y \geq n_u^2 \ , \tag{7.128}$$

Treat the general case, introduce weighting matrices $\mathbf{W}$ and $\mathbf{Z}$, and define the estimate $\hat{\boldsymbol{\vartheta}}$ as the minimizing element of

$$V(\boldsymbol{\vartheta}) = \text{tr}(\mathbf{W}\mathbf{f}(\boldsymbol{\vartheta})\mathbf{Z}\mathbf{f}^T(\boldsymbol{\vartheta})) \ . \tag{7.129}$$

Assume that $\mathbf{f}$ is written columnwise as

$$\mathbf{f} = \left(\mathbf{f}_1 \ \dots \ \mathbf{f}_{n_y}\right) . \tag{7.130}$$

Then the criterion (7.129) can be written as

$$\begin{aligned} V &= \operatorname{tr}\left[\begin{pmatrix} \mathbf{f}_1^T \\ \vdots \\ \mathbf{f}_{n_y}^T \end{pmatrix} \mathbf{W} \left(\mathbf{f}_1 \ \dots \ \mathbf{f}_{n_y}\right) \mathbf{Z}\right] . \\ &= \sum_i \sum_j \mathbf{Z}_{i,j} \mathbf{f}_j^T \mathbf{W} \mathbf{f}_i . \end{aligned} \tag{7.131}$$

An alternative way would be to first vectorize the matrix $\mathbf{f}$. Then set

$$\overline{\mathbf{f}} = \operatorname{vec}(\mathbf{f}) = \begin{pmatrix} \mathbf{f}_1 \\ \mathbf{f}_2 \\ \vdots \\ \mathbf{f}_{n_y} \end{pmatrix} , \tag{7.132}$$

and consider the criterion

$$\overline{V} = \overline{\mathbf{f}}^T \overline{\mathbf{W}} \overline{\mathbf{f}} = \sum_i \sum_j \mathbf{f}_i^T \mathbf{W}_{i,j} \mathbf{f}_j , \tag{7.133}$$

where

$$\overline{\mathbf{W}} = \begin{pmatrix} \mathbf{W}_{1,1} & \mathbf{W}_{1,2} & \dots & \mathbf{W}_{1,n_y} \\ \vdots & & & \\ \mathbf{W}_{n_y,1} & \dots & & \mathbf{W}_{n_y,n_y} \end{pmatrix} . \tag{7.134}$$

This approach will give the same criterion as in (7.131), provided

$$\mathbf{W}_{i,j} = \mathbf{Z}_{i,j} \mathbf{W} , \tag{7.135}$$

that is

$$\overline{\mathbf{W}} = \mathbf{Z} \otimes \mathbf{W} . \tag{7.136}$$

The two criteria, (7.129) and (7.133), *are hence equivalent* as long as the condition (7.136) holds. Both forms have specific advantages.

- The form $V(\boldsymbol{\theta}, \boldsymbol{\rho})$ in (7.129) is convenient when deriving numerically efficient algorithms for the minimization. More aspects on how to carry out the minimization of $V(\boldsymbol{\theta}, \boldsymbol{\rho})$ in (7.129) are given in Appendix 13.A.

- The form (7.133) where $\overline{\mathbf{f}}$ is regarded as a function of the total parameter vector $\boldsymbol{\vartheta}$, see (5.68), is the convenient choice when carrying out a statistical analysis of the parameter estimate, as shown in Chap. 14, see Sect. 14.4.3.

This section is concluded by giving some comments around the parameterization. Distinguish between two cases, for which the output noise $\tilde{\mathbf{y}}(t)$ is white or correlated, respectively:

- When the output noise is white (i.e., Assumption **AN3c** applies), the noise parameter vector $\boldsymbol{\rho}$ is chosen as in (5.66). The model integers p_y and p_u are nonnegative and subject to the condition (7.128).
- When the output noise is correlated, the noise parameter vector $\boldsymbol{\rho}$ is taken as in (5.67). Further, in this case one must choose $p_y = 0$, and p_u must be taken to fulfill $p_u n_y \geq n_u$.

How to modify the above analysis when treating canonical forms for MIMO systems is described in Appendix 7.A.2.

7.5 Extensions and Bibliographical Notes

The bias-compensation principle has also been used in other contexts than EIV. It is applied for order determination in Xin et al. (1996) and for estimation of a time-delay in Zheng and Feng (1990). It is extended to modeling nonlinear systems in Vajk and Hetthéssy (2003). Handling the bias caused in closed-loop identification is treated in Zheng and Feng (1995).

The bias-compensation principle for standard system identification with an arbitrary process disturbance but no input measurement noise has been treated using the BELS principle in Feng and Zheng (1991) and Zheng (2002b).

7.5.1 BELS

There are also alternative BELS methods described in the literature. BELS approaches using prefilters of the input are launched in Zheng and Feng (1989) and Zheng (1999d). The idea is to use a polynomial prefilter $F(q^{-1})$ and define the filtered input

$$\overline{u}(t) = \frac{1}{F(q^{-1})} u(t) \ .$$

Then in addition to computing the LS estimate in the ARX model

$$A(q^{-1})y(t) = B(q^{-1})u(t) + \varepsilon(t) \ , \tag{7.137}$$

one also considers the LS estimation of

$$A(q^{-1})y(t) = \beta(q^{-1})\overline{u}(t) + \varepsilon(t) , \tag{7.138}$$

where in some ideal situation $\beta(q^{-1}) = B(q^{-1})F(q^{-1})$. By comparing the LS estimates of (7.137) and (7.138) and the corresponding minimal values of the loss function, it is possible to derive equations for determining the unknown noise variances λ_y and λ_u.

Other approaches for deriving the required two equations to be used in addition to (7.28) have been proposed by Wada et al. (1990), Zheng (1999c), Jia et al. (2001), Ikenoue et al. (2005) and Ikenoue and Wada (2015). A related analysis for the case of noisy observations of an AR process appear in Jin et al. (2001), Jia et al. (2003).

Combining ideas of BELS and Yule–Walker equations appear in Diversi (2009).

Alternative, and simple, ways to derive the BELS estimates for EIV problems are presented in Hong et al. (2007b).

There are several further papers deriving BELS type of methods:

- Use of over-parameterization (one A or one B parameter added) is treated in Zheng (1998a, 1999c).
- BELS methods for cases where output noise $\tilde{y}$ may be an ARMA process appear in Zheng and Feng (1992), Zheng (1999a, 2002a).
- Analysis of convergence over iterations for BELS algorithms is given in Söderström et al. (2005).
- The paper Diversi (2013) treats BELS for mutually correlated noises.
- The case of an ARX system combined with the input measurements corrupted by white noise

$$\begin{aligned} y(t) &= \frac{B_0(q^{-1})}{A_0(q^{-1})}u_0(t) + \frac{1}{A_0(q^{-1})}e(t) \\ u(t) &= u_0(t) + \tilde{u}(t), \quad \mathsf{E}\left\{\tilde{u}^2(t)\right\} = \lambda_u \end{aligned} \tag{7.139}$$

  is considered in Liu and Zhu (2015). A bias-compensation scheme is derived from estimating $\boldsymbol{\theta}$ and λ_u. The bias-compensated normal equations are used along with a minimization of a criterion that depends on fourth-order moments of the data. An extension is considered in Liu and Zhu (2016, 2017). The general linear model of the form

$$\begin{aligned} y(t) &= \frac{B_0(q^{-1})}{A_0(q^{-1})}u_0(t) + \frac{C_0(q^{-1})}{D_0(q^{-1})}e(t) \\ u(t) &= u_0(t) + \tilde{u}(t), \quad \mathsf{E}\left\{\tilde{u}^2(t)\right\} = \lambda_u \end{aligned} \tag{7.140}$$

  with white input noise is considered. The proposed method consists of two steps. In the first step a high-order ARX model is fitted; that is, the model (7.139) is applied. In the second step a frequency-weighted model reduction takes place.

Special Case with One Degree of Freedom

When the noise variance ratio is known, or there is only one noise source, the identification problem simplifies considerably. There are many papers treating this special case. See also the treatment in Sect. 3.3.

One BELS-related method considering the case of output noise only includes extra delayed measurements and was suggested in Zheng (2000b).

The paper Jia et al. (2011) contains a general framework for describing many types of BELS algorithms with no input noise and generally correlated output noise.

Other general techniques for the case of one degree of freedom include Söderström (2013), Stoica et al. (1995b).

Papers treating BELS for the case of no input noise present include Sagara and Wada (1977), Stoica and Söderström (1982), Zheng and Feng (1990, 1995), Feng and Zheng (1991), Stoica et al. (1995b), Zheng (2002b).

The paper Vajk and Hetthéssy (2003) describes extension of BELS to handle nonlinear models for the case of one degree of freedom, that is, in line with the Koopmans–Levin estimator, Koopmans (1937).

The paper Vajk (2008) considers some extensions to the Koopmans-Levin algorithm (which assumes known noise variances). See also Vajk and Hetthéssy (2005, 2006), Vajk (2005).

Papers treating the case of known noise variance ratio include Levin (1964), Aoki and Yue (1970), Guidorzi (1975, 1981), Fernando and Nicholson (1985).

Relation to IV

BELS with 'one degree of freedom' can be interpreted as an IV method, see Gilson and Van den Hof (2001), Söderström et al. (1999), Garnier et al. (2000).

Relations between BELS and IV are discussed in Jia et al. (2002). The non-EIV situation is treated, and it is shown that the two classes of method are equivalent under certain conditions.

7.5.2 The Frisch Scheme

The roots of Frisch scheme appear in Frisch (1934), with reference to static problems; the extension to dynamic systems was proposed in Beghelli et al. (1990).

One of the first criteria that has shown remarkable robustness properties was originally proposed in Beghelli et al. (1993) and further in Diversi et al. (2004). This criterion relies on the shift properties of time-invariant systems and is based on rank deficiency conditions of the noise-free covariance matrix. For further treatments of Example 7.2, see Diversi et al. (2006).

The idea Frisch scheme variant described in Example 7.3 appeared in Diversi et al. (2003b).

Frisch schemes extended to handle diagonal bilinear models appear in Larkowski et al. (2009c). Frisch schemes for multivariable systems are treated in Diversi and Guidorzi (2009). The Frisch scheme combined with YW equations (a way to define the governing equations) for MIMO models is proposed and analyzed in Diversi and Guidorzi (2010). Identification of MIMO models is also presented in Guidorzi (1996). The extension of the Frisch scheme to multivariable systems is also treated in Diversi and Guidorzi (2017). In particular, it is described how the identification procedure relies on the association of the EIV models with directions in the noise space. A set of high-order Yule–Walker equations are used.

The paper Beghelli et al. (1997) deals with the Frisch scheme formulated in the frequency domain. Another frequency domain approach, somewhat inspired by the Frisch scheme, is developed in Söderström et al. (2003). Among more recent work using a frequency domain formulation of the Frisch scheme are Soverini and Söderström (2014a, 2015a).

A modified Frisch scheme that can handle correlated output noise was suggested in Söderström (2007b).

The paper Castaldi et al. (1995) deals with ARX models with additional measurement noise on the input and output sides. A related and more general paper handling noisy ARX and ARARX models is Diversi et al. (2010).

Some general comparison of the Frisch method and other alternative EIV identification methods appears in Söderström et al. (2002) and Soverini and Söderström (2000).

7.5.3 GIVE

The GIVE method was introduced in Söderström (2011). A somewhat related method was presented in Mahata (2007). The approach in Mahata (2007) is more general though. The vector $\mathbf{z}(t)$ is then not formed from a small set of delayed inputs and outputs, but both backward filtering and forward filtering of the data are exploited when forming $\mathbf{z}(t)$. The extended compensated least squares method, see Example 7.9, was presented in Ekman et al. (2006). Another early paper on using the principle of bias-compensation in the IV equations is Thil et al. (2008b). For further treatments, see also Larkowski et al. (2008b, c).

The paper Hong and Söderström (2009) demonstrates the equivalence between several GIVE cases.

The GIVE method for the MIMO case is first treated in Söderström (2012). The paper Diversi and Guidorzi (2012) compares covariances for equation errors using small lags in the MIMO case, similar to the treatment in Diversi et al. (2003b) for the SISO case.

It is shown in Diversi and Guidorzi (2013) how certain GIVE estimates can be reformulated as a quadratic eigenvalue problem, which in turn can be handled as a generalized eigenvalue problem. The paper Diversi et al. (2012) presents a GIVE and inspired Söderström et al. (2014a) to show how many GIVE cases can be extended to handle the case of correlated input and output noise.

For some alternative methods related to compensating the IV equations, see Ikenoue et al. (2008), Jun and Bernstein (2006). Some further studies of various bias-compensating approaches for system identification appear in Zhang et al. (2015), Diversi and Soverini (2015), and Arablouei et al. (2014).

Appendix

7.A Further Details

7.A.1 *Proof of Lemma 7.1*

Using Lemma A.3 in Söderström and Stoica (1989), the matrix $\mathbf{R}_{\bar{\bar{\varphi}}}$ in (7.56) is positive semidefinite with one eigenvalue in the origin precisely when the matrix

$$\overline{\mathbf{R}}_{\bar{\bar{\varphi}}} \stackrel{\Delta}{=} \hat{\mathbf{R}}_{\overline{\varphi}_y} - \hat{\lambda}_y \mathbf{I}_{n_a+1} - \hat{\mathbf{R}}_{\overline{\varphi}_y \varphi_u} \left(\hat{\mathbf{R}}_{\varphi_u} - \hat{\lambda}_u \mathbf{I}_{n_b} \right)^{-1} \hat{\mathbf{R}}_{\varphi_u \overline{\varphi}_y}$$

is positive semidefinite with one eigenvalue in the origin. However, by construction

$$\lambda_{\min}(\overline{\mathbf{R}}_{\bar{\bar{\varphi}}}) = -\hat{\lambda}_y + \lambda_{\min} \left(\hat{\mathbf{R}}_{\overline{\varphi}_y} - \hat{\mathbf{R}}_{\overline{\varphi}_y \varphi_u} \left(\hat{\mathbf{R}}_{\varphi_u} - \hat{\lambda}_u \mathbf{I}_{n_b} \right)^{-1} \hat{\mathbf{R}}_{\varphi_u \overline{\varphi}_y} \right) = 0 ,$$

where (7.55) is used. Hence, the smallest eigenvalue of $\overline{\mathbf{R}}_{\bar{\bar{\varphi}}}$ is in the origin. There is a need to check that the variance estimate $\hat{\lambda}_y$ is nonnegative. However,

$$\begin{aligned}
&\hat{\lambda}_y \geq 0 \Leftrightarrow \\
&\hat{\mathbf{R}}_{\overline{\varphi}_y} - \hat{\mathbf{R}}_{\overline{\varphi}_y \varphi_u} \left(\hat{\mathbf{R}}_{\varphi_u} - \hat{\lambda}_u \mathbf{I}_{n_b} \right)^{-1} \hat{\mathbf{R}}_{\varphi_u \overline{\varphi}_y} \text{ positive semidefinite } \Leftrightarrow \\
&\begin{pmatrix} \hat{\mathbf{R}}_{\overline{\varphi}_y} & \hat{\mathbf{R}}_{\overline{\varphi}_y \varphi_u} \\ \hat{\mathbf{R}}_{\varphi_u \overline{\varphi}_y} & \left(\hat{\mathbf{R}}_{\varphi_u} - \hat{\lambda}_u \mathbf{I}_{n_b} \right) \end{pmatrix} \text{ positive semidefinite } \Leftrightarrow \\
&\hat{\mathbf{R}}_{\varphi_u} - \hat{\lambda}_u \mathbf{I}_{n_b} - \hat{\mathbf{R}}_{\varphi_u \overline{\varphi}_y} \hat{\mathbf{R}}_{\overline{\varphi}_y}^{-1} \hat{\mathbf{R}}_{\overline{\varphi}_y \varphi_u} \text{ positive semidefinite ,}
\end{aligned}$$

which is true due to the given bound (7.54) on $\hat{\lambda}_u$. This completes the proof.

7.A.2 *Algorithm for the Canonical Form*

How would the algorithm in Sect. 7.4.3 modify for the alternative parameterization (5.69)? In this case it seems convenient to use the parameter vector $\overline{\boldsymbol{\theta}}$ rather than the matrix $\boldsymbol{\Theta}$ and to write the loss function $V(\overline{\boldsymbol{\theta}}, \boldsymbol{\rho})$ in the form (7.133).

It holds here

$$\mathbf{f} = \mathbf{f}(\overline{\boldsymbol{\theta}}, \rho) = \mathrm{vec}\left(\frac{1}{N}\sum_{t=1}^{N} \mathbf{z}(t)\mathbf{y}^T(t) - \frac{1}{N}\sum_{t=1}^{N} \mathbf{z}(t)\overline{\boldsymbol{\theta}}^T\overline{\boldsymbol{\Phi}}^T(t) - \mathbf{R}_{\tilde{\mathbf{z}}\tilde{\mathbf{y}}}(\rho) - \mathsf{E}\left\{\tilde{\mathbf{z}}(t)\overline{\boldsymbol{\theta}}^T\tilde{\overline{\boldsymbol{\Phi}}}^T(t)\right\}\right)$$

$$= \frac{1}{N}\sum_{t=1}^{N} \mathbf{y}(t) \otimes \mathbf{z}(t) - \left(\frac{1}{N}\sum_{t=1}^{N} \overline{\boldsymbol{\Phi}}^T(t) \otimes \mathbf{z}(t)\right)\overline{\boldsymbol{\theta}} - \mathrm{vec}\left(\mathbf{R}_{\tilde{\mathbf{z}}\tilde{\mathbf{y}}}(\rho)\right) - \mathsf{E}\left\{\tilde{\overline{\boldsymbol{\Phi}}}(t) \otimes \tilde{\mathbf{z}}(t)\right\}\overline{\boldsymbol{\theta}}$$

$$\stackrel{\Delta}{=} \bar{\mathbf{g}}(\rho) - \bar{\mathbf{H}}(\rho)\overline{\boldsymbol{\theta}}\ , \tag{7.141}$$

where Lemma A.3 was used.

Next minimize

$$V(\overline{\boldsymbol{\theta}}, \rho) = \bar{\mathbf{f}}^T\overline{\mathbf{W}}\bar{\mathbf{f}} = \left[\bar{\mathbf{g}}^T(\rho) - \overline{\boldsymbol{\theta}}^T\bar{\mathbf{H}}^T(\overline{\rho})\right]^T \overline{\mathbf{W}}\left[\bar{\mathbf{g}}(\rho) - \bar{\mathbf{H}}(\overline{\rho})\overline{\boldsymbol{\theta}}\right] \tag{7.142}$$

with respect to $\overline{\boldsymbol{\theta}}$ subject to the constraint

$$\mathbf{S}\overline{\boldsymbol{\theta}} = \mathbf{0}\ . \tag{7.143}$$

Using Lagrangian multipliers, the augmented criterion becomes (omitting the argument $\overline{\rho}$ for convenience)

$$L(\overline{\boldsymbol{\theta}}, \rho) = \frac{1}{2}\left[\bar{\mathbf{g}}^T - \overline{\boldsymbol{\theta}}^T\bar{\mathbf{H}}^T\right]\overline{\mathbf{W}}\left[\bar{\mathbf{g}} - \bar{\mathbf{H}}\overline{\boldsymbol{\theta}}\right] + \boldsymbol{\lambda}^T\mathbf{S}\overline{\boldsymbol{\theta}}\ . \tag{7.144}$$

Setting the derivatives of L with respect to $\overline{\boldsymbol{\theta}}$ and $\boldsymbol{\lambda}$ to zero leads directly to

$$\begin{pmatrix} \bar{\mathbf{H}}^T\overline{\mathbf{W}}\bar{\mathbf{H}} & \mathbf{S}^T \\ \mathbf{S} & \mathbf{0} \end{pmatrix}\begin{pmatrix} \overline{\boldsymbol{\theta}} \\ \boldsymbol{\lambda} \end{pmatrix} = \begin{pmatrix} \bar{\mathbf{H}}^T\overline{\mathbf{W}}\bar{\mathbf{g}} \\ \mathbf{0} \end{pmatrix}\ . \tag{7.145}$$

Then it is straightforward to compute

$$V_2(\rho) = V(\overline{\boldsymbol{\theta}}(\rho), \rho)\ , \tag{7.146}$$

where $\overline{\boldsymbol{\theta}}(\rho)$ is obtained from the solution to (7.145).

Chapter 8
Covariance Matching

Among the different methods for errors-in-variables problems there are several ones that are based on a number of covariances of the noisy input–output data. The methods are typically based on the compensated least squares principle. In many cases, the equations are bilinear in the unknown parameters. Several examples were given in Chap. 7.

An alternative method is proposed and analyzed in this chapter. The starting point is not the normal equations of least squares, but a small set of covariance elements of the measured data. These covariance elements are compared to what can be expected from the underlying model, and this idea leads to equations defining the parameter estimates. The noise-free input is assumed to be a stationary process, but no specific type of model for its statistical properties is used.

The chapter is organized as follows. In the next section the basic ideas of covariance matching are described. The formal description of the method appears in Sect. 8.2. Some details are given in Sect. 8.3, while user choices of the method are summarized and discussed in Sect. 8.4. It turns out that the covariance matrix approach is closely linked to confirmatory factor analysis (CFA), which was considered for static models in Sect. 2.2. Details for how CFA can be used for dynamic models are given in Sect. 8.5. The chapter is concluded in Sect. 8.6 with extensions and bibliographical notes.

8.1 The Basic Idea of Covariance Matching

To describe the basic idea of covariance matching (CM), consider a first-order system

$$y_0(t) + ay_0(t-1) = bu_0(t-1) , \tag{8.1}$$

T. Söderström, *Errors-in-Variables Methods in System Identification*, Communications and Control Engineering,
https://doi.org/10.1007/978-3-319-75001-9_8

and assume that it is identified with GIVE with the instrumental variable vector

$$\mathbf{z}(t) = \begin{pmatrix} y(t) \\ y(t-1) \\ u(t-1) \\ u(t-2) \end{pmatrix} \tag{8.2}$$

for the case of white input and white output noise. The equations for determining the four unknowns $a,\ b,\ \lambda_y$, and λ_u are then

$$\begin{cases} \hat{r}_y(0) + a\hat{r}_y(1) - b\hat{r}_{yu}(1) = \lambda_y \ , \\ \hat{r}_y(1) + a\hat{r}_y(0) - b\hat{r}_{yu}(0) = a\lambda_y \ , \\ \hat{r}_{yu}(1) + a\hat{r}_{yu}(0) - b\hat{r}_u(0) = -b\lambda_u \ , \\ \hat{r}_{yu}(2) + a\hat{r}_{yu}(1) - b\hat{r}_u(1) = 0 \ . \end{cases} \tag{8.3}$$

The information in the data is first condensed into the covariance elements $\hat{r}_y(0)$, $\hat{r}_y(1),\ \hat{r}_{yu}(0),\ \hat{r}_{yu}(1),\ \hat{r}_{yu}(2),\ \hat{r}_u(0),\ \hat{r}_u(1)$. It is then of interest to explore if there are other ways to infer the parameters from these covariance elements. If so, is there any way that gives particularly accurate estimates?

To cope with this issue, introduce the signal

$$z_0(t) = \frac{1}{1 + aq^{-1}} u_0(t) \ . \tag{8.4}$$

Consider the asymptotic case ($N \to \infty$) and use

$$\begin{aligned} y(t) &= y_0(t) + \tilde{y}(t) = bz_0(t-1) + \tilde{y}(t) \ , \\ u(t) &= u_0(t) + \tilde{u}(t) = z_0(t) + az_0(t-1) + \tilde{u}(t) \ . \end{aligned} \tag{8.5}$$

Set

$$R_k = \mathsf{E}\left\{z_0(t+k)z_0(t)\right\}, \quad k \geq 0 \ . \tag{8.6}$$

One finds by straightforward calculations

$$\begin{aligned} r_y(0) &= b^2 R_0 + \lambda_y \ , \\ r_y(1) &= b^2 R_1 \ , \\ r_{yu}(0) &= bR_1 + abR_0 \ , \\ r_{yu}(1) &= bR_0 + abR_1 \ , \\ r_{yu}(2) &= bR_1 + abR_2 \ , \\ r_u(0) &= (1 + a^2)R_0 + 2aR_1 + \lambda_u \ , \\ r_u(1) &= aR_0 + (1 + a^2)R_1 + aR_2 \ . \end{aligned} \tag{8.7}$$

The elements in the left-hand sides of (8.7) can all be estimated from data. The seven equations in (8.7) contain seven unknowns $(a, b, \lambda_y, \lambda_u, R_0, R_1, R_2)$. Unless there is a particular interest in determining λ_y and λ_u, one can exclude the equations for

$r_y(0)$ and $r_u(0)$ and thereby reduce the number of equations and unknowns to five. To solve this system of equations is the covariance matching approach.

Note that

- The coefficients R_0, R_1, R_2 are auxiliary unknowns.
- The parameterization implied by (8.4) and (8.5) allows an arbitrary type of stationary process for the noise-free input $u_0(t)$.
- In principle it is possible to derive additional equations by exploiting also other covariance elements. A so obtained system of equations may be solved in least squares sense.

The system of equations (dropping those involving λ_y and λ_u) can be written as

$$\begin{pmatrix} r_y(1) \\ r_{yu}(0) \\ r_{yu}(1) \\ r_{yu}(2) \\ r_u(1) \end{pmatrix} = \begin{pmatrix} 0 & b^2 & 0 \\ ab & b & 0 \\ b & ab & 0 \\ 0 & b & ab \\ a & 1+a^2 & a \end{pmatrix} \begin{pmatrix} R_0 \\ R_1 \\ R_2 \end{pmatrix} . \tag{8.8}$$

In the next section the CM approach is introduced and described for the general case.

8.2 The Covariance Matching Method

Using the general EIV setup laid out in Chaps. 3 and 5, define

$$z_0(t) = \frac{1}{A(q^{-1})} u_0(t) . \tag{8.9}$$

This means that both $u_0(t)$ and $y_0(t)$ can be expressed in terms of $z_0(t)$ and that

$$u(t) = u_0(t) + \tilde{u}(t) = A(q^{-1}) z_0(t) + \tilde{u}(t) , \tag{8.10}$$

$$y(t) = y_0(t) + \tilde{y}(t) = B(q^{-1}) z_0(t) + \tilde{y}(t) . \tag{8.11}$$

It is a key issue that both $u_0(t)$ and $y_0(t)$ are obtained by finite impulse response (FIR) filtering (and not IIR filtering) of $z_0(t)$, as in (8.10) and (8.11).

Using (8.10) and (8.11) the covariance function of the measured input $u(t)$ can be written as

$$r_u(\tau) = \mathrm{E}\{u(t+\tau)u(t)\} = \sum_{i=0}^{n_a} \sum_{j=0}^{n_a} a_i a_j r_0(\tau - i + j) \tag{8.12}$$

for $\tau > 0$, where $a_0 = 1$ and

$$r_0(\tau) = \mathrm{E}\{z_0(t+\tau)z_0(t)\} \, . \tag{8.13}$$

Similarly,

$$r_y(\tau) = \mathrm{E}\{y(t+\tau)y(t)\} \sum_{i=1}^{n_b}\sum_{j=1}^{n_b} b_i b_j r_0(\tau - i + j) \tag{8.14}$$

for $\tau > 0$, and

$$r_{yu}(\tau) = \mathrm{E}\{y(t+\tau)u(t)\} = \sum_{i=1}^{n_b}\sum_{j=0}^{n_a} b_i a_j r_0(\tau - i + j) \, . \tag{8.15}$$

for arbitrary values of τ.

In the covariance matching approach, the idea is to regard $\boldsymbol{\theta}$ and the covariance elements $\{r_0(\tau)\}$ as unknowns that are fitted to some specified input–output covariance elements, as in (8.12), (8.14), and (8.15). The approach thus means that first the covariance vectors

$$\mathbf{r}_y = \begin{pmatrix} r_y(1) \\ \vdots \\ r_y(p_y) \end{pmatrix}, \quad \mathbf{r}_u = \begin{pmatrix} r_u(1) \\ \vdots \\ r_u(p_u) \end{pmatrix}, \quad \mathbf{r}_{yu} = \begin{pmatrix} r_{yu}(p_1) \\ \vdots \\ r_{yu}(p_2) \end{pmatrix} \tag{8.16}$$

are estimated. Here, the integers p_y, p_u, p_1, and p_2 are treated as user parameters. The choice of these parameters is discussed in Sect. 8.4 below. When the output is assumed to be corrupted by white output noise, one can choose $p_y \geq 1$. In the extended case of correlated output noise, no vector $\mathbf{r}_y$ at all should be used. As the input is assumed to be subject to white measurement noise, the condition $p_u \geq 1$ should apply.

To keep the situation general one may use the convention that $p_y = 0$ means that no vector $\mathbf{r}_y$ is utilized. The choice $p_u = 0$ can be used in an analogous way. The integers p_1 and p_2 can be chosen more freely, but must by construction fulfill $p_1 \leq p_2$. It is a common experience that covariance elements tend to be more informative when the lag is small. Thus, it is beneficial to include the covariance element $r_{yu}(0)$. Note that (8.15) holds also for $\tau = 0$. Assume hence for convenience that

$$p_1 \leq 0 \leq p_2 \, . \tag{8.17}$$

Now introduce a number of notational conventions. The covariance vector of the noise-free signal $z_0(t)$ is introduced as

$$\mathbf{r}_z = \begin{pmatrix} r_0(0) \\ \vdots \\ r_0(k) \end{pmatrix}, \tag{8.18}$$

where the maximal lag k is to be specified later. Further, introduce the conventions

$$a_i = \begin{cases} 1 & \text{if } i = 0 \,, \\ 0 & \text{if } i < 0, i > n_a \,, \end{cases} \tag{8.19}$$

$$b_i = 0 \quad \text{if } i < 1, i > n_b \,. \tag{8.20}$$

The vector of known covariance elements to be utilized is taken as

$$\mathbf{r} \overset{\Delta}{=} \begin{pmatrix} \mathbf{r}_y \\ \mathbf{r}_u \\ \mathbf{r}_{yu} \end{pmatrix} . \tag{8.21}$$

From (8.12), (8.14), and (8.15) it is possible to write

$$\mathbf{r} = \mathbf{F}(\boldsymbol{\theta})\mathbf{r}_z = \begin{pmatrix} \mathbf{F}_y(\boldsymbol{\theta}) \\ \mathbf{F}_u(\boldsymbol{\theta}) \\ \mathbf{F}_{yu}(\boldsymbol{\theta}) \end{pmatrix} \mathbf{r}_z \,. \tag{8.22}$$

It will soon be described how to express $\mathbf{F}(\boldsymbol{\theta})$ for arbitrary values of $\boldsymbol{\theta}$. The matrix elements are quadratic functions in $\boldsymbol{\theta}$. See Sect. 8.3 for details.

Remark 8.1 A related but somewhat different approach has been previously described, for example, in Mossberg (2008), Söderström et al. (2006). In those references it was assumed that the noise-free input can be described as an ARMA model, cf. Assumption **AI4**. The ARMA parameters were estimated together with the unknown system parameters. This differs from the approach here where it is not specified any particular type of model for the statistical properties of $u_0(t)$. ■

The identification approach is to first estimate $\mathbf{r}$ from data, which is a straightforward task, cf. (8.25) below. Then $\boldsymbol{\theta}$ and $\mathbf{r}_z$ are determined by considering the system of equations given by (8.22). Here, of course, $\boldsymbol{\theta}$ is the interesting quantity to determine, while the elements of $\mathbf{r}_z$ have the role of being auxiliary variables.

Based on the system of equations implied by (8.22), the estimator

$$\{\hat{\boldsymbol{\theta}}, \hat{\mathbf{r}}_z\} = \arg\min_{\boldsymbol{\theta}, \mathbf{r}_z} \; J(\boldsymbol{\theta}, \mathbf{r}_z) \,, \tag{8.23}$$

with

$$J(\boldsymbol{\theta}, \mathbf{r}_z) = ||\hat{\mathbf{r}} - \mathbf{F}(\boldsymbol{\theta})\mathbf{r}_z||^2_{\mathbf{W}} \,, \tag{8.24}$$

can be defined for determining $\boldsymbol{\theta}$ and $\mathbf{r}_z$. Here $\mathbf{W}$ is a symmetric and positive definite weighting matrix, and $\hat{\mathbf{r}}$ is defined as in (8.21) but with the covariance and cross-covariance function elements replaced by estimates. For example, one can take a typical element of $\hat{\mathbf{r}}$ as

$$\hat{r}_y(\tau) = \frac{1}{N} \sum_{t=1}^{N} y(t+\tau)y(t) \tag{8.25}$$

but also other normalizations, such as replacing $1/N$ by $1/(N-\tau)$, are possible. From the separable least squares problem (8.23) and (8.24)

$$\hat{\mathbf{r}}_z = \left(\mathbf{F}^T(\boldsymbol{\theta})\mathbf{W}\mathbf{F}(\boldsymbol{\theta})\right)^{-1}\mathbf{F}^T(\boldsymbol{\theta})\mathbf{W}\hat{\mathbf{r}} \, , \tag{8.26}$$

and finally

$$\hat{\boldsymbol{\theta}} = \arg\min_{\boldsymbol{\theta}} V(\boldsymbol{\theta}) \tag{8.27}$$

$$V(\boldsymbol{\theta}) = \hat{\mathbf{r}}^T\mathbf{W}\hat{\mathbf{r}} - \hat{\mathbf{r}}^T\mathbf{W}\mathbf{F}(\boldsymbol{\theta})\left(\mathbf{F}^T(\boldsymbol{\theta})\mathbf{W}\mathbf{F}(\boldsymbol{\theta})\right)^{-1}\mathbf{F}^T(\boldsymbol{\theta})\mathbf{W}\hat{\mathbf{r}} \, . \tag{8.28}$$

A number of remarks on the implementation aspects are now in order.

Remark 8.2 The optimization problem in (8.27) is a variable projection problem; see Appendix A.1.4. It can be solved with standard optimization algorithms. It is though possible in this case to provide explicit expressions for the gradient, as the dependence of $\mathbf{F}(\boldsymbol{\theta})$ on $\boldsymbol{\theta}$ has a relatively straightforward structure. Such algorithms are discussed further in Sect. 13.4. ■

Remark 8.3 Note that for the algorithm (8.27) and (8.28) to work, it is necessary that the matrix $\mathbf{F}(\boldsymbol{\theta})$ has full column rank. This issue is further discussed in Appendix 8.A.1. ■

Remark 8.4 In a practical implementation one should use a QR factorization of $\mathbf{F}(\boldsymbol{\theta})$ to ensure that the corresponding projection in (8.28) is guaranteed to stay nonnegative definite. ■

Remark 8.5 It seems to be an open issue whether or not the loss function defined in (8.28) can have multiple local minima. In numerical studies so far reported problems with local minima have not been encountered. ■

Remark 8.6 The framework presented above can be extended to include $r_y(0)$ and $r_u(0)$. In that case, write

$$\hat{\mathbf{r}}_2 = \begin{pmatrix} r_y(0) \\ r_u(0) \end{pmatrix} = \begin{pmatrix} \mathbf{F}_2(\boldsymbol{\theta}) & \mathbf{I} \end{pmatrix} \begin{pmatrix} \mathbf{r}_z \\ \boldsymbol{\rho} \end{pmatrix} , \tag{8.29}$$

$$\boldsymbol{\rho} = \begin{pmatrix} \lambda_y & \lambda_u \end{pmatrix}^T . \tag{8.30}$$

The estimate $\hat{\boldsymbol{\theta}}$ will then be the same as the one obtained in (8.27). Note that in (8.29) two more equations and two more unknowns are added as compared to (8.27). Use of (8.12) and (8.14) for $\tau = 0$ would not lead to other estimates of $\boldsymbol{\theta}$, as there is no new information about $\boldsymbol{\theta}$ to infer. Still, such a consideration may have its value, when comparing the covariance matching approach to other methods such as GIVE. Recall that GIVE exploits the covariances $r_y(0)$ and $r_u(0)$ when determining the parameter vector $\boldsymbol{\theta}$. More details are given in Söderström et al. (2009). ■

So far in the book some different methods are presented, which are all based on a finite-dimensional $\hat{\mathbf{r}}$ as a starting point. A few general comments can be made on the relations between the approaches.

1. **The Covariance Matching (CM) Approach**. A quite general form of the noise-free input is allowed as long as it is a stationary process. The underlying equations lead to an optimization of order $n_a + n_b$. The CM method is hence somewhat more computationally demanding than the GIVE methods. There is a specific parameterization of the spectrum of the noise-free input that allows arbitrary spectra and still leads to a low-dimensional optimization problem. The spectrum of the noise-free input is parameterized via the vector $\mathbf{r}_z$ in (8.18), which describes the influence from an arbitrary, stationary process $u_0(t)$. Note that the vector $\mathbf{r}_z$ has a low dimension due to the specific structure of the problem. As will be shown later in Example 8.1, CM gives in general much more accurate estimates than GIVE. In CM, a more efficient use of the information in $\hat{\mathbf{r}}$ is made, as compared to GIVE.
2. For **GIVE methods** such as **bias-eliminating least squares (BELS)**, Zheng (1998a, 2002a), and **the Frisch scheme**, Beghelli et al. (1990), Diversi et al. (2003b), the underlying equations are relatively simple and lead to a low-dimensional (say dimension 2) optimization problem. There is no modeling at all of the noise-free input. An arbitrary form of the noise-free input is allowed.

8.3 Expressions for the Covariance Elements

To show how the matrix $\mathbf{F}(\boldsymbol{\theta})$ in (8.22) explicitly depends on $\boldsymbol{\theta}$, start with $\mathbf{F}_y(\boldsymbol{\theta})$. For this purpose, it is convenient to introduce

$$\beta_j = \sum_{i=1}^{n_b} b_i b_{i+j} \ . \tag{8.31}$$

It follows from (8.20) that

$$\beta_j = 0 \text{ if either } j \leq -n_b \text{ or } j \geq n_b \ . \tag{8.32}$$

Then write (8.14) as

$$r_y(\tau) = \sum_{j=-n_b+1}^{n_b-1} \beta_j r_0(\tau + j) \ , \tag{8.33}$$

and therefore

$$\mathbf{r}_y = \begin{pmatrix} r_y(1) \\ \vdots \\ r_y(p_y) \end{pmatrix} = \begin{pmatrix} \beta_{-n_b+1} & \cdots & \beta_{n_b-1} & & \mathbf{0} \\ & \ddots & & \ddots & \\ \mathbf{0} & & \beta_{-n_b+1} & \cdots & \beta_{n_b-1} \end{pmatrix} \begin{pmatrix} r_0(2-n_b) \\ \vdots \\ r_0(p_y+n_b-1) \end{pmatrix}$$

$$= \begin{pmatrix} \beta_{-n_b+1} & \cdots & \beta_{n_b-1} & & \mathbf{0} \\ & \ddots & & \ddots & \\ \mathbf{0} & & \beta_{-n_b+1} & \cdots & \beta_{n_b-1} \end{pmatrix} \begin{pmatrix} \mathbf{0} & & 1 & & \\ & \cdot^{\cdot^{\cdot}} & & & \\ 1 & & \mathbf{0} & & \\ & \ddots & & & \\ & & & \ddots & \\ \mathbf{0} & & & & 1 \end{pmatrix} \begin{pmatrix} r_0(0) \\ \vdots \\ r_0(p_y+n_b-1) \end{pmatrix}$$

$$\triangleq \mathbf{T}_y(\boldsymbol{\theta})\mathbf{P}_y\mathbf{r}_z \ . \tag{8.34}$$

Here, the matrix $\mathbf{T}_y(\boldsymbol{\theta})$ has dimension $p_y \times (p_y + 2n_b - 2)$ and the dimension of $\mathbf{P}_y$ is $(p_y + 2n_b - 2) \times (p_y + n_b)$. In the matrix $\mathbf{P}_y$ all elements are zero, except one element being equal to one in each row. In the first row, the element in column $n_b - 1$ is equal to one. It now holds

$$\mathbf{F}_y(\boldsymbol{\theta}) = \mathbf{T}_y(\boldsymbol{\theta})\mathbf{P}_y \ . \tag{8.35}$$

The minimal dimension of $\mathbf{r}_z$ in (8.22) may be different for $\mathbf{r}_y$, $\mathbf{r}_u$, and $\mathbf{r}_{yu}$. To handle this, set

$$\mathbf{F}_y(\boldsymbol{\theta}) = \begin{pmatrix} \mathbf{T}_y(\boldsymbol{\theta})\mathbf{P}_y & \mathbf{0} \end{pmatrix} \ . \tag{8.36}$$

to account for the possibly different lengths of $\mathbf{r}_z$ compatible with the vectors $\mathbf{r}_y$, $\mathbf{r}_u$, and $\mathbf{r}_{yu}$. The null matrix in (8.36) has $k - p_y - n_b + 1$ columns.

Next consider the matrix $\mathbf{F}_u(\boldsymbol{\theta})$. The procedure is analogous to the one above. First introduce

$$\alpha_j = \sum_{i=0}^{n_a} a_i a_{i+j} \ . \tag{8.37}$$

It follows from (8.19) that

$$\alpha_j = 0 \text{ if either } j < -n_a \text{ or } j > n_a \ . \tag{8.38}$$

Then write (8.12) as

$$r_u(\tau) = \sum_{j=-n_a}^{n_a} \alpha_j r_0(\tau + j) \ , \tag{8.39}$$

and therefore

$$\mathbf{r}_u = \begin{pmatrix} r_u(1) \\ \vdots \\ r_u(p_u) \end{pmatrix} = \begin{pmatrix} \alpha_{-n_a} & \dots & \alpha_{n_a} & & \mathbf{0} \\ & \ddots & & \ddots & \\ \mathbf{0} & & \alpha_{-n_a} & \dots & \alpha_{n_a} \end{pmatrix} \begin{pmatrix} r_0(1-n_a) \\ \vdots \\ r_0(p_u+n_a) \end{pmatrix}$$

$$= \begin{pmatrix} \alpha_{-n_a} & \dots & \alpha_{n_a} & & \mathbf{0} \\ & \ddots & & \ddots & \\ \mathbf{0} & & \alpha_{-n_a} & \dots & \alpha_{n_a} \end{pmatrix} \begin{pmatrix} \mathbf{0} & 1 & & \\ & \cdot^{\cdot^{\cdot}} & & \\ 1 & \mathbf{0} & & \\ & \ddots & & \\ & & \ddots & \\ \mathbf{0} & & & 1 \end{pmatrix} \begin{pmatrix} r_0(0) \\ \vdots \\ r_0(p_u+n_a) \end{pmatrix}$$

$$\stackrel{\Delta}{=} \mathbf{T}_u(\boldsymbol{\theta})\mathbf{P}_u\mathbf{r}_z \ . \tag{8.40}$$

Here, the matrix $\mathbf{T}_u(\boldsymbol{\theta})$ has dimension $p_u \times (p_u + 2n_a)$ and the dimension of $\mathbf{P}_u$ is $(p_u + 2n_a) \times (p_u + n_a + 1)$. In the matrix $\mathbf{P}_u$ all elements are zero, except one element being equal to one in each row. In the first row, the element in column n_a is equal to one. It now holds

$$\mathbf{F}_u(\boldsymbol{\theta}) = \mathbf{T}_u(\boldsymbol{\theta})\mathbf{P}_u \ , \tag{8.41}$$

or more generally

$$\mathbf{F}_u(\boldsymbol{\theta}) = \begin{pmatrix} \mathbf{T}_u(\boldsymbol{\theta})\mathbf{P}_u & \mathbf{0} \end{pmatrix} \tag{8.42}$$

to account for the possibly different lengths of $\mathbf{r}_z$ compatible with the vectors $\mathbf{r}_y$, $\mathbf{r}_u$, and $\mathbf{r}_{yu}$. The null matrix in (8.42) has $k - p_u - n_a$ columns.

Finally, consider the matrix $\mathbf{F}_{yu}(\boldsymbol{\theta})$. The procedure follows the same line of logic as before. Introduce

$$\gamma_j = \sum_{i=1}^{n_b} b_i a_{i+j} \ . \tag{8.43}$$

It follows from (8.19) and (8.20) that

$$\gamma_j = 0 \text{ if either } j < -n_b \text{ or } j \geq n_a \ . \tag{8.44}$$

One can then write (8.15) as

$$r_{yu}(\tau) = \sum_{j=-n_b}^{n_a-1} \gamma_j r_0(\tau + j) \ , \tag{8.45}$$

and therefore

$$\mathbf{r}_{yu} = \begin{pmatrix} r_{yu}(p_1) \\ \vdots \\ r_{yu}(p_2) \end{pmatrix} = \begin{pmatrix} \gamma_{-n_b} & \cdots & \gamma_{n_a-1} & & \mathbf{0} \\ & \ddots & & \ddots & \\ \mathbf{0} & & \gamma_{-n_b} & \cdots & \gamma_{n_a-1} \end{pmatrix} \begin{pmatrix} r_0(p_1 - n_b) \\ \vdots \\ r_0(p_2 + n_a - 1) \end{pmatrix}$$

$$= \begin{pmatrix} \gamma_{-n_b} & \cdots & \gamma_{n_a-1} & & \mathbf{0} \\ & \ddots & & \ddots & \\ \mathbf{0} & & \gamma_{-n_b} & \cdots & \gamma_{n_a-1} \end{pmatrix} \begin{pmatrix} \mathbf{0} & & 1 \\ & \cdot^{\cdot^{\cdot}} & \\ 1 & & \mathbf{0} \\ & \ddots & \\ \mathbf{0} & & 1 \end{pmatrix} \begin{pmatrix} r_0(0) \\ \vdots \\ r_0(k_{yu}) \end{pmatrix}$$

$$\stackrel{\Delta}{=} \mathbf{T}_{yu}(\boldsymbol{\theta})\mathbf{P}_{yu}\mathbf{r}_z \ . \tag{8.46}$$

The lag k_{yu} in (8.46) is given by

$$k_{yu} = \max(-p_1 + n_b, p_2 + n_a - 1) \ . \tag{8.47}$$

In (8.46), the matrix $\mathbf{T}_{yu}(\boldsymbol{\theta})$ is $(-p_1 + p_2 + 1) \times (p_2 - p_1 + n_b + n_a)$ and the dimension of $\mathbf{P}_{yu}$ is $(p_2 - p_1 + n_b + n_a) \times (k_{yu} + 1)$. In the matrix $\mathbf{P}_{yu}$ all elements are zero, except one element being equal to one in each row. In the first row, the element in column $n_b - p_1 + 1$ is equal to one. Depending on the values of p_1 and p_2 it may happen that this element appears in the rightmost column.

If (8.17) is relaxed to

$$p_1 \leq p_2 \ , \tag{8.48}$$

then for the case $p_1 > 0$, Eq. (8.46) still applies, but $\mathbf{P}_{yu}$ has then to be modified to

$$\mathbf{P}_{yu} = \begin{pmatrix} \mathbf{0} & 1 & & \\ & & \ddots & \\ & & & 1 \end{pmatrix} , \tag{8.49}$$

where the nonzero element of the first row appears at position $p_1 + 1$. It now holds

$$\mathbf{F}_{yu}(\boldsymbol{\theta}) = \mathbf{T}_{yu}(\boldsymbol{\theta})\mathbf{P}_{yu} \ , \tag{8.50}$$

or

$$\mathbf{F}_{yu}(\boldsymbol{\theta}) = \begin{pmatrix} \mathbf{T}_{yu}(\boldsymbol{\theta})\mathbf{P}_{yu} & \mathbf{0} \end{pmatrix} \tag{8.51}$$

to account for the possibly different lengths of $\mathbf{r}_0$ compatible with the vectors $\mathbf{r}_y$, $\mathbf{r}_u$, and $\mathbf{r}_{yu}$. The null matrix in (8.51) has $k - k_{yu}$ columns.

8.4 User Choices in the Algorithm

In this section some aspects on the user choices in the proposed algorithm are discussed.

8.4.1 General Aspects

It may be worthwhile to emphasize the 'structure' of the problem, and let it be defined as the quadruple of integers

$$\mathbf{p} = \left(p_y \; p_u \; p_1 \; p_2 \right)^T . \tag{8.52}$$

Note that the choice of structure is certainly in the hands of the user. How should $\mathbf{p}$ be chosen? What aspects should be considered in that choice? Below some ideas are briefly discussed.

1. The choice of $\mathbf{p}$ must of course be such that a sufficient number of equations are obtained. This leads to some specific necessary conditions; see (8.57) below.
2. It is by experience often an advantage to use covariance elements with small τ arguments. The main reason is that then the estimated covariances deviate only a little from their true values. For 'large' values of τ, the accuracy of an estimated covariance can be quite low.
3. One should avoid making the elements of $\mathbf{p}$ too large, in the sense that $\hat{\mathbf{r}}$ with many elements would then increase the dimension of the problem. However, the essential optimization problem will be of order $n_a + n_b$, and the dimension of $\mathbf{r}_z$ is really of no particular concern. Though, the quantities in the loss function $V(\boldsymbol{\theta})$ in (8.28) have higher dimension when $\mathbf{p}$ has large elements. One should also avoid choosing the elements of $\mathbf{p}$ too small. In such situations the accuracy of the estimates may be relatively poor. Further, it is more likely to experience numerical troubles in the optimization of $V(\boldsymbol{\theta})$ in (8.28).
4. If more than the minimal number of equations is formed, one may consider active weighting in the loss function (8.28) by choosing $\mathbf{W} \neq \mathbf{I}$. An optimal weighting is

$$\mathbf{W} = \mathbf{R}^{-1} , \tag{8.53}$$

   where $\mathbf{R}$ is the covariance matrix of the vector $\hat{\mathbf{r}}$; see (14.135).
5. If one expects the measurement noise in y, or in both u and y to be correlated, the appropriate choices in $\mathbf{p}$ include to take $p_y = 0$, or $p_y = p_u = 0$, respectively.

8.4.2 Compatibility and Identifiability Conditions

There are at least two questions of relevance for the properties of the problem of solving (8.22), namely *compatibility* and *identifiability*.

In order to be able to solve the system of equations given in (8.22), it is necessary that there are at least as many equations as unknowns. This consideration leads to the *compatibility condition*

$$\dim(\mathbf{r}_y) + \dim(\mathbf{r}_u) + \dim(\mathbf{r}_{yu}) \geq \dim(\boldsymbol{\theta}) + \dim(\mathbf{r}_z) \,, \tag{8.54}$$

which is easily rewritten into

$$p_y + p_u + (p_2 - p_1 + 1) \geq n_a + n_b + k + 1 \,. \tag{8.55}$$

The value of the integer k is to be chosen as the maximum number that appears in the lags of $r_0(\tau)$ in the expressions (8.34), (8.40), and (8.46). This turns out to give

$$k = \max(p_y + n_b - 1, \, p_u + n_a, \, \max(-p_1 + n_b, \, p_2 + n_a - 1)) \,. \tag{8.56}$$

Therefore, the compatibility condition (8.55) becomes

$$\begin{aligned} &p_y + p_u + p_2 - p_1 - n_a - n_b \geq \\ &\max(p_y + n_b - 1, \, p_u + n_a, \, -p_1 + n_b, \, p_2 + n_a - 1) \,. \end{aligned} \tag{8.57}$$

One may ask if there is a natural choice of the user parameters p_y, p_u, p_1, and p_2 to satisfy the condition (8.57). One can see that if p_y, p_u, and p_2 are all chosen large enough the condition will be fulfilled. On the other hand, large values of these integers will imply that the dimension of $\hat{\mathbf{r}}$ is large, and that the computational load is increased.

Example 8.1 The parameter vector $\boldsymbol{\theta}$ in (8.22) contains $n_a + n_b$ parameters. This may be taken as a motivation of using a 'horizon' of $n_a + n_b$, meaning that covariance lags τ are chosen so that $|\tau| \leq n_a + n_b$. This idea leads to the choices

$$p_1 = 0, \quad p_u = p_y = p_2 = n_a + n_b \,. \tag{8.58}$$

With these choices one gets for (8.57):

$$\begin{aligned} \text{LHS} &= p_y + p_u + p_2 - p_1 - n_a - n_b \\ &= 2n_a + 2n_b \,, \\ \text{RHS} &= \max(n_a + 2n_b - 1, 2n_a + n_b, n_b, 2n_a + n_b - 1) \\ &= \max(n_a + 2n_b - 1, 2n_a + n_b) < \text{LHS} \,, \end{aligned}$$

so the necessary compatibility condition (8.57) is satisfied. ■

Example 8.2 Consider the case when one wants to allow correlated noise on both the input and output sides. Then no elements of $\mathbf{r}_y$ and $\mathbf{r}_u$ should be used in $\mathbf{r}$. This means that the user makes the choices

$$p_y = 0, \quad p_u = 0 , \tag{8.59}$$

which just is a convention to state that only the covariance vector $\mathbf{r}_{yu}$ is used in $\mathbf{r}$. For this particular case, it turns out that the condition (8.57) becomes

$$p_2 - p_1 - n_a - n_b \geq \max(n_b - 1, n_a, -p_1 + n_b, p_2 + n_a - 1) , \tag{8.60}$$

which turns out to be equivalent to

$$p_1 \leq -2n_a - n_b + 1 , \tag{8.61}$$
$$p_2 \geq n_a + 2n_b . \tag{8.62}$$

As some covariance elements $r_{yu}(\tau)$ have large lags (τ) as compared to white measurement noise and the use of $\mathbf{r}_y$ and $\mathbf{r}_u$, potentially a degraded accuracy may be achieved. ■

Remark 8.7 Note that (8.57) can be seen as a necessary condition for identifiability; when it is violated, there is no hope that one can determine $\boldsymbol{\theta}$ uniquely for the given set of covariance elements. ■

Example 8.3 In this example the covariance matching approach is considered and compared to the behavior of the GIVE estimate. The user parameters of the CM approach were chosen as

$$p_y = 2, \, p_u = 2, \, p_1 = -2, \, p_2 = 2 .$$

This means that almost the same covariance elements $\{r_y(\tau), r_u(\tau), r_{yu}(\tau)\}$ are used in both GIVE and CM. The difference is that for CM also the element $r_{yu}(-2)$ is used as well, just in order to get identifiability for the CM method.

The obtained results using Monte Carlo simulations with the standard data sets are displayed in Table 8.1.

One can see that the CM estimates are generally much more accurate than those obtained by using GIVE. This means that with the CM approach the information available in the estimated covariance functions of the data is exploited in a more efficient way. One should recall though that the optimization problem becomes more complex in the CM case, in the sense that for CM the optimization is over $n_a + n_b$ variables, while it is over 2 variables in the GIVE case. This shows up in the computation times. While they are quite modest, they are on the average about 5 times larger for CM compared to GIVE in this example. ■

Example 8.4 To consider CM for a case when more covariance elements than in Example 8.3 are used, let here

Table 8.1 Covariance matching (CM) estimates for the standard data sets

Parameter	True value	CM		GIVE	
		Mean	Standard dev.	Mean	Standard dev.
a_1	−1.5	−1.496	0.038	−1.528	0.085
a_2	0.7	0.698	0.026	0.709	0.040
b_1	2.0	1.982	0.268	3.57	2.82
b_2	1.0	1.047	0.513	−0.69	3.30
λ_y	10.0	11.02	1.34	8.37	4.51
λ_u	1.81	1.83	0.32	2.00	0.64

Table 8.2 Covariance matching (CM) estimates for some different values of p using the standard data sets. The arithmetic mean values are given with standard deviations within parentheses

Parameter	a_1	a_2	b_1	b_2	λ_y	λ_u
p True	−1.5	0.7	2.0	1.0	10.0	1.81
1	−1.489	0.692	1.994	1.087	10.96	1.92
	(0.020)	(0.014)	(0.28)	(0.38)	(1.24)	(0.22)
2	−1.494	0.696	2.008	1.039	11.09	1.86
	(0.014)	(0.010)	(0.17)	(0.24)	(1.34)	(0.17)
3	−1.495	0.697	2.005	1.028	11.16	1.83
	(0.013)	(0.010)	(0.17)	(0.22)	(1.39)	(0.17)
4	−1.497	0.698	2.005	1.019	11.18	1.81
	(0.013)	(0.010)	(0.17)	(0.22)	(1.41)	(0.18)
5	−1.497	0.698	2.010	1.006	11.12	1.81
	(0.014)	(0.011)	(0.20)	(0.25)	(1.42)	(0.19)

$$p_y = 2 + p, \ p_u = 1 + p, \ p_1 = -1 - p, \ p_2 = 2 + p, \quad p = 1, \ldots, 5 . \tag{8.63}$$

The numerical results obtained are summarized in Table 8.2. The RMS values are plotted versus p in Fig. 8.1.

The results in Table 8.2 are all good, and one can see some modest improvement when p is increased from 1. The smallest RMS value is obtained for $p = 3$, while almost the same value is also obtained for $p = 2$ and $p = 4$. Increasing to $p = 5$ gives a slight deterioration. ■

A further illustration of the influence of p is given in Example 14.5 where the Monte Carlo simulations are compared to a theoretical analysis.

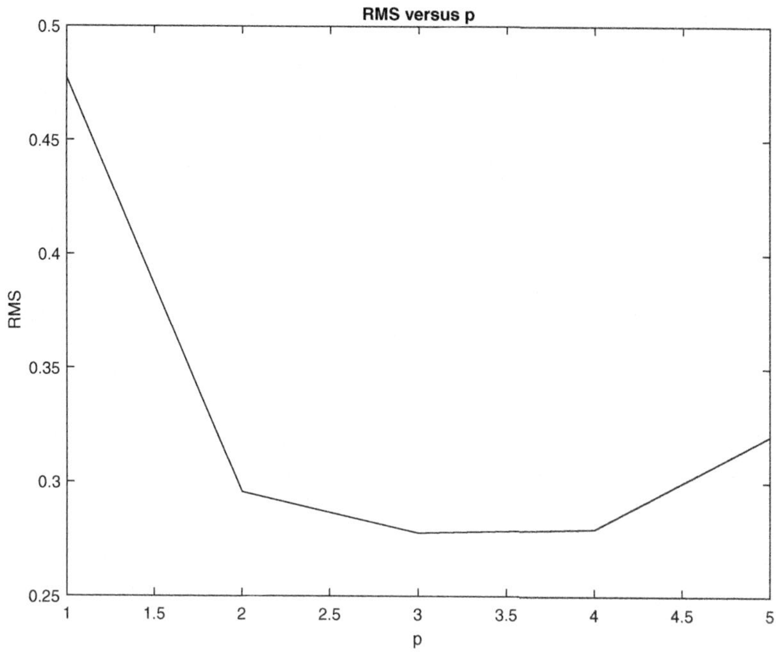

Fig. 8.1 RMS values versus p for the CM estimates

8.5 Applying Confirmatory Factor Analysis Modeling for EIV Identification

The confirmatory factor analysis (CFA) technique was introduced in Sect. 2.2. In this approach model covariance matrices are fitted to empirical covariances. Thus CFA, which is based on a prespecified set of equations, has potential connections to the covariance matching method of Sect. 8.2. However, it is not trivial how to apply the ideas of CFA for treating the EIV problem in the dynamic case.

To get a feasible set of equations (meaning that the number of free parameters is at most n_e, (2.45)), it turns out to be convenient (and perhaps even more or less necessary) to use the parameterization trick in (8.9).

Recall that the CFA model (2.27) reads

$$\mathbf{y} = \mathbf{A}(\boldsymbol{\vartheta})\boldsymbol{\eta} + \boldsymbol{\varepsilon} \tag{8.64}$$

Now make the attempt to set

$$\mathbf{y} = \big(y(t) \ \ldots \ y(t - n_a - p_y) \ u(t-1) \ \ldots \ u(t - n_b - p_u) \big)^T \ , \tag{8.65}$$

$$\boldsymbol{\eta} = \big(z_0(t) \ \ldots \ z_0(t-k) \big)^T \ , \quad k = n_a + n_b + \max(p_y, p_u) \ . \tag{8.66}$$

Here, p_y and p_u are nonnegative integers that are to be chosen by the user. For this case it apparently holds

$$n = \dim(\mathbf{y}) = n_a + n_b + 1 + p_y + p_u, \quad m = \dim(\boldsymbol{\eta}) = k + 1 . \tag{8.67}$$

The number of 'initial' unknowns (neglecting the $\mathbf{r}_z$ elements) is

$$n_\theta = n_a + n_b + 2 . \tag{8.68}$$

Further,

$$\mathbf{y} - \boldsymbol{\varepsilon} = \begin{pmatrix} 0 \; b_1 \; \dots \; b_{n_b} & & \mathbf{0} \\ & \ddots & \ddots \\ \mathbf{0} & & b_1 \; \dots \; b_{n_b} \\ 1 \; \dots \; a_{n_a} & & \mathbf{0} \\ & \ddots & \ddots \\ & \ddots & \ddots \\ \mathbf{0} & & 1 \; \dots \; a_{n_a} \end{pmatrix} \boldsymbol{\eta} \triangleq \mathbf{A}(\boldsymbol{\vartheta})\boldsymbol{\eta} , \tag{8.69}$$

and it also holds

$$\mathbf{R}_{\boldsymbol{\varepsilon}} = \begin{pmatrix} \lambda_y \mathbf{I}_{n_a+p_y+1} & \mathbf{0} \\ \mathbf{0} & \lambda_u \mathbf{I}_{n_b+p_u} \end{pmatrix} . \tag{8.70}$$

To parameterize the Toeplitz matrix

$$\mathbf{R}_{\eta}(\boldsymbol{\vartheta}) = \mathsf{E}\left\{\boldsymbol{\eta}\boldsymbol{\eta}^T\right\} = \begin{pmatrix} r_z(0) & \dots & r_z(k) \\ \vdots & \ddots & \vdots \\ r_z(k) & \dots & r_z(0) \end{pmatrix} \tag{8.71}$$

requires that the first column elements are used as parameters, i.e.

$$n_z = k + 1 \tag{8.72}$$

additional parameters are used.

Note that in this case, when applying CFA to EIV modeling, the vector $\mathbf{x}(\boldsymbol{\vartheta})$ in (2.52) is precisely the same as $\mathbf{r}$ in (8.21).

It is interesting to compare the CM and CFA approaches, as both methods are based on a finite and modest number of covariance elements.

Some initial comments are as follows:

1. CM can be seen as the more general approach, as there is large flexibility in which covariance elements to include. To have the same set of covariances in both approaches, then for fixed values of p_y and p_u, one has to make the specific choices
$$p_1 = 1 - n_a - p_y, \quad p_2 = n_b + p_u \tag{8.73}$$

for CM.
2. CFA, on the other hand, is a more general methodology in the sense that it covers a large set of situations, and is not limited to EIV identification of dynamic models.
3. The criterion for CM is quadratic in the estimated covariances $\hat{r}(\tau)$. On the other hand the criterion V_1 for CFA is linear in these estimated (sample) covariances. Hence, the criteria *cannot* be equivalent.

The identifiability properties of the CFA approach can be examined as follows. The number of equations becomes here, taking the Toeplitz structure into account

$$\begin{aligned} r_y \text{ elements } &: \; n_a + p_y + 1 \\ r_u \text{ elements } &: \; n_b + p_u \\ r_{yu} \text{ elements } &: \; n_a + p_y + n_b + p_u \end{aligned} \tag{8.74}$$

and hence

$$n_e = 2(n_a + p_y + n_b + p_u) + 1 \; . \tag{8.75}$$

The compatibility condition becomes now

$$\begin{aligned} & n_e - n_\theta - n_z \geq 0 \\ & 2(n_a + p_y + n_b + p_u) - (n_a + n_b + 2) - (k + 1) \geq 0 \\ & 2(p_y + p_u) - 2 - \max(p_y, p_u) \geq 0 \\ & p_y + p_u \geq 2 \; . \end{aligned} \tag{8.76}$$

Note that the condition (8.76) is the same identifiability condition that applies for the CM approach; see Sect. 8.4.2. (Recall that p_u has different meanings for the two approaches: The maximal lag of $r_u(\tau)$ that appears in the CFA approach is $n_b + p_u - 1$, not $n_b + p_u$.)

8.6 Extensions and Bibliographical Notes

The use of the transformation (8.9) has also been proposed in Polderman and Willems (1998). See also Diversi et al. (2007).

Note that similar analysis appears also in many other fields for estimation problems leading to optimization of a loss function of the form (8.24). The key aspect is the formulation of the estimation problem as a separable nonlinear least squares problem, which is handled with a variable projection algorithm. This approach is quite general. Sensor array processing, Haykin (1991b), Viberg and Ottersten (1991), and estimation of material functions, Mossberg et al. (2001), Mahata et al. (2003), are just some examples among others, Mahata and Söderström (2004). In the present case $\hat{\mathbf{r}}$ consists of sampled covariance elements formed from the data, while in some other applications, the correspondence to $\hat{\mathbf{r}}$ is formed directly from the data.

The use of Markov chain (MC) models will lead to covariance functions of the same form as for a finite-order state space model. It has hence direct couplings to the CM approach. How to carry out identification for MC models is discussed in Kennedy and Chung (1992, 1996). See also Krusevac et al. (2006), Pulford et al. (1995), Sadeghi et al. (2008).

A CM version where in addition to $\boldsymbol{\theta}$, the parameters of an ARMA model of the noise-free input are estimated, is proposed in Mossberg (2007b).

A continuous-time version of the CM method is analyzed in Mossberg (2007a). The asymptotic covariance matrix of the parameter estimates is derived.

Extension of the CM method to multivariable systems is described in Mossberg and Söderström (2012).

The idea of covariance matching (although that particular terminology was not used) for estimating ARMA model parameters without additional measurement noise in the observations appeared in Stoica and Nehorai (1986), Porat and Friedlander (1986). In terms of this chapter, only $\mathbf{r}_y$ was used in (8.21). It was shown that when p_y grows to infinity, and an optimal weighting is used, the estimate of the ARMA parameters becomes statistically efficient; that is the Cramér-Rao bound is achieved.

Confirmatory Factor Analysis

More explicit details about how the structural equation modeling (SEM) can be applied to the CM formulation appeared recently in Kreiberg et al. (2016). A preliminary analysis appeared in Kreiberg et al. (2013). In particular both the attempt (8.65) and (8.66) and an alternative formulation are considered and analyzed.

Some attempts to extend the CFA methodology to general stochastic state space model appear in Otter (1986).

Appendix

8.A Further Details

8.A.1 The Rank Condition

As noted in Sect. 8.3, see Remark 8.3, it is necessary for the proposed covariance matching algorithm to work that the matrix $\mathbf{F}(\boldsymbol{\theta})$ has full rank. This condition will be examined in some detail in this section.

First note that if $\mathbf{F}(\boldsymbol{\theta})$ is rank-deficient, this will be easily detected in the optimization of (8.28). Appropriate actions, such as modifying the structure $\mathbf{p}$ (8.52), can then be taken. As will be seen in Sect. 8.A.2 an alternative is to use a modified parameterization.

The issue is illustrated by means of an example.

Example 8.5 Consider the case when correlated output noise is allowed and therefore

$$p_y = 0 \tag{8.77}$$

is chosen. For convenience only, the additional choices

$$n = n_a = n_b, \; p_1 = 0, \; p_2 = p_u = p \tag{8.78}$$

are made. Set

$$A(z) = z^n + a_1 z^{n-1} + \cdots + a_n \;, \tag{8.79}$$

$$B(z) = b_1 z^{n-1} + \cdots + b_n \;, \tag{8.80}$$

$$\alpha(z) = \sum_{i=0}^{2n} \alpha_{-n+i} z^{2n-i} = z^n A(z) A(z^{-1}) \;, \tag{8.81}$$

$$\gamma(z) = \sum_{i=0}^{2n-1} \gamma_{-n+i} z^{2n-1-i} = z^n A(z) B(z^{-1}) \;. \tag{8.82}$$

One then finds from (8.40) and (8.46),

$$\mathbf{r} = \begin{pmatrix} r_u(1) \\ \vdots \\ r_u(p) \\ r_{yu}(0) \\ \vdots \\ r_{yu}(p) \end{pmatrix} = \begin{pmatrix} \alpha_{-n} & \alpha_{-n+1} & \cdots & \alpha_n & & \mathbf{0} \\ \mathbf{0} & \ddots & & & \ddots & \\ & & \alpha_{-n} & \cdots & & \alpha_n \\ \gamma_{-n} & \gamma_{-n+1} & \cdots & \gamma_{n-1} & & \mathbf{0} \\ \mathbf{0} & \ddots & & & \ddots & \\ & & \gamma_{-n} & \cdots & & \gamma_{n-1} \end{pmatrix} \begin{pmatrix} r_0(1-n) \\ \vdots \\ r_0(p+n) \end{pmatrix} . \tag{8.83}$$

The matrix $\mathbf{F}(\boldsymbol{\theta})$ can therefore be written as

$$\mathbf{F}(\boldsymbol{\theta}) = \mathbf{T}(\boldsymbol{\theta})\mathbf{P} \;, \tag{8.84}$$

$$\mathbf{T}(\boldsymbol{\theta}) = \begin{pmatrix} \alpha_{-n} & \alpha_{-n+1} & \cdots & \alpha_n & & \mathbf{0} \\ \mathbf{0} & \ddots & & & \ddots & \\ & & \alpha_{-n} & \cdots & & \alpha_n \\ \gamma_{-n} & \gamma_{-n+1} & \cdots & \gamma_{n-1} & & \mathbf{0} \\ \mathbf{0} & \ddots & & & \ddots & \\ & & \gamma_{-n} & \cdots & & \gamma_{n-1} \end{pmatrix} , \tag{8.85}$$

$$\mathbf{P} = \begin{pmatrix} \mathbf{0} & 1 & & \\ & \iddots & & \\ 1 & \mathbf{0} & & \\ & \ddots & & \\ & & \ddots & \\ \mathbf{0} & & & 1 \end{pmatrix} . \tag{8.86}$$

Here, $\mathbf{T}(\boldsymbol{\theta})$ is a generalized Sylvester matrix, of dimension $(2p+1) \times (p+2n)$, while the matrix $\mathbf{P}$ has dimensions $(p+2n) \times (p+n+1)$ and has full column rank $p+n+1$. The polynomials $\alpha(z)$ and $\gamma(z)$ have precisely n common factors, which is clear from (8.81) and (8.82) (assume that the polynomials $A(z)$ and $B(z)$ are co-prime, as usual!). Then it follows from the theory of generalized Sylvester matrices, see for example Lemma A.30 of Söderström and Stoica (1989), that

$$\text{rank}(\mathbf{T}(\boldsymbol{\theta})) = 2n + (2n-1) + (p-2n+1) - n = p+n . \tag{8.87}$$

and

$$\begin{aligned} \text{rank}(\mathbf{F}(\boldsymbol{\theta})) &= \text{rank}(\mathbf{T}(\boldsymbol{\theta})\mathbf{P}) \\ &\leq \text{rank}(\mathbf{T}(\boldsymbol{\theta})) < p+n+1 = \dim(\mathbf{r}_z) . \end{aligned} \tag{8.88}$$

Hence, the matrix $\mathbf{F}(\boldsymbol{\theta})$ is rank-deficient, and this is true independent of how n and p in (8.78) are related.

Furthermore, for the case treated in this example, the identifiability condition (8.57) can be simplified to

$$p \geq 3n , \tag{8.89}$$

which certainly is satisfied for some parameter combinations. ■

Remark 8.8 An alternative (and more explicit) way of reasoning to arrive at (8.87) goes as follows. First recall that it holds

$$\dim(\mathcal{N}(\mathbf{F}^T(\boldsymbol{\theta}))) = 2p + 1 - \text{rank}(\mathbf{F}(\boldsymbol{\theta})) . \tag{8.90}$$

Then examine this null space, and let an arbitrary vector in it be written as

$$\mathbf{x} = \begin{pmatrix} x_1 & \dots & x_p & y_1 & \dots & y_{p+1} \end{pmatrix}^T . \tag{8.91}$$

One finds that $\mathbf{F}(\boldsymbol{\theta})^T \mathbf{x} = 0$ can be equivalently written as the polynomial equation

$$X(z)\alpha(z) + Y(z)\gamma(z) = 0 , \tag{8.92}$$

where

$$X(z) = x_1 z^{p-1} + \cdots + x_p \ ,$$
$$Y(z) = y_1 z^p + \cdots + y_{p+1} \ ,$$

are introduced as unknown polynomials. In this case (8.92) implies

$$X(z)z^n A(z)A(z^{-1}) + Y(z)z^n A(z)B(z^{-1}) = 0 \ , \tag{8.93}$$

and one can cancel the factor $A(z)$. This means that (8.93) is effectively a linear system of equations with

$$\begin{aligned} \#\,\text{equations} &= 1 + \deg\,(\text{LHS})\text{ of }(8.93) = p + n \ , \\ \#\,\text{unknowns} &= \dim(\mathbf{x}) = 2p + 1 \ . \end{aligned} \tag{8.94}$$

Hence the number of degrees of freedom in the solution, which is the same as the dimension of the null space $\mathcal{N}^T(\mathbf{F}(\boldsymbol{\theta}))$, becomes

$$\dim(\mathcal{N}(\mathbf{F}^T(\boldsymbol{\theta}))) = \#\,\text{unknowns} \ - \#\,\text{equations} \ = p - n + 1 \ , \tag{8.95}$$

which using (8.90) leads to

$$\begin{aligned} \text{rank}(\mathbf{T}(\boldsymbol{\theta})) &= (2p+1) - \dim(\mathcal{N}(\mathbf{F}^T(\boldsymbol{\theta}))) \\ &= (2p+1) - (p - n + 1) \\ &= p + n \ , \end{aligned} \tag{8.96}$$

which is the same result as in (8.87). ■

One might perhaps believe from the example above that it generally (or often) is true that $\mathbf{F}(\boldsymbol{\theta})$ is rank-deficient. The following example shows that this is *not* so.

Example 8.6 Consider the simple case given by

$$n_a = n_b = 1, \ p_y = 0, \ p_u = 2, \ p_1 = -2, \ p_2 = 2 \ . \tag{8.97}$$

For this case, it holds

$$\alpha_0 = 1 + a^2, \alpha_1 = \alpha_{-1} = a, \gamma_{-1} = b, \gamma_0 = ab \ . \tag{8.98}$$

Further, the basic equation (8.22) becomes

$$\begin{pmatrix} r_u(1) \\ r_u(2) \\ r_{yu}(-2) \\ r_{yu}(-1) \\ r_{yu}(0) \\ r_{yu}(1) \\ r_{yu}(2) \end{pmatrix} = \begin{pmatrix} 0 & \alpha_{-1} & \alpha_0 & \alpha_1 & 0 & 0 \\ 0 & 0 & \alpha_{-1} & \alpha_0 & \alpha_1 & 0 \\ \gamma_{-1} & \gamma_0 & 0 & 0 & 0 & 0 \\ 0 & \gamma_{-1} & \gamma_0 & 0 & 0 & 0 \\ 0 & 0 & \gamma_{-1} & \gamma_0 & 0 & 0 \\ 0 & 0 & 0 & \gamma_{-1} & \gamma_0 & 0 \\ 0 & 0 & 0 & 0 & \gamma_{-1} & \gamma_0 \end{pmatrix} \begin{pmatrix} r_0(-2) \\ r_0(-1) \\ r_0(0) \\ r_0(1) \\ r_0(2) \\ r_0(3) \end{pmatrix}$$

$$= \begin{pmatrix} \alpha_0 & 2\alpha_1 & 0 & 0 \\ \alpha_1 & \alpha_0 & \alpha_1 & 0 \\ 0 & \gamma_0 & \gamma_{-1} & 0 \\ \gamma_0 & \gamma_{-1} & 0 & 0 \\ \gamma_{-1} & \gamma_0 & 0 & 0 \\ 0 & \gamma_{-1} & \gamma_0 & 0 \\ 0 & 0 & \gamma_{-1} & \gamma_0 \end{pmatrix} \begin{pmatrix} r_0(0) \\ r_0(1) \\ r_0(2) \\ r_0(3) \end{pmatrix}$$

$$= \mathbf{F}(\boldsymbol{\theta})\mathbf{r}_0 , \tag{8.99}$$

and it is verified by inspection that the matrix $\mathbf{F}(\boldsymbol{\theta})$ has full column rank (equal to 4) in this case. ■

To summarize this section, it has been seen by means of examples that the matrix $\mathbf{F}(\boldsymbol{\theta})$ may or may not be rank-deficient. Which alternative that applies depends in a rather involved way on the structure (8.52) of the problem.

8.A.2 An Alternative Parameterization

It may still happen that the parameter vector $\boldsymbol{\theta}$ can be uniquely identified even if the matrix $\mathbf{T}(\boldsymbol{\theta})$ in (8.84) should be rank-deficient, by using a somewhat modified approach involving an alternative parameterization resulting from a factorization. Next this situation is analyzed for the case $p_y = 0$, a situation that must apply if correlated output noise should be allowed.

First introduce some additional notations. Let the dimension of $\mathbf{F}(\boldsymbol{\theta})$ be $n_r \times n_z$, the rank of $\mathbf{F}(\boldsymbol{\theta})$ be n_f, and the length of $\boldsymbol{\theta}$ be n_θ. For the equation

$$\mathbf{r} = \mathbf{F}(\boldsymbol{\theta})\mathbf{r}_z \tag{8.100}$$

to be compatible, it must hold that, cf (8.57)

$$n_r \geq n_\theta + n_z . \tag{8.101}$$

The case to consider and discuss is when $\mathbf{F}$ is rank-deficient, that is when

$$n_f < n_z . \tag{8.102}$$

Assume that $\mathbf{F}$ can be factorized as

$$\mathbf{F}(\boldsymbol{\theta}) = \mathbf{F}_1(\boldsymbol{\theta})\mathbf{F}_2(\boldsymbol{\theta}) , \tag{8.103}$$

where $\mathbf{F}_1(\boldsymbol{\theta})$ is $n_r \times n_f$ and $\mathbf{F}_2(\boldsymbol{\theta})$ is $n_f \times n_z$. As rank$(\mathbf{F}) = n_f$, it must then hold that $\mathbf{F}_1(\boldsymbol{\theta})$ has full column rank equal to n_f. Equation (8.100) can now be transformed into

$$\mathbf{r} = \mathbf{F}_1(\boldsymbol{\theta})\mathbf{g} , \tag{8.104}$$

$$\mathbf{g} = \mathbf{F}_2(\boldsymbol{\theta})\mathbf{r}_z . \tag{8.105}$$

Consider first (8.104) only, and treat the vector $\mathbf{g}$, which has length n_f, as consisting of auxiliary unknowns. It turns out that this equation is overdetermined, as

$$\dim(\mathbf{r}) = n_r \geq n_\theta + n_z > n_\theta + n_f = \dim(\boldsymbol{\theta}) + \dim(\mathbf{g}) . \tag{8.106}$$

As $\mathbf{F}_1$ has full rank, one can expect that (8.104) should lead to a sound least squares solution. Once, $\boldsymbol{\theta}$ and $\mathbf{g}$ are found, (8.105) can be treated with $\mathbf{r}_z$ as unknowns. This system is under-determined, and $\mathbf{r}_z$ cannot be determined uniquely. Still, it is $\boldsymbol{\theta}$ that is of primary interest to estimate.

All in all, this consideration indicates that when $\mathbf{F}(\boldsymbol{\theta})$ is rank-deficient, it may still be possible to determine $\boldsymbol{\theta}$ uniquely, but not $\mathbf{r}_z$. However, the approach laid out in the end of Sect. 8.3 for the algorithm has to be modified, as it breaks down when $\mathbf{F}(\boldsymbol{\theta})$ looses rank.

The analysis was based on the assumption that the factorization (8.103) is possible. Next examine this issue in some examples.

Example 8.7 Recalling the calculations made in Example 8.5, it is sufficient to show how to factorize $\mathbf{T}(\boldsymbol{\theta})$ as

$$\mathbf{T}(\boldsymbol{\theta}) = \mathbf{T}_1(\boldsymbol{\theta})\mathbf{T}_2(\boldsymbol{\theta}) , \tag{8.107}$$

where $\mathbf{T}_1(\boldsymbol{\theta})$ is $(2p+1) \times (p+n)$ and of rank $p+n$. However, some calculations reveal that

$$\mathbf{T}(\boldsymbol{\theta}) = \begin{pmatrix} \alpha_{-n} & \alpha_{-n+1} & \dots & \alpha_n & & \mathbf{0} \\ \mathbf{0} & \ddots & & & \ddots & \\ & & \alpha_{-n} & \dots & & \alpha_n \\ \gamma_{-n} & \gamma_{-n+1} & \dots & \gamma_{n-1} & & \mathbf{0} \\ \mathbf{0} & \ddots & & & \ddots & \\ & & \gamma_{-n} & \dots & & \gamma_{n-1} \end{pmatrix}$$

$$= \begin{pmatrix} a_n & a_{n-1} & \dots & 1 & & \mathbf{0} \\ & \ddots & & & & \\ \mathbf{0} & & a_n & \dots & & 1 \\ b_n & \dots & & b_1 & & 0 \\ & \ddots & & & \ddots & \\ \mathbf{0} & & b_n & \dots & & b_1 \end{pmatrix} \begin{pmatrix} 1 & a_1 & \dots & a_n & & \mathbf{0} \\ & \ddots & & & \ddots & \\ \mathbf{0} & & 1 & a_1 & \dots & a_n \end{pmatrix}$$

$$\stackrel{\Delta}{=} \mathbf{T}_1(\boldsymbol{\theta})\mathbf{T}_2(\boldsymbol{\theta}) \,. \tag{8.108}$$

The matrix $\mathbf{T}_1(\boldsymbol{\theta})$ is a generalized Sylvester matrix; see Sect. A.1.5. It has dimension $(2p+1) \times (p+n)$ and is of rank $p+n$. The matrix $\mathbf{T}_2(\boldsymbol{\theta})$ has dimension $(p+n) \times (2n+p)$ and has also rank $p+n$. Thus a factorization of the form (8.103) is found that leads to identifiability of $\boldsymbol{\theta}$. ■

It may be of interest to consider also another type of factorization that is closely related to (8.108).

Example 8.8 Reconsider the case in Example 8.5. Recall that

$$u(t) = \sum_{i=1}^{n} b_i z(t-i) + \tilde{u}(t) \,, \tag{8.109}$$

$$y(t) = \sum_{i=0}^{n} a_i z(t-i) + \tilde{y}(t) \,. \tag{8.110}$$

Then one finds that

$$\mathbf{r}_u = \begin{pmatrix} r_u(1) \\ \vdots \\ r_u(p) \end{pmatrix} = \sum_{i=1}^{n} b_i \mathsf{E} \left\{ \begin{pmatrix} z(t+1-i) \\ \vdots \\ z(t+p-i) \end{pmatrix} u(t) \right\}$$

$$= \begin{pmatrix} b_n & \dots & b_1 & & \mathbf{0} \\ & \ddots & & \ddots & \\ \mathbf{0} & & b_n & \dots & b_1 \end{pmatrix} \begin{pmatrix} r_{zu}(1-n) \\ \vdots \\ r_{zu}(p-1) \end{pmatrix}, \tag{8.111}$$

$$\mathbf{r}_{yu} = \begin{pmatrix} r_{yu}(0) \\ \vdots \\ r_{yu}(p) \end{pmatrix} = \sum_{i=0}^{n} a_i \mathsf{E} \left\{ \begin{pmatrix} z(t-i) \\ \vdots \\ z(t+p-i) \end{pmatrix} u(t) \right\}$$

$$= \begin{pmatrix} a_n & \dots & a_1 & 1 & & \mathbf{0} \\ & \ddots & & & \ddots & \\ \mathbf{0} & & a_n & \dots & a_1 & 1 \end{pmatrix} \begin{pmatrix} r_{zu}(-n) \\ \vdots \\ r_{zu}(p) \end{pmatrix}. \tag{8.112}$$

Hence it holds

$$
\mathbf{r} = \begin{pmatrix} \mathbf{r}_u \\ \mathbf{r}_{yu} \end{pmatrix}
= \begin{pmatrix}
0 & b_n & \dots & b_1 & 0 & \mathbf{0} \\
 & \ddots & & & \ddots & \\
\mathbf{0} & & 0 & b_n & \dots & b_1 \; 0 \\
a_n & a_{n-1} & \dots & & 1 & \mathbf{0} \\
 & \ddots & & & \ddots & \\
\mathbf{0} & & a_n & \dots & & 1
\end{pmatrix}
\begin{pmatrix} r_{zu}(-n) \\ \vdots \\ r_{zu}(p) \end{pmatrix} . \tag{8.113}
$$

However, now note that the matrix in (8.113) is a Sylvester matrix and of dimension $(2p + 1) \times (p + n + 1)$. It has rank $n + p + 1$, not $n + p$, assuming $a_n \neq 0$. This is in contrast to $\mathbf{T}(\boldsymbol{\theta})$ in (8.108). As the matrix in (8.113) has full column rank, the parameterization there can be used to estimate $\boldsymbol{\theta}$ with the algorithm (8.27) and (8.28). ■

Remark 8.9 The case $p_y = 0$ is further generalized in Söderström et al. (2009), extending the idea of Example 8.8. ■

Chapter 9
Prediction Error and Maximum Likelihood Methods

The chapter is organized as follows. Some basic ideas are first given in Sect. 9.1. The prediction error and the maximum likelihood (ML) methods can be derived and handled either in the time domain (see Sect. 9.2) or in the frequency domain (see Sect. 9.3). A frequency domain ML method where the noise variance ratio is known and all samples of the noise-free input u_0 are estimated is presented in Sect. 9.4. An extended version of the ML approach in the frequency domain, where also a marginal likelihood function is utilized, which allows the noise variance ratio to be unknown is described in Sect. 9.5. Section 9.6 contains extensions and bibliographical notes.

9.1 The Basic Ideas

The maximum likelihood and prediction error methods are based on considering the probability function of the observed data as a function of the unknown parameters. The parameter values that maximize this pdf is taken as the estimates, cf. Convention A.7.

There are a number of possibilities how to arrange this idea in practice.

- An important aspect is how to regard the unmeasured input signal $u_0(t)$. One option is to regard it as a stochastic process, whose properties are described with a modest number of parameters. This way is followed in Sects. 9.2 and 9.3. Another option is to regard all the sample values $u_0(1), \ldots, u_0(N)$ as auxiliary unknowns. This approach is considered in Sects. 9.4 and 9.5.
- A second important aspect is whether to formulate the estimation problem in the time domain or in the frequency domain. While a time domain formulation may be more in line with the derived GIVE and CM methods in Chaps. 7 and 8, a frequency domain formulation has the advantage that convolutions appear as

T. Söderström, *Errors-in-Variables Methods in System Identification*, Communications and Control Engineering,
https://doi.org/10.1007/978-3-319-75001-9_9

simple multiplications. In both domains appropriate actions can be included to take care of initial values and effects of transients in the data.

A quite general treatment of identification of systems where the input is uncertain is given in Risuleo et al. (2017c). See also Risuleo (2016). The input is then assumed not fully known but some partial knowledge of its behavior is available. This is modeled by assuming $u_0(t)$ as well as the system weighting function g to be Gaussian processes, cf. Rasmussen and Williams (2006),

$$\begin{pmatrix} u_0 \\ g \end{pmatrix} \sim \mathscr{GP}\left(\begin{pmatrix} \mu_{u_0}(\cdot;\eta_1) \\ \mu_g(\cdot;\eta_2) \end{pmatrix}, \begin{pmatrix} K_{u_0}(\cdot;\eta_1) & K_{gu_0}^T(\cdot;\eta_2,\eta_1) \\ K_{gu_0}(\cdot;\eta_2,\eta_1) & K_g(\cdot;\eta_2) \end{pmatrix}\right), \tag{9.1}$$

where the mean values μ_{u_0} and μ_g as well as the covariance functions K_{u_0}, K_g and K_{gu_0} are parameterized with some parameter vectors η_1 and η_2. These parameters are called hyperparameters.

Using this type of description of the unknown quantities lead naturally to a Bayesian framework. It is also clearly related to the use of kernel methods and regularization in system identification, see Pillonetto et al. (2014).

The general formulation (9.1) allows for many possibilities.

- One option is to assume that $u_0(t)$ is a Gaussian distributed ARMA process,

$$D(q^{-1})u_0(t) = C(q^{-1})e(t) \tag{9.2}$$

as in Assumption **AI4**. Then one chooses $\mu_{u_0} = 0$, and let K_{u_0} be the Teoplitz matrix formed by expressing the covariance function of $u_0(t)$. In this case the parameter vector η_1 becomes

$$\eta_1 = \begin{pmatrix} c_1 \dots c_{n_c} \; d_1 \dots d_{n_d} \; \lambda_e \end{pmatrix}^T . \tag{9.3}$$

Further, one would then regard the system transfer function operator to take the form $B(q^{-1})/A(q^{-1})$. This would be achieved by taking

$$\eta_2 = \begin{pmatrix} a_1 \dots a_{n_a} \; b_1 \dots b_{n_b} \end{pmatrix}^T , \tag{9.4}$$

and let μ_g be the weighting function associated with B/A, and $K_g = 0$. This option is followed in Sect. 9.2.
- Another option is to consider all samples of $u_0(t)$ as unknowns, and one would then take

$$\mu_{u_0}(\cdot;\eta_1) = \eta_1, \quad K_{u_0}(\cdot;\eta_1) = 0 . \tag{9.5}$$

This approach leads to nonidentifiability unless some further assumption is induced. This was illustrated in Sect. 3.4.2. It will be used and analyzed in Sect. 9.4 for the case when the noise variance ratio $r = \lambda_y/\lambda_u$ is known.

Still another approach of using the maximum likelihood method is described in Sect. 9.5 see also Zhang et al. (2013). The samples of the nonmeasured input $u_0(t)$ are estimated. In contrast to the previous mentioned approach, both the likelihood function and a marginal likelihood function are used. This is shown to give identifiability. It is convenient to carry out the calculations in the frequency domain.

The optimization problem to be solved when applying the ML technique can often be quite complex as the problem contains latent variables. When taking the general description (9.1) and using an empirical Bayesian approach the marginal distribution of the data is to be maximized. One possibility for organizing the calculations is to apply the expectation-maximimization (EM) method. This is an iterative procedure that alternates between computation of the expected value of the joint likelihood of the data, the unknown system and the input (the E step), and the maximimization of this expected value with respect to the unknown parameters (the M step). The procedure results in a sequence of local approximations of the likelihood function.

Remark 9.1 The Bayesian approach can be applied also for treating identification of nonlinear dynamics. The numerical procedure then often involve Markov Chain Monte Carlo (MCMC) methods and particle filters, Risuleo et al. (2017c). ■

9.2 Time Domain Formulation

The prediction error method (PEM) is based on the idea that a major goal of the identification task is to be able to *predict* the future behavior. PEM is a standard technique that is well studied, see for example Ljung (1999), Söderström and Stoica (1989). In many cases (but not typically so for EIV problems!) it coincides with the maximum likelihood (ML) method, and it then enjoys many strong statistical properties such as consistency and statistical efficiency.

A common ingredient for PEM is that a model parameterization has to be given in (or transformed into) innovations form. Then it is straightforward to compute the mean-square optimal one-step prediction errors. The parameter estimates are determined so that the size of these prediction errors becomes as small as possible. In the multivariable case the situation is more complex, as the effect of the disturbances can be parameterized in many more ways.

When using the PEM approach for the errors–in–variables model of Fig. 4.1 the model is regarded as a multivariable system with both $y(t)$ and $u(t)$ as components. It is called the joint-input output approach in Söderström (1981). Of crucial importance for this approach is the assumption **AI4**, *i.e.*, the noise–free input $u_0(t)$ is characterized by a rational spectrum. The signal $u_0(t)$ can thus be represented by its innovations form, described as an ARMA process of the type (3.8):

$$u_0(t) = \frac{C(q^{-1})}{D(q^{-1})} e(t) , \tag{9.6}$$

where $e(t)$ is a white noise with variance λ_e. In this way, the whole errors–in–variables model

$$\begin{aligned} u(t) &= u_0(t) + \tilde{u}(t) , \\ y(t) &= B(q^{-1})/A(q^{-1})\, u_0(t) + \tilde{y}(t) , \\ u_0(t) &= C(q^{-1})/D(q^{-1})\ e(t) , \end{aligned} \tag{9.7}$$

can be considered as a system with a two-dimensional measurement vector $\boldsymbol{\zeta}(t) = \begin{pmatrix} y(t) & u(t) \end{pmatrix}^T$ and three mutually uncorrelated white noise sources $e(t)$, $\tilde{u}(t)$ and $\tilde{y}(t)$:

$$\boldsymbol{\zeta}(t) = \begin{pmatrix} y(t) \\ u(t) \end{pmatrix} = \begin{pmatrix} \frac{B(q^{-1})C(q^{-1})}{A(q^{-1})D(q^{-1})} & 1\ 0 \\ \frac{C(q^{-1})}{D(q^{-1})} & 0\ 1 \end{pmatrix} \begin{pmatrix} e(t) \\ \tilde{y}(t) \\ \tilde{u}(t) \end{pmatrix} . \tag{9.8}$$

Thus the model $C(q^{-1})/D(q^{-1})$ of the undisturbed input is a part of the errors–in–variables representation and its coefficients must be estimated together with the parameters of $A(q^{-1})$ and $B(q^{-1})$.

The model (9.8) can be transformed to a general state space model, as described in Sect. 5.6, see Example 5.2. Consider a linear multivariable system given on state space form as

$$\mathbf{x}(t+1) = \mathbf{F}(\boldsymbol{\vartheta})\mathbf{x}(t) + \mathbf{v}_1(t) , \tag{9.9}$$

$$\boldsymbol{\zeta}(t) = \mathbf{C}(\boldsymbol{\vartheta})\mathbf{x}(t) + \mathbf{v}_2(t) , \tag{9.10}$$

$$\mathsf{E}\left\{\begin{pmatrix} \mathbf{v}_1(t) \\ \mathbf{v}_2(t) \end{pmatrix}\right\} = \mathbf{0} , \tag{9.11}$$

$$\mathsf{E}\left\{\begin{pmatrix} \mathbf{v}_1(t) \\ \mathbf{v}_2(t) \end{pmatrix} (\mathbf{v}_1^T(s)\ \ \mathbf{v}_2^T(s))\right\} = \begin{pmatrix} \mathbf{R}_1(\boldsymbol{\vartheta}) & \mathbf{R}_{12}(\boldsymbol{\vartheta}) \\ \mathbf{R}_{12}^T(\boldsymbol{\vartheta}) & \mathbf{R}_2(\boldsymbol{\vartheta}) \end{pmatrix} \delta_{t,s} , \tag{9.12}$$

where the total parameter vector $\boldsymbol{\vartheta}$ is

$$\boldsymbol{\vartheta} = \left(\boldsymbol{\theta}^T\ \boldsymbol{\rho}^T\right)^T . \tag{9.13}$$

In this case $\boldsymbol{\theta} = \left(a_1 \ldots a_{n_a}\ b_1 \ldots b_{n_b}\right)^T$ as before, while now $\boldsymbol{\rho}$ contains both the measurement noise variances λ_y and λ_u, as well as the parameters of the noise-free input model, that is $c_1, \ldots, c_{n_c}, d_1, \ldots, d_{n_d}, \lambda_e$.

The noise sequences $\{\mathbf{v}_1(t)\}$ and $\{\mathbf{v}_2(t)\} = \{(\tilde{y}(t)\ \tilde{u}(t))^T\}$ are assumed to be Gaussian distributed. The system matrices $\mathbf{F}(\boldsymbol{\vartheta})$, $\mathbf{C}(\boldsymbol{\vartheta})$, $\mathbf{R}_1(\boldsymbol{\vartheta})$, $\mathbf{R}_{12}(\boldsymbol{\vartheta})$ and $\mathbf{R}_2(\boldsymbol{\vartheta})$ will be smooth functions of the parameter vector $\boldsymbol{\vartheta}$. When there is no risk of confusion, the argument $\boldsymbol{\vartheta}$ will be suppressed in the following for simplicity.

In what follows consider prediction error methods (PEMs), Ljung (1999), Söderström and Stoica (1989) for this type of model. Hence application to errors-in-variables modeling will be included when an pertinent model parameterization

(9.9)–(9.12) is chosen. The maximum likelihood (ML) method will be a special case, as explained below.

Considering prediction error methods for estimating the parameter vector $\boldsymbol{\vartheta}$, it will be necessary to transform the state space model (9.9) into innovations form, obtained from the Kalman filter, see for example, Söderström (2002). Rewriting the innovations form to show how the prediction errors $\boldsymbol{\varepsilon}(t, \boldsymbol{\vartheta}) = \boldsymbol{\zeta}(t) - \hat{\boldsymbol{\zeta}}(t|t-1; \boldsymbol{\vartheta})$ depend on the data and the model matrices gives

$$\hat{\mathbf{x}}(t+1|t) = (\mathbf{F} - \mathbf{KC})\hat{\mathbf{x}}(t|t-1) + \mathbf{K}\boldsymbol{\zeta}(t) \,, \tag{9.14}$$

$$\boldsymbol{\varepsilon}(t, \boldsymbol{\vartheta}) = \boldsymbol{\zeta}(t) - \mathbf{C}\hat{\mathbf{x}}(t|t-1) \,, \tag{9.15}$$

where

$$\mathbf{K} = [\mathbf{FPC}^T + \mathbf{R}_{12}][\mathbf{CPC}^T + \mathbf{R}_2]^{-1} \,, \tag{9.16}$$

$$\mathbf{P} = \mathbf{FPF}^T + \mathbf{R}_1 - [\mathbf{FPC}^T + \mathbf{R}_{12}][\mathbf{CPC}^T + \mathbf{R}_2]^{-1}[\mathbf{CPF}^T + \mathbf{R}_{12}^T] \,. \tag{9.17}$$

In addition, these calculations provide an expression for the covariance matrix $\mathbf{Q}$ of the prediction errors:

$$\mathbf{Q}(\boldsymbol{\vartheta}) = \mathsf{E}\left\{\boldsymbol{\varepsilon}(t, \boldsymbol{\vartheta})\boldsymbol{\varepsilon}^T(t, \boldsymbol{\vartheta})\right\} = \mathbf{CPC}^T + \mathbf{R}_2 \,. \tag{9.18}$$

Often one may only be interested in estimating $\boldsymbol{\theta}$. With the prediction error approach, the augmented vector $\boldsymbol{\vartheta}$, (9.13), needs to be estimated. Then the noise parameter vector $\boldsymbol{\rho}$ can be regarded as a vector of nuisance parameters.

Remark 9.2 Alternatively one can use polynomial calculations, see Kučera (1979), to find the innovations model of the form

$$\boldsymbol{\zeta}(t) = \mathbf{S}(q^{-1}; \boldsymbol{\vartheta})\boldsymbol{\varepsilon}(t), \;\; \mathsf{E}\left\{\boldsymbol{\varepsilon}(t)\boldsymbol{\varepsilon}^T(t)\right\} = \mathbf{Q}(\boldsymbol{\vartheta}) \,. \tag{9.19}$$

The relation between the state space model (9.9)–(9.12) and the innovations form (9.19) is given in Example 5.2. ■

When identifying a model of the system, the parameter vector $\boldsymbol{\vartheta}$ will be estimated from a data sequence $\{\boldsymbol{\zeta}(t)\}_{t=1}^N$. This is done by minimizing a criterion function:

$$\hat{\boldsymbol{\vartheta}}_{\text{PEM}} = \arg\min_{\boldsymbol{\vartheta}} V_N(\boldsymbol{\vartheta}) \,. \tag{9.20}$$

Assume that $V_N(\boldsymbol{\vartheta})$ is a (sufficiently) smooth function of $\boldsymbol{\vartheta}$. In addition, assume that $V_N(\boldsymbol{\vartheta})$ converges (uniformly on compact subsets) as $N \to \infty$. This makes $\hat{\boldsymbol{\vartheta}}_{\text{PEM}}$ consistent. Hence

$$V_N(\boldsymbol{\vartheta}) \to V_\infty(\boldsymbol{\vartheta}), \;\; N \to \infty \,, \tag{9.21}$$

$$\hat{\boldsymbol{\vartheta}}_{\text{PEM}} \to \boldsymbol{\vartheta}_0, \;\; N \to \infty \,. \tag{9.22}$$

There is no unique way of defining a PEM criterion, but all alternatives must penalize the sequence $\{\boldsymbol{\varepsilon}(t, \boldsymbol{\vartheta})\}_{t=1}^{N}$. A few typical cases are given in the following examples.

Example 9.1 Take

$$V_N(\boldsymbol{\vartheta}) = \det\left(\frac{1}{N}\sum_{t=1}^{N} \boldsymbol{\varepsilon}(t, \boldsymbol{\vartheta})\boldsymbol{\varepsilon}^T(t, \boldsymbol{\vartheta})\right) . \tag{9.23}$$

Assume the data to be Gaussian and that the innovation matrix $\mathbf{Q}(\boldsymbol{\vartheta})$ in (9.19) is fully parameterized, that is all its elements are assumed to be unknown. The log-likelihood function can then be written as

$$\begin{aligned}\log L &= -\frac{N}{2}\log\left(\det(\mathbf{Q})\right) - \frac{1}{2}\sum_{t=1}^{N} \boldsymbol{\varepsilon}^T(t, \boldsymbol{\theta})\mathbf{Q}^{-1}\boldsymbol{\varepsilon}(t, \boldsymbol{\theta}) \\ &= -\frac{N}{2}\log\left(\det(\mathbf{Q})\right) - \frac{N}{2}\mathrm{tr}\left(\mathbf{Q}^{-1}\frac{1}{N}\sum_{t=1}^{N} \boldsymbol{\varepsilon}(t, \boldsymbol{\theta})\boldsymbol{\varepsilon}^T(t, \boldsymbol{\theta})\right) \\ &\triangleq -V_0(\boldsymbol{\theta}, \mathbf{Q}) .\end{aligned} \tag{9.24}$$

Applying now the analysis of Appendix 2.A.2 it follows that

$$\min_{\mathbf{Q}} V_0(\boldsymbol{\theta}, \mathbf{Q}) = \det\left(\mathbf{Q}^{-1}\frac{1}{N}\sum_{t=1}^{N} \boldsymbol{\varepsilon}(t, \boldsymbol{\theta})\boldsymbol{\varepsilon}^T(t, \boldsymbol{\theta})\right) . \tag{9.25}$$

This gives a rationale for the choice (9.23). ■

Example 9.2 Choose

$$V_N(\boldsymbol{\vartheta}) = \frac{1}{N}\sum_{t=1}^{N} \ell(\boldsymbol{\varepsilon}(t, \boldsymbol{\vartheta}), \boldsymbol{\vartheta}, t) . \tag{9.26}$$

The scalar function $\ell(\boldsymbol{\varepsilon}, \boldsymbol{\vartheta}, t)$ in (9.26) can in turn be chosen in several ways. It should always be minimized with respect to $\boldsymbol{\varepsilon}$ for $\boldsymbol{\varepsilon} = \mathbf{0}$. Further, $\boldsymbol{\vartheta} = \boldsymbol{\vartheta}_0$ should minimize $\mathsf{E}\{\ell(\boldsymbol{\varepsilon}(t, \boldsymbol{\vartheta}), \boldsymbol{\vartheta}, t)\}$ with respect to $\boldsymbol{\vartheta}$. These conditions imply that

$$\mathsf{E}\{\ell_{\boldsymbol{\varepsilon}}(\boldsymbol{\varepsilon}, \boldsymbol{\vartheta}, t)\} = \mathbf{0}, \quad \mathsf{E}\{\ell_{\boldsymbol{\vartheta}}(\boldsymbol{\varepsilon}, \boldsymbol{\vartheta}, t)\} = \mathbf{0} \tag{9.27}$$

hold for $\boldsymbol{\vartheta} = \boldsymbol{\vartheta}_0$, where $\ell_{\boldsymbol{\varepsilon}} = \frac{\partial \ell}{\partial \boldsymbol{\varepsilon}}, \quad \ell_{\boldsymbol{\vartheta}} = \frac{\partial \ell}{\partial \boldsymbol{\vartheta}}$. ■

Example 9.3 A very important special case of Example 9.2 is to choose

$$\ell(\boldsymbol{\varepsilon}, \boldsymbol{\vartheta}, t) = \frac{1}{2}\log\det\mathbf{Q}(\boldsymbol{\vartheta}) + \frac{1}{2}\boldsymbol{\varepsilon}^T(t, \boldsymbol{\vartheta})\mathbf{Q}^{-1}(\boldsymbol{\vartheta})\boldsymbol{\varepsilon}(t, \boldsymbol{\vartheta}) . \tag{9.28}$$

This corresponds to the maximum likelihood estimator, in case the data are Gaussian distributed, that is Assumptions **AN2, AI3** apply. ■

Example 9.4 Choose

$$V_N(\vartheta) = \frac{1}{N} \sum_{t=1}^{N} \boldsymbol{\varepsilon}^T(t, \vartheta) \mathbf{W} \boldsymbol{\varepsilon}(t, \vartheta) , \tag{9.29}$$

where $\mathbf{W}$ is a fixed, positive definite weighting matrix. ■

Consider for a moment the general case of a state space model given as (9.9)–(9.12). In many practical cases the user may have some a priori information about appropriate parameterization, as illustrated in Sect. 5.6. In fact, the prediction errors are invariant to a joint scaling of the noise covariance matrices $\mathbf{R}_1(\vartheta)$, $\mathbf{R}_{12}(\vartheta)$, $\mathbf{R}_2(\vartheta)$. This means that if these matrices are substituted by $r\mathbf{R}_1$, $r\mathbf{R}_{12}$, and $r\mathbf{R}_2$, respectively, the same prediction error will result. Hence such a scaling parameter r should not be included in the parameter vector, if only the prediction errors themselves are parameterized. Nevertheless, it can be estimated but this has to be done as a separate step. The trick when r is to be determined also, is to utilize the minimum value of the criterion

$$\min_{\vartheta} V_N(\vartheta)$$

as a further piece of information from which r can be estimated. For details see Ljung (1999), Söderström and Stoica (1989) and Söderström (2006).

Below some general comments on the PEM approach are given.

- In Söderström (1979, 1981) it was proved that the prediction error method applied to the model structure (9.8) yields consistent parameter estimates under quite mild conditions. See also Theorem 4.1 for some general results on identifiability.
- In the optimization of the loss function, at every iteration a spectral factorization (for example by solving the Riccati equation (9.17)) must be performed in order to find the innovations $\boldsymbol{\varepsilon}(t)$ used in the criterion function.
- The procedure may fail to give good results if only poor initial parameter estimates are available.
- This technique can work with rather arbitrarily correlated noises. For such cases an extended system representation by including also models for the noises is needed.
- The general prediction error approach described above is extended to the case of a continuous-time modeling of the input model and the system dynamics in Söderström et al. (2006). After appropriate sampling of the continuous-time stochastic processes, Söderström (2002), a general state space formalism as in (9.9)–(9.12) is obtained, and the same principle as for discrete-time modeling is applied.
- In Bottegal et al. (2017) an alternative way of organizing the minimization is presented. The formulation is done via the impulse responses of B/A and C/D. The optimization is carried out by applying a variant of the expectation-maximization (EM) method.

Table 9.1 Prediction error estimates for the standard data sets

Parameter	True value	PEM		CM	
		Mean	Standard dev.	Mean	Standard dev.
a_1	−1.5	−1.502	0.014	−1.496	0.038
a_2	0.7	0.701	0.011	0.698	0.026
b_1	2.0	1.934	0.137	1.982	0.268
b_2	1.0	1.017	0.165	1.047	0.513
c_1	0.7	0.066	0.230	–	–
d_1	−1.0	−1.113	0.137	–	–
d_2	0.2	0.292	0.118	–	–
λ_y	10.0	9.25	1.55	11.02	1.34
λ_u	1.81	1.65	0.15	1.83	0.32
λ_v	1.0	3.03	4.93	–	–

Example 9.5 Use of the prediction error method is considered, where the criterion described in Example 9.1 is used. The numerical results are displayed in Table 9.1. The standard data sets are used for a Monte Carlo simulation study. One can note that the PEM estimates are much more accurate than the CM estimates. For the PEM estimates, the user exploits indeed the knowledge that, in this case, the noise-free input $u_0(t)$ is an ARMA(1,1) process. ■

Example 9.6 Use of the maximum likelihood method is considered, where the criterion described in Example 9.3 is used. The numerical results are displayed in Table 9.2. The standard data sets are used for a Monte Carlo simulation study. It clearly appears that the ML estimates are superior in quality to the PEM estimates of Example 9.5. The estimates of c_1 stick out as not being accurate at all. The values presented in the tables are indeed those obtained in the simulations, and not due to particular outlier realizations. The expected accuracy of the ML estimates are shown in Sect. 14.6 to be given by the Cramér–Rao lower bound (CRB). An algorithm to compute CRB is presented in Sect. 14.8.2. The numerical values obtained from that algorithm describe quite well the behavior in the tables. In particular, the expected error in c_1 is one magnitude larger than expected the errors of the other parameters. ■

9.3 Frequency Domain Formulation

The time domain formulation of the PEM given in Sect. 9.2 is for a SISO system with known input equal to the minimization problem

Table 9.2 Maximum likelihood estimates for the standard data sets

Parameter	True value	ML		PEM	
		Mean	Standard dev.	Mean	Standard dev.
a_1	−1.5	−1.497	0.012	−1.502	0.014
a_2	0.7	0.698	0.010	0.701	0.011
b_1	2.0	1.990	0.125	1.934	0.137
b_2	1.0	1.028	0.165	1.017	0.165
c_1	0.7	0.030	0.165	0.066	0.230
d_1	−1.0	−1.188	0.098	−1.113	0.137
d_2	0.2	0.355	0.086	0.292	0.118
λ_y	10.0	10.07	0.66	9.25	1.55
λ_u	1.81	1.70	0.10	1.65	0.15
λ_v	1.0	1.76	0.25	3.03	4.93

$$\hat{\vartheta} = \min_{\vartheta} \frac{1}{N} \sum_{t=1}^{N} \varepsilon^2(t, \vartheta) . \tag{9.30}$$

By using Parseval's relation, this criterion can be transformed to the frequency domain. To cope with the details, consider the prediction error sequence $\varepsilon(1), \dots \varepsilon(N)$ as given. Introduce the discrete Fourier transform (DFT) of this sequence as

$$\varepsilon_k = \frac{1}{\sqrt{N}} \sum_{t=1}^{N} \mathrm{e}^{-\mathrm{i}tk\frac{2\pi}{N}} \varepsilon(t), \quad k = 0, \dots, N-1 . \tag{9.31}$$

Note that the sequence $\{\varepsilon_k\}$ is complex-valued. Then, by Parseval's relation is holds

$$\sum_{t=1}^{N} \varepsilon^2(t) = \sum_{k=0}^{N-1} |\varepsilon_k|^2 . \tag{9.32}$$

Hence it is possible to use frequency domain data by taking DFT of the measurements and for any relevant value of the parameter vector ϑ compute the prediction errors $\varepsilon_0(\vartheta), \dots, \varepsilon_{N-1}(\vartheta)$, and determine the estimate as

$$\hat{\vartheta} = \min_{\vartheta} \sum_{k=0}^{N-1} |\varepsilon_k(\vartheta)|^2 . \tag{9.33}$$

Referring to (9.32) it seems that (9.30) and (9.33) are exactly equivalent.

However, this is not exactly so without some further specifications, see Agüero et al. (2010), Pintelon and Schoukens (2007). The two estimates (9.30) and (9.33) are shown in Agüero et al. (2010) to be fully equivalent provided that initial and final values conditions are considered in compatible ways for the two estimators. The effect of initial values of the criterion in (9.30) is known to have only a marginal influence for large data sets. Hence, if no specific consideration of initial and final values is done, the two estimators are asymptotically equivalent when $N \to \infty$. A more precise way of handling the effects of initial and final values is to apply the frequency domain model presented in Sect. 5.5. For examples of how this can be done, see Sects. 10.5 and 9.4.

To make use of some statistical properties of the DFT of a time series, Lemma A.13 in Appendix A is useful.

The result of Lemma A.13 can be formulated in words as follows. Asymptotically, as the number of data points tends to infinity, the frequency domain data, say $Y_0, \ldots, Y_{N-1}$ are uncorrelated for different frequencies. Further, the distribution is circular symmetric in the sense that $\mathsf{E}\left\{Y_j^2\right\} = 0$, meaning that the real and imaginary parts of Y_j are uncorrelated and have both the same variance, namely $\frac{1}{2}\phi(\frac{2\pi}{N}j)$.

Next derive the maximum likelihood criterion in the frequency domain using the asymptotic expressions in Lemma A.13, following the ideas of Pintelon and Schoukens (2001), Pintelon and Schoukens (2007).

Let $\mathbf{Z}_k,\ k = 0, \ldots, N-1$ denote the DFT of the data $\boldsymbol{\zeta}(t), t = 1, \ldots, N$. It follows from Lemma A.13 that $\mathbf{Z}_j$ behaves as a circular complex-valued white noise sequence with covariance matrix

$$\mathbf{Q}_j = \mathsf{E}\left\{\mathbf{Z}_j\mathbf{Z}_j^*\right\} = \begin{pmatrix} \phi_y(\omega_j) & \phi_{yu}(\omega_j) \\ \phi_{uy}(\omega_j) & \phi_u(\omega_j) \end{pmatrix} = \begin{pmatrix} G_jG_j^*\phi_j + \lambda_y & G_j\phi_j \\ G_j^*\phi_j & \phi_j + \lambda_u \end{pmatrix}, \tag{9.34}$$

where the conventions

$$\omega_j = \frac{2\pi}{N}j,\ \ j = 0, \ldots, N-1\ , \tag{9.35}$$

$$G_j = G(\mathrm{e}^{-\mathrm{i}\omega_j}),\ \ \phi_j = \phi_{u_0}(\omega_j) \tag{9.36}$$

are used.

Assume further that the data are Gaussian distributed (Assumptions **AN2**, **AI3** apply). Then, up to a $\boldsymbol{\vartheta}$ independent constant, the negative log-likelihood function is given by, see Sect. A.3.2,

$$V(\boldsymbol{\vartheta}) = \sum_{j=0}^{N-1}\left[\mathbf{Z}_j^*\mathbf{Q}_j^{-1}\mathbf{Z}_j + \log\det(\mathbf{Q}_j)\right]. \tag{9.37}$$

By direct calculations, using (9.34),

$$\det(\mathbf{Q}_j) = \lambda_y \lambda_u + \phi_{u_0}(\omega_j)\left[\lambda_y + |G(e^{-i\omega_j})|^2 \lambda_u\right], \tag{9.38}$$

$$\begin{aligned}\mathbf{Z}_j^* \mathbf{Q}_j^{-1} \mathbf{Z}_j &= \frac{1}{\det(\mathbf{Q}_j)} \begin{pmatrix} Y_j^* & U_j^* \end{pmatrix} \\ &\times \begin{pmatrix} \phi_{u_0}(\omega_j) + \lambda_u & -G(e^{-i\omega_j})\phi_{u_0}(\omega_j) \\ -G^*(e^{-i\omega_j})\phi_{u_0}(\omega_j) & G(e^{-i\omega_j})G^*(e^{-i\omega_j})\phi_{u_0}(\omega_j) + \lambda_y \end{pmatrix} \begin{pmatrix} Y_j \\ U_j \end{pmatrix} \\ &= \frac{1}{\det(\mathbf{Q}_j)} \left[\phi_{u_0}(\omega_j)|Y_j - G(e^{-i\omega_j})U_j|^2 + \lambda_u |Y_j|^2 + \lambda_y |U_j|^2\right]. \end{aligned} \tag{9.39}$$

Remark 9.3 The expressions hold asymptotically, cf Lemma A.13. An exact (and a bit more involved) expression that also includes the effects of transients is presented in Pintelon and Schoukens (2007). It is also possible to extend the analysis to cases where the input and output noises, $\tilde{u}(t)$ and $\tilde{y}(t)$, are correlated. ■

Remark 9.4 As it stands, the frequency domain formulation (9.37) of the maximum likelihood criterion can be seen as an alternative formulation of the time domain version in Sect. 9.2. ■

9.4 A Frequency Domain Maximum Likelihood Method

Consider again the frequency domain model (5.35).

Assume that the measurement noise is white on both the input and the output side (Assumption **AN3c** applies) and that the noise variance ratio

$$r = \lambda_y / \lambda_u \tag{9.40}$$

is known. See Assumption **AN4**.

The frequency domain model can be written as (5.35),

$$A(e^{-i\omega_k})\, Y_0(\omega_k) = B(e^{-i\omega_k})\, U_0(\omega_k) + T(e^{-i\omega_k}). \tag{9.41}$$

Introduce also the following notations

$$\mathbf{U} = \begin{pmatrix} U_0, & U_1, & \cdots, & U_M \end{pmatrix}^T, \tag{9.42}$$

$$\mathbf{Y} = \begin{pmatrix} Y_0, & Y_1, & \cdots, & Y_M \end{pmatrix}^T, \tag{9.43}$$

and let $\mathbf{U}_0$, $\mathbf{Y}_0$ be the correspondence to (9.42) and (9.43) for the noise–free data. Above, M denotes the number of frequencies to be used, and $M \leq N$ must apply. It follows from Appendix A.2 that the choice $M = N/2$ gives no loss of information, but other (lower) values may be used as a way of lowpass filtering the data.

Further, introduce also (for $k = 1, \ldots, M$)

$$A_k = A(e^{-i\omega_k}),\ B_k = B(e^{-i\omega_k}),\ T_k = T(e^{-i\omega_k}) , \tag{9.44}$$
$$G_k = B(e^{-i\omega_k})/A(e^{-i\omega_k}) , \tag{9.45}$$
$$H_k = T(e^{-i\omega_k})/A(e^{-i\omega_k}) , \tag{9.46}$$

and

$$\mathbf{G} = \operatorname{diag}\left(G_0,\ G_1, \ldots, G_M\right) , \tag{9.47}$$
$$\mathbf{H} = \left(H_0,\ H_1, \cdots , H_M\right)^T . \tag{9.48}$$

For the noise–free data and an arbitrary frequency ω_k relation (9.41) holds, i.e.

$$A_k Y_k^0 = B_k U_k^0 + T_k . \tag{9.49}$$

For all the frequencies, the system model can be written as

$$\mathbf{Y}_0 = \mathbf{G}\mathbf{U}_0 + \mathbf{H} . \tag{9.50}$$

In the noisy case it holds instead

$$\mathbf{Y} = \mathbf{G}\mathbf{U}_0 + \mathbf{H} + \tilde{\mathbf{Y}} , \tag{9.51}$$
$$\mathbf{U} = \mathbf{U}_0 + \tilde{\mathbf{U}} , \tag{9.52}$$

where the noise vectors $\tilde{\mathbf{Y}}$ and $\tilde{\mathbf{U}}$ are defined in similar way to (9.42) and (9.43). Because of Assumptions **AN3c** and **AN4**, they are jointly complex Gaussian distributed. In fact,

$$\begin{pmatrix} \mathbf{Y} \\ \mathbf{U} \end{pmatrix} \sim \mathscr{CN}\left(\begin{pmatrix} \mathbf{G}\mathbf{U}_0 + \mathbf{H} \\ \mathbf{U}_0 \end{pmatrix}, \begin{pmatrix} r\lambda_u \mathbf{I}_M & \mathbf{0} \\ \mathbf{0} & \lambda_u \mathbf{I}_M \end{pmatrix}\right) . \tag{9.53}$$

where the identity matrices in (9.53) are of dimension $M \times M$. It is assumed that the realization $\mathbf{U}_0$ of the noise-free input is a deterministic sequence to be estimated.

The unknown quantities in the model (9.53) are hence:

- The parameter vector $\boldsymbol{\theta}$, which enters in $\mathbf{G}$ and $\mathbf{H}$,
- The input noise variance λ_u,
- The auxiliary parameter vector $\boldsymbol{\tau}$, which enters in $\mathbf{H}$. In particular, note that $\mathbf{H}$ depends *linearly* on $\boldsymbol{\tau}$.
- The noise-free data vector $\mathbf{U}_0$.

Remark 9.5 The model (9.53) is valid for any set of frequency values, even for M smaller than $N/2$. ■

9.4.1 The Frequency Domain ML Estimator

Using the model (9.53), one can directly set up the likelihood function, cf. Söderström (2002). Use the pdf of complex-valued circular Gaussian variables, see Appendix A.3.2. The result will be

$$\begin{aligned}\log(L) &= -(\mathbf{Y} - \mathbf{G}\mathbf{U}_0 - \mathbf{H})^*(\mathbf{Y} - \mathbf{G}\mathbf{U}_0 - \mathbf{H})/(r\lambda_u) \\ &\quad -(\mathbf{U} - \mathbf{U}_0)^*(\mathbf{U} - \mathbf{U}_0)/\lambda_u \\ &\quad - \log(\det(r\lambda_u\mathbf{I})) - \log(\det(\lambda_u\mathbf{I}))\ , \end{aligned} \tag{9.54}$$

where L is the likelihood function. The maximum likelihood estimates of the unknowns $\boldsymbol{\theta},\ \lambda_u,\ \boldsymbol{\tau}$ and $\mathbf{U}_0$ are

$$\left(\hat{\boldsymbol{\theta}}, \hat{\lambda}_u, \hat{\boldsymbol{\tau}}, \hat{\mathbf{U}}_0\right) = \arg \max_{\boldsymbol{\theta}, \lambda_u, \boldsymbol{\tau}, \mathbf{U}_0} \log(L)\ . \tag{9.55}$$

To derive the ML estimator, one has to find a way to express the maximum point of the likelihood function, with respect to all the unknowns. Fortunately, this is a separable nonlinear least squares problem, which allows the maximization to be done with respect to the different unknowns to be treated in separate steps. Next the details are developed.

9.4.2 Maximization with Respect to λ_u

First consider how the likelihood function L depends on the unknown λ_u. Rewrite (9.54) as

$$\log(L) = -V_1/\lambda_u - \log(r^M \lambda_u^M) - \log(\lambda_u^M)\ , \tag{9.56}$$

$$\begin{aligned} V_1(\boldsymbol{\theta}, \mathbf{U}_0, \boldsymbol{\tau}) &\stackrel{\Delta}{=} (\mathbf{Y} - \mathbf{G}\mathbf{U}_0 - \mathbf{H})^*(\mathbf{Y} - \mathbf{G}\mathbf{U}_0 - \mathbf{H})/r \\ &\quad + (\mathbf{U} - \mathbf{U}_0)^*(\mathbf{U} - \mathbf{U}_0)\ . \end{aligned} \tag{9.57}$$

Maximization of L with respect to λ_u is now straightforward. Direct differentiation gives

$$0 = \frac{\partial \log(L)}{\partial \lambda_u} = V_1/\lambda_u^2 - 2M/\lambda_u\ , \tag{9.58}$$

and hence the maximizing argument for λ_u is

$$\hat{\lambda}_u = \frac{1}{2M} \min_{\boldsymbol{\theta}, \mathbf{U}_0, \boldsymbol{\vartheta}} V_1 = \frac{1}{2M} V_1(\hat{\boldsymbol{\theta}}, \hat{\mathbf{U}}_0, \hat{\boldsymbol{\vartheta}})\ . \tag{9.59}$$

The conclusion so far is that to maximize the likelihood function, one needs to find the minimum point of the criterion V_1 with respect to $\boldsymbol{\theta}$, $\mathbf{U}_0$, $\boldsymbol{\tau}$.

9.4.3 Minimization with Respect to $\mathbf{U}_0$

Note that the loss function $V_1(\boldsymbol{\theta}, \mathbf{U}_0, \boldsymbol{\tau})$ is quadratic in $\mathbf{U}_0$, see (9.57). Hence this minimization is straightforward. It holds

$$\begin{aligned} V_1(\boldsymbol{\theta}, \mathbf{U}_0, \boldsymbol{\tau}) &= \mathbf{U}_0^* \left[\mathbf{G}^*\mathbf{G}/r + \mathbf{I}\right] \mathbf{U}_0 - \mathbf{U}_0^* \left[\mathbf{G}^*(\mathbf{Y} - \mathbf{H})/r + \mathbf{U}\right] \\ &\quad - \left[(\mathbf{Y}^* - \mathbf{H}^*)\mathbf{G}/r + \mathbf{U}^*\right] \mathbf{U}_0 + (\mathbf{Y}^* - \mathbf{H}^*)(\mathbf{Y} - \mathbf{H})/r + \mathbf{U}^*\mathbf{U} , \end{aligned} \tag{9.60}$$

so the minimizing argument is

$$\hat{\mathbf{U}}_0 = \left[\hat{\mathbf{G}}^*\hat{\mathbf{G}}/r + \mathbf{I}\right]^{-1} \left[\hat{\mathbf{G}}^*(\mathbf{Y} - \hat{\mathbf{H}})/r + \mathbf{U}\right] , \tag{9.61}$$

and the minimal value of the loss function is

$$\begin{aligned} V_2(\boldsymbol{\theta}, \boldsymbol{\tau}) &\stackrel{\Delta}{=} \min_{\mathbf{U}_0} V_1(\boldsymbol{\theta}, \mathbf{U}_0, \boldsymbol{\tau}) \\ &= (\mathbf{Y}^* - \mathbf{H}^*)(\mathbf{Y} - \mathbf{H})/r + \mathbf{U}^*\mathbf{U} \\ &\quad - \left[(\mathbf{Y}^* - \mathbf{H}^*)\mathbf{G}/r + \mathbf{U}^*\right] \left[\mathbf{G}^*\mathbf{G}/r + \mathbf{I}\right]^{-1} \left[\mathbf{G}^*(\mathbf{Y} - \mathbf{H})/r + \mathbf{U}\right] . \end{aligned} \tag{9.62}$$

It also follows that, see (9.59),

$$\hat{\lambda}_u = \frac{1}{2M} V_2(\hat{\boldsymbol{\theta}}, \hat{\boldsymbol{\tau}}) . \tag{9.63}$$

9.4.4 Minimization with Respect to $\boldsymbol{\tau}$

Next proceed and minimize $V_2(\boldsymbol{\theta}, \boldsymbol{\tau})$ with respect to $\boldsymbol{\tau}$. The key observation is then to note that V_2 is a quadratic function of $\mathbf{H}$, and that $\mathbf{H}$ depends linearly on $\boldsymbol{\tau}$. From (10.53), (10.54), (9.46) and (9.48) one can write

$$\mathbf{H} = \mathbf{K}\boldsymbol{\tau} , \tag{9.64}$$

$$\mathbf{K} = \begin{pmatrix} \frac{1}{A_0} \mathrm{e}^{-\mathrm{i}\omega_0(\mathrm{n}-1)} & \dots & \frac{1}{A_0} \\ \vdots & & \vdots \\ \frac{1}{A_{M-1}} \mathrm{e}^{-\mathrm{i}\omega_{M-1}(\mathrm{n}-1)} & \dots & \frac{1}{A_{M-1}} \end{pmatrix} \stackrel{\Delta}{=} \begin{pmatrix} \mathbf{k}_0 \\ \vdots \\ \mathbf{k}_{M-1} \end{pmatrix} . \tag{9.65}$$

Inserting (9.64) into (9.62) gives

$$
\begin{aligned}
V_2(\boldsymbol{\theta}, \boldsymbol{\tau}) = {} & \boldsymbol{\tau}^T \left(\mathbf{K}^*\mathbf{K}/r - \mathbf{K}^*\mathbf{G}\rho \left[\mathbf{G}^*\mathbf{G}/r + \mathbf{I}\right]^{-1} \mathbf{G}^*\mathbf{K}/r\right) \boldsymbol{\tau} \\
& + \boldsymbol{\tau}^T \left(-\mathbf{K}^*\mathbf{Y}/r + \mathbf{K}^*\mathbf{G}/r \left[\mathbf{G}^*\mathbf{G}/r + \mathbf{I}\right]^{-1} (\mathbf{G}^*\mathbf{Y}/r + \mathbf{U})\right) \\
& + \left((\mathbf{Y}^*\mathbf{G}/r + \mathbf{U}^*) \left[\mathbf{G}^*\mathbf{G}/r + \mathbf{I}\right]^{-1} \mathbf{G}^*\mathbf{K}/r - \mathbf{Y}^*\mathbf{K}/r\right) \boldsymbol{\tau} \\
& + \mathbf{Y}^*\mathbf{Y}/r + \mathbf{U}^*\mathbf{U} - (\mathbf{Y}^*\mathbf{G}/r + \mathbf{U}^*) \left[\mathbf{G}^*\mathbf{G}/r + \mathbf{I}\right]^{-1} (\mathbf{G}^*\mathbf{Y}/r + \mathbf{U}) \ .
\end{aligned}
\tag{9.66}
$$

It is seen that the argument $\boldsymbol{\tau}$ that minimizes $V_2(\boldsymbol{\theta}, \boldsymbol{\tau})$ is given by

$$
\begin{aligned}
\hat{\boldsymbol{\tau}} = {} & -\left(\hat{\mathbf{K}}^*\hat{\mathbf{K}}/r - \hat{\mathbf{K}}^*\hat{\mathbf{G}}/r \left[\hat{\mathbf{G}}^*\hat{\mathbf{G}}/r + \mathbf{I}\right]^{-1} \hat{\mathbf{G}}^*\hat{\mathbf{K}}/r\right)^{-1} \\
& \times \left(- \hat{\mathbf{K}}^*\mathbf{Y}/r + \hat{\mathbf{K}}^*\hat{\mathbf{G}}/r \left[\hat{\mathbf{G}}^*\hat{\mathbf{G}}/r + \mathbf{I}\right]^{-1} \left(\hat{\mathbf{G}}^*\mathbf{Y}/r + \mathbf{U}\right)\right) ,
\end{aligned}
\tag{9.67}
$$

while the minimal value is

$$
\min_{\boldsymbol{\theta}, \boldsymbol{\tau}} V_2(\boldsymbol{\theta}, \boldsymbol{\tau}) = V_2(\hat{\boldsymbol{\theta}}, \hat{\boldsymbol{\tau}}) = M V_3(\hat{\boldsymbol{\theta}}) \ , \tag{9.68}
$$

$$
\hat{\boldsymbol{\theta}} = \arg\min_{\boldsymbol{\theta}} V_3(\boldsymbol{\theta}) \ , \tag{9.69}
$$

$$
\begin{aligned}
V_3(\boldsymbol{\theta}) \stackrel{\Delta}{=} {} & \frac{1}{M} \left[\mathbf{Y}^*\mathbf{Y}/r + \mathbf{U}^*\mathbf{U}\right] \\
& - \frac{1}{M} (\mathbf{Y}^*\mathbf{G}/r + \mathbf{U}^*) \left[\mathbf{G}^*\mathbf{G}/r + \mathbf{I}\right]^{-1} (\mathbf{G}^*\mathbf{Y}/r + \mathbf{U}) \\
& - \frac{1}{M} \left((\mathbf{Y}^*\mathbf{G}/r + \mathbf{U}^*) \left[\mathbf{G}^*\mathbf{G}/r + \mathbf{I}\right]^{-1} \mathbf{G}^*\mathbf{K}/r - \mathbf{Y}^*\mathbf{K}/r\right) \\
& \times \left(\mathbf{K}^*\mathbf{K}/r - \mathbf{K}^*\mathbf{G}/r \left[\mathbf{G}^*\mathbf{G}/r + \mathbf{I}\right]^{-1} \mathbf{G}^*\mathbf{K}/r\right)^{-1} \\
& \times \left(-\mathbf{K}^*\mathbf{Y}/r + \mathbf{K}^*\mathbf{G}/r \left[\mathbf{G}^*\mathbf{G}/r + \mathbf{I}\right]^{-1} (\mathbf{G}^*\mathbf{Y}/r + \mathbf{U})\right) \ .
\end{aligned}
\tag{9.70}
$$

Also, cf. (9.63),

$$
\hat{\lambda}_u = \frac{1}{2} V_3(\hat{\boldsymbol{\theta}}) \ . \tag{9.71}
$$

Note that the normalization factor M is inserted in (9.68), (9.70) for convenience in the following analysis only.

9.4.5 Minimization with Respect to $\boldsymbol{\theta}$

The result of the calculations so far is that the ML estimate of $\boldsymbol{\theta}$ can be obtained by minimizing the function $V_3(\boldsymbol{\theta})$, in which none of the other originally introduced unknowns appear. In this subsection the expression for the loss function $V_3(\boldsymbol{\theta})$ is rewritten using frequency domain formulas. It is important to recall that the matrix $\mathbf{G}$ is diagonal, see (9.47). It holds, with all summations being over $\sum_0^{M-1}$,

$$
\begin{aligned}
V_3(\boldsymbol{\theta}) &= \frac{1}{M}\left[\frac{1}{r}\sum_j |Y_j|^2 + \sum_j |U_j|^2\right] - \frac{1}{M}\sum_j \frac{|G_j^* Y_j/r + U_j|^2}{G_j^* G_j/r + 1} \\
&-\frac{1}{M}\left(\sum_j \frac{Y_j^* G_j/r + U_j^*}{G_j^* G_j/r + 1} G_j^* \mathbf{k}_j/r - \sum_j Y_j^* \mathbf{k}_j/r\right) \\
&\times \left(\frac{1}{r}\sum_j \mathbf{k}_j^* \mathbf{k}_j - \sum_j \frac{\mathbf{k}_j^* G_j}{r}\frac{G_j^* \mathbf{k}_j}{r}\frac{1}{G_j G_j^*/r + 1}\right)^{-1} \\
&\times \left(\sum_j -\mathbf{k}_j^* Y_j/r + \sum_j \frac{\mathbf{k}_j^* G_j}{r}\frac{G_j^* Y_j/r + U_j}{G_j G_j^*/r + 1}\right) .
\end{aligned} \tag{9.72}
$$

The result (9.72) can be further rewritten as

$$
\begin{aligned}
V_3(\boldsymbol{\theta}) &= \frac{1}{M}\sum_j \frac{|Y_j|^2 (G_j G_j^*/r + 1) + |U_j|^2 r (G_j G_j^*/r + 1)}{r(G_j G_j^*/r + 1)} \\
&- \frac{1}{M}\sum_j \frac{r|G_j^* Y_j/r + U_j|^2}{r(G_j G_j^*/r + 1)} - \frac{1}{M}\left(\sum_j \frac{(Y_j^* G_j/r + U_j^*)G_j^* - Y_j^*(G_j G_j^*/r + 1)}{r(G_j G_j^*/r + 1)}\mathbf{k}_j\right) \\
&\times \left(\sum_j \mathbf{k}_j \mathbf{k}_j^* \frac{r(G_j G_j^*/r + 1) - G_j G_j^*}{r^2 (G_j G_j^*/r + 1)}\right)^{-1} \\
&\times \left(\sum_j \mathbf{k}_j^* \frac{G_j (G_j^* Y_j/r + U_j) - Y_j (G_j G_j^*/r + 1)}{r(G_j G_j^*/r + 1)}\right) \\
&= \frac{1}{M}\sum_j \frac{Y_j Y_j^* + U_j U_j^* G_j G_j^* - G_j^* Y_j U_j^* - G_j Y_j^* U_j}{G_j G_j^* + r} \\
&- \frac{1}{M}\left(\sum_j \frac{U_j^* G_j^* - Y_j^*}{G_j G_j^* + r}\mathbf{k}_j\right)\left(\sum_j \mathbf{k}_j^* \mathbf{k}_j \frac{1}{G_j G_j^* + r}\right)^{-1}\left(\sum_j \mathbf{k}_j^* \frac{G_j U_j - Y_j}{G_j G_j^* + r}\right) .
\end{aligned} \tag{9.73}
$$

The first part of (9.73) can be rewritten as

$$
V_{31}(\boldsymbol{\theta}) = \frac{1}{M}\sum_{j=0}^{M-1} \frac{|Y_j - B_j/A_j U_j|^2}{|B_j/A_j|^2 + r} = \frac{1}{M}\sum_{j=0}^{M-1} \frac{|A_j Y_j - B_j U_j|^2}{r|A_j|^2 + |B_j|^2} . \tag{9.74}
$$

Remark 9.6 $V_{31}(\boldsymbol{\theta})$ turns out to be precisely the form of the ML criterion in the frequency domain used as the basis for the SML method, see Schoukens et al. (1997) and Sect. 12.3. The SML method assumes periodic data, and includes then also estimates of the noise variances, replacing r. The second part of (9.73) corresponds to the effects of the transients, in the computation of the DFT, when the data are finite and not periodic. One would expect the impact of the second term to be small compared to that of the first term. ∎

Remark 9.7 One can in fact rewrite the loss function $V_2(\boldsymbol{\theta}, \boldsymbol{\tau})$ as given by (9.66) into a form using the polynomial formalism, following the principles used in this subsection. The result turns out to be

$$V_2(\boldsymbol{\theta}, \boldsymbol{\tau}) = \frac{1}{M} \sum_{i=0}^{M-1} \frac{|A_j Y_j - T_j - B_j U_j|^2}{r|A_j|^2 + |B_j|^2} \tag{9.75}$$

which can be compared to $V_3(\boldsymbol{\theta})$ in (9.73) and $V_{31}(\boldsymbol{\theta})$ in (9.74). The expression of $V_2(\boldsymbol{\theta}, \boldsymbol{\tau})$ in (9.75) looks neater than that in (9.73), but on the hand it contains $3n$ independent parameters instead of $2n$. ∎

9.4.6 The ML Algorithm

The general minimization problem (9.55) with respect to $\boldsymbol{\theta}, \lambda_u, \boldsymbol{\tau}, \mathbf{U}_0$ has been reduced in dimension to optimization over only $\boldsymbol{\theta}$:

$$\hat{\boldsymbol{\theta}} = \arg\min_{\theta} V_3(\boldsymbol{\theta}) \tag{9.76}$$

with $V_3(\boldsymbol{\theta})$ given by (9.70) or equivalently (9.73). The minimization in (9.76) has to be carried out using a numerical search algorithm, and will then require an iterative procedure. As for similar optimization problems, this is NP hard, and there is a potential risk for converging to local false minima. It makes sense to start the iterations over $\boldsymbol{\theta}$ using some consistent estimate, for example obtained with the generalized instrumental variable method.

Once $\hat{\boldsymbol{\theta}}$ is found, the remaining parameters are found as, cf. (9.67), (9.61), (9.71),

$$\begin{aligned} \hat{\boldsymbol{\tau}} &= -\left(\hat{\mathbf{K}}^*\hat{\mathbf{K}}/r - \hat{\mathbf{K}}^*\hat{\mathbf{G}}r\left[\hat{\mathbf{G}}^*\hat{\mathbf{G}}/r + \mathbf{I}\right]^{-1}\hat{\mathbf{G}}^*\hat{\mathbf{K}}/r\right)^{-1} \\ &\quad \times \left(-\hat{\mathbf{K}}^*\mathbf{Y}/r + \hat{\mathbf{K}}^*\hat{\mathbf{G}}/r\left[\hat{\mathbf{G}}^*\hat{\mathbf{G}}/r + \mathbf{I}\right]^{-1}\left(\hat{\mathbf{G}}^*\mathbf{Y}/r + \mathbf{U}\right)\right), \end{aligned} \tag{9.77}$$

$$\hat{\mathbf{U}}_0 = \left[\hat{\mathbf{G}}^*\hat{\mathbf{G}}/r + \mathbf{I}\right]^{-1}\left[\hat{\mathbf{G}}^*(\mathbf{Y} - \hat{\mathbf{H}})/r + \mathbf{U}\right], \tag{9.78}$$

$$\hat{\lambda}_u = \frac{1}{2} V_3(\hat{\boldsymbol{\theta}}). \tag{9.79}$$

Remark 9.8 The ML estimation problem formulated in (9.54) turns out to have the same algebraic form as the EIV problem for a static linear system where all the latent variables (corresponding to $\mathbf{U}_0$ here) are estimated. Also for such cases it is known that the estimated parameters $\hat{\boldsymbol{\theta}}$ are consistent, but the estimated noise variance (corresponding to λ_u) is asymptotically off by 50% in (9.79), cf. Lindley (1947) and (2.142). For this reason, in order to achieve consistency the estimate of λ_u must be modified by doubling the result of (9.79),

$$\hat{\lambda}_u = V_3(\hat{\boldsymbol{\theta}}) . \tag{9.80}$$

■

Remark 9.9 In many other estimation problems, maximum likelihood estimates are consistent, in contrast to (9.79). A particular aspect here is that as $\mathbf{U}_0$ is included as unknown to be estimated, the number of estimates grows as $\sim N$ when the data length N increases. This is in contrast to many 'standard' ML problems, where the number of unknown parameters stay fixed. It is also noted in Zhang et al. (2013) that the derivatives of the likelihood function with respect to the noise variances are non-zero when evaluated for the true parameters. This fact indicates that the noise statistical properties cannot be consistently estimated using the ML approach. ■

9.5 An Extended Frequency Domain ML Method

It was explained in Sect. 9.4 that the maximum likelihood method, treating the noise-free input realization, as additional unknown nuisance parameters to be estimated, would give consistent parameter estimates only if the noise variance ratio $r = \lambda_y/\lambda_u$ is known. In practice, the assumption that r is known can be quite restrictive.

Here an extension of the ML approach is presented; see Zhang et al. (2013) for further details. The main advantage of this extension is that the parameter r is estimated as well.

The general setup of Sect. 9.4 is used. To simplify the exposition, the treatment of transient effects based on the vector $\boldsymbol{\vartheta}$ is neglected. Recall that the likelihood function L can be written,

$$\begin{aligned} L &= L(\mathbf{Y}, \mathbf{U}|\boldsymbol{\theta}, \mathbf{U}_0, r, \lambda_u) \\ &= \frac{1}{r^N \lambda_u^{2N}} \exp\left(\sum_k \left(-\frac{|Y_k - G_k(\boldsymbol{\theta}) U_{0,k}|^2}{r\lambda_u} - \frac{|U_k - U_{0,k}|^2}{\lambda_u} \right) \right) . \end{aligned} \tag{9.81}$$

The maximization of L with respect to the noise-free input $\mathbf{U}_0 = \left(U_{0,0}, \ldots U_{0,N-1}\right)^T$ is achieved for

$$\hat{U}_{0,k} = \frac{rU_k + G_k^* Y_k}{r + |G_k|^2} . \tag{9.82}$$

It was derived in Sect. 9.4 that combining (9.81) and (9.82), the ML estimate of $\boldsymbol{\theta}$ is obtained as the minimizing element of

$$V_{31}(\boldsymbol{\theta}, r) = \frac{1}{N} \sum_{k=0}^{N-1} \frac{|Y_k - G_k(\boldsymbol{\theta}) U_k|^2}{r + |G_k(\boldsymbol{\theta})|^2} , \tag{9.83}$$

where though r was assumed to be known.

The idea here is to consider, in addition to the likelihood function $L = L(\mathbf{Y}, \mathbf{U}|\boldsymbol{\theta}, \mathrm{U}_0, r, \lambda_u)$ in (9.81), also a marginal likelihood function, where the dependence on the noise-free input is integrated out. To this aim, consider therefore

$$\begin{aligned} L_2(\mathbf{Y}, \mathbf{U}|\boldsymbol{\theta}, r, \lambda_u) &= \int L(\mathbf{Y}, \mathbf{U}|\boldsymbol{\theta}, \mathbf{U}_0, r, \lambda_u) d\mathbf{U}_0 \\ &= \int \frac{1}{r^N \lambda_u^N} \exp\left(-\frac{\sum_k |Y_k - G_k(\boldsymbol{\theta}) U_{0,k}|^2}{r\lambda_u} \right) \\ &\quad \times \frac{1}{\lambda_u^N} \exp\left(-\frac{\sum_k |U_k - U_{0,k}|^2}{\lambda_u} \right) d\mathbf{U}_0 \,. \end{aligned} \tag{9.84}$$

Set

$$N_k = r + |G_k(\boldsymbol{\theta})|^2 \,. \tag{9.85}$$

The negative exponential part in (9.84) can now be written as, suppressing the dependence of $G_k(\boldsymbol{\theta})$ on $\boldsymbol{\theta}$,

$$\begin{aligned} &\frac{1}{r\lambda_u} \sum_k |Y_k - G_k U_{0,k}|^2 + \frac{1}{\lambda_u} \sum_k |U_k - U_{0,k}|^2 \\ &= \frac{1}{r\lambda_u} \sum_k \left[\left(Y_k - G_k U_{0,k}\right)^* \left(Y_k - G_k U_{0,k}\right) + r \left(U_k - U_{0,k}\right)^* \left(U_k - U_{0,k}\right) \right] \\ &= \frac{1}{r\lambda_u} \sum_k \left[U_{0,k}^* U_{0,k} N_k + U_{0,k}^* \left(-G_k^* Y_k - r U_k \right) \right. \\ &\quad \left. + U_{0,k} \left(-G_k Y_k^* - r U_k^* \right) + \left(Y_k^* Y_k + r U_k^* U_k \right) \right] \\ &= \frac{1}{r\lambda_u} \sum_k \left[N_k (U_{0,k} - \hat{U}_{0,k})^* (U_{0,k} - \hat{U}_{0,k}) + \left(Y_k^* Y_k + r U_k^* U_k \right) - \hat{U}_{0,k}^* \hat{U}_{0,k} N_k \right] . \end{aligned} \tag{9.86}$$

Next noting that for a scalar complex Gaussian variable x with mean μ it holds, cf (A.88),

$$\int \exp\left(-\frac{|x - \mu|^2}{\sigma} \right) dx = \pi\sigma , \tag{9.87}$$

one finds

$$L_2(\mathbf{Y}, \mathbf{U}|\boldsymbol{\theta}, r, \lambda_u) = \frac{1}{r^N \lambda_u^{2N}} \pi^N \frac{r^N \lambda_u^N}{\prod_k N_k} \times \exp\left(-\frac{1}{r\lambda_u} \sum_k \left[\left(Y_k^* Y_k + r U_k^* U_k\right) - \hat{U}_{0,k}^* \hat{U}_{0,k} N_k \right] \right) . \tag{9.88}$$

The terms in the sum in the exponential part of (9.88) can be rewritten as, cf. (9.82),

$$\begin{aligned} & Y_k^* Y_k + r U_k^* U_k - \hat{U}_{0,k}^* \hat{U}_{0,k} N_k \\ &= Y_k^* Y_k + r U_k^* U_k - \frac{1}{N_k} \left(r U_k^* + G_k Y_k^* \right) \left(r U_k + G_k^* Y_k \right) \\ &= \frac{1}{N_k} \left[Y_k^* Y_k r + r |G_k|^2 U_k^* U_k - r U_k^* G_k^* Y_k - r U_k G_k Y_k^* \right] \\ &= \frac{r}{N_k} |Y_k - G_k U_k|^2 . \end{aligned} \tag{9.89}$$

Thus,

$$L_2(\mathbf{Y}, \mathbf{U}|\boldsymbol{\theta}, r, \lambda_u) = \frac{\pi^N}{\lambda_u^N} \frac{1}{\prod_k (r + |G_k(\boldsymbol{\theta})|^2} \times \exp\left(-\frac{1}{\lambda_u} \sum_k \frac{|Y_k - G_k(\boldsymbol{\theta}) U_k|^2}{r + |G_k(\boldsymbol{\theta})|^2} \right) . \tag{9.90}$$

To maximize L_2 is obviously equivalent to minimize

$$-\frac{1}{N} \log L_2(\mathbf{Y}, \mathbf{U}|\boldsymbol{\theta}, r, \lambda_u) = -\log(\pi) + \log(\lambda_u) + \frac{1}{N} \sum_k \log\left(r + |G_k(\boldsymbol{\theta})|^2 \right) + \frac{1}{\lambda_u} V_{31}(\boldsymbol{\theta}, r) . \tag{9.91}$$

Minimization with respect to λ_u is straightforward as

$$\frac{\partial}{\partial \lambda_u} (-\log(L_2)) = \frac{1}{\lambda_u} - \frac{1}{\lambda_u^2} V_{31}(\boldsymbol{\theta}, r) = 0 \tag{9.92}$$

leads to

$$\hat{\lambda}_u = V_{31}(\hat{\boldsymbol{\theta}}, \hat{r}) . \tag{9.93}$$

Note that this result differs from (9.79) as here the factor $1/2$ (which prevents consistency) *does not appear*.

Further, minimization with respect to r leads to

$$0 = \frac{\partial}{\partial r}(-\log(L_2)) = \frac{1}{N}\sum_k \frac{1}{r + |G_k|^2} - \frac{1}{\lambda_u}\frac{1}{N}\sum_k \frac{|Y_k - G_k U_k|^2}{\left(r + |G_k|^2\right)^2}$$
$$= \frac{1}{N}\sum_k \frac{1}{r + |G_k|^2}\left(1 - \frac{1}{\lambda_u}\frac{|Y_k - G_k U_k|^2}{r + |G_k|^2}\right). \quad (9.94)$$

Consider now the asymptotic case, $N \to \infty$, and let $\boldsymbol{\theta} = \boldsymbol{\theta}_0$. Then

$$0 = \lim_{N\to\infty} \frac{\partial}{\partial r}(-\log(L_2))$$
$$= \lim_{N\to\infty} \frac{1}{N}\sum_k \frac{1}{r + |G_k|^2}\left(1 - \frac{\lambda_{u,0}}{\lambda_u}\frac{r_0 + |G_k|^2}{r + |G_k|^2}\right). \quad (9.95)$$

It is clear that the true values $r = r_0$, $\lambda_u = \lambda_{u,0}$ form a solution to (9.92) and (9.95).

The extended ML estimate is now defined as the solution to the following system of equations, with $\boldsymbol{\theta}$, r, λ_u as unknowns:

$$\frac{\partial V_{31}}{\partial \boldsymbol{\theta}}(\boldsymbol{\theta}, r) = \mathbf{0}, \quad (9.96)$$
$$\frac{\partial}{\partial r}\log L_2(\boldsymbol{\theta}, r, \lambda_u) = 0, \quad (9.97)$$
$$\frac{\partial}{\partial \lambda_u}\log L_2(\boldsymbol{\theta}, r, \lambda_u) = 0. \quad (9.98)$$

The above analysis has shown that in the asymptotic case ($N \to \infty$), the true parameters $\boldsymbol{\theta}_0$, r_0, $\lambda_{u,0}$ are obtained as a solution. In other words, the method gives consistent parameter estimates, subject to identifiability.

The consistency of the estimate follow from the following considerations, Zhang and Pintelon (2017). Let $\boldsymbol{\rho}$ contain the noise variances, or equivalently the vector composed of r and λ_u. Then it is shown that

$$\frac{\partial}{\partial \boldsymbol{\theta}}\mathsf{E}\{V_{31}(\boldsymbol{\theta}, \boldsymbol{\rho})\} = \mathbf{0}, \quad (9.99)$$
$$\frac{\partial}{\partial \boldsymbol{\rho}}\mathsf{E}\{V_{31}(\boldsymbol{\theta}, \boldsymbol{\rho})\} \neq \mathbf{0}, \quad (9.100)$$
$$\frac{\partial}{\partial \boldsymbol{\theta}}\mathsf{E}\{V_2(\boldsymbol{\theta}, \boldsymbol{\rho})\} \neq \mathbf{0}, \quad (9.101)$$
$$\frac{\partial}{\partial \boldsymbol{\rho}}\mathsf{E}\{V_2(\boldsymbol{\theta}, \boldsymbol{\rho})\} = \mathbf{0}. \quad (9.102)$$

Using (9.99) and (9.102) and omitting the expectation operator is precisely the extended ML estimate (9.96)–(9.98).

Remark 9.10 The system of equations (9.96)–(9.98) can be implemented as an optimization problem:

$$\left(\hat{\boldsymbol{\theta}}, \hat{r}, \hat{\lambda}_u\right) = \arg\min_{\boldsymbol{\theta}, r, \lambda_u} V_e(\boldsymbol{\theta}, r, \lambda_u) , \tag{9.103}$$

$$V_e(\boldsymbol{\theta}, r, \lambda_u) \stackrel{\Delta}{=} \| \frac{\partial V_{31}}{\partial \boldsymbol{\theta}} \|^2 + \left(\frac{\partial}{\partial r} \log(L_2)\right)^2 \tag{9.104}$$

$$\hat{\lambda}_u = V_{31}(\hat{\boldsymbol{\theta}}, \hat{r}) . \tag{9.105}$$

■

Possibly some weights can be included in (9.104) so as to make the terms having the same magnitude. This may speed up the convergence when searching numerically for the minimization in (9.103).

9.6 Extensions and Bibliographical Notes

General Aspects on Unknown Inputs (Sect. 9.1)

For other treatments of system identification with (partly) unknown inputs, see Linder (2017), Linder and Enqvist (2017b), Risuleo et al. (2017a), Risuleo et al. (2017c). Kernel-based estimation is considered in Risuleo et al. (2015). The impulse response of the system is modelled as a Gaussian process. An empirical Bayes method is used to estimate the posterior distribution of the impulse response.

PEM and ML in the Time Domain (Sect. 9.2)

For a general description of prediction error and maximum likelihood methods, see for example the textbooks Ljung (1999) and Söderström and Stoica (1989). Some additional general aspects related to estimation in state space models are given in Sandell and Yared (1978) and McKelvey (1995).

In Chen and Yang (2005) a recursive ELS (extended least squares) algorithm, Ljung and Söderström (1983), is applied, and consistency of the estimates is proved. The model is assumed to have white measurement noise on the input and output sides, that is Assumption **AN3c** is applied, and the noise-free input is modelled as an ARMA process, Assumption **AI4**.

The so called dynamic shock error model is used in Ghosh (1989). Consistency of a maximum likelihood estimator for this model parameterization is examined.

Some general comparisons of the ML estimates with other identification methods for the EIV setting appear in Soverini and Söderström (2000) and Söderström et al. (2002).

How to apply PEM and ML for the EIV problem was early explained in Söderström (1981), while the corresponding case of using a continuous-time model appeared in Söderström et al. (2006). A similar approach to the one described in Sect. 9.3 is proposed in Guo et al. (2017b). See also Wu et al. (2016). The character of the noise-free input is described by a state space model, which corresponds to describing it with an ARMAX model driven by a known signal. The system dynamics is assumed to be an ARX model. The identification problem is essentially defined as a maximum likelihood problem. It is though splitted so that the estimation of the noise-free input is carried out with a Kalman smoother, while the estimation of the system parameters is implemented using the EM algorithm.

Frequency Domain ML (Sect. 9.3)

Frequency domain descriptions of the ML estimate are given in Pintelon and Schoukens (2007) and Pintelon and Schoukens (2012b). The possibility to work with non-parametric noise model for frequency domain ML estimation without affecting the asymptotic properties is described in Schoukens et al. (1999a).

The relations between time domain and frequency domain descriptions of the ML estimates are analyzed in a clear way in Agüero et al. (2010).

ML for Known Noise Variance Ratio (Sect. 9.4)

ML identification of EIV models is considered in Diversi et al. (2007) for the case when input and output noise are both white, and their variance ratio is known. The maximum likelihood method for known noise variance ratio was described and analyzed in Soverini and Söderström (2014b) and Söderström and Soverini (2017).

The identification method proposed in Vajk and Hetthéssy (2008) is essentially an ML estimator for known noise variance ratio, combined with a covariance matching criteria, in order to deliver consistent parameter estimates.

Extended ML Approach (Sect. 9.5)

The extended frequency domain ML method was proposed in Zhang et al. (2013). Extension to the multivariable case appears in Zhang and Pintelon (2017).

A variant of the extended ML approach is considered in Zhang and Pintelon (2018), where it is assumed that the frequency response function and the system leakage term can be locally very well approximated by a polynomial of low degree.

Implementation Aspects and Extensions to Nonlinear Dynamics

The ML estimate for Gaussian distributed data is given by minimization of (9.28). An interesting alternative numerical implementation by using the EM (expectation maximization) principle is provided in Bottegal et al. (2017), Risuleo (2016).

The expectation-maximization (EM) algorithm for successive approximation of the likelihood function is described in Dempster et al. (1977). More examples of the EM method can be found, for example, in Guo et al. (2017a), Agüero et al. (2017).

Nonlinear dynamics can be considered by generalizing (9.9)–(9.12) into a nonlinear form, and then applying state estimation. This is a large field. For some examples of identification of nonlinear systems given in state space form, see Ninness et al. (2010), Schön et al. (2011), Patwardhan et al. (2012), Svensson and Schön (2017).

Chapter 10
Frequency Domain Methods

The chapter is organized as follows. The next section gives some introductory remarks. Section 10.2 presents some nonparametric methods. A frequency domain version of the Frisch scheme is treated in Sect. 10.3. A parametric method based on multivariable ARMA model for $\boldsymbol{\zeta}(t) = (y(t)\ u(t))^T$ is described in Sect. 10.4. Section 10.5 presents a generalized instrumental variable estimator (GIVE) method based on the frequency domain model (5.35) introduced in Chap. 5. The chapter ends with Sect. 10.6 which contains some extensions and bibliographical notes.

10.1 Introduction

Almost all the methods described in the previous chapters are based on time domain techniques. It is, however, also possible to work in the frequency domain. Then, typically, as a first step the spectrum of the observations is determined, or the discrete Fourier transforms of the recorded signals are computed. Note that most of the methods treated so far are based solely on the second-order moments of the data. The information available in the covariances is also present in the spectrum, and that is a reason why frequency domain methods offer an alternative approach.

A general treatment of frequency domain estimators for the EIV problem is surveyed in Pintelon et al. (1994), Pintelon and Schoukens (2001). For many of these methods it is assumed that the noise-free input $u_0(t)$ is periodic. More details for such cases are provided in Chap. 12.

An H^∞ estimate is proposed in Agüero et al. (2005), see also (4.20). The idea is to not impose additional assumptions, but to accept that there is a lack of identifiability. There is indeed one degree of freedom in the set of possible models that are compatible with the first- and second-order statistics of the measured data; see also Sect. 4.1. The estimate in Agüero et al. (2005) is based on choosing this degree of freedom such that the H^∞ norm of the estimation error is minimized.

T. Söderström, *Errors-in-Variables Methods in System Identification*, Communications and Control Engineering,
https://doi.org/10.1007/978-3-319-75001-9_10

To be more specific in describing the general problem, apply Assumption **AN3c** so both input and output noises are white. Using the notations

$$\zeta(t) = \begin{pmatrix} y(t) \\ u(t) \end{pmatrix}, \qquad \zeta_0(t) = \begin{pmatrix} y_0(t) \\ u_0(t) \end{pmatrix}, \tag{10.1}$$

for the input–output data, the spectral density of the input–output data satisfies; see (4.2),

$$\boldsymbol{\Phi}_\zeta(\omega) = \begin{pmatrix} G(e^{i\omega}) \\ 1 \end{pmatrix} \begin{pmatrix} G^*(e^{i\omega}) & 1 \end{pmatrix} \phi_{u_0}(\omega) + \begin{pmatrix} \lambda_y & 0 \\ 0 & \lambda_u \end{pmatrix}, \qquad \forall \omega \,. \tag{10.2}$$

If the spectral density matrix $\boldsymbol{\Phi}_\zeta$ is known, and an appropriate constant diagonal matrix is subtracted, then one would get a rank 1 matrix, corresponding to the first term of (10.2), for all frequencies ω. In case the decomposition (10.2) can be carried out, the first term would easily lead to estimates of the transfer function $G(e^{i\omega})$ and the true input spectrum $\phi_{u_0}(\omega)$.

The setup may seem restrictive. It is crucial that the measurement noise is temporarily white. However, it would not be complicated to allow a correlation $\mathsf{E}\{\tilde{u}(t)\tilde{y}(t)\} \neq 0$. Further, one can allow a feedback from $y_0(t)$ to $u_0(t)$ as long as no noise is entered inside the feedback loop. Presence of a feedback will lead to a changed model for the unperturbed input $u_0(t)$.

Remark 10.1 The methods in this chapter are all described for the discrete-time case, and the frequency varies from 0 to π. As already indicated in Remark 5.3 it is possible to extend the methods to handle continuous-time data by replacing $e^{i\omega}$ by $i\omega$ and letting ω be unbounded. ∎

10.2 Nonparametric Methods

The idea here is similar to that of the Frisch scheme, with the important difference that here the full spectral information in the data, that is (10.2), is considered for all frequencies and not only the covariance matrix (7.56).

The basic steps of the approach are the following:

1. Estimate the spectral density matrix $\boldsymbol{\Phi}_\zeta(\omega)$. This can be done in a number of ways, as will be outlined in Sect. 10.2.1.
2. Using the estimate $\hat{\boldsymbol{\Phi}}_\zeta(\omega)$ for a suitable number of frequencies, determine the unknown transfer function $G(e^{i\omega})$ and the noise variances λ_y and λ_u. Possibly also the spectrum $\phi_{u_0}(\omega)$ will be determined. Again, this can be done in different ways, and examples appear in Sect. 10.2.2.

A key ingredient in the total procedure is the relation (10.2). Here, the spectrum is written as the sum of one rank-deficient term and one constant, diagonal term. This is very similar to, for example, estimation problems in sensor array processing and to factor analysis.

10.2.1 Estimating the Spectrum of the Observed Signals

Spectrum estimation is a rich field, and many methods have been proposed and analyzed. For some overviews see the books (Stoica and Moses 2005; Kay 1988). Some possibilities are the following:

1. A first tentative method would be to compute the periodogram of the data $\boldsymbol{\zeta}(t)$ as an estimate of the spectrum $\boldsymbol{\Phi}_{\zeta}(\omega)$. The periodogram is commonly used in general and is very easy to compute, but it is inaccurate as its variance does not decay with the number of data points. In the present situation, it has a drawback that rules it out as a reasonable alternative. The periodogram of the vector observations $\boldsymbol{\zeta}(1), \ldots, \boldsymbol{\zeta}(N)$ is given by

$$\hat{\boldsymbol{\Phi}}_{\zeta}(\omega) = \frac{1}{N}\left[\sum_{t=1}^{N} \boldsymbol{\zeta}(t)e^{-\mathrm{i}\omega_t}\right]\left[\sum_{s=1}^{N} \boldsymbol{\zeta}^T(s)e^{\mathrm{i}\omega_s}\right] \tag{10.3}$$

and is by construction a rank-one matrix. Hence a decomposition as in (10.2) is not possible.
2. A better alternative is to use any other 'standard' spectrum estimator, such as a weighted periodogram, Capon's method, or Welsch method.
3. A further alternative is to use a parametric estimator. Under weak conditions, see Sect. 10.4, the observations can be modeled as a multivariable ARMA process. It therefore can make sense to estimate an ARMA model of the form

$$\mathbf{M}(q^{-1})\boldsymbol{\zeta}(t) = \mathbf{N}(q^{-1})\boldsymbol{\varepsilon}(t), \quad \mathsf{E}\left\{\boldsymbol{\varepsilon}(t)\boldsymbol{\varepsilon}^T(s)\right\} = \boldsymbol{\Lambda}\delta_{t,s}\ , \tag{10.4}$$

where

$$\mathbf{M}(q^{-1}) = \mathbf{I} + \mathbf{M}_1 q^{-1} + \cdots + \mathbf{M}_K q^{-K}\ , \tag{10.5}$$

$$\mathbf{N}(q^{-1}) = \mathbf{I} + \mathbf{N}_1 q^{-1} + \cdots + \mathbf{N}_K q^{-K}\ . \tag{10.6}$$

For the time being regard (10.4) as an unstructured black-box model of the observations. Once the matrix polynomials $\mathbf{M}(q^{-1})$ and $\mathbf{N}(q^{-1})$ and the innovations covariance matrix $\boldsymbol{\Lambda}$ are estimated, the spectrum is estimated as

$$\hat{\boldsymbol{\Phi}}_{\zeta}(\omega) = \hat{\mathbf{M}}^{-1}(\mathrm{e}^{-\mathrm{i}\omega})\hat{\mathbf{N}}(\mathrm{e}^{-\mathrm{i}\omega})\hat{\boldsymbol{\Lambda}}\hat{\mathbf{N}}^T(\mathrm{e}^{\mathrm{i}\omega})\hat{\mathbf{M}}^{-T}(\mathrm{e}^{\mathrm{i}\omega})\ . \tag{10.7}$$

There are ways to impose certain structures of the matrix polynomials; see Sect. 10.4. Among the many possible alternatives to estimate the ARMA coefficients in (10.4), here a well-known procedure, previously used in Durbin (1959), Mayne and Firoozan (1982), is adopted. The idea is to first estimate the innovation process $\{\boldsymbol{\varepsilon}(t)\}_{t=1}^{N}$ and then take it as a known second input, thus using a linear regression model to estimate $\mathbf{M}(q^{-1})$ and $\mathbf{N}(q^{-1})$. In order to estimate the innovation $\boldsymbol{\varepsilon}(t)$, fit an AR model of large order L to the data $\boldsymbol{\zeta}(t)$, i.e.,

$$\left[\mathbf{I} + \mathbf{A}_1 q^{-1} + \cdots + \mathbf{A}_L q^{-L}\right] \boldsymbol{\zeta}(t) = \boldsymbol{\varepsilon}(t) . \tag{10.8}$$

The model order L should be sufficiently large, in order to ensure a good approximation of the ARMA model (10.4) by the AR model (10.8). The AR parameters $\{\mathbf{A}_k\}_{k=1}^{L}$ are estimated using least squares. Introduce

$$\boldsymbol{\Theta}_1 = \left(\mathbf{A}_1 \ \ldots \ \mathbf{A}_L\right) , \tag{10.9}$$

$$\mathbf{Z}_1(t) = \left(-\boldsymbol{\zeta}^T(t-1) \ \ldots \ -\boldsymbol{\zeta}^T(t-L)\right)^T . \tag{10.10}$$

Then

$$\begin{aligned} \hat{\boldsymbol{\Theta}}_1 &= \arg\min_{\boldsymbol{\Theta}_1} \sum_{t=1}^{N} \|\boldsymbol{\zeta}(t) - \boldsymbol{\Theta}_1 \mathbf{Z}_1(t)\|^2 \\ &= \left[\sum_{t=1}^{N} \boldsymbol{\zeta}(t) \mathbf{Z}_1^T(t)\right] \left[\sum_{t=1}^{N} \mathbf{Z}_1(t) \mathbf{Z}_1^T(t)\right]^{-1} . \end{aligned} \tag{10.11}$$

Hence, an estimate of the innovations $\boldsymbol{\varepsilon}(t)$ and their covariance matrix $\boldsymbol{\Lambda}$ are obtained as

$$\hat{\boldsymbol{\varepsilon}}(t) = \boldsymbol{\zeta}(t) - \hat{\boldsymbol{\Theta}}_1 \mathbf{Z}_1(t), \quad \hat{\boldsymbol{\Lambda}} = \frac{1}{N} \sum_{t=1}^{N} \hat{\boldsymbol{\varepsilon}}(t) \hat{\boldsymbol{\varepsilon}}^T(t) . \tag{10.12}$$

Next, consider $\hat{\boldsymbol{\varepsilon}}(t)$ as an additional and *known* input and estimate the coefficients of the ARMA model (10.4). Using least squares it is straightforward to estimate

$$\boldsymbol{\Theta}_2 = \left(\mathbf{M}_1 \ \ldots \ \mathbf{M}_K \ \mathbf{N}_1 \ \ldots \ \mathbf{N}_K\right) , \tag{10.13}$$

where the regressor matrix will be composed of $\{\boldsymbol{\zeta}^T(t-i)\}_{i=1}^{K}$ and $\{\hat{\boldsymbol{\varepsilon}}^T(t-i)\}_{i=1}^{K}$:

$$\mathbf{Z}_2(t) = \left(-\boldsymbol{\zeta}^T(t-1) \cdots -\boldsymbol{\zeta}^T(t-K) \ \hat{\boldsymbol{\varepsilon}}^T(t-1) \cdots \hat{\boldsymbol{\varepsilon}}^T(t-K)\right)^T . \tag{10.14}$$

The least squares estimate is

$$\hat{\boldsymbol{\Theta}}_2 = \left[\sum_{t=1}^{N} \left(\boldsymbol{\zeta}(t) - \hat{\boldsymbol{\varepsilon}}(t)\right) \mathbf{Z}_2^T(t)\right] \left[\sum_{t=1}^{N} \mathbf{Z}_2(t) \mathbf{Z}_2^T(t)\right]^{-1} . \tag{10.15}$$

The estimate is to be modified, if some structure on the ARMA model (10.4) is imposed, as exemplified in Sect. 10.4.

10.2.2 Estimating the Transfer Function $G(e^{i\omega})$

Now turn to the problem of how to estimate $G(e^{i\omega})$. Apply the basic relation (10.2) for a number of frequency values and regard the spectrum

$$\boldsymbol{\Phi}_{\zeta}(\omega) = \begin{pmatrix} \phi_y(\omega) & \phi_{yu}(\omega) \\ \phi_{uy}(\omega) & \phi_u(\omega) \end{pmatrix} \tag{10.16}$$

as known.

As the first term in (10.2) is singular, its determinant is zero and hence it must hold for each frequency $\omega_k,\ k = 1, 2, \ldots$ that

$$[\phi_y(\omega_k) - \lambda_y][\phi_u(\omega_k) - \lambda_u] - |\phi_{yu}(\omega_k)|^2 = 0\ . \tag{10.17}$$

By utilizing (10.17) for a number of frequencies, the unknown noise variances λ_u and λ_y can be determined. Once that is done, the (1,2) and (2,2) elements of (10.2) give that the transfer function is finally to be estimated as

$$\hat{G}(e^{i\omega_k}) = \phi_{yu}(\omega_k)/[\phi_u(\omega_k) - \hat{\lambda}_u]\ . \tag{10.18}$$

As the basic idea is to exploit the properties of singular matrices, note the similarity with the Frisch scheme methods treated in Sect. 7.3. There a certain covariance matrix of the data is singular, while here it is a spectrum. For this reason, the idea exploited here is sometimes referred to as a Frisch scheme in the frequency domain.

10.2.3 Estimating the Noise Variances

There are several possibilities for how the noise variances are estimated.

1. A computationally simple approach, treated in Söderström et al. (2002), Söderström et al. (2003) is to apply (10.17) as a linear regression with $\lambda_y,\ \lambda_u,\ \lambda_y\lambda_u$ as three unknowns, to derive an estimate of the noise variances. This just means that (10.17) is rewritten as a *linear* regression model

$$\phi_y(\omega_k)\phi_u(\omega_k) - |\phi_{yu}(\omega_k)|^2 = \begin{pmatrix} \phi_y(\omega_k) & \phi_u(\omega_k) & -1 \end{pmatrix} \begin{pmatrix} x_1 \\ x_2 \\ x_3 \end{pmatrix}\ . \tag{10.19}$$

Then a least squares fit is applied using k as a running index. The estimates are obtained as

$$\hat{\lambda}_u = \hat{x}_1 , \quad \hat{\lambda}_y = \hat{x}_2 . \tag{10.20}$$

When proceeding like this, there is no guarantee that the estimated noise variances are positive (this is a price to be paid for using a very simple procedure). Should this happens, the estimated noise variances need to be adjusted to be positive, or at least zero, before using (10.18) for determining the transfer function.

2. A somewhat more complicated approach would be to proceed as above, but using a *nonlinear* regression model. Then it holds

$$\phi_y(\omega_k)\phi_u(\omega_k) - |\phi_{yu}(\omega_k)|^2 = \begin{pmatrix} \phi_y(\omega_k) & \phi_u(\omega_k) & -1 \end{pmatrix} \begin{pmatrix} x_1 \\ x_2 \\ x_1 x_2 \end{pmatrix} . \tag{10.21}$$

One may also impose the constraints that x_1 and x_2 are positive. Estimating the noise variances, (10.20), using nonlinear least squares can be expected to lead to more accurate estimates than for the simpler scheme with a linear regression described above.

3. A more complex variant possibly leading to more accurate estimates is to impose the constraints of the form

$$0 \le \lambda_u \le \lambda_u^{\max} , \qquad 0 \le \lambda_y \le \lambda_y^{\max} , \tag{10.22}$$

where $\lambda_y^{\max}$ is obtained by posing $\lambda_u = 0$ in (10.17) and minimizing the resulting λ_y over all the considered frequencies, i.e.,

$$\lambda_y^{\max} = \min_k \left\{ \phi_y(\omega_k) - |\phi_{yu}(\omega_k)|^2 / \phi_u(\omega_k) \right\} . \tag{10.23}$$

Similarly $\lambda_u^{\max}$ is obtained by posing $\lambda_y = 0$ in Eq. (10.17), i.e.

$$\lambda_u^{\max} = \min_k \left\{ \phi_u(\omega_k) - |\phi_{yu}(\omega_k)|^2 / \phi_y(\omega_k) \right\} . \tag{10.24}$$

In practice the upper bounds $\lambda_y^{\max}$ and $\lambda_u^{\max}$ given by (10.23) and (10.24) have to be computed from the *estimated* spectrum, resulting in modified bounds, say $\hat{\lambda}_y^{\max}$ and $\hat{\lambda}_u^{\max}$.

4. In Castaldi et al. (2002) still another Frisch domain approach is used. The problem is formulated as a bilinear matrix inequality. Set

$$\begin{pmatrix} x_1 \\ x_2 \\ x_3 \end{pmatrix} = \begin{pmatrix} \lambda_y \\ \lambda_u \\ \lambda_y \lambda_u \end{pmatrix} . \tag{10.25}$$

The following criterion

$$\ell(x_1, x_2, x_3, \varepsilon) = \sum_k \big[x_3 - \phi_y(\omega_k)x_2 - \phi_u(\omega_k)x_1 \\ +\phi_y(\omega_k)\phi_u\omega_k) - |\phi_{yu}(\omega_k)|^2\big] + \varepsilon \tag{10.26}$$

is minimized with respect to $x_1,\ x_2,\ x_3,\ \varepsilon$ under the constraints

$$\boldsymbol{\Phi}_\zeta - \begin{pmatrix} x_1 0 \\ 0 x_2 \end{pmatrix} \geq 0\,, \tag{10.27}$$

$$x_1 x_2 \geq x_3 - \varepsilon\,, \tag{10.28}$$

$$-x_1 x_2 \geq -x_3 - \varepsilon\,, \tag{10.29}$$

$$x_1 \geq 0\,, \tag{10.30}$$

$$x_2 \geq 0\,, \tag{10.31}$$

$$\varepsilon \geq 0\,. \tag{10.32}$$

10.2.4 An Estimate of θ

If wanted, one can use the nonparametric estimate of $G(e^{-\mathrm{i}\omega})$ also to derive a parametric estimates of the parameter vector

$$\boldsymbol{\theta} = \begin{pmatrix} a_1 \ldots a_{n_a}\ b_1 \ldots b_{n_b} \end{pmatrix}^T\,. \tag{10.33}$$

One simple way to estimate $\boldsymbol{\theta}$ is to minimize the weighted spectral distance between $B(\mathrm{e}^{-\mathrm{i}\omega}, \boldsymbol{\theta})$ and $A(\mathrm{e}^{-\mathrm{i}\omega}, \boldsymbol{\theta})\hat{G}(\mathrm{e}^{-\mathrm{i}\omega})$, where $\hat{G}(\mathrm{e}^{-\mathrm{i}\omega})$ is given by (10.18), i.e.,

$$\hat{\boldsymbol{\theta}} = \arg\min_{\boldsymbol{\theta}} \sum_{k=0}^{M-1} W_k |B(\mathrm{e}^{-\mathrm{i}\omega}, \boldsymbol{\theta}) - A(\mathrm{e}^{-\mathrm{i}\omega}, \boldsymbol{\theta})\hat{G}(\mathrm{e}^{-\mathrm{i}\omega_k})|^2\,, \tag{10.34}$$

where $\{\omega_k\}_{k=0}^{M-1} \subset [0, \pi]$ are user defined discrete frequency points and W_k is a user chosen weight corresponding to frequency ω_k. The cost function is quadratic in $\boldsymbol{\theta}$ and can be optimized analytically using least squares.

Since the transfer function estimate given by (10.18) is expected to be more accurate at the frequencies for which the data SNR is large, a natural choice of the weight W_k in (10.34) while deriving the parametric estimate would be

$$W_k = \hat{\phi}_{u_0}(\omega_k)\,. \tag{10.35}$$

10.3 A Frisch Scheme-Based Method in the Frequency Domain

The following method is proposed in Tanaka and Katayama (2006). The starting point is the input–output spectrum, see (10.16), $\boldsymbol{\Phi}_{\zeta}(\omega)$. In practice an estimate of the spectrum will be used. For the standard Frisch solution the second term in the decomposition (10.2) is forced to be diagonal (and constant, if the measurement noise is white).

The approach in Tanaka and Katayama (2006) is instead based on the assumption that the noise level in the measured spectrum is bounded. Further, it is assumed that a PEM approach on the input–output data would lead to a model of the form

$$y(t) = G(q^{-1})u(t) + H(q^{-1})e(t) , \tag{10.36}$$

where $H(q^{-1})$ is an error model for frequency weighting and given as some prespecified information. Thus, the corresponding prediction error will be

$$\varepsilon(t) = H^{-1}(q^{-1}) \left(1 \ -G(q^{-1}) \right) \boldsymbol{\zeta}(t) , \tag{10.37}$$

where, as before, $\boldsymbol{\zeta}(t) = (y(t) \ u(t))^T$ denotes the output–input data at time t. Denote the spectrum of $\varepsilon(t)$ by ϕ_{ε}. The transfer function $G(q^{-1})$ is to be estimated under the constraint

$$\phi_{\varepsilon} < \gamma^2 , \tag{10.38}$$

where γ^2 is a prespecified bound. The estimation problem is then to find *all* $G(q^{-1})$ that are compatible with (10.2) and the constraint (10.38). The constraint can be transformed into

$$\left(1 \ -G(e^{i\omega}) \right) \boldsymbol{\Phi}_{\zeta}(\omega) \begin{pmatrix} 1 \\ -G^*(e^{i\omega}) \end{pmatrix} \leq \gamma^2 H(e^{i\omega}) H^*(e^{i\omega}) . \tag{10.39}$$

Rather than looking for a unique estimate of G one seeks in this approach to parameterize all models which explains the joint input–output spectrum. As an intermediate step one identifies the innovations form of the spectrum using a subspace identification algorithm (Van Overschee et al. 1997).

10.4 A Parametric Method

In this section a parametric model is developed for the observation vector $\boldsymbol{\zeta}(t)$. It is still assumed that the measurement noise is white for both the input and the output (that is, Assumption **AN3c** applies). Further, assume that the noise-free input can be modeled as an ARMA process; see Assumption **AI4**

$$D(q^{-1})u_0(t) = C(q^{-1})e(t), \quad \mathsf{E}e^2(t) = \lambda_e . \tag{10.40}$$

It will be shown that the observed vector $\boldsymbol{\zeta}(t)$ admits a *structured* multivariable ARMA model.

As a starting point, set

$$\mathbf{M}(q^{-1}) = \begin{pmatrix} A(q^{-1}) & -B(q^{-1}) \\ 0 & D(q^{-1}) \end{pmatrix} . \tag{10.41}$$

One can easily show that

$$\begin{aligned} \mathbf{M}(q^{-1})\boldsymbol{\zeta}(t) &= \begin{pmatrix} A(q^{-1})y(t) - B(q^{-1})u(t) \\ D(q^{-1})u(t) \end{pmatrix} \\ &= \begin{pmatrix} A(q^{-1})\tilde{y}(t) - B(q^{-1})\tilde{u}(t) \\ C(q^{-1})e(t) + D(q^{-1})\tilde{u}(t) \end{pmatrix} . \end{aligned} \tag{10.42}$$

From (10.42) one can see that $\mathbf{M}(q^{-1})\boldsymbol{\zeta}(t)$ is clearly a moving average process in the sense that its covariance function vanishes after K lags, where

$$K = \max\,(n_a, n_b, n_c, n_d) .$$

Therefore, $\boldsymbol{\zeta}(t)$ admits the two-dimensional ARMA representation

$$\mathbf{M}(q^{-1})\boldsymbol{\zeta}(t) = \mathbf{N}(q^{-1})\boldsymbol{\varepsilon}(t) . \tag{10.43}$$

where $\mathbf{N}(q^{-1})$ is a monic, 2×2 matrix polynomial of order K

$$\mathbf{N}(q^{-1}) = \mathbf{I} + \mathbf{N}_1 q^{-1} + \cdots + \mathbf{N}_K q^{-K} . \tag{10.44}$$

Further, in (10.43) $\boldsymbol{\varepsilon}(t)$ is the two-dimensional innovation of $\boldsymbol{\zeta}(t)$, a zero mean white noise with covariance matrix $\boldsymbol{\Lambda}$.

The ARMA representation of a multi-dimensional process is non-unique in general; see Gevers and Wertz (1984), but the structure (10.41) of the polynomial $\mathbf{M}(q^{-1})$ makes it possible to derive also the parametric estimates of the polynomials $A(q^{-1})$, $B(q^{-1})$, $C(q^{-1})$, and $D(q^{-1})$ along with the noise variances λ_y, λ_u, and λ_e from $\mathbf{M}(q^{-1})$, $\mathbf{N}(q^{-1})$, and $\boldsymbol{\Lambda}$. Note that parametric estimates of $A(q^{-1})$, $B(q^{-1})$, and $D(q^{-1})$ can be obtained from the estimate of $\mathbf{M}(q^{-1})$ using (10.41). Using the identified ARMA model it is possible to estimate also the other relevant parameters. Introduce the vector process $\mathbf{v}(t)$ and its correlation sequence, as well as their estimates

$$\mathbf{v(t)} = \mathbf{M(q^{-1})}\boldsymbol{\zeta}\mathbf{(t)}, \quad \mathbf{R}_v(\tau) = \mathsf{E}\left\{\mathbf{v}(t)\mathbf{v}^T(t-\tau)\right\} , \tag{10.45}$$

$$\hat{\mathbf{v}}(t) = \hat{\mathbf{M}}(q^{-1})\zeta(t) \,, \quad \hat{\mathbf{R}}_v(\tau) = \frac{1}{N}\sum_{t=1}^{N} \hat{\mathbf{v}}(t)\hat{\mathbf{v}}^T(t-\tau) \,. \tag{10.46}$$

To explore how $C(q^{-1})$, λ_u, λ_y, and λ_e can be determined, examine $\mathbf{R}_v(\tau)$ elementwise.

Using (10.42), it is straightforward to see that

$$\mathbf{R}_v(\tau)_{11} = \lambda_y \sum_{i=0}^{n_a-\tau} a_i a_{i+\tau} + \lambda_u \sum_{i=1}^{n_b-\tau} b_i b_{i+\tau}, \quad 0 \le \tau \le \max\,(n_a, n_b) \,, \tag{10.47}$$

$$\mathbf{R}_v(\tau)_{12} = -\lambda_u \sum_{i=0}^{\max\,(n_d, n_b-\tau)} d_i b_{i+\tau}, \quad 0 \le \tau \le n_b \,, \tag{10.48}$$

$$\mathbf{R}_v(\tau)_{21} = -\lambda_u \sum_{i=1}^{\max\,(n_b, n_d-\tau)} b_i d_{i+\tau}, \quad 0 \le \tau \le n_d - 1 \,, \tag{10.49}$$

where the conventions $b_0 = 0$ and $a_0 = d_0 = 1$ are used. Note that estimates of $A(q^{-1})$, $B(q^{-1})$, and $D(q^{-1})$ are already known. Hence (10.47), (10.48), and (10.49) can be used for different values of $\tau \le K$ to get an over determined system of linear equations in λ_y and λ_u. This system can be solved to obtain the estimates of the unknowns. Then finally use the remaining element of $\mathbf{R}_v(\tau)$ to estimate $C(q^{-1})$. It holds

$$\begin{aligned}\mathbf{R}_v(\tau)_{22} = \mathsf{E}\big\{&\big[C(q^{-1})e(t) + D(q^{-1})\tilde{u}(t)\big] \\ &\times \big[C(q^{-1})e(t-\tau) + D(q^{-1})\tilde{u}(t-\tau)\big]\big\} \,,\end{aligned} \tag{10.50}$$

$$\sum_{\tau=-n_c}^{n_c} \mathbf{R}_v(\tau)_{22} q^{-\tau} - \lambda_u D(q^{-1})D(q) = \lambda_e C(q^{-1})C(q) \,. \tag{10.51}$$

The polynomial at the left-hand side of (10.51) can be factorized to obtain $\hat{C}(q^{-1})$. Since $\hat{\mathbf{R}}_v(\tau)_{22}$ and $\hat{\lambda}_u$ deviate somewhat from their true values, it may happen that the left-hand side cannot be exactly factorized. This problem is encountered precisely when the left-hand side is not positive definite.

As an essential basis for the described method is to first estimate, and then explore, the spectral factor $\mathbf{M}^{-1}\mathbf{N}$, the method will be called the spectral factor method in what follows.

Example 10.1 In this example the spectral factor estimator is considered and applied to the standard data sets. The results are displayed in Table 10.1.

The accuracy in terms of standard deviations varies greatly between different parameters. An heuristic explanation is as follows. In a first step the spectral factor is

Table 10.1 Spectral factor estimates for the standard data sets

Parameter	True value	Spec fact		GIVE	
		Mean	Standard dev.	Mean	Standard dev.
a_1	−1.5	−1.494	0.014	−1.528	0.085
a_2	0.7	0.696	0.011	0.709	0.040
b_1	2.0	1.599	0.485	3.57	2.82
b_2	1.0	1.353	0.419	−0.69	3.30
c_1	0.7	0.968	0.022	–	–
d_1	−1.0	−0.757	0.152	–	–
d_2	0.2	0.014	0.119	–	–
λ_y	10.0	10.07	0.78	8.37	4.51
λ_u	1.81	1.96	0.32	2.00	0.64
λ_v	1.0	19.7	9.88	–	–

estimated. This comes with some error or uncertainty. From the spectral factor one extracts rather directly the estimates of A, B, and D. At a later stage, one uses these estimates, with their uncertainties, to extract also C and λ_v. This procedure leads to large uncertainties in the estimates of C and λ_v.

In summary, the spectral factor estimate is superior to the basic GIVE estimate and gives a similar performance as GIVE used with some additional equations (say, $p \geq 2$ in Example 7.11). ■

10.5 A Frequency Domain GIVE Method

10.5.1 Description

The starting point is the frequency domain model (5.35). For a general frequency ω it holds

$$A(\mathrm{e}^{-\mathrm{i}\omega})Y_0(\omega) = B(\mathrm{e}^{-\mathrm{i}\omega})U_0(\omega) + T(\mathrm{e}^{-\mathrm{i}\omega}) \ . \tag{10.52}$$

where $T(z^{-1})$ is a polynomial of order $n-1$

$$T(z^{-1}) = \tau_0 + \tau_1 z^{-1} + \cdots + \tau_{n-1} z^{-n+1}, \quad n = \max(n_a, n_b) \ , \tag{10.53}$$

that takes into account the effects of the initial and final conditions of the experiment. For the subsequent analysis introduce also the parameter vector

$$\boldsymbol{\tau} = \begin{pmatrix} \tau_0 & \dots & \tau_{n-1} \end{pmatrix}^T . \tag{10.54}$$

Equation (10.52) can be equivalently written as

$$\boldsymbol{\varphi}_0^H(\omega)\boldsymbol{\Theta} = 0 \,, \tag{10.55}$$

where the 'frequency domain regressor vector' and the extended parameter vector are, respectively,

$$\begin{aligned} \boldsymbol{\varphi}_0^H(\omega) = \big(Y_0(\omega) \ \mathrm{e}^{-\mathrm{i}\omega}Y_0(\omega) \quad \ldots \ \mathrm{e}^{-\mathrm{i}n\omega}Y_0(\omega) \quad -\mathrm{e}^{-\mathrm{i}\omega}U_0(\omega) \\ \ldots \ -\mathrm{e}^{-\mathrm{i}n\omega}U_0(\omega) \ -1 \ \ldots \ -\mathrm{e}^{-\mathrm{i}(n-1)\omega}\big) \,, \end{aligned} \tag{10.56}$$

$$\boldsymbol{\Theta} = \left(1 \ a_1 \ \ldots \ a_n \ \ b_1 \ \ldots \ b_n \ \boldsymbol{\tau}^T\right)^T . \tag{10.57}$$

Frequency domain data are assumed available for

$$\omega_k = \frac{2\pi}{N}k, \quad k = 0, \ldots, N-1 \,. \tag{10.58}$$

Introduce further

$$\boldsymbol{\Phi}_0 = \begin{pmatrix} \boldsymbol{\varphi}_0^H(\omega_0) \\ \vdots \\ \boldsymbol{\varphi}_0^H(\omega_{N-1}) \end{pmatrix} . \tag{10.59}$$

It follows from (10.55) that

$$\boldsymbol{\Phi}_0\boldsymbol{\Theta} = \mathbf{0} \,. \tag{10.60}$$

Set also

$$\boldsymbol{\Sigma}_0 = \frac{1}{N}\boldsymbol{\Phi}_0^H\boldsymbol{\Phi}_0 = \frac{1}{N}\sum_{k=0}^{N-1}\boldsymbol{\varphi}_0(\omega_k)\boldsymbol{\varphi}_0^H(\omega_k) \,, \tag{10.61}$$

$$\tilde{\boldsymbol{\Sigma}} = \begin{pmatrix} \lambda_y\mathbf{I}_{n+1} & \mathbf{0} & \mathbf{0} \\ \mathbf{0} & \lambda_u\mathbf{I}_n & \mathbf{0} \\ \mathbf{0} & \mathbf{0} & \mathbf{0}_n \end{pmatrix} . \tag{10.62}$$

Then the normal equations related to the extended parameter vector $\boldsymbol{\Theta}$ (10.57) can be expressed in the frequency domain, as

$$\boldsymbol{\Sigma}_0\boldsymbol{\Theta} = \mathbf{0} \,. \tag{10.63}$$

Next introduce for the noise-disturbed signals the symbols $\boldsymbol{\varphi}(\omega)$, $\boldsymbol{\Phi}$, and $\boldsymbol{\Sigma}$ fully analogously to (10.56), (10.59), (10.61).

In the limiting case, when $N \to \infty$, the *compensated normal equations* can be written as

$$\left(\boldsymbol{\Sigma} - \tilde{\boldsymbol{\Sigma}}\right)\boldsymbol{\Theta} = \mathbf{0}\,. \tag{10.64}$$

This system of equations contains $3n + 1$ equations and has $3n + 2$ unknowns. Therefore some further equations are needed to define an estimate of $\boldsymbol{\Theta}$.

Introduce the 'generalized IV vector' $\mathbf{Z}^H(\omega)$. It should contain at least the first $2n+1$ elements of $\boldsymbol{\varphi}(\omega)$ (which correspond to $y(t), \ldots y(t-n), u(t), \ldots u(t-n)$ in the time domain). The other elements, which must be at least $n+1$ in number, are typically delayed (noisy) inputs or delayed (noisy) outputs.

Using the above notations leads to the GIVE equations:

$$\left[\frac{1}{N}\sum_{k=0}^{N-1}\mathbf{Z}(\omega_k)\boldsymbol{\varphi}^H(\omega_k) - \begin{pmatrix}\tilde{\boldsymbol{\Sigma}}_0\\ \mathbf{0}\end{pmatrix}\right]\boldsymbol{\Theta} = \mathbf{0}\,. \tag{10.65}$$

where

$$\tilde{\boldsymbol{\Sigma}}_0 = \begin{pmatrix}\lambda_y\mathbf{I}_{n+1} & \mathbf{0} & \mathbf{0}\\ \mathbf{0} & \lambda_u\mathbf{I}_n & \mathbf{0}\end{pmatrix} \tag{10.66}$$

The system (10.65) of equations thus has at least as many equations as unknowns. When the number of equations is larger than minimal, the system (10.65) has to be solved in a weighted least squares sense. In this way, the resulting estimated $\hat{\boldsymbol{\Theta}}$ is uniquely defined.

10.5.2 Some Comments

In what sense does the frequency domain GIVE correspond to the time domain GIVE described in Sect. 7.4? Cf (7.96) and (10.65). Essentially, the defining equations are the same. The reason is that the generic element of the first matrix in (10.65) can be written, using Parseval's relation, as

$$\frac{1}{N}\sum_{k=0}^{N-1}X_1(\omega_k)X_2^H(\omega_k) = \frac{1}{N}\sum_{t=0}^{N-1}x_1(t)x_2(t) \to \mathsf{E}\left\{x_1(t)x_2(t)\right\}\,. \tag{10.67}$$

Note the exact equality in (10.67). Further, the precise numerical algorithm applied to solve (10.65) has *no* importance when the *properties* of the solution are considered, for example the asymptotic covariance matrix of the parameter estimates. What matters is that the solution is found, not the recipe how to arrive to it!

Nevertheless, the choice of numerical algorithm is important, for example, to guarantee that an appropriate solution is found. Different algorithms can also have different convergence speed.

In the time domain the typical element in the first matrix in (10.65) has the form $\frac{1}{N}\sum_{t=0}^{N-1}x_1(t)x_2(t)$. If $x_1(t)$ consists of a delayed signal, some terms in the sum are naturally to be skipped. This is a simple way to treat the effect of the initial value of the dynamics. One may choose to handle this issue such that all elements in the matrix $\sum_{t=1}^{N} \mathbf{z}(t)\boldsymbol{\varphi}^T(t)$ are summed over the same set of t-values.

For the frequency domain formulation, one is instead using the model (10.52), which contains some explicit auxiliary parameters $(\tau_0, \ldots, \tau_{n-1})$ to account for all effects of the unknown initial values and final values. Thereby it also takes care of leakage effects.

10.5.3 Some Analysis

Consider the equations (10.65) that define the GIVE estimate in the frequency domain. One may write the system of equations as

$$\left[\hat{\mathbf{R}} - \begin{pmatrix} \mathbf{J}(\hat{\rho}) & \mathbf{0} \\ \mathbf{0} & \mathbf{0} \end{pmatrix}\right] \hat{\boldsymbol{\Theta}} = \mathbf{0} , \tag{10.68}$$

where

$$\hat{\boldsymbol{\Theta}} = \begin{pmatrix} \hat{\boldsymbol{\theta}} \\ \hat{\boldsymbol{\tau}} \end{pmatrix} , \tag{10.69}$$

$$\hat{\mathbf{R}} = \frac{1}{N} \sum_{k=0}^{N-1} \mathbf{Z}(\omega_k)\boldsymbol{\varphi}^H(\omega_k) , \tag{10.70}$$

$$\mathbf{J}(\hat{\rho}) = \begin{pmatrix} \hat{\lambda}_y \mathbf{I}_{n+1} & \mathbf{0} \\ \mathbf{0} & \hat{\lambda}_u \mathbf{I}_n \end{pmatrix} , \tag{10.71}$$

$$\hat{\rho} = \begin{pmatrix} \hat{\lambda}_y \\ \hat{\lambda}_u \end{pmatrix} . \tag{10.72}$$

Remark 10.2 Referring to Parseval's relation (10.67) the time domain GIVE estimate of the same parameters is given by the solution to (10.68) but deleting the last n columns of the matrix and the last n elements of the vector $\hat{\boldsymbol{\Theta}}$. It is clear that the two estimates (7.96) and (10.65) are not identical. *The difference is not due to what domain (time of frequency) that is used, but that the model* (10.52) *takes explicitly initial and final data effects into account.* Indeed, it may be somewhat misleading to talk about frequency domain and time domain versions of GIVE. An alternative description would be whether or not the leakage effect (that is, the term $T(\mathrm{e}^{-\mathrm{i}\omega})$ in (10.52)) is taken into account. ■

The elements of the last n columns of the matrix in (10.65) turns to zero as $N \to \infty$. As the defining equations for GIVE in the frequency domain and in the time domain, see (7.96) and (10.65), respectively, are asymptotically the same, it is fairly natural to expect that the GIVE estimates will be asymptotically the equal.

More precisely,

$$\hat{\boldsymbol{\theta}}_{\text{TD}} - \boldsymbol{\theta}_0 = \mathbf{0}(1/\sqrt{N}) , \tag{10.73}$$

$$\hat{\boldsymbol{\theta}}_{\text{FD}} - \boldsymbol{\theta}_0 = \mathbf{0}(1/\sqrt{N}) , \tag{10.74}$$

$$\hat{\boldsymbol{\theta}}_{\text{TD}} - \hat{\boldsymbol{\theta}}_{\text{FD}} = \mathbf{0}(1/N) . \tag{10.75}$$

Here TD and FD stands for time domain and frequency domain, respectively. The right-hand sides of (10.73)–(10.75) refer to the magnitude of the standard deviation of the left-hand sides, respectively. See Chap. 14 for further details and an explicit derivation of (10.73) and characterizing its right-hand side.

Example 10.2 This example considers the frequency domain GIVE method and compares it to the associated time domain GIVE method. The standard data sets are considered. The system was subject to Monte Carlo simulations in the usual way. Note though that different types of data lengths were considered, from quite short (only 50 of the total 1000 data points in each run) to the full data lengths. GIVE was applied with a non-minimal size of the instrumental vector, more precisely

$$\mathbf{z}(t) = (y(t) \ldots y(t-5)\, u(t) \ldots u(t-5))^T . \tag{10.76}$$

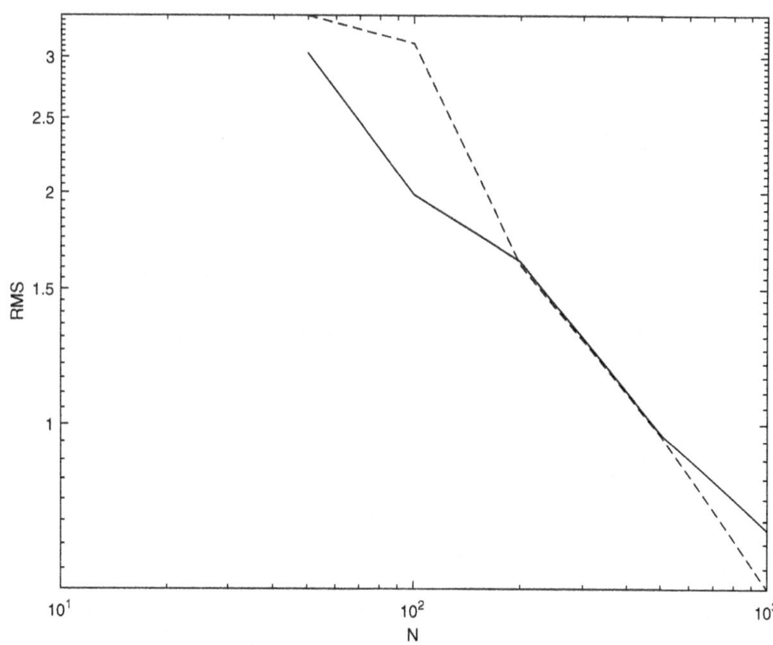

Fig. 10.1 RMS values for GIVE estimates versus the data length N. Solid line: RMS values for the frequency domain GIVE, which estimates the leakage term. Dashed line: RMS values for the time domain GIVE, in which the leakage term is neglected

The behavior was evaluated using the root-mean-square (RMS) values of the parameter estimates. The results are displayed in Fig. 10.1.

As expected by theory, the FD and TD results are very similar for the long data lengths. For short data records, there is a significant difference though, and then the FD estimates are certainly more accurate than the TD estimates. ■

10.6 Extensions and Bibliographical Notes

For a general background on frequency domain techniques in system identification, see Pintelon and Schoukens (2001), Pintelon and Schoukens (2012b).

The paper Heath (2005) treats the variance of a modified empirical transfer function estimator (ETFE) in the frequency domain, for an EIV situation.

The paper Liu and Zhu (2017) generalizes an earlier developed identification method (called ASYM) to handle the EIV case. As a first step a high-order ARX model is fitted to the data, and then a model reduction is applied using frequency domain tools.

The paper Carvajal et al. (2012) is based on frequency domain ML and a restricted set of frequencies.

A frequency domain method to estimate transfer functions from noisy measurements, including missing data, is presented in Ugryomova et al. (2015). The method can be applied to EIV problems, in case there is a known reference signal available.

A frequency domain method for estimating linear parameter-varying systems in an EIV setting is considered in Goos et al. (2017). It is assumed that a scheduling signal is known.

Chapter 11
Total Least Squares

The chapter is organized as follows. The total least squares (TLS) problem is introduced in Sect. 11.1, and ways to compute its solution are found in Sect. 11.2. How to apply TLS to system identification is presented in Sects. 11.3 and 11.4. Consistency analysis follows in Sect. 11.5. Extensions and bibliographical comments appear in Sect. 11.6.

11.1 The Total Least Squares Problem

The total least squares method is in principle a method for treating overdetermined linear systems of equations where both the coefficient matrix and the right-hand side are subject to errors.

Consider the overdetermined system of equations

$$\mathbf{A}\mathbf{x} \approx \mathbf{b} \ , \tag{11.1}$$

where it is assumed that the $m \times n$ matrix $\mathbf{A}$ has full column rank n. The least squares solution to (11.1) is

$$\mathbf{x}_{\text{LS}} = (\mathbf{A}^T\mathbf{A})^{-1}\mathbf{A}^T\mathbf{b} \ , \tag{11.2}$$

and can also be formulated as the solution to the following optimization problem

$$\min \parallel \Delta\mathbf{b} \parallel^2 \ \text{subject to} \ \mathbf{A}\mathbf{x}_{\text{LS}} = \mathbf{b} + \Delta\mathbf{b} \ . \tag{11.3}$$

The total least squares (TLS) problem is different and takes into account also uncertainties in the coefficient matrix $\mathbf{A}$. The TLS problem has some roots in orthogonal regression, Adcock (1877, 1878).

T. Söderström, *Errors-in-Variables Methods in System Identification*, Communications and Control Engineering,
https://doi.org/10.1007/978-3-319-75001-9_11

The TLS problem can be formulated as, compare (11.3),

$$\min \parallel (\Delta\mathbf{A}\ \Delta\mathbf{b}) \parallel_{\mathrm{F}}^2 \text{ subject to } (\mathbf{A} + \Delta\mathbf{A})\mathbf{x}_{\mathrm{TLS}} = \mathbf{b} + \Delta\mathbf{b}\,. \tag{11.4}$$

Note that in (11.4) it is important to use the Frobenius norm, while for (11.3) the Frobenius norm and the common 2-norm coincide as **b** is a vector.

11.2 Computation of the TLS Estimate

This section describes how the solution $\mathbf{x}_{\mathrm{TLS}}$ to the TLS problem can be found. Set

$$\mathbf{C} = \begin{pmatrix} -\mathbf{b} & \mathbf{A} \end{pmatrix}\,. \tag{11.5}$$

The following result applies.

Lemma 11.1 *Let the smallest singular value of the matrix* **C** *be* σ*, and the associated right singular vector be denoted by* **u**.

1. *Assume that the first component of* **u** *is nonzero. Then the TLS solution is*

$$\mathbf{x}_{\mathrm{TLS}} = \begin{pmatrix} u_2 \\ \vdots \\ u_{n+1} \end{pmatrix} \frac{1}{u_1}\,. \tag{11.6}$$

2. *In case there are more than one singular value equal to* σ*, the solution to the TLS problem is not unique, but (11.6) can be applied using any right singular vector associated to* σ.

Proof See Appendix 11.A.1. ■

Remark 11.1 It holds generically that $u_1 \neq 0$. For the particular case when $u_1 = 0$, there is no guarantee that a TLS solution exists at all; see Van Huffel and Vandewalle (1991). ■

Remark 11.2 If in (11.4) the Frobenius norm is replaced by the spectral norm, the solution is still given by Lemma 11.1. This is so because the low-rank approximation of **C** obtained via the truncated SVD is optimal in both the Frobenius and the spectral norms. ■

Remark 11.3 It holds

$$\mathbf{x}_{\mathrm{TLS}} = \left(\mathbf{A}^T\mathbf{A} - \sigma^2\mathbf{I}\right)^{-1} \mathbf{A}^T\mathbf{b}\,. \tag{11.7}$$

Proof See Appendix 11.A.1. ■

The result (11.7) means in a sense that $\mathbf{x}_{\mathrm{TLS}}$ is larger than $\mathbf{x}_{\mathrm{LS}}$, cf. (11.2). Compare with the findings in Fig. 2.4 and in (2.95) for the scalar static case.

11.3 Using the TLS Estimate for System Identification

How is TLS to be applied to the errors-in-variables problem? Consider a simple example of applying TLS to system identification. Let the system dynamics be given by (5.6):

$$y(t) = \boldsymbol{\varphi}^T(t)\boldsymbol{\theta} \; . \tag{11.8}$$

Repeating this equation for various values of t gives

$$\begin{pmatrix} \boldsymbol{\varphi}^T(n+1) \\ \vdots \\ \boldsymbol{\varphi}^T(N) \end{pmatrix} \boldsymbol{\theta} = \begin{pmatrix} y(n+1) \\ \vdots \\ y(N) \end{pmatrix} , \tag{11.9}$$

where

$$n = \max(n_a, n_b) \; . \tag{11.10}$$

The reason why elements with $t \leq n$ are not included in (11.9) is that some input-output elements of $\boldsymbol{\varphi}(t)$ may be missing for $t \leq n$. Equation (11.9) can be written more explicitly as

$$\boldsymbol{\Phi}\boldsymbol{\theta} = \mathbf{Y} \; , \tag{11.11}$$

where

$$\boldsymbol{\Phi} = \begin{pmatrix} -y(n) & \ldots & -y(1-n_a+n) & u(n) & \ldots & u(1-n_b+n) \\ -y(n+1) & \ldots & -y(2-n_a+n) & u(n+1) & \ldots & \\ \vdots & \ddots & & \vdots & \ddots & \\ \vdots & & & \vdots & & \\ -y(N-1) & \ldots & -y(N-n_a) & u(N-1) & \ldots & u(N-n_b) \end{pmatrix} , \tag{11.12}$$

$$\mathbf{Y} = \begin{pmatrix} y(n+1) \\ \vdots \\ y(N) \end{pmatrix} . \tag{11.13}$$

The TLS principle can now be applied to the overdetermined system of equations in (11.11).

A generalization to the MIMO case is considered in Markovsky et al. (2005b).

11.4 The Structured Total Least Squares (STLS) Estimate

A potential weakness when applying the TLS solution to the system of equations (11.11) is that the matrix $\boldsymbol{\Phi}$ has a lot of structure that is not exploited. In fact, the data matrix

$$\begin{pmatrix} -\mathbf{Y} & \boldsymbol{\Phi} \end{pmatrix}$$

is a block Toeplitz matrix. When computing a TLS solution such structures are not taken into account, and the modified matrix

$$\begin{pmatrix} -\mathbf{Y} - \Delta\mathbf{Y} & \boldsymbol{\Phi} + \Delta\boldsymbol{\Phi} \end{pmatrix}$$

is no longer block Toeplitz.

In order to take such structural aspects into account, several extensions of the basic TLS problem have been suggested and introduced in the literature. One possibility of how to introduce a structured TLS (STLS) when the data matrix depends linearly on some unknown disturbances is as now described.

Assume, as is the case in (11.12), (11.13), that the matrix $\mathbf{C}$, see (11.5), depends linearly on a number of unknown parameters, and set

$$\mathbf{C} = \mathbf{C}_0 + \sum_{i=1}^{K} \mathbf{C}_i \eta_i \, , \tag{11.14}$$

where $\{\mathbf{C}_i\}_{i=1}^{K}$ are known matrices, and $\{\eta_i\}_{i=1}^{K}$ are unknown parameters. The modified system of equations are

$$\left(\mathbf{C} - \sum_{i=1}^{K} \mathbf{C}_i \eta_i \right) \bar{\mathbf{x}} = \mathbf{0} \, , \tag{11.15}$$

where

$$\bar{\mathbf{x}} = \begin{pmatrix} 1 \\ \mathbf{x} \end{pmatrix} \, . \tag{11.16}$$

Now seek, except for $\mathbf{x}$ also for the unknown vector

$$\boldsymbol{\eta} = \begin{pmatrix} \eta_1 & \dots & \eta_K \end{pmatrix}^T \tag{11.17}$$

which should be chosen so as to make the modification $-\sum_{i=1}^{K} \mathbf{C}_i \eta_i$ of $\mathbf{C}$ as small as possible. To be more specific, the criterion

$$V = \| \sum_{i=1}^{K} \mathbf{C}_i \eta_i \|_F^2 \tag{11.18}$$

is to be minimized. The rationale of the chosen criterion (11.18) is that small values of $\{\eta_i\}$ mean a small modification of $\mathbf{C}$. The criterion V can be rewritten as follows:

$$\begin{aligned} V &= \operatorname{tr}\left(\left(\sum_{i=1}^{K} \mathbf{C}_i \eta_i\right)\left(\sum_{j=1}^{K} \mathbf{C}_j^T \eta_j\right)\right) \\ &= \sum_{i=1}^{K}\sum_{j=1}^{K} \eta_i \eta_j \operatorname{tr}\left(\mathbf{C}_i \mathbf{C}_j^T\right) \\ &\stackrel{\Delta}{=} \boldsymbol{\eta}^T \mathbf{Q} \boldsymbol{\eta} , \end{aligned} \tag{11.19}$$

where the weighting matrix $\mathbf{Q}$ is given elementwise by

$$\mathbf{Q}_{ij} = \operatorname{tr}\left(\mathbf{C}_i \mathbf{C}_j^T\right), \quad i, j = 1, \ldots, K . \tag{11.20}$$

By construction, cf (11.18), $V \geq 0$ holds. Hence the matrix $\mathbf{Q}$ is certainly nonnegative definite. What is required for $\mathbf{Q}$ to be positive definite? The answer is that it must then hold $V > 0$ as soon as $\boldsymbol{\eta} \neq \mathbf{0}$. The requirement can also be written as

$$\sum_{i=1}^{K} \mathbf{C}_i \boldsymbol{\eta}_i = \mathbf{0} \quad \Rightarrow \boldsymbol{\eta}_i = 0 \; \forall i ,$$

which means that $\{\mathbf{C}_i\}_{i=1}^K$ are linearly independent. This is a most reasonable assumption in (11.14). The condition ensures that any deviation $\mathbf{C} - \mathbf{C}_0$ has a *unique* representation in the form $\sum_i \mathbf{C}_i \boldsymbol{\eta}_i$.

To minimize the criterion V with respect to $\boldsymbol{\eta}$ and $\mathbf{x}$ under the constraint (11.15) leads to a specific optimization problem. Note that all terms in (11.15) depend linearly on $\mathbf{x}$ and linearly on $\boldsymbol{\eta}$. It is though a bilinear function, as it does not depend linearly on both vectors simultaneously. It is indeed possible to write (11.15) in the form

$$\mathbf{f}(\mathbf{x}) = \mathbf{F}(\mathbf{x})\boldsymbol{\eta} . \tag{11.21}$$

Using this formulation, the optimization problem can as a matter of fact be reformulated as a unconstrained minimization problem with respect to $\mathbf{x}$ only. This is in fact application of the variable projection principle to the separable least squares problem.

Lemma 11.2 *Consider the optimization problem of minimizing (11.19) with respect to* $\mathbf{x}$ *and* $\boldsymbol{\eta}$ *under the constraint (11.21). The solution can be arranged as*

$$\hat{\mathbf{x}} = \arg\min_{\mathbf{x}} \overline{V}(\mathbf{x}) \,, \tag{11.22}$$

$$\overline{V}(\mathbf{x}) = \min_{\eta} V(\eta) = \mathbf{f}^T(\mathbf{x}) \left(\mathbf{F}(\mathbf{x}) \mathbf{Q}^{-1} \mathbf{F}^T(\mathbf{x}) \right)^{-1} \mathbf{f}(\mathbf{x}) \,, \tag{11.23}$$

$$\hat{\eta} = \mathbf{Q}^{-1} \mathbf{F}^T(\hat{\mathbf{x}}) \left(\mathbf{F}(\hat{\mathbf{x}}) \mathbf{Q}^{-1} \mathbf{F}^T(\hat{\mathbf{x}}) \right)^{-1} \mathbf{f}(\hat{\mathbf{x}}) \,. \tag{11.24}$$

Proof See Appendix 11.A.3. ■

Remark 11.4 Lemma 11.2 is a special case of Theorem 2.3 in Markovsky et al. (2005a), where a general (matrix **B**) affine structured TLS problem is considered. ■

The consistency properties of the TLS and STLS estimates of $\boldsymbol{\theta}$ are presented below in Sect. 11.5. It is shown that to guarantee consistency of $\hat{\boldsymbol{\theta}}$ it is necessary to assume that the noise variance ratio $r = \lambda_y/\lambda_u$ is known. This means that when using TLS, Assumption **AN4** is necessary for consistency.

11.5 Analysis of the TLS Estimate

11.5.1 General Aspects

If the errors in the various **A** and **b** coefficients are independent and identically distributed, the TLS solution to the problem (11.1) coincides with the maximum likelihood estimate, Gleser (1981). Further, both consistency and the asymptotic distribution of the estimates are examined under the assumption that the errors in all matrix elements of ($\Delta\mathbf{A}$ $\Delta\mathbf{b}$) are independent and identically distributed.

Unfortunately, the consistency result is not of much help, when TLS is applied for identification of dynamic systems. The assumption that all elements of the matrix perturbations are independent is not fulfilled.

A similar paper is Kukush and Tsaregorodtsev (2016) which treats both consistency and asymptotic distribution under the condition that the error matrices $\tilde{\mathbf{A}}$ and $\tilde{\mathbf{B}}$ have a covariance structure that is known up to a scalar factor.

11.5.2 Analysis of the TLS Estimate in an EIV Setting

Consider the TLS solution to the system of equations

$$\mathbf{Y} = \boldsymbol{\Phi}\boldsymbol{\theta} \,, \tag{11.25}$$

where $\mathbf{Y}$ and $\boldsymbol{\Phi}$ are given by (11.12), (11.13) and set

$$\hat{\overline{\boldsymbol{\theta}}}_{\text{TLS}} = \begin{pmatrix} 1 \\ \hat{\boldsymbol{\theta}}_{\text{TLS}} \end{pmatrix} . \tag{11.26}$$

The TLS solution (11.26) is characterized as the right singular vector associated with the smallest singular value for the data matrix

$$\left(-\mathbf{Y} \ \boldsymbol{\Phi} \right) ,$$

that is, it is the eigenvector associated to the smallest eigenvalue of

$$\left(\begin{pmatrix} -\mathbf{Y}^T \\ \boldsymbol{\Phi}^T \end{pmatrix} \left(-\mathbf{Y} \ \boldsymbol{\Phi} \right) \right) . \tag{11.27}$$

To examine consistency properties, consider the asymptotic case $N \to \infty$ and normalize (11.27) with $1/N$. This means that $\hat{\overline{\boldsymbol{\theta}}}_{\text{TLS}}$ is the eigenvector associated with

$$\mathbf{R}_{\overline{\varphi}} = \begin{pmatrix} r_y(0) & -\mathbf{r}_{y\varphi} \\ -\mathbf{r}_{\varphi y} & \mathbf{R}_{\varphi} \end{pmatrix} . \tag{11.28}$$

Assume that both the input noise and the output noise are white (Assumption **AN3c**). It then holds

$$\mathbf{R}_{\overline{\varphi}} = \mathbf{R}_{\overline{\varphi}_0} + \underbrace{\begin{pmatrix} \lambda_y \mathbf{I}_{n_a+1} & \mathbf{0} \\ \mathbf{0} & \lambda_u \mathbf{I}_{n_b} \end{pmatrix}}_{\mathbf{J}} , \tag{11.29}$$

$$\mathbf{R}_{\overline{\varphi}_0} \overline{\boldsymbol{\theta}}_0 = \mathbf{0} . \tag{11.30}$$

Let σ denote the smallest eigenvalue of $\mathbf{R}_{\overline{\varphi}}$. Then it holds

$$\mathbf{R}_{\overline{\varphi}} \hat{\overline{\boldsymbol{\theta}}}_{\text{TLS}} = \sigma \hat{\overline{\boldsymbol{\theta}}}_{\text{TLS}} . \tag{11.31}$$

The estimate is consistent if $\hat{\boldsymbol{\theta}}_{\text{TLS}} = \boldsymbol{\theta}_0$, which turns out to be equivalent to

$$\begin{aligned} & \mathbf{R}_{\overline{\varphi}} \overline{\boldsymbol{\theta}}_0 = \sigma \overline{\boldsymbol{\theta}}_0 \\ \Rightarrow\ & \mathbf{J} \overline{\boldsymbol{\theta}}_0 = \sigma \overline{\boldsymbol{\theta}}_0 \\ \Rightarrow\ & \lambda_y \overline{\mathbf{a}} = \sigma \overline{\mathbf{a}}, \quad \lambda_u \mathbf{b} = \sigma \mathbf{b} , \end{aligned} \tag{11.32}$$

cf. (5.11). Apparently, this can only hold if

$$\lambda_y = \lambda_u , \tag{11.33}$$

for which case it also follows that $\sigma = \lambda_u$. Of course, (11.33) holds if the input noise and output noise variances are the same. One should though rather interpret (11.33) as an identifiability condition that the ratio $r = \lambda_y / \lambda_u$ is known; that is Assumption **AN4** applies. When this ratio is known, it is straightforward to scale the recorded signals so that the input and the output are perturbed by white noise to identical extents (of the same variance), and then the condition (11.33) is met.

Note that the consistency analysis here, leading to the condition (11.33), is in perfect agreement with the previous findings in Chap. 2, where the static case was considered and analyzed.

11.5.3 Analysis of the STLS Estimate in an EIV Setting

The statistical properties of the solution to a structured TLS problem is considered in several papers, for example Kukush et al. (2005). A quite general TLS situation is examined. In the analysis, however, it is assumed that the covariance structure may vary from row to row, but that the total covariance structure is known up to a scalar factor. A related analysis for a very general TLS situation is given in Kukush et al. (2005), again assuming that the covariance matrix of the errors is known up to a scaling factor. When the results are applied to identification of a dynamic system, the assumptions imply that the input and output noise variances are assumed to be the same ($\lambda_u = \lambda_y$). This corresponds in the GIVE framework in Sect. 7.4 to the simpler, but not so realistic, case of a one degree of freedom problem.

Can identifiability be achieved in the more general case when both the noise variances λ_y and λ_u are unknown? The experience from Chap. 2 and Sect. 11.5.2 indicates that this is hardly the case. For the STLS approach to the identification problem of (11.11) the $\boldsymbol{\eta}$ variables will correspond to estimates $\hat{\tilde{y}}(t),\ \hat{\tilde{u}}(t),\ t = 1, \ldots, N$ of the noises. To estimate them is essentially the same as estimating the noise-free input signal, $u_0(t),\ t = 1, \ldots, N$. It was found in Sect. 3.4.2 that identifiability is lost in the general case (λ_y and λ_u both unknown) for such a situation.

A bit more detailed analysis goes as follows. In this case $\mathbf{x}$ in (11.22)–(11.24) is the same as $\boldsymbol{\theta}$. Further, $\mathbf{f}(\mathbf{x})$ in (11.21) is

$$\mathbf{f}(\mathbf{x}) = \begin{pmatrix} \varepsilon(1, \boldsymbol{\theta}) \\ \vdots \\ \varepsilon(N, \boldsymbol{\theta}) \end{pmatrix} , \tag{11.34}$$

where

$$\varepsilon(t, \boldsymbol{\theta}) = A(q^{-1})y(t) - B(q^{-1})u(t) . \tag{11.35}$$

As $\boldsymbol{\eta}$ consists of the estimated noise contributions,

$$\boldsymbol{\eta} = \begin{pmatrix} \hat{\tilde{y}}(1) \ldots \hat{\tilde{y}}(N)\ \hat{\tilde{u}}(1) \ldots \hat{\tilde{u}}(N) \end{pmatrix}^T , \tag{11.36}$$

it holds that $K = 2N$, and each $\mathbf{C}_i$ is a block Toeplitz matrix with all elements equal to zero except for one subdiagonal having elements equal to one. As a consequence, (11.20) implies

$$\mathbf{Q}_{ij} = N\delta_{i,j} \ . \tag{11.37}$$

Further, $\mathbf{F}(\mathbf{x})$ in (11.21) will be a block Toeplitz matrix of the form

$$\mathbf{F}(\mathbf{x}) = \begin{pmatrix} a_{n_a} & \dots & 1 & \mathbf{0} & -b_{n_b} & \dots & -b_1 & \mathbf{0} & \\ & \ddots & & & & \ddots & & & \\ \mathbf{0} & & a_{n_a} & \dots 1 & \mathbf{0} & & -b_{n_b} & \dots & -b_1 \end{pmatrix} . \tag{11.38}$$

Thus the loss function $\overline{V}(\mathbf{x})$ in (11.23) can be written in the asymptotic case ($N \to \infty$) as

$$\overline{\overline{V}}(\boldsymbol{\theta}) = \lim_{N\to\infty} \frac{1}{N} \mathbf{f}^T(\mathbf{x}) \left(\mathbf{F}(\mathbf{x})\mathbf{F}^T(\mathbf{x})\right)^{-1} \mathbf{f}(\mathbf{x}) \stackrel{\Delta}{=} \mathbf{f}^T(\mathbf{x})\mathbf{G}(\mathbf{x})\mathbf{f}(\mathbf{x}) \ . \tag{11.39}$$

Using (11.34) one finds that

$$\overline{\overline{V}}(\boldsymbol{\theta}) = \lim_{N\to\infty} \sum_{t,s=1}^{N} \varepsilon(t,\boldsymbol{\theta})\varepsilon(s,\boldsymbol{\theta})\mathbf{G}_{t,s}(\boldsymbol{\theta}) \ . \tag{11.40}$$

Assume that the data satisfies the general assumptions, and hence

$$y(t) = \tilde{y}(t) + B_0(q^{-1})z(t) \ , \tag{11.41}$$

$$u(t) = \tilde{u}(t) + A_0(q^{-1})z(t) \ , \tag{11.42}$$

$$z(t) = \frac{1}{A_0(q^{-1})} u_0(t) \ , \tag{11.43}$$

$$\begin{aligned} \varepsilon(t,\boldsymbol{\theta}) &= A(q^{-1})\tilde{y}(t) - B(q^{-1})\tilde{u}(t) \\ &\quad + \left[A(q^{-1})B_0(q^{-1}) - A_0(q^{-1})B(q^{-1})\right] z(t) \\ &\stackrel{\Delta}{=} \varepsilon_y(t,\boldsymbol{\theta}) + \varepsilon_u(t,\boldsymbol{\theta}) + \varepsilon_z(t,\boldsymbol{\theta}) \ . \end{aligned} \tag{11.44}$$

Note that the three different terms in (11.44) are independent. Therefore one can write

$$\begin{aligned} \overline{\overline{V}}(\boldsymbol{\theta}) &= \lim_{N\to\infty} \frac{1}{N} \sum_{t,s=1}^{N} \varepsilon_y(t,\boldsymbol{\theta})\varepsilon_y(s,\boldsymbol{\theta})\mathbf{G}_{t,s}(\boldsymbol{\theta}) \\ &\quad + \lim_{N\to\infty} \frac{1}{N} \sum_{t,s=1}^{N} \varepsilon_u(t,\boldsymbol{\theta})\varepsilon_u(s,\boldsymbol{\theta})\mathbf{G}_{t,s}(\boldsymbol{\theta}) \\ &\quad + \lim_{N\to\infty} \frac{1}{N} \sum_{t,s=1}^{N} \varepsilon_z(t,\boldsymbol{\theta})\varepsilon_z(s,\boldsymbol{\theta})\mathbf{G}_{t,s}(\boldsymbol{\theta}) \\ &\stackrel{\Delta}{=} V_y(\boldsymbol{\theta}) + V_u(\boldsymbol{\theta}) + V_z(\boldsymbol{\theta}) \ . \end{aligned} \tag{11.45}$$

Table 11.1 Total least squares estimates for the standard data sets

Parameter	True value	TLS		LS	
		Mean	Standard dev.	Mean	Standard dev.
a_1	−1.5	0.664	8.73	−1.414	0.015
a_2	0.7	−0.57	5.25	0.619	0.014
b_1	2.0	−29.3	129	1.599	0.103
b_2	1.0	46.4	184	1.361	0.129

Table 11.2 Scaled total least squares estimates for the standard data sets

Parameter	a_1	a_2	b_1	b_2
True value	−1.5	0.7	2.0	1.0
Scale factor				
0.1	5.21 (40.0)	−3.23 (21.8)	−98.9 (585)	146 (839)
0.5	1.306 (11.2)	−0.96 (6.42)	−39.3 (175)	60.5 (251)
1	0.664 (8.73)	−0.57 (5.25)	−29.3 (129)	46.4 (184)
2	−1.494 (1.060)	0.689 (0.635)	2.644 (15.7)	0.414 (22.4)
$\sqrt{\frac{\lambda_y}{\lambda_u}} \approx 2.35$	−1.499 (0.029)	0.699 (0.019)	2.030 (0.414)	0.980 (0.579)
5	−1.572 (0.012)	0.754 (0.009)	1.573 (0.133)	0.884 (0.164)
10	−1.587 (0.011)	0.765 (0.009)	1.496 (0.118)	0.849 (0.140)

Due to its construction, $V_z(\boldsymbol{\theta})$ will be minimized with respect to $\boldsymbol{\theta}$ for $\boldsymbol{\theta} = \boldsymbol{\theta}_0$. However, the terms $V_y(\boldsymbol{\theta})$ and $V_u(\boldsymbol{\theta})$ do not depend on the true parameter vector $\boldsymbol{\theta}_0$, and thus $\boldsymbol{\theta}_0$ cannot be the global minimum point of $\overline{\overline{V}}(\boldsymbol{\theta})$. Therefore identifiability is not achieved.

Example 11.1 The basic TLS estimator was applied to the standard data sets. The outcome is displayed in Table 11.1, where a comparison to the LS estimates is also provided.

The obtained TLS estimates are quite lousy. For about 10 of the 100 realizations the estimation errors are very large, which indicates that the TLS estimate is certainly less robust than the LS estimate. In contrast, for LS there is a bias, but the standard deviations are small. ■

Example 11.2 What would happen with the TLS estimate if another scaling of the data is used? In Table 11.2 the input is multiplied with the scale factor before the

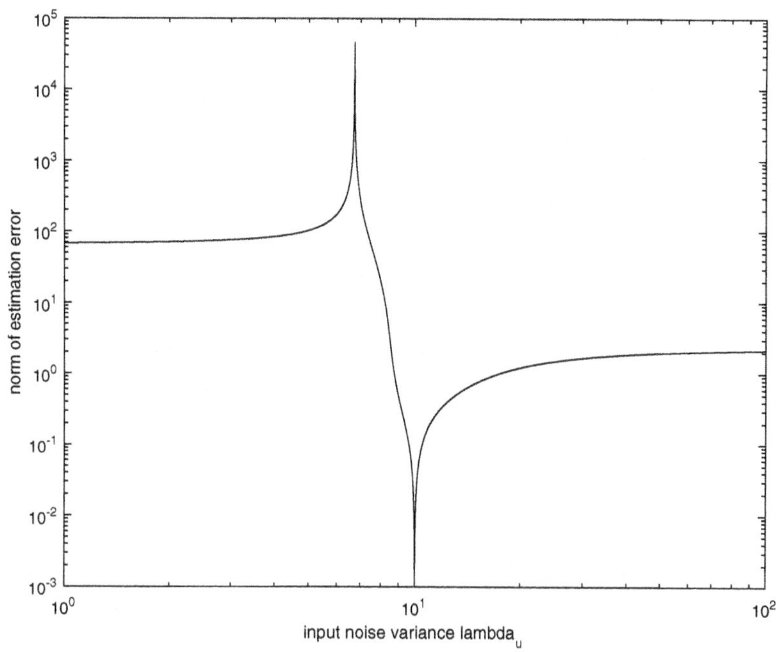

Fig. 11.1 Scaled TLS estimates, asymptotic case. Plot of the norm of the parameter estimation error $\| \hat{\boldsymbol{\theta}} - \boldsymbol{\theta}_0 \|$ as a function of the input noise variance λ_u. Note the logarithmic scales

TLS estimator is applied, and then the B-parameters are rescaled so that the model applies to the original data. It is apparently crucial to use an appropriate scale factor. In particular, if the scale factor (which should be regarded as a user parameter) is chosen too small, the result can be disastrous. Note that the value 2.35 corresponds to equal noise variances on the (scaled) input and the output. Hence this is the ideal scale factor, and in the fact the value that must be chosen for the estimates to be consistent. The results are reasonable only when the scale factor is close to the optimal value $\sqrt{\lambda_y / \lambda_u}$. ■

Example 11.3 In Fig. 11.1 the quality of the TLS estimate is shown as a function of the amount of input noise. The output noise variance was fixed to $\lambda_y = 10$. The asymptotic case is considered, so expected values of covariance elements are used rather than a Monte Carlo simulation. The input noise variance λ_u was varied, and its impact on the norm of the estimation error $\| \hat{\boldsymbol{\theta}} - \boldsymbol{\theta}_0 \|$ was studied. It is clearly seen that the estimation error becomes small just in a small region around the case of equal variances, that is, when $\lambda_u \approx \lambda_y = 10$. ■

11.6 Extensions and Bibliographical Notes

The total least squares approach to identification of errors-in-variables systems has been used in a number of contexts, and there are many publications in the field.

TLS as such has been given a detailed treatment in Golub and Van Loan (1980) and Van Huffel and Vandewalle (1991). Some further general treatments of TLS and its variants are given in Paige and Strakoš (2002), Paige and Strakoš (2006), De Moor and Vandewalle (1990), and De Moor (1993).

There are several connections between TLS and the errors-in-variables problems, as manifested in the workshop proceedings Van Huffel (1997), Van Huffel and Lemmerling (2002). An overview of TLS problems, where EIV identification is just an example is given in Markovsky and Van Huffel (2007).

The papers Adcock (1877) and Adcock (1878) are recognized to be among the first ones to talk about measurement error models. Adcok's least squares method is now known as orthogonal regression in statistics and is indeed total least squares in its simplest form (assuming equal variances in both variables).

TLS with Modified Criterion or with Constraints

Use of structured TLS (STLS), for example imposing Toeplitz or Hankel structures, with a focus on algorithmic aspects and computational efficiency is studied in Markovsky et al. (2004), Markovsky et al. (2005b). The problem is typically reformulated as a nonlinear optimization problem. Other publications on STLS variants include Lemmerling and Van Huffel (2002), Abatzoglou and Mendel (1991), De Moor (1994), Ben Rosen et al. (1996). In Beck (2007) a structured TLS problem is treated, where the uncertainty in the matrix $\mathbf{A}$ takes the form $\Delta\mathbf{A} = \mathbf{D}\Delta\mathbf{E}\mathbf{F}$, where $\mathbf{D}$ and $\mathbf{F}$ are given matrices.

There are several iterative quadratic maximum likelihood (IQML) algorithms for solving the STLS problem, Lemmerling et al. (2001). The classical IQML algorithm provides only a suboptimal solution, while there are other versions of IQML that give the optimal solution. For other illustrations of IQML leading to non-consistent estimates; see, e.g., Stoica et al. (1997).

An equivalence. between STLS and a constrained TLS problem is demonstrated in Lemmerling et al. (1996)

In Markovsky et al. (2006b) the TLS problem is generalized to elementwise weighting of the errors ($\Delta\mathbf{A}\ \Delta\mathbf{b}$) again leading to a nonlinear optimization. Consistency is examined, but only for restricted cases such as when the errors are uncorrelated between different rows, which does not apply to the model (11.12), (11.13).

The work in Schaffrin and Felus (2009) examines the TLS problem with linear and quadratic constraints.

The case when some elements of the matrix $\mathbf{A}$ are not subject to noise or uncertainties can be treated using the structured TLS algorithm. An alternative approach is to convert the problem to a nonlinear Gauss–Markov model; see Shi et al. (2015) and Xu et al. (2012).

The papers Markovsky (2008), Markovsky (2014) consider structured low-rank approximation and its applications, while several estimation problems related to low-rank approximation are surveyed in Markovsky (2009).

The case when some elements of the matrix $\mathbf{A}$ are not subject to noise or uncertainties can be treated using the structured TLS algorithm. An alternative approach is to convert the problem to a nonlinear Gauss-Markov model; see Shi et al. (2015) and Xu et al. (2012).

An extension to. the standard TLS formulation by adding linear observations, and taking nonlinear adjustments into account, is proposed in Pan et al. (2015)

The standard TLS problems and its generalizations to take internal structures of the data into account, structured TLS and constrained TLS, are designed to take care of measurements errors in the data. The situation covered in Yeredor (2001) concerns the case when there is also some modeling error to cope with. An 'extended least squares' (XLS) criterion is proposed that makes a trade off between minimizing the effect of measurement errors and modeling errors. The underlying model is written in the form

$$\mathbf{g}(\mathbf{x}, \boldsymbol{\theta}) \approx \mathbf{0} \ , \tag{11.46}$$

and the XLS optimization problem is formulated as

$$\min_{\hat{\mathbf{x}}, \hat{\boldsymbol{\theta}}} \left[\mathbf{g}^T(\hat{\mathbf{x}}, \hat{\boldsymbol{\theta}}) \mathbf{W}_g \mathbf{g}(\hat{\mathbf{x}}, \hat{\boldsymbol{\theta}}) + (\mathbf{x} - \hat{\mathbf{x}})^T \mathbf{W}_{\mathbf{x}} (\mathbf{x} - \hat{\mathbf{x}}) \right] , \tag{11.47}$$

where $\mathbf{W}_g$ and $\mathbf{W}_x$ are user-specified weighting matrices, and $\mathbf{x}$ some noisy observations of $\hat{\mathbf{x}}$. As an illustration, the XLS problem is applied to the estimation of parameters in an autoregressive process observed with measurement noise.

There are several papers, see Rhode et al. (2014) and Arablouei et al. (2015) for example, that deal with recursive implementations of TLS.

Applications of TLS to System Identification

The TLS principle can be applied not only to the direct data matrices as in (11.12), (11.13), but also to IV equations such as (6.14):

$$\hat{\mathbf{R}}_{z\varphi} \hat{\boldsymbol{\theta}} = \hat{r}_{\mathbf{z}y} \ . \tag{11.48}$$

An alternative approach for exploiting (11.48) is to solve it in a total least squares sense, Van Huffel and Vandewalle (1991). An extensive comparison of IV and TLS based on numerical examples is presented in Van Huffel and Vandewalle (1989). A theoretical justification that TLS and a weighted IV approach can give estimates with equal asymptotic distribution is given in Söderström and Mahata (2002). In this specific case note that the relation means that the matrix $\hat{\mathbf{R}}_{z\overline{\varphi}}$ is close to singular. Let $\hat{\mathbf{w}}$ denote its right singular vector, associated with its smallest singular value. Then, except for a normalization, $\hat{\mathbf{w}}$ should coincide with $\hat{\overline{\boldsymbol{\theta}}}$. Using MATLAB notations, this leads to the estimate

$$\hat{\boldsymbol{\theta}}_{\mathrm{TLS}} = \hat{\mathbf{w}}_{2:n_a+n_b+1} / \hat{w}_1 \ . \tag{11.49}$$

For the case when $\dim \mathbf{z}(t) = n_a + n_b$ the estimator $\hat{\boldsymbol{\theta}}_{\text{IV}}$ and $\hat{\boldsymbol{\theta}}_{\text{TLS}}$ coincide. This is easily seen as follows. First note that in this case $\hat{\mathbf{R}}_{z\overline{\varphi}}$ is of dimension $(n_a + n_b) \times (n_a + n_b + 1)$ and has hence a null space of dimension at least equal to one. The smallest singular value is zero, and the associated singular vector turns out to be $(1 \ \hat{\boldsymbol{\theta}}_{\text{IV}}^T)^T$:

$$\hat{\mathbf{R}}_{z\overline{\varphi}} \begin{pmatrix} 1 \\ \hat{\boldsymbol{\theta}}_{\text{IV}} \end{pmatrix} = \begin{pmatrix} \hat{\mathbf{r}} \ \hat{\mathbf{R}}_{z\varphi} \end{pmatrix} \begin{pmatrix} 1 \\ -\hat{\mathbf{R}}_{z\varphi}^{-1}\hat{\mathbf{r}} \end{pmatrix} = \mathbf{0} \,. \tag{11.50}$$

When $\dim \mathbf{z}(t) > n_a + n_b$ holds, the two estimators $\hat{\boldsymbol{\theta}}_{\text{TLS}}$ and $\hat{\boldsymbol{\theta}}_{\text{IV}}$ do differ.

A simple TLS approach for identifying dynamic systems, assuming $\lambda_u = \lambda_y$, (or that the variance ratio is known so that the data first can be rescaled) is presented in Cadzow and Solomon (1986).

In De Moor and Roorda (1994) STLS is applied for system identification and treated as a special case of L_2-optimal approximation.

Approaches based on TLS and different regularizations for identification of sparse systems (models with several zero coefficients) have been proposed by Lim and Pang (2016a), Lim and Pang (2016b), Lim (2017).

The papers Roorda and Heij (1995), Roorda (1995) apply TLS ideas in a much more general setting, using the behavioral approach for modeling and approximating dynamic systems. Primarily no noise consideration is made. The idea is to impose a dynamic model which is fulfilled by the noise-free input–output data $\{\zeta_0(t)\}, t = 1, \dots N$. Within the class of such generated noise-free data time series, one seeks the one that minimizes

$$\| \mathbf{Z} - \mathbf{Z}_0 \| \tag{11.51}$$

with respect to possible $\mathbf{Z}_0$. Here, $\mathbf{Z}$ denotes the full sequence (or time series) of measured values of $z(t)$, and $\mathbf{Z}_0$ is the corresponding quantity for the noise-free model. No specific distinction is made between inputs and outputs. The methodology is called 'global total least squares,' and various characteristics for the optimal model are presented. It should be noted that in this context, 'optimal' means the model with the minimal l_2 distance from the observed time series. The approach can be used also for multivariable systems, with more than one input $u(t)$ and more than one output $y(t)$. As one is not specifying any noise characteristics, and input and output errors are given equal weights in the criterion (11.51), one may possibly say that the method fits the noise assumption **AN3c** combined with **AN4** with equal input and output noise variances. However, it is fair to underline that the approach primarily concerns the best approximation of the given time series without any specification of the noise properties. In fact, the analysis of this approach carried out in Heij and Scherrer (1999) concerns the limit estimate as N goes to infinity. It is shown that one then gets the optimal approximation, that is the best approximation using infinitely long time series. (N.B. The concept of 'consistency' is here not used in the standard statistical sense.)

The book Markovsky et al. (2006c) on the behavioral approach is written in the same spirit as Roorda's work. This approach focuses on a representation-free setting, thus avoiding specific model forms. See also Markovsky (2015). Also the paper Maurya et al. (2016) follows the behavioral approach. It essentially solves the EIV problem as a structured TLS problem by using a principle component analysis framework. Another paper dealing with PCA when handling noisy data is Narasimhan and Shah (2008).

An extension of TLS where one performs a minimization of a weighted combination of $\sum_t \hat{\tilde{u}}^2(t)$, $\sum_t \hat{\tilde{y}}^2(t)$, and $\sum_t \varepsilon^2(t)$, where $\hat{\tilde{u}}(t)$ is an estimate of the input noise, $\hat{\tilde{y}}(t)$ an estimate of the output noise, and $\varepsilon(t)$ is an estimate of the equation error, is presented in Lemmerling and De Moor (2001).

A rather general setting for EIV application is considered in Kukush et al. (2005). Equal noise levels are assumed, so $\lambda_u = \lambda_y$ and multivariable systems are treated.

Other Aspects of TLS

A constrained TLS, used for harmonics estimation (sine waves in noise), is proposed by Abatzoglou (1989).

The paper Fuchs and Maria (2007) deals with the order estimation problem, using matrix perturbation analysis.

For an extension of the Eckart–Young theorem (Lemma 11.3), see Golub et al. (1987).

The paper Markovsky et al. (2006a) is based on clustering (which essentially means that there are data from two different experiments, where the noise parts are uncorrelated, while the noise-free data is correlated between the experiments).

The papers Dreesen et al. (2012a, b), Batselier et al. (2012) show how many problems in system identification and system theory can be considered and solved using multivariate polynomials. See also Dreesen and De Moor (2009), Batselier et al. (2014).

Appendix

11.A Further Details

11.A.1 The Eckart–Young–Mirsky Lemma

There is a neat result, due to Eckart and Young (1936) and Mirsky (1960), on the optimal low-rank approximation of a given matrix. It will be useful when deriving the solution to the TLS problem.

Lemma 11.3 *Consider an* $m \times n$ *matrix* $\mathbf{C}$ *with* $m \geq n$. *Let* $\mathbf{C}$ *have a singular value decomposition*

$$\mathbf{C} = \begin{pmatrix} \mathbf{U}_1 & \mathbf{U}_2 \end{pmatrix} \begin{pmatrix} \mathbf{\Sigma}_1 & \mathbf{0} \\ \mathbf{0} & \mathbf{\Sigma}_2 \end{pmatrix} \begin{pmatrix} \mathbf{V}_1^T \\ \mathbf{V}_2^T \end{pmatrix} , \tag{11.52}$$

where $\boldsymbol{\Sigma}_1$ is an $r \times r$ matrix containing the r largest singular values, and the other matrices have compatible dimensions.

The matrix $\hat{\mathbf{C}}$ defined as

$$\hat{\mathbf{C}} = \arg \min_{\text{rank}\hat{\mathbf{C}}=r} \| \mathbf{C} - \hat{\mathbf{C}} \|_F^2 , \tag{11.53}$$

is given by

$$\hat{\mathbf{C}} = \mathbf{U}_1 \boldsymbol{\Sigma}_1 \mathbf{V}_1^T . \tag{11.54}$$

Further,

$$\| \mathbf{C} - \hat{\mathbf{C}} \|_F^2 = \| \boldsymbol{\Sigma}_2 \|_F^2 = \text{tr}(\boldsymbol{\Sigma}_2^T \boldsymbol{\Sigma}_2) . \tag{11.55}$$

Proof See Eckart and Young (1936). ∎

11.A.2 Characterization of the TLS Solution

11.A.2.1 Proof of Lemma 11.1

Set

$$\mathbf{u} = \begin{pmatrix} 1 \\ \mathbf{x}_{\text{TLS}} \end{pmatrix} .$$

One needs to find $\Delta\mathbf{C}$ with minimal norm, such that

$$(\mathbf{C} + \Delta\mathbf{C})\mathbf{u} = \mathbf{0} .$$

By applying Lemma 11.3, one gets $r = n - 1$ and

$$\Delta\mathbf{C} = -\mathbf{U}_2 \boldsymbol{\Sigma}_2 \mathbf{V}_2^T , \quad \mathbf{u} = \mathbf{V}_2 , \quad \sigma = \boldsymbol{\Sigma}_2 ,$$

and the lemma follows directly.

11.A.2.2 Proof of Remark 11.3

First establish $\mathbf{V}_1^T \mathbf{U} = \mathbf{0}$ and

$$\begin{aligned} \mathbf{C}^T \mathbf{C}\mathbf{u} &= \left(\mathbf{V}_1 \boldsymbol{\Sigma}_1 \mathbf{U}_1^T + \mathbf{V}_2 \boldsymbol{\Sigma}_2 \mathbf{U}_2^T\right) \left(\mathbf{U}_1 \boldsymbol{\Sigma}_1 \mathbf{V}_1^T + \mathbf{U}_2 \boldsymbol{\Sigma}_2 \mathbf{V}_2^T\right) \mathbf{u} \\ &= \mathbf{V}_2 \boldsymbol{\Sigma}_2^2 = \sigma^2 \mathbf{u} . \end{aligned}$$

Using the definition (11.5) of $\mathbf{C}$ it thus holds

$$\begin{pmatrix} -\mathbf{b}^T \\ \mathbf{A}^T \end{pmatrix} \begin{pmatrix} -\mathbf{b} & \mathbf{A} \end{pmatrix} \begin{pmatrix} 1 \\ \mathbf{x}_{\text{TLS}} \end{pmatrix} = \sigma^2 \begin{pmatrix} 1 \\ \mathbf{x}_{\text{TLS}} \end{pmatrix} .$$

Spelling out the lower part of this equation gives

$$-\mathbf{A}^T\mathbf{b} + \mathbf{A}^T\mathbf{A}\mathbf{x}_{\text{TLS}} = \sigma^2 \mathbf{x}_{\text{TLS}} ,$$

which proves (11.7).

11.A.3 Proof of Lemma 11.2

To examine the optimization problem, introduce the Lagrange multiplier vector $\boldsymbol{\lambda}$ and the Lagrange function

$$L(\mathbf{x}, \eta) = \eta^T \mathbf{Q} \eta + \boldsymbol{\lambda}^T \left(\mathbf{f}(\mathbf{x}) - \mathbf{F}(\mathbf{x})\eta\right) . \tag{11.56}$$

Setting the gradient of L with respect to η to zero leads to

$$2\mathbf{Q}\eta - \mathbf{F}^T(\mathbf{x})\boldsymbol{\lambda} = \mathbf{0} \Rightarrow \eta = \frac{1}{2}\mathbf{Q}^{-1}\mathbf{F}^T(\mathbf{x})\boldsymbol{\lambda} . \tag{11.57}$$

Considering the constraint (11.21) (or setting the gradient of L with respect to $\boldsymbol{\lambda}$ to zero) leads to

$$\mathbf{f}(\mathbf{x}) = \mathbf{F}(\mathbf{x})\eta = \frac{1}{2}\mathbf{F}(\mathbf{x})\mathbf{Q}^{-1}\mathbf{F}^T(\mathbf{x})\boldsymbol{\lambda} \Rightarrow \boldsymbol{\lambda} = 2\left(\mathbf{F}(\mathbf{x})\mathbf{Q}^{-1}\mathbf{F}^T(\mathbf{x})\right)^{-1}\mathbf{f}(\mathbf{x}) . \tag{11.58}$$

Therefore

$$\eta = \mathbf{Q}^{-1}\mathbf{F}^T(\mathbf{x})\left(\mathbf{F}(\mathbf{x})\mathbf{Q}^{-1}\mathbf{F}^T(\mathbf{x})\right)^{-1}\mathbf{f}(\mathbf{x}) , \tag{11.59}$$

which always satisfies the constraint (11.21). Furthermore, simple algebra shows that the minimal value of the loss function (with respect η) becomes

$$\overline{V}(\mathbf{x}) = \min_{\eta} V(\eta) = \mathbf{f}^T(\mathbf{x})\left(\mathbf{F}(\mathbf{x})\mathbf{Q}^{-1}\mathbf{F}^T(\mathbf{x})\right)^{-1}\mathbf{f}(\mathbf{x}) , \tag{11.60}$$

which shows (11.23) and completes the proof.

Chapter 12
Methods for Periodic Data

This chapter is organized as follows. After an introduction in Sect. 12.1, use of instrumental variable methods is presented in Sect. 12.2. The so-called sample maximum likelihood (SML) method is described in Sect. 12.3, while extensions and bibliographical notes appear in Sect. 12.4.

12.1 Introduction

The methods described in this chapter are tied to periodic data. However, also the case when there is more than one experiment is included. Should the unperturbed input signal, $u_0(t)$, be the same in all (two or more) experiments, and the experiments are of equal length, then concatenating the measurements will indeed produce periodic data.

To simplify the description in this chapter, it is generally assumed here that transient effects can be neglected.

Recall from Sect. 4.3, see Examples 4.5 and 4.6, that under mild conditions the system will be identifiable if the noise-free input signal $u_0(t)$ is periodic.

Example 12.1 A straightforward way to handle multiple experiments using time domain data is the following, which leads to an instrumental variable estimator.

Consider the linear regression model

$$y(t) = \boldsymbol{\varphi}^T(t)\boldsymbol{\theta} + \varepsilon(t) , \tag{12.1}$$

where $\varepsilon(t)$ denotes the equation error. Assume that more than one data set is available, so that

$$y^{(i)}(t) = \boldsymbol{\varphi}^{(i)T}(t)\boldsymbol{\theta} + \varepsilon^{(i)}(t), \quad i = 1, 2, \ldots \tag{12.2}$$

T. Söderström, *Errors-in-Variables Methods in System Identification*, Communications and Control Engineering,
https://doi.org/10.1007/978-3-319-75001-9_12

The true parameter vector fits perfectly the models when undisturbed data are used:

$$y_0^{(i)}(t) = \boldsymbol{\varphi}_0^{(i)T}(t)\boldsymbol{\theta}_0 \,. \tag{12.3}$$

Assume now that **AE2b** applies, that is, the noise is independent in the different data sets, and the unperturbed regressor vector $\boldsymbol{\varphi}_0^{(i)}(t)$ is (well) correlated in the different data sets.

Using two data sets, one then gets

$$\begin{aligned} & \mathsf{E}\left\{\boldsymbol{\varphi}^{(1)}(t)\boldsymbol{\varphi}^{(2)T}(t)\right\}\boldsymbol{\theta}_0 - \mathsf{E}\left\{\boldsymbol{\varphi}^{(1)}(t)y^{(2}(t))\right\} \\ & = \mathsf{E}\left\{\boldsymbol{\varphi}_0^{(1)}(t)\boldsymbol{\varphi}_0^{(2)T}(t)\right\}\boldsymbol{\theta}_0 - \mathsf{E}\left\{\boldsymbol{\varphi}_0^{(1)}(t)y_0^{(2)}(t)\right\} = \mathbf{0} \,. \end{aligned} \tag{12.4}$$

Assume that the matrix $\mathsf{E}\left\{\boldsymbol{\varphi}_0^{(1)}(t)\boldsymbol{\varphi}_0^{(2)T}(t)\right\}$ is non-singular. This is partly a condition on the inputs being persistently exciting. It is also a condition on sufficient correlation between the data sets. The consequence is that from *two* data sets, it is possible to derive a consistent parameter estimator as

$$\left[\frac{1}{N}\sum_{t=1}^{N}\boldsymbol{\varphi}^{(1)}(t)\boldsymbol{\varphi}^{(2)T}(t)\right]\hat{\boldsymbol{\theta}} = \left[\frac{1}{N}\sum_{t=1}^{N}\boldsymbol{\varphi}^{(1)}(t)y^{(2)}(t)\right] \,. \tag{12.5}$$

This is indeed an instrumental variable estimator.

It is though also possible to apply the estimator (12.5) in other situations, and the 'experiments' can be allowed to be overlapping, as long as the basic assumptions are satisfied. In particular two subsets of a single experiment may be used, as illustrated in Fig. 12.1. ■

More extended versions of the idea of Example 12.1 including the use of optimal weightings will be dealt with in Sect. 12.2.

In what follows in this chapter the following additional assumptions and notations will be used.

AI5. The noise-free signal $u_0(t)$ is a periodic function. The length of the period is denoted N. It is assumed that M periods of the data $u(t)$, $y(t)$ are available. Hence the total data length is NM. In each period $u_0(t)$ is a stationary process.

AI6. The measurement noise signals $\tilde{u}(t)$ and $\tilde{y}(t)$ are uncorrelated with the noise-free input $u_0(s)$ for all t and s. Further, the measurement noise signals within different periods are uncorrelated.

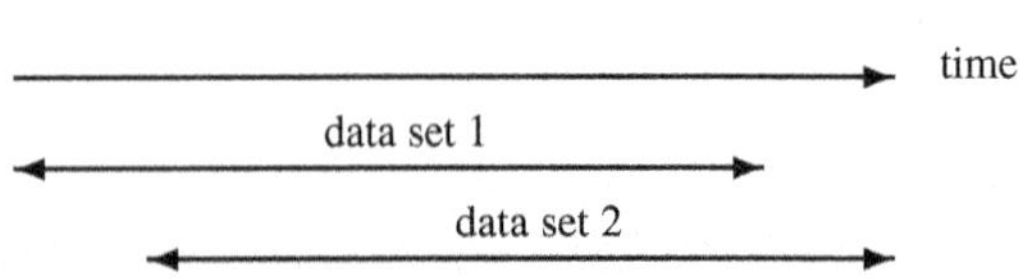

Fig. 12.1 Split of a total data set into two overlapping ones

12.2 Using Instrumental Variable Estimation

12.2.1 Introduction

First introduce some more detailed notation, where the periodicity of the noise-free data $u_0(t)$, $y_0(t)$ is exploited. In period j (where $1 \leq j \leq M$), write the regressor vector as

$$\varphi_j(t) = \varphi_0(t) + \tilde{\varphi}_j(t) \qquad t = 1, \ldots, N \ , \tag{12.6}$$

where $\varphi_0(t)$ contains the noise-free data (which due to Assumption **AI5** does not depend on j), and $\tilde{\varphi}_j(t)$ denotes the noise contribution. Similarly, $\mathbf{z}_j(t)$ (for $1 \leq j \leq M$) will denote the instrumental vector for period j.

Note that as a consequence of the above, in this chapter the following conventions are used:

- N denotes the number of data within each period
- M denotes the number of periods

Hence the total amount of data is NM. When the asymptotic case is considered ($NM \to \infty$), this means in this context that $M \to \infty$.

Now proceed and consider two examples of instrumental variable vectors.

Example 12.2 For an IV method, it is natural to let the vector $\mathbf{z}_j(t)$ be formed by regressors other than $\varphi_j(t)$. A simple choice of a basic IV is to take

$$\mathbf{z}_j(t) = \begin{cases} \varphi_{j+1}(t) & j = 1, \ldots, M-1 \ , \\ \varphi_1(t) & j = M \ . \end{cases} \tag{12.7}$$

■

Example 12.3 For an extended IV approach, consider an overdetermined instrumental variable vector as

$$\mathbf{z}_j(t) = \left(\varphi_{j+1}^T(t) \ \ldots \ \varphi_{j+L}^T(t) \right)^T \ , \tag{12.8}$$

where the convention

$$\varphi_{M+k}(t) = \varphi_k(t) \ , \quad k \geq 1 \tag{12.9}$$

is used, and L is a positive and finite integer. The particular choice $L = 1$ corresponds precisely to the IV variant of Example 12.2. ■

Remark 12.1 The IV variants in Examples 12.2 and 12.3 do not rely on periodic data. The key assumption (which is weaker than periodicity) is rather that the noise records in different segments are mutually uncorrelated. ■

For the further treatment, it will be convenient to introduce the following matrices

$$\begin{aligned} \boldsymbol{\Phi}(t) &= \left(\boldsymbol{\varphi}_1(t) \dots \boldsymbol{\varphi}_M(t)\right) , \\ \mathbf{Z}(t) &= \left(\mathbf{z}_1(t) \dots \mathbf{z}_M(t)\right) , \end{aligned} \qquad t = 1, \dots, N , \tag{12.10}$$

and the vector

$$\mathbf{Y}(t) = \begin{pmatrix} y_1(t) \\ \vdots \\ y_M(t) \end{pmatrix} , \qquad t = 1, \dots, N , \tag{12.11}$$

where $y_j(t)$ is the output at time t within period j:

$$y_j(t) = y(t + (j-1)N) , \quad \begin{cases} t = 1, \dots, N . \\ j = 1, \dots, M . \end{cases} \tag{12.12}$$

The basic IV Eq. (6.14) can be compactly written as

$$\left[\sum_{t=1}^{N} \mathbf{Z}(t)\boldsymbol{\Phi}^T(t)\right] \boldsymbol{\theta} \cong \left[\sum_{t=1}^{N} \mathbf{Z}(t)Y(t)\right] . \tag{12.13}$$

Write the general extended IV estimate as, see (6.16),

$$\hat{\boldsymbol{\theta}}_{\text{EIV}} = (\hat{\mathbf{R}}^T \mathbf{W} \hat{\mathbf{R}})^{-1} \hat{\mathbf{R}}^T \mathbf{W} \hat{\mathbf{r}} , \tag{12.14}$$

where now

$$\hat{\mathbf{R}} = \frac{1}{NM} \sum_{t=1}^{N} \mathbf{Z}(t)\boldsymbol{\Phi}^T(t) = \frac{1}{NM} \sum_{j=1}^{M} \sum_{t=1}^{N} \mathbf{z}_j(t)\boldsymbol{\varphi}_j^T(t) , \tag{12.15}$$

$$\hat{\mathbf{r}} = \frac{1}{NM} \sum_{t=1}^{N} \mathbf{Z}(t)\mathbf{Y}(t) = \frac{1}{NM} \sum_{j=1}^{M} \sum_{t=1}^{N} \mathbf{z}_j(t) y_j(t) . \tag{12.16}$$

Note that the basic IV estimate (12.13) is a special case of the more general expression (12.14). It is applicable when $\mathbf{R}$ is square, that is, $\dim(\mathbf{z}_j) = \dim(\boldsymbol{\varphi}_j)$.

12.2.2 Consistency Analysis

When analyzing the general estimate (12.14), it is worth noticing that the underlying model equation

$$\mathbf{Y}(t) = \boldsymbol{\Phi}^T(t)\boldsymbol{\theta} \;, \tag{12.17}$$

cf (5.69) takes the form of a multivariable instrumental variable model (Söderström and Stoica (1989), p. 262). This observation is useful when examining the statistical properties of the estimate (12.14).

In the analysis, assume that the model structure captures the true dynamics. More precisely, assume that there is a true parameter vector $\boldsymbol{\theta}_0$ such that

$$y(t) = \boldsymbol{\varphi}^T(t)\boldsymbol{\theta}_0 + v(t) \;, \tag{12.18}$$

where

$$v(t) = A_0(q^{-1})\tilde{y}(t) - B_0(q^{-1})\tilde{u}(t) \;. \tag{12.19}$$

Introduce the vector $\mathbf{V}(t)$ similarly to $\mathbf{Y}(t)$, (12.11). The relation (12.18) now implies

$$\mathbf{Y}(t) = \boldsymbol{\Phi}^T(t)\boldsymbol{\theta}_0 + \mathbf{V}(t) \;. \tag{12.20}$$

Inserting (12.20) into (12.14), and combining this with (12.16) gives

$$\hat{\boldsymbol{\theta}}_{\text{EIV}} - \boldsymbol{\theta}_0 = (\hat{\mathbf{R}}^T\mathbf{W}\hat{\mathbf{R}})^{-1}\hat{\mathbf{R}}^T\mathbf{W}\hat{\tilde{\mathbf{r}}} \;, \tag{12.21}$$

where

$$\hat{\tilde{\mathbf{r}}} = \frac{1}{NM}\sum_{t=1}^{N}\mathbf{Z}(t)\mathbf{V}(t) \;. \tag{12.22}$$

To analyze consistency (*i.e.*, to examine whether $\lim_{M\to\infty}\hat{\boldsymbol{\theta}}_{\text{EIV}} = \boldsymbol{\theta}_0$), it follows from (12.21) and standard conditions, see Chap. 6 and Söderström and Stoica (1983), Söderström and Stoica (1989) that consistency is guaranteed if

$$\lim_{M\to\infty}\hat{\mathbf{R}} \stackrel{\Delta}{=} \mathbf{R} \tag{12.23}$$

has full column rank and

$$\lim_{M\to\infty}\hat{\tilde{\mathbf{r}}} = \mathbf{0} \;. \tag{12.24}$$

To examine these conditions further, note that the left-hand side of (12.24) can be written

$$\lim_{M\to\infty}\hat{\tilde{\mathbf{r}}} = \lim_{M\to\infty}\frac{1}{NM}\sum_{j=1}^{M}\sum_{t=1}^{N}\mathbf{z}_j(t)v_j(t) = \frac{1}{N}\sum_{j=1}^{N}\mathsf{E}\left\{\mathbf{z}_j(t)v_j(t)\right\} \;. \tag{12.25}$$

The following result holds true.

Lemma 12.1 *Consider the IV estimate (12.14). Assume that* $\mathbf{z}_j(t)$ *and* $v_j(t)$ *are uncorrelated for all* j. *Also assume that*

$$\mathbf{R}_0 = \frac{1}{N}\sum_{t=1}^{N} \varphi_0(t)\varphi_0^T(t) \tag{12.26}$$

is positive definite. Then the estimate (12.14) is consistent for the IV estimators in Examples 12.2 and 12.3.

Proof First note that (12.24) follows directly from (12.25) and the assumptions.
Next evaluate the matrix **R** in (12.23). It holds

$$\begin{aligned}
\mathbf{R} &= \lim_{M\to\infty} \frac{1}{NM}\sum_{j=1}^{M}\sum_{t=1}^{N} \mathbf{z}_j(t)[\varphi_0^T(t) + \tilde{\varphi}_j^T(t)] \\
&= \mathsf{E}\left\{\frac{1}{N}\sum_{t=1}^{N} \mathbf{z}_j(t)\varphi_0^T(t)\right\} + \mathsf{E}\left\{\frac{1}{N}\sum_{t=1}^{N} \mathbf{z}_j(t)\tilde{\varphi}_j^T(t)\right\} \\
&= \mathsf{E}\left\{\frac{1}{N}\sum_{t=1}^{N} \mathbf{z}_j(t)\varphi_0^T(t)\right\} ,
\end{aligned} \tag{12.27}$$

as $\mathbf{z}_j(t)$ is uncorrelated with the noise $\tilde{\varphi}_j(t)$ during period j.
In order to evaluate **R** first note that

$$\mathsf{E}\left\{\frac{1}{N}\sum_{t=1}^{N} \varphi_j(t)\varphi_0^T(t)\right\} = \frac{1}{N}\sum_{t=1}^{N} \varphi_0(t)\varphi_0^T(t) + \mathsf{E}\left\{\frac{1}{N}\sum_{t=1}^{N} \tilde{\varphi}_j(t)\varphi_0^T(t)\right\} = \mathbf{R}_0 . \tag{12.28}$$

Next evaluate the limiting matrix **R** in (12.27), for the instrumental vectors introduced in Examples 12.2 and 12.3, see Eq. (12.7) and (12.8).
For Example 12.2 one gets directly

$$\mathbf{R} = \mathbf{R}_0 , \tag{12.29}$$

which is non-singular and of full rank since $\mathbf{R}_0$ is positive definite.
For the Extended IV of Example 12.3, one gets

$$\begin{aligned}
\mathbf{R} &= \mathsf{E}\left\{\frac{1}{N}\sum_{t=1}^{N} \begin{pmatrix} \varphi_0(t) + \tilde{\varphi}_{j+1}(t) \\ \vdots \\ \varphi_0(t) + \tilde{\varphi}_{j+L}(t) \end{pmatrix} \varphi_0^T(t)\right\} \\
&= \begin{pmatrix} \mathbf{R}_0 \\ \vdots \\ \mathbf{R}_0 \end{pmatrix} = \mathbf{g}_L \otimes \mathbf{R}_0 .
\end{aligned} \tag{12.30}$$

where $\mathbf{g}_L = (1 \ \dots \ 1)^T$ has dimension $L \times 1$, and $\otimes$ denotes Kronecker product, see Sect. A.1.2. Apparently, the matrix $\mathbf{R}$ is of full rank in (12.30). ■

Remark 12.2 The condition on $\mathbf{R}_0$ in (12.26) to be positive definite is a condition on persistent excitation. ■

12.3 The Sample Maximum Likelihood Method

For the frequency domain maximum likelihood method in Schoukens et al. (1997), see also Pintelon and Schoukens (2001), it is assumed that at least four periods of the data (or four independent experiments) are available. This assumption can be somewhat relaxed, see Sect. 12.4. The data are first preprocessed to estimate the noise covariance functions (or equivalently, the frequency-dependent variances of the DFT of the input and output noise). Note that the method can be extended to handle also cases where the input noise $\tilde{u}(t)$ is correlated with the output noise $\tilde{y}(t)$. Once the noise covariance functions can be regarded as known, it basically remains to estimate the transfer function $G(\mathrm{e}^{-\mathrm{i}\omega})$ which can be done in several ways. The resulting estimates of $G(\mathrm{e}^{-\mathrm{i}\omega})$ will be consistent.

The SML method to be described here is certainly related to the frequency domain ML method developed in Sect. 9.3. The SML differs though in some respects:

- The excitation is assumed to be periodic.
- The samples of the noise-free input are estimated.
- Nonparametric modles of the input and output noise are allowed.

12.3.1 A Frequency Domain ML Method

Consider the situation in Sect. 9.3, but assume that the noise-free input values, $u_0(1), \dots, u_o(N)$, are auxiliary deterministic values to be determined together with the parameter vector. In the development assume for the moment that the noise statistics are known.

Similar to the treatment in (9.34) let $\mathbf{Z}_k,\ k = 0, \dots, N-1$, denote the DFT of the measured input–output data. In this case note that $\{\mathbf{Z}_k\}$ are still asymptotically circular symmetric complex Gaussian distributed, but now with a nonzero mean value

$$\mathsf{E}\left\{\mathbf{Z}_j\right\} = \overline{\mathbf{G}}_j U_{0j} \ , \tag{12.31}$$

where

$$\overline{\mathbf{G}}_j = \begin{pmatrix} G_j \\ 1 \end{pmatrix} \tag{12.32}$$

and where $\{U_{0j}\}_{j=0}^{N-1}$ is the DFT sequence of the noise-free input $u_0(t)$. Here,

$$\omega_j = j\frac{2\pi}{N}, \quad j = 0, \ldots, N-1 , \tag{12.33}$$

and

$$G_j = \frac{B_j}{A_j} = \frac{B(e^{-i\omega_j})}{A(e^{-i\omega_j})} . \tag{12.34}$$

Further, the data have a covariance matrix now being

$$\mathbf{Q}_j = \text{cov}\left(\mathbf{Z}_j\right) = \begin{pmatrix} \lambda_y & 0 \\ 0 & \lambda_u \end{pmatrix} . \tag{12.35}$$

The expression (12.35) applies when both the input noise and the output noise are white (Assumption **AN3c** is fulfilled), but it is straightforward to generalize to arbitrary correlation of both $\tilde{u}(t)$ and $\tilde{y}(t)$, as well as to allow these two signals to be mutually correlated. Compare also (12.31) and (12.35) to (9.53).

The negative log-likelihood function can therefore, up to a $\boldsymbol{\theta}$-independent constant, be written as follows:

$$L = \sum_{j=0}^{N-1} \left[\left(\mathbf{Z}_j^* - \overline{\mathbf{G}}_j^* U_{0j}^*\right) \mathbf{Q}_j^{-1} \left(\mathbf{Z}_j - \overline{\mathbf{G}}_j U_{0j}\right) + \log(\det(\mathbf{Q}_j)) \right] . \tag{12.36}$$

It is straightforward to minimize L with respect to U_{0j} leading first to

$$U_{0j} = \frac{\overline{\mathbf{G}}_j^* \mathbf{Q}_j^{-1} \mathbf{Z}_j}{\overline{\mathbf{G}}_j^* \mathbf{Q}_j^{-1} \overline{\mathbf{G}}_j} , \tag{12.37}$$

and then to a concentrated loss function of $\boldsymbol{\theta}$:

$$\begin{aligned} V_{\text{FML}}(\boldsymbol{\theta}) &= \min_{\{U_{0j}\}} L \\ &= \sum_{j=0}^{N-1} \left[\mathbf{Z}_j^* \mathbf{Q}_j^{-1} \mathbf{Z}_j - \frac{\mathbf{Z}_j^* \mathbf{Q}_j^{-1} \overline{\mathbf{G}}_j \overline{\mathbf{G}}_j^* \mathbf{Q}_j^{-1} \mathbf{Z}_j}{\overline{\mathbf{G}}_j^* \mathbf{Q}_j^{-1} \overline{\mathbf{G}}_j} + \log(\det(\mathbf{Q}_j)) \right] . \end{aligned} \tag{12.38}$$

Next specialize to white input and output noise (Assumption **AN3c**), cf (12.35), and set

$$\mathbf{Z}_j = \begin{pmatrix} Y_j \\ U_j \end{pmatrix} . \tag{12.39}$$

Then (12.35) applies and the loss function (12.38) can be simplified to, neglecting a constant term,

$$\begin{aligned}
V_{\mathrm{FML}}(\boldsymbol{\theta}) &= \sum_{j=0}^{N-1} \Big[Y_j^* Y_j/\lambda_y + U_j^* U_j/\lambda_u \\
&\quad - \frac{\left(Y_j^* G_j/\lambda_y + U_j^*/\lambda_u\right)\left(G_j^* Y_j/\lambda_y + U_j/\lambda_u\right)}{G_j^* G_j/\lambda_y + 1/\lambda_u} \Big] \\
&= \sum_{j=0}^{N-1} \frac{1}{\lambda_y \lambda_u (G_j^* G_j/\lambda_y + 1/\lambda_u)} \\
&\quad \times \big[(G_j^* G_j/\lambda_y + 1/\lambda_u)\left(\lambda_u Y_j^* Y_j + \lambda_y U_j^* U_j\right) \\
&\quad - \left(Y_j^* G_j \lambda_u + U_j^* \lambda_y\right)\left(G_j^* Y_j/\lambda_y + U_j/\lambda_u\right) \big] \\
&= \sum_{j=0}^{N-1} \frac{1}{G_j^* G_j \lambda_u + \lambda_y} \left(Y_j^* - U_j^* G_j^*\right)\left(Y_j - G_j U_j\right) .
\end{aligned} \tag{12.40}$$

It follows directly that an alternative form of the loss function $V_{\mathrm{FML}}(\boldsymbol{\theta})$ is

$$V_{\mathrm{FML}}(\boldsymbol{\theta}) = \sum_{j=0}^{N-1} \frac{\left(A_j Y_j - B_j U_j\right)^* \left(A_j Y_j - B_j U_j\right)}{B_j B_j^* \lambda_u + A_j A_j^* \lambda_y} = \sum_{j=0}^{N-1} \frac{|A_j Y_j - B_j U_j|^2}{B_j B_j^* \lambda_u + A_j A_j^* \lambda_y} . \tag{12.41}$$

Remark 12.3 As mentioned earlier, the approach leading to (12.41) can be generalized to an arbitrary correlation structure. If instead of Assumption **AN3c** allow

$$\mathbf{Q}_j = \boldsymbol{\Phi}_{\tilde{\zeta}}(\omega_j) = \begin{pmatrix} \phi_{\tilde{y}}(\omega_j) & \phi_{\tilde{y}\tilde{u}}(\omega_j) \\ \phi_{\tilde{u}\tilde{y}}(\omega_j) & \phi_{\tilde{u}}(\omega_j) \end{pmatrix} , \tag{12.42}$$

it turns out that some calculations lead to

$$\begin{aligned}
V_{\mathrm{FML}}(\boldsymbol{\theta}) &= \sum_{j=0}^{N-1} \frac{|A_j Y_j - B_j U_j|^2}{\phi_{\tilde{u}}(\omega_j)|B_j|^2 + \phi_{\tilde{y}}(\omega_j)|A_j|^2 - \phi_{\tilde{y}\tilde{u}}(\omega_j) A_j B_j^* - \phi_{\tilde{u}\tilde{y}}(\omega_j) A_j^* B_j} \\
&= \sum_{j=0}^{N-1} \frac{|A_j Y_j - B_j U_j|^2}{\phi_{\tilde{u}}(\omega_j)|B_j|^2 + \phi_{\tilde{y}}(\omega_j)|A_j|^2 - 2\,\mathrm{Real}\left(\phi_{\tilde{y}\tilde{u}}(\omega_j) A_j B_j^*\right)} .
\end{aligned} \tag{12.43}$$

■

12.3.2 The SML Method

The ML criterion in the frequency domain for *known* noise statistics can be written as in (12.41) and (12.43). Note that there is no need to assume that A is stable as long as the system has stationary input and output signals, e.g., an unstable plant captured in a stabilizing feedback loop is allowed. If the noise (co)variances are known a priori, it is straightforward to minimize the cost function to get good estimates.

However, knowing exactly the noise model is not realistic in most practical cases. Then one has to consider the (co)variances of the noises as additional parameters which should also be estimated from the data. This is though not feasible for a general case, cf. the findings in Chap. 4.

For periodic data the situation is much more favorable from an identifiability point of view. The noise statistics may be estimated separately, even in a nonparametric way. This is a main idea of the sample maximum likelihood method, SML, Schoukens et al. (1997).

Let $\bar{U}(\omega_k)$ and $\bar{Y}(\omega_k)$ denote the sample mean (over the periods) of the input and the output,

$$\bar{U}(\omega_k) = \frac{1}{M}\sum_{l=1}^{M} U_l(\omega_k), \ \ \bar{Y}(\omega_k) = \frac{1}{M}\sum_{l=1}^{M} Y_l(\omega_k) \, . \tag{12.44}$$

Introduce the estimated sample (co)variances $\hat{\sigma}_{\tilde{u}}^2(\omega_k)$, $\hat{\sigma}_{\tilde{y}}^2(\omega_k)$, and $\hat{\sigma}_{\tilde{y}\tilde{u}}(\omega_k)$ as

$$\begin{aligned} \hat{\sigma}_{\tilde{u}}^2(\omega_k) &= \tfrac{1}{M-1}\sum_{l=1}^{M} |U_l(\omega_k) - \bar{U}(\omega_k)|^2 \, , \\ \hat{\sigma}_{\tilde{y}}^2(\omega_k) &= \tfrac{1}{M-1}\sum_{l=1}^{M} |Y_l(\omega_k) - \bar{Y}(\omega_k)|^2 \, , \\ \hat{\sigma}_{\tilde{y}\tilde{u}}(\omega_k) &= \tfrac{1}{M-1}\sum_{l=1}^{M} \left(Y_l(\omega_k) - \bar{Y}(\omega_k)\right)\left(U_l(\omega_k) - \bar{U}(\omega_k)\right)^* \, , \end{aligned} \tag{12.45}$$

Once the noise statistics is determined in the above way, the parameter vector $\boldsymbol{\theta}$ is estimated by minimizing

$$V_{\text{SML}}(\boldsymbol{\theta}) = \frac{1}{N}\sum_{k=0}^{N-1} \frac{|B(\mathrm{e}^{-\mathrm{i}\omega_k}, \boldsymbol{\theta})\bar{U}(\omega_k) - A(\mathrm{e}^{-\mathrm{i}\omega_k}, \boldsymbol{\theta})\bar{Y}(\omega_k)|^2}{W(\omega_k, \boldsymbol{\theta})} \, , \tag{12.46}$$

with

$$\begin{aligned} W(\omega_k, \boldsymbol{\theta}) &= \hat{\sigma}_{\tilde{u}}^2(\omega_k)|B(\mathrm{e}^{-\mathrm{i}\omega_k}, \boldsymbol{\theta})|^2 + \hat{\sigma}_{\tilde{y}}^2(\omega_k)|A(\mathrm{e}^{-\mathrm{i}\omega_k}, \boldsymbol{\theta})|^2 \\ &\quad -2\,\text{Real}\left(\hat{\sigma}_{\tilde{y}\tilde{u}}(\omega_k)A(\mathrm{e}^{-\mathrm{i}\omega_k}, \boldsymbol{\theta})B(\mathrm{e}^{\mathrm{i}\omega_k}, \boldsymbol{\theta})\right) \, . \end{aligned} \tag{12.47}$$

The cost function (12.46) is thus an approximation of $W_{\text{FML}}(\boldsymbol{\theta})$ in (12.43) by replacing the exact covariances of the noise by their sample estimates. The major advantage of this approach is that the plant parameters remain as the only unknowns to be estimated, which leads to a low dimension of the nonlinear optimization problem.

Example 12.4 Some identification methods were tested on periodic data. The system and the noise levels were the same as in the standard data sets. See Sect. 3.2. In each Monte Carlo run, the noise-free input was a realization of the standard data ARMA process of length $N = 1000$ and repeated $M = 10$ times. Thus the data length becomes $NM = 10\,000$ for each realization. Three different methods were tested:

- The covariance matching method using

$$p_y = 2, \ p_u = 2, \ p_1 = -2, \ p_2 = 2$$

  as in Example 8.3. The fact that the data are periodic was not exploited for this method.
- The IV method given by (12.8) with $L = M - 1$. Hence the instruments for a specific period consist of the regressors of all other periods. The periodicity of the data is hence exploited.
- The SML method implemented using the knowledge that $\tilde{u}(t)$ and $\tilde{y}(t)$ are white and mutually uncorrelated. The minimization of V_{SML} was initialized using a least squares estimate computed from the frequency domain data.

The numerical results are summarized in Table 12.1.

One can make the following observations:

- Covariance matching gives quite reasonable estimates. The standard deviations are about a factor $\sqrt{10} \approx 3.1$ smaller than in Example 8.3, which is expected as the data length is increased with a factor $10\,000/1000 = 10$.
- The IV approach, taken periodicity of the data into account, gives a considerably improvement as compared to the CM method. There is indeed a very substantial improvement as compared to IV applied without taking the periodicity into account, see Example 6.3.
- The SML method gives clearly the best result here. Even if it requires the longest computation time (with the used implementations) of the three methods in this example, it should be the natural choice when data are known to be periodic. ■

Table 12.1 CM, IV and SML estimates for periodic data

Parameter	True value	CM		IV		SML	
		Mean	Standard dev.	Mean	Standard dev.	Mean	Standard dev.
a_1	−1.5	−1.4998	0.012	−1.4993	0.005	−1.4993	0.004
a_2	0.7	0.6999	0.008	0.6995	0.004	0.6995	0.003
b_1	2.0	1.9953	0.070	1.9964	0.054	1.9955	0.040
b_2	1.0	1.0050	0.140	1.0087	0.073	1.0069	0.050

12.4 Extensions and Bibliographical Notes

A general EIV setup, treated in the frequency domain appears in Pintelon et al. (1994). The situation of a non-parametric noise model, $\tilde{u}$ and $\tilde{y}$ white, but possibly cross-correlated, and using a small number of experiments (hence periodic data) is described in Schoukens et al. (1997). See also Pintelon and Schoukens (2001, 2012b).

A special case is mentioned in Forssell et al. (1999), Schoukens et al. (1997), Guillaume et al. (1995), where some specific measurement situations allow for repeated experiments where the noise-free regressor vector remain the same; that is, $\boldsymbol{\varphi}_0^{(i)}(t)$ does not vary from one experiment to another. The estimator scheme in Forssell et al. (1999) is based on total least squares, TLS.

Also the method described in Markovsky et al. (2006a) is based on two experiments, although formally presented as that there is a clustering in time of the data, so that during the whole experiment the statistical properties of the noise-free input, such as its spectrum $\phi_{u_0}(\omega)$, change at some point of time. The same type of idea is discussed in Wald (1940) under the name of grouping. A sufficient condition for consistency is given. It basically means that the data can be divided into two groups where the input properties are sufficiently different as $N \to \infty$. Unfortunately, this condition will not hold when the noise-free input signal is a stationary stochastic process.

The paper Kukush et al. (2007) considers the case when there is a time point in the data where the dynamics change. This is essentially the same as treating data from two different experiments.

Vandersteen et al. (2009) treat frequency domain techniques for EIV identification. Periodic data (very few periods are needed) are used for a boot-strapped TLS technique.

The paper Markovsky and Pintelon (2015) deals with identification of EIV systems from multiple experiments. A behavioral approach is used. The effects of initial conditions of each experiment are eliminated analytically using a variable projection approach, so that the remaining optimization is done only over the system parameters.

Frequency domain techniques for estimating nonparametric frequency response models with some periodic excitation are described in Pintelon et al. (2013a), Louarroudi et al. (2014), Hostettler et al. (2016).

Instrumental Variable

The analysis in Sect. 12.2 expands the ideas of Söderström and Hong (2005).

The principle to use instrumental variable estimation and cross-correlation between the different experiments has earlier been reported by van den Bos (1992) and Pintelon et al. (1994).

Sample Maximum Likelihood

The SML methods for periodic data, or rather from a small set or experiments, is presented in Schoukens et al. (1997, 1999b).

An extension of the SML method, for the EIV situation, using data from overlapping sub-records is presented in Barbé et al. (2007, 2008, 2012).

In Pintelon and Schoukens (2012a) the authors extend SML to work under weaker assumptions. The noise-free input is assumed to be periodic (two periods are enough to give identifiability). The analysis is based on asymptotic theory, in the sense that it is assumed that the considered subintervals become large.

The SML technique is extended to identify slowly time-varying systems from one experiment, when excited with a multi-sine input in Lataire and Pintelon (2009, 2011). The system dynamics is modeled by a difference equation whose coefficients are written as polynomials in time. The fitted noise model is nonparametric and can cope with arbitrarily correlated measurement noise.

The SML method is applied to parameter estimation on partial differential equation models in van Berkel et al. (2014).

The paper Söderström et al. (2010) aims to give a comparison between time domain ML (see Sect. 9.2) and SML in terms of accuracy.

Periodic Excitation

Sometimes it is possible to fully or partly design the input signal. In an EIV setting, this can correspond to that an external signal can be chosen, but its relation to the true input signal may be unknown or nonlinear or both. See Fig. 3.3 for illustration. When using periodic excitation, frequency domain considerations are often used. A survey of perturbation signals in the frequency domain are given in Godfrey et al. (2005). The use of periodic excitation combined with nonparametric estimation methods is described in Schoukens et al. (2009). Applications to various areas in mechanical and electrical engineering can be found, for example in Oomen et al. (2014), Criens et al. (2015) and Pierre et al. (2012).

Chapter 13
Algorithmic Properties

The chapter is organized in the following way. Section 13.1 contains a general discussion highlighting the difference between definitions of estimates on one hand and numerical algorithms to compute them on the other. Section 13.2 discusses typical user choices. General properties, such as uniqueness, consistency, and convergence are treated in Sect. 13.3. Section 13.4 covers aspects of variable projection algorithms, while Sect. 13.5 treats handling of overdetermined systems of equations. Algorithms where the parameter estimates are updated recursively in time are dealt with in Sect. 13.6. The chapter ends with Sect. 13.7 which gives some extensions and bibliographical notes.

13.1 Introduction

Many of the estimators treated so far can be phrased in a quite general form as

$$\mathbf{f}(\hat{\boldsymbol{\vartheta}}) = \mathbf{0} \ , \tag{13.1}$$

where it is not specified exactly how the vector $\mathbf{f}$ depends on data. For many methods, such as generalized instrumental variable (GIVE) and covariance matching (CM), the function depends on a small set of covariance elements, so (13.1) may then be written as

$$\mathbf{f}(\hat{\boldsymbol{\vartheta}}, \hat{\mathbf{r}}_N) = \mathbf{0} \ . \tag{13.2}$$

For prediction error methods (including the maximum likelihood method), the estimate $\hat{\boldsymbol{\vartheta}}$ is defined as the minimizing argument of a loss function

$$\hat{\boldsymbol{\vartheta}} = \arg\min_{\boldsymbol{\vartheta}} V_N(\boldsymbol{\vartheta}) \ , \tag{13.3}$$

T. Söderström, *Errors-in-Variables Methods in System Identification*, Communications and Control Engineering,
https://doi.org/10.1007/978-3-319-75001-9_13

where V_N depends on the data in a more complicated way than in (13.2). In fact, V_N would depend on $O(N)$ number of covariance elements, and thus the formulation (13.2) is not pertinent to use in this case.

The PEM estimate (13.3) may be written as the solution to the equation for stationary points,

$$V_N'(\hat{\boldsymbol{\vartheta}}) = \mathbf{0} \tag{13.4}$$

which has a form similar to (13.1). Note though, that (13.4) is not exactly equivalent to (13.3), as other stationary points than the global minimum point are solutions to (13.4) but differ, of course, from $\hat{\boldsymbol{\vartheta}}$ as defined in (13.3).

It is important to distinguish between

- *an estimate*, such as the solution to a set of nonlinear equations in (13.1), on one hand, and
- *a numerical algorithm* to solve these equations on the other hand.

As the equations are nonlinear in one form or another, the numerical algorithm is bound to be iterative in some way. Further, for a given set of equations, there may be more than one pertinent algorithm available for solving the equations, and these algorithms may have different numerical properties, say in terms of computational complexity, and convergence of iterations, etc. To analyze the statistical properties of the estimate means to explore the properties of the estimation error $\hat{\boldsymbol{\vartheta}} - \boldsymbol{\vartheta}_0$, and then it does not matter in what particular computational way the defining equations (13.1) have been solved.

Sometimes algorithms for solving the defining equations are derived by rewriting (13.2) as

$$\hat{\boldsymbol{\vartheta}} = \mathbf{g}(\hat{\boldsymbol{\vartheta}}, \hat{\mathbf{r}}_N) \ . \tag{13.5}$$

Examples for when this may be feasible occur for BELS; see Sect. 7.2. Of course, a simple alternative is to define

$$\mathbf{g}(\boldsymbol{\vartheta}, \hat{\mathbf{r}}_N) = \mathbf{f}(\boldsymbol{\vartheta}, \hat{\mathbf{r}}_N) - \boldsymbol{\vartheta} \ . \tag{13.6}$$

An iterative scheme for solving (13.5) may then be derived as

$$\hat{\boldsymbol{\vartheta}}^{(i)} = \mathbf{g}(\hat{\boldsymbol{\vartheta}}^{(i-1)}, \hat{\mathbf{r}}_N) \ , \tag{13.7}$$

where i is the iteration number.

An algorithm of the type (13.7) can be proven to converge if it is a contraction mapping, meaning that $\| \mathbf{g} \| < 1$.

13.2 Algorithmic User Choices

In previous chapters several different types of identification methods were introduced. There are many possible varieties of these methods, though, and their underlying user choices. In a general setting one may describe the situation as follows, as far as the 'design' or 'description' of such methods goes.

- As a first step, before dealing with the algorithmic aspects, the user has to provide the structure of a noise model. More specifically, the type of input and output noise is to be provided. Two natural choices here are Assumption **AN3c** (both $\tilde{u}(t)$ and $\tilde{y}(t)$ are white noise, with unknown variances) or Assumption **AN3b** ($\tilde{u}(t)$ is white noise with unknown variance, and $\tilde{y}(t)$ is an ARMA process with unknown parameters).
- A second step is to set up the underlying equations. In the GIVE case, the set of equations are (7.28)–(7.30). While the normal equations for a standard least squares estimation typically are kept, the other equations may be substituted. It is also possible to use more equations than unknowns. Due to the nature of the problem, the set of equations will always be nonlinear, although they have a considerable amount of structural properties. For the CM method, this step consists of choosing the structural integers p_y, p_u, p_1, p_2; see Sect. 8.4.1. For the prediction error method, there is a choice of the specific criterion to be minimized; see Examples 9.1–9.4 for illustration.
- The third step to consider is the numerical algorithm to use for solving the system of equations. For example, in the standard BELS case there are at least two such schemes, which turned out to have different convergence properties. Details are presented in Appendix 13.A.1.

Once an estimation method is defined along the lines just outlined, it is possible to make an analysis of its properties.

13.3 Some General Concepts

One can analyze the estimate and a possible algorithm in several different respects. Below follows a short discussion of some general questions.

1. **Uniqueness**. This is basically an identifiability issue. It concerns whether or not there is a unique solution to the defining set of equations (13.2). In this context it is reasonable to treat the asymptotic case when $N \to \infty$. As it has been assumed generally that adequate parameterizations are used, the true parameter vector $\boldsymbol{\vartheta}_0$ is always a solution to (13.2), as $\mathbf{f}(\boldsymbol{\vartheta}_0, \mathbf{r}_\infty) = \mathbf{0}$. The issue is whether or not the implication
$$\mathbf{f}(\hat{\boldsymbol{\vartheta}}, \mathbf{r}_\infty) = \mathbf{0} \Rightarrow \hat{\boldsymbol{\vartheta}} = \boldsymbol{\vartheta} \tag{13.8}$$

holds. For the prediction error method, the issue of uniqueness concerns whether the asymptotic loss function $V_\infty(\boldsymbol{\vartheta})$ has a unique *global* minimum point or not.

By a simple linearization it follows that the estimate is *locally unique* if the gradient matrix $\partial\mathbf{f}/\partial\boldsymbol{\vartheta}$ has full rank, as close to $\boldsymbol{\vartheta}_0$ it holds

$$\mathbf{0} = \mathbf{f}(\hat{\boldsymbol{\vartheta}}, \mathbf{r}_\infty) \approx \mathbf{f}(\boldsymbol{\vartheta}_0, \mathbf{r}_\infty) + \frac{\partial \mathbf{f}}{\partial \boldsymbol{\vartheta}} \left(\hat{\boldsymbol{\vartheta}} - \boldsymbol{\vartheta}_0\right) . \tag{13.9}$$

2. **Consistency**. This concept concerns whether or not the estimate for finite but large N becomes arbitrarily close to the true parameter vector. In mathematical form, the issue is about the question

$$\hat{\boldsymbol{\vartheta}}_N \to \boldsymbol{\vartheta}_0, \quad N \to \infty\ ? \tag{13.10}$$

The estimate is indeed consistent, if it can be shown that the function $\mathbf{f}(\boldsymbol{\vartheta}, \mathbf{r})$ converges *uniformly* over compact intervals. See for example Ljung (1999), Söderström (1974). This is typically the case for all the methods treated in this book.

3. **Convergence**. This concept concerns a property of a computational algorithm, rather than of an estimator as such. Note that two different algorithms both aimed to solve a given set of equations may very well have different convergence properties. As an example, consider an iterative algorithm of the form (13.7). It is convergent if

$$\hat{\boldsymbol{\vartheta}}_N^{(i)} \to \hat{\boldsymbol{\vartheta}}_N, \quad \text{as } i \to \infty . \tag{13.11}$$

To find out the convergence properties of (13.7) one can proceed as follows. The *local convergence* properties are obtained by linearizing the iterations (13.7) around the desired estimate. One then has

$$\hat{\boldsymbol{\vartheta}}_N^{(i)} - \hat{\boldsymbol{\vartheta}} = \mathbf{g}(\hat{\boldsymbol{\vartheta}}_N^{(i-1)}, \hat{\mathbf{r}}_N) - \mathbf{g}(\hat{\boldsymbol{\vartheta}}_N, \hat{\mathbf{r}}_N) \approx \mathbf{G}\left(\hat{\boldsymbol{\vartheta}}_N^{(i-1)} - \hat{\boldsymbol{\vartheta}}_N\right) , \tag{13.12}$$

where the square matrix $\mathbf{G}$ is

$$\mathbf{G} = \frac{\partial \mathbf{g}(\boldsymbol{\vartheta}, \mathbf{r})}{\partial \boldsymbol{\vartheta}}_{|\boldsymbol{\vartheta}=\boldsymbol{\vartheta}_0} . \tag{13.13}$$

It may then be sufficient to consider the matrix $\mathbf{G}$ for the asymptotic case when $N \to \infty$. It follows that a *sufficient condition for local convergence* is that the matrix $\mathbf{G}$ has all eigenvalues inside the unit circle. The closer the eigenvalues are clustered around the origin, the faster the convergence.

As illustration of identifiability analysis, consider the covariance matching approach of Chap. 8.

Example 13.1 Assume that the compatibility condition (8.57) is fulfilled. Is the system then identifiable in the sense that the underlying system of equations (8.22) has a unique solution with respect to $\boldsymbol{\theta}$ and $\mathbf{r}_z$?

A partial analysis is as follows. Check the solutions $\hat{\boldsymbol{\theta}}$ and $\hat{\mathbf{r}}_z$ of the system of nonlinear equations

$$\mathbf{F}(\boldsymbol{\theta})\mathbf{r}_z = \mathbf{F}(\hat{\boldsymbol{\theta}})\hat{\mathbf{r}}_z \, . \tag{13.14}$$

Note that the global minimum points of the criterion $V(\boldsymbol{\theta})$ in (8.28) do satisfy (when $N \to \infty$)

$$V(\hat{\boldsymbol{\theta}}) = 0 \Rightarrow J(\hat{\boldsymbol{\theta}}, \hat{\mathbf{r}}_z) = 0 \Rightarrow \text{ (13.14)} \, . \tag{13.15}$$

Hence the solutions to (8.22) are also the global minimum points of the criterion $V(\boldsymbol{\theta})$ in (8.28), and vice versa.

Assume for convenience that $\mathbf{F}(\hat{\boldsymbol{\theta}})$ has full column rank. Then it follows that ($\mathbf{F}^{\dagger} = (\mathbf{F}^T\mathbf{F})^{-1}\mathbf{F}^T$ denotes the pseudo-inverse of $\mathbf{F}$),

$$\begin{aligned}\text{(13.14)} &\Rightarrow \hat{\mathbf{r}}_z = \mathbf{F}^{\dagger}(\hat{\boldsymbol{\theta}})\mathbf{F}(\boldsymbol{\theta})\mathbf{r}_z \\ &\Rightarrow \mathbf{F}(\boldsymbol{\theta})\mathbf{r}_z - \mathbf{F}(\hat{\boldsymbol{\theta}})\mathbf{F}^{\dagger}(\hat{\boldsymbol{\theta}})\mathbf{F}(\boldsymbol{\theta})\mathbf{r}_z = \mathbf{0} \\ &\Rightarrow \left[\mathbf{I} - \mathbf{F}(\hat{\boldsymbol{\theta}})\mathbf{F}^{\dagger}(\hat{\boldsymbol{\theta}})\right]\mathbf{F}(\boldsymbol{\theta})\mathbf{r}_z = \mathbf{0} \, . \end{aligned} \tag{13.16}$$

Consider the generic case; that is, assume (13.16) to hold for an arbitrary and randomly chosen vector $\mathbf{r}_z$. In such a case it must hold

$$\mathbf{F}^T(\boldsymbol{\theta})\left[\mathbf{I} - \mathbf{F}(\hat{\boldsymbol{\theta}})\mathbf{F}^{\dagger}(\hat{\boldsymbol{\theta}})\right]\mathbf{F}(\boldsymbol{\theta}) = \mathbf{0} \, , \tag{13.17}$$

or yet

$$\mathbf{F}^T(\boldsymbol{\theta})\mathbf{F}(\boldsymbol{\theta}) = \mathbf{F}^T(\boldsymbol{\theta})\mathbf{F}(\hat{\boldsymbol{\theta}})\left(\mathbf{F}^T(\hat{\boldsymbol{\theta}}))\mathbf{F}(\hat{\boldsymbol{\theta}})\right)^{-1}\mathbf{F}^T(\hat{\boldsymbol{\theta}}))\mathbf{F}(\boldsymbol{\theta}) \, . \tag{13.18}$$

Let $\mathbf{F}(\boldsymbol{\theta})$ have n_z columns. According to Lemma A.3 of Söderström and Stoica (1989), it then follows from (13.18) that

$$\text{rank}\left(\begin{pmatrix}\mathbf{F}^T(\boldsymbol{\theta}) \\ \mathbf{F}^T(\hat{\boldsymbol{\theta}}))\end{pmatrix}\begin{pmatrix}\mathbf{F}(\boldsymbol{\theta}) & \mathbf{F}(\hat{\boldsymbol{\theta}}))\end{pmatrix}\right) = n_z \, , \tag{13.19}$$

and next

$$\mathbf{F}^T(\hat{\boldsymbol{\theta}})\left[\mathbf{I} - \mathbf{F}(\boldsymbol{\theta})\mathbf{F}^{\dagger}(\boldsymbol{\theta})\right]\mathbf{F}(\hat{\boldsymbol{\theta}}) = \mathbf{0} \, . \tag{13.20}$$

This condition can also be phrased as $\mathbf{F}(\boldsymbol{\theta})$ and $\mathbf{F}(\hat{\boldsymbol{\theta}})$ have the same range space. Now, (13.20) can be interpreted and reformulated as a system of nonlinear equations in the components of the unknown vector $\hat{\boldsymbol{\theta}}$. As $\mathbf{F}(\hat{\boldsymbol{\theta}})$ depends quadratically on $\hat{\boldsymbol{\theta}}$, the equations are polynomials of degree four in the components of $\hat{\boldsymbol{\theta}}$. In general, this system of nonlinear equations is overdetermined (there are often many more equations than unknowns). It seems therefore reasonable to conjecture that the only solution to (13.17) is the trivial and 'desirable' one, $\hat{\boldsymbol{\theta}} = \boldsymbol{\theta}$. The numerical experiences reported in Söderström et al. (2009) point in the same direction. ■

13.4 Variable Projection Algorithms

Both GIVE (see Chap. 7) and covariance matching (CM) methods (see Chap. 8) lead to so-called separable nonlinear least squares problems. These are computational formulations where variable projection algorithms can be applied; see Sect. A.1.4. In brief notations, a loss function of the form

$$V(\boldsymbol{\theta}, \mathbf{x}) = \| \mathbf{b} - \mathbf{A}(\boldsymbol{\theta})\mathbf{x} \|_{\mathbf{W}}^2 \tag{13.21}$$

is to be minimized with respect to the two vectors $\boldsymbol{\theta}$ and $\mathbf{x}$. As V depends quadratically on $\mathbf{x}$ it is easy to minimize it with respect to $\mathbf{x}$, and then get a concentrated loss function, depending only on $\boldsymbol{\theta}$:

$$\overline{V}(\boldsymbol{\theta}) = \min_{\mathbf{x}} V(\boldsymbol{\theta}, \mathbf{x}) = V(\boldsymbol{\theta}, \mathbf{x}(\boldsymbol{\theta})) \,, \tag{13.22}$$

$$\mathbf{x}(\boldsymbol{\theta}) = \left(\mathbf{A}^T(\boldsymbol{\theta})\mathbf{W}\mathbf{A}(\boldsymbol{\theta})\right)^{-1} \mathbf{A}^T(\boldsymbol{\theta})\mathbf{W}\mathbf{b} \,, \tag{13.23}$$

leading to

$$\overline{V}(\boldsymbol{\theta}) = \mathbf{b}^T \left(\mathbf{W} - \mathbf{W}\mathbf{A}(\boldsymbol{\theta}) \left(\mathbf{A}^T(\boldsymbol{\theta})\mathbf{W}\mathbf{A}(\boldsymbol{\theta})\right)^{-1} \mathbf{A}^T(\boldsymbol{\theta})\mathbf{W}\right) \mathbf{b} \,. \tag{13.24}$$

When $\mathbf{W} = \mathbf{I}$, the matrix in (13.24) becomes $\mathbf{I} - \mathbf{A}\left(\mathbf{A}^T\mathbf{A}\right)^{-1}\mathbf{A}^T$, which is the orthogonal projection onto the range space of the matrix $\mathbf{A}$. As $\mathbf{A} = \mathbf{A}(\boldsymbol{\theta})$ depends on the variable $\boldsymbol{\theta}$, the projection varies with $\boldsymbol{\theta}$, and the name 'variable projection algorithm' is often used.

A major advantage of reformulating the optimization problem as (13.24) instead of (13.21) is that the dimension of the free variables is reduced from $\dim(\boldsymbol{\theta}) + \dim(\mathbf{x})$ to $\dim(\boldsymbol{\theta})$. As demonstrated in Appendices 2.A.2 and 13.A.2 there may be other advantages as well. In particular, for BELS, better convergence properties are obtained when a variable projection algorithm is used, as compared to a standard method.

Remark 13.1 For the GIVE method in the multivariable case, one can choose between different problem formulations, as discussed in Sect. 7.4.3. In Appendix 13.A it is shown how to conveniently arrange a variable projection algorithm in the multivariable case. ■

13.5 Handling Overdetermined Systems of Equations

Another particular algorithmic aspect deserves to be discussed. Sometimes the system of equations (13.1) is overdetermined. One possibility is then to introduce a weighting as already discussed. There are some aspects on how it may be chosen.

- It is possible to optimize the covariance matrix of the parameter estimates; see Remark 14.10 in Chap. 14 for details. There is indeed a lower bound on the covariance matrix.
- A more heuristically weighting is, of course, also possible. The weighting should reflect the confidence one can have in the different equations. It is a rule of thumb that equations that correspond to high time lags are not so reliable, and should be given smaller weights than those with small time lags. For example, for GIVE equations elements in $\mathbf{z}(t)$ that correspond to a significantly delayed output or input should be avoided or given low weights.
- Another possibility is to directly treat some of the equations as fully exact, and to minimize the norm of the remaining ones. Assume, for example, that the equations are written as

$$\mathbf{f}(\boldsymbol{\vartheta}) = \begin{pmatrix} \mathbf{f}_1(\boldsymbol{\vartheta}) \\ \mathbf{f}_2(\boldsymbol{\vartheta}) \end{pmatrix} \approx \mathbf{0} , \tag{13.25}$$

where

$$\dim(\mathbf{f}_1) < \dim(\boldsymbol{\vartheta}) < \dim(\mathbf{f}) . \tag{13.26}$$

Consider the case when it is desired that $\mathbf{f}_1(\boldsymbol{\vartheta}) = \mathbf{0}$ holds exactly (noting that this gives too few equations to determine $\boldsymbol{\vartheta}$), and then to minimize $\| \mathbf{f}_2(\boldsymbol{\vartheta}) \|$ under this constraint. This can be handled practically in different ways as will now be described.

Remark 13.2 In Appendix 13.C it is shown how the CFA estimator developed in Sect. 8.5 can be formulated as a variable projection algorithm. ■

The following two algorithms can be used for handling, analyzing, and describing the problem of letting some of the equations hold exactly.

Algorithm 13.1 The problem may be treated as

$$\begin{aligned} &\min_{\boldsymbol{\vartheta}} \| \mathbf{f}_2(\boldsymbol{\vartheta}) \|^2_{\mathbf{W}_2} \\ &\text{such that } \mathbf{f}_1(\boldsymbol{\vartheta}) = \mathbf{0} . \end{aligned} \tag{13.27}$$

The problem (13.27) may be treated using Lagrange multipliers by introducing

$$L(\boldsymbol{\vartheta}, \boldsymbol{\lambda}) = \mathbf{f}_2^T(\boldsymbol{\vartheta})\mathbf{W}_2\mathbf{f}_2(\boldsymbol{\vartheta}) + \boldsymbol{\lambda}^T \mathbf{f}_1(\boldsymbol{\vartheta})^T \mathbf{f}_1(\boldsymbol{\vartheta}) . \tag{13.28}$$

For the minimum point the following equations must hold

$$\begin{cases} \mathbf{f}_1(\boldsymbol{\vartheta}) = \mathbf{0}, \\ \mathbf{f}_2^T(\boldsymbol{\vartheta})\mathbf{W}_2\frac{\partial \mathbf{f}_2}{\partial \boldsymbol{\vartheta}} + \boldsymbol{\lambda}^T\frac{\partial \mathbf{f}_1}{\partial \boldsymbol{\vartheta}} = \mathbf{0} . \end{cases} \tag{13.29}$$

Apparently (13.29) is a nonlinear system with $\dim(\mathbf{f}_1) + \dim(\boldsymbol{\vartheta})$ equations and the same number of unknowns. ■

Algorithm 13.2 Another approach to the problem (13.27) is to use the general weighting and set

$$\mathbf{W} = \begin{pmatrix} \alpha\mathbf{W}_1 & \mathbf{0} \\ \mathbf{0} & \mathbf{W}_2 \end{pmatrix} \tag{13.30}$$

with $\mathbf{W}_1$ positive definite and α a scalar that will tend to infinity. Then the loss function $\| \mathbf{f}(\boldsymbol{\vartheta}) \|_{\mathbf{W}}^2$ can be written as

$$\begin{aligned} \bar{V}(\boldsymbol{\vartheta}) &= \alpha \| \mathbf{f}_1(\boldsymbol{\vartheta}) \|_{\mathbf{W}_1}^2 + \| \mathbf{f}_2(\boldsymbol{\vartheta}) \|_{\mathbf{W}_2}^2 \\ &= \alpha\mathbf{f}_1^T(\boldsymbol{\vartheta})\mathbf{W}_1\mathbf{f}_1(\boldsymbol{\vartheta}) + \mathbf{f}_2^T(\boldsymbol{\vartheta})\mathbf{W}_2\mathbf{f}_2(\boldsymbol{\vartheta}) . \end{aligned} \tag{13.31}$$

The equations for finding the minimum point of $\bar{V}(\boldsymbol{\vartheta})$ will in this algorithm be

$$\mathbf{0} = \alpha\mathbf{f}_1^T(\boldsymbol{\vartheta})\mathbf{W}_1\frac{\partial \mathbf{f}_1}{\partial \boldsymbol{\vartheta}} + \mathbf{f}_2^T(\boldsymbol{\vartheta})\mathbf{W}_2\frac{\partial \mathbf{f}_2}{\partial \boldsymbol{\vartheta}} . \tag{13.32}$$

■

A comparison of (13.29) and (13.32) is given in Appendix 13.B.

13.6 Recursive Algorithms

13.6.1 General Aspects

In some situations it is of interest to arrange the computational algorithm recursively in time. There is a relative rich literature on recursive and adaptive algorithms for general estimation and identification problems; see for example Ljung and Söderström (1983), Haykin (1991a), Sayed (2003). In an EIV context, such algorithms can be useful when the model is applied for fault detection. Then changes in the underlying system should show up as changes in the identified model, and proper actions may be taken based on the current model.

The basic idea of a recursive algorithm is to let the current estimate $\hat{\boldsymbol{\vartheta}}(t)$ at time t depend on all the data achieved at that time, that is, Y^t, U^t. The recursive estimator describes how the estimate $\hat{\boldsymbol{\vartheta}}(t)$ is updated to $\hat{\boldsymbol{\vartheta}}(t+1)$ once the new measurements $\boldsymbol{\zeta}(t+1)$ are obtained. In most cases, it is also required that the computational complexity is limited, so that the algorithm contains a known bounded amount of flops (floating point operations) per time step.

For the methods GIVE and CM it was previously found that the estimate $\hat{\boldsymbol{\vartheta}}$ is determined from a small set of covariances collected in a vector $\hat{\mathbf{r}}_t$. Symbolically, this can be written as, cf. (13.2)

$$\mathbf{f}(\hat{\boldsymbol{\vartheta}}(t), \hat{\mathbf{r}}_t) = \mathbf{0} . \tag{13.33}$$

In an off-line or batch setting, the main computational load lies often in computing the vector $\hat{\mathbf{r}}_t$. This will typically require $O(tn)$ flops, where $n = \dim(\boldsymbol{\vartheta})$, cf. Eq. (13.36) below. In case (13.33) is a linear system of equations, then to compute a solution with respect to $\boldsymbol{\vartheta}$ would require another $O(n^3)$ flops. The amount of computations is larger in the nonlinear case. Still, if $t \gg n$, it is the computation of the covariances that accounts for the main computational load.

In the non-EIV case, recursive algorithms are mostly constructed to have complexity $O(n^2)$ rather than $O(n^3)$ per time step. In certain cases, it is even possible to derive algorithms that has as low complexity as $O(n)$, Ljung and Söderström (1983). Some aspects to consider when constructing and choosing a recursive algorithm include

- The computational load per time step
- The transient behavior
- The convergence properties

A simple and general approach for recursive algorithms can be derived for cases when the equations defining the parameter estimates are formulated in the form (13.5):

$$\hat{\boldsymbol{\vartheta}} = \mathbf{g}(\hat{\boldsymbol{\vartheta}}, \hat{\mathbf{r}}_t) . \tag{13.34}$$

It is then a possibility to update $\hat{\mathbf{r}}_t$ recursively, and to substitute the nonlinear system of equations (13.34) with one update per time step as follows:

$$\hat{\boldsymbol{\vartheta}}(t+1) = \mathbf{g}(\hat{\boldsymbol{\vartheta}}(t), \hat{\mathbf{r}}_{t+1}) . \tag{13.35}$$

The updating of the covariance vector $\hat{\mathbf{r}}_t$ is straightforward. This vector (or an arbitrary part of it) can be written as

$$\hat{\mathbf{r}}_t = \frac{1}{t} \sum_{s=1}^{t} \boldsymbol{\varphi}(s) y(s) . \tag{13.36}$$

A recursion is obtained by rewriting (13.36) as

$$\begin{aligned} \hat{\mathbf{r}}_{t+1} &= \frac{1}{t+1} \sum_{s=1}^{t+1} \boldsymbol{\varphi}(s) y(t) = \frac{1}{t+1} \left[\boldsymbol{\varphi}(t+1) y(t+1) + t \hat{\mathbf{r}}_t \right] \\ &= \hat{\mathbf{r}}_t + \frac{1}{t+1} \left[\boldsymbol{\varphi}(t+1) y(t+1) - \hat{\mathbf{r}}_t \right] . \end{aligned} \tag{13.37}$$

Together, (13.35) and (13.37) form a recursive algorithm.

Next some more specific details are given for how recursive algorithms may be constructed for some of the main estimation methods presented in earlier chapters of this book.

As a background on recursive least squares estimation, consider the least squares estimate

$$\hat{\boldsymbol{\theta}}(t) = \left(\sum_{s=1}^{t} \boldsymbol{\varphi}(s)\boldsymbol{\varphi}^T(s) \right)^{-1} \left(\sum_{s=1}^{t} \boldsymbol{\varphi}(s)y(s) \right) . \tag{13.38}$$

It is well known, see Ljung and Söderström (1983) for example, that the estimate (13.38) can be arranged recursively as

$$\hat{\boldsymbol{\theta}}(t) = \hat{\boldsymbol{\theta}}(t-1) + \mathbf{K}(t) \left[y(t) - \boldsymbol{\varphi}^T(t)\hat{\boldsymbol{\theta}}(t-1) \right] , \tag{13.39}$$

$$\mathbf{K}(t) = \frac{\mathbf{P}(t-1)\boldsymbol{\varphi}(t)}{1 + \boldsymbol{\varphi}^T(t)\mathbf{P}(t-1)\boldsymbol{\varphi}(t)} = \mathbf{P}(t)\boldsymbol{\varphi}(t) , \tag{13.40}$$

$$\mathbf{P}(t) = \mathbf{P}(t-1) - \frac{\mathbf{P}(t-1)\boldsymbol{\varphi}(t)\boldsymbol{\varphi}^T(t)\mathbf{P}(t-1)}{1 + \boldsymbol{\varphi}^T(t)\mathbf{P}(t-1)\boldsymbol{\varphi}(t)} . \tag{13.41}$$

This algorithm has computational complexity $O(n^2)$ rather than $O(n^3)$, which means that the number of floating point operations (flops) per time step is of the magnitude $O(n^2)$.

A modified scheme (often derived using stochastic approximation) runs as

$$\hat{\boldsymbol{\theta}}(t) = \hat{\boldsymbol{\theta}}(t-1) + \gamma(t)\mathbf{K}(t) \left[y(t) - \boldsymbol{\varphi}^T(t)\hat{\boldsymbol{\theta}}(t-1) \right] , \tag{13.42}$$

$$\mathbf{K}(t) = \mathbf{R}^{-1}(t)\boldsymbol{\varphi}(t) , \tag{13.43}$$

$$\mathbf{R}(t) = \mathbf{R}(t-1) + \gamma(t) \left[\boldsymbol{\varphi}(t)\boldsymbol{\varphi}^T(t) - \mathbf{R}(t-1) \right] . \tag{13.44}$$

Here, $\gamma(t)$ is a gain sequence, that decays with time. As it stands the algorithm (13.42)–(13.44) has computational complexity $O(n^3)$. Note though, that it can be rewritten after some algebraic manipulations as

$$\hat{\boldsymbol{\theta}}(t) = \hat{\boldsymbol{\theta}}(t-1) + \mathbf{K}(t) \left[y(t) - \boldsymbol{\varphi}^T(t)\hat{\boldsymbol{\theta}}(t-1) \right] , \tag{13.45}$$

$$\mathbf{K}(t) = \frac{\mathbf{P}(t-1)\boldsymbol{\varphi}(t)}{\lambda(t) + \boldsymbol{\varphi}^T(t)\mathbf{P}(t-1)\boldsymbol{\varphi}(t)} , \tag{13.46}$$

$$\mathbf{P}(t) = \left[\mathbf{P}(t-1) - \frac{\mathbf{P}(t-1)\boldsymbol{\varphi}(t)\boldsymbol{\varphi}^T(t)\mathbf{P}(t-1)}{\lambda(t) + \boldsymbol{\varphi}^T(t)\mathbf{P}(t-1)\boldsymbol{\varphi}(t)} \right] \frac{1}{\lambda(t)} . \tag{13.47}$$

provided the time-varying forgetting factor $\lambda(t)$ and the gain sequence $\gamma(t)$ are related as

$$\lambda(t) = \frac{\gamma(t-1)}{\gamma(t)} (1 - \gamma(t)) . \tag{13.48}$$

In particular, one combination that fulfills (13.48) is the constant values $\lambda(t) = 1 - \gamma,\ \gamma(t) = \gamma$. Another possibility is $\gamma(t) = 1/t$ leading to $\lambda(t) \equiv 1$. The computational complexity of (13.45)–(13.47) is $O(n^2)$.

13.6.2 Recursive Version of the GIVE Estimate

First recall, see Sect. 7.4 and Eq. (7.96), that the GIVE estimates $\hat{\boldsymbol{\theta}}$, $\hat{\boldsymbol{\rho}}$ can be written as the solution to

$$\frac{1}{t}\sum_{s=1}^{t} \mathbf{z}(s)\left[y(s) - \boldsymbol{\varphi}^T(s)\boldsymbol{\theta}\right] - \mathbf{h}(\boldsymbol{\rho}, \boldsymbol{\theta}) = \mathbf{0}\ , \tag{13.49}$$

where

$$\mathbf{h}(\boldsymbol{\rho}, \boldsymbol{\theta}) = \mathbf{r}_{\tilde{\mathbf{z}}\tilde{y}}(\boldsymbol{\rho}) - \mathbf{R}_{\tilde{\mathbf{z}}\tilde{\boldsymbol{\varphi}}}(\boldsymbol{\rho})\boldsymbol{\theta}\ . \tag{13.50}$$

Introduce now

$$\mathbf{R}(t) = \frac{1}{t}\sum_{s=1}^{t} \mathbf{z}(s)\boldsymbol{\varphi}^T(s)\ , \tag{13.51}$$

$$\mathbf{b}(t) = \mathbf{b}(t, \boldsymbol{\rho}, \boldsymbol{\theta}) = \left[\frac{1}{t}\sum_{s=1}^{t} \mathbf{z}(s)y(s)\right] - \mathbf{h}(\boldsymbol{\rho}, \boldsymbol{\theta})\ . \tag{13.52}$$

Equation (13.49) is then equivalent to

$$\mathbf{R}(t)\boldsymbol{\theta} = \mathbf{b}(t, \boldsymbol{\rho}, \boldsymbol{\theta})\ . \tag{13.53}$$

A recursive estimator of $\hat{\boldsymbol{\theta}}(t)$ will be based on recursion of $\mathbf{R}(t)$ and $\mathbf{b}(t, \boldsymbol{\rho}, \boldsymbol{\theta})$.

The rectangular matrix $\mathbf{R}(t)$ can easily be computed in a recursive fashion:

$$\begin{aligned}\mathbf{R}(t+1) &= \frac{1}{t+1}\left[\mathbf{z}(t+1)\boldsymbol{\varphi}^T(t+1) + t\mathbf{R}(t)\right] \\ &= \mathbf{R}(t) + \frac{1}{t+1}\left[\mathbf{z}(t+1)\boldsymbol{\varphi}^T(t+1) - \mathbf{R}(t)\right]\ .\end{aligned} \tag{13.54}$$

Similarly, for fixed arguments $\boldsymbol{\rho}$ and $\boldsymbol{\theta}$,

$$\begin{aligned}\mathbf{b}(t+1) &= \frac{1}{t+1}\left[\mathbf{z}(t+1)y(t+1) + t\left(\mathbf{b}(t) + \mathbf{h}(\boldsymbol{\rho}, \boldsymbol{\theta})\right)\right] - \mathbf{h}(\boldsymbol{\rho}, \boldsymbol{\theta}) \\ &= \mathbf{b}(t) + \frac{1}{t+1}\left[\mathbf{z}(t+1)y(t+1) - \mathbf{b}(t) - \mathbf{h}(\boldsymbol{\rho}, \boldsymbol{\theta})\right]\ .\end{aligned} \tag{13.55}$$

A recursive updating of $\hat{\boldsymbol{\theta}}(t)$ can then be taken as (13.54) combined with

$$\hat{\boldsymbol{\theta}}(t+1) = \left[\mathbf{R}^T(t+1)\mathbf{R}(t+1)\right]^{-1} \mathbf{R}^T(t+1)\mathbf{b}(t+1, \hat{\boldsymbol{\rho}}(t), \hat{\boldsymbol{\theta}}(t)) \ . \tag{13.56}$$

It remains to derive an updating equation for the estimate $\hat{\boldsymbol{\rho}}(t)$ of the noise variances.
It holds in this case

$$\mathbf{h}(\boldsymbol{\rho}, \boldsymbol{\theta}) = \begin{pmatrix} \lambda_y \bar{\mathbf{a}} \\ \mathbf{0}_{p_y \times 1} \\ -\lambda_u \mathbf{b} \\ \mathbf{0}_{p_u \times 1} \end{pmatrix} \stackrel{\Delta}{=} \underbrace{\begin{pmatrix} \bar{\mathbf{a}} & \mathbf{0} \\ \mathbf{0} & \mathbf{0} \\ \mathbf{0} & -\mathbf{b} \\ \mathbf{0} & \mathbf{0} \end{pmatrix}}_{\mathbf{J}(\boldsymbol{\theta})} \boldsymbol{\rho} \ . \tag{13.57}$$

Set further

$$\mathbf{q}(t) = \frac{1}{t} \sum_{s=1}^{t} \mathbf{z}(s) y(s) \ . \tag{13.58}$$

Similarly to (13.54) one gets the recursion

$$\mathbf{q}(t+1) = \mathbf{q}(t) + \frac{1}{t+1} \left[\mathbf{z}(t+1)y(t+1) - \mathbf{q}(t)\right] \ . \tag{13.59}$$

Further, from (13.57) a possible recursion for $\hat{\boldsymbol{\rho}}(t)$ is

$$\begin{aligned} \hat{\boldsymbol{\rho}}(t) &= \mathbf{J}^{\dagger}(\hat{\boldsymbol{\theta}}(t)) \left[\frac{1}{t} \sum_{s=1}^{t} \mathbf{z}(s) \left[y(s) - \boldsymbol{\varphi}^T(s) \hat{\boldsymbol{\theta}}(t) \right] \right] \\ &= \mathbf{J}^{\dagger}(\hat{\boldsymbol{\theta}}(t)) \left[\mathbf{q}(t) - \mathbf{R}(t) \hat{\boldsymbol{\theta}}(t) \right] , \end{aligned} \tag{13.60}$$

where $\mathbf{J}^{\dagger}(\boldsymbol{\theta}) = \left[\mathbf{J}^T(\boldsymbol{\theta})\mathbf{J}(\boldsymbol{\theta})\right]^{-1} \mathbf{J}^T(\boldsymbol{\theta})$ is the pseudo-inverse of $\mathbf{J}(\boldsymbol{\theta})$. Note that in (13.60) the pseudo-inverse is included in the updating, and that the updating of $\mathbf{R}(t)$ itself includes a rank-one modification; see (13.54). How to arrange this with $O(n^2)$ instead of $O(n^3)$ flops is developed for the so-called overdetermined recursive instrumental variable estimator; see Friedlander (1984) or Söderström and Stoica (1989).

13.6.3 *Recursive Version of the Covariance Matching Estimate*

For the CM approach, introduced in Chap. 8, it holds

$$\hat{\boldsymbol{\theta}}(t) = \arg\min_{\boldsymbol{\theta}} V_t(\boldsymbol{\theta}) \;, \tag{13.61}$$

where

$$V_t(\boldsymbol{\theta}) = \min_{\mathbf{r}_z} \frac{1}{2} \parallel \hat{\mathbf{r}}_t - \mathbf{F}(\boldsymbol{\theta})\mathbf{r}_z \parallel^2_{\mathbf{W}} \;. \tag{13.62}$$

One possibility for deriving a recursive estimator is to at each time step take one (approximate) Gauss–Newton step for minimizing the criterion. Conceptually, the algorithm would then be written as

$$\hat{\boldsymbol{\theta}}(t+1) = \hat{\boldsymbol{\theta}}(t) - \alpha_{t+1} \left[V_t^{''}(\hat{\boldsymbol{\vartheta}}(t)) \right]^{-1} \left(V_t^{'}(\hat{\boldsymbol{\vartheta}}(t)) \right)^T \;. \tag{13.63}$$

Here, the step length α_{t+1} should ideally be taken as 1, but a smaller value may be used to ensure a good transient behavior, and to avoid divergence. How to find the gradient and the Hessian of the loss function is shown in Chap. 14, from which one can make use of some results, cf. (14.132), (14.133). In the evaluations several approximations are made. The expressions for the Hessian are valid in the asymptotic case ($t \to \infty$), and for $\boldsymbol{\theta}$ equal to the true parameter vector. Similarly, the expression used for the gradient $V_t^{'}$ is valid for the true parameter vector. These expressions are reasonable, at least to describe the behavior for large data sets, and when convergence of the parameter estimates to the true values has almost occurred. Further, using these approximations does indeed simplify the expressions.

One can write

$$V_t^{''}(\hat{\boldsymbol{\vartheta}}(t)) \approx \mathbf{S}^T \mathbf{P} \mathbf{S} \;, \tag{13.64}$$

$$\left(V_t^{'}(\hat{\boldsymbol{\vartheta}}(t)) \right)^T \approx -\mathbf{S}^T \mathbf{P} \tilde{\mathbf{r}}_t \;, \tag{13.65}$$

$$\mathbf{S} = \left(\frac{\partial \mathbf{F}(\boldsymbol{\theta})}{\partial \theta_1} \cdots \frac{\partial \mathbf{F}(\boldsymbol{\theta})}{\partial \theta_{na+nb}} \right) \;, \tag{13.66}$$

$$\mathbf{P} = \mathbf{W} - \mathbf{W}\mathbf{F} \left(\mathbf{F}^T \mathbf{W} \mathbf{F} \right)^{-1} \mathbf{F}^T \mathbf{W} \;, \tag{13.67}$$

$$\tilde{\mathbf{r}}_t = \hat{\mathbf{r}}_t - \mathbf{r} \;. \tag{13.68}$$

One way to approximate $\tilde{\mathbf{r}}_t$ is as follows. First, to approximate $\hat{\mathbf{r}}_t$ set

$$\mathbf{q}(t) = \frac{1}{t} \sum_{s=1}^{t} \boldsymbol{\varphi}(s) y(s) \;. \tag{13.69}$$

In case several types of covariance elements are included in the vector $\hat{\mathbf{r}}_t$, then the different separate parts can be written as in (13.69). The quantity $\mathbf{q}(t)$ can be updated as before, using

$$\mathbf{q}(t+1) = \mathbf{q}(t) + \frac{1}{t+1} \left[\boldsymbol{\varphi}(t+1) y(t+1) - \mathbf{q}(t) \right] \;. \tag{13.70}$$

Then using (8.26), one may take

$$\tilde{\mathbf{r}}_t = \mathbf{q}(t) - \mathbf{F}(\hat{\boldsymbol{\theta}}) \left(\mathbf{F}^T(\hat{\boldsymbol{\theta}})\mathbf{W}\mathbf{F}(\hat{\boldsymbol{\theta}})\right)^{-1} \mathbf{F}^T(\hat{\boldsymbol{\theta}})\mathbf{W}\mathbf{q}(t)_{\hat{\theta}=\hat{\theta}(t)} \,. \tag{13.71}$$

The total recursive CM algorithm thus consists of (13.63), (13.64), (13.65), and (13.71).

13.6.4 Recursive Version of the Maximum Likelihood Estimate

For the maximum likelihood estimator, derived in Sect. 9.2 it is in principle straightforward to use a general machinery for deriving a recursive estimator, Ljung and Söderström (1983). First note that the overall model can be written in state space form as, see (9.9)–(9.12),

$$\mathbf{x}(t+1) = \mathbf{A}_0(\boldsymbol{\vartheta})\mathbf{x}(t) + \mathbf{v}(t) \,, \tag{13.72}$$

$$\boldsymbol{\zeta}(t) = \mathbf{C}_0(\boldsymbol{\vartheta})\mathbf{x}(t) + \mathbf{e}(t) \,, \tag{13.73}$$

$$\operatorname{cov}\begin{pmatrix} \mathbf{v}(t) \\ \mathbf{e}(t) \end{pmatrix} = \begin{pmatrix} \mathbf{R}_1(\boldsymbol{\vartheta}) & \mathbf{R}_{12}(\boldsymbol{\vartheta}) \\ \mathbf{R}_{21}(\boldsymbol{\vartheta}) & \mathbf{R}_2(\boldsymbol{\vartheta}) \end{pmatrix} . \tag{13.74}$$

Further, the estimate $\hat{\boldsymbol{\vartheta}}(t)$ minimizes

$$V_t(\boldsymbol{\vartheta}) = \sum_{s=1}^{t} \ell(\boldsymbol{\varepsilon}(s,\boldsymbol{\vartheta}), \boldsymbol{\vartheta}) \,, \tag{13.75}$$

where $\boldsymbol{\varepsilon}(t,\boldsymbol{\vartheta})$ is the one-step prediction error of $\boldsymbol{\zeta}(t)$ in (13.72), (13.73). The criterion $\ell(\boldsymbol{\varepsilon},\boldsymbol{\vartheta})$ is

$$\ell(\boldsymbol{\varepsilon},\boldsymbol{\vartheta}) = \frac{1}{2}\left[\boldsymbol{\varepsilon}^T\mathbf{Q}(\boldsymbol{\vartheta})^{-1}\boldsymbol{\varepsilon} + \log\det(\mathbf{Q}(\boldsymbol{\vartheta}))\right] \,, \tag{13.76}$$

where $\mathbf{Q}(\boldsymbol{\vartheta})$ is the covariance matrix of the prediction errors. See Eqs. (9.28) and (9.18). To arrive at (13.76) the model (13.72), (13.73) is extended and rewritten to describe how both the one-step prediction $\hat{\boldsymbol{\zeta}}(t|\boldsymbol{\theta})$ and its gradient $\boldsymbol{\psi}(t,\boldsymbol{\vartheta})$ with respect to $\boldsymbol{\vartheta}$ are computed by filtering the data $\boldsymbol{\zeta}(t)$.

$$\boldsymbol{\xi}(t+1,\boldsymbol{\vartheta}) = \mathbf{A}(\boldsymbol{\vartheta})\boldsymbol{\xi}(t,\boldsymbol{\vartheta}) + \mathbf{B}(\boldsymbol{\vartheta})\boldsymbol{\zeta}(t) \,, \tag{13.77}$$

$$\begin{pmatrix} \hat{\boldsymbol{\zeta}}(t|\boldsymbol{\vartheta}) \\ \operatorname{vec}(\boldsymbol{\psi}(t,\boldsymbol{\vartheta})) \end{pmatrix} = \mathbf{C}(\boldsymbol{\vartheta})\boldsymbol{\xi}(t,\boldsymbol{\vartheta}) \,. \tag{13.78}$$

From this general formulation, a recursive estimator can be constructed as follows; see Ljung and Söderström (1983),

$$\boldsymbol{\varepsilon}(t) = \boldsymbol{\zeta}(t) - \hat{\boldsymbol{\zeta}}(t) \,, \tag{13.79}$$

$$\hat{\boldsymbol{\Lambda}}(t) = \hat{\boldsymbol{\Lambda}}(t-1) + \gamma(t)\left[\boldsymbol{\varepsilon}(t)\boldsymbol{\varepsilon}^T(t) - \hat{\boldsymbol{\Lambda}}(t-1)\right] \,, \tag{13.80}$$

$$\mathbf{R}(t) = \mathbf{R}(t-1) + \gamma(t)\left[\boldsymbol{\psi}(t)\hat{\boldsymbol{\Lambda}}^{-1}(t)\boldsymbol{\psi}^T(t) - \mathbf{R}(t-1)\right] \,, \tag{13.81}$$

$$\boldsymbol{\xi}(t+1) = \mathbf{A}(\hat{\boldsymbol{\vartheta}}(t))\boldsymbol{\xi}(t) + \mathbf{B}(\hat{\boldsymbol{\vartheta}}(t))\boldsymbol{\zeta}(t) \,, \tag{13.82}$$

$$\begin{pmatrix} \hat{\boldsymbol{\zeta}}(t) \\ \mathrm{vec}(\boldsymbol{\psi}(t)) \end{pmatrix} = \mathbf{C}(\hat{\boldsymbol{\vartheta}}(t-1))\boldsymbol{\xi}(t) \,, \tag{13.83}$$

$$\begin{aligned} \hat{\boldsymbol{\vartheta}}(t) &= \hat{\boldsymbol{\vartheta}}(t-1) + \gamma(t)\mathbf{R}^{-1}(t) \\ &\times \left[-\ell_{\boldsymbol{\vartheta}}^T(t, \hat{\boldsymbol{\vartheta}}(t-1), \boldsymbol{\varepsilon}(t)) + \boldsymbol{\psi}(t)\ell_{\boldsymbol{\varepsilon}}^T(t, \hat{\boldsymbol{\vartheta}}(t-1), \boldsymbol{\varepsilon}(t))\right] \,. \end{aligned} \tag{13.84}$$

In (13.84) the gradient $\ell_{\boldsymbol{\vartheta}}$ of the momentary loss function

$$\ell(\boldsymbol{\varepsilon}, \boldsymbol{\vartheta}) = \frac{1}{2}\boldsymbol{\varepsilon}^T\mathbf{Q}^{-1}\boldsymbol{\varepsilon} + \frac{1}{2}\log\det\mathbf{Q} \tag{13.85}$$

is needed. Straightforward differentiation gives that the gradient is given componentwise as

$$\ell_{\vartheta_i} = \frac{\partial \ell(\boldsymbol{\varepsilon}, \boldsymbol{\vartheta})}{\partial \vartheta_i} = -\frac{1}{2}\boldsymbol{\varepsilon}^T\mathbf{Q}^{-1}\frac{\partial \mathbf{Q}}{\partial \vartheta_i}\mathbf{Q}^{-1}\boldsymbol{\varepsilon} + \frac{1}{2}\mathrm{tr}\left(\mathbf{Q}^{-1}\frac{\partial \mathbf{Q}}{\partial \vartheta_i}\right) \,. \tag{13.86}$$

See also Sect. 14.8.2 and Corollary 14.2 for similar types of calculations.

13.7 Extensions and Bibliographical Notes

Computational Efficiency

Potentially, an expectation–maximization (EM) form of the ML estimator could be used to compute the EIV estimate. Aspects on how to compute the ML estimator for a general state space model using the EM algorithm are described in Wills et al. (2010), Shumway and Stoffer (1982), Gibson and Ninness (2005). See also Schön et al. (2011) and Ninness et al. (2010).

Variable Projection Algorithms

There are indeed many problems, for example in system identification, signal processing, and control, that lead to separable nonlinear least squares problems of the type (13.21), Golub and Pereyra (1973), Golub and Pereyra (2003).

One example within system identification concerns modeling pseudo-linear regressions (PLR) (sometimes called extended least squares); see for example Ljung and Söderström (1983), Goodwin and Sin (1984). PLR modeling is usually carried out in recursive form, but can also be applied in an batch form, Stoica et al. (1985).

More recent work that can be interpreted as PLR modeling includes Ding et al. (2006), Ding and Chen (2005a, b). As a distinction to the separable nonlinear least squares problem treated here, for PLR modeling the correspondence to (8.24) has a vector $\mathbf{r}_z$ of a length proportional to the number of data, N. This makes the estimation problem and its analysis quite different from the case treated here. Furthermore, in Ding et al. (2006), Ding and Chen (2005a, b) the primary interest lies in estimating $\mathbf{r}_z$ rather than $\boldsymbol{\theta}$.

Recursive Algorithms

The thesis Linden (2009) treats recursive Frisch schemes and EIV filtering problems.

Other papers dealing with computationally efficient implementations of recursive BELS and Frisch estimators include Zheng and Feng (1989), Wada et al. (1990), Feng and Zheng (1991), Linden et al. (2008a, b), Linden and Burnham (2009), Linden et al. (2009, 2007, 2012), Arablouei et al. (2014).

Recursive Frisch methods are also treated in Larkowski et al. (2009a). For further examples of recursive algorithms for implementing the Frisch scheme estimator; see Fan and Luo (2010), Fan and Lo (2013).

Recursive algorithms for dynamic shock error models (a special case of EIV models) appear in Krishnamurthy (1994).

The paper Song and Chen (2008) considers recursive identification in the case the noise-free input is white noise, while the observation noise on the input and output sides are both ARMA processes. Consistency is established.

Recursive algorithms for multivariable EIV models are treated in Chen (2007), under the assumption of white noise on the input and output sides, and the noise-free input being an ARMA process.

Appendix

13.A Algorithmic Aspects of the GIVE Estimate

This appendix describes some properties of GIVE algorithms. Some general aspects are treated in Sect. 13.A.1, while specifics for variable projection algorithms follow in Sect. 13.A.2.

Part of the analysis was carried out in detail for the BELS case in Söderström et al. (2005).

13.A.1 General Aspects

As described in Sect. 7.4, the general set of equations for determining $\boldsymbol{\theta}$ and $\boldsymbol{\rho}$ is given by (7.96). Ideally one should solve the possibly overdetermined system of equations

$$\mathbf{f}(\boldsymbol{\vartheta}) = \hat{\mathbf{r}}_{z\varepsilon}(\boldsymbol{\theta}) - \mathbf{r}_{z\varepsilon}(\boldsymbol{\theta}, \boldsymbol{\rho}) \approx \mathbf{0} \tag{13.87}$$

with respect to $\boldsymbol{\theta}$ and $\boldsymbol{\rho}$.

Consider the variant with a fixed weighting matrix $\mathbf{W}$, that does not depend on $\boldsymbol{\theta}$. Thus the Frisch variant which compares the covariance function for the residuals is not included in the present discussion. It turns out that the function $\mathbf{f}(\boldsymbol{\theta}, \boldsymbol{\rho})$ is bilinear: It is linear in $\boldsymbol{\theta}$ and linear in $\boldsymbol{\rho}$, but not linear simultaneously in both vectors. In early BELS algorithms, Zheng and Feng (1989), this property was exploited in the following way.

Algorithm 13.3 Assume that the system of equations is not overdetermined. Write the Eq. (13.87) symbolically as

$$\mathbf{A}_1(\boldsymbol{\rho})\boldsymbol{\theta} = \mathbf{b}_1(\boldsymbol{\rho}) \, , \tag{13.88}$$

$$\mathbf{A}_2(\boldsymbol{\theta})\boldsymbol{\rho} = \mathbf{b}_2(\boldsymbol{\theta}) \, , \tag{13.89}$$

where $\mathbf{A}_1(\boldsymbol{\theta})$ is a square matrix of dimension $\dim(\boldsymbol{\theta})$, and $\mathbf{A}_2(\boldsymbol{\theta})$ is another square matrix of dimension $\dim(\boldsymbol{\rho})$. Then iterating successively between solving the two linear systems of equations above can be written as follows

1. Start with an initial guess, $\boldsymbol{\theta}^{(0)}$.
2. For $i = 1, \ldots,$ repeat until convergence

   2a. Solve $\mathbf{A}_2(\boldsymbol{\theta}^{(i-1)})\boldsymbol{\rho} = \mathbf{b}_2(\boldsymbol{\theta}^{(i-1)})$ to get $\boldsymbol{\rho}^{(i)}$.
   2b. Solve $\mathbf{A}_1(\boldsymbol{\rho}^{(i)})\boldsymbol{\theta} = \mathbf{b}_1(\boldsymbol{\rho}^{(i)})$ to get $\boldsymbol{\theta}^{(i)}$.

■

After linearizing these iterations around the true parameter vector, one may write

$$\boldsymbol{\theta}^{(i)} - \boldsymbol{\theta}_0 = \mathbf{G}\left[\boldsymbol{\theta}^{(i-1)} - \boldsymbol{\theta}_0\right] \, . \tag{13.90}$$

The (local) convergence properties of the iterations are determined by the eigenvalues of the matrix $\mathbf{G}$. Such a convergence analysis is undertaken in Söderström et al. (2005). It turns out that:

1. Local convergence takes always place ($\mathbf{G}$ has all eigenvalues inside the unit circle) if the signal-to-noise ratio, SNR, on the input and output sides is large. There, it is also proved that the matrix $\mathbf{G}$ has always one eigenvalue equal to zero. In case $\mathsf{E}\left\{u_0^2(t)\right\}$ becomes large, the eigenvalues of $\mathbf{G}$ all satisfy

$$\lambda_j(\mathbf{G}) = O(1/\mathsf{E}\left\{u_0^2(t)\right\}) \, . \tag{13.91}$$

   Unfortunately, this is of somewhat limited practical value, as for large SNR the bias introduced by the least squares method may be insignificant anyway.
2. There are indeed cases, with low SNR, where the matrix $\mathbf{G}$ has eigenvalues outside the unit circle. Then Algorithm 13.3 will not produce the desired solution (although the solution does exist) to the Eq. (13.87).

In Söderström et al. (2005) it was proposed to preferably solve the Eq. (13.87) using a variable projection algorithm, cf. Sect. A.1.4. This principle is outlined below.

Algorithm 13.4 Write the Eq. (13.87) as

$$\begin{aligned} \mathbf{0} = \mathbf{f}(\boldsymbol{\theta}, \boldsymbol{\rho}) &= \hat{\mathbf{r}}_{\mathbf{z}y} - \hat{\mathbf{R}}_{\mathbf{z}\boldsymbol{\varphi}}\boldsymbol{\theta} - \mathbf{r}_{\tilde{\mathbf{z}}\tilde{y}}(\boldsymbol{\rho}) - \mathbf{R}_{\tilde{\mathbf{z}}\tilde{\boldsymbol{\varphi}}}\boldsymbol{\theta} \\ &\stackrel{\Delta}{=} \mathbf{g}(\boldsymbol{\rho}) - \mathbf{F}(\boldsymbol{\rho})\boldsymbol{\theta} \ . \end{aligned} \tag{13.92}$$

Then the loss function to be minimized can be written as

$$\| \mathbf{f}(\boldsymbol{\theta}, \boldsymbol{\rho}) \|_{\mathbf{W}}^2 = \left[\mathbf{g}(\boldsymbol{\rho}) - \mathbf{F}(\boldsymbol{\rho})\boldsymbol{\theta}\right]^T \mathbf{W} \left[\mathbf{g}(\boldsymbol{\rho}) - \mathbf{F}(\boldsymbol{\rho})\boldsymbol{\theta}\right] \ . \tag{13.93}$$

To minimize the criterion with respect to $\boldsymbol{\theta}$ for any fixed $\boldsymbol{\rho}$ is easy:

$$\boldsymbol{\theta} = \boldsymbol{\theta}(\boldsymbol{\rho}) = \left(\mathbf{F}^T(\boldsymbol{\rho})\mathbf{W}\mathbf{F}(\boldsymbol{\rho})\right)^{-1} \mathbf{F}^T(\boldsymbol{\rho})\mathbf{W}\mathbf{g}(\boldsymbol{\rho}) \ . \tag{13.94}$$

Inserting (13.94) into (13.93) gives the concentrated loss function

$$\begin{aligned} V(\boldsymbol{\rho}) &= \min_{\boldsymbol{\theta}} \| \mathbf{f}(\boldsymbol{\theta}, \boldsymbol{\rho}) \|_{\mathbf{W}}^2 \\ &= \mathbf{g}^T(\boldsymbol{\rho})\mathbf{W}\mathbf{g}(\boldsymbol{\rho}) - \mathbf{g}^T(\boldsymbol{\rho})\mathbf{W}\mathbf{F}(\boldsymbol{\rho}) \left(\mathbf{F}^T(\boldsymbol{\rho})\mathbf{W}\mathbf{F}(\boldsymbol{\rho})\right)^{-1} \mathbf{F}^T(\boldsymbol{\rho})\mathbf{W}\mathbf{g}(\boldsymbol{\rho}) \ . \end{aligned} \tag{13.95}$$

Minimization of $V(\boldsymbol{\rho})$ in (13.95) has to be carried out using some numerical search algorithm.

One advantage of minimizing $V(\boldsymbol{\rho})$ numerically instead of $\| f(\boldsymbol{\theta}, \boldsymbol{\rho}) \|_{\mathbf{W}}^2$ is, as reported in Söderström et al. (2005), that this algorithm is much more robust than Algorithm 13.3 and no particular numerical problems have been reported. ■

When using Algorithm 13.4 to solve the equations at least two theoretical and practical questions arise.

1. Does the loss function $V(\boldsymbol{\rho})$ have a unique global minimum point? Every solution to the Eq. (13.87) would correspond to a global minimum point giving V its theoretical minimal value equal to zero.
2. Does the loss function $V(\boldsymbol{\rho})$ have any 'false' local minimum points? (with 'false' is here understood points such that $\boldsymbol{\rho} \neq \boldsymbol{\rho}_0$). If this happens to be the case, there is a potential risk that the numerical search procedure is stuck in a local minimum point not corresponding to the global minimum.

The two above questions are largely open and still unanswered. As mentioned before the reported experience indicates that in practice one should not expect convergence problems. Still, the optimization problem is certainly nonlinear and has a good deal of structure, so it would not be a surprise if false minima can sometimes exist. A further aspect is that in practice one would prefer not to consider any arbitrary vector $\boldsymbol{\rho}$ but only such ones that correspond to positive noise variances (and to a positive definite covariance sequence for $r_{\tilde{y}}(\tau)$ when $\tilde{y}(t)$ is treated as a correlated noise).

To exemplify the above reasoning of the two questions, consider the following example.

Example 13.2 Consider the very simple case (the detailed calculations will easily become messy with more advanced examples!)

$$y_0(t) = bu_0(t) , \tag{13.96}$$

and set

$$\mathbf{z}(t) = \begin{pmatrix} y(t) \\ u(t) \\ u(t-1) \end{pmatrix}, \quad \varphi(t) = u(t) , \tag{13.97}$$

$$r_0 = \mathsf{E}\left\{u_0^2(t)\right\}, \quad r_1 = \mathsf{E}\left\{u_0(t)u_0(t-1)\right\} . \tag{13.98}$$

Let the true noise variances be denoted as

$$\lambda_u^0 = \mathsf{E}\left\{\tilde{u}^2(t)\right\}, \quad \lambda_y^0 = \mathsf{E}\left\{\tilde{y}^2(t)\right\} , \tag{13.99}$$

and set for future calculations

$$\xi = \lambda_y^0 - \lambda_y, \quad \eta = \lambda_u^0 - \lambda_u . \tag{13.100}$$

In this case it holds (when $N \to \infty$)

$$\mathbf{g} = \mathbf{g}(\xi) = \mathbf{r}_{\mathbf{z}y} - \mathbf{r}_{\tilde{\mathbf{z}}\tilde{y}}(\boldsymbol{\rho}) = \begin{pmatrix} b^2 r_0 + \xi \\ b r_0 \\ b r_1 \end{pmatrix} , \tag{13.101}$$

$$\mathbf{F} = \mathbf{F}(\eta) = \mathbf{r}_{\mathbf{z}\varphi} - \mathbf{r}_{\tilde{\mathbf{z}}\tilde{\varphi}}(\boldsymbol{\rho}) = \begin{pmatrix} b r_0 \\ r_0 + \eta \\ r_1 \end{pmatrix} . \tag{13.102}$$

First consider *global minima*, or equivalently, all solutions to

$$\mathbf{g}(\xi) - \mathbf{F}(\eta)\hat{b} = 0 , \tag{13.103}$$

where the unknowns are $\hat{b}$, ξ, and η. In order to have identifiability one must require that $r_1 \neq 0, \ b \neq 0$. The last component of (13.103) then gives $\hat{b} = b$. Then the first and the second components give directly $\xi = 0, \ \eta = 0$. Hence in this specific example, there is only one solution to the GIVE equation.

Next consider the *local minima* of V, (13.95). For this examination choose for convenience the weighting (scaling)

$$\mathbf{W} = \begin{pmatrix} 1 & 0 & 0 \\ 0 & b^2 & 0 \\ 0 & 0 & b^2 \end{pmatrix} . \tag{13.104}$$

Use (13.101), (13.102) and express the loss function as

$$V(\xi, \eta) = \mathbf{g}^T(\xi)\mathbf{W}\mathbf{g}(\xi) - \frac{\left[\mathbf{g}^T(\xi)\mathbf{W}\mathbf{F}(\eta)\right]^2}{\mathbf{F}^T(\eta)\mathbf{W}\mathbf{F}(\eta)} . \tag{13.105}$$

The stationary points of $V(\xi, \eta)$ are the solutions to

$$V'_\xi = 0 \Rightarrow \mathbf{g}^T(\xi)\mathbf{W}\mathbf{g}_\xi(\xi) - \frac{\mathbf{g}^T(\xi)\mathbf{W}\mathbf{F}(\eta)}{\mathbf{F}^T(\eta)\mathbf{W}\mathbf{F}(\eta)}\mathbf{F}^T(\eta)\mathbf{W}\mathbf{g}_\xi(\xi) = 0 , \tag{13.106}$$

$$\begin{aligned} V'_\eta = 0 \Rightarrow & -\frac{\mathbf{g}^T(\xi)\mathbf{W}\mathbf{F}(\eta)}{\mathbf{F}^T(\eta)\mathbf{W}\mathbf{F}(\eta)}\mathbf{g}^T(\xi)\mathbf{W}\mathbf{F}_\eta(\eta) \\ & + \frac{\left[\mathbf{g}^T(\xi)\mathbf{W}\mathbf{F}(\eta)\right]^2}{\left[\mathbf{F}^T(\eta)\mathbf{W}\mathbf{F}(\eta)\right]^2}\mathbf{F}^T(\eta)\mathbf{W}\mathbf{F}_\eta(\eta) = 0 . \end{aligned} \tag{13.107}$$

Furthermore,

$$\mathbf{g}_\xi(\xi) = \frac{\partial \mathbf{g}(\xi)}{\partial \xi} = \begin{pmatrix} 1 \\ 0 \\ 0 \end{pmatrix} , \quad \mathbf{F}_\eta(\eta) = \frac{\partial \mathbf{F}(\eta)}{\partial \eta} = \begin{pmatrix} 0 \\ 1 \\ 0 \end{pmatrix} . \tag{13.108}$$

Next, (13.106) is simplified:

$$\begin{aligned} V'_\xi = 0 \Rightarrow & \left[b^2 r_0^2 + b^2(r_0 + \eta)^2 + b^2 r_1^2\right](b^2 r_0 + \xi) \\ & - \left[b r_0(b^2 r_0 + \xi) + b^3 r_0(r_0 + \eta) + b^3 r_1^2\right] b r_0 = 0 \\ \Rightarrow & \left[(r_0 + \eta)^2 + r_1^2\right]\xi - b^2 r_0 \eta (r_0 + \eta) = 0 \\ \Rightarrow & \xi = \frac{b^2 r_0 (r_0 + \eta)}{(r_0 + \eta)^2 + r_1^2}\eta . \end{aligned} \tag{13.109}$$

Simplification of (13.107) leads to

$$\begin{aligned} V'_\eta = 0 \Rightarrow & -\left[b^2 r_0^2 + b^2(r_0 + \eta)^2 + b^2 r_1^2\right] b^3 r_0 \\ & + \left[b r_0(b^2 r_0 + \xi) + b^3 r_0(r_0 + \eta) + b^3 r_1^2\right] b^2(r_0 + \eta) = 0 \\ \Rightarrow & -b^2 r_0(r_0^2 + r_1^2) + (r_0 + \eta)\left[b^2(r_0^2 + r_1^2) + r_0 \xi\right] = 0 \\ \Rightarrow & \eta b^2(r_0^2 + r_1^2) + r_0 \xi (r_0 + \eta) = 0 . \end{aligned} \tag{13.110}$$

Inserting (13.109) into (13.110) leads now to an equation in only η:

$$\eta b^2(r_0^2 + r_1^2)\left[(r_0+\eta)^2 + r_1^2\right] + r_0(r_0+\eta)b^2 r_0\eta(r_0+\eta) = 0\,. \tag{13.111}$$

Obviously $\eta = 0$ is a solution (which corresponds to the true parameter values, $\hat{\boldsymbol{\vartheta}} = \boldsymbol{\vartheta}_0$). Cancelling a factor $b^2\eta$ in (13.111) leads to

$$(r_0^2 + r_1^2)\left[(r_0+\eta)^2 + r_1^2\right] + r_0^2(r_0+\eta)^2 = 0\,. \tag{13.112}$$

However, all terms in this equation must be positive, and hence $\eta = 0$ is the only solution to (13.111). This fact means that the loss function $V(\xi, \eta)$ has a *unique* stationary point, namely $\xi = 0,\ \eta = 0$. ■

13.A.2 Use of a Variable Projection Algorithm for MIMO Systems

In this section some details are presented for a variable projection algorithm applied for the GIVE in the multivariable case. For general aspects on such algorithms; see Sect. A.1.4.

Noting that the matrix $\mathbf{f}$, (7.125), is an affine transformation of the system parameter matrix $\boldsymbol{\Theta}$, one can, by simplifying notations, write it as

$$\begin{aligned}\mathbf{f}(\boldsymbol{\vartheta}) &= \hat{\mathbf{r}}_{\mathbf{zy}} - \hat{\mathbf{R}}_{\mathbf{z}\varphi}\boldsymbol{\Theta} - \mathbf{r}_{\tilde{\mathbf{z}}\tilde{\mathbf{y}}}(\boldsymbol{\rho}) + \mathbf{R}_{\tilde{\mathbf{z}}\tilde{\varphi}}(\boldsymbol{\rho})\boldsymbol{\Theta} \\ &\stackrel{\Delta}{=} \mathbf{G} - \mathbf{H}\boldsymbol{\Theta}\,,\end{aligned} \tag{13.113}$$

where

$$\mathbf{G} = \mathbf{G}(\boldsymbol{\rho}) = \hat{\mathbf{r}}_{\mathbf{zy}} - \mathbf{r}_{\tilde{\mathbf{z}}\tilde{\mathbf{y}}}(\boldsymbol{\rho})\,, \tag{13.114}$$

$$\mathbf{H} = \mathbf{H}(\boldsymbol{\rho}) = \hat{\mathbf{R}}_{\mathbf{z}\varphi} - \mathbf{R}_{\tilde{\mathbf{z}}\tilde{\varphi}}(\boldsymbol{\rho})\,. \tag{13.115}$$

Here $\boldsymbol{\rho}$ is the noise parameter vector; see (5.66) or (5.67). Note that $\mathbf{f}(\boldsymbol{\vartheta})$ is linear in the parameter matrix $\boldsymbol{\Theta}$. This can be exploited to simplify the minimization problem. The parameter estimate is defined as

$$(\hat{\boldsymbol{\Theta}}, \hat{\boldsymbol{\rho}}) \stackrel{\Delta}{=} \arg\min_{\boldsymbol{\vartheta}} V(\boldsymbol{\Theta}, \boldsymbol{\rho})\,, \tag{13.116}$$

and can be arranged by first minimizing V with respect to $\boldsymbol{\Theta}$. Set

$$\hat{\boldsymbol{\Theta}}(\boldsymbol{\rho}) = \arg\min_{\boldsymbol{\Theta}}\ V(\boldsymbol{\Theta}, \boldsymbol{\rho})\,, \tag{13.117}$$

$$\hat{\boldsymbol{\rho}} = \arg\min_{\boldsymbol{\rho}} V_2(\boldsymbol{\rho}) \stackrel{\Delta}{=} \arg\min_{\boldsymbol{\rho}} V(\hat{\boldsymbol{\Theta}}(\boldsymbol{\rho}), \boldsymbol{\rho})\,. \tag{13.118}$$

This means that $V_2(\boldsymbol{\rho})$ is a concentrated loss function. The minimization in (13.117) is simple, as V depends quadratically on $\boldsymbol{\Theta}$. The minimization problem in (13.118) is simpler than that in the original formulation (13.116), as the number of unknown variables is significantly reduced.

In what follows some details for the optimization problem (13.116) are given. Write the criterion $V(\boldsymbol{\Theta}, \boldsymbol{\rho})$ as, see (5.71)

$$V(\boldsymbol{\Theta}, \boldsymbol{\rho}) = \mathrm{tr}\left[\mathbf{W}(\mathbf{G} - \mathbf{H}\boldsymbol{\Theta})\mathbf{Z}(\mathbf{G} - \mathbf{H}\boldsymbol{\Theta})^T\right] . \tag{13.119}$$

Here let, by using general notations, $\mathbf{H}$ be an $k \times m$ matrix, $\boldsymbol{\Theta}$ an $m \times n$ matrix, $\mathbf{G}$ an $k \times n$ matrix, $\mathbf{Z}$ an $n \times n$ matrix, and $\mathbf{W}$ an $k \times k$ matrix. Further, introduce $\mathbf{e}_i$ as an m-dimensional unit vector, and $\mathbf{f}_j$ as an n-dimensional unit vector. Then one can write

$$\frac{\partial}{\partial \boldsymbol{\Theta}_{i,j}} \boldsymbol{\Theta} = \mathbf{e}_i \mathbf{f}_j^T . \tag{13.120}$$

Now find the minimal value of $V(\boldsymbol{\Theta}, \boldsymbol{\rho})$ and the minimizing argument when the matrix $\boldsymbol{\Theta}$ is varied. Direct differentiation gives

$$\begin{aligned} 0 = \frac{\partial V}{\partial \boldsymbol{\Theta}_{i,j}} &= -2\mathrm{tr}\left[\mathbf{W}\mathbf{H}\mathbf{e}_i\mathbf{f}_j^T\mathbf{Z}(\mathbf{G} - \mathbf{H}\boldsymbol{\Theta})^T\right] \\ &= -2\mathbf{f}_j^T\mathbf{Z}(-\boldsymbol{\Theta}^T\mathbf{H}^T + \mathbf{G}^T)\mathbf{W}\mathbf{H}\mathbf{e}_i \quad \forall i, j . \end{aligned} \tag{13.121}$$

Hence one can conclude

$$\mathbf{Z}(\boldsymbol{\Theta}^T\mathbf{H}^T - \mathbf{G}^T)\mathbf{W}\mathbf{H} = \mathbf{0} ,$$

and therefore

$$\hat{\boldsymbol{\Theta}} = \left(\mathbf{H}^T\mathbf{W}\mathbf{H}\right)^{-1}\mathbf{H}^T\mathbf{W}\mathbf{G} . \tag{13.122}$$

Note that $\hat{\boldsymbol{\Theta}}$ does not depend on the weighting matrix $\mathbf{Z}$. The minimal value $V_2(\boldsymbol{\rho})$ of the criterion is easily found to be

$$\begin{aligned} V_2(\boldsymbol{\rho}) = \min_{\boldsymbol{\Theta}} V(\boldsymbol{\Theta}, \boldsymbol{\rho}) &= \mathrm{tr}\left[\mathbf{W}\left\{\mathbf{I} - \mathbf{H}\left(\mathbf{H}^T\mathbf{W}\mathbf{H}\right)^{-1}\mathbf{H}^T\mathbf{W}\right\}\mathbf{G}\mathbf{Z}\mathbf{G}^T\right. \\ &\quad \left. \times \left\{\mathbf{I} - \mathbf{H}\left(\mathbf{H}^T\mathbf{W}\mathbf{H}\right)^{-1}\mathbf{H}^T\mathbf{W}\right\}^T\right] \\ &= \mathrm{tr}\left[\left\{\mathbf{I} - \mathbf{W}\mathbf{H}\left(\mathbf{H}^T\mathbf{W}\mathbf{H}\right)^{-1}\mathbf{H}^T\right\}\mathbf{W}\right. \\ &\quad \left. \times \left\{\mathbf{I} - \mathbf{H}\left(\mathbf{H}^T\mathbf{W}\mathbf{H}\right)^{-1}\mathbf{H}^T\mathbf{W}\right\}\mathbf{G}\mathbf{Z}\mathbf{G}^T\right] \\ &= \mathrm{tr}\left[\left\{\mathbf{W} - \mathbf{W}\mathbf{H}\left(\mathbf{H}^T\mathbf{W}\mathbf{H}\right)^{-1}\mathbf{H}^T\mathbf{W}\right\}\mathbf{G}\mathbf{Z}\mathbf{G}^T\right] . \end{aligned} \tag{13.123}$$

It is illustrative to derive the results (13.122), (13.123) using completion of squares as an alternative technique. To this aim, first set

$$\boldsymbol{\Theta}^* = \left(\mathbf{H}^T\mathbf{W}\mathbf{H}\right)^{-1}\mathbf{H}^T\mathbf{W}\mathbf{G}\,. \tag{13.124}$$

Next the criterion (13.119) can be rewritten as

$$\begin{aligned} V(\boldsymbol{\Theta}, \rho) &= \mathrm{tr}\left[\mathbf{W}\left(\mathbf{G} - \mathbf{H}\boldsymbol{\Theta} + \mathbf{H}\boldsymbol{\Theta}^* - \mathbf{H}\boldsymbol{\Theta}^*\right)\mathbf{Z}\left(\mathbf{G} - \mathbf{H}\boldsymbol{\Theta} + \mathbf{H}\boldsymbol{\Theta}^* - \mathbf{H}\boldsymbol{\Theta}^*\right)^T\right] \\ &= \mathrm{tr}\left[\mathbf{W}\mathbf{H}(\boldsymbol{\Theta} - \boldsymbol{\Theta}^*)\mathbf{Z}(\boldsymbol{\Theta} - \boldsymbol{\Theta}^*)^T\mathbf{H}^T\right] \\ &\quad + \mathrm{tr}\left[\mathbf{W}(\mathbf{H}\boldsymbol{\Theta}^* - \mathbf{G})\mathbf{Z}(\mathbf{H}\boldsymbol{\Theta}^* - \mathbf{G})^T\right] \\ &\quad + 2\mathrm{tr}\left[\mathbf{W}\mathbf{H}(\boldsymbol{\Theta} - \boldsymbol{\Theta}^*)\mathbf{Z}(\mathbf{H}\boldsymbol{\Theta}^* - \mathbf{G})^T\right]\,. \end{aligned} \tag{13.125}$$

Here, the last term can be evaluated as

$$\begin{aligned} & 2\mathrm{tr}\left[\mathbf{W}\mathbf{H}(\boldsymbol{\Theta} - \boldsymbol{\Theta}^*)\mathbf{Z}(\mathbf{H}\boldsymbol{\Theta}^* - \mathbf{G})^T\right] \\ &= 2\mathrm{tr}\left[\left\{\mathbf{G}^T\mathbf{W}\mathbf{H}\left(\mathbf{H}^T\mathbf{W}\mathbf{H}\right)^{-1}\mathbf{H}^T - \mathbf{G}^T\right\}\mathbf{W}\mathbf{H}(\boldsymbol{\Theta} - \boldsymbol{\Theta}^*)\mathbf{Z}\right] \\ &= 2\mathrm{tr}\left[\mathbf{G}^T\left\{\mathbf{W}\mathbf{H}\left(\mathbf{H}^T\mathbf{W}\mathbf{H}\right)^{-1}\mathbf{H}^T\mathbf{W}\mathbf{H} - \mathbf{W}\mathbf{H}\right\}(\boldsymbol{\Theta} - \boldsymbol{\Theta}^*)\mathbf{Z}\right] \\ &= 0\,. \end{aligned}$$

From this and (13.125) it follows directly that $V(\boldsymbol{\Theta}, \rho)$ is minimized with respect to $\boldsymbol{\Theta}$ for $\boldsymbol{\Theta} = \boldsymbol{\Theta}^*$.

13.B Handling Overdetermined Systems of Equations

This section treats the situation where some of the generalized IV equations are required to hold exactly and others approximately for the parameter estimates. The Algorithms 13.1 and 13.2 will be compared.

For this aim set

$$\mathbf{f}(\boldsymbol{\vartheta}) = \begin{pmatrix} \mathbf{f}_1(\boldsymbol{\vartheta}) \\ \mathbf{f}_2(\boldsymbol{\vartheta}) \end{pmatrix}, \tag{13.126}$$

$$\mathbf{S} = \frac{\partial \mathbf{f}}{\partial \boldsymbol{\vartheta}} = \begin{pmatrix} \frac{\partial \mathbf{f}_1}{\partial \boldsymbol{\vartheta}} \\ \\ \frac{\partial \mathbf{f}_2}{\partial \boldsymbol{\vartheta}} \end{pmatrix} \triangleq \begin{pmatrix} \mathbf{F}_1 \\ \mathbf{F}_2 \end{pmatrix}, \tag{13.127}$$

where $\mathbf{F}_1$ is $n_1 \times n$ and of rank n_1, and $\mathbf{S}$ has rank n. The matrix $\mathbf{S}$ is a form of sensitivity matrix.

The equations governing the parameter estimates are for Algorithm 13.1

$$\begin{aligned} &\mathbf{f}_1(\boldsymbol{\vartheta}) = \mathbf{0} \,, \\ &\mathbf{f}_2^T(\boldsymbol{\vartheta})\mathbf{W}_2\mathbf{F}_2(\boldsymbol{\vartheta}) + \boldsymbol{\lambda}^T\mathbf{F}_1(\boldsymbol{\vartheta}) = \mathbf{0} \,, \end{aligned} \tag{13.128}$$

while Algorithm 13.2 leads to

$$\mathbf{0} = \alpha\mathbf{f}_1^T(\boldsymbol{\vartheta})\mathbf{W}_1\mathbf{F}_1(\boldsymbol{\vartheta}) + \mathbf{f}_2^T(\boldsymbol{\vartheta})\mathbf{W}_2\mathbf{F}_2(\boldsymbol{\vartheta}) \,, \tag{13.129}$$

where α should be chosen large.

The sensitivity matrix $\mathbf{S} = \frac{\partial \mathbf{f}}{\partial \boldsymbol{\vartheta}}$ appears also in (14.90). Next it will be shown that the factor that matters for the covariance matrix of the parameter estimates is

$$\mathbf{G} \triangleq \left(\mathbf{S}^T\mathbf{W}\mathbf{S}\right)^{-1}\mathbf{S}^T\mathbf{W} \,. \tag{13.130}$$

Equations (13.128) and (13.129) are compatible if one sets

$$\boldsymbol{\lambda}^T = \alpha\mathbf{f}_1^T(\boldsymbol{\vartheta})\mathbf{W}_1 \,. \tag{13.131}$$

It then makes sense that

$$\lim_{\alpha\to\infty} \mathbf{f}_1(\boldsymbol{\vartheta}) = \mathbf{0} \,, \tag{13.132}$$

as $\boldsymbol{\lambda}$ does not depend on α.

Using (13.129) and the associated weighting

$$\mathbf{W} = \begin{pmatrix} \alpha\mathbf{W}_1 & \mathbf{0} \\ \mathbf{0} & \mathbf{W}_2 \end{pmatrix} \tag{13.133}$$

in the expression (13.130) gives

$$\mathbf{G} = \mathbf{G}(\alpha) = \left(\alpha\mathbf{F}_1^T\mathbf{W}_1\mathbf{F}_1 + \mathbf{F}_2^T\mathbf{W}_2\mathbf{F}_2\right)^{-1}\left(\alpha\mathbf{F}_1^T\mathbf{W}_1 \;\; \mathbf{F}_2^T\mathbf{W}_2\right) \,. \tag{13.134}$$

It is of interest to examine the limit of $\mathbf{G}(\alpha)$ when α tends to infinity. Due to the nature of the problem, one expects that the limit exists and that it is independent of $\mathbf{W}_1$. Note that both terms in the inverse appearing in (13.134) can be singular, and thus the matrix inversion lemma cannot be applied in a standard fashion to examine the convergence.

Before examining the limit using an algebraic approach, consider the stationary points obtained in (13.128) for Algorithm 13.1. After linearization, these equations may be written as

$$\begin{aligned} &\mathbf{F}_1\tilde{\boldsymbol{\vartheta}} + \mathbf{y}_1 = \mathbf{0} \,, \\ &\mathbf{F}_2^T\mathbf{W}_2\mathbf{F}_2\tilde{\boldsymbol{\vartheta}} + \mathbf{F}_1^T\tilde{\boldsymbol{\lambda}} + \mathbf{F}_2^T\mathbf{W}_2\mathbf{y}_2 = \mathbf{0} \,. \end{aligned} \tag{13.135}$$

From this equation one can find the parameter error $\tilde{\boldsymbol{\vartheta}}$ as

$$\tilde{\boldsymbol{\vartheta}} = -\left(\mathbf{I}\ \mathbf{0}\right)\begin{pmatrix} \mathbf{F}_1 & \mathbf{0} \\ \mathbf{F}_2^T\mathbf{W}_2\mathbf{F}_2 & \mathbf{F}_1^T \end{pmatrix}^{-1}\begin{pmatrix} \mathbf{I} & \mathbf{0} \\ \mathbf{0} & \mathbf{F}_2^T\mathbf{W}_2 \end{pmatrix}\begin{pmatrix} \mathbf{y}_1 \\ \mathbf{y}_2 \end{pmatrix}, \tag{13.136}$$

and one would therefore expect that the matrix in front of the $\mathbf{y}$'s relates to $\lim_{\alpha\to\infty}\mathbf{G}(\alpha)$. It will now be shown that this is indeed the case.

Lemma 13.1 *Under the dimension and rank assumptions of* $\mathbf{F}_1$ *and* $\mathbf{F}_2$ *it holds for any fixed* α *that*

$$\begin{aligned} \mathbf{G}(\alpha) &= \left(\alpha\mathbf{F}_1^T\mathbf{W}_1\mathbf{F}_1 + \mathbf{F}_2^T\mathbf{W}_2\mathbf{F}_2\right)^{-1}\left(\alpha\mathbf{F}_1^T\mathbf{W}_1\ \mathbf{F}_2^T\mathbf{W}_2\right) \\ &= \left(\mathbf{I}\ \mathbf{0}\right)\begin{pmatrix} \mathbf{F}_1 & -\mathbf{W}_1^{-1}/\alpha \\ \mathbf{F}_2^T\mathbf{W}_2\mathbf{F}_2 & \mathbf{F}_1^T \end{pmatrix}^{-1}\begin{pmatrix} \mathbf{I} & \mathbf{0} \\ \mathbf{0} & \mathbf{F}_2^T\mathbf{W}_2 \end{pmatrix}. \end{aligned} \tag{13.137}$$

Proof The right-hand side of (13.137) can be evaluated as

$$\begin{aligned} \text{RHS} &= \left(\mathbf{I}\ \mathbf{0}\right)\left(\begin{pmatrix} \mathbf{F}_1 & -\mathbf{W}_1^{-1}/\alpha \\ \mathbf{F}_2^T\mathbf{W}_2\mathbf{F}_2 & \mathbf{F}_1^T \end{pmatrix}\begin{pmatrix} \mathbf{0} & \mathbf{I} \\ \mathbf{I} & \mathbf{0} \end{pmatrix}\begin{pmatrix} \mathbf{0} & \mathbf{I} \\ \mathbf{I} & \mathbf{0} \end{pmatrix}\right)^{-1}\begin{pmatrix} \mathbf{I} & \mathbf{0} \\ \mathbf{0} & \mathbf{F}_2^T\mathbf{W}_2 \end{pmatrix} \\ &= \left(\mathbf{I}\ \mathbf{0}\right)\begin{pmatrix} \mathbf{0} & \mathbf{I} \\ \mathbf{I} & \mathbf{0} \end{pmatrix}\begin{pmatrix} -\mathbf{W}_1^{-1}/\alpha & \mathbf{F}_1 \\ \mathbf{F}_1^T & \mathbf{F}_2^T\mathbf{W}_2\mathbf{F}_2 \end{pmatrix}^{-1}\begin{pmatrix} \mathbf{I} & \mathbf{0} \\ \mathbf{0} & \mathbf{F}_2^T\mathbf{W}_2 \end{pmatrix} \\ &= \left(\mathbf{0}\ \mathbf{I}\right)\left[\begin{pmatrix} -\alpha\mathbf{W}_1 & \mathbf{0} \\ \mathbf{0} & \mathbf{0} \end{pmatrix} + \begin{pmatrix} \alpha\mathbf{W}_1\mathbf{F}_1 \\ \mathbf{I} \end{pmatrix}\right. \\ &\quad \left.\times\left(\alpha\mathbf{F}_1^T\mathbf{W}_1\mathbf{F}_1 + \mathbf{F}_2^T\mathbf{W}_2\mathbf{F}_2\right)^{-1}\left(\alpha\mathbf{F}_1^T\mathbf{W}_1\ \mathbf{I}\right)\right]\begin{pmatrix} \mathbf{I} & \mathbf{0} \\ \mathbf{0} & \mathbf{F}_2^T\mathbf{W}_2 \end{pmatrix} \\ &= \left(\alpha\mathbf{F}_1^T\mathbf{W}_1\mathbf{F}_1 + \mathbf{F}_2^T\mathbf{W}_2\mathbf{F}_2\right)^{-1}\left(\alpha\mathbf{F}_1^T\mathbf{W}_1\ \mathbf{F}_2^T\mathbf{W}_2\right) \\ &= \text{LHS}\,. \end{aligned}$$

■

Remark 13.3 When $\alpha \to \infty$ it holds

$$\lim_{\alpha\to\infty}\mathbf{G}(\alpha) = \left(\mathbf{I}\ \mathbf{0}\right)\begin{pmatrix} \mathbf{F}_1 & \mathbf{0} \\ \mathbf{F}_2^T\mathbf{W}_2\mathbf{F}_2 & \mathbf{F}_1^T \end{pmatrix}^{-1}\begin{pmatrix} \mathbf{I} & \mathbf{0} \\ \mathbf{0} & \mathbf{F}_2^T\mathbf{W}_2 \end{pmatrix}, \tag{13.138}$$

which obviously does not depend on $\mathbf{W}_1$. Further, the right-hand side of (13.138) is precisely the matrix appearing in (13.136). ■

To express the inverse in (13.138) is relatively complicated in the general case. First rewrite $\mathbf{G}$ by making some block permutations,

$$\mathbf{G} = (\mathbf{I}\ \mathbf{0})\left[\begin{pmatrix}\mathbf{0} & \mathbf{I}\\ \mathbf{I} & \mathbf{0}\end{pmatrix}\begin{pmatrix}\mathbf{0} & \mathbf{I}\\ \mathbf{I} & \mathbf{0}\end{pmatrix}\begin{pmatrix}\mathbf{F}_1 & \mathbf{0}\\ \mathbf{F}_2^T\mathbf{W}_2\mathbf{F}_2 & \mathbf{F}_1^T\end{pmatrix}\right.$$
$$\left.\times\begin{pmatrix}\mathbf{0} & \mathbf{I}\\ \mathbf{I} & \mathbf{0}\end{pmatrix}\begin{pmatrix}\mathbf{0} & \mathbf{I}\\ \mathbf{I} & \mathbf{0}\end{pmatrix}\right]^{-1}\begin{pmatrix}\mathbf{I} & \mathbf{0}\\ \mathbf{0} & \mathbf{F}_2^T\mathbf{W}_2\end{pmatrix}$$
$$= (\mathbf{0}\ \mathbf{I})\begin{pmatrix}\mathbf{F}_1^T & \mathbf{F}_2^T\mathbf{W}_2\mathbf{F}_2\\ \mathbf{0} & \mathbf{F}_1\end{pmatrix}^{-1}\begin{pmatrix}\mathbf{0} & \mathbf{F}_2^T\mathbf{W}_2\\ \mathbf{I} & \mathbf{0}\end{pmatrix}. \tag{13.139}$$

To proceed, the inverse in (13.139) needs to be rewritten. For that aim apply Lemma A.6 to the matrix in (13.139). It is needed to verify that the matrix

$$\mathbf{D}_0 \stackrel{\Delta}{=} \mathbf{P} + \mathbf{P}_\perp\mathbf{F}_2^T\mathbf{W}_2\mathbf{F}_2\mathbf{P}_\perp\ , \tag{13.140}$$

is non-singular, where

$$\mathbf{P} = \mathbf{F}_1\mathbf{F}_1^\dagger,\quad \mathbf{P}_\perp = \mathbf{I} - \mathbf{P}\ . \tag{13.141}$$

Clearly, by construction $\mathbf{D}_0$ is symmetric and nonnegative definite. Further,

$$\begin{aligned}\mathbf{x}^T\mathbf{D}_0\mathbf{x} = \mathbf{0},\ &\Rightarrow \mathbf{x}^T\left[\mathbf{P} + \mathbf{P}_\perp\mathbf{F}_2^T\mathbf{W}_2\mathbf{F}_2\mathbf{P}_\perp\right]\mathbf{x} = \mathbf{0}\ ,\\ &\Rightarrow \mathbf{x}^T\mathbf{P}\mathbf{x} = 0,\ \ \mathbf{F}_2\mathbf{P}_\perp\mathbf{x} = \mathbf{0}\ ,\\ &\Rightarrow \mathbf{F}_1\mathbf{x} = \mathbf{0},\ \ \mathbf{F}_2\left[\mathbf{I} - \mathbf{F}_1^T\left(\mathbf{F}_1\mathbf{F}_1^T\right)^{-1}\mathbf{F}_1\right]\mathbf{x} = \mathbf{0}\ ,\\ &\Rightarrow \mathbf{F}_1\mathbf{x} = \mathbf{0},\ \ \mathbf{F}_2\mathbf{x} = \mathbf{0},\ \ \Rightarrow \mathbf{F}\mathbf{x} = \mathbf{0}\ .\end{aligned}$$

Thus $\mathbf{D}_0$ is non-singular. Then according to Lemma A.6, (13.139) can be expressed as

$$\mathbf{G} = (\mathbf{0}\ \mathbf{I})\begin{pmatrix}\mathbf{H}_{11} & \mathbf{H}_{12}\\ \mathbf{H}_{21} & \mathbf{H}_{22}\end{pmatrix}\begin{pmatrix}\mathbf{0} & \mathbf{F}_2^T\mathbf{W}_2\\ \mathbf{I} & \mathbf{0}\end{pmatrix}, \tag{13.142}$$

where

$$\begin{aligned}\mathbf{H}_{11} &= \left(\mathbf{F}_1\mathbf{F}_1^T\right)^{-1}\mathbf{F}_1(\mathbf{I} - \mathbf{F}_2^T\mathbf{W}_2\mathbf{F}_2\mathbf{D})\ ,\\ \mathbf{H}_{12} &= -\left(\mathbf{F}_1\mathbf{F}_1^T\right)^{-1}\mathbf{F}_1\mathbf{F}_2^T\mathbf{W}_2\mathbf{F}_2(\mathbf{I} - \mathbf{D}\mathbf{F}_2^T\mathbf{W}_2\mathbf{F}_2)\mathbf{F}_1^T\left(\mathbf{F}_1\mathbf{F}_1^T\right)^{-1}\ ,\\ \mathbf{H}_{21} &= \mathbf{D}\ ,\\ \mathbf{H}_{22} &= (\mathbf{I} - \mathbf{D}\mathbf{F}_2^T\mathbf{W}_2\mathbf{F}_2)\mathbf{F}_1^T\left(\mathbf{F}_1\mathbf{F}_1^T\right)^{-1}\ ,\\ \mathbf{D} &= \mathbf{D}_0^{-1}\mathbf{P}_\perp\ .\end{aligned}$$

Straightforward multiplications in (13.142) then lead to

$$\mathbf{G} = \left((\mathbf{I} - \mathbf{D}\mathbf{F}_2^T\mathbf{W}_2\mathbf{F}_2)\mathbf{F}_1^T\left(\mathbf{F}_1\mathbf{F}_1^T\right)^{-1}\ \ \mathbf{D}\mathbf{F}_2^T\mathbf{W}_2\right). \tag{13.143}$$

For the particular case when rank $\mathbf{F}_2 = n$ (which requires dim $\mathbf{f}_2 \geq n$) a simpler expression is possible:

Lemma 13.2 *Consider the expression*

$$\mathbf{G}(\alpha) = \left[\alpha \mathbf{F}_1^T \mathbf{W}_1 \mathbf{F}_1 + \mathbf{F}_2^T \mathbf{W}_2 \mathbf{F}_2\right]^{-1} \left(\alpha \mathbf{F}_1^T \mathbf{W}_1 \ \mathbf{F}_2^T \mathbf{W}_2\right) , \tag{13.144}$$

where $\mathbf{F}_1$ *is an* $n_1 \times n$ *matrix of rank* n_1, $\mathbf{F}_2$ *an* $n_2 \times n$ *matrix of rank* n, $\mathbf{W}_1$ *an* $n_1 \times n_1$ *matrix,* $\mathbf{W}_2$ *an* $n_2 \times n_2$ *matrix, and* $\mathbf{W}_1$ *and* $\mathbf{W}_2$ *are positive definite. Set*

$$\mathbf{H} = \mathbf{F}_2^T \mathbf{W}_2 \mathbf{F}_2 . \tag{13.145}$$

Then it holds that the following limit exists and

$$\lim_{\alpha \to \infty} \mathbf{G}(\alpha) \stackrel{\Delta}{=} \left(\mathbf{G}_1 \ \mathbf{G}_2\right) , \tag{13.146}$$

$$\mathbf{G}_1 = \mathbf{H}^{-1}\mathbf{F}_1^T \left(\mathbf{F}_1 \mathbf{H}^{-1} \mathbf{F}_1^T\right)^{-1} , \tag{13.147}$$

$$\mathbf{G}_2 = \left[\mathbf{I} - \mathbf{H}^{-1}\mathbf{F}_1^T \left(\mathbf{F}_1 \mathbf{H}^{-1} \mathbf{F}_1^T\right)^{-1} \mathbf{F}_1\right] \mathbf{H}^{-1} \mathbf{F}_2^T \mathbf{W}_2 . \tag{13.148}$$

Proof Note that matrix $\mathbf{H}$ by construction is invertible. Using the matrix inversion lemma

$$\left[\alpha \mathbf{F}_1^T \mathbf{W}_1 \mathbf{F}_1 + \mathbf{F}_2^T \mathbf{W}_2 \mathbf{F}_2\right]^{-1} = \mathbf{H}^{-1} - \mathbf{H}^{-1}\mathbf{F}_1^T \left(\frac{1}{\alpha}\mathbf{W}_1^{-1} + \mathbf{F}_1 \mathbf{H}^{-1} \mathbf{F}_1^T\right)^{-1} \mathbf{F}_1 \mathbf{H}^{-1} . \tag{13.149}$$

Thus

$$\begin{aligned}
\mathbf{G}_2 &= \lim_{\alpha \to \infty} \left[\mathbf{H}^{-1} - \mathbf{H}^{-1}\mathbf{F}_1^T \left(\frac{1}{\alpha}\mathbf{W}_1^{-1} + \mathbf{F}_1 \mathbf{H}^{-1} \mathbf{F}_1^T\right)^{-1} \mathbf{F}_1 \mathbf{H}^{-1}\right] \mathbf{F}_2^T \mathbf{W}_2 \\
&= \mathbf{H}^{-1}\mathbf{F}_2^T \mathbf{W}_2 - \mathbf{H}^{-1}\mathbf{F}_1^T \left(\mathbf{F}_1 \mathbf{H}^{-1} \mathbf{F}_1^T\right)^{-1} \mathbf{F}_1 \mathbf{H}^{-1} \mathbf{F}_2^T \mathbf{W}_2 \\
&= \left[\mathbf{I} - \mathbf{H}^{-1}\mathbf{F}_1^T \left(\mathbf{F}_1 \mathbf{H}^{-1} \mathbf{F}_1^T\right)^{-1} \mathbf{F}_1\right] \mathbf{H}^{-1} \mathbf{F}_2^T \mathbf{W}_2 .
\end{aligned} \tag{13.150}$$

Similarly,

$$\begin{aligned}
\mathbf{G}_1 &= \lim_{\alpha \to \infty} \left[\alpha \mathbf{H}^{-1}\mathbf{F}_1^T \mathbf{W}_1 - \alpha \mathbf{H}^{-1}\mathbf{F}_1^T \left(\frac{1}{\alpha}\mathbf{W}_1^{-1} + \mathbf{F}_1 \mathbf{H}^{-1} \mathbf{F}_1^T\right)^{-1} \mathbf{F}_1 \mathbf{H}^{-1} \mathbf{F}_1^T \mathbf{W}_1\right] \\
&= \lim_{\alpha \to \infty} \left[\alpha \mathbf{H}^{-1}\mathbf{F}_1^T \left(\frac{1}{\alpha}\mathbf{W}_1^{-1} + \mathbf{F}_1 \mathbf{H}^{-1} \mathbf{F}_1^T\right)^{-1} \right. \\
&\quad \times \left. \left(\frac{1}{\alpha}\mathbf{W}_1^{-1} + \mathbf{F}_1 \mathbf{H}^{-1} \mathbf{F}_1^T - \mathbf{F}_1 \mathbf{H}^{-1} \mathbf{F}_1^T\right) \mathbf{W}_1\right]
\end{aligned}$$

$$= \lim_{\alpha\to\infty} \mathbf{H}^{-1}\mathbf{F}_1^T \left(\frac{1}{\alpha}\mathbf{W}_1^{-1} + \mathbf{F}_1\mathbf{H}^{-1}\mathbf{F}_1^T\right)^{-1}$$

$$= \mathbf{H}^{-1}\mathbf{F}_1^T \left(\mathbf{F}_1\mathbf{H}^{-1}\mathbf{F}_1^T\right)^{-1} . \tag{13.151}$$

■

13.C Algorithmic Aspects of CFA-Based Estimators

The loss function $V_2(\boldsymbol{\vartheta})$, (2.49), turns out to depend quadratically on some of the parameters. When applied to dynamic models as in Sect. 8.5, this can be exploited to treat the minimization problem as a separable nonlinear least squares problem.

It is convenient to split the parameter vector $\boldsymbol{\vartheta}$ into two parts:

$$\boldsymbol{\vartheta} = \begin{pmatrix} \boldsymbol{\theta} \\ \boldsymbol{\rho} \end{pmatrix}, \quad \boldsymbol{\theta} = \begin{pmatrix} a_1 \\ \vdots \\ a_{n_a} \\ b_1 \\ \vdots \\ b_{n_b} \end{pmatrix}, \quad \boldsymbol{\rho} = \begin{pmatrix} \lambda_y \\ \lambda_u \\ r_z(0) \\ \vdots \\ r_z(k) \end{pmatrix} . \tag{13.152}$$

Next exploit that $V_2(\boldsymbol{\vartheta})$ depends quadratically on $\boldsymbol{\rho}$. For this purpose, write the covariance matrix $\mathbf{R}(\boldsymbol{\vartheta})$ in the form

$$\mathbf{R} = \sum_{j=1}^{k+3} \rho_j \mathbf{J}_j . \tag{13.153}$$

Specifically, the coefficient matrices $\mathbf{J}_j$ become:

$$\mathbf{J}_1 = \begin{pmatrix} \mathbf{I}_{n_a+p_y+1} & \mathbf{0} \\ \mathbf{0} & \mathbf{0} \end{pmatrix} , \tag{13.154}$$

$$\mathbf{J}_2 = \begin{pmatrix} \mathbf{0} & \mathbf{0} \\ \mathbf{0} & \mathbf{I}_{n_b+p_u} \end{pmatrix} , \tag{13.155}$$

$$\mathbf{J}_3 = \boldsymbol{\Gamma}\mathbf{I}_{k+1}\boldsymbol{\Gamma}^T , \tag{13.156}$$

$$\mathbf{J}_j = \boldsymbol{\Gamma} \begin{pmatrix} \mathbf{0} & 1 & & \\ 1 & & \ddots & \\ & \ddots & & 1 \\ & & 1 & \end{pmatrix} \boldsymbol{\Gamma}^T, \quad j = 4, \ldots, k+3 . \tag{13.157}$$

Note that $\boldsymbol{\Gamma}$ depends on $\boldsymbol{\theta}$, and hence the coefficient matrices $\mathbf{J}_j$, $j = 3, \ldots k+3$ are also functions of $\boldsymbol{\theta}$.

The loss function can now be written as, where the dependence on $\boldsymbol{\rho}$ is emphasized,

$$\begin{aligned} V_2(\boldsymbol{\vartheta}) &= \mathrm{tr}\left[\mathbf{Q}_1\left(\hat{\mathbf{R}} - \sum_j \rho_j \mathbf{J}_j\right)\mathbf{Q}_2\left(\hat{\mathbf{R}} - \sum_\ell \rho_\ell \mathbf{J}_\ell\right)\right] \\ &= \sum_j \sum_\ell \rho_j \rho_\ell \mathrm{tr}\left[\mathbf{Q}_1\mathbf{J}_j\mathbf{Q}_2\mathbf{J}_\ell\right] - 2\sum_j \rho_j \mathrm{tr}\left[\mathbf{Q}_1\mathbf{J}_j\mathbf{Q}_2\hat{\mathbf{R}}\right] + \mathrm{tr}\left[\mathbf{Q}_1\hat{\mathbf{R}}\mathbf{Q}_2\hat{\mathbf{R}}\right]. \end{aligned} \tag{13.158}$$

It is straightforward to minimize (13.158) with respect to $\boldsymbol{\rho}_j$:

$$0 = \sum_\ell \rho_\ell 2\mathrm{tr}\left[\mathbf{Q}_1\mathbf{J}_j\mathbf{Q}_2\mathbf{J}_\ell\right] - 2\mathrm{tr}\left[\mathbf{Q}_1\mathbf{J}_j\mathbf{Q}_2\hat{\mathbf{R}}\right], \quad j = 1, \ldots, m = k+3\,. \tag{13.159}$$

This is easily expressed as a system of linear equations:

$$\begin{pmatrix} \mathrm{tr}\,(\mathbf{Q}_1\mathbf{J}_1\mathbf{Q}_2\mathbf{J}_1) & \ldots & \mathrm{tr}\,(\mathbf{Q}_1\mathbf{J}_1\mathbf{Q}_2\mathbf{J}_m) \\ \vdots & & \\ \mathrm{tr}\,(\mathbf{Q}_1\mathbf{J}_m\mathbf{Q}_2\mathbf{J}_1) & \ldots & \mathrm{tr}\,(\mathbf{Q}_1\mathbf{J}_m\mathbf{Q}_2\mathbf{J}_m) \end{pmatrix} \begin{pmatrix} \rho_1 \\ \vdots \\ \rho_m \end{pmatrix} = \begin{pmatrix} \mathrm{tr}\left(\mathbf{Q}_1\mathbf{J}_1\mathbf{Q}_2\hat{\mathbf{R}}\right) \\ \vdots \\ \mathrm{tr}\left(\mathbf{Q}_1\mathbf{J}_m\mathbf{Q}_2\hat{\mathbf{R}}\right) \end{pmatrix}, \tag{13.160}$$

which can be compactly written as

$$\mathbf{A}(\boldsymbol{\theta})\rho = \mathbf{b}(\boldsymbol{\theta})\,. \tag{13.161}$$

As $\{\mathbf{J}_k\}$ depends on $\boldsymbol{\theta}$, so will $\mathbf{A}$ and $\mathbf{b}$.

The loss function (13.158) can now be written as

$$V_2(\boldsymbol{\vartheta}) = \boldsymbol{\rho}^T\mathbf{A}\boldsymbol{\rho} - 2\boldsymbol{\rho}^T\mathbf{b} + \mathrm{tr}\left[\mathbf{Q}_1\hat{\mathbf{R}}\mathbf{Q}_2\hat{\mathbf{R}}\right]. \tag{13.162}$$

By minimizing over $\boldsymbol{\rho}$ one gets the concentrated loss function, which depends on $\boldsymbol{\theta}$ only, as

$$\begin{aligned} \bar{V}_2(\boldsymbol{\theta}) &= \min_{\rho} V_2(\boldsymbol{\rho}) \\ &= \mathrm{tr}\left[\mathbf{Q}_1\hat{\mathbf{R}}\mathbf{Q}_2\hat{\mathbf{R}}\right] - \mathbf{b}^T(\boldsymbol{\theta})\mathbf{A}^{-1}(\boldsymbol{\theta})\mathbf{b}(\boldsymbol{\theta})\,. \end{aligned} \tag{13.163}$$

To minimize the criterion $\bar{V}_2(\boldsymbol{\theta})$ a numerical search method has to be applied.

Chapter 14
Asymptotic Distributions

Asymptotic distribution of the parameter estimates includes the covariance matrix, which can be taken as a measure of accuracy. Several aspects on the asymptotic distributions are treated in this chapter. First, some general considerations are provided in Sect. 14.1. In Sect. 14.2 general CFA methods are treated. Then in the following sections, the different estimation methods introduced in the book are analyzed, and explicit expressions for the covariance matrices are given. The distributions for IV estimators is the topic of Sect. 14.3, while Sect. 14.4 gives similar results for the GIVE. Covariance matching methods are covered in Sect. 14.5, and prediction error methods in Sect. 14.6. Results for periodic data are presented in Sect. 14.7. An analysis of the Cramér–Rao lower bound (CRB) for the frequency domain ML problem is given in Sect. 14.8. Section 14.9 treats extensions and bibliographical notes.

14.1 Background and General Considerations

To analyze at what rate a consistent estimate approaches the true value as the number of data points grows to infinity, consider the asymptotic covariance matrix of the parameter estimate. Often, this matrix will be regarded here as a measure of the accuracy of the estimate.

Consider in particular the general formulation (13.2) of an estimate:

$$\mathbf{f}(\hat{\boldsymbol{\vartheta}}_N, \hat{\mathbf{r}}_N) = \mathbf{0} . \tag{14.1}$$

In the limiting case ($N \to \infty$), the true parameter vector $\boldsymbol{\vartheta}_0$ is the desired solution, so

$$\mathbf{f}(\boldsymbol{\vartheta}_0, \mathbf{r}) = \mathbf{0} . \tag{14.2}$$

T. Söderström, *Errors-in-Variables Methods in System Identification*, Communications and Control Engineering,
https://doi.org/10.1007/978-3-319-75001-9_14

By subtracting (14.2) from (14.1) and making a series expansion one finds that for small deviations it holds

$$\mathbf{f}_{\vartheta}\left(\hat{\boldsymbol{\vartheta}}_N - \boldsymbol{\vartheta}_0\right) + \mathbf{f}_{\mathbf{r}}\left(\hat{\mathbf{r}}_N - \mathbf{r}_0\right) = \mathbf{0}\,, \tag{14.3}$$

where

$$\mathbf{f}_{\vartheta} = \frac{\partial \mathbf{f}}{\partial \boldsymbol{\vartheta}}, \quad \mathbf{f}_{\mathbf{r}} = \frac{\partial \mathbf{f}}{\partial \mathbf{r}} \tag{14.4}$$

evaluated for the true parameter values. Under weak assumptions $\hat{\mathbf{r}}_N$ converges to $\mathbf{r}_0$ at a rate $O(1/\sqrt{N})$, in the sense that the limit

$$\mathbf{R} = \lim_{N\to\infty} N\mathsf{E}\left\{(\hat{\mathbf{r}} - \mathbf{r})(\hat{\mathbf{r}} - \mathbf{r})^T\right\} \tag{14.5}$$

exists and is finite. Next assume that $\mathbf{f}_{\vartheta}$, which is a 'tall' or a square matrix, has full column rank. If Eq. (14.1) is solved in an unweighted least squares sense, this applies also to (14.3), and it follows that (up to higher-order terms)

$$\sqrt{N}\left(\hat{\boldsymbol{\vartheta}}_N - \boldsymbol{\vartheta}_0\right) = -\mathbf{f}_{\vartheta}^{\dagger}\mathbf{f}_{\mathbf{r}}\sqrt{N}\left(\hat{\mathbf{r}}_N - \mathbf{r}\right)\,. \tag{14.6}$$

This implies that the asymptotic normalized covariance matrix of the parameter estimates $\hat{\boldsymbol{\vartheta}}_N$ can be computed as

$$\begin{aligned}\mathbf{P} &\triangleq \lim_{N\to\infty} N\mathrm{cov}\left(\hat{\boldsymbol{\vartheta}}_N\right) = \lim_{N\to\infty} N\mathsf{E}\left\{\left(\hat{\boldsymbol{\vartheta}}_N - \boldsymbol{\vartheta}_0\right)\left(\hat{\boldsymbol{\vartheta}}_N - \boldsymbol{\vartheta}_0\right)^T\right\} \\ &= \mathbf{f}_{\vartheta}^{\dagger}\mathbf{f}_{\mathbf{r}}\mathbf{R}\left(\mathbf{f}_{\vartheta}^{\dagger}\mathbf{f}_{\mathbf{r}}\right)^T\,. \end{aligned} \tag{14.7}$$

The above style of analysis is applicable for estimates that can be written in the form (14.1). This does not mean that it is always the most convenient way to carry out the analysis.

For prediction error methods, the estimate cannot be written in the form (14.1), but is determined as the global minimum point of a loss function. A similar analysis as above can be applied though, by linearizing the gradient of the loss function $V_N(\boldsymbol{\vartheta})$. It holds, when the estimate $\hat{\boldsymbol{\vartheta}}$ is close to the true value $\boldsymbol{\vartheta}_0$, neglecting higher-order terms:

$$\begin{aligned}\mathbf{0} = V_N'(\hat{\boldsymbol{\vartheta}}_N) &\approx V_N'(\boldsymbol{\vartheta}_0) + V_N''(\boldsymbol{\vartheta}_0)\left(\hat{\boldsymbol{\vartheta}}_N - \boldsymbol{\vartheta}_0\right) \\ &\approx V_N'(\boldsymbol{\vartheta}_0) + V_\infty''(\boldsymbol{\vartheta}_0)\left(\hat{\boldsymbol{\vartheta}}_N - \boldsymbol{\vartheta}_0\right)\,. \end{aligned} \tag{14.8}$$

Here $V_\infty(\boldsymbol{\vartheta}) = \lim_{N\to\infty}(\boldsymbol{\vartheta})$, and $V_\infty''(\boldsymbol{\vartheta}_0)$ denotes its Hessian evaluated for $\boldsymbol{\vartheta} = \boldsymbol{\vartheta}_0$.

Similar to (14.5) the term $V_N'(\boldsymbol{\vartheta}_0)$ is of order $O(1/\sqrt{N})$, so one may in this case write

$$\mathbf{P} = \lim_{N\to\infty} N\text{cov}\left(\hat{\boldsymbol{\vartheta}}_N\right) = \lim_{N\to\infty} N\mathsf{E}\left\{\left(\hat{\boldsymbol{\vartheta}}_N - \boldsymbol{\vartheta}_0\right)\left(\hat{\boldsymbol{\vartheta}}_N - \boldsymbol{\vartheta}_0\right)^T\right\}$$
$$= [V''_\infty(\boldsymbol{\vartheta}_0)]^{-1}\left[\lim_{N\to\infty} N\mathsf{E}\left\{\left(V'_N(\boldsymbol{\vartheta}_0)\right)^T V'_N(\boldsymbol{\vartheta}_0)\right\}\right][V''_\infty(\boldsymbol{\vartheta}_0)]^{-1} . \quad (14.9)$$

In some cases, notably for IV and GIVE, it is possible to utilize the linearization argument in a slightly different fashion to arrive at expressions of the form (again after neglecting higher-order terms)

$$\hat{\boldsymbol{\vartheta}}_N - \boldsymbol{\vartheta}_0 = \mathbf{F}\frac{1}{N}\sum_{t=1}^{N}\mathbf{x}(t)v(t) , \quad (14.10)$$

where $\mathbf{F}$ is constant matrix, and $\mathbf{x}(t)$ and $v(t)$ are stationary stochastic processes. It turns out that the right hand side of (14.10) is of order $O(1/\sqrt{N})$ so an expression for the asymptotic normalized covariance matrix, $\lim_{N\to\infty} N\text{cov}(\hat{\boldsymbol{\vartheta}})$, can be derived.

Remark 14.1 It is important to note the general assumption in this analysis. The asymptotic case is considered, and it is assumed that the estimate is consistent, so that the estimate $\hat{\boldsymbol{\theta}}$ is close to its true value. The aim of the analysis is to obtain a measure of the size of this deviation, expressed in form of the covariance matrix of the estimation error. For technical reasons, a normalized case (the covariance matrix is multiplied with N) is considered for the asymptotic case ($N \to \infty$). Failing to realize this basic assumption, can no doubt lead to misunderstandings and wrong conclusions, as exemplified in Fan and Luo (2010). ■

The purpose of this chapter is to analyze the covariance matrix of the parameter error $\hat{\boldsymbol{\vartheta}} - \boldsymbol{\vartheta}_0$. Trivially, the error will depend on the disturbances, and it will therefore be necessary to introduce some assumptions about the statistical properties of the disturbances. To this aim, note from the general assumption **AN1** that the disturbance $v(t)$ in (6.8) is a stationary stochastic process, and introduce an innovations description

$$v(t) = H(q^{-1})e(t), \quad \mathsf{E}\{e(t)e(s)\} = \lambda\delta_{t,s}, \quad H(q^{-1}) = 1 + \sum_{k=1}^{\infty} h_k q^{-k} . \quad (14.11)$$

Note that this is equivalent to a spectral factorization of the disturbance spectrum $\phi_v(\omega)$. Assume that coefficients h_k decay at an exponential rate; that is, there is a constant C and a number α, $0 \le \alpha < 1$, such that

$$|h_k| \le C\alpha^k, \forall k \ge 0 . \quad (14.12)$$

Example 14.1 Assume that both $\tilde{u}(t)$ and $\tilde{y}(t)$ are white noise, so Assumption **AN3c** applies. Then the spectrum $\phi_v(\omega)$ is easily obtained as

$$\begin{aligned}\phi_v(\omega) &= |A(e^{i\omega})|^2\lambda_y + |B(e^{i\omega})|^2\lambda_u \\ &= |H(e^{i\omega})|^2\lambda \,,\end{aligned} \tag{14.13}$$

which also demonstrates how $H(q^{-1})$ and λ are obtained by spectral factorization. In this particular case, $H(q^{-1})$ will be a FIR filter, with order $\max(n_a, n_b - 1)$.

Should instead the output noise $\tilde{y}(t)$ be an ARMA process (**AN3b** applies), say

$$\tilde{y}(t) = F(q^{-1})e_y(t) \,, \tag{14.14}$$

then (14.13) modifies to

$$\begin{aligned}\phi_v(\omega) &= |A(e^{i\omega})|^2|F(e^{i\omega})|^2\lambda_y + |B(e^{i\omega})|^2\lambda_u \\ &= |H(e^{i\omega})|^2\lambda \,.\end{aligned} \tag{14.15}$$

■

14.2 CFA-Based Parameter Estimates

The aim in this section is to investigate the accuracy of the parameter estimates for general CFA estimators. In particular the normalized parameter error $\sqrt{N}\left(\hat{\boldsymbol{\vartheta}}_N - \boldsymbol{\vartheta}_0\right)$ and its covariance matrix in case N is large are examined.

First note that $\hat{\boldsymbol{\vartheta}}_N \to \boldsymbol{\vartheta}_0$, as $N \to \infty$. Next make a series expansion of the gradient of the loss function as in (14.8) leading to, in the usual way, the asymptotic Gaussian distribution and the asymptotic (for large N) covariance matrix of the parameter estimates, Ljung (1999), Söderström and Stoica (1989). For CFA-based estimation, one gets, precisely as in (14.9), the expression

$$\sqrt{N}\left(\hat{\boldsymbol{\vartheta}}_N - \boldsymbol{\vartheta}_0\right) \xrightarrow{\text{dist}} \mathscr{N}(\mathbf{0}, \mathbf{P}) \,, \tag{14.16}$$

$$\mathbf{P} = \left[V''_\infty(\boldsymbol{\vartheta}_0)\right]^{-1}\left[\lim_{N\to\infty} N\mathsf{E}\left\{\left(V'_N(\boldsymbol{\vartheta}_0)\right)^T V'_N(\boldsymbol{\vartheta}_0)\right\}\right]\left[V''_\infty(\boldsymbol{\vartheta}_0)\right]^{-1} . \tag{14.17}$$

Next one has to find the matrices in (14.17). This is done by successive differentiation. Each of the possible loss functions described in Sect. 2.2.2 will be treated below.

For general purposes, introduce the notations

$$\mathbf{R}_i = \frac{\partial}{\partial \boldsymbol{\vartheta}_i}\mathbf{R}(\boldsymbol{\vartheta}), \quad \mathbf{R}_{ij} = \frac{\partial^2}{\partial \boldsymbol{\vartheta}_i \partial \boldsymbol{\vartheta}_j}\mathbf{R}(\boldsymbol{\vartheta}) \,, \tag{14.18}$$

$$\mathbf{R}_{\vartheta} = \left(\text{vec}(\mathbf{R}_1) \ \ldots \ \text{vec}(\mathbf{R}_{n_\vartheta})\right) \,, \tag{14.19}$$

$$\mathbf{C}_r = \lim_{N\to\infty} N\mathsf{E}\left\{\tilde{\mathbf{r}}(\boldsymbol{\vartheta})\tilde{\mathbf{r}}^T(\boldsymbol{\vartheta})\right\} . \tag{14.20}$$

A derivation of expressions for $\mathbf{C}_r$ is found in Sect. 14.2.5 and Appendix 14.A.

Using the relation (2.52) it follows that

$$\mathbf{R}_{\vartheta} = -\mathbf{F}\mathbf{X}_{\vartheta} , \tag{14.21}$$

$$\mathbf{X}_{\vartheta} = \left(\frac{\partial \mathbf{x}}{\partial \vartheta_1} \cdots \frac{\partial \mathbf{x}}{\partial \vartheta_{n_\vartheta}} \right) . \tag{14.22}$$

Further, write the sample covariance matrix $\hat{\mathbf{R}}_N$ as

$$\hat{\mathbf{R}}_N = \frac{1}{N}\sum_{t=1}^{N} \varphi(t)\varphi^T(t) , \tag{14.23}$$

and let $\mathbf{e}_\mu$ denote the μth unit vector.

Expressions for $\mathbf{P}$ using different criteria in the estimation are described in the following subsections.

14.2.1 The ML Criterion $V_1(\boldsymbol{\vartheta})$

For $V_1(\boldsymbol{\vartheta})$, the criterion to minimize is, see (2.48),

$$V_N(\boldsymbol{\vartheta}) = \mathrm{tr}\left(\hat{\mathbf{R}}_N\mathbf{R}^{-1}(\boldsymbol{\vartheta})\right) + \log(\det \mathbf{R}(\boldsymbol{\vartheta})) . \tag{14.24}$$

Using (14.24) one gets by differentiation

$$\frac{\partial}{\partial \vartheta_i} V_N(\boldsymbol{\vartheta}) = \mathrm{tr}\left(-\hat{\mathbf{R}}_N\mathbf{R}^{-1}(\boldsymbol{\vartheta})\mathbf{R}_i(\boldsymbol{\vartheta})\mathbf{R}^{-1}(\boldsymbol{\vartheta})\right) + \mathrm{tr}\left(\mathbf{R}^{-1}(\boldsymbol{\vartheta})\mathbf{R}_i(\boldsymbol{\vartheta})\right) , \tag{14.25}$$

$$\begin{aligned}\frac{\partial^2}{\partial \vartheta_i \partial \vartheta_j} V_N(\boldsymbol{\vartheta}) = \mathrm{tr}\Big(&\hat{\mathbf{R}}_N\mathbf{R}^{-1}(\boldsymbol{\vartheta})\mathbf{R}_j(\boldsymbol{\vartheta})\mathbf{R}^{-1}(\boldsymbol{\vartheta})\mathbf{R}_i(\boldsymbol{\vartheta})\mathbf{R}^{-1}(\boldsymbol{\vartheta}) \\ &-\hat{\mathbf{R}}_N\mathbf{R}^{-1}(\boldsymbol{\vartheta})\mathbf{R}_{ij}(\boldsymbol{\vartheta})\mathbf{R}^{-1}(\boldsymbol{\vartheta}) \\ &+\hat{\mathbf{R}}_N\mathbf{R}^{-1}(\boldsymbol{\vartheta})\mathbf{R}_i(\boldsymbol{\vartheta})\mathbf{R}^{-1}(\boldsymbol{\vartheta})\mathbf{R}_j(\boldsymbol{\vartheta})\mathbf{R}^{-1}(\boldsymbol{\vartheta})\Big) \\ &+\mathrm{tr}\Big(-\mathbf{R}^{-1}(\boldsymbol{\vartheta})\mathbf{R}_j(\boldsymbol{\vartheta})\mathbf{R}^{-1}(\boldsymbol{\vartheta})\mathbf{R}_i(\boldsymbol{\vartheta}) + \mathbf{R}^{-1}(\boldsymbol{\vartheta})\mathbf{R}_{ij}(\boldsymbol{\vartheta})\Big) ,\end{aligned} \tag{14.26}$$

$$\begin{aligned}\frac{\partial^2}{\partial \vartheta_i \partial \vartheta_j} V_\infty(\boldsymbol{\vartheta}_0) = \mathrm{tr}\Big(&2\mathbf{R}_j(\boldsymbol{\vartheta})\mathbf{R}^{-1}(\boldsymbol{\vartheta})\mathbf{R}_i(\boldsymbol{\vartheta})\mathbf{R}^{-1}(\boldsymbol{\vartheta}) - \mathbf{R}_{ij}(\boldsymbol{\vartheta})\mathbf{R}^{-1}(\boldsymbol{\vartheta}) \\ &- \mathbf{R}_i(\boldsymbol{\vartheta})\mathbf{R}^{-1}(\boldsymbol{\vartheta})\mathbf{R}_j(\boldsymbol{\vartheta})\mathbf{R}^{-1}(\boldsymbol{\vartheta}) + \mathbf{R}^{-1}(\boldsymbol{\vartheta})\mathbf{R}_{ij}(\boldsymbol{\vartheta})\Big)_{|\boldsymbol{\vartheta}=\boldsymbol{\vartheta}_0}\end{aligned}$$

$$= \text{tr}\left(\mathbf{R}_j(\boldsymbol{\vartheta}_0)\mathbf{R}^{-1}(\boldsymbol{\vartheta}_0)\mathbf{R}_i(\boldsymbol{\vartheta}_0)\mathbf{R}^{-1}(\boldsymbol{\vartheta}_0)\right) . \tag{14.27}$$

Using the convention (14.19), it follows for the criterion $V_1(\boldsymbol{\vartheta})$ that

$$V_{1\infty}^{''}(\boldsymbol{\vartheta}_0) = \mathbf{R}_{\vartheta}^T(\mathbf{R}^{-1} \otimes \mathbf{R}^{-1})\mathbf{R}_{\vartheta} . \tag{14.28}$$

Denote the covariance matrix $\mathbf{P}$ by $\mathbf{P}_1$ for this method. So far, it is found that $\mathbf{P}_1$ is given by (14.17), and the matrices involved are given elementwise as specified by (14.25) and (14.28).

Set

$$\mathbf{S}_i = \mathbf{R}^{-1}\mathbf{R}_i\mathbf{R}^{-1} . \tag{14.29}$$

In the calculation below, skip the argument $\boldsymbol{\vartheta} = \boldsymbol{\vartheta}_0$ for convenience. One can write, using (14.24),

$$\begin{aligned}
(\mathbf{P}_1)_{ij} &\stackrel{\Delta}{=} \lim_{N\to\infty} N\mathsf{E}\left\{V_{Ni}^{'}V_{Nj}^{'}\right\} \\
&= \mathsf{E}\left\{\left[\text{tr}\left(-\hat{\mathbf{R}}_N\mathbf{R}^{-1}\mathbf{R}_i\mathbf{R}^{-1}\right) + \text{tr}\left(\mathbf{R}^{-1}\mathbf{R}_i\right)\right]\right. \\
&\quad \left.\times\left[\text{tr}\left(-\hat{\mathbf{R}}_N\mathbf{R}^{-1}\mathbf{R}_j\mathbf{R}^{-1}\right) + \text{tr}\left(\mathbf{R}^{-1}\mathbf{R}_j\right)\right]\right\} \\
&= \mathsf{E}\lim_{N\to\infty} N\left\{\left[\text{tr}\left(-\hat{\mathbf{R}}_N\mathbf{S}_i + \mathbf{R}\mathbf{S}_i\right)\right]\left[\text{tr}\left(-\hat{\mathbf{R}}_N\mathbf{S}_j + \mathbf{R}\mathbf{S}_j\right)\right]\right\} \\
&= \mathsf{E}\lim_{N\to\infty} N\left\{(\text{vec}(\mathbf{S}_i))^T\text{vec}(\hat{\mathbf{R}}_N - \mathbf{R})\left(\text{vec}(\hat{\mathbf{R}}_N - \mathbf{R})\right)^T(\text{vec}(\mathbf{S}_j))\right\} \\
&= (\text{vec}(\mathbf{S}_i))^T\mathbf{C}_r(\text{vec}(\mathbf{S}_j)) .
\end{aligned} \tag{14.30}$$

Further, from (14.19) and (14.29) and Lemma A.2

$$\text{vec}(\mathbf{S}_i) = \left(\mathbf{R}^{-1} \otimes \mathbf{R}^{-1}\right)\text{vec}(\mathbf{R}_i) , \tag{14.31}$$

$$\left(\text{vec}(\mathbf{S}_1) \ldots \text{vec}(\mathbf{S}_{n_\vartheta})\right) = \left(\mathbf{R}^{-1} \otimes \mathbf{R}^{-1}\right)\mathbf{R}_{\vartheta} . \tag{14.32}$$

Combining (14.31) and (14.32) leads finally to

$$\mathbf{P}_1 = \mathbf{R}_{\vartheta}^T\left(\mathbf{R}^{-1} \otimes \mathbf{R}^{-1}\right)\mathbf{C}_r\left(\mathbf{R}^{-1} \otimes \mathbf{R}^{-1}\right)\mathbf{R}_{\vartheta} . \tag{14.33}$$

Recall that $\mathbf{C}_r$ was defined in (14.20).

As the use of V_1 is the maximum likelihood estimate *for static problems*, it is known to be statistically efficient for such cases, with $\mathbf{P}_1$ equal to the inverse of the Fisher information matrix, cf Cramér (1946b).

Expressions for $\mathbf{P}_1$ for such cases are given and discussed in Jöreskog and Goldberger (1975), Jöreskog (1972), Jöreskog (1967), Chan and Mak (1984).

14.2.2 The Criterion $V_2(\boldsymbol{\vartheta})$

The criterion $V_2(\boldsymbol{\vartheta})$ was introduced in (2.49). To find $\mathbf{P}_2$ for general weightings $\mathbf{W}_1$ and $\mathbf{W}_2$ one needs V'_{2Ni} and $V''_{2\infty ij}$. Some comments were given in Chap. 2 on how to select these weightings. It holds

$$V'_{2Ni} = \operatorname{tr}\left(-\mathbf{R}_i\mathbf{W}_1(\hat{\mathbf{R}}_N - \mathbf{R})\mathbf{W}_2 - (\hat{\mathbf{R}}_N - \mathbf{R})\mathbf{W}_1\mathbf{R}_i\mathbf{W}_2\right) , \tag{14.34}$$

$$\left(V''_{2\infty}\right)_{ij} = \operatorname{tr}\left(\mathbf{R}_i\mathbf{W}_1\mathbf{R}_j\mathbf{W}_2 + \mathbf{R}_j\mathbf{W}_1\mathbf{R}_i\mathbf{W}_2\right) . \tag{14.35}$$

Using the convention (14.19), it follows that

$$V''_{2\infty} = \mathbf{R}_{\vartheta}^T\left(\mathbf{W}_2 \otimes \mathbf{W}_1 + \mathbf{W}_1 \otimes \mathbf{W}_2\right)\mathbf{R}_{\vartheta} . \tag{14.36}$$

Next, introduce the notations

$$\bar{\mathbf{R}}_{i21} = \mathbf{W}_2\mathbf{R}_i\mathbf{W}_1, \quad \bar{\mathbf{R}}_{i12} = \mathbf{W}_1\mathbf{R}_i\mathbf{W}_2, \quad \bar{\mathbf{R}}_i = \bar{\mathbf{R}}_{i21} + \bar{\mathbf{R}}_{i12} . \tag{14.37}$$

It then holds

$$\begin{aligned} V'_{2Ni} &= -\operatorname{tr}\left(\left(\hat{\mathbf{R}}_N - \mathbf{R}\right)\bar{\mathbf{R}}_i\right) \\ &= -\left(\operatorname{vec}(\bar{\mathbf{R}}_i)\right)^T \operatorname{vec}\left(\hat{\mathbf{R}}_N - \mathbf{R}\right) , \end{aligned} \tag{14.38}$$

$$\operatorname{vec}(\bar{\mathbf{R}}_i) = \left(\mathbf{W}_2 \otimes \mathbf{W}_1 + \mathbf{W}_1 \otimes \mathbf{W}_2\right)\operatorname{vec}(\mathbf{R}_i) . \tag{14.39}$$

Paralleling the developments in Sect. 14.2.1, one finds that

$$\begin{aligned} \mathbf{P}_2 &= \lim_{N\to\infty} N\mathsf{E}\left\{\left(V'_{2N}\right)^T V'_{2N}\right\} \\ &= \mathbf{R}_{\vartheta}^T\left(\mathbf{W}_2 \otimes \mathbf{W}_1 + \mathbf{W}_1 \otimes \mathbf{W}_2\right)\mathbf{C}_r\left(\mathbf{W}_2 \otimes \mathbf{W}_1 + \mathbf{W}_1 \otimes \mathbf{W}_2\right)\mathbf{R}_{\vartheta} . \end{aligned} \tag{14.40}$$

14.2.3 The Criterion $V_3(\boldsymbol{\vartheta})$

The criterion $V_3(\boldsymbol{\vartheta})$ was introduced in (2.51). Here one can use (2.50) to write

$$V'_{3N} = -2\mathbf{R}_{\vartheta}^T\mathbf{W}\tilde{\mathbf{r}}_N(\boldsymbol{\vartheta}) , \tag{14.41}$$

$$V''_{3\infty} = 2\mathbf{R}_{\vartheta}^T\mathbf{W}\mathbf{R}_{\vartheta} , \tag{14.42}$$

$$\tilde{\mathbf{r}}_N(\boldsymbol{\vartheta}) = \operatorname{vec}(\hat{\mathbf{R}}(\boldsymbol{\vartheta})) - \operatorname{vec}(\mathbf{R}(\boldsymbol{\vartheta}))$$

$$= \text{vec}\left(\frac{1}{N}\sum_{t=1}^{N}\varphi(t)\varphi^T(t)\right) - \text{vec}(\mathbf{R}) . \tag{14.43}$$

Then, similar to the analysis in Sect. 14.2.2,

$$\mathbf{P}_3 = \left(\mathbf{R}_\vartheta^T\mathbf{W}\mathbf{R}_\vartheta\right)^{-1}\mathbf{R}_\vartheta^T\mathbf{W}\mathbf{C}_r\mathbf{W}\mathbf{R}_\vartheta\left(\mathbf{R}_\vartheta^T\mathbf{W}\mathbf{R}_\vartheta\right)^{-1} . \tag{14.44}$$

14.2.4 Comparisons

The analysis in the previous subsections shows that the covariance matrix of the parameter estimates can be written in a general form, namely

$$\mathbf{P} = \left(\mathbf{R}_\vartheta^T\mathbf{W}\mathbf{R}_\vartheta\right)^{-1}\mathbf{R}_\vartheta^T\mathbf{W}\mathbf{C}_r\mathbf{W}\mathbf{R}_\vartheta\left(\mathbf{R}_\vartheta^T\mathbf{W}\mathbf{R}_\vartheta\right)^{-1} . \tag{14.45}$$

The expression (14.45) applies to the different criteria as follows:

- For V_1, set

$$\mathbf{W} = \mathbf{R}^{-1} \otimes \mathbf{R}^{-1} . \tag{14.46}$$

- For V_2, set

$$\mathbf{W} = \mathbf{W}_1 \otimes \mathbf{W}_2 + \mathbf{W}_2 \otimes \mathbf{W}_1 . \tag{14.47}$$

- For V_3, $\mathbf{W}$ is the user-chosen weighting matrix.
- For V_{3c}, set

$$\mathbf{W} = \mathbf{F}^T\mathbf{W}_c\mathbf{F} . \tag{14.48}$$

A specific conclusion at this point is that when the constraint (2.52) applies, the weighting $\mathbf{W}$ influences the parameter accuracy only through $\mathbf{W}_c$ as specified in (14.48).

One may also rephrase these results as

$$\mathbf{P}_1 = \mathbf{P}_2(\mathbf{R}^{-1}, \mathbf{R}^{-1}) , \tag{14.49}$$

$$\mathbf{P}_3(\mathbf{W}_1 \otimes \mathbf{W}_2) = \mathbf{P}_2(\mathbf{W}_1, \mathbf{W}_2) , \tag{14.50}$$

$$\mathbf{P}_3(\mathbf{F}^T\mathbf{W}_c\mathbf{F}) = \mathbf{P}_{3c}(\mathbf{W}_c) . \tag{14.51}$$

The relation (14.49) also shows that this choice of weighting gives asymptotically (as $\hat{\mathbf{R}} \to \mathbf{R},\ N \to \infty$)

$$\mathbf{P}_2(\mathbf{W}_1 = \hat{\mathbf{R}}^{-1}, \mathbf{W}_2 = \hat{\mathbf{R}}^{-1}) = \mathbf{P}_1 , \tag{14.52}$$

where $\mathbf{P}_1$ denotes the covariance matrix obtained by minimizing the criterion (14.24). Indeed, the particular weighting in (14.52) gives the smallest possible covariance matrix $\mathbf{P}_2$.

14.2.5 The Matrix $\mathbf{C}_r$

Next consider the matrix $\mathbf{C}_r$, introduced in (14.20) for Gaussian distributed data.

Lemma 14.1 *Let the general assumptions as well as Assumptions* **AI3**, **AI4**, *and* **AN3c** *apply. Then it holds*

$$\mathbf{C}_r = \sum_{\tau=-\infty}^{\infty} \left[\mathbf{R}_\varphi(\tau) \otimes \mathbf{R}_\varphi(\tau) + \left(\mathbf{I} \otimes \mathbf{R}_\varphi(\tau)\right) \begin{pmatrix} \mathbf{e}_1\mathbf{e}_1^T & \ldots & \mathbf{e}_{n_\varphi}\mathbf{e}_1^T \\ \vdots & & \\ \mathbf{e}_1\mathbf{e}_{n_\varphi}^T & \ldots & \mathbf{e}_{n_\varphi}\mathbf{e}_{n_\varphi}^T \end{pmatrix} \left(\mathbf{I} \otimes \mathbf{R}_\varphi(\tau)\right) \right]$$

$$\stackrel{\Delta}{=} \sum_{\tau=-\infty}^{\infty} \left[\mathbf{R}_\varphi(\tau) \otimes \mathbf{R}_\varphi(\tau) + \left(\mathbf{I} \otimes \mathbf{R}_\varphi(\tau)\right) \mathbf{K} \left(\mathbf{I} \otimes \mathbf{R}_\varphi(\tau)\right) \right] . \tag{14.53}$$

The matrix **K** *is a permutation matrix, with eigenvalues in* ± 1, *and it is generally indefinite.*

Proof See Appendix 14.A.1. ∎

Remark 14.2 The result means that the matrix $\mathbf{C}_r$ can be written as a block matrix, where the μ, ν block ($\mu, \nu = 1, \ldots, n_\varphi$) can be written as follows

$$(\mathbf{C}_r)_{\mu,\nu} = \sum_{\tau=-\infty}^{\infty} \left[r_{\varphi_\mu \varphi_\nu}(\tau) \mathbf{R}_\varphi(\tau) + \mathbf{R}_\varphi(\tau) \mathbf{e}_\nu \mathbf{e}_\mu^T \mathbf{R}_\varphi(\tau) \right] . \tag{14.54}$$

∎

Remark 14.3 In the static case, where $\mathbf{R}_\varphi(\tau) = \mathbf{0}$ for any $\tau \neq 0$, a further simplification applies. Although the expression for $\mathbf{C}_r$ as such does not simplify, the expression (14.45) when used for the criterion V_1 leads to

$$\mathbf{P}_1 = \left(\mathbf{R}_\vartheta^T (\mathbf{R}^{-1} \otimes \mathbf{R}^{-1}) \mathbf{R}_\vartheta\right)^{-1} . \tag{14.55}$$

A simple proof is based on the fact that V_1 in this particular case is indeed the maximum likelihood estimator and that the Cramér–Rao bound is achieved. In particular, the equality in (A.131) applies, which leads to

$$\mathbf{R}_\vartheta^T \mathbf{W} \mathbf{C}_r \mathbf{W} \mathbf{R}_\vartheta = \mathbf{R}_\vartheta^T \mathbf{W} \mathbf{R}_\vartheta . \tag{14.56}$$

Note though that (14.53) *does not simplify to*

$$\mathbf{C}_r = (\mathbf{R} \otimes \mathbf{R}) . \tag{14.57}$$

Indeed, if $\mathbf{W} = \mathbf{R}^{-1} \otimes \mathbf{R}^{-1}$ in (14.45) and (14.57) would hold, then (14.55) can be derived. However, when evaluating $\mathbf{C}_r$ from (14.53), it is only the first sum that simplifies to (14.57). ■

14.2.6 A Lower Bound on the Parameter Covariance Matrix

Recall the general expression (14.45) for the covariance matrix $\mathbf{P}$:

$$\mathbf{P} = \left[\mathbf{R}_\vartheta^T \mathbf{W} \mathbf{R}_\vartheta\right]^{-1} \mathbf{R}_\vartheta^T \mathbf{W} \mathbf{C}_r \mathbf{W} \mathbf{R}_\vartheta \left[\mathbf{R}_\vartheta^T \mathbf{W} \mathbf{R}_\vartheta\right]^{-1} . \tag{14.58}$$

Now simplify, and assume $\mathbf{C}_r$ to be non-singular. One can then apply Lemma A.2 and get a lower bound:

$$\mathbf{P} \geq \left[\mathbf{R}_\vartheta^T \mathbf{C}_r^{-1} \mathbf{R}_\vartheta\right]^{-1} . \tag{14.59}$$

A sufficient, *but not necessary*, condition for equality to hold in (14.59) is

$$\mathbf{W} = \mathbf{C}_r^{-1} . \tag{14.60}$$

However, *the above analysis does not apply to the standard CFA problem*. The reason is that $\mathbf{C}_r$ is singular. In fact, cf. (2.52), (14.19), (14.20)

$$\mathbf{R}_\vartheta = -\mathbf{F}\mathbf{X}_\vartheta , \tag{14.61}$$

$$\mathbf{C}_r = \mathbf{F}\mathbf{C}_x\mathbf{F}^T , \tag{14.62}$$

where though $\mathbf{C}_x$ is non-singular. Using this one finds, cf. (2.56),

$$\begin{aligned}
\mathbf{P} &= \left(\mathbf{X}_\vartheta^T \mathbf{F}^T \mathbf{W} \mathbf{F} \mathbf{X}_\vartheta\right)^{-1} \mathbf{X}_\vartheta^T \mathbf{F}^T \mathbf{W} \mathbf{F} \mathbf{C}_x \mathbf{F}^T \mathbf{W} \mathbf{F} \mathbf{X}_\vartheta \left(\mathbf{X}_\vartheta^T \mathbf{F}^T \mathbf{W} \mathbf{F} \mathbf{X}_\vartheta\right)^{-1} \\
&= \left(\mathbf{X}_\vartheta^T \mathbf{W}_\mathrm{c} \mathbf{X}_\vartheta\right)^{-1} \mathbf{X}_\vartheta^T \mathbf{W}_\mathrm{c} \mathbf{C}_x \mathbf{W}_\mathrm{c} \mathbf{X}_\vartheta \left(\mathbf{X}_\vartheta^T \mathbf{W}_\mathrm{c} \mathbf{X}_\vartheta\right)^{-1} \\
&\geq \left(\mathbf{X}_\vartheta^T \mathbf{C}_x^{-1} \mathbf{X}_\vartheta\right)^{-1} \stackrel{\Delta}{=} \mathbf{P}_\mathrm{opt}
\end{aligned} \tag{14.63}$$

with equality, for example, for

$$\mathbf{W}_\mathrm{c} = \mathbf{C}_x^{-1} . \tag{14.64}$$

The matrix $\mathbf{P}_\mathrm{opt}$ in (14.63) is a lower bound of $\mathbf{P}$. Due to (14.64) it is achievable for the criteria V_3 and $V_{3\mathrm{c}}$ with appropriate weightings. It is so far an open issue if and when it can be achieved for the criteria V_1 and V_2.

14.3 Instrumental Variable Methods

14.3.1 The Basic IV Estimator

Consider the basic IV estimator (6.14). The normalized parameter error can be written as

$$\sqrt{N}\left(\hat{\boldsymbol{\theta}}_{\mathrm{IV}} - \boldsymbol{\theta}_0\right) = \left[\frac{1}{N}\sum_{t=1}^{N} \mathbf{z}(t)\boldsymbol{\varphi}^T(t)\right]^{-1} \left[\frac{1}{\sqrt{N}}\sum_{t=1}^{N} \mathbf{z}(t)v(t)\right] . \tag{14.65}$$

Under weak assumptions, the underlying signals are ergodic, Ljung (1999), Söderström and Stoica (1989), and the normalized sum converges to its expected value, cf. (6.20)

$$\frac{1}{N}\sum_{t=1}^{N} \mathbf{z}(t)\boldsymbol{\varphi}^T(t) \rightarrow \mathsf{E}\left\{\mathbf{z}(t)\boldsymbol{\varphi}^T(t)\right\} \triangleq \mathbf{R}, \quad N \rightarrow \infty . \tag{14.66}$$

Further, one can show that the second factor in (14.65) is asymptotically Gaussian distributed in the sense, see Lemma A.14

$$\frac{1}{\sqrt{N}}\sum_{t=1}^{N} \mathbf{z}(t)v(t) \xrightarrow{\text{dist}} \mathscr{N}(\mathbf{0}, \mathbf{C}), \quad N \rightarrow \infty , \tag{14.67}$$

$$\mathbf{C} = \lim_{N\rightarrow\infty} \mathsf{E}\left\{\left[\frac{1}{\sqrt{N}}\sum_{t=1}^{N} \mathbf{z}(t)v(t)\right]\left[\frac{1}{\sqrt{N}}\sum_{s=1}^{N} \mathbf{z}(s)v(s)\right]^T\right\} . \tag{14.68}$$

Compare also the general analysis in Sect. 14.1. Expressions for the covariance matrix $\mathbf{C}$ in various situations will be given below. It then follows from Slutsky's theorem, see Lemma A.15, that the normalized parameter error is also asymptotically Gaussian distributed, Söderström and Stoica (1989), as

$$\sqrt{N}\left(\hat{\boldsymbol{\theta}}_{\mathrm{IV}} - \boldsymbol{\theta}_0\right) \xrightarrow{\text{dist}} \mathscr{N}(\mathbf{0}, \mathbf{P}_{\mathrm{IV}}), \quad \mathbf{P}_{\mathrm{IV}} = \mathbf{R}^{-1}\mathbf{C}\mathbf{R}^{-T} . \tag{14.69}$$

Remark 14.4 Recall the consistency condition (6.21). If $\mathbf{R}$ is almost singular (more precisely, $\mathbf{R}$ has a large condition number), the estimate $\hat{\boldsymbol{\theta}}_{\mathrm{IV}}$ is almost not consistent. In (14.69) this shows up in that $\mathbf{R}^{-1}$ and hence $\mathbf{P}_{\mathrm{IV}}$ both will have large elements. ■

To characterize the matrix $\mathbf{C}$ in (14.69), the following result is of interest.

Lemma 14.2 *Assume that*

1. $\mathbf{z}(t)$ *and* $v(s)$ *are jointly Gaussian, (cf. Assumptions* **AN2** *and* **AI3***)*
2. *The signals* $\mathbf{z}(t)$ *and* $v(t)$ *are (at least) partly independent in the sense*

$$\mathsf{E}\left\{\mathbf{z}(t)v(s)\right\} = 0 \text{ if either } t \leq s \text{ or } t \geq s \ . \tag{14.70}$$

Let $v(t)$ *have the innovations form*

$$v(t) = H(q^{-1})v_0(t), \quad \mathsf{E}\left\{v_0(t)v_0(s)\right\} = \lambda\delta_{t,s} \ , \tag{14.71}$$

Then

$$\mathbf{C} = \lambda\mathsf{E}\left\{[H(q^{-1})\mathbf{z}(t)][H(q^{-1})\mathbf{z}(t)]^T\right\} \ . \tag{14.72}$$

Proof See Appendix 14.B. ■

Remark 14.5 The condition (14.70) is fairly general. A main point of IV estimators is that they should work for quite general disturbances. It is therefore hard to construct realistic conditions on the instrumental vector and the disturbances that are substantially weaker than (14.70). ■

14.3.2 Extensions

Consider now various extensions of the simple and basic case treated above.

Lemma 14.3 *Consider the extended IV estimator (6.16). Then (14.69) generalizes to*

$$\sqrt{N}(\hat{\boldsymbol{\theta}}_{\text{IV}} - \boldsymbol{\theta}_0) \xrightarrow{\text{dist}} \mathscr{N}(\mathbf{0}, \mathbf{P}_{\text{IV}}) \ , \tag{14.73}$$

with the covariance matrix $\mathbf{P}_{\text{IV}}$ *now given by*

$$\mathbf{P}_{\text{IV}} = (\mathbf{R}^T\mathbf{W}\mathbf{R})^{-1}\mathbf{R}^T\mathbf{W}\mathbf{C}\mathbf{W}\mathbf{R}(\mathbf{R}^T\mathbf{W}\mathbf{R})^{-1} \ , \tag{14.74}$$

$$\mathbf{C} = \lambda\mathsf{E}\left\{[F(q^{-1})H(q^{-1})\mathbf{z}(t)][F(q^{-1})H(q^{-1})\mathbf{z}(t)]^T\right\} \ . \tag{14.75}$$

Proof For a derivation, see Söderström and Stoica (1983), Söderström and Stoica (1989). ■

Lemma 14.4 *The results of Lemma 14.3 can be generalized to the MIMO case. The result is as in (14.73), (14.74), but (14.75) has to be modified. In this case, write the innovations form for the disturbances as*

$$\mathbf{v}(t) = \mathbf{H}(q^{-1})\mathbf{e}(t), \quad \mathbf{H}(0) = \mathbf{I}, \quad \mathsf{E}\left\{\mathbf{e}(t)\mathbf{e}^T(s)\right\} = \boldsymbol{\Lambda}\delta_{t,s} \ . \tag{14.76}$$

Further, introduce matrix coefficients $\{\mathbf{K}_i\}_{i=0}^{\infty}$ *by*

$$\sum_{i=0}^{\infty} \mathbf{K}_i z^i = \mathbf{F}(z)\mathbf{H}(z) . \tag{14.77}$$

It holds in this case

$$\mathbf{C} = \mathsf{E} \left\{ \left[\sum_{i=0}^{\infty} \mathbf{Z}(t+i)\mathbf{K}_i \right] \boldsymbol{\Lambda} \left[\sum_{j=0}^{\infty} \mathbf{K}_j^T \mathbf{Z}^T(t+j) \right] \right\} . \tag{14.78}$$

Proof A derivation of the result appears in Söderström and Stoica (1989). ■

Remark 14.6 For the IV with weighted subspace fitting, see Sect. 6.6, it holds that

$$\sqrt{N}(\hat{\boldsymbol{\theta}}_{\mathrm{IV}} - \boldsymbol{\theta}_0) \xrightarrow{\text{dist}} \mathscr{N}(\mathbf{0}, \mathbf{P}_{\mathrm{IVSF}}) . \tag{14.79}$$

An explicit but somewhat involved expression for the covariance matrix $\mathbf{P}_{\mathrm{IVSF}}$ is provided in Cedervall and Stoica (1996). ■

14.3.3 Optimal IV

By user choices, such as $F(q^{-1})$, $\mathbf{z}(t)$ and $\mathbf{W}$, the covariance matrix $\mathbf{P}_{\mathrm{IV}}$ in (14.74) can be affected. Here it is discussed *how to choose* these variables so that the covariance matrix $\mathbf{P}_{\mathrm{IV}}$ of the parameter estimates becomes as small as possible.

First there is a result on choosing the weighting matrix $\mathbf{W}$ optimally for a given instrumental vector $\mathbf{z}(t)$ and a fixed prefilter $F(q^{-1})$.

Lemma 14.5 *Consider the covariance matrix* $\mathbf{P}_{\mathrm{IV}} = \mathbf{P}_{\mathrm{IV}}(\mathbf{W})$ *given by (14.74), where the dependence on* $\mathbf{W}$ *is highlighted. Then it holds*

$$\mathbf{P}(\mathbf{W}) \geq \mathbf{P}(\mathbf{C}^{-1}) \tag{14.80}$$

in the sense that $\mathbf{P}(\mathbf{W}) \geq \mathbf{P}(\mathbf{C}^{-1})$ *is a nonnegative definite matrix.*

Proof Direct application of Lemma A.1. ■

Remark 14.7 The condition $\mathbf{W} = \mathbf{C}^{-1}$ is sufficient, but not always necessary to get optimal accuracy. An example in a particular IV situation, where even the choice of no weighting ($\mathbf{W} = \mathbf{I}$) gives optimal accuracy, is given in Söderström and Hong (2005) and in Sect. 14.7. ■

Further optimization is described by the following result.

Lemma 14.6 *Consider the general extended IV estimator, with optimal weighting. Then the covariance matrix* $\mathbf{P}_{\text{IV}}$ *has a lower bound*

$$\mathbf{P}_{\text{IV}} \geq \lambda \mathsf{E}\left\{[H^{-1}(q^{-1})\varphi_0(t)][H^{-1}(q^{-1})\varphi_0^T(t)]\right\}^{-1} \stackrel{\Delta}{=} \mathbf{P}_{\text{IV}}^{\text{opt}} . \tag{14.81}$$

The lower bound is achieved if

$$F(q^{-1}) = H^{-1}(q^{-1}), \quad \mathbf{z}(t) = H^{-1}(q^{-1})\varphi_0(t), \quad (\mathbf{W} \text{ irrelevant}) . \tag{14.82}$$

In (14.81), (14.82) $\varphi_0(t)$ *denotes the 'noise–free' part of* $\varphi(t)$

$$\varphi_0(t) = (-y_0(t-1) \ldots - y_0(t-n_a) \quad u_0(t-1) \ldots u_0(t-n_b))^T . \tag{14.83}$$

Proof See Appendix 14.B.2. ■

Remark 14.8 The choices (14.82) are 'ideal,' as they rely on knowledge of unknown quantities. Certainly, (14.81) gives a lower bound. It is possible to get asymptotic accuracy, $\mathbf{P}_{\text{IV}}^{\text{opt}}$, in practice by using a few steps of an iterative algorithm, see Söderström and Stoica (1989). ■

Example 14.2 The weighted instrumental variable estimator considered in Example 6.2 is reconsidered and repeated. Now also the theoretical standard deviations of the estimates are included and compared to the numerical results obtained by the Monte Carlo simulations.

The numerical findings are displayed in Table 14.1.

As for most other estimators, the estimates of A are much more accurate than the estimates of B. There is in this example some improvement by using optimal weighting. The theory does not describe the simulations well for small p ($p = 0$). This is possibly due to that the asymptotic theory does not really apply here. Expressed differently, the estimates are in this case not robust to imperfections of the data. The theoretical results for optimal weighting imply that there is not much to be gained by choosing $p > 2$.

A closer examination of the results revealed a few outlier realizations, which was noticeable particularly in the B estimates. This can be viewed as a lack of robustness.

For example, for $p = 1$ with no weighting ($\mathbf{W} = \mathbf{I}$) 8 troublesome realizations were found. When these realizations were removed, the statistics of the remaining 92 was much more reasonable, as illustrated in Table 14.2. It can be seen that the estimate of the A parameters is almost the same as before when using all 100 realizations, while the quality of the B estimates improves considerably when the outlier realizations are not considered. ■

Table 14.1 Instrumental variable estimates for increasing length of the instrumental variable vector using the standard data sets. The arithmetic mean values are given with standard deviations within parentheses. The theoretical standard deviations are given within brackets

Parameter		a_1	a_2	b_1	b_2
True value		−1.5	0.7	2.0	1.0
p	weight				
0	I	−1.440	0.670	−0.510	3.599
		(0.534)	(0.302)	(27.86)	(26.20)
		[0.048]	[0.029]	[2.83]	[2.57]
0	opt	−1.440	0.670	−0.510	3.5999
		(0.534)	(0.302)	(27.86)	(26.20)
		[0.048]	[0.029]	[2.83]	[2.57]
1	I	−1.493	0.697	1.510	1.468
		(0.042)	(0.029)	(4.56)	(3.77)
		[0.030]	[0.017]	[2.35]	[2.06]
	opt	−1.493	0.697	1.510	1.468
		(0.042)	(0.029)	(4.56)	(3.77)
		[0.030]	[0.017]	[2.28]	[2.01]
2	I	−1.495	0.698	1.734	1.271
		(0.028)	(0.021)	(2.89)	(2.33)
		[0.020]	[0.013]	[2.01]	[1.72]
	opt	−1.493	0.696	1.964	1.106
		(0.036)	(0.024)	(3.43)	(2.83)
		[0.020]	[0.012]	[1.76]	[1.52]
3	I	−1.499	0.700	1.726	1.236
		(0.020)	(0.016)	(2.18)	(1.75)
		[0.017]	[0.012]	[1.85]	[1.57]
	opt	−1.494	0.696	1.848	1.169
		(0.026)	(0.019)	(2.82)	(2.32)
		[0.017]	[0.012]	[1.62]	[1.38]
4	I	−1.500	0.700	1.862	1.122
		(0.017)	(0.014)	(1.63)	(1.32)
		[0.017]	[0.011]	[1.82]	[1.53]
	opt	−1.493	0.695	1.832	1.196
		(0.024)	(0.017)	(2.38)	(1.97)
		[0.016]	[0.011]	[1.51]	[1.28]
5	I	−1.500	0.701	1.835	1.138
		(0.017)	(0.014)	(1.54)	(1.23)
		[0.017]	[0.011]	[1.89]	[1.58]
	opt	−1.491	0.694	1.728	1.291
		(0.023)	(0.018)	(2.24)	(1.81)
		[0.016]	[0.011]	[1.48]	[1.26]

Table 14.2 Instrumental variable estimates for $p = 1$, no weighting, using the standard data sets. The arithmetic mean values are given with standard deviations within parentheses. The theoretical standard deviations are given within brackets

Parameter	a_1	a_2	b_1	b_2
True value	−1.5	0.7	2.0	1.0
All realizations	−1.493	0.697	1.510	1.468
	(0.042)	(0.029)	(4.56)	(3.77)
	[0.030]	[0.017]	[2.35]	[2.06]
Best 92 realizations	−1.492	0.695	2.051	1.026
	(0.042)	(0.027)	(3.24)	(2.73)
	[0.030]	[0.017]	[2.35]	[2.06]

14.4 Generalized Instrumental Variable Methods

14.4.1 The SISO Case

The generalized instrumental variable (GIVE) method was defined in (7.98) by

$$\hat{\boldsymbol{\vartheta}} = \arg\min_{\boldsymbol{\vartheta}} \parallel \mathbf{f}(\boldsymbol{\vartheta}) \parallel^2_{\mathbf{W}} , \tag{14.84}$$

where $\mathbf{W}$ is a weighting matrix, that may depend on $\boldsymbol{\vartheta}$, and the function $\mathbf{f}(\boldsymbol{\vartheta})$ is given by

$$\mathbf{f}(\boldsymbol{\vartheta}) = \frac{1}{N}\sum_{t=1}^{N} \mathbf{z}(t, \boldsymbol{\theta})\varepsilon(t, \boldsymbol{\theta}) - \mathbf{r}_{\mathbf{z}\varepsilon}(\boldsymbol{\theta}, \boldsymbol{\rho}) . \tag{14.85}$$

This formulation of GIVE applies to SISO systems. Analysis of the multivariable case appears in Sect. 14.4.3.

To derive the asymptotic distribution of the general estimate $\hat{\boldsymbol{\vartheta}}$ given by (14.84), first utilize that $\hat{\boldsymbol{\vartheta}}$ is assumed to be consistent, as $\mathbf{f}(\boldsymbol{\vartheta}_0) \to \mathbf{0}$, when $N \to \infty$. Linearizing $\mathbf{f}(\boldsymbol{\vartheta})$ around $\boldsymbol{\vartheta} = \boldsymbol{\vartheta}_0$ leads to

$$\begin{aligned} \mathbf{f}(\hat{\boldsymbol{\vartheta}}) &= \frac{1}{N}\sum_{t=1}^{N} \mathbf{z}(t, \hat{\boldsymbol{\theta}})\varepsilon(t, \hat{\boldsymbol{\theta}}) - \mathbf{r}_{z\varepsilon}(\hat{\boldsymbol{\theta}}, \hat{\rho}) \\ &\approx \frac{1}{N}\sum_{t=1}^{N} \mathbf{z}(t, \hat{\boldsymbol{\theta}}) \left[\boldsymbol{\varphi}_0^T(t)\boldsymbol{\theta}_0 + \tilde{y}(t) - \boldsymbol{\varphi}^T(t)\left(\hat{\boldsymbol{\theta}} - \boldsymbol{\theta}_0\right) - \boldsymbol{\varphi}^T(t)\boldsymbol{\theta}_0 \right] \\ &\quad -\mathbf{r}_{\mathbf{z}\varepsilon}(\boldsymbol{\theta}_0, \boldsymbol{\rho}_0) - \mathbf{r}_{\boldsymbol{\theta}}(\boldsymbol{\theta}_0, \boldsymbol{\rho}_0)\left(\hat{\boldsymbol{\theta}} - \boldsymbol{\theta}_0\right) - \mathbf{r}_{\rho}(\boldsymbol{\theta}_0, \boldsymbol{\rho}_0)\left(\hat{\boldsymbol{\rho}} - \boldsymbol{\rho}_0\right) , \end{aligned} \tag{14.86}$$

where

$$\mathbf{r}_{\boldsymbol{\theta}}(\boldsymbol{\theta}, \boldsymbol{\rho}) = \frac{\partial \mathbf{r}_{\mathbf{z}\varepsilon}}{\partial \boldsymbol{\theta}}(\boldsymbol{\theta}, \boldsymbol{\rho}), \quad \mathbf{r}_{\rho}(\boldsymbol{\theta}, \boldsymbol{\rho}) = \frac{\partial \mathbf{r}_{\mathbf{z}\varepsilon}}{\partial \boldsymbol{\rho}}(\boldsymbol{\theta}, \boldsymbol{\rho}) . \tag{14.87}$$

Hence,

$$\mathbf{f}(\hat{\boldsymbol{\vartheta}}) \approx \frac{1}{N} \sum_{t=1}^{N} \mathbf{z}(t, \hat{\boldsymbol{\theta}}) \left[\tilde{y}(t) - \tilde{\boldsymbol{\varphi}}^T(t) \boldsymbol{\theta}_0 \right] - \mathbf{r}_{\mathbf{z}\varepsilon}(\boldsymbol{\theta}_0, \boldsymbol{\rho}_0)$$
$$-\mathbf{R}_{\mathbf{z}\varphi} \left(\hat{\boldsymbol{\theta}} - \boldsymbol{\theta}_0 \right) - \mathbf{r}_{\theta}(\boldsymbol{\theta}_0, \boldsymbol{\rho}_0) \left(\hat{\boldsymbol{\theta}} - \boldsymbol{\theta}_0 \right) - \mathbf{r}_{\rho}(\boldsymbol{\theta}_0, \boldsymbol{\rho}_0) \left(\hat{\boldsymbol{\rho}} - \boldsymbol{\rho}_0 \right) . \qquad (14.88)$$

From this one can see that (when $\hat{\boldsymbol{\vartheta}}$ is close to $\boldsymbol{\vartheta}_0$)

$$\mathbf{f}(\hat{\boldsymbol{\vartheta}}) \approx \frac{1}{N} \sum_{t=1}^{N} \mathbf{z}(t, \boldsymbol{\theta}_0) \varepsilon(t, \boldsymbol{\theta}_0) - \mathsf{E} \left\{ \mathbf{z}(t, \boldsymbol{\theta}_0) \varepsilon(t, \boldsymbol{\theta}_0) \right\}$$
$$- \left(\mathbf{R}_{\mathbf{z}\varphi} + \mathbf{r}_{\theta}(\boldsymbol{\theta}_0, \boldsymbol{\rho}_0) \;\; \mathbf{r}_{\rho}(\boldsymbol{\theta}_0, \boldsymbol{\rho}_0) \right) \begin{pmatrix} \hat{\boldsymbol{\theta}} - \boldsymbol{\theta}_0 \\ \hat{\boldsymbol{\rho}} - \boldsymbol{\rho}_0 \end{pmatrix} . \qquad (14.89)$$

Introduce the notation

$$\mathbf{S} \stackrel{\Delta}{=} \frac{\partial \mathbf{f}}{\partial \boldsymbol{\vartheta}} = - \left(\mathbf{R}_{\mathbf{z}\varphi} + \mathbf{r}_{\theta}(\boldsymbol{\theta}_0, \boldsymbol{\rho}_0) \;\; \mathbf{r}_{\rho}(\boldsymbol{\theta}_0, \boldsymbol{\rho}_0) \right) , \qquad (14.90)$$

and assume **S** to have full column rank. Further details of the sensitivity matrix **S** are given in Appendix 14.C.1.

The minimizing argument in (14.84) must fulfill

$$\mathbf{0} = \mathbf{f}^T(\hat{\boldsymbol{\vartheta}}) \mathbf{W} \frac{\partial \mathbf{f}}{\partial \boldsymbol{\vartheta}}(\hat{\boldsymbol{\vartheta}}) . \qquad (14.91)$$

From (14.89) one may write (for $\hat{\boldsymbol{\vartheta}} \approx \boldsymbol{\vartheta}_0$)

$$\mathbf{0} = \mathbf{S}^T \mathbf{W} \mathbf{f}(\hat{\boldsymbol{\vartheta}}) \approx \mathbf{S}^T \mathbf{W} \mathbf{S} \left(\hat{\boldsymbol{\vartheta}} - \boldsymbol{\vartheta}_0 \right) + \mathbf{S}^T \mathbf{W} \mathbf{f}(\boldsymbol{\vartheta}_0) , \qquad (14.92)$$

where

$$\mathbf{f}(\boldsymbol{\vartheta}_0) = \frac{1}{N} \sum_{t=1}^{N} \mathbf{z}(t, \boldsymbol{\theta}_0) \varepsilon(t, \boldsymbol{\theta}_0) - \mathsf{E} \left\{ \mathbf{z}(t, \boldsymbol{\theta}_0) \varepsilon(t, \boldsymbol{\theta}_0) \right\} \qquad (14.93)$$

has zero mean. Hence from (14.92)

$$\hat{\boldsymbol{\vartheta}} - \boldsymbol{\vartheta}_0 \approx - \left(\mathbf{S}^T \mathbf{W} \mathbf{S} \right)^{-1} \mathbf{S}^T \mathbf{W} \mathbf{f}(\boldsymbol{\vartheta}_0) . \qquad (14.94)$$

Then invoke the central limit theorem, see Sect. A.3.4 or Söderström and Stoica (1989), to conclude that asymptotically in N,

$$\sqrt{N} \left(\hat{\boldsymbol{\vartheta}} - \boldsymbol{\vartheta}_0 \right) \xrightarrow{\text{dist}} \mathscr{N}(\mathbf{0}, \mathbf{P}_{\text{GIVE}}) , \qquad (14.95)$$

where the covariance matrix $\mathbf{P}_{\text{GIVE}}$ is given by

$$\mathbf{P}_{\text{GIVE}} \stackrel{\Delta}{=} \left(\mathbf{S}^T\mathbf{W}\mathbf{S}\right)^{-1}\mathbf{S}^T\mathbf{W}\mathbf{C}\mathbf{W}\mathbf{S}\left(\mathbf{S}^T\mathbf{W}\mathbf{S}\right)^{-1} , \tag{14.96}$$

and

$$\mathbf{C} \stackrel{\Delta}{=} \lim_{N\to\infty} N\text{cov}\left(\tilde{\mathbf{r}}_{\mathbf{z}\varepsilon}\right) , \tag{14.97}$$

$$\tilde{\mathbf{r}}_{\mathbf{z}\varepsilon} = \frac{1}{N}\sum_{t=1}^{N}\mathbf{z}(t,\boldsymbol{\theta}_0)\varepsilon(t,\boldsymbol{\theta}_0) - \mathsf{E}\left\{\mathbf{z}(t,\boldsymbol{\theta}_0)\varepsilon(t,\boldsymbol{\theta}_0)\right\} = \mathbf{f}(\boldsymbol{\vartheta}_0) . \tag{14.98}$$

Remark 14.9 In the non-over-determined case, $\mathbf{S}$ is a square matrix, and so is $\mathbf{W}$. Then (14.96) simplifies to

$$\mathbf{P}_{\text{GIVE}} = \mathbf{S}^{-1}\mathbf{C}\mathbf{S}^{-T} . \tag{14.99}$$

■

Remark 14.10 As for traditional IV methods, cf Lemma 14.5, the covariance matrix $\mathbf{P}_{\text{GIVE}}$ is minimized with respect to the weighting $\mathbf{W}$ if $\mathbf{W} = \mathbf{C}^{-1}$, and

$$\min_{\mathbf{W}} \mathbf{P}_{\text{GIVE}} = \left(\mathbf{S}^T\mathbf{C}^{-1}\mathbf{S}\right)^{-1} . \tag{14.100}$$

■

So far it has been assumed that the weighting matrix $\mathbf{W}$ is fixed. In some cases, see Example 7.8 and (7.122), this is not the case. Then the matrix $\mathbf{W}$ depends on $\hat{\boldsymbol{\theta}}$. In such cases the calculations and a first-order approximation still apply, but $\mathbf{W}$ is to be interpreted as

$$\mathbf{W}(\boldsymbol{\theta}) = \mathbf{M}^T(\boldsymbol{\theta})\mathbf{M}(\boldsymbol{\theta}) . \tag{14.101}$$

Indeed, taking derivatives of (14.84) leads to

$$\mathbf{0} = \mathbf{f}^T(\hat{\boldsymbol{\vartheta}})\mathbf{W}(\hat{\boldsymbol{\vartheta}})\frac{\partial\mathbf{f}(\hat{\boldsymbol{\vartheta}})}{\partial\hat{\boldsymbol{\vartheta}}} + \frac{\partial\mathbf{f}^T(\hat{\boldsymbol{\vartheta}})}{\partial\hat{\boldsymbol{\vartheta}}}\mathbf{W}(\hat{\boldsymbol{\vartheta}})\mathbf{f}(\hat{\boldsymbol{\vartheta}}) + \mathbf{f}^T(\hat{\boldsymbol{\vartheta}})\frac{\partial\mathbf{W}(\hat{\boldsymbol{\vartheta}})}{\partial\hat{\boldsymbol{\vartheta}}}\mathbf{f}(\hat{\boldsymbol{\vartheta}}) , \tag{14.102}$$

where $\frac{\partial\mathbf{W}(\hat{\boldsymbol{\vartheta}})}{\partial\hat{\boldsymbol{\vartheta}}}$ is a symbolically written term (this middle part of the third term is in fact a tensor, not a matrix). As $\mathbf{f} \to \mathbf{0}$, the last term in (14.102) can though be neglected in comparison with the other two, and (14.96) still applies with $\mathbf{W} = \mathbf{W}(\boldsymbol{\vartheta}_0)$.

14.4.2 Evaluation of the Matrix C

Next evaluate the matrix $\mathbf{C}$ as given by (14.97).

Consider first the case of Gaussian distributed data. It then follows from Lemma A.11 that

$$\mathbf{C} = \sum_{\tau=-\infty}^{\infty} \left[\mathbf{R}_{\mathbf{z}}(\tau) r_{\varepsilon}(\tau) + \mathbf{r}_{\mathbf{z}\varepsilon}(\tau) \mathbf{r}_{\mathbf{z}\varepsilon}^{T}(-\tau) \right] . \tag{14.103}$$

The matrix $\mathbf{C}$ in (14.103) contains an infinite sum over τ. In turns out that if the output noise $\tilde{y}(t)$ is white, the sum is finite, and only a few of the terms are nonzero. However, if $\tilde{y}(t)$ is correlated, this is no longer the case. The matrix $\mathbf{C}$ can in fact be computed by taking expectations of some other, related, processes, and the explicit summation of τ can be completely avoided. How such computations can be arranged is shown in Appendix 14.C.2.

14.4.2.1 Non-Gaussian Distributed Data

When the assumption of Gaussian distributed noise-free signals (**AI3**, **AN2**) does not hold, it turns out that the correct expression for $\mathbf{C}$ is not (14.103), but instead

$$\mathbf{C} = \sum_{\tau=-\infty}^{\infty} \left[\mathbf{R}_{\mathbf{z}}(\tau) r_{\varepsilon}(\tau) + \mathbf{r}_{\mathbf{z}\varepsilon}(\tau) \mathbf{r}_{\mathbf{z}\varepsilon}^{T}(-\tau) \right] + \mathbf{C}^{\mathrm{NG}} , \tag{14.104}$$

where $\mathbf{C}^{\mathrm{NG}}$ is a term expressing an additional contribution. This matrix has the following property.

Lemma 14.7 *Assume that the data are non-Gaussian. Let the measurement noise fulfill*

$$\mathsf{E}\{\tilde{u}(t)\} = 0, \quad \mathsf{E}\{\tilde{u}^2(t)\} = \lambda_u, \quad \mathsf{E}\{\tilde{u}^4(t)\} = \mu_u , \tag{14.105}$$
$$\tilde{y}(t) = F(q^{-1}) e_y(t) , \tag{14.106}$$
$$\mathsf{E}\left\{e_y(t)\right\} = 0, \quad \mathsf{E}\left\{e_y^2(t)\right\} = \lambda_y, \quad \mathsf{E}\left\{e_y^4(t)\right\} = \mu_y , \tag{14.107}$$

Then (14.104) holds with

$$\mathbf{C}^{\mathrm{NG}} = \begin{pmatrix} \mathbf{C}_{11}^{\mathrm{NG}} & \mathbf{0}_{(n_a+1)\times p_y} & \mathbf{0}_{(n_a+1)\times n_b} & \mathbf{0}_{(n_a+1)\times p_u} \\ \mathbf{0}_{p_y\times(n_a+1)} & \mathbf{0}_{p_y\times p_y} & \mathbf{0}_{p_y\times n_b} & \mathbf{0}_{p_y\times p_u} \\ \mathbf{0}_{n_b\times(n_a+1)} & \mathbf{0}_{n_b\times p_y} & \mathbf{C}_{33}^{\mathrm{NG}} & \mathbf{0}_{n_b\times p_u} \\ \mathbf{0}_{p_u\times(n_a+1)} & \mathbf{0}_{p_u\times p_y} & \mathbf{0}_{p_u\times n_b} & \mathbf{0}_{p_u\times p_u} \end{pmatrix} . \tag{14.108}$$

If $\tilde{y}(t)$ is white noise (Assumption **AN3c** *applies), then*

$$\mathbf{C}_{11}^{\mathrm{NG}} = (\mu_y - 3\lambda_y^2)\bar{\mathbf{a}}\bar{\mathbf{a}}^T , \quad \mathbf{C}_{33}^{\mathrm{NG}} = (\mu_u - 3\lambda_u^2)\mathbf{b}\mathbf{b}^T . \tag{14.109}$$

where

$$\bar{\mathbf{a}} = \begin{pmatrix} 1 \\ a_1 \\ \vdots \\ a_{n_a} \end{pmatrix}, \quad \mathbf{b} = \begin{pmatrix} b_1 \\ \vdots \\ b_{n_b} \end{pmatrix}, \tag{14.110}$$

If $\tilde{y}(t)$ is correlated noise (Assumption **AN3b** *applies), then*

$$p_y = 0, \tag{14.111}$$

and

$$\mathbf{C}_{11}^{\mathrm{NG}} = (\mu_y - 3\lambda_y^2)\mathbf{r}_x\mathbf{r}_x^T, \quad \mathbf{C}_{33}^{\mathrm{NG}} = (\mu_u - 3\lambda_u^2)\mathbf{b}\mathbf{b}^T, \tag{14.112}$$

$$\mathbf{r}_x = \begin{pmatrix} \mathscr{A} & \mathbf{0}_{(n_a+1)\times 1} \end{pmatrix} \boldsymbol{\rho}, \tag{14.113}$$

where the matrix $\mathscr{A}$ is given by

$$\mathscr{A} = \begin{pmatrix} 1 & a_1 & \dots & a_{n_a} \\ a_1 & a_2 & \iddots & \\ \vdots & \iddots & & \\ a_{n_a} & & & \mathbf{0} \end{pmatrix} + \begin{pmatrix} 0 & & & & \\ 0 & 1 & & & \mathbf{0} \\ \vdots & a_1 & \ddots & & \\ & \vdots & \ddots & \ddots & \\ 0 & a_{n_a-1} & \dots & a_1 & 1 \end{pmatrix}, \tag{14.114}$$

and $\boldsymbol{\rho}$ is given by (5.27).

Proof See Appendix 14.C.3. ■

14.4.3 The MIMO Case

The parameter estimate $\hat{\boldsymbol{\vartheta}}$ is introduced as the minimizing element of the loss function $\overline{V}(\boldsymbol{\vartheta})$, see (7.133). Following the analysis leading to (14.9) it now holds for large enough N

$$\operatorname{cov}\left(\sqrt{N}(\hat{\boldsymbol{\vartheta}}_N - \boldsymbol{\vartheta}_0)\right) = \mathbf{P}_1^{-1}\mathbf{P}_2\mathbf{P}_1^{-1}, \tag{14.115}$$

$$\mathbf{P}_1 = \frac{\partial^2 V_\infty}{\partial \boldsymbol{\vartheta}^2}(\boldsymbol{\vartheta}_0), \tag{14.116}$$

$$\mathbf{P}_2 = \lim_{N\to\infty} N \mathsf{E}\left\{ \left(\frac{\partial \overline{V}_N}{\partial \boldsymbol{\vartheta}}(\boldsymbol{\vartheta}_0) \right)^T \frac{\partial \overline{V}_N}{\partial \boldsymbol{\vartheta}}(\boldsymbol{\vartheta}_0) \right\}. \tag{14.117}$$

Introduce the matrix

$$\mathbf{F} = \left(\mathbf{F}_\theta \ \mathbf{F}_\rho \right) \triangleq \frac{\partial \bar{\mathbf{f}}_\infty}{\partial \boldsymbol{\vartheta}}_{|\boldsymbol{\vartheta}=\boldsymbol{\vartheta}_0} . \tag{14.118}$$

Straightforward differentiation using the form (7.133) leads to

$$\frac{\partial \overline{V}_N}{\partial \boldsymbol{\vartheta}}(\boldsymbol{\vartheta}_0) = 2\bar{\mathbf{f}}^T \overline{\mathbf{W}} \mathbf{F} . \tag{14.119}$$

As $\overline{\mathbf{W}}$ and $\mathbf{F}$ in (14.119) are non-stochastic matrices, it follows that

$$\mathbf{P}_2 = 4\mathbf{F}^T \overline{\mathbf{W}} \mathbf{Q} \overline{\mathbf{W}} \mathbf{F} , \tag{14.120}$$

$$\mathbf{C} = \lim_{N\to\infty} N \mathsf{E} \left\{ \bar{\mathbf{f}}_N \bar{\mathbf{f}}_N^T \right\} . \tag{14.121}$$

To proceed, expressions for the matrices $\mathbf{F}$, $\mathbf{P}_1$, $\mathbf{C}$ are needed.

The matrix $\mathbf{P}_1$ is relatively easy. Taking derivatives in (14.119) leads, in symbolic form, to

$$\frac{\partial^2 \overline{V}_N}{\partial \boldsymbol{\vartheta}^2} = 2 \frac{\partial \bar{\mathbf{f}}^T}{\partial \boldsymbol{\vartheta}} \overline{\mathbf{W}} \mathbf{F} + 2\bar{\mathbf{f}}^T \frac{\partial \overline{\mathbf{W}}}{\partial \boldsymbol{\vartheta}} \mathbf{F} + 2\bar{\mathbf{f}}^T \overline{\mathbf{W}} \frac{\partial \mathbf{F}}{\partial \boldsymbol{\vartheta}} . \tag{14.122}$$

The expression is written in symbolic form, as many of the terms are tensors, not matrices. Note, however, that in the case studied $N \to \infty$ applies, and then the factor $\bar{\mathbf{f}}$ vanishes. Utilizing this fact gives the expression

$$\mathbf{P}_1 = 2\mathbf{F}^T \overline{\mathbf{W}} \mathbf{F} . \tag{14.123}$$

Combining (14.115), (14.120), (14.123) gives

$$\lim_{N\to\infty} N \operatorname{cov} \left(\hat{\boldsymbol{\vartheta}}_N - \boldsymbol{\vartheta}_0 \right) = \left(\mathbf{F}^T \overline{\mathbf{W}} \mathbf{F} \right)^{-1} \mathbf{F}^T \overline{\mathbf{W}} \mathbf{C} \overline{\mathbf{W}} \mathbf{F} \left(\mathbf{F}^T \overline{\mathbf{W}} \mathbf{F} \right)^{-1} . \tag{14.124}$$

Proceed by treating the matrix $\mathbf{C}$.

Lemma 14.8 *Assume that the data are Gaussian distributed (so Assumptions* **AN2, AI3** *apply). Then the matrix* **C** *(14.121) can be written as*

$$\mathbf{C} = \sum_{\tau=-\infty}^{\infty} \left[\mathbf{R}_\varepsilon(\tau) \otimes \mathbf{R}_z(\tau) + \operatorname{vec}\left(\mathbf{R}_{z\varepsilon}(\tau)\right) \operatorname{vec}\left(\mathbf{R}_{z\varepsilon}^T(-\tau)\right) \right] , \tag{14.125}$$

where $\otimes$ denotes Kronecker product, see Sect. A.1.2.

Proof See Appendix 14.C.4. ∎

Further details of how to compute the sensitivity matrix $\mathbf{F}$, (14.118) in the MIMO case are given in Söderström (2012).

As noted earlier, see Sect. 5.7, identifying canonical forms can be done by considering optimization problems under certain linear constraints. How to analyze the model accuracy for such cases is outlined in Appendix 14.D.

Example 14.3 The weighted GIVE estimator considered in Example 7.11 is reconsidered. Now also the theoretical standard deviations of the estimates are included and compared to the numerical results obtained by the Monte Carlo simulations.

The numerical findings are displayed in Table 14.3.

A closer examination revealed a few outlier realizations, which was noticeable particularly in the B estimates. This can be viewed as a lack of robustness. For example, for $p = 1$ with no weighting ($\mathbf{W} = \mathbf{I}$) two troublesome realizations were found. When these realizations were removed, the statistics of the remaining 98 was much more reasonable, as illustrated in Table 14.4.

Some observations:

- The theory explains the simulation outcome reasonably for $p \geq 3$.
- There is in this example some improvement by using optimal weighting, even if it often is minor.
- The theoretical results for optimal weighting imply that very little is to be gained by choosing $p > 2$.

■

14.5 Covariance Matching Methods

For an analysis using the confirmatory factor analysis approach (cf. Sect. 8.5) see Sect. 14.2.

14.5.1 Asymptotic Distribution

When analyzing the asymptotic distribution of the covariance matching estimates, the general expression (14.9) will be applied.

Now derive the asymptotic covariance matrix of the joint unknown parameter vector

$$\eta = \begin{pmatrix} \theta \\ \mathbf{r}_z \end{pmatrix} . \tag{14.126}$$

Introduce

$$\mathbf{L} = \begin{pmatrix} \mathbf{S} \ \mathbf{F} \end{pmatrix} , \tag{14.127}$$

$$\mathbf{S} = \begin{pmatrix} \frac{\partial \mathbf{F}}{\partial \theta_1} \mathbf{r_z} \ \dots \ \frac{\partial \mathbf{F}}{\partial \theta_{na+nb}} \mathbf{r_z} \end{pmatrix} . \tag{14.128}$$

Table 14.3 GIVE estimates for increasing length of the instrumental variable vector using the standard data sets. The arithmetic mean values are given with standard deviations within parentheses. The theoretical standard deviations are given within brackets

Parameter		a_1	a_2	b_1	b_2	λ_y	λ_u
True value		−1.5	0.7	2.0	1.0	10.0	1.81
p	weight						
1	I	−1.528	0.709	3.568	−0.689	8.37	2.00
		(0.086)	(0.040)	(2.82)	(3.30)	(4.51)	(0.64)
		[0.016]	[0.012]	[0.27]	[0.31]	[0.74]	[0.19]
	opt	−1.496	0.696	2.160	0.877	10.82	1.93
		(0.018)	(0.012)	(0.36)	(0.42)	(1.21)	(0.25)
		[0.016]	[0.012]	[0.27]	[0.31]	[0.73]	[0.19]
2	I	−1.500	0.698	2.360	0.652	10.56	1.97
		(0.018)	(0.011)	(1.01)	(1.08)	(1.77)	(0.23)
		[0.017]	[0.012]	[0.43]	[0.48]	[0.79]	[0.24]
	opt	−1.496	0.697	2.089	0.942	10.96	1.89
		(0.012)	(0.010)	(0.24)	(0.28)	(1.26)	(0.19)
		[0.015]	[0.011]	[0.25]	[0.29]	[0.64]	[0.16]
3	I	−1.498	0.698	2.213	0.807	10.80	1.95
		(0.013)	(0.010)	(0.33)	(0.37)	(1.17)	(0.22)
		[0.015]	[0.011]	[0.34]	[0.38]	[0.75]	[0.21]
	opt	−1.498	0.697	2.124	0.896	10.99	1.89
		(0.012)	(0.009)	(0.26)	(0.30)	(1.28)	(0.20)
		[0.014]	[0.011]	[0.24]	[0.28]	[0.63]	[0.16]
4	I	−1.499	0.698	2.225	0.793	10.80	1.95
		(0.014)	(0.010)	(0.40)	(0.44)	(1.16)	(0.26)
		[0.016]	[0.011]	[0.36]	[0.41]	[0.73]	[0.22]
	opt	−1.498	0.697	2.141	0.879	10.97	1.90
		(0.012)	(0.009)	(0.26)	(0.29)	(1.26)	(0.20)
		[0.014]	[0.011]	[0.24]	[0.28]	[0.63]	[0.16]
5	I	−1.499	0.698	2.208	0.811	10.82	1.94
		(0.015)	(0.010)	(0.46)	(0.51)	(1.20)	(0.28)
		[0.017]	[0.011]	[0.45]	[0.50]	[0.75]	[0.28]
	opt	−1.498	0.697	2.139	0.880	10.97	1.90
		(0.012)	(0.009)	(0.25)	(0.28)	(1.27)	(0.20)
		[0.014]	[0.011]	[0.24]	[0.28]	[0.63]	[0.16]

Table 14.4 GIVE estimates with $p = 1$, no weighting ($\mathbf{W} = \mathbf{I}$), using the standard data sets. The arithmetic mean values are given with standard deviations within parentheses. The theoretical standard deviations are given within brackets

Parameter	a_1	a_2	b_1	b_2	λ_y	λ_u
True value	−1.5	0.7	2.0	1.0	10.0	1.81
All realizations	−1.528	0.709	3.568	−0.689	8.37	2.00
	(0.086)	(0.040)	(2.82)	(3.30)	(4.51)	(0.64)
	[0.016]	[0.012]	[0.27]	[0.31]	[0.74]	[0.19]
Best 98 realizations	−1.498	0.698	2.227	0.790	10.76	1.96
	(0.016)	(0.011)	(0.41)	(0.47)	(1.19)	(0.22)
	[0.016]	[0.012]	[0.27]	[0.31]	[0.74]	[0.19]

For estimates being close to the true values, write

$$\begin{aligned}\hat{\mathbf{r}} - \mathbf{F}(\hat{\boldsymbol{\theta}})\hat{\mathbf{r}}_z &= \hat{\mathbf{r}} - \mathbf{r} - \sum_j \mathbf{F}_j \mathbf{r}_z(\hat{\boldsymbol{\theta}}_j - \boldsymbol{\theta}_j) - \mathbf{F}(\hat{\mathbf{r}}_z - \mathbf{r}_z) + O(\| \hat{\boldsymbol{\eta}} - \boldsymbol{\eta} \|^2) \\ &\approx (\hat{\mathbf{r}} - \mathbf{r}) - \mathbf{S}(\hat{\boldsymbol{\theta}} - \boldsymbol{\theta}) - \mathbf{F}(\hat{\mathbf{r}}_z - \mathbf{r}_z) \\ &= (\hat{\mathbf{r}} - \mathbf{r}) - \mathbf{L}(\hat{\boldsymbol{\eta}} - \boldsymbol{\eta}) \, . \end{aligned} \tag{14.129}$$

Note that the asymptotic case (N large enough) is studied, and the estimate $\hat{\boldsymbol{\eta}}$ converges to its true value $\boldsymbol{\eta}$ as $N \to \infty$. Hence it is not restrictive for the current study to assume $\hat{\boldsymbol{\eta}} - \boldsymbol{\eta}$ to be small.

By writing the estimate $\hat{\boldsymbol{\eta}}$ as

$$\hat{\boldsymbol{\eta}} = \arg\min_{\boldsymbol{\eta}} J(\boldsymbol{\eta}) \tag{14.130}$$

with $J(\boldsymbol{\eta})$ defined in (8.24), one finds

$$\lim_{N\to\infty} N\mathrm{cov}\left(\hat{\boldsymbol{\eta}}\right) = \left(J''(\boldsymbol{\eta}_0)\right)^{-1} \lim_{N\to\infty} N\mathrm{cov}\left(J'(\boldsymbol{\eta}_0)\right)\left(J''(\boldsymbol{\eta}_0)\right)^{-1} , \tag{14.131}$$

where

$$J'(\boldsymbol{\eta}_0) = -2\left(\hat{\mathbf{r}} - \mathbf{r}\right)^T \mathbf{W}\mathbf{L} \, , \tag{14.132}$$

$$J''(\boldsymbol{\eta}_0) = 2\mathbf{L}^T\mathbf{W}\mathbf{L} \, . \tag{14.133}$$

Now introduce the notations

$$\begin{aligned}\mathbf{P}_\eta &\stackrel{\Delta}{=} \lim_{N\to\infty} N\mathrm{cov}\left(\hat{\boldsymbol{\eta}}\right) \\ &= \left(\mathbf{L}^T\mathbf{W}\mathbf{L}\right)^{-1}\mathbf{L}^T\mathbf{W}\mathbf{R}\mathbf{W}\mathbf{L}\left(\mathbf{L}^T\mathbf{W}\mathbf{L}\right)^{-1} , \end{aligned} \tag{14.134}$$

$$\mathbf{R} = \lim_{N\to\infty} N\mathsf{E}\{(\hat{\mathbf{r}} - \mathbf{r})(\hat{\mathbf{r}} - \mathbf{r})^T\} \, . \tag{14.135}$$

Apparently, it holds

$$\mathbf{P}_\theta = \begin{pmatrix} \mathbf{I} & \mathbf{0} \end{pmatrix} \mathbf{P}_\eta \begin{pmatrix} \mathbf{I} \\ \mathbf{0} \end{pmatrix} . \tag{14.136}$$

To proceed, use the notation

$$\begin{pmatrix} \bar{\mathbf{W}}_{11} & \bar{\mathbf{W}}_{12} \\ \bar{\mathbf{W}}_{21} & \bar{\mathbf{W}}_{22} \end{pmatrix} \stackrel{\Delta}{=} \mathbf{L}^T\mathbf{W}\mathbf{L} = \begin{pmatrix} \mathbf{S}^T\mathbf{W}\mathbf{S} & \mathbf{S}^T\mathbf{W}\mathbf{F} \\ \mathbf{F}^T\mathbf{W}\mathbf{S} & \mathbf{F}^T\mathbf{W}\mathbf{F} \end{pmatrix} . \tag{14.137}$$

Therefore it holds

$$\begin{aligned} \left(\mathbf{L}^T\mathbf{W}\mathbf{L}\right)^{-1} \begin{pmatrix} \mathbf{I} \\ \mathbf{0} \end{pmatrix} &= \begin{pmatrix} \bar{\mathbf{W}}_{11} & \bar{\mathbf{W}}_{12} \\ \bar{\mathbf{W}}_{21} & \bar{\mathbf{W}}_{22} \end{pmatrix}^{-1} \begin{pmatrix} \mathbf{I} \\ \mathbf{0} \end{pmatrix} \\ &= \left[\begin{pmatrix} \mathbf{0} & \mathbf{0} \\ \mathbf{0} & \bar{\mathbf{W}}_{22}^{-1} \end{pmatrix} + \begin{pmatrix} \mathbf{I} \\ -\bar{\mathbf{W}}_{22}^{-1}\bar{\mathbf{W}}_{21} \end{pmatrix} \right. \\ &\quad \left. \times \left(\bar{\mathbf{W}}_{11} - \bar{\mathbf{W}}_{12}\bar{\mathbf{W}}_{22}^{-1}\bar{\mathbf{W}}_{21}\right)^{-1} \left(\mathbf{I} \; -\bar{\mathbf{W}}_{12}\bar{\mathbf{W}}_{22}^{-1}\right) \right] \begin{pmatrix} \mathbf{I} \\ \mathbf{0} \end{pmatrix} \\ &= \begin{pmatrix} \mathbf{I} \\ -\bar{\mathbf{W}}_{22}^{-1}\bar{\mathbf{W}}_{21} \end{pmatrix} \left(\bar{\mathbf{W}}_{11} - \bar{\mathbf{W}}_{12}\bar{\mathbf{W}}_{22}^{-1}\bar{\mathbf{W}}_{21}\right)^{-1} , \end{aligned} \tag{14.138}$$

and

$$\begin{aligned} \mathbf{W}\mathbf{L}\left(\mathbf{L}^T\mathbf{W}\mathbf{L}\right)^{-1} \begin{pmatrix} \mathbf{I} \\ \mathbf{0} \end{pmatrix} &= \begin{pmatrix} \mathbf{W}\mathbf{S} & \mathbf{W}\mathbf{F} \end{pmatrix} \begin{pmatrix} \mathbf{I} \\ -\bar{\mathbf{W}}_{22}^{-1}\bar{\mathbf{W}}_{21} \end{pmatrix} \left(\bar{\mathbf{W}}_{11} - \bar{\mathbf{W}}_{12}\bar{\mathbf{W}}_{22}^{-1}\bar{\mathbf{W}}_{21}\right)^{-1} \\ &= \left(\mathbf{W}\mathbf{S} - \mathbf{W}\mathbf{F}\left(\mathbf{F}^T\mathbf{W}\mathbf{F}\right)^{-1}\mathbf{F}^T\mathbf{W}\mathbf{S}\right)\left(\mathbf{S}^T\mathbf{M}\mathbf{S}\right)^{-1} \\ &= \mathbf{M}\mathbf{S}\left(\mathbf{S}^T\mathbf{M}\mathbf{S}\right)^{-1} , \end{aligned} \tag{14.139}$$

where

$$\mathbf{M} = \mathbf{W} - \mathbf{W}\mathbf{F}\left(\mathbf{F}^T\mathbf{W}\mathbf{F}\right)^{-1}\mathbf{F}^T\mathbf{W} . \tag{14.140}$$

Further, it is a standard result, see Söderström and Stoica (1989), that expressions such as (14.134) are minimized with respect to $\mathbf{W}$ with the particular choice $\mathbf{W} = \mathbf{R}^{-1}$. This specific choice of $\mathbf{W}$ minimizes not only $\mathbf{P}_\theta$ but also the full matrix $\mathbf{P}_\eta$.

To sum up, the following theorem has been proved.

Theorem 14.1 *The asymptotic normalized covariance matrix of the parameter estimates fulfills*

$$\mathbf{P}_\theta \stackrel{\Delta}{=} \lim_{N\to\infty} N\operatorname{cov}(\hat{\theta} - \theta_0) = [\mathbf{S}^\mathrm{T}\mathbf{M}\mathbf{S}]^{-1}\mathbf{S}^\mathrm{T}\mathbf{M}\mathbf{R}\mathbf{M}\mathbf{S}[\mathbf{S}^\mathrm{T}\mathbf{M}\mathbf{S}]^{-1} , \tag{14.141}$$

where

$$\mathbf{M} = \mathbf{W} - \mathbf{W}\mathbf{F}\left(\mathbf{F}^T\mathbf{W}\mathbf{F}\right)^{-1}\mathbf{F}^T\mathbf{W} . \tag{14.142}$$

Remark 14.11 The result can be extended to the case introduced in Remark 8.6, where also λ_y and λ_u are estimated and $r_y(0)$ and $r_u(0)$ exploited. This extension is presented in Appendix 14.E.1, see also Söderström et al. (2014c). ∎

14.5.2 *The Matrix* R

In this section it will be described how the matrix **R** given in (14.135) can be computed. Needless to say, it requires that the true system is known, as well as the statistical properties of the measurement noises and the unperturbed input signal.

The full vector $\hat{\mathbf{r}}$ is partitioned according to (8.21) as

$$\hat{\mathbf{r}} = \begin{pmatrix} \hat{\mathbf{r}}_y \\ \hat{\mathbf{r}}_u \\ \hat{\mathbf{r}}_{yu} \end{pmatrix} . \tag{14.143}$$

The normalized covariance matrix **R** has a corresponding partitioning as

$$\mathbf{R} = \begin{pmatrix} \mathbf{R}_{11} & \mathbf{R}_{12} & \mathbf{R}_{13} \\ \mathbf{R}_{21} & \mathbf{R}_{22} & \mathbf{R}_{23} \\ \mathbf{R}_{31} & \mathbf{R}_{32} & \mathbf{R}_{33} \end{pmatrix} . \tag{14.144}$$

The following result applies.

Theorem 14.2 *Let the general assumptions as well as Assumptions* **AI3**, **AI4**, *and* **AN3c** *apply. Then the elements of the matrix* **R** *can be written as follows. For* $\mu, \nu = 0, \dots p_y$,

$$(\mathbf{R}_{11})_{\mu\nu} = \lambda_y^2 \left[\delta_{\mu,\nu} + \delta_{\mu,0}\delta_{\nu,0}\right] + 2\lambda_y \left[r_{y_0}(\mu-\nu) + r_{y_0}(\mu+\nu)\right] + \beta_{\mu-\nu} + \beta_{\mu+\nu} \,, \tag{14.145}$$

where

$$\beta_k = \mathsf{E}\left\{\left[\frac{B^2(q^{-1})}{A^2(q^{-1})}\frac{C^2(q^{-1})}{D^2(q^{-1})}e(t+k)\right]\left[\frac{B^2(q^{-1})}{A^2(q^{-1})}\frac{C^2(q^{-1})}{D^2(q^{-1})}e(t)\right]\right\} . \tag{14.146}$$

For $\mu = 0, \dots p_y$, $\nu = 0, \dots p_u$,

$$(\mathbf{R}_{12})_{\mu\nu} = \beta^{(2)}_{-\mu+\nu} + \beta^{(2)}_{\mu+\nu} \,, \tag{14.147}$$

where

$$\beta_k^{(2)} = \mathsf{E}\left\{\left[\frac{B(q^{-1})}{A(q^{-1})}\frac{C^2(q^{-1})}{D^2(q^{-1})}e(t+k)\right]\left[\frac{B(q^{-1})}{A(q^{-1})}\frac{C^2(q^{-1})}{D^2(q^{-1})}e(t)\right]\right\} . \tag{14.148}$$

For $\mu = 0, \dots p_y, \nu = p_1, \dots p_2,$

$$(\mathbf{R}_{13})_{\mu\nu} = \lambda_y[r_{yu}(-\mu+\nu) + r_{yu}(\mu+\nu)] + \beta^{(3)}_{\nu-\mu} + \beta^{(3)}_{\mu+\nu} \,, \tag{14.149}$$

where

$$\beta^{(3)}_k = \mathsf{E}\left\{\left[\frac{B^2(q^{-1})}{A^2(q^{-1})}\frac{C^2(q^{-1})}{D^2(q^{-1})}e(t+k)\right]\left[\frac{B(q^{-1})}{A(q^{-1})}\frac{C^2(q^{-1})}{D^2(q^{-1})}e(t)\right]\right\} . \tag{14.150}$$

For the block $\mathbf{R}_{22}$ *it holds* $(0 \le \mu \le p_u,\ 0 \le \nu \le p_u)$

$$(\mathbf{R}_{22})_{\mu\nu} = \lambda_u^2\left[\delta_{\mu,\nu} + \delta_{\mu,0}\delta_{\nu,0}\right] + 2\lambda_u\left[r_{u_0}(\mu-\nu) + r_{u_0}(\mu+\nu)\right] + \beta^{(4)}_{\mu-\nu} + \beta^{(4)}_{\mu+\nu} \,, \tag{14.151}$$

where

$$\beta^{(4)}_k = \mathsf{E}\left\{\left[\frac{C^2(q^{-1})}{D^2(q^{-1})}e(t+k)\right]\left[\frac{C^2(q^{-1})}{D^2(q^{-1})}e(t)\right]\right\} . \tag{14.152}$$

For $\mu = 0, \dots p_u, \nu = p_1, \dots p_2,$ *it holds*

$$(\mathbf{R}_{23})_{\mu\nu} = \lambda_u\left[r_{yu}(\nu-\mu) + r_{yu}(\nu+\mu)\right] + \beta^{(5)}_{-\mu+\nu} + \beta^{(5)}_{\mu+\nu} \,, \tag{14.153}$$

where

$$\beta^{(5)}_k = \mathsf{E}\left\{\left[\frac{B(q^{-1})}{A(q^{-1})}\frac{C^2(q^{-1})}{D^2(q^{-1})}e(t+k)\right]\left[\frac{C^2(q^{-1})}{D^2(q^{-1})}e(t)\right]\right\} . \tag{14.154}$$

Finally, for $\mu, \nu = p_1, \dots p_2,$

$$\begin{aligned}(\mathbf{R}_{33})_{\mu\nu} &= \lambda_y\lambda_u\delta_{\mu,\nu} + \lambda_u r_{y_0}(\mu-\nu) + \lambda_y r_{u_0}(\nu-\mu) \\ &\quad + \beta^{(6)}_{-\mu+\nu} + \gamma^{(6)}_{\mu+\nu} \,,\end{aligned} \tag{14.155}$$

where

$$\begin{aligned}\beta^{(6)}_k &= \mathsf{E}\left\{\left[\frac{B(q^{-1})}{A(q^{-1})}\frac{C^2(q^{-1})}{D^2(q^{-1})}e(t+k)\right]\right. \\ &\quad \times\left.\left[\frac{B(q^{-1})}{A(q^{-1})}\frac{C^2(q^{-1})}{D^2(q^{-1})}e(t)\right]\right\} ,\end{aligned} \tag{14.156}$$

$$\gamma^{(6)}_k = \mathsf{E}\left\{\left[\frac{B^2(q^{-1})}{A^2(q^{-1})}\frac{C^2(q^{-1})}{D^2(q^{-1})}e(t+k)\right]\left[\frac{C^2(q^{-1})}{D^2(q^{-1})}e(t)\right]\right\} . \tag{14.157}$$

Proof See Appendix 14.E.2. ■

What happens when the data are not Gaussian distributed? The answer is given in the following lemma.

Lemma 14.9 *Assume that $u_0(t)$ is non-Gaussian, but the measurement noise $\tilde{u}(t)$, $\tilde{y}(t)$ is Gaussian. Then the matrix* **R** *can be written as*

$$\mathbf{R} = \mathbf{R}^{\mathrm{G}} + \mathbf{R}^{\mathrm{NG}} , \tag{14.158}$$

where the 'Gaussian part' $\mathbf{R}^{\mathrm{G}}$ is given as in Theorem 14.2. The non-Gaussian part $\mathbf{R}^{\mathrm{NG}}$ turns out to be

$$\mathbf{R}^{\mathrm{NG}} = (\mu - 3)\mathbf{r}_0\mathbf{r}_0^T , \tag{14.159}$$

where

$$\mu = \frac{\mathsf{E}\left\{e^4(t)\right\}}{\left(\mathsf{E}\left\{e^2(t)\right\}\right)^2} , \tag{14.160}$$

and $\mathbf{r}_0$ is the part of the vector **r** *that is composed solely of the noise-free input–output data.*

Proof See Appendix 14.E.3. ■

Example 14.4 The weighted CM estimator considered in Example 8.4 is reconsidered and repeated. Now also the theoretical standard deviations of the estimates are included and compared to the numerical results obtained by the Monte Carlo simulations.

The numerical findings are displayed in Table 14.5. The user parameter p was introduced in (8.63).

The theory does not describe the simulations well for small p. Probably the asymptotic theory does not fully apply here. In other words, the estimates lack some robustness.

One observation is that the optimally weighted CM estimate seems more sensitive to unexpected errors ('outlier realizations') than the unweighted CM estimate. This is noticeable for the cases $p = 3$ and $p = 5$. As a further investigation of this behavior, the results for these values of p were repeated when one troublesome realization was removed. The results, which are presented in Table 14.6, confirm that the remaining realizations lead to reasonable statistics.

The results in Table 14.7 concern the case with $p = 1$ and no weighting ($\mathbf{W} = \mathbf{I}$). Comparison is made using all 100 realizations or removing one realization. It is clear that the estimates are improved, in particular the estimates of the B parameters, when the single realization with 'bad data' is removed.

Some further observations:

- There is in this example only a marginal improvement by using optimal weighting.
- The theory explains the outcome reasonably for $p \geq 4$.
- The theoretical results for optimal weighting imply there is here no point in choosing $p > 2$.

■

Table 14.5 CM estimates for increasing length of the covariance vector using the standard data sets. The arithmetic mean values are given with standard deviations within parentheses. The theoretical standard deviations are given within brackets

Parameter		a_1	a_2	b_1	b_2	λ_y	λ_u
True value		−1.5	0.7	2.0	1.0	10.0	1.81
p	weight						
1	I	−1.489	0.692	1.994	1.087	10.96	1.92
		(0.020)	(0.014)	(0.28)	(0.38)	(1.24)	(0.22)
		[0.015]	[0.011]	[0.14]	[0.19]	[0.24]	[0.22]
	opt	−1.488	0.692	1.980	1.101	11.01	1.10
		(0.018)	(0.013)	(0.17)	(0.25)	(1.24)	(0.23)
		[0.015]	[0.011]	[0.14]	[0.19]	[0.24]	[0.22]
2	I	−1.494	0.696	2.008	1.039	11.09	1.86
		(0.014)	(0.010)	(0.17)	(0.24)	(1.34)	(0.17)
		[0.014]	[0.011]	[0.13]	[0.17]	[0.27]	[0.21]
	opt	−1.491	0.694	1.978	1.085	10.98	1.85
		(0.014)	(0.011)	(0.16)	(0.20)	(1.24)	(0.18)
		[0.013]	[0.010]	[0.12]	[0.16]	[0.24]	[0.21]
3	I	−1.495	0.697	2.005	1.028	11.16	1.83
		(0.013)	(0.010)	(0.17)	(0.22)	(1.39)	(0.17)
		[0.014]	[0.011]	[0.14]	[0.19]	[0.30]	[0.21]
	opt	−1.480	0.684	1.866	1.220	10.84	1.82
		(0.013)	(0.011)	(1.04)	(1.30)	(1.84)	(0.20)
		[0.013]	[0.010]	[0.12]	[0.16]	[0.23]	[0.20]
4	I	−1.497	0.698	2.005	1.019	11.18	1.81
		(0.013)	(0.010)	(0.17)	(0.22)	(1.41)	(0.18)
		[0.015]	[0.011]	[0.16]	[0.22]	[0.32]	[0.22]
	opt	−1.493	0.696	1.961	1.097	10.97	1.81
		(0.013)	(0.011)	(0.17)	(0.21)	(1.23)	(0.18)
		[0.012]	[0.009]	[0.11]	[0.16]	[0.23]	[0.20]
5	I	−1.497	0.698	2.010	1.006	11.20	1.81
		(0.014)	(0.011)	(0.20)	(0.25)	(1.42)	(0.19)
		[0.016]	[0.012]	[0.18]	[0.25]	[0.33]	[0.22]
	opt	−1.480	0.685	1.850	1.230	10.81	1.81
		(0.013)	(0.011)	(1.12)	(1.38)	(1.96)	(0.19)
		[0.012]	[0.009]	[0.11]	[0.16]	[0.63]	[0.20]

Example 14.5 As a further illustration of the CM estimates, also another system than the standard one was used, in order to show that the use of optimal weighting can lead to more substantial improvements.

Table 14.6 CM estimates using the standard data sets. Comparison with removing one realization. In all cases, optimal weighting is used. The arithmetic mean values are given with standard deviations within parentheses. The theoretical standard deviations are given within brackets

Parameter	a_1	a_2	b_1	b_2	λ_y	λ_u
True value	−1.5	0.7	2.0	1.0	10.0	1.81
p # real.						
3 100	−1.480	0.684	1.866	1.220	10.84	1.83
	(0.013)	(0.011)	(1.04)	(1.30)	(1.84)	(0.20)
	[0.014]	[0.011]	[0.14]	[0.19]	[0.30]	[0.21]
99	−1.492	0.695	1.968	1.092	10.97	1.82
	(0.013)	(0.011)	(0.17)	(0.22)	(1.24)	(0.20)
	[0.014]	[0.011]	[0.14]	[0.19]	[0.30]	[0.21]
5 100	−1.480	0.685	1.850	1.230	10.81	1.81
	(0.013)	(0.011)	(1.12)	(1.38)	(1.96)	(0.19)
	[0.016]	[0.012]	[0.18]	[0.25]	[0.33]	[0.22]
99	−1.493	0.696	1.961	1.094	10.96	1.81
	(0.014)	(0.011)	(0.17)	(0.22)	(1.23)	(0.19)
	[0.016]	[0.012]	[0.18]	[0.25]	[0.33]	[0.22]

Table 14.7 CM estimates using the standard data sets. Comparison with removing one realization. No weighting ($\mathbf{W} = \mathbf{I}$) is used. The arithmetic mean values are given with standard deviations within parentheses. The theoretical standard deviations are given within brackets

Parameter	a_1	a_2	b_1	b_2	λ_y	λ_u
True value	−1.5	0.7	2.0	1.0	10.0	1.81
p # real.						
1 100	−1.489	0.692	1.994	1.087	10.96	1.92
	(0.020)	(0.014)	(0.28)	(0.38)	(1.24)	(0.22)
	[0.015]	[0.011]	[0.14]	[0.19]	[0.24]	[0.22]
99	−1.490	0.693	2.014	1.062	10.96	1.92
	(0.018)	(0.013)	(0.20)	(0.29)	(1.24)	(0.26)
	[0.015]	[0.011]	[0.14]	[0.19]	[0.24]	[0.22]

The considered system is of second order and given by

$$\begin{aligned}
&y_0(t) - 1.5y_0(t-1) + 0.7y_0(t-2) = 1.0u_0(t-1) + 0.5u_0(t-2)\,, \\
&u_0(t) - 0.5u_0(t-1) = v(t) + 0.7v(t-1)\,, \\
&\mathsf{E}\left\{v^2(t)\right\} = 1, \quad \lambda_y = 1.64, \quad \lambda_u = 0.83\,,
\end{aligned} \tag{14.161}$$

which leads to the signal-to-noise ratio of 6 dB on the input side and 20 dB on the output side.

Table 14.8 CM estimates for increasing length of the covariance vector using the data sets for the system (14.161). The arithmetic mean values are given with standard deviations within parentheses. The theoretical standard deviations are given within brackets

Parameter		a_1	a_2	b_1	b_2	λ_y	λ_u
True value		−1.5	0.7	1.0	0.5	1.64	0.83
p	weight						
1	I	−1.494	0.695	1.003	0.521	1.720	0.855
		(0.019)	(0.015)	(0.063)	(0.089)	(0.138)	(0.113)
		[0.017]	[0.015]	[0.060]	[0.080]	[0.089]	[0.137]
	opt	−1.494	0.695	1.002	0.521	1.720	0.852
		(0.019)	(0.016)	(0.063)	(0.091)	(0.139)	(0.114)
		[0.017]	[0.015]	[0.059]	[0.080]	[0.089]	[0.136]
2	I	−1.495	0.696	1.000	0.519	1.727	0.850
		(0.014)	(0.012)	(0.060)	(0.075)	(0.140)	(0.082)
		[0.015]	[0.012]	[0.056]	[0.070]	[0.087]	[0.124]
	opt	−1.495	0.696	0.999	0.520	1.723	0.844
		(0.015)	(0.013)	(0.060)	(0.073)	(0.137)	(0.086)
		[0.014]	[0.012]	[0.053]	[0.068]	[0.084]	[0.115]
3	I	−1.495	0.697	0.999	0.518	1.735	0.844
		(0.014)	(0.012)	(0.068)	(0.081)	(0.146)	(0.084)
		[0.015]	[0.013]	[0.063]	[0.080]	[0.096]	[0.125]
	opt	−1.495	0.698	0.997	0.521	1.724	0.838
		(0.014)	(0.013)	(0.061)	(0.073)	(0.138)	(0.088)
		[0.013]	[0.011]	[0.051]	[0.068]	[0.084]	[0.106]
4	I	−1.496	0.697	1.000	0.514	1.743	0.838
		(0.015)	(0.013)	(0.080)	(0.096)	(0.157)	(0.094)
		[0.016]	[0.013]	[0.074]	[0.095]	[0.107]	[0.131]
	opt	−1.496	0.697	0.994	0.524	1.722	0.831
		(0.014)	(0.013)	(0.062)	(0.074)	(0.137)	(0.086)
		[0.013]	[0.011]	[0.049]	[0.066]	[0.083]	[0.102]
5	I	−1.497	0.698	1.002	0.510	1.748	0.836
		(0.016)	(0.013)	(0.093)	(0.116)	(0.164)	(0.104)
		[0.017]	[0.014]	[0.085]	[0.112]	[0.117]	[0.139]
	opt	−1.496	0.698	0.992	0.524	1.721	0.826
		(0.015)	(0.013)	(0.063)	(0.073)	(0.137)	(0.090)
		[0.012]	[0.011]	[0.048]	[0.066]	[0.083]	[0.100]

The CM method was applied to 100 realizations, each of length $N = 1000$, with

$$p_y = 2 + p, \;\; p_u = 1 + p, \, p_1 = -1 - p, \, p_2 = 2 + p, \quad p = 1, \ldots, 5 \,. \quad (14.162)$$

The results obtained are summarized in Table 14.8.

Some observations:

- No significant problems with outlier realizations were observed.
- There is much better agreement between MC simulations and theory as compared to Example 14.4.
- There is now also a larger (relative) difference using between the cases of no and optimal weighting, respectively.

■

14.6 The Maximum Likelihood Method

14.6.1 The Prediction Error Method

The class of prediction error methods, which includes ML methods as a special case, was introduced in Sect. 9.2.

For the prediction error method it holds

$$\sqrt{N}(\hat{\boldsymbol{\vartheta}}_{\text{PEM}} - \boldsymbol{\vartheta}_0) \xrightarrow{\text{dist}} \mathscr{N}(\mathbf{0}, \mathbf{P}_{\text{PEM}}) , \tag{14.163}$$

with

$$\mathbf{P}_{\text{PEM}} = [V''_{\infty}(\boldsymbol{\vartheta}_0)]^{-1}\mathbf{P}_0[V''_{\infty}(\boldsymbol{\vartheta}_0)]^{-1} , \tag{14.164}$$

$$\mathbf{P}_0 = \lim_{N\to\infty} N\mathsf{E}\left\{V'_N(\boldsymbol{\vartheta}_0)^T V'_N(\boldsymbol{\vartheta}_0)\right\} . \tag{14.165}$$

cf (14.9). In (14.164), $V''_{\infty}(\boldsymbol{\vartheta}_0)$ denotes the Hessian matrix of $V_{\infty}(\boldsymbol{\vartheta}) = \lim_{N\to\infty} V_N(\boldsymbol{\vartheta})$, evaluated for $\boldsymbol{\vartheta} = \boldsymbol{\vartheta}_0$. Due to the assumption on identifiability, one can assume that $V''_{\infty}(\boldsymbol{\vartheta}_0)$ is indeed invertible.

To find expressions for the covariance matrix $\mathbf{P}_{\text{PEM}}$ (14.164), $V''_{\infty}(\boldsymbol{\vartheta}_0)$ and $\mathbf{P}_0$ are needed. Introduce the notation

$$\boldsymbol{\psi}(t, \boldsymbol{\vartheta}) = -\left(\frac{\partial \boldsymbol{\varepsilon}(t, \boldsymbol{\vartheta})}{\partial \boldsymbol{\vartheta}}\right)^T , \tag{14.166}$$

which is an $(n_{\vartheta} \times n_y)$-dimensional matrix ($\dim \boldsymbol{\vartheta} = n_{\vartheta}$, $\dim \mathbf{y} = n_y$).

The following results, that all relate to Examples 9.1–9.4, are useful. Let $\mathbf{Q} = \mathbf{Q}(\boldsymbol{\vartheta})$ denote the covariance matrix of the true innovations, see (9.19).

Below the superscript, such as (1) in $\mathbf{P}^{(1)}_{\text{PEM}}$ refers to the associate loss function, that is, V_1 in this example.

Lemma 14.10 *Consider the criterion function (9.23) in Example 9.1. Then it holds*

$$\mathbf{P}^{(1)}_{\text{PEM}} = \left[\mathsf{E}\left\{\boldsymbol{\psi}^T(t)\mathbf{Q}^{-1}(\boldsymbol{\vartheta}_0)\boldsymbol{\psi}(t)\right\}\right]^{-1} . \tag{14.167}$$

Proof See Ljung (1999) or Söderström and Stoica (1989). ■

Lemma 14.11 *Consider the criterion function (9.26) in Example 9.2. Then it holds*

$$V''_{\infty}(\boldsymbol{\vartheta}_0) = \mathsf{E}\left\{\boldsymbol{\psi}(t)\ell_{\boldsymbol{\varepsilon}\boldsymbol{\varepsilon}}\boldsymbol{\psi}^T(t) + \ell_{\boldsymbol{\vartheta}\boldsymbol{\vartheta}}\right\} , \tag{14.168}$$

$$\mathbf{P}_0 = \mathsf{E}\left\{\boldsymbol{\psi}(t)\mathsf{E}\left\{\ell_{\boldsymbol{\varepsilon}}^T\ell_{\boldsymbol{\varepsilon}}\right\}\boldsymbol{\psi}^T(t)\right\} + \mathsf{E}\left\{\ell_{\boldsymbol{\vartheta}}^T\ell_{\boldsymbol{\vartheta}}\right\} , \tag{14.169}$$

$$\mathbf{P}_{\text{PEM}}^{(2)} = \left[V''_{\infty}(\boldsymbol{\vartheta}_0)\right]^{-1}\mathbf{P}_0\left[V''_{\infty}(\boldsymbol{\vartheta}_0)\right]^{-1} . \tag{14.170}$$

where

$$\ell_{\varepsilon} = \frac{\partial \ell}{\partial \varepsilon}, \quad \ell_{\boldsymbol{\vartheta}} = \frac{\partial \ell}{\partial \boldsymbol{\vartheta}}, \quad \ell_{\varepsilon\varepsilon} = \frac{\partial^2 \ell}{\partial \varepsilon^2}, \quad \ell_{\boldsymbol{\vartheta}\boldsymbol{\vartheta}} = \frac{\partial^2 \ell}{\partial \boldsymbol{\vartheta}^2} . \tag{14.171}$$

Proof See Appendix 14.F.1. ■

Corollary 14.1 *For the criterion (9.29) it holds that*

$$V''_{\infty}(\boldsymbol{\vartheta}_0) = \mathsf{E}\left\{\boldsymbol{\psi}(t)\mathbf{W}\boldsymbol{\psi}^T(t)\right\} , \tag{14.172}$$

$$\mathbf{P}_0 = \mathsf{E}\left\{\boldsymbol{\psi}(t)\mathbf{WQW}\boldsymbol{\psi}^T(t)\right\} , \tag{14.173}$$

$$\mathbf{P}_{\text{PEM}}^{(4)} = \left[\mathsf{E}\left\{\boldsymbol{\psi}(t)\mathbf{W}\boldsymbol{\psi}^T(t)\right\}\right]^{-1}\mathsf{E}\left\{\boldsymbol{\psi}(t)\mathbf{WQW}\boldsymbol{\psi}^T(t)\right\}\left[\mathsf{E}\left\{\boldsymbol{\psi}(t)\mathbf{W}\boldsymbol{\psi}^T(t)\right\}\right]^{-1} . \tag{14.174}$$

Proof In this case, one finds directly

$$\ell_{\boldsymbol{\varepsilon}} = 2\boldsymbol{\varepsilon}^T\mathbf{W}, \quad \ell_{\boldsymbol{\varepsilon}\boldsymbol{\varepsilon}} = 2\mathbf{W}, \quad \ell_{\boldsymbol{\vartheta}} = \mathbf{0}, \quad \ell_{\boldsymbol{\vartheta}\boldsymbol{\vartheta}} = \mathbf{0} ,$$

and

$$\mathsf{E}\left\{\ell_{\boldsymbol{\varepsilon}}^T\ell_{\boldsymbol{\varepsilon}}\right\} = 4\mathsf{E}\left\{\mathbf{W}\boldsymbol{\varepsilon}\boldsymbol{\varepsilon}^T\mathbf{W}\right\} = 4\mathbf{WQW} ,$$

and the result follows. Note that the result is well-known, see Ljung (1999), Söderström and Stoica (1989). ■

Remark 14.12 As shown in Ljung (1999), Söderström and Stoica (1989), use of V in (9.23) gives always at least as good accuracy as V in (9.29), in the sense

$$\mathbf{P}_{\text{PEM}}^{(1)} \leq \mathbf{P}_{\text{PEM}}^{(4)}(\mathbf{W}) \quad \forall\, \mathbf{W} . \tag{14.175}$$

Furthermore, the covariance matrix $\mathbf{P}_{\text{PEM}}^{(4)}(\mathbf{W})$ can be minimized with respect to the weighting matrix $\mathbf{W}$. The optimal choice of $\mathbf{W}$ is

$$\mathbf{W} = \mathbf{Q}^{-1} , \tag{14.176}$$

which follows from Lemma A.1. This choice cannot be used immediately, as $\mathbf{Q}$ is not known prior to the identification. It may be implemented using a two-step procedure: First apply PEM with an arbitrary $\mathbf{W}$, such as $\mathbf{W} = \mathbf{I}$. Then estimate $\mathbf{Q}$ as the sample covariance matrix of the residuals $\hat{\mathbf{Q}} = \frac{1}{N}\sum_t \boldsymbol{\varepsilon}(t, \hat{\boldsymbol{\theta}})\boldsymbol{\varepsilon}^T(t, \hat{\boldsymbol{\theta}})$ and use $\mathbf{W} = \hat{\mathbf{Q}}^{-1}$ in a second PEM estimation. ■

Corollary 14.2 *For the criterion in (9.28), it holds*

$$V''_{\infty}(\boldsymbol{\vartheta}_0) = \mathbf{P}_0 = \mathsf{E}\left\{\boldsymbol{\psi}(t)\mathbf{Q}^{-1}\boldsymbol{\psi}(t)\right\} + \overline{\mathbf{P}}\,, \tag{14.177}$$

where

$$\overline{\mathbf{P}}_{ij} = \frac{1}{2}\mathrm{tr}[\mathbf{Q}^{-1}\mathbf{Q}_i\mathbf{Q}^{-1}\mathbf{Q}_j]\,, \tag{14.178}$$

and

$$\mathbf{Q}_i = \frac{\partial \mathbf{Q}}{\partial \boldsymbol{\theta}_i}\,. \tag{14.179}$$

Proof See Appendix 14.F.1. ■

Remark 14.13 The covariance matrix $\mathbf{P}_0$ in Corollary 14.2 is in this case also the covariance matrix of the maximum likelihood estimate for Gaussian distributed data and equal to the Cramer–Rao bound. For arbitrary distributions, it gives a smaller covariance matrix than the other options considered here, due to the term $\overline{\mathbf{P}}$. It holds

$$\mathbf{P}_{\mathrm{CRB}} = \mathbf{P}_{\mathrm{ML}} = \mathbf{P}^{(3)}_{\mathrm{PEM}} \leq \mathbf{P}^{(1)}_{\mathrm{PEM}} \leq \mathbf{P}^{(4)}_{\mathrm{PEM}}(\mathbf{W})\,. \tag{14.180}$$

■

The EIV problem considered here is a case where a prediction error method and the maximum likelihood method do differ in general. In this case the ML estimate will be more accurate than the PEM estimate, in the sense that it gives a smaller covariance matrix for the parameter errors. The explanation for the difference is that the innovations covariance matrix $\mathbf{Q}(\boldsymbol{\vartheta})$ and the prediction errors $\{\boldsymbol{\varepsilon}(t, \boldsymbol{\vartheta})\}$ are parameterized with joint parameters. In contrast, in 'standard identification' problems $\mathbf{Q}$ and $\boldsymbol{\varepsilon}(t)$ are parameterized using *separate* parameters. In such a situation, the loss function (9.26) with ℓ given by (9.28) can first be minimized with respect to $\mathbf{Q}$ and the obtained concentrated criterion turns out to be equivalent to the criterion (9.23).

A consequence is that the CRB for Gaussian distributed data can be computed using the results in Corollary 14.2. How to organize the calculations for a generally parameterized state space model is described in Sect. 14.8.2.

Remark 14.14 An alternative way of computing the CRB than the one presented in Sect. 14.8.2 is presented in Hjalmarsson et al. (2011). The asymptotic covariance matrix will in general depend on the system parameters and the noise variances in a quite complicated way. However, in Hjalmarsson et al. (2011) it is also shown, by theoretical calculations, that the information matrix is increasing with the input noise variance, in the sense that $\frac{\partial}{\partial \lambda_u}\mathbf{P}^{-1}_{\mathrm{CRB}}$ is a non-positive definite matrix. ■

14.6.2 The Maximum Likelihood Method in the Frequency Domain

The ML estimate in question was described in Sect. 9.4. One can derive the asymptotic covariance matrix of the estimate $\hat{\boldsymbol{\theta}}$ by using the general line of logic outlined in Sect. 14.1 leading to

$$\sqrt{N}\left(\hat{\boldsymbol{\theta}} - \boldsymbol{\theta}_0\right) \stackrel{N\to\infty}{\longrightarrow} \mathscr{N}(\mathbf{0}, \mathbf{P}_\theta) , \tag{14.181}$$

with

$$\mathbf{P}_\theta = \mathbf{P}_1^{-1}\mathbf{P}_2\mathbf{P}_1^{-1} , \tag{14.182}$$

$$\mathbf{P}_1 = \lim_{N\to\infty} V_{31}''(\boldsymbol{\theta}) , \tag{14.183}$$

$$\mathbf{P}_2 = \lim_{N\to\infty} \mathsf{E}\left\{\left(\sqrt{N}V_{31}'(\boldsymbol{\theta})\right)^T \left(\sqrt{N}V_{31}'(\boldsymbol{\theta})\right)\right\} . \tag{14.184}$$

(It is assumed that $M = N$.) The purpose of the following calculations is to derive explicit expressions for the matrices $\mathbf{P}_1$ and $\mathbf{P}_2$.

First introduce the polynomial F as stable but non-monic from the spectral factorization

$$FF^* = rAA^* + BB^* . \tag{14.185}$$

Then from (9.74) and Parseval's relation it follows

$$V_{31}(\boldsymbol{\theta}) = \frac{1}{N}\sum_{t=1}^{N}\left[\frac{A(q^{-1})}{F(q^{-1})}y(t) - \frac{B(q^{-1})}{F(q^{-1})}u(t)\right]^2 . \tag{14.186}$$

For shortness write

$$\varepsilon(t) = \frac{A}{F}y(t) - \frac{B}{F}u(t) . \tag{14.187}$$

Set

$$\boldsymbol{\psi}(t) = \varepsilon'(t, \boldsymbol{\theta}_0) = \frac{\partial\varepsilon}{\partial\boldsymbol{\theta}}(t, \boldsymbol{\theta})_{|\boldsymbol{\theta}=\boldsymbol{\theta}_0} . \tag{14.188}$$

As $\boldsymbol{\theta}_0$ is a minimum point of $V_{31}(\boldsymbol{\theta})$ (when $N \to \infty$), it must hold

$$\mathsf{E}\left\{\varepsilon(t, \boldsymbol{\theta}_0)\boldsymbol{\psi}(t)\right\} = \mathbf{0} . \tag{14.189}$$

One gets

$$\lim_{N\to\infty} V_{31}(\boldsymbol{\theta}) = \mathsf{E}\left\{\varepsilon^2(t, \boldsymbol{\theta})\right\} , \tag{14.190}$$

$$\lim_{N\to\infty} V_{31}'(\boldsymbol{\theta}_0) = 2\mathsf{E}\left\{\varepsilon(t, \boldsymbol{\theta}_0)\boldsymbol{\psi}(t)\right\} , \tag{14.191}$$

$$\mathbf{P}_1 = \lim_{N\to\infty} V_{31}''(\boldsymbol{\theta}_0) = 2\mathsf{E}\left\{\varepsilon(t,\boldsymbol{\theta}_0)\varepsilon''(t,\boldsymbol{\theta}_0)\right\} + 2\mathsf{E}\left\{\boldsymbol{\psi}(t)\boldsymbol{\psi}^T(t)\right\}, \quad (14.192)$$

$$\begin{aligned}\mathbf{P}_2 &= \lim_{N\to\infty} N\frac{4}{N^2}\mathsf{E}\left\{\sum_{t=0}^{N-1}\sum_{s=0}^{N-1}\varepsilon(t)\boldsymbol{\psi}(t)\varepsilon(s)\boldsymbol{\psi}^T(s)\right\} \\ &= \lim_{N\to\infty}\frac{4}{N}\sum_{t=0}^{N-1}\sum_{s=0}^{N-1}\Bigg[\underbrace{\mathsf{E}\left\{\varepsilon(t)\boldsymbol{\psi}(t)\right\}}_{=0}\underbrace{\mathsf{E}\left\{\varepsilon(s)\boldsymbol{\psi}^T(s)\right\}}_{=0} \\ &\quad + \mathsf{E}\left\{\varepsilon(t)\varepsilon(s)\right\}\mathsf{E}\left\{\boldsymbol{\psi}(t)\boldsymbol{\psi}^T(s)\right\} + \mathsf{E}\left\{\boldsymbol{\psi}(t)\varepsilon(s)\right\}\mathsf{E}\left\{\varepsilon(t)\boldsymbol{\psi}^T(s)\right\}\Bigg].\end{aligned} \quad (14.193)$$

In evaluating (14.193) Lemma A.9 about the properties of the expectation $\mathsf{E}\{x_1x_2x_3x_4\}$ for jointly Gaussian variables is used.

To proceed, the different terms in (14.193) need to be considered in some detail.

For $\boldsymbol{\theta} = \boldsymbol{\theta}_0$, the residuals $\varepsilon(t,\boldsymbol{\theta}_0)$ are indeed white noise and have a constant spectrum. Therefore, the relation

$$\mathsf{E}\left\{\varepsilon(t)\varepsilon(s)\right\} = \lambda_u^0\delta_{t,s} \quad (14.194)$$

follows. Indeed, to spell out details of an explicit proof, let $\tau \geq 0$. It holds

$$\varepsilon(t,\boldsymbol{\theta}_0) = \frac{A}{F}\tilde{y}(t) - \frac{B}{F}\tilde{u}(t), \quad (14.195)$$

and

$$\begin{aligned}\mathsf{E}\left\{\varepsilon(t+\tau,\boldsymbol{\theta}_0)\varepsilon(t,\boldsymbol{\theta}_0)\right\} &= \frac{1}{2\pi}\int_{-\pi}^{\pi}\left[\frac{AA^*r\lambda_u^0}{FF^*} + \frac{BB^*\lambda_u^0}{FF^*}\right]\mathrm{e}^{\mathrm{i}\omega\tau}d\omega \\ &= \lambda_u^0\frac{1}{2\pi i}\oint z^{\tau}\frac{dz}{z} = \lambda_u^0\delta_{\tau,0},\end{aligned} \quad (14.196)$$

which means that $\varepsilon(t)$ behaves as white noise.

Using (14.194) simplifies the second terms in (14.193).

The double sum of the third terms in (14.193) can be rewritten as

$$\begin{aligned}&\lim_{N\to\infty}\frac{4}{N}\sum_{t=0}^{N-1}\sum_{s=0}^{N-1}\left(\mathsf{E}\left\{\varepsilon(s)\boldsymbol{\psi}(t)\right\}\right)\left(\mathsf{E}\left\{\varepsilon(t)\boldsymbol{\psi}^T(s)\right\}\right) \\ &= \lim_{N\to\infty}\frac{4}{N}\sum_{\tau=-N+1}^{N-1}\left(N-|\tau|\right)\left(\mathsf{E}\left\{\varepsilon(t+\tau)\boldsymbol{\psi}(t)\right\}\right)\left(\mathsf{E}\left\{\varepsilon(t-\tau)\boldsymbol{\psi}^T(t)\right\}\right) \\ &= 4\sum_{\tau=-\infty}^{\infty}\left(\mathsf{E}\left\{\varepsilon(t+\tau)\boldsymbol{\psi}(t)\right\}\right)\left(\mathsf{E}\left\{\varepsilon(t-\tau)\boldsymbol{\psi}^T(t)\right\}\right),\end{aligned} \quad (14.197)$$

assuming that the cross-covariance $\mathsf{E}\{\varepsilon(t+\tau)\boldsymbol{\psi}(t)\}$ is exponentially bounded as a function of τ.

The expressions for $\mathbf{P}_1$ and $\mathbf{P}_2$ therefore run as follows, cf. (14.192), (14.193).

$$\mathbf{P}_1 = 2\mathsf{E}\left\{\boldsymbol{\psi}(t)\boldsymbol{\psi}^T(t)\right\} + 2\mathsf{E}\left\{\varepsilon(t,\boldsymbol{\theta}_0)\varepsilon''(t,\boldsymbol{\theta}_0)\right\} , \tag{14.198}$$

$$\mathbf{P}_2 = 4\lambda_u^0\mathsf{E}\left\{\boldsymbol{\psi}(t)\boldsymbol{\psi}^T(t)\right\} + 4\sum_{\tau=-\infty}^{\infty}\left(\mathsf{E}\left\{\varepsilon(t+\tau)\boldsymbol{\psi}(t)\right\}\right)\left(\mathsf{E}\left\{\varepsilon(t-\tau)\boldsymbol{\psi}^T(t)\right\}\right) . \tag{14.199}$$

It is clear from the previous analysis that for the EIV case all terms in (14.198), (14.199) must be taken into account, and none can be assumed to be identically zero. It remains to develop some algorithm for how to compute these additional terms.

Algorithmic details, based on the development in this section, for how to compute $\mathbf{P}_\theta$ are given in Appendix 14.F.2.

Remark 14.15 Equation (14.189) happens to be the same type of expression as for the ML estimation of ARMA processes with all polynomial coefficients as unknowns. However, in the ARMA case $\varepsilon(t,\boldsymbol{\theta}_0)$ is white noise, and the derivative $\partial\varepsilon(t,\boldsymbol{\theta})/\partial\boldsymbol{\theta}$ depends automatically *only on past values* of the measured data, and thereby on past values of $\varepsilon(t,\boldsymbol{\theta})$. It turns out that this *does not apply* in the EIV case. As a consequence, the expressions for $\mathbf{P}_1$ and $\mathbf{P}_2$ become more complicated, with no terms being zero in (14.198), (14.199). ■

Finally, one can derive the asymptotic covariance matrix of the extended vector $\left(\boldsymbol{\theta}^T \ \lambda_u\right)^T$. Referring to the analysis for the pure ARMA case, the result appears to be, *mutatis mutandis*,

$$\sqrt{N}\left(\begin{pmatrix}\hat{\boldsymbol{\theta}}\\ \hat{\lambda}_u\end{pmatrix} - \begin{pmatrix}\boldsymbol{\theta}_0\\ \lambda_u^0\end{pmatrix}\right) \stackrel{N\to\infty}{\longrightarrow} \mathscr{N}\left(\begin{pmatrix}\mathbf{0}\\ \mathbf{0}\end{pmatrix}, \begin{pmatrix}\mathbf{P}_\theta & \mathbf{0}\\ \mathbf{0} & 2\lambda_u^2\end{pmatrix}\right), \tag{14.200}$$

cf. Ljung (1999), Söderström and Stoica (1989).

14.6.3 The Extended Maximum Likelihood in the Frequency Domain

The estimate is given by Eqs. (9.96)–(9.98). These equations can be written as

$$\mathbf{f}_1(\hat{\boldsymbol{\theta}},\hat{r}) = \mathbf{0} , \tag{14.201}$$

$$\mathbf{f}_2(\hat{\boldsymbol{\theta}},\hat{r},\hat{\lambda}_u) = 0 , \tag{14.202}$$

$$\mathbf{f}_3(\hat{\boldsymbol{\theta}},\hat{r},\hat{\lambda}_u) \stackrel{\Delta}{=} \hat{\lambda}_u - V_{31}(\hat{\boldsymbol{\theta}},\hat{r}) = 0 , \tag{14.203}$$

or summarized as

$$\mathbf{f}_N(\hat{\boldsymbol{\theta}}, \hat{r}, \hat{\lambda}_u) = \mathbf{0} \,, \tag{14.204}$$

where the dependence on the sample length is marked with a subscript N. As the true parameter values form a solution in the asymptotic case, linearization leads to (higher-order terms are neglected by assuming N is large enough)

$$\mathbf{0} = \mathbf{f}_N(\hat{\boldsymbol{\theta}}, \hat{r}, \hat{\lambda}_u) = \mathbf{f}_N(\boldsymbol{\theta}_0, r_0, \lambda_{u,0}) + \mathbf{F} \begin{pmatrix} \hat{\boldsymbol{\theta}} - \boldsymbol{\theta}_0 \\ \hat{r} - r_0 \\ \hat{\lambda}_u - \lambda_{u,0} \end{pmatrix} , \tag{14.205}$$

where

$$\mathbf{F} = \lim_{N\to\infty} \begin{pmatrix} \frac{\partial \mathbf{f}_1}{\partial \boldsymbol{\theta}} & \frac{\partial \mathbf{f}_1}{\partial r} & \frac{\partial \mathbf{f}_1}{\partial \lambda_u} \\ \frac{\partial \mathbf{f}_2}{\partial \boldsymbol{\theta}} & \frac{\partial \mathbf{f}_2}{\partial r} & \frac{\partial \mathbf{f}_2}{\partial \lambda_u} \\ \frac{\partial \mathbf{f}_3}{\partial \boldsymbol{\theta}} & \frac{\partial \mathbf{f}_3}{\partial r} & \frac{\partial \mathbf{f}_3}{\partial \lambda_u} \end{pmatrix} . \tag{14.206}$$

The asymptotic normalized covariance matrix of the parameter estimates is then obtained from (14.205) as

$$\mathbf{P} = \lim_{N\to\infty} N \text{cov} \begin{pmatrix} \hat{\boldsymbol{\theta}} - \boldsymbol{\theta}_0 \\ \hat{r} - r_0 \\ \hat{\lambda}_u - \lambda_{u,0} \end{pmatrix} = \mathbf{F}^{-1}\mathbf{H}\mathbf{F}^{-T} , \tag{14.207}$$

$$\mathbf{H} = \lim_{N\to\infty} N \text{cov} \begin{pmatrix} \mathbf{f}_1(\boldsymbol{\theta}_0, r_0) \\ \mathbf{f}_2(\boldsymbol{\theta}_0, r_0, \lambda_{u,0}) \\ \mathbf{f}_3(\boldsymbol{\theta}_0, r_0, \lambda_{u,0}) \end{pmatrix} . \tag{14.208}$$

To characterize **P** one then needs an algorithm for how to compute the matrices **F** and **H**. This is done in Appendix 14.F.3.

14.7 Methods for Periodic Data

14.7.1 The General Case

When the unperturbed input is periodic, a common way the estimation problem is treated is to average over the M periods. In this way one gets a new data set, where the data length is N (not NM as for the original data series). The effect is also that the variance of the measurement noise decreases with a factor M, both on the input side and on the output side. *Using the averaged data in this way is not necessarily statistically efficient.* The true Cramér–Rao lower bound (CRB) is not achieved with such an estimator, but is lower. The reason can be explained as follows. Let the measured data series be a long vector, partitioned into M blocks each of size N,

$$\mathbf{Z} = \left(\mathbf{Z}_1^T \ \mathbf{Z}_2^T \ \ldots \ \mathbf{Z}_M^T\right)^T . \tag{14.209}$$

Then make a linear (non-singular) transformation of the full data vector as

$$\mathbf{X} = \begin{pmatrix} \frac{1}{M}\sum_{k=1}^{M} \mathbf{Z}_k \\ \mathbf{Z}_1 - \mathbf{Z}_2 \\ \vdots \\ \mathbf{Z}_{M-1} - \mathbf{Z}_M \end{pmatrix} \stackrel{\Delta}{=} \begin{pmatrix} \mathbf{X}_1 \\ \mathbf{X}_2 \\ \vdots \\ \mathbf{X}_M \end{pmatrix} . \tag{14.210}$$

To compute the CRB from $\mathbf{Z}$ is the same as to compute the CRB from $\mathbf{X}$. However, in the simplified form when only $\mathbf{X}_1$ is used for computing the CRB, the remaining part of the data is neglected. The parts $\mathbf{X}_2, \ldots, \mathbf{X}_M$ do not depend on the noise-free input, but on the noise statistics (say the variance λ_u of the input noise $\tilde{u}(t)$ and the variance λ_y of the output noise $\tilde{y}(t)$). As the CRB of the system parameters (i.e., the A and B coefficients) and the noise parameters is **not** block diagonal, it will be beneficial from the accuracy point of view, to make use of also the remaining data $\mathbf{X}_2, \ldots, \mathbf{X}_M$. What can be gained if the data $\mathbf{X}_2, \ldots, \mathbf{X}_M$ are in fact used?

To analyze this question, it is useful to use the Slepian–Bang formula, see Lemma A.18, for the CRB. It holds for Gaussian distributed data that the information matrix is

$$\mathbf{J}_{j,k} = \frac{1}{2}\mathrm{tr}\left(\mathbf{R}^{-1}\mathbf{R}_j\mathbf{R}^{-1}\mathbf{R}_k\right), \quad \mathbf{R}_j = \frac{\partial \mathbf{R}}{\partial \boldsymbol{\theta}_j} , \tag{14.211}$$

and write the covariance matrix of the full data vector $\mathbf{X}$ as

$$\mathbf{R} = \mathsf{E}\{\mathbf{X}\mathbf{X}^T\} = \begin{pmatrix} \mathbf{R}_{11} & \mathbf{R}_{12} \\ \mathbf{R}_{21} & \mathbf{R}_{22} \end{pmatrix} , \tag{14.212}$$

where $\mathbf{R}_{11}$ corresponds to the data part $\mathbf{X}_1$, while the block $\mathbf{R}_{22}$ is associated with $\mathbf{X}_2, \ldots, \mathbf{X}_M$. It is easy to see that $\mathbf{R}_{12} = \mathbf{0}$, and for large λ_v

$$\mathbf{R} = \begin{pmatrix} O(\lambda_v) & \mathbf{0} \\ \mathbf{0} & O(1) \end{pmatrix} , \tag{14.213}$$

where λ_v is the variance of the noise-free data.

To make this analysis more explicit, write the data blocks in (14.209) as

$$\mathbf{Z}_k = \mathbf{Z}_0 + \tilde{\mathbf{Z}}_k, \quad k = 1, \ldots, M , \tag{14.214}$$

where $\mathbf{Z}_0$ denotes the effect of the noise-free input, and the noise contributions $\{\tilde{\mathbf{Z}}_k\}_{k=1}^{M}$ are assumed to be uncorrelated between different periods. Introduce the notations

$$\mathbf{R}_0 = \text{cov}(\mathbf{Z}_0), \quad \tilde{\mathbf{R}} = \text{cov}(\tilde{\mathbf{Z}}_k) . \tag{14.215}$$

Using the full data vector $\mathbf{X}$ from (14.210) leads to

$$\mathbf{R} = \begin{pmatrix} \mathbf{R}_0 + \frac{1}{M}\tilde{\mathbf{R}} & \mathbf{0} \\ \mathbf{0} & \mathbf{J}_0 \otimes \tilde{\mathbf{R}} \end{pmatrix} \stackrel{\Delta}{=} \begin{pmatrix} \mathbf{R}_1 & 0 \\ 0 & \mathbf{R}_2 \end{pmatrix} , \tag{14.216}$$

where $\otimes$ denotes Kronecker product and

$$\mathbf{J}_0 = 2\mathbf{I}_{M-1} - \begin{pmatrix} 0 & 1 & & \\ 1 & 0 & \ddots & \\ & \ddots & 0 & 1 \\ & & 1 & 0 \end{pmatrix} . \tag{14.217}$$

As $\mathbf{R}$ in (14.216) is block diagonal, one finds directly that taking the block $\mathbf{R}_2$ into account means that an additional term in the Fisher information matrix appears. Indeed, from (14.211)

$$\begin{aligned} \mathbf{J}_{j,k} &= \frac{1}{2}\text{tr}\left[\mathbf{R}^{-1}\mathbf{R}_j\mathbf{R}^{-1}\mathbf{R}_k\right] \\ &= \frac{1}{2}\text{tr}\left[\mathbf{R}_1^{-1}\mathbf{R}_{1,j}\mathbf{R}_1^{-1}\mathbf{R}_{1,k}\right] + \frac{1}{2}\text{tr}\left[\mathbf{R}_2^{-1}\mathbf{R}_{2,j}\mathbf{R}_2^{-1}\mathbf{R}_{2,k}\right] . \end{aligned} \tag{14.218}$$

The first term in the RHS of (14.218) is the contribution when only $\mathbf{X}_1$ is used. The second term which gives a strict increase to the information matrix can be expressed more explicitly as

$$\begin{aligned} \tilde{\mathbf{J}}_{j,k} &\stackrel{\Delta}{=} \frac{1}{2}\text{tr}\left[\mathbf{R}_2^{-1}\mathbf{R}_{2,j}\mathbf{R}_2^{-1}\mathbf{R}_{2,k}\right] \\ &= \frac{1}{2}\text{tr}\left[(\mathbf{J}_0 \otimes \tilde{\mathbf{R}})^{-1}(\mathbf{J}_0 \otimes \tilde{\mathbf{R}}_j)(\mathbf{J}_0 \otimes \tilde{\mathbf{R}})^{-1}(\mathbf{J}_0 \otimes \tilde{\mathbf{R}}_k)\right] \\ &= \frac{1}{2}\text{tr}\left[\mathbf{I}_{M-1} \otimes \left(\tilde{\mathbf{R}}^{-1}\tilde{\mathbf{R}}_j\tilde{\mathbf{R}}^{-1}\tilde{\mathbf{R}}_k\right)\right] \\ &= \frac{M-1}{2}\text{tr}\left[\tilde{\mathbf{R}}^{-1}\tilde{\mathbf{R}}_j\tilde{\mathbf{R}}^{-1}\tilde{\mathbf{R}}_k\right] . \end{aligned} \tag{14.219}$$

In particular, when both $\tilde{y}(t)$ and $\tilde{u}(t)$ are white noise,

$$\tilde{\mathbf{R}} = \begin{pmatrix} \lambda_y \mathbf{I}_N & \mathbf{0} \\ \mathbf{0} & \lambda_u \mathbf{I}_N \end{pmatrix} . \tag{14.220}$$

Then it is straightforward to derive

$$\tilde{\mathbf{J}}_{\lambda_y,\lambda_y} = \frac{(M-1)N}{2\lambda_y^2}, \quad \tilde{\mathbf{J}}_{\lambda_u,\lambda_u} = \frac{(M-1)N}{2\lambda_u^2} , \tag{14.221}$$

while all other elements of $\tilde{\mathbf{J}}$ are zero. Note that the first term of (14.218) is $O(\lambda_v)$, while obviously the second term is $O(1)$ and is hence almost negligible for large λ_v. Large values of λ_v correspond to large SNR values at both input and output sides. For such cases there is thus not much information lost if the identification is based on only the data averaged over the periods (i.e., only $\mathbf{X}_1$ in (14.210) is used).

14.7.2 Instrumental Variable

Now return to the specific IV estimators treated in Sect. 12.2. Here an accuracy analysis is provided.

Lemma 14.12 *The estimate $\hat{\boldsymbol{\theta}}_{\text{EIV}}$ of (12.14) is asymptotically Gaussian distributed as*

$$\sqrt{MN}(\hat{\boldsymbol{\theta}}_{\text{EIV}} - \boldsymbol{\theta}_0) \xrightarrow{\text{dist}} \mathcal{N}(\mathbf{0}, \mathbf{P}_{\text{EIV}}) , \tag{14.222}$$

where

$$\mathbf{P}_{\text{EIV}} = \mathbf{P}_{\text{EIV}}(\mathbf{W}) = (\mathbf{R}^T\mathbf{W}\mathbf{R})^{-1}\mathbf{R}^T\mathbf{W}\mathbf{C}_{\text{EIV}}\mathbf{W}\mathbf{R}(\mathbf{R}^T\mathbf{W}\mathbf{R})^{-1} , \tag{14.223}$$

$$\mathbf{C}_{\text{EIV}} = \lambda \mathsf{E}\left\{\left[H(q^{-1})\mathbf{Z}(t)\right]\left[H(q^{-1})\mathbf{Z}(t)\right]^T\right\} , \tag{14.224}$$

and $H(q^{-1})$ is given by (14.11).

Proof See Appendix 14.G.1. ■

Remark 14.16 The covariance matrix $\mathbf{P}_{\text{EIV}}$ in (14.223) apparently depends on the weighting matrix $\mathbf{W}$. It follows from Lemma 14.5 that

$$\mathbf{P}_{\text{EIV}} \geq \mathbf{P}_{\text{EIV}}^{\text{opt}} , \tag{14.225}$$

meaning that the difference $\mathbf{P}_{\text{EIV}} - \mathbf{P}_{\text{EIV}}^{\text{opt}}$ is nonnegative definite, where

$$\mathbf{P}_{\text{EIV}}^{\text{opt}} = (\mathbf{R}^T\mathbf{C}_{\text{EIV}}^{-1}\mathbf{R})^{-1} . \tag{14.226}$$

Further, equality holds in (14.225) if

$$\mathbf{W} = \mathbf{C}_{\text{EIV}}^{-1} . \tag{14.227}$$

These statements follow directly from Lemma A.1, cf. Lemma 14.5. ■

Now examine the role of weighting for the IV methods of Example 12.3.

Lemma 14.13 *Consider the instrumental vectors chosen in the extended IV method of Example 12.3. It holds that*

$$\mathbf{C}_{\text{EIV}} = \lambda \mathbf{g}_L \mathbf{g}_L^T \otimes \mathbf{C}_1 + \lambda \mathbf{I}_L \otimes \mathbf{C}_2 , \tag{14.228}$$

where $\mathbf{g}_L$ *is a vector of dimension L with all elements equal to one, and*

$$\begin{aligned} \mathbf{C}_1 &= \mathsf{E}\left\{ \left[H(q^{-1})\varphi_0(t)\right]\left[H(q^{-1})\varphi_0(t)\right]^T \right\} , \\ \mathbf{C}_2 &= E\left\{ \left[H(q^{-1})\tilde{\varphi}_j(t)\right]\left[H(q^{-1})\tilde{\varphi}_j(t)\right]^T \right\} . \end{aligned} \tag{14.229}$$

Furthermore, the optimal covariance matrix (14.226) becomes in this case

$$\mathbf{P}_{\text{EIV}}^{\text{opt}} = \lambda \mathbf{R}_0^{-1} \left(\mathbf{C}_1 + \frac{\mathbf{C}_2}{L} \right) \mathbf{R}_0^{-1} . \tag{14.230}$$

Proof See Appendix 14.G.2. ■

Next examine whether the lower bound can be achieved for other weighting matrices $\mathbf{W}$. It is interesting to evaluate the possible degradation of using $\mathbf{W} = \mathbf{I}$.

Corollary 14.3 *Due to the general form of the covariance matrix* $\mathbf{P}_{\text{EIV}}$ *in (14.223), it follows that*

$$\mathbf{P}_{\text{EIV}}(\mathbf{I}) = \left(\mathbf{R}^T\mathbf{R}\right)^{-1} \mathbf{R}^T \mathbf{C}_{\text{EIV}} \mathbf{R} \left(\mathbf{R}^T\mathbf{R}\right)^{-1} = \lambda \mathbf{R}_0^{-1} \left(\mathbf{C}_1 + \frac{\mathbf{C}_2}{L} \right) \mathbf{R}_0^{-1} . \tag{14.231}$$

Proof See Appendix 14.G.3. ■

Remark 14.17 The important consequence of Corollary 14.3 is that for this particular IV estimator, optimal weighting and no weighting both lead to the same covariance matrix of the parameter estimates! Recall that in Lemma A.1, Eq. (A.8) is only a sufficient, not a necessary, condition for equality in (A.7). For the estimation problem at hand, there is hence no need to apply any particular weighting, as optimal accuracy is readily obtained by the choice $\mathbf{W} = \mathbf{I}$. ■

Remark 14.18 For the basic IV method of Example 12.2, the covariance matrix of $\hat{\boldsymbol{\theta}}$ is:

$$\mathbf{P}_{\text{IV}} = \lambda \mathbf{R}_0^{-1} \left(\mathbf{C}_1 + \mathbf{C}_2 \right) \mathbf{R}_0^{-1} . \tag{14.232}$$

Proof This follows directly from the previous analysis by setting $L = 1$. ■

Remark 14.19 From the explicit expressions (14.230), (14.231), (14.232) one can easily reestablish

$$\mathbf{P}_{\text{IV}} \geq \mathbf{P}_{\text{EIV}} = \mathbf{P}_{\text{EIV}}^{\text{opt}} . \tag{14.233}$$

Further, the noise contribution in $\mathbf{P}_{\text{EIV}}$, as compared to $\mathbf{P}_{\text{IV}}$, is reduced by a factor $1/L$. ■

Remark 14.20 As the EIV estimate gives the same covariance matrix of the parameter estimates with no weighting ($\mathbf{W} = \mathbf{I}$) as with optimal weighting, one may ask if there are ways to get better accuracy than that expressed in (14.230) for any IV method. As in EIV all possible vectors are used as instruments, this can hardly be expected to hold. A more detailed algebraic analysis is carried out in Söderström and Hong (2005), Söderström and Hong (2004), where no smaller covariance matrices were found in a still wider class of IV estimators. ■

14.7.3 The FML and SML Estimates

For the frequency domain ML (FML) estimate, see Sect. 12.3, the following result applies. Recall from (12.40) that the criterion for the FML loss function can be written as

$$V_N(\boldsymbol{\theta}) = \frac{2\pi}{N} \sum_{j=0}^{N-1} \frac{1}{G_j^* G_j \lambda_u + \lambda_y} \left(Y_j^* - U_j^* G_j^*\right)\left(Y_j - G_j U_j\right) , \tag{14.234}$$

where the normalization factor $2\pi/N$ is included for convenience only. (It facilitates asymptotic expressions when $N \to \infty$.)

Following the general procedure in (14.9), the asymptotic covariance matrix of the estimate $\hat{\boldsymbol{\theta}}$ can be evaluated as

$$\mathbf{P}_{\text{FML}} = \left[V''_{\infty}\right]^{-1} \lim_{N\to\infty} N \mathsf{E}\left\{(V'_N)^T V'_N\right\} \left[V''_{\infty}\right]^{-1} , \tag{14.235}$$

where the derivatives are evaluated for the true parameter vector.

Lemma 14.14 *The asymptotic covariance matrix for the FML method is given by*

$$\mathbf{P}_{\text{FML}} = \mathbf{M}_1^{-1} \left(\mathbf{M}_1 + \mathbf{M}_2\right) \mathbf{M}_1^{-1} , \tag{14.236}$$

$$(\mathbf{M}_1)_{k,\ell} = 2 \int_{-\pi}^{\pi} \frac{\phi_{u_0}}{\left(\lambda_y + GG^*\lambda_u\right)} \frac{\partial G^*}{\partial \boldsymbol{\theta}_k} \frac{\partial G}{\partial \boldsymbol{\theta}_\ell} \mathrm{d}\omega , \tag{14.237}$$

$$(\mathbf{M}_2)_{k,\ell} = 2 \int_{-\pi}^{\pi} \frac{\lambda_u \lambda_y}{\left(\lambda_y + GG^*\lambda_u\right)^2} \frac{\partial G^*}{\partial \boldsymbol{\theta}_k} \frac{\partial G}{\partial \boldsymbol{\theta}_\ell} \mathrm{d}\omega . \tag{14.238}$$

Proof See Appendix 14.G.4. ■

To find yet alternative forms of $\mathbf{M}_1$ and $\mathbf{M}_2$, introduce the spectral factorizations

$$\phi_{u_0}(\omega) = H(\mathrm{e}^{\mathrm{i}\omega}) H(\mathrm{e}^{-\mathrm{i}\omega}) \lambda_v , \tag{14.239}$$

$$\lambda_y + G(\mathrm{e}^{\mathrm{i}\omega}) G(\mathrm{e}^{-\mathrm{i}\omega}) \lambda_u = F(\mathrm{e}^{\mathrm{i}\omega}) F(\mathrm{e}^{-\mathrm{i}\omega}) \lambda_\varepsilon , \tag{14.240}$$

where $H(q^{-1})$ and $F(q^{-1})$ are restricted to be monic, stable, and having stable inverses. Then it follows that the matrices in (14.236) can alternatively be written as

$$\mathbf{M}_1 = \frac{\lambda_v}{\lambda_\varepsilon}\mathsf{E}\left\{\frac{H(q^{-1}}{F(q^{-1})}\left(\frac{\partial G(q^{-1})}{\partial \boldsymbol{\theta}}\right)^T v_0(t)\,\frac{H(q^{-1})}{F(q^{-1})}\left(\frac{\partial G(q^{-1})}{\partial \boldsymbol{\theta}}\right)v_0(t)\right\}, \tag{14.241}$$

$$\mathbf{M}_2 = \frac{\lambda_u\lambda_y}{\lambda_\varepsilon^2}\mathsf{E}\left\{\frac{1}{F^2(q^{-1})}\left(\frac{\partial G(q^{-1})}{\partial \boldsymbol{\theta}}\right)^T v_0(t)\,\frac{1}{F^2(q^{-1})}\left(\frac{\partial G(q^{-1})}{\partial \boldsymbol{\theta}}\right)v_0(t)\right\}, \tag{14.242}$$

where $v_0(t)$ denotes white noise of unit variance.

The SML estimator is not an exact ML estimator. However, it was proven in Schoukens et al. (1997) that the estimator is consistent if the number of experiments $M \geq 4$. Further, it was also shown that for $M \geq 6$ the covariance matrix of the estimates of $\boldsymbol{\theta}$ satisfies (asymptotically for large N)

$$\mathbf{P}_{\text{SML}} = \frac{M-2}{M-3}\mathbf{P}_{\text{FML}}\,. \tag{14.243}$$

Here, $\mathbf{P}_{\text{FML}}$ denotes the covariance matrix for the frequency domain maximum likelihood method, assuming that the noise variances are fully known, see (14.236).

Remark 14.21 The factor $(M-2)/(M-3)$ gives a loss of efficiency. However, also for small values of M, this factor is rather close to 1. ■

14.8 The Cramér–Rao Lower Bound for Maximum Likelihood Problems

14.8.1 Introduction

For consistent and unbiased estimates there exists a lower bound on the covariance matrix of the parameter estimates. This bound is the celebrated Cramér–Rao lower bound (CRB), see Appendix A.3.5 for some general expressions. It is further considered in this section.

In Sect. 14.8.2 an algorithm is derived for computing the CRB for a general state space model. Section 14.8.3 is devoted to the particular situation of complex data using also the unknown input samples as unknowns, that is, the identification problem considered in Sect. 9.4. Finally, some comparisons are given in Sect. 14.8.4.

14.8.2 *Algorithm for Computing the CRB for Arbitrary State Space Models*

For Gaussian distributed data the covariance matrix of the parameter estimates is given by Corollary 14.2.

14.8.2.1 Model Description

Consider a linear multivariable system given on state space form as

$$\begin{aligned} \mathbf{x}(t+1) &= \mathbf{F}(\boldsymbol{\theta})\mathbf{x}(t) + \mathbf{v}(t) , \\ \mathbf{y}(t) &= \mathbf{C}(\boldsymbol{\theta})\mathbf{x}(t) + \mathbf{e}(t) , \end{aligned} \tag{14.244}$$

$$\mathsf{E}\left\{\begin{pmatrix} \mathbf{v}(t) \\ \mathbf{e}(t) \end{pmatrix}\right\} = \mathbf{0} , \tag{14.245}$$

$$\mathsf{E}\left\{\begin{pmatrix} \mathbf{v}(t) \\ \mathbf{e}(t) \end{pmatrix} (\mathbf{v}^T(s) \ \mathbf{e}^T(s))\right\} = \begin{pmatrix} \mathbf{R}_1(\boldsymbol{\theta}) & \mathbf{R}_{12}(\boldsymbol{\theta}) \\ \mathbf{R}_{12}^T(\boldsymbol{\theta}) & \mathbf{R}_2(\boldsymbol{\theta}) \end{pmatrix} \delta_{t,s} . \tag{14.246}$$

The noise sequences $\{\mathbf{v}(t)\}$ and $\{\mathbf{e}(t)\}$ are assumed to be Gaussian distributed. The system matrices $\mathbf{F}(\boldsymbol{\theta})$, $\mathbf{C}(\boldsymbol{\theta})$, $\mathbf{R}_1(\boldsymbol{\theta})$, $\mathbf{R}_{12}(\boldsymbol{\theta})$ and $\mathbf{R}_2(\boldsymbol{\theta})$ are assumed to be smooth functions of the parameter vector $\boldsymbol{\theta}$. When there is no risk of confusion, the argument $\boldsymbol{\theta}$ will be suppressed for simplicity.

Further consider the case of an identifiable model, that is, there is a 'true' and unique parameter vector $\boldsymbol{\theta}_0$ that describes the exact dynamics. This assumption introduces some limitations on how the system matrices $\mathbf{F}(\boldsymbol{\theta})$, $\mathbf{C}(\boldsymbol{\theta})$, $\mathbf{R}_1(\boldsymbol{\theta})$, $\mathbf{R}_{12}(\boldsymbol{\theta})$, $\mathbf{R}_2(\boldsymbol{\theta})$ can be parameterized. The model (14.244) can easily be complemented with a known input signal. For convenience only, drop such input terms here and focus on the effects of the noise.

It will be necessary to transform the state space model (14.244) into innovations form, obtained from the Kalman filter, Sect. 9.2:

$$\hat{\mathbf{x}}(t+1|t) = (\mathbf{F} - \mathbf{K}\mathbf{C})\hat{\mathbf{x}}(t|t-1) + \mathbf{K}\mathbf{y}(t) , \tag{14.247}$$

$$\boldsymbol{\varepsilon}(t, \boldsymbol{\theta}) = \mathbf{y}(t) - \mathbf{C}\hat{\mathbf{x}}(t|t-1) , \tag{14.248}$$

where

$$\mathbf{K} = [\mathbf{F}\mathbf{P}\mathbf{C}^T + \mathbf{R}_{12}][\mathbf{C}\mathbf{P}\mathbf{C}^T + \mathbf{R}_2]^{-1} , \tag{14.249}$$

$$\mathbf{P} = \mathbf{F}\mathbf{P}\mathbf{F}^T + \mathbf{R}_1 - [\mathbf{F}\mathbf{P}\mathbf{C}^T + \mathbf{R}_{12}][\mathbf{C}\mathbf{P}\mathbf{C}^T + \mathbf{R}_2]^{-1}[\mathbf{C}\mathbf{P}\mathbf{F}^T + \mathbf{R}_{12}^T] , \tag{14.250}$$

$$\mathbf{Q} = \mathsf{E}\left\{\varepsilon(t, \boldsymbol{\theta})\varepsilon^T(t, \boldsymbol{\theta})\right\} = \mathbf{C}\mathbf{P}\mathbf{C}^T + \mathbf{R}_2 . \tag{14.251}$$

14.8.2.2 Sensitivity Matrices

What is needed to calculate in order to find expressions for the covariance matrix of the parameter estimates? Set

$$\boldsymbol{\psi}(t) = \begin{pmatrix} \boldsymbol{\psi}_1^T(t) \\ \vdots \\ \boldsymbol{\psi}_{n_\theta}(t) \end{pmatrix} . \tag{14.252}$$

To compute the CRB apparently, one needs

- $\mathsf{E}\left\{\boldsymbol{\psi}_i(t)\boldsymbol{\psi}_j^T(t)\right\}, \ i, j = 1, \ldots, n_\theta$.
- $\mathbf{Q}_i$.

These quantities can be derived from sensitivity derivatives of the optimal predictor, (14.247), (14.248), and the Riccati equation (14.250). In what follows let an index i denote the derivative with respect to $\boldsymbol{\theta}_i$ (rather than the ith component).

Start by deriving $\mathbf{P}_i$. Differentiating the Riccati equation gives

$$\begin{aligned}
\mathbf{P}_i &= \mathbf{F}_i\mathbf{P}\mathbf{F}^T + \mathbf{F}\mathbf{P}_i\mathbf{F}^T + \mathbf{F}\mathbf{P}\mathbf{F}_i^T + \mathbf{R}_{1i} \\
&\quad -(\mathbf{F}_i\mathbf{P}\mathbf{C}^T + \mathbf{F}\mathbf{P}_i\mathbf{C}^T + \mathbf{F}\mathbf{P}\mathbf{C}_i^T + \mathbf{R}_{12i})\mathbf{K}^T \\
&\quad +\mathbf{K}(\mathbf{C}_i\mathbf{P}\mathbf{C}^T + \mathbf{C}\mathbf{P}_i\mathbf{C}^T + \mathbf{C}\mathbf{P}\mathbf{C}_i^T + \mathbf{R}_{2i})\mathbf{K}^T \\
&\quad -\mathbf{K}(\mathbf{C}_i\mathbf{P}\mathbf{F}^T + \mathbf{C}\mathbf{P}_i\mathbf{F}^T + \mathbf{C}\mathbf{P}\mathbf{F}_i^T + \mathbf{R}_{12i}^T) \\
&= (\mathbf{F} - \mathbf{K}\mathbf{C})\mathbf{P}_i(\mathbf{F} - \mathbf{K}\mathbf{C})^T \\
&\quad +(\mathbf{R}_{1i} - \mathbf{K}\mathbf{R}_{12i}^T - \mathbf{R}_{12i}\mathbf{K}^T + \mathbf{K}\mathbf{R}_{2i}\mathbf{K}^T) \\
&\quad +(\mathbf{F}_i - \mathbf{K}\mathbf{C}_i)\mathbf{P}(\mathbf{F} - \mathbf{K}\mathbf{C})^T + (\mathbf{F} - \mathbf{K}\mathbf{C})\mathbf{P}(\mathbf{F}_i - \mathbf{K}\mathbf{C}_i)^T .
\end{aligned} \tag{14.253}$$

This is a (linear) Lyapunov equation in $\mathbf{P}_i$ that is easy to solve numerically.

The sensitivity of $\mathbf{Q}$ and $\mathbf{K}$ is easily related to $\mathbf{P}_i$:

$$\mathbf{Q}_i = \mathbf{C}_i\mathbf{P}\mathbf{C}^T + \mathbf{C}\mathbf{P}_i\mathbf{C}^T + \mathbf{C}\mathbf{P}\mathbf{C}_i^T + \mathbf{R}_{2i} , \tag{14.254}$$

$$\begin{aligned}
\mathbf{K}_i &= (\mathbf{F}_i\mathbf{P}\mathbf{C}^T + \mathbf{F}\mathbf{P}_i\mathbf{C}^T + \mathbf{F}\mathbf{P}\mathbf{C}_i^T + \mathbf{R}_{12i})\mathbf{Q}^{-1} - \mathbf{K}\mathbf{Q}_i\mathbf{Q}^{-1} \\
&= (\mathbf{F}_i - \mathbf{K}\mathbf{C}_i)\mathbf{P}\mathbf{C}^T\mathbf{Q}^{-1} + (\mathbf{F} - \mathbf{K}\mathbf{C})\mathbf{P}_i\mathbf{C}^T\mathbf{Q}^{-1} \\
&\quad +\mathbf{R}_{12i}\mathbf{Q}^{-1} + (\mathbf{F} - \mathbf{K}\mathbf{C})\mathbf{P}\mathbf{C}_i^T\mathbf{Q}^{-1} - \mathbf{K}\mathbf{R}_{2i}\mathbf{Q}^{-1} .
\end{aligned} \tag{14.255}$$

Next differentiate the optimal predictor (14.247), (14.248):

$$\begin{aligned}
\boldsymbol{\psi}_i(t) = \boldsymbol{\varepsilon}_i(t) &= -\mathbf{C}_i\hat{\mathbf{x}}(t|t-1) - \mathbf{C}\hat{\mathbf{x}}(t|t-1)_i , \\
\hat{\mathbf{x}}(t+1|t)_i &= (\mathbf{F}_i - \mathbf{K}_i\mathbf{C} - \mathbf{K}\mathbf{C}_i)\hat{\mathbf{x}}(t|t-1) \\
&\quad +(\mathbf{F} - \mathbf{K}\mathbf{C})\hat{\mathbf{x}}(t|t-1)_i + \mathbf{K}_i\mathbf{y}(t) .
\end{aligned} \tag{14.256}$$

One is now in a position to form an augmented state space model for computing $\boldsymbol{\psi}_i(t)$. Introduce the notations

$$\overline{\mathbf{F}} = \begin{pmatrix} \mathbf{F}_1 \\ \vdots \\ \mathbf{F}_{n\theta} \end{pmatrix}, \ \overline{\mathbf{C}} = \begin{pmatrix} \mathbf{C}_1 \\ \vdots \\ \mathbf{C}_{n\theta} \end{pmatrix}, \ \overline{\mathbf{K}} = \begin{pmatrix} \mathbf{K}_1 \\ \vdots \\ \mathbf{K}_{n\theta} \end{pmatrix}, \tag{14.257}$$

$$\mathscr{F} = \begin{pmatrix} \mathbf{F} & \mathbf{0}_{n_x \times n_x} & \mathbf{0}_{n_x \times n_\theta n_x} \\ \mathbf{KC} & \mathbf{F} - \mathbf{KC} & \mathbf{0}_{n_x \times n_\theta n_x} \\ \overline{\mathbf{K}}\mathbf{C} & \overline{\mathbf{F}} - \overline{\mathbf{K}}\mathbf{C} - \mathbf{K}\overline{\mathbf{C}} & \mathbf{I}_{n_\theta} \otimes (\mathbf{F} - \mathbf{KC}) \end{pmatrix}. \tag{14.258}$$

Then

$$\begin{pmatrix} \mathbf{x}(t+1) \\ \hat{\mathbf{x}}(t+1) \\ \hat{\mathbf{x}}(t+1)_1 \\ \hat{\mathbf{x}}(t+1)_2 \\ \vdots \\ \hat{\mathbf{x}}(t+1)_{n_\theta} \end{pmatrix} = \mathscr{F} \begin{pmatrix} \mathbf{x}(t) \\ \hat{\mathbf{x}}(t) \\ \hat{\mathbf{x}}(t)_1 \\ \hat{\mathbf{x}}(t)_2 \\ \vdots \\ \hat{\mathbf{x}}(t)_{n_\theta} \end{pmatrix} + \begin{pmatrix} \mathbf{I} & \mathbf{0} \\ \mathbf{0} & \mathbf{K} \\ \mathbf{0} & \overline{\mathbf{K}} \end{pmatrix} \begin{pmatrix} \mathbf{v}(t) \\ \mathbf{e}(t) \end{pmatrix}, \tag{14.259}$$

$$\begin{pmatrix} \boldsymbol{\psi}_1(t) \\ \vdots \\ \boldsymbol{\psi}_{n_\theta}(t) \end{pmatrix} = \begin{pmatrix} \mathbf{0} & -\overline{\mathbf{C}} & -\mathbf{I}_{n_\theta} \otimes \mathbf{C} \end{pmatrix} \begin{pmatrix} x(t) \\ \hat{\mathbf{x}}(t) \\ \hat{\mathbf{x}}(t)_1 \\ \hat{\mathbf{x}}(t)_2 \\ \vdots \\ \hat{\mathbf{x}}(t)_{n_\theta} \end{pmatrix}. \tag{14.260}$$

The covariance matrix of the augmented state vector in (14.259) can easily be found by solving a Lyapunov equation.

To sum up, the following generic algorithm has thus been derived.

Algorithm 14.1 The matrices $\mathbf{F}, \ \mathbf{C}, \ \mathbf{R}_1, \ \mathbf{R}_{12}, \ \mathbf{R}_2$ and $\mathbf{F}_i, \ \mathbf{C}_i, \ \mathbf{R}_{1i}, \ \mathbf{R}_{12i}, \ \mathbf{R}_{2i}$ for $i = 1, \ldots, n_\theta$ are given.

1. Solve the Riccati equation (14.250) to get $\mathbf{P}, \ \mathbf{K}, \ \mathbf{Q}$.
2. For $i = 1, \ldots, m$: Solve the Lyapunov equation (14.253) to get $\mathbf{P}_i$. Compute $\mathbf{K}_i$ from (14.255) and $\mathbf{Q}_i$ from (14.254). Form the corresponding block rows of the state space model (14.259), (14.260).
3. Denoting the state space model (14.259), (14.260) in brief as

$$\bar{\mathbf{x}}(t+1) = \mathscr{F}\bar{\mathbf{x}}(t) + \mathscr{K}\bar{\mathbf{v}}(t), \quad \boldsymbol{\psi}(t) = \mathscr{C}\bar{\mathbf{x}}(t),$$

solve the Lyapunov equation

$$\mathscr{P} = \mathscr{F}\mathscr{P}\mathscr{F}^T + \mathscr{K} \operatorname{cov}(\bar{\mathbf{v}}(t)) \mathscr{K}^T \tag{14.261}$$

to get

$$\mathsf{E}\left\{\boldsymbol{\psi}(t)\boldsymbol{\psi}^T(t)\right\} = \mathscr{C}\mathscr{P}\mathscr{C}^T .$$

4. The covariance matrix of the parameter estimates is composed of two kinds of terms. One term, such as $\overline{\mathbf{P}}$ in (14.178), depends solely on $\mathbf{Q}$ and $\{\mathbf{Q}_i\}$ and is straightforward to form. The other possible term is of the form $\mathsf{E}\left\{\boldsymbol{\psi}(t)\mathbf{S}\boldsymbol{\psi}^T(t)\right\}$, where $\mathbf{S}$ is a known weighting matrix. This term is easily formed from $\mathsf{E}\left\{\boldsymbol{\psi}(t)\boldsymbol{\psi}^T(t)\right\}$ as

$$\begin{aligned}[\mathsf{E}\left\{\boldsymbol{\psi}(t)\mathbf{S}\boldsymbol{\psi}^T(t)\right\}]_{i,j} &= \mathsf{E}\left\{\boldsymbol{\psi}_i^T(t)\mathbf{S}\boldsymbol{\psi}_j(t)\right\} = \mathsf{E}\left\{\operatorname{tr}\left\{\boldsymbol{\psi}_i^T(t)\mathbf{S}\boldsymbol{\psi}_j(t)\right\}\right. \\ &= \operatorname{tr}\left[\mathsf{E}\left\{\boldsymbol{\psi}_j(t)\boldsymbol{\psi}_i^T(t)\right\}\mathbf{S}\right] . \end{aligned} \tag{14.262}$$

■

Remark 14.22 As pointed out earlier the prediction errors are invariant to a scaling of $\mathbf{R}_1$, $\mathbf{R}_{12}$, and $\mathbf{R}_2$. Note that this is not the case for the ML estimator criterion defined by (9.28). If the $\mathbf{R}$ matrices are multiplied by r, so is $\mathbf{Q}$, and the ML criterion (9.28) will in fact be effected. This means that the scaling factor can be conveniently included in the parameter vector when the ML criterion is applied, in contrast to other PEM cases.

How to modify and complement Algorithm 14.1 to also compute the covariance matrix of prediction error estimates is described in Söderström (2006). A key trick is to exploit also the minimal value of the identification criterion in order to determine the scaling factor r. ■

14.8.3 The Cramér–Rao Lower Bound for the Frequency Domain Maximum Likelihood Problem

Consider the ML problem defined in the frequency domain for estimating $\mathbf{U}_0$, $\boldsymbol{\theta}$, λ_u and possibly r. See Sect. 9.4.

The Cramér–Rao bound, which gives a lower bound for the covariance matrix of an estimated $\mathbf{x}$ vector for any unbiased estimator, is given by the inverse of the Fisher information matrix. The information matrix can be found as, see (A.132),

$$\mathbf{J} = \mathsf{E}\left\{\left(\frac{\partial}{\partial \mathbf{x}}(-\log L)\right)^T\left(\frac{\partial}{\partial \mathbf{x}}(-\log L)\right)\right\} . \tag{14.263}$$

Consider the ML problem defined in the frequency domain for estimating $\mathbf{U}_0$, $\boldsymbol{\theta}$, λ_u, and possibly r. As the information matrix is defined for real-valued parameters, it makes sense to consider here

$$\mathbf{x} = \left(\mathbf{U}_{0,R}^T\ \mathbf{U}_{0,I}^T\ \boldsymbol{\theta}^T\ \lambda_u\ r\right)^T \tag{14.264}$$

as the total parameter vector to be estimated, where $\mathbf{U}_{0,R}$ is the real part of $\mathbf{U}_0$ and $\mathbf{U}_{0,I}$ the imaginary part.

Use now the notations of Sect. 14.6.2. In particular, note that $\mathbf{Y}$ and $\mathbf{U}$ are column vectors and $\mathbf{G}$ a diagonal matrix. Further set

$$G_k^{\mu} = \frac{\partial}{\partial \boldsymbol{\theta}_{\mu}} G_k, \quad \mathbf{G}_{\mu} = \frac{\partial}{\partial \boldsymbol{\theta}_{\mu}} \mathbf{G} . \tag{14.265}$$

Write in the following also the other variables with a real part and an imaginary part as

$$\mathbf{Y} = \mathbf{Y}_R + \mathrm{i}\mathbf{Y}_I, \quad \mathbf{G} = \mathbf{G}_R + \mathrm{i}\mathbf{G}_I , \tag{14.266}$$

and similarly for $\mathbf{U}$, $\tilde{\mathbf{Y}}$, and $\tilde{\mathbf{U}}$. Further, recall that $\bar{X}$ denotes the complex conjugate of X.

Special care has to be exercised when handling the complex-valued variables. The derivation here is patterned after Stoica and Nehorai (1991) which handles a related problem in array signal processing.

From (9.81) it follows that the likelihood function is in this case

$$\begin{aligned} L(\mathbf{Y}, \mathbf{U}) = & \frac{1}{(2\pi)^{2N} \left(\frac{\lambda_u}{2}\right)^N \left(\frac{r\lambda_u}{2}\right)^N} \\ & \times \exp\left(-\frac{1}{r\lambda_u} (\mathbf{Y} - \mathbf{G}\mathbf{U}_0)^* (\mathbf{Y} - \mathbf{G}\mathbf{U}_0) - \frac{1}{\lambda_u} (\mathbf{U} - \mathbf{U}_0)^* (\mathbf{U} - \mathbf{U}_0) \right) , \end{aligned} \tag{14.267}$$

and hence

$$\begin{aligned} \log(L) = & \ \mathrm{const} - N \log(\lambda_u) - N \log(r\lambda_u) \\ & - \frac{1}{r\lambda_u} (\mathbf{Y} - \mathbf{G}\mathbf{U}_0)^* (\mathbf{Y} - \mathbf{G}\mathbf{U}_0) - \frac{1}{\lambda_u} (\mathbf{U} - \mathbf{U}_0)^* (\mathbf{U} - \mathbf{U}_0) . \end{aligned} \tag{14.268}$$

It is straightforward to find the derivatives of $\log(L)$ with respect to the components of the vector $\mathbf{x}$, (14.264). The results will be as follows:

$$\begin{aligned} \frac{\partial \log(L)}{\partial \mathbf{U}_{0,R}} & = -\frac{1}{r\lambda_u} \left(\mathbf{G}^* \tilde{\mathbf{Y}} + \bar{\mathbf{G}} \bar{\tilde{\mathbf{Y}}} \right) - \frac{1}{\lambda_u} \left(\tilde{\mathbf{U}} + \bar{\tilde{\mathbf{U}}} \right) \\ & = -\frac{2}{r\lambda_u} \mathrm{Real}\left(\mathbf{G}^* \tilde{\mathbf{Y}} \right) - \frac{2}{\lambda_u} \mathrm{Real}\left(\tilde{\mathbf{U}} \right) , \end{aligned} \tag{14.269}$$

$$\begin{aligned} \frac{\partial \log(L)}{\partial \mathbf{U}_{0,I}} & = -\frac{1}{r\lambda_u} \left(-\mathrm{i}\mathbf{G}^* \tilde{\mathbf{Y}} + \mathrm{i}\mathbf{G}^T \bar{\tilde{\mathbf{Y}}} \right) - \frac{1}{\lambda_u} \left(-\mathrm{i}\tilde{\mathbf{U}} + \mathrm{i}\bar{\tilde{\mathbf{U}}} \right) \\ & = -\frac{2}{r\lambda_u} \mathrm{Imag}\left(\mathbf{G}^* \tilde{\mathbf{Y}} \right) - \frac{2}{\lambda_u} \mathrm{Imag}\left(\tilde{\mathbf{U}} \right) , \end{aligned} \tag{14.270}$$

$$\frac{\partial \log(L)}{\partial \boldsymbol{\theta}_\mu} = \frac{2}{r\lambda_u} \text{Real}\left(\mathbf{U}_0^* \mathbf{G}_\mu^* \tilde{\mathbf{Y}}\right) , \tag{14.271}$$

$$\frac{\partial \log(L)}{\partial \lambda_u} = -\frac{N}{\lambda_u} - \frac{N}{r\lambda_u} r + \frac{1}{r\lambda_u^2} \tilde{\mathbf{Y}}^* \tilde{\mathbf{Y}} + \frac{1}{\lambda_u^2} \tilde{\mathbf{U}}^* \tilde{\mathbf{U}} , \tag{14.272}$$

$$\frac{\partial \log(L)}{\partial r} = -\frac{N}{r} + \frac{1}{r^2 \lambda_u} \tilde{\mathbf{Y}}^* \tilde{\mathbf{Y}} . \tag{14.273}$$

Next one can note that all odd moments of Gaussian variables (real- or complex-valued) with zero mean will automatically vanish. Further, the expressions (14.269)–(14.271) are all linear in $\tilde{\mathbf{Y}}$ and $\tilde{\mathbf{U}}$, while the expressions (14.272), (14.273) depend on even powers only. The implication is that these two groups of expressions will be uncorrelated, and the associated blocks of the information matrix will be zero.

To proceed, the following results will be useful. They are verified by direct computations. Let $\mathbf{x}$ and $\mathbf{y}$ be two arbitrary complex-valued vectors. Then it holds

$$\text{Real}(\mathbf{x})\text{Real}(\mathbf{y}^T) = \frac{1}{2}\left[\text{Real}(\mathbf{x}\mathbf{y}^T) + \text{Real}(\mathbf{x}\mathbf{y}^*)\right] , \tag{14.274}$$

$$\text{Imag}(\mathbf{x})\text{Imag}(\mathbf{y}^T) = -\frac{1}{2}\left[\text{Real}(\mathbf{x}\mathbf{y}^T) - \text{Real}(\mathbf{x}\mathbf{y}^*)\right] , \tag{14.275}$$

$$\text{Real}(\mathbf{x})\text{Imag}(\mathbf{y}^T) = \frac{1}{2}\left[\text{Imag}(\mathbf{x}\mathbf{y}^T) - \text{Imag}(\mathbf{x}\mathbf{y}^*)\right] . \tag{14.276}$$

For the second-order moments it holds

$$\mathsf{E}\left\{\tilde{\mathbf{U}}\tilde{\mathbf{U}}^*\right\} = \lambda_u \mathbf{I}_N, \quad \mathsf{E}\left\{\tilde{\mathbf{U}}\tilde{\mathbf{U}}^T\right\} = \mathbf{0}, \quad \mathsf{E}\left\{\tilde{\mathbf{U}}^*\tilde{\mathbf{U}}\right\} = N\lambda_u \tag{14.277}$$

with obvious modifications if the output noise is considered instead.

Corresponding to the decomposition (14.264) of the parameter vector, the information matrix (14.263) has the block structure

$$\mathbf{J} = \begin{pmatrix} \mathbf{J}_{1,1} & \mathbf{J}_{1,2} & \mathbf{J}_{1,3} & \mathbf{0} & \mathbf{0} \\ \mathbf{J}_{2,1} & \mathbf{J}_{2,2} & \mathbf{J}_{2,3} & \mathbf{0} & \mathbf{0} \\ \mathbf{J}_{3,1} & \mathbf{0} & \mathbf{J}_{3,3} & \mathbf{0} & \mathbf{0} \\ \mathbf{0} & \mathbf{0} & \mathbf{0} & \mathbf{J}_{4,4} & \mathbf{J}_{4,5} \\ \mathbf{0} & \mathbf{0} & \mathbf{0} & \mathbf{J}_{5,4} & \mathbf{J}_{5,5} \end{pmatrix} . \tag{14.278}$$

In Appendix 14.H the following expressions are derived for the blocks of (14.278):

$$\mathbf{J}_{1,1} = \frac{2}{r\lambda_u} \text{Real}\left(\mathbf{G}^*\mathbf{G}\right) + \frac{2}{\lambda_u}\mathbf{I} , \tag{14.279}$$

$$\mathbf{J}_{1,2} = \mathbf{0} , \tag{14.280}$$

$$\left(\mathbf{J}_{1,3}\right)_\mu = -\frac{2}{r\lambda_u} \text{Real}\left(\mathbf{G}^*\mathbf{G}_\mu \mathbf{U}_0\right) , \tag{14.281}$$

$$\mathbf{J}_{2,2} = \frac{2}{r\lambda_u}\text{Real}\left(\mathbf{G}^*\mathbf{G}\right) + \frac{2}{\lambda_u}\mathbf{I}\,, \tag{14.282}$$

$$\left(\mathbf{J}_{2,3}\right)_\mu = \frac{2}{r\lambda_u}\text{Imag}\left(\mathbf{G}^*\mathbf{G}_\mu\mathbf{U}_0\right)\,, \tag{14.283}$$

$$\left(\mathbf{J}_{3,3}\right)_{\mu,\nu} = \frac{2}{r\lambda_u}\text{Real}\left(\mathbf{U}_0^*\mathbf{G}_\mu^*\mathbf{G}_\nu\mathbf{U}_0\right)\,, \tag{14.284}$$

$$\mathbf{J}_{4,4} = \frac{2N}{\lambda_u^2}\,, \tag{14.285}$$

$$\mathbf{J}_{4,5} = \frac{N}{r\lambda_u}\,, \tag{14.286}$$

$$\mathbf{J}_{5,5} = \frac{N}{r^2}\,. \tag{14.287}$$

Now some comments and observations are in order:

- In case the noise variance ratio r happens to be known, all the calculations above still hold, if the last row and the last column of the information matrix $\mathbf{J}$ are dropped.
- All the blocks $\mathbf{J}_{1,4}, \mathbf{J}_{1,5}, \mathbf{J}_{2,4}, \mathbf{J}_{2,5}, \mathbf{J}_{3,4}, \mathbf{J}_{3,5}$ vanish. This implies in particular that the best achievable accuracy of the parameter estimate $\hat{\boldsymbol{\theta}}$ does not depend on whether or not r is known.
- Considering only the lower right corner of the information matrix gives

$$\begin{aligned}\begin{pmatrix} \mathbf{J}_{4,4} & \mathbf{J}_{4,5} \\ \mathbf{J}_{5,4} & \mathbf{J}_{5,5} \end{pmatrix}^{-1} &= \begin{pmatrix} 2N/\lambda_u^2 & N/(r\lambda_u) \\ N/(r\lambda_u) & N/r^2 \end{pmatrix}^{-1} \\ &= \frac{1}{N}\begin{pmatrix} \lambda_u^2 & -r\lambda_u \\ -r\lambda_u & 2r^2 \end{pmatrix}\,. \end{aligned} \tag{14.288}$$

This means in particular:

– If r is unknown then for any unbiased estimator

$$\text{var}(\hat{\lambda}_u) \geq \frac{\lambda_u^2}{N}\,, \tag{14.289}$$

– while if r is known, then instead

$$\text{var}(\hat{\lambda}_u) \geq \mathbf{J}_{4,4}^{-1} = \frac{\lambda_u^2}{2N}\,. \tag{14.290}$$

It follows that the noise level (the variance λ_u) can be estimated more accurately when the noise variance ratio r is known. This is in contrast to the accuracy of the system parameter estimates.

Remark 14.23 Assume the parameter vector would change from (14.264) to

$$\mathbf{x} = \left(\mathbf{U}_{0,R}^T \; \mathbf{U}_{0,I}^T \; \boldsymbol{\theta}^T \; \lambda_u \; \lambda_y\right)^T . \tag{14.291}$$

The overall block structure (14.278) still applies. The upper left corner is of course unchanged. The lower right corner turns out to be

$$\begin{pmatrix} \mathbf{J}_{4,4} & \mathbf{J}_{4,5} \\ \mathbf{J}_{5,4} & \mathbf{J}_{5,5} \end{pmatrix} = N \begin{pmatrix} 1/\lambda_u^2 & 0 \\ 0 & 1/\lambda_y^2 \end{pmatrix} . \tag{14.292}$$

■

The block $\mathbf{J}_{3,3}$ can be further elaborated. Set

$$G_k = \frac{B_k}{A_k} , \tag{14.293}$$

leading to

$$G_k^\nu = \frac{A_k B_k^\nu - A_k^\nu B_k}{A_k^2} , \tag{14.294}$$

and therefore

$$\left(\mathbf{J}_{3,3}\right)_{\mu,\nu} = \frac{2}{r\lambda_u} \sum_k \mathsf{E}\left\{\left(\frac{A_k B_k^\mu - A_k^\mu B_k}{A_k^2} U_{0,k}\right)^* \left(\frac{A_k B_k^\nu - A_k^\nu B_k}{A_k^2} U_{0,k}\right)\right\} . \tag{14.295}$$

By Parseval's relation, one then has from (14.295)

$$\lim_{N\to\infty} \frac{1}{N} \left(\mathbf{J}_{3,3}\right)_{\mu,\nu} = \frac{2}{r\lambda_u} \mathsf{E}\left\{\left[\frac{AB^\mu - A^\mu B}{A^2} u_0(t)\right]\left[\frac{AB^\nu - A^\nu B}{A^2} u_0(t)\right]\right\} , \tag{14.296}$$

and

$$\lim_{N\to\infty} \frac{1}{N} \mathbf{J}_{3,3} = \frac{2}{r\lambda_u} \mathcal{S}(-B, A) \mathsf{E}\left\{\boldsymbol{\varphi}(t)\boldsymbol{\varphi}^T(t)\right\} \mathcal{S}^T(-B, A) , \tag{14.297}$$

where

$$\boldsymbol{\varphi}(t) = \frac{1}{A^2(q^{-1})} \begin{pmatrix} u_0(t-1) \\ \vdots \\ u_0(t-n_a-n_b) \end{pmatrix} , \tag{14.298}$$

and the Sylvester matrix $\mathscr{S}(-B, A)$ of dimension $(n_a + n_b) \times (n_a + n_b)$ is given by

$$\mathscr{S}(-B, A) = \begin{pmatrix} 0 & -b_1 & \dots & b_{n_b} & & \mathbf{0} \\ & \ddots & & & \ddots & \\ & \mathbf{0} & & -b_1 & \dots & b_{n_b} \\ 1 & a_1 & \dots & a_{n_a} & & \mathbf{0} \\ & \ddots & & & \ddots & \\ & \mathbf{0} & 1 & a_1 & \dots & a_{n_a} \end{pmatrix} . \tag{14.299}$$

The total CRB of the parameter estimate $\hat{\boldsymbol{\theta}}$ is

$$\begin{aligned} \text{CRB}(\hat{\boldsymbol{\theta}}) &= \begin{pmatrix} \mathbf{0} & \mathbf{0} & \mathbf{I} \end{pmatrix} \begin{pmatrix} \mathbf{J}_{1,1} & \mathbf{J}_{1,2} & \mathbf{J}_{1,3} \\ \mathbf{J}_{2,1} & \mathbf{J}_{2,2} & \mathbf{J}_{2,3} \\ \mathbf{J}_{3,1} & \mathbf{J}_{3,2} & \mathbf{J}_{3,3} \end{pmatrix}^{-1} \begin{pmatrix} \mathbf{0} \\ \mathbf{0} \\ \mathbf{I} \end{pmatrix} \\ &= \left(\mathbf{J}_{3,3} - \mathbf{X}\right)^{-1} , \end{aligned} \tag{14.300}$$

where

$$\mathbf{X} = \begin{pmatrix} \mathbf{J}_{3,1} & \mathbf{J}_{3,2} \end{pmatrix} \begin{pmatrix} \mathbf{J}_{1,1} & \mathbf{J}_{1,2} \\ \mathbf{J}_{2,1} & \mathbf{J}_{2,2} \end{pmatrix}^{-1} \begin{pmatrix} \mathbf{J}_{1,3} \\ \mathbf{J}_{2,3} \end{pmatrix} . \tag{14.301}$$

It is shown in Appendix 14.H that

$$\lim_{N \to \infty} \frac{1}{N} \mathbf{X} = \frac{2}{r\lambda_u} \mathscr{S}(-B, A) \mathsf{E} \left\{ \boldsymbol{\psi}(t) \boldsymbol{\psi}^T(t) \right\} \mathscr{S}^T(-B, A) , \tag{14.302}$$

$$\boldsymbol{\psi}(t) = \frac{B(q^{-1})}{A^2(q^{-1})F(q^{-1})} \begin{pmatrix} u_0(t-1) \\ \vdots \\ u_0(t - n_a - n_b) \end{pmatrix} , \tag{14.303}$$

and

$$\left[\lim_{N \to \infty} \frac{1}{N} \left(\mathbf{J}_{3,3} - \mathbf{X} \right) \right]^{-1} = \frac{\lambda_u}{2} \mathscr{S}^{-T}(-B, A) \mathsf{E} \left\{ \boldsymbol{\psi}_2(t) \boldsymbol{\psi}_2^T(t) \right\} \mathscr{S}^{-1}(-B, A) , \tag{14.304}$$

$$\boldsymbol{\psi}_2(t) = \frac{1}{A(q^{-1})F(q^{-1})} \begin{pmatrix} u_0(t-1) \\ \vdots \\ u_0(t - n_a - n_b) \end{pmatrix} . \tag{14.305}$$

Table 14.9 Cramér–Rao lower bounds for the standard system. CRB_1 refers to modeling the noise-free input as an ARMA process. CRB_2 refers to estimating the realization of the noise-free input

Parameter	CRB_1	CRB_2
a_1	0.0093	0.0066
a_2	0.0076	0.0054
b_1	0.0716	0.0506
b_2	0.0966	0.0683
λ_y	0.0028	0.0010
λ_u	0.0021	0.0010

14.8.4 Numerical Illustration of the Cramér–Rao Lower Bounds

Two different Cramér–Rao lower bounds were derived and presented in the text. They apply under different circumstances, but are compared numerically in the following example.

The first bound is based on estimating the character of the noise-free input $u_0(t)$ as an ARMA process. The details for the CRB were developed in Sect. 14.8.2.

The second bound is based on estimating the full realization of the noise-free input $u_0(t)$, with calculations performed in the frequency domain. The values of $u_0(t), t = 1, \ldots, N$ are thus treated as unknown deterministic parameters to be estimated. For details see Sect. 14.8.3.

Example 14.6 The two CRBs were computed for the standard system used in the previous numerical examples. Table 14.9 presents the corresponding standard deviations of the parameters that appear for both cases.

It is clear from the table that the situation corresponding to CRB_2 gives lower values. A closer investigation of details shows in fact for this example that the difference matrix CRB_1 - CRB_2 is indeed positive definite.

The bound CRB_1 depends in a rather intricate way on the different noise variances λ_y, λ_u, and λ_v. For a certain asymptotic case, namely when the signal-to-noise ratio grows without bounds on both input and output sides, a more explicit analysis is possible. Technically, this is achieved by letting $u_0(t)$ be an ARMA process,

$$u_0(t) = H(q^{-1})v(t), \quad \mathsf{E}\{v(t)v(s)\} = \lambda_v \delta_{t,s} , \tag{14.306}$$

and considering the case $\lambda_v \to \infty$.

Let in the following CRB denote the bound on the covariance matrix of $\hat{\boldsymbol{\theta}}$, i.e., only the estimated A and B parameters are considered.

It is derived in Söderström et al. (2010)

$$\text{CRB}_1 = \frac{1}{\lambda_v} \mathcal{S}^{-T}(-B, A) \mathbf{M}_0^{-1} \mathcal{S}^{-1}(-B, A) , \tag{14.307}$$

$$\mathbf{M}_0 = \frac{1}{\lambda_u} \text{cov} \left(\frac{H(q^{-1})}{A(q^{-1})F(q^{-1})} \begin{pmatrix} e(t-1) \\ \vdots \\ e(t - n_a - n_b) \end{pmatrix} \right) , \tag{14.308}$$

$$FF^* = \frac{\lambda_y}{\lambda_u} AA^* + BB^* . \tag{14.309}$$

Here $e(t)$ is white noise of unit variance, and $\mathcal{S}(-B, A)$ denotes the Sylvester matrix as in (14.299).

The developments in Sect. 14.8. lead in the asymptotic case when $\lambda_v \to \infty$,

$$\text{CRB}_2 = 0.5\text{CRB}_1 . \tag{14.310}$$

■

14.9 Extensions and Bibliographical Notes

Confirmatory Factor Analysis, CFA (Sect. 14.2*)*

The paper Huwang et al. (2009) treats the asymptotic distribution of the estimates in the static case and develops robust tests based on that distribution. See also Fuller (1980) and Schneeweiss (1976).

Alternative analysis of CFA estimates is given in Browne (1974) and Browne (1984). Emphasis is given on effects for the case that the data are not Gaussian distributed.

The GIVE Method (Sect. 14.4*)*

More details in the analysis of the GIVE method can be found in Söderström (2012). Accuracy analysis of BELS is treated in Hong et al. (2006), Hong et al. (2007a). Accuracy analysis of the Frisch scheme appeared in Söderström (2005).

Covariance Matching Methods (Sect. 14.5*)*

The section giving the accuracy analysis of CM follows the ideas developed in Söderström and Mossberg (2011). A similar analysis for continuous-time models can be found in Söderström et al. (2012) and Söderström et al. (2013).

Prediction Error and Maximum Likelihood Methods (Sect. 14.6*)*

Some aspects on the accuracy of MIMO models obtained by ML estimation are provided in Agüero et al. (2012).

The analysis of ML in the frequency domain (Sect. 14.6.2) summarizes the developments in Soverini and Söderström (2014b), Söderström and Soverini (2016).

Periodic Data (Sect. 14.7*)*

The paper Pintelon and Hong (2007) derives the asymptotic accuracy for SML estimates, using nonparametric noise models. Comparisons of TML and SML appear in Hong et al. (2008) and Söderström et al. (2010). The methods work under different assumptions and are shown to give rather similar results.

Cramér–Rao Lower Bound (Sect. 14.8*)*

Analysis of the CRB when both the number of observations and the number of free parameters grow without bounds is given in Stoica and Li (1995), See also Stoica and Ng (1998), Stoica and Larsson (2001), Stoica et al. (2001), Li et al. (2008), Moore et al. (2007).

Further details on the computation of CRB in Sect. 14.8.2 can be found in Söderström (2006).

An alternative way to compute the CRB in general under the assumption of Gaussian distributed data is to utilize Whittle's formula, see Lemma A.17. This idea is exploited in Karlsson et al. (2000), where it is applied for the EIV model of Fig. 4.1, with the noise-free input being an ARMA process, and the measurement noises $\tilde{y}(t)$, $\tilde{u}(t)$ both being white (Assumption **AN3e**). The algorithm is converted to a number of polynomial operations and comprises both linear Diophantine equations and some spectral factorizations. Both approaches can be extended to cover cases when the measurement noises are modeled as ARMA processes.

Appendix

14.A Asymptotic Distribution of CFA Estimates

14.A.1 Proof of Lemma 14.1

First note that $\mathbf{C}_r$ can be written as a block matrix, where an arbitrary partition $(\mu, \nu = 1, \ldots, n_\varphi)$ can be written as follows

$$
\begin{aligned}
(\mathbf{C}_r)_{\mu,\nu} &= \lim_{N\to\infty} N \mathsf{E}\left\{\left[\frac{1}{N}\sum_{t=1}^{N}\boldsymbol{\varphi}(t)\boldsymbol{\varphi}_{\mu}(t) - \mathbf{R}\mathbf{e}_{\mu}\right]\left[\frac{1}{N}\sum_{s=1}^{N}\boldsymbol{\varphi}^T(s)\boldsymbol{\varphi}_{\nu}(s) - \mathbf{e}_{\nu}^T\mathbf{R}\right]\right\} \\
&= \lim_{N\to\infty} N\frac{1}{N^2}\sum_{t=1}^{N}\sum_{s=1}^{N}\left[r_{\varphi_\mu\varphi_\nu}(t-s)\mathbf{R}_{\varphi}(t-s) + r_{\varphi\varphi_\nu}(t-s)\mathbf{r}_{\varphi_\mu\varphi}(t-s)\right] \\
&= \lim_{N\to\infty}\frac{1}{N}\sum_{\tau=-N}^{N}(N-|\tau|)\left[r_{\varphi_\mu\varphi_\nu}(\tau)\mathbf{R}_{\varphi}(\tau) + r_{\varphi\varphi_\nu}(\tau)\mathbf{r}^T_{\varphi_\mu\varphi}(\tau)\right] \\
&= \sum_{\tau=-\infty}^{\infty}\left[r_{\varphi_\mu\varphi_\nu}(\tau)\mathbf{R}_{\varphi}(\tau) + r_{\varphi\varphi_\nu}(\tau)\mathbf{r}_{\varphi_\mu\varphi}(\tau)\right] \\
&= \sum_{\tau=-\infty}^{\infty}\left[r_{\varphi_\mu\varphi_\nu}(\tau)\mathbf{R}_{\varphi}(\tau) + \mathbf{R}_{\varphi}(\tau)\mathbf{e}_{\nu}\mathbf{e}_{\mu}^T\mathbf{R}_{\varphi}(\tau)\right] .
\end{aligned}
\tag{14.311}
$$

This means that the total matrix $\mathbf{C}_r$ can be written as

$$
\begin{aligned}
\mathbf{C}_r &= \sum_{\tau=-\infty}^{\infty}\left[\mathbf{R}_{\varphi}(\tau)\otimes\mathbf{R}_{\varphi}(\tau) + \left(\mathbf{I}\otimes\mathbf{R}_{\varphi}(\tau)\right)\begin{pmatrix}\mathbf{e}_1\mathbf{e}_1^T & \dots & \mathbf{e}_{n_\varphi}\mathbf{e}_1^T \\ \vdots & & \\ \mathbf{e}_1\mathbf{e}_{n_\varphi}^T & \dots & \mathbf{e}_{n_\varphi}\mathbf{e}_{n_\varphi}^T\end{pmatrix}\left(\mathbf{I}\otimes\mathbf{R}_{\varphi}(\tau)\right)\right] \\
&\stackrel{\Delta}{=} \sum_{\tau=-\infty}^{\infty}\left[\mathbf{R}_{\varphi}(\tau)\otimes\mathbf{R}_{\varphi}(\tau) + \left(\mathbf{I}\otimes\mathbf{R}_{\varphi}(\tau)\right)\mathbf{K}\left(\mathbf{I}\otimes\mathbf{R}_{\varphi}(\tau)\right)\right] ,
\end{aligned}
\tag{14.312}
$$

which is (14.53).

14.A.2 Evaluation of $\mathbf{R}_{\varphi}(\tau)$

To proceed it is relevant to evaluate the covariance function $\mathbf{R}_{\varphi}(\tau)$. Here it is done when CFA is applied to dynamic models as in Sect. 8.5. Then it is useful to write

$$
\begin{aligned}
\boldsymbol{\varphi}(t) &= \boldsymbol{\varphi}_0(t) + \tilde{\boldsymbol{\varphi}}(t) \\
&= \boldsymbol{\Gamma}\mathbf{z}(t) + \tilde{\boldsymbol{\varphi}}(t) \\
&= \boldsymbol{\Gamma}\begin{pmatrix} z(t) \\ \vdots \\ z(t-k)\end{pmatrix} + \begin{pmatrix}\tilde{y}(t) \\ \vdots \\ \tilde{y}(t-n_a-p_y) \\ \tilde{u}(t-1) \\ \vdots \\ \tilde{u}(t-n_b-p_u)\end{pmatrix} .
\end{aligned}
\tag{14.313}
$$

The two terms in (14.313) are uncorrelated.

Next, introduce the notation $\mathbf{J}_m$ as the 'shift matrix' of dimension $m \times m$,

$$\mathbf{J}_m = \begin{pmatrix} 0\,1\,0 \ldots & \\ 0\,0\,1 & \mathbf{0} \\ \vdots & \\ \mathbf{0} & 1 \\ \mathbf{0} & 0 \end{pmatrix} . \tag{14.314}$$

With some abuse of notation, set also

$$\mathbf{J}_m^{-1} = \mathbf{J}_m^T, \quad \mathbf{J}_m^{-k} = \left(\mathbf{J}_m^T\right)^k . \tag{14.315}$$

One can then write

$$\mathbf{R}_{\tilde{\varphi}}(\tau) = \begin{pmatrix} \lambda_y \mathbf{J}_{n_a+p_y+1}^{\tau} & \mathbf{0} \\ \mathbf{0} & \lambda_u \mathbf{J}_{n_b+p_u}^{\tau} \end{pmatrix} . \tag{14.316}$$

Note that

- Due to the convention (14.315) the result (14.316) holds for all (positive as well as negative) values of τ.
- The expression in (14.316) will in fact be zero, if

$$|\tau| > n_a + p_y + 1 \text{ and } |\tau| > n_b + p_u , \tag{14.317}$$

that is, as soon as

$$|\tau| > \min(n_a + p_y + 1, n_b + p_u) \stackrel{\Delta}{=} \tau_0 . \tag{14.318}$$

Therefore

$$\mathbf{R}_{\varphi}(\tau) = \boldsymbol{\Gamma} \mathbf{R}_{\mathbf{z}}(\tau) \boldsymbol{\Gamma}^T + \mathbf{R}_{\tilde{\varphi}}(\tau) . \tag{14.319}$$

It also holds

$$\begin{aligned} \mathbf{R}_{\mathbf{z}}(\tau) &= \sum_{i=1}^{k+1} \sum_{j=1}^{k+1} \mathbf{e}_i \mathbf{e}_j^T \left(\mathbf{R}_{\mathbf{z}}(\tau)\right)_{ij} \\ &= \sum_i \sum_j \mathbf{e}_i \mathbf{e}_j^T \mathsf{E}\left\{z(t-i+1+\tau) z(t-j+1)\right\} \\ &= \sum_i \sum_j \mathbf{e}_i \mathbf{e}_j^T r_z(\tau - i + j) . \end{aligned} \tag{14.320}$$

Using (14.311), (14.318), and (14.320) one can now write

$$
\begin{aligned}
(\mathbf{C}_r)_{\mu\nu} &= \sum_{\tau=-\infty}^{\infty} \left[\mathbf{e}_\mu^T \mathbf{R}_{\boldsymbol{\varphi}}(\tau)\mathbf{e}_\nu \mathbf{R}_{\boldsymbol{\varphi}}(\tau) + \mathbf{R}_{\boldsymbol{\varphi}}(\tau)\mathbf{e}_\nu \mathbf{e}_\mu^T \mathbf{R}_{\boldsymbol{\varphi}}(\tau)\right] \\
&= \sum_{\tau=-\tau_0}^{\tau_0} \left[\mathbf{e}_\mu^T \tilde{\mathbf{R}}_{\boldsymbol{\varphi}}(\tau)\mathbf{e}_\nu \tilde{\mathbf{R}}_{\boldsymbol{\varphi}}(\tau) + \tilde{\mathbf{R}}_{\boldsymbol{\varphi}}(\tau)\mathbf{e}_\nu \mathbf{e}_\mu^T \tilde{\mathbf{R}}_{\boldsymbol{\varphi}}(\tau)\right] \\
&+ \sum_{\tau=-\tau_0}^{\tau_0} \Bigg[\mathbf{e}_\mu^T \tilde{\mathbf{R}}_{\boldsymbol{\varphi}}(\tau)\mathbf{e}_\nu \boldsymbol{\Gamma} \sum_i \sum_j \mathbf{e}_i \mathbf{e}_j^T r_z(\tau - i + j)\boldsymbol{\Gamma}^T \\
&+\mathbf{e}_\mu^T \boldsymbol{\Gamma} \sum_i \sum_j \mathbf{e}_i \mathbf{e}_j^T r_z(\tau - i + j)\boldsymbol{\Gamma}^T \mathbf{e}_\nu \tilde{\mathbf{R}}_{\boldsymbol{\varphi}}(\tau) \\
&+\tilde{\mathbf{R}}_{\boldsymbol{\varphi}}(\tau)\mathbf{e}_\nu \mathbf{e}_\mu^T \boldsymbol{\Gamma} \sum_i \sum_j \mathbf{e}_i \mathbf{e}_j^T r_z(\tau - i + j)\boldsymbol{\Gamma}^T \\
&+\boldsymbol{\Gamma} \sum_i \sum_j \mathbf{e}_i \mathbf{e}_j^T r_z(\tau - i + j)\boldsymbol{\Gamma}^T \mathbf{e}_\nu \mathbf{e}_\mu^T \tilde{\mathbf{R}}_{\boldsymbol{\varphi}}(\tau) \Bigg] \\
&+ \sum_{\tau=-\infty}^{\infty} \Bigg[\mathbf{e}_\mu^T \boldsymbol{\Gamma} \sum_i \sum_j \mathbf{e}_i \mathbf{e}_j^T r_z(\tau - i + j)\boldsymbol{\Gamma}^T \mathbf{e}_\nu \boldsymbol{\Gamma} \sum_k \sum_\ell \mathbf{e}_k \mathbf{e}_\ell^T r_z(\tau - k + \ell)\boldsymbol{\Gamma}^T \\
&+\boldsymbol{\Gamma} \sum_i \sum_j \mathbf{e}_i \mathbf{e}_j^T r_z(\tau - i + j)\boldsymbol{\Gamma}^T \mathbf{e}_\nu \mathbf{e}_\mu^T \boldsymbol{\Gamma} \sum_k \sum_\ell \mathbf{e}_k \mathbf{e}_\ell^T r_z(\tau - k + \ell)\boldsymbol{\Gamma}^T \Bigg] \\
&= \sum_{\tau=-\tau_0}^{\tau_0} \left[\mathbf{e}_\mu^T \tilde{\mathbf{R}}_{\boldsymbol{\varphi}}(\tau)\mathbf{e}_\nu \tilde{\mathbf{R}}_{\boldsymbol{\varphi}}(\tau) + \tilde{\mathbf{R}}_{\boldsymbol{\varphi}}(\tau)\mathbf{e}_\nu \mathbf{e}_\mu^T \tilde{\mathbf{R}}_{\boldsymbol{\varphi}}(\tau)\right] \\
&+ \sum_{\tau=-\tau_0}^{\tau_0} \Bigg[\mathbf{e}_\mu^T \tilde{\mathbf{R}}_{\boldsymbol{\varphi}}(\tau)\mathbf{e}_\nu \boldsymbol{\Gamma} \sum_i \sum_j \mathbf{e}_i \mathbf{e}_j^T r_z(\tau - i + j)\boldsymbol{\Gamma}^T \\
&+\sum_i \sum_j \left(\boldsymbol{\Gamma}_{\mu i} \boldsymbol{\Gamma}_{\nu j} r_z(\tau - i + j)\tilde{\mathbf{R}}_{\boldsymbol{\varphi}}(\tau) + \tilde{\mathbf{R}}_{\boldsymbol{\varphi}}(\tau)\mathbf{e}_\nu \boldsymbol{\Gamma}_{\mu i} \mathbf{e}_j^T r_z(\tau - i + j)\boldsymbol{\Gamma}^T\right) \\
&+\boldsymbol{\Gamma} \sum_i \sum_j \mathbf{e}_i \boldsymbol{\Gamma}_{\nu j} r_z(\tau - i + j)\mathbf{e}_\mu^T \tilde{\mathbf{R}}_{\boldsymbol{\varphi}}(\tau) \Bigg] \\
&+\sum_i \sum_j \boldsymbol{\Gamma}_{\mu i} \boldsymbol{\Gamma}_{\nu j} \boldsymbol{\Gamma} \sum_k \sum_\ell \mathbf{e}_k \mathbf{e}_\ell^T \boldsymbol{\Gamma}^T \sum_{\tau=-\infty}^{\infty} r_z(\tau - i + j) r_z(\tau - k + \ell) \\
&+\sum_i \sum_j \sum_k \sum_\ell \boldsymbol{\Gamma} \mathbf{e}_i \boldsymbol{\Gamma}_{\nu j} \boldsymbol{\Gamma}_{\mu k} \mathbf{e}_\ell^T \boldsymbol{\Gamma}^T \sum_{\tau=-\infty}^{\infty} r_z(\tau - i + j) r_z(\tau - k + \ell) . \qquad (14.321)
\end{aligned}
$$

To proceed it would be convenient to have an algorithm for computing sums of the form

$$\alpha_i \stackrel{\Delta}{=} \sum_{\tau=-\infty}^{\infty} r_z(\tau+i)r_z(\tau) \tag{14.322}$$

for an arbitrary (fixed) integer i.

This can be done using Lemma A.10. Assume that the noise-free input is an ARMA process

$$u_0(t) = \lambda_v \frac{C(q^{-1})}{D(q^{-1})} v(t) , \tag{14.323}$$

where $v(t)$ is white noise of unit variance. Then it holds

$$z(t) = \frac{1}{A(q^{-1})} u_0(t) = \lambda_v \frac{C(q^{-1})}{A(q^{-1})D(q^{-1})} v(t) , \tag{14.324}$$

and it follows that

$$\alpha_i = \mathsf{E}\{s(t+i)s(t)\} = r_s(i) , \tag{14.325}$$

where

$$s(t) = \lambda_v^2 \frac{C^2(q^{-1})}{A^2(q^{-1})D^2(q^{-1})} v(t) . \tag{14.326}$$

14.B Asymptotic Distribution for IV Estimates

14.B.1 *Proof of Lemma 14.2*

Using the assumption of joint Gaussian distribution one can apply the general rule for product of Gaussian variables, see Lemma A.9,

$$\mathsf{E}\{x_1x_2x_3x_4\} = \mathsf{E}\{x_1x_2\}\,\mathsf{E}\{x_3x_4\} + \mathsf{E}\{x_1x_3\}\,\mathsf{E}\{x_2x_4\} + \mathsf{E}\{x_1x_4\}\,\mathsf{E}\{x_2x_3\} . \tag{14.327}$$

Using the result (14.327) in (14.68) leads to

$$\begin{aligned} \mathbf{C} &= \lim_{N\to\infty} \frac{1}{N} \sum_{t=1}^{N} \sum_{s=1}^{N} \Big\{ \left[\mathsf{E}\left\{\mathbf{z}(t)\mathbf{z}^T(s)\right\}\right] \left[\mathsf{E}\left\{v(t)v(s)\right\}\right] \\ &\quad + \left[\mathsf{E}\left\{\mathbf{z}(t)v(s)\right\}\right] \left[\mathsf{E}\left\{\mathbf{z}^T(s)v(t)\right\}\right] \Big\} \\ &= \lim_{N\to\infty} \sum_{\tau=-N}^{N} \left(1 - \frac{|\tau|}{N}\right) \left\{\mathbf{R}_{\mathbf{z}}(\tau)r_v(\tau) + \mathbf{r}_{\mathbf{z}v}(\tau)\mathbf{r}_{\mathbf{z}v}^T(-\tau)\right\} . \end{aligned} \tag{14.328}$$

The assumption (14.70) implies that

$$\mathbf{r}_{\mathbf{z}v}(\tau)\mathbf{r}_{\mathbf{z}v}^T(-\tau) = \mathbf{0} \quad \forall\, \tau \tag{14.329}$$

as at least one of the factors is zero.

Now use the conventions

$$h_0 = 1, \quad h_i = 0 \quad \text{for } i < 0\,. \tag{14.330}$$

Recall that the covariance function $r_v(\tau)$ decays exponentially with τ. Therefore it holds

$$\| \lim_{N\to\infty} \frac{1}{N} \sum_{\tau=-N}^{N} |\tau| \mathbf{R}_{\mathbf{z}}(\tau) r_v(\tau) \| \leq \lim_{N\to\infty} \frac{2}{N} \sum_{\tau=0}^{N} \tau C \alpha^{\tau} = 0 \tag{14.331}$$

for some $|\alpha| < 1$. Using this result, one gets from (14.328)

$$\begin{aligned}
\mathbf{C} &= \sum_{\tau=-\infty}^{\infty} \mathbf{R}_{\mathbf{z}}(\tau) r_v(\tau) = \sum_{\tau=-\infty}^{\infty} \left[\mathbf{R}_{\mathbf{z}}(\tau) \lambda \sum_{i=0}^{\infty} h_i h_{i+\tau} \right] \\
&= \lambda \sum_{\tau=-\infty}^{\infty} \sum_{i=-\infty}^{\infty} h_i h_{i+\tau} \mathsf{E} \left\{ \mathbf{z}(t+\tau)\mathbf{z}^T(t) \right\} \\
&= \lambda \sum_{\tau=-\infty}^{\infty} \sum_{i=-\infty}^{\infty} h_i h_{i+\tau} \mathsf{E} \left\{ \mathbf{z}(t-i)\mathbf{z}^T(t-i-\tau) \right\} \\
&= \lambda \mathsf{E} \left\{ \sum_{i=-\infty}^{\infty} h_i \mathbf{z}(t-i) \sum_{\tau=-\infty}^{\infty} h_{i+\tau} \mathbf{z}^T(t-i-\tau) \right\} \\
&= \lambda \mathsf{E} \left\{ \left(\sum_{i=-\infty}^{\infty} h_i \mathbf{z}(t-i) \right) \left(\sum_{k=-\infty}^{\infty} h_k \mathbf{z}^T(t-k) \right) \right\} \\
&= \lambda \mathsf{E} \left\{ \left[H(q^{-1})\mathbf{z}(t) \right] \left[H(q^{-1})\mathbf{z}(t) \right]^T \right\} ,
\end{aligned} \tag{14.332}$$

which is (14.72).

14.B.2 Proof of Lemma 14.6

Use the definition (14.77) of $K(z)$ and introduce the notations

$$\boldsymbol{\alpha}(t) = \mathbf{R}^T \mathbf{W} \sum_{i=0}^{\infty} K_i \mathbf{z}(t+i) \,, \tag{14.333}$$

$$\boldsymbol{\beta}(t) = H^{-1}(q^{-1}) \boldsymbol{\varphi}_0(t) \,. \tag{14.334}$$

Then it holds

$$\begin{aligned} \mathbf{R}^T \mathbf{W} \mathbf{R} &= \mathbf{R}^T \mathbf{W} \mathsf{E} \left\{ \mathbf{z}(t) F(q^{-1}) \boldsymbol{\varphi}_0^T(t) \right\} \\ &= \mathbf{R}^T \mathbf{W} \mathsf{E} \left\{ \mathbf{z}(t) K(q^{-1}) H^{-1}(q^{-1}) \boldsymbol{\varphi}_0^T(t) \right\} \\ &= \mathbf{R}^T \mathbf{W} \mathsf{E} \left\{ \mathbf{z}(t) \sum_{i=0}^{\infty} K_i H^{-1}(q^{-1}) \boldsymbol{\varphi}_0^T(t-i) \right\} \\ &= \mathsf{E} \left\{ \boldsymbol{\alpha}(t) \boldsymbol{\beta}^T(t) \right\} \,. \end{aligned} \tag{14.335}$$

Using (14.78)
leads to

$$\lambda \mathbf{R}^T \mathbf{W} \mathbf{C} \mathbf{W} \mathbf{R} = \mathsf{E} \left\{ \boldsymbol{\alpha}(t) \boldsymbol{\alpha}^T(t) \right\} \,. \tag{14.336}$$

The stated inequality (14.81) then reads

$$\begin{aligned} \mathbf{P}_{\mathrm{IV}} &= \lambda \left(\mathsf{E} \left\{ \boldsymbol{\alpha}(t) \boldsymbol{\beta}^T(t) \right\} \right)^{-1} \left(\mathsf{E} \left\{ \boldsymbol{\alpha}(t) \boldsymbol{\alpha}^T(t) \right\} \right) \left(\mathsf{E} \left\{ \boldsymbol{\beta}(t) \boldsymbol{\alpha}^T(t) \right\} \right)^{-1} \\ &\geq \lambda \left(\mathsf{E} \left\{ \boldsymbol{\beta}(t) \boldsymbol{\beta}^T(t) \right\} \right)^{-1} \,. \end{aligned} \tag{14.337}$$

Now, (14.337) is equivalent to

$$\left(\mathsf{E} \left\{ \boldsymbol{\beta}(t) \boldsymbol{\alpha}^T(t) \right\} \right) \left(\mathsf{E} \left\{ \boldsymbol{\alpha}(t) \boldsymbol{\alpha}^T(t) \right\} \right)^{-1} \left(\mathsf{E} \left\{ \boldsymbol{\alpha}(t) \boldsymbol{\beta}^T(t) \right\} \right) \leq \left(\mathsf{E} \left\{ \boldsymbol{\beta}(t) \boldsymbol{\beta}^T(t) \right\} \right) \,, \tag{14.338}$$

which follows from the theory of partitioned matrices, cf Lemma A.4 in Söderström and Stoica (1989), as

$$\mathsf{E} \left\{ \begin{pmatrix} \boldsymbol{\beta}(t) \boldsymbol{\beta}^T(t) & \boldsymbol{\beta}(t) \boldsymbol{\alpha}^T(t) \\ \boldsymbol{\alpha}(t) \boldsymbol{\beta}^T(y) & \boldsymbol{\alpha}(t) \boldsymbol{\alpha}^T(t) \end{pmatrix} \right\} = \mathsf{E} \left\{ \begin{pmatrix} \boldsymbol{\beta}(t) \\ \boldsymbol{\alpha}(t) \end{pmatrix} \begin{pmatrix} \boldsymbol{\beta}^T(t) & \boldsymbol{\alpha}^T(t) \end{pmatrix} \right\} \geq \mathbf{0} \,. \tag{14.339}$$

Further, for the specific choice

$$\mathbf{z}(t) = H^{-1}(q^{-1}) \boldsymbol{\varphi}_0(t), \quad F(q^{-1}) = H^{-1}(q^{-1}) \,, \tag{14.340}$$

(**W** has then no influence) it holds that

$$\mathbf{R} = \mathsf{E} \left\{ [H^{-1}(q^{-1}) \boldsymbol{\varphi}_0(t)][H^{-1}(q^{-1}) \boldsymbol{\varphi}_0^T(t)] \right\} = \mathsf{E} \left\{ \boldsymbol{\beta}(t) \boldsymbol{\beta}^T(t) \right\} = \mathbf{C} \,, \tag{14.341}$$

from which equality in (14.81) follows.

14.C Asymptotic Distribution for GIVE

14.C.1 The Sensitivity Matrix S *for the SISO Case*

In this section the sensitivity matrix **S** will be analyzed in more detail. In addition, expressions for the covariance function $r_\varepsilon(\tau)$ and the cross-covariance functions $\mathbf{r}_{\mathbf{z}\varepsilon}(\tau)$ are provided.

Use generally the conventions

$$\begin{aligned} a_i &= 0, \text{ if } i < 0 \text{ or } i > n_a \,, \\ a_0 &= 1 \,, \\ b_i &= 0, \text{ if } i < 1 \text{ or } i > n_b \,. \end{aligned} \tag{14.342}$$

The two cases of white and correlated output noises are treated in the next two subsections, respectively.

14.C.1.1 Expressions for White Output Noise

Using the conventions (14.342), the auto-covariance function of the residuals is easily found to be

$$r_\varepsilon(\tau) = \lambda_y \sum_i a_i a_{i+\tau} + \lambda_u \sum_i b_i b_{i+\tau} \,. \tag{14.343}$$

Similarly, the covariance vector $\mathbf{r}_{\mathbf{z}\varepsilon}(\tau) = \mathsf{E}\left\{\mathbf{z}(t+\tau)\varepsilon(t)\right\}$ is calculated as follows. As the generalized IV vector is given by (7.97), it holds

$$\mathbf{r}_{\mathbf{z}\varepsilon}(\tau) = \mathsf{E}\left\{\begin{pmatrix} y(t) \\ \vdots \\ y(t-n_a-p_y) \\ u(t-1) \\ \vdots \\ u(t-n_b-p_u) \end{pmatrix} \right.$$

$$\left. \times \left[\sum_i a_i \tilde{y}(t-\tau-i) - \sum_i b_i \tilde{u}(t-\tau-i)\right]\right\}$$

$$= \begin{pmatrix} \lambda_y \begin{pmatrix} a_{-\tau} \\ \vdots \\ a_{n_a+p_y-\tau} \end{pmatrix} \\ -\lambda_u \begin{pmatrix} b_{1-\tau} \\ \vdots \\ b_{n_b+p_u-\tau} \end{pmatrix} \end{pmatrix} . \tag{14.344}$$

The general expression of the matrix **S**, (14.90), reads

$$\mathbf{S} = -\left(\mathbf{R}_{\mathbf{z}\varphi} + \mathbf{r}_\theta \ \mathbf{r}_\rho\right) . \tag{14.345}$$

When the output noise is white, the noise parameter vector is

$$\rho = \begin{pmatrix} \lambda_y \\ \lambda_u \end{pmatrix} , \tag{14.346}$$

and using (14.344) as well as $\bar{\mathbf{a}} = \left(1 \ a_1 \ \ldots \ a_{n_a}\right)^T$, $\mathbf{b} = \left(b_1 \ \ldots \ b_{n_b}\right)^T$

$$\mathbf{r}(\boldsymbol{\theta}, \boldsymbol{\rho}) = \mathbf{r}_{\mathbf{z}\varepsilon}(0) = \begin{pmatrix} \lambda_y \bar{\mathbf{a}} \\ \mathbf{0}_{p_y \times 1} \\ -\lambda_u \mathbf{b} \\ \mathbf{0}_{p_u \times 1} \end{pmatrix} . \tag{14.347}$$

It follows from (14.347) that

$$\mathbf{r}_\theta = \frac{\partial \mathbf{r}_{\mathbf{z}\varepsilon}}{\partial \boldsymbol{\theta}} = \begin{pmatrix} \mathbf{0}_{1 \times (n_a+n_b)} \\ \lambda_y \mathbf{I}_{n_a} \ \mathbf{0}_{n_a \times n_b} \\ \mathbf{0}_{p_y \times (n_a+n_b)} \\ \mathbf{0}_{n_b \times n_a} \ -\lambda_u \mathbf{I}_{n_b} \\ \mathbf{0}_{p_u \times (n_a+n_b)} \end{pmatrix} , \tag{14.348}$$

and

$$\mathbf{r}_\rho = \frac{\partial \mathbf{r}_{\mathbf{z}\varepsilon}}{\partial \boldsymbol{\rho}} = \begin{pmatrix} \bar{\mathbf{a}} \ \mathbf{0}_{(n_a+1) \times 1} \\ \mathbf{0}_{p_y \times 2} \\ \mathbf{0}_{n_b \times 1} \ -\mathbf{b} \\ \mathbf{0}_{p_u \times 2} \end{pmatrix} . \tag{14.349}$$

A closer evaluation of the first part of **S** gives

$$\mathbf{R}_{\mathbf{z}\varphi} + \mathbf{r}_\theta = \mathbf{R}_{\mathbf{z}_0\varphi_0} + \mathbf{R}_{\tilde{\mathbf{z}}\tilde{\varphi}} + \mathbf{r}_\theta = \mathbf{R}_{\mathbf{z}_0\varphi_0} , \tag{14.350}$$

as $\mathbf{R}_{\tilde{\mathbf{z}}\tilde{\varphi}} + \mathbf{r}_\theta = \mathbf{0}$, using, say, (14.347).

14.C.1.2 Expressions for Correlated Output Noise

Now consider the case of correlated output noise. The instrumental vector $\mathbf{z}(t)$ is in this case taken as in (7.97) with $p_y = 0$; that is,

$$\mathbf{z}(t) = \begin{pmatrix} y(t) \\ \vdots \\ y(t-n_a) \\ u(t-1) \\ \vdots \\ u(t-n_b-p_u) \end{pmatrix} . \tag{14.351}$$

The noise parameter vector $\boldsymbol{\rho}$ is given by, see (5.27),

$$\boldsymbol{\rho} = \begin{pmatrix} r_{\tilde{y}}(0) \\ \vdots \\ r_{\tilde{y}}(n_a) \\ \lambda_u \end{pmatrix} . \tag{14.352}$$

The vector $\mathbf{r}_{\mathbf{z}\varepsilon}(\boldsymbol{\theta})$ becomes

$$\begin{aligned} \mathbf{r}_{\mathbf{z}\varepsilon}(\boldsymbol{\theta}) &= \mathsf{E}\left\{\left[\mathbf{z}_0(t) + \tilde{\mathbf{z}}(t)\right]\left[\varepsilon_0(t,\boldsymbol{\theta}) + \tilde{\varepsilon}(t,\boldsymbol{\theta})\right]\right\} \\ &= \mathbf{r}_{\mathbf{z}_0\varepsilon_0}(\boldsymbol{\theta}) + \mathbf{r}_{\tilde{\mathbf{z}}\tilde{\varepsilon}}(\boldsymbol{\theta}) . \end{aligned} \tag{14.353}$$

Note that in particular $\varepsilon_0(t, \boldsymbol{\theta}_0) = 0$.

The first part of the matrix $\mathbf{S}$, (14.90), will now be

$$\begin{aligned} \mathbf{R}_{\mathbf{z}\varphi} + \mathbf{r}_{\theta}\Big|_{\theta=\theta_0} &= \mathsf{E}\left\{\left[\mathbf{z}_0(t) + \tilde{\mathbf{z}}(t)\right]\left[\boldsymbol{\varphi}_0^T(t) + \tilde{\boldsymbol{\varphi}}^T(t)\right]\right\} \\ &\quad -\mathsf{E}\left\{\left[\mathbf{z}_0(t) + \tilde{\mathbf{z}}(t)\right]\left[\tilde{\boldsymbol{\varphi}}^T(t)\right]\right\} \\ &= \mathsf{E}\left\{\mathbf{z}_0(t)\boldsymbol{\varphi}_0^T(t)\right\} = \mathbf{R}_{\mathbf{z}_0\varphi_0} . \end{aligned} \tag{14.354}$$

This is the same relation as in (14.350).

Furthermore, the vector $\mathbf{r}_{\mathbf{z}\varepsilon}(\boldsymbol{\theta}_0)$ is given by

$$\mathbf{r}_{\mathbf{z}\varepsilon}(\boldsymbol{\theta}_0) = \mathsf{E}\left\{ \begin{pmatrix} y(t) \\ \vdots \\ y(t-n_a) \\ u(t-1) \\ \vdots \\ u(t-n_b-p_u) \end{pmatrix} \left[\sum_i a_i \tilde{y}(t-i) - \sum_i b_i \tilde{u}(t-i) \right]_{\theta=\theta_0} \right\}$$

$$= \begin{pmatrix} \mathscr{A} & \mathbf{0}_{(n_a+1)\times 1} \\ \mathbf{0}_{n_b\times(n_a+1)} & -\mathbf{b} \\ \mathbf{0}_{p_u\times(n_a+1)} & \mathbf{0}_{p_u\times 1} \end{pmatrix} \begin{pmatrix} r_{\tilde{y}}(0) \\ \vdots \\ r_{\tilde{y}}(n_a) \\ \lambda_u \end{pmatrix} . \tag{14.355}$$

The upper left part $\mathscr{A}$ of the matrix in (14.355) is an $(n_a+1) \times (n_a+1)$ matrix, given in (14.114).

Trivially, the derivative $\mathbf{r}_\rho$ is precisely the matrix appearing in (14.355).

14.C.2 Computation of the Matrix C

An efficient tool for computing the elements of the matrix $\mathbf{C}$ without using explicit summation over τ is presented here, for the case when both $\tilde{y}(t)$ and the noise-free input $u_0(t)$ are ARMA processes. Details for how to compute such sums of products of covariances are described in Lemma A.10.

When applying Lemma A.10 for evaluation of $\mathbf{C}$, it is convenient to first decompose the variables as

$$\mathbf{z}(t) = \mathbf{z}_0(t) + \tilde{\mathbf{z}}(t), \quad \varepsilon(t) = \varepsilon_y(t) + \varepsilon_u(t) , \tag{14.356}$$

$$\varepsilon_y(t) = A(q^{-1})\tilde{y}(t), \quad \varepsilon_u(t) = B(q^{-1})\tilde{u}(t) . \tag{14.357}$$

It then holds

$$\begin{aligned} \mathbf{C} = \sum_{\tau=-\infty}^{\infty} &\left[\left(\mathbf{R}_{\mathbf{z}_0}(\tau) + \mathbf{R}_{\tilde{\mathbf{z}}}(\tau)\right) \left(r_{\varepsilon_y}(\tau) + r_{\varepsilon_u}(\tau)\right) \right. \\ &\left. + \left(\mathbf{r}_{\tilde{\mathbf{z}}\varepsilon_y}(\tau) + \mathbf{r}_{\tilde{\mathbf{z}}\varepsilon_u}(\tau)\right) \left(\mathbf{r}_{\varepsilon_y\tilde{\mathbf{z}}}(\tau) + \mathbf{r}_{\varepsilon_u\tilde{\mathbf{z}}}(\tau)\right) \right] . \end{aligned} \tag{14.358}$$

Introduce also the following model assumptions and notations, where $e_0(t)$ and $v(t)$ are white noise signals,

$$\tilde{y}(t) = F(q^{-1})e_0(t), \quad \mathsf{E}\left\{e_0^2(t)\right\} = \lambda_y , \tag{14.359}$$

$$u_0(t) = H(q^{-1})v(t), \quad \mathsf{E}\left\{v^2(t)\right\} = \lambda_v , \tag{14.360}$$

$$G(q^{-1}) = \frac{B(q^{-1})}{A(q^{-1})} . \tag{14.361}$$

Further observe:

1. When the output noise $\tilde{y}(t)$ is white, it holds

$$F(q^{-1}) = 1 . \tag{14.362}$$

2. When the output noise is correlated, it holds

$$p_y = 0\,. \tag{14.363}$$

It will be convenient though to derive expressions for the general case.

The evaluation of **C** has to be done blockwise. Set

$$\mathbf{C} = \begin{pmatrix} \mathbf{C}_{11} & \mathbf{C}_{12} \\ \mathbf{C}_{21} & \mathbf{C}_{22} \end{pmatrix}, \tag{14.364}$$

where the block $\mathbf{C}_{11}$ has dimension $n_a + p_y + 1$, and the block $\mathbf{C}_{22}$ has dimension $n_b + p_u$. The three blocks in (14.364) can be evaluated using repeatedly Lemma A.10.

14.C.2.1 Evaluation of the Block $\mathbf{C}_{11}$

Consider the μ, ν element where it is practical to number the elements such that $\mu, \nu = 0, \ldots n_a + p_y$. From (14.358)

$$\begin{aligned}(\mathbf{C}_{11})_{\mu,\nu} = & \sum_{\tau=-\infty}^{\infty} \left[\left(r_{y_0}(\tau - \mu + \nu) + r_{\tilde{y}}(\tau - \mu + \nu) \right) \left(r_{\varepsilon_y}(\tau) + r_{\varepsilon_u}(\tau) \right) \right. \\ & \left. + r_{\tilde{y}\varepsilon_y}(\tau - \mu) r_{\varepsilon_y \tilde{y}}(\tau + \nu) \right]\,. \end{aligned} \tag{14.365}$$

Note that due to the assumptions and notions above

$$\begin{aligned} y_0(t) &= G(q^{-1})H(q^{-1})v(t), \ \tilde{y}(t) = F(q^{-1})e_0(t)\,, \\ \varepsilon_y(t) &= A(q^{-1})F(q^{-1})e_0(t), \ \varepsilon_u(t) = B(q^{-1})\tilde{u}(t)\,. \end{aligned} \tag{14.366}$$

Using Lemma A.10, one now gets, with $e(t)$ being white noise of unit variance,

$$\begin{aligned}(\mathbf{C}_{11})_{\mu,\nu} = & \lambda_v \lambda_y \mathsf{E}\left\{[q^{-\nu}G(q^{-1})H(q^{-1})A(q^{-1})F(q^{-1})e(t)] \right.\\ & \left. \times [q^{-\mu}G(q^{-1})H(q^{-1})A(q^{-1})F(q^{-1})e(t)]\right\} \\ & + \lambda_v \lambda_u \mathsf{E}\left\{[q^{-\nu}G(q^{-1})H(q^{-1})B(q^{-1})e(t)] \right.\\ & \left. \times [q^{-\mu}G(q^{-1})H(q^{-1})B(q^{-1})e(t)]\right\} \\ & + \lambda_y^2 \mathsf{E}\left\{[q^{-\nu}A(q^{-1})F^2(q^{-1})e(t)][q^{-\mu}A(q^{-1})F^2(q^{-1})e(t)]\right\} \\ & + \lambda_y \lambda_u \mathsf{E}\left\{[q^{-\nu}F(q^{-1})B(q^{-1})e(t)][q^{-\mu}F(q^{-1})B(q^{-1})e(t)]\right\} \\ & + \lambda_y^2 \mathsf{E}\left\{[A^2(q^{-1})F^2(q^{-1})e(t)][q^{-\mu-\nu}F^2(q^{-1})e(t)]\right\}\,. \end{aligned} \tag{14.367}$$

14.C.2.2 Evaluation of the Block $\mathbf{C}_{22}$

Consider the μ, ν element where $\mu, \nu = 1, \ldots n_b + p_u$. From (14.358)

$$
\begin{aligned}
(\mathbf{C}_{22})_{\mu,\nu} = \sum_{\tau=-\infty}^{\infty} & \left[\left(r_{u_0}(\tau-\mu+\nu) + r_{\tilde{u}}(\tau-\mu+\nu) \right) \left(r_{\varepsilon_y}(\tau) + r_{\varepsilon_u}(\tau) \right) \right. \\
& \left. + r_{\tilde{u}\varepsilon_u}(\tau-\mu) r_{\varepsilon_u \tilde{u}}(\tau+\nu) \right] \\
= {} & \lambda_\nu \lambda_u \mathsf{E}\left\{ [q^{-\nu} B(q^{-1}) H(q^{-1}) e(t)][q^{-\mu} B(q^{-1}) H(q^{-1}) e(t)] \right\} \\
& + \lambda_\nu \lambda_y \mathsf{E}\left\{ [q^{-\nu} A(q^{-1}) F(q^{-1}) H(q^{-1}) e(t)] \right. \\
& \left. \times [q^{-\mu} A(q^{-1}) F(q^{-1}) H(q^{-1}) e(t)] \right\} \\
& + \lambda_u^2 \mathsf{E}\left\{ [q^{-\nu} B(q^{-1}) e(t)][q^{-\mu} B(q^{-1}) e(t)] \right\} \\
& + \lambda_y \lambda_u \mathsf{E}\left\{ [q^{-\nu} A(q^{-1}) F(q^{-1}) e(t)][q^{-\mu} A(q^{-1}) F(q^{-1}) e(t)] \right\} \\
& + \lambda_u^2 \mathsf{E}\left\{ [B^2(q^{-1}) e(t)][q^{-\mu-\nu} e(t)] \right\} . \qquad (14.368)
\end{aligned}
$$

14.C.2.3 Evaluation of the Block $\mathbf{C}_{12}$

Consider the μ, ν element where $\mu = 0, \ldots, n_a + p_y, \ \nu = 1, \ldots n_b + p_u$. From (14.358)

$$
\begin{aligned}
(\mathbf{C}_{12})_{\mu,\nu} = \sum_{\tau=-\infty}^{\infty} & \left[r_{y_0 u_0}(\tau-\mu+\nu) \left(r_{\varepsilon_y}(\tau) + r_{\varepsilon_u}(\tau) \right) \right. \\
& \left. + r_{\tilde{y}\varepsilon_y}(\tau-\mu) r_{\varepsilon_u \tilde{u}}(\tau+\nu) \right] \\
= {} & \lambda_\nu \lambda_y \mathsf{E}\left\{ [q^{-\nu} H(q^{-1}) A(q^{-1}) F(q^{-1}) e(t)] \right. \\
& \left. \times [q^{-\mu} G(q^{-1}) H(q^{-1}) A(q^{-1}) F(q^{-1}) e(t)] \right\} \\
& + \lambda_\nu \lambda_u \mathsf{E}\left\{ [q^{-\nu} H(q^{-1}) B(q^{-1}) e(t)][q^{-\mu} G(q^{-1}) H(q^{-1}) B(q^{-1}) e(t)] \right\} \\
& + \lambda_y \lambda_u \mathsf{E}\left\{ [A(q^{-1}) B(q^{-1}) F(q^{-1}) e(t)][q^{-\mu-\nu} F(q^{-1}) e(t)] \right\} . \qquad (14.369)
\end{aligned}
$$

14.C.3 Non-Gaussian Distributed Data. Proof of Lemma 14.7

It follows by construction that an arbitrary element of $\mathbf{C}$ in (14.97) has the form of $\mathbf{C}$ in (A.80). One thus can write $\mathbf{C}$ as in (14.104) where the elements of the 'non-Gaussian contribution' $\mathbf{C}^{\mathrm{NG}}$ are given by the last term in (A.80). Reflecting the structure of the vector $\mathbf{z}(t)$ it is found that $\mathbf{C}^{\mathrm{NG}}$ will always have the form given by (14.108).

Consider first the block $\mathbf{C}_{33}^{\mathrm{NG}}$. For its μ, ν element (with $\mu, \nu = 1, \ldots, n_b$), apply Lemma A.12 using

$$
\begin{aligned}
x_1(t) &= \tilde{u}(t-\mu), \quad x_2(t) = B(q^{-1}) \tilde{u}(t) , \\
x_3(t) &= \tilde{u}(t-\nu), \quad x_4(t) = B(q^{-1}) \tilde{u}(t) ,
\end{aligned}
\qquad (14.370)
$$

and therefore the expressions for $\mathbf{C}_{33}^{\mathrm{NG}}$ in (14.109) and (14.112) follow.

Next consider the block $\mathbf{C}_{11}^{\mathrm{NG}}$. For its μ, ν element ($\mu, \nu = 0, \dots n_a$), apply Lemma A.12 using

$$\begin{aligned} x_1(t) &= \tilde{y}(t-\mu), \quad x_2(t) = A(q^{-1})\tilde{y}(t) , \\ x_3(t) &= \tilde{y}(t-\nu), \quad x_4(t) = A(q^{-1})\tilde{y}(t) . \end{aligned} \tag{14.371}$$

If $\tilde{y}(t)$ is white noise, the expression (14.109) for $\mathbf{C}_{11}^{\mathrm{NG}}$ is obtained. In the more general case of $\tilde{y}(t)$ being correlated noise, one gets instead the expression in (14.112) that holds with

$$\begin{aligned} \mathbf{r}_x &= \mathsf{E}\left\{ \begin{pmatrix} \tilde{y}(t) \\ \vdots \\ \tilde{y}(t-n_a) \end{pmatrix} \sum_{i=0}^{n_a} a_i \tilde{y}(t-i) \right\} \\ &= \mathscr{A} \begin{pmatrix} r_{\tilde{y}}(0) \\ \vdots \\ r_{\tilde{y}}(n_a) \end{pmatrix} , \end{aligned} \tag{14.372}$$

with $\mathscr{A}$ given by (14.114). This proves (14.113).

14.C.4 Proof of Lemma 14.8

First note that by construction, see (7.130), and (7.132) it holds

$$\overline{\mathbf{f}} = \hat{\mathbf{r}}_{\boldsymbol{\varepsilon}\otimes\mathbf{z}}(\boldsymbol{\theta}) - \mathbf{r}_{\boldsymbol{\varepsilon}\otimes\mathbf{z}}(\boldsymbol{\theta}, \boldsymbol{\rho}) . \tag{14.373}$$

The definition (14.121) of $\mathbf{C}$ gives

$$\begin{aligned} \mathbf{C} &= \lim_{N\to\infty} N\mathsf{E}\left\{ \overline{\mathbf{f}}\,\overline{\mathbf{f}}^T \right\} \\ &= \lim_{N\to\infty} N\mathsf{E}\left\{ \left(\frac{1}{N}\sum_{t=1}^{N} \boldsymbol{\varepsilon}(t)\otimes\mathbf{z}(t) \right) \left(\frac{1}{N}\sum_{s=1}^{N} \boldsymbol{\varepsilon}^T(s)\otimes\mathbf{z}^T(s) \right) - \mathbf{r}_{\boldsymbol{\varepsilon}\otimes\mathbf{z}}\mathbf{r}_{\boldsymbol{\varepsilon}\otimes\mathbf{z}}^T \right\} . \end{aligned} \tag{14.374}$$

The arguments $\boldsymbol{\theta}$ and $\boldsymbol{\rho}$ were omitted for simplicity.

Consider now an arbitrary block, say block j, k of $\mathbf{C}$, and apply the assumption on Gaussian distributed data, using Lemma A.9

$$
\begin{aligned}
(\mathbf{C})_{j,k} &= \lim_{N\to\infty} N \left(\frac{1}{N^2} \sum_{t=1}^{N} \sum_{s=1}^{N} \left[\mathbf{r}_{\varepsilon_j \mathbf{z}} \mathbf{r}_{\varepsilon_k \mathbf{z}}^T + \mathsf{E}\left\{\varepsilon_j(t)\varepsilon_k(s)\right\} \mathsf{E}\left\{\mathbf{z}(t)\mathbf{z}^T(s)\right\} \right.\right. \\
&\quad \left.\left. + \mathsf{E}\left\{\varepsilon_k(s)\mathbf{z}(t)\right\} \mathsf{E}\left\{\varepsilon_j(t)\mathbf{z}^T(s)\right\} \right] - \mathbf{r}_{\varepsilon_j \mathbf{z}} \mathbf{r}_{\varepsilon_k \mathbf{z}}^T \right) \\
&= \lim_{N\to\infty} N \frac{1}{N^2} \sum_{\tau=-N}^{N} (N - |\tau|) \left[r_{\varepsilon_j \varepsilon_k}(\tau) \mathbf{R}_{\mathbf{z}}(\tau) + \mathbf{r}_{\mathbf{z}\varepsilon_k}(\tau) \mathbf{r}_{\varepsilon_j \mathbf{z}}(\tau) \right] \\
&= \sum_{\tau=-\infty}^{\infty} \left[r_{\varepsilon_j \varepsilon_k}(\tau) \mathbf{R}_{\mathbf{z}}(\tau) + \mathbf{r}_{\mathbf{z}\varepsilon_k}(\tau) \mathbf{r}_{\varepsilon_j \mathbf{z}}(\tau) \right] . \qquad (14.375)
\end{aligned}
$$

The last equality in (14.375) follows as in the scalar IV case, see Sect. 14.B.1, using arguments that all the involved covariance functions do decay exponentially as $|\tau|$ increases to infinity.

Going back to the whole matrix $\mathbf{C}$ it turns out from (14.375) that it can be written as

$$
\begin{aligned}
\mathbf{C} &= \sum_{\tau=-\infty}^{\infty} \left[\mathbf{R}_{\boldsymbol{\varepsilon}}(\tau) \otimes \mathbf{R}_{\mathbf{z}}(\tau) + \begin{pmatrix} \mathbf{r}_{\mathbf{z}\varepsilon_1}(\tau) \\ \vdots \\ \mathbf{r}_{\mathbf{z}\varepsilon_{ny}}(\tau) \end{pmatrix} \begin{pmatrix} \mathbf{r}_{\mathbf{z}\varepsilon_1}^T(-\tau) & \dots & \mathbf{r}_{\mathbf{z}\varepsilon_{ny}}^T(-\tau) \end{pmatrix} \right] \\
&= \sum_{\tau=-\infty}^{\infty} \left[\mathbf{R}_{\boldsymbol{\varepsilon}}(\tau) \otimes \mathbf{R}_{\mathbf{z}}(\tau) + \operatorname{vec}\left(\mathbf{R}_{\mathbf{z}\boldsymbol{\varepsilon}}(\tau)\right) \left(\operatorname{vec}\left(\mathbf{R}_{\mathbf{z}\boldsymbol{\varepsilon}}(-\tau)\right) \right)^T \right] , \qquad (14.376)
\end{aligned}
$$

which completes the proof.

14.D Asymptotic Accuracy for Models Obtained under Linear Constraints

Consider the problem of minimizing

$$V_N(\boldsymbol{\vartheta}) = \frac{1}{2} \mathbf{f}_N^T(\boldsymbol{\vartheta}) \mathbf{f}_N(\boldsymbol{\vartheta}) \qquad (14.377)$$

under the constraint

$$\mathbf{g}_N(\boldsymbol{\vartheta}) = \mathbf{0} . \qquad (14.378)$$

Here, $\mathbf{f}$ and $\mathbf{g}$ are smooth functions of the parameter vector $\boldsymbol{\vartheta}$. They depend also on the number of data points, N, and converge uniformly to $\mathbf{f}_\infty$ and $\mathbf{g}_\infty$, respectively, as $N \to \infty$. Further, in the limiting case for the true parameter vector it holds

$$\mathbf{f}_\infty(\boldsymbol{\vartheta}_0) = \mathbf{0}, \quad \mathbf{g}_\infty(\boldsymbol{\vartheta}_0) = \mathbf{0} . \qquad (14.379)$$

The solution to the optimization problem will be denoted by $\hat{\boldsymbol{\vartheta}}_N$. The issue to be considered here is how to express the estimation error $\hat{\boldsymbol{\vartheta}}_N - \boldsymbol{\vartheta}_0$ (for large enough values of N) in terms of $\mathbf{f}_N$ and $\mathbf{g}_N$.

To treat the general optimization problem, introduce the Lagrangian

$$L(\boldsymbol{\vartheta}, \boldsymbol{\lambda}) = \frac{1}{2}\mathbf{f}_N^T(\boldsymbol{\vartheta})\mathbf{f}_N(\boldsymbol{\vartheta}) + \boldsymbol{\lambda}^T\mathbf{g}_N(\boldsymbol{\vartheta}) \ . \tag{14.380}$$

Setting the gradient of L to zero gives the necessary conditions for the optimum to be

$$\mathbf{0} = \mathbf{f}_N^T(\hat{\boldsymbol{\vartheta}})\frac{\partial \mathbf{f}_N(\hat{\boldsymbol{\vartheta}}_N)}{\partial \boldsymbol{\vartheta}} + \boldsymbol{\lambda}^T\frac{\partial \mathbf{g}_N(\hat{\boldsymbol{\vartheta}}_N)}{\partial \boldsymbol{\vartheta}} \ , \tag{14.381}$$

$$\mathbf{0} = \mathbf{g}_N(\hat{\boldsymbol{\vartheta}}_N) \ . \tag{14.382}$$

To this aim introduce the sensitivity matrices

$$\mathbf{F} = \frac{\partial \mathbf{f}(\boldsymbol{\vartheta})}{\partial \boldsymbol{\vartheta}}|_{\boldsymbol{\vartheta}=\boldsymbol{\vartheta}_0} \ , \tag{14.383}$$

$$\mathbf{G} = \frac{\partial \mathbf{g}(\boldsymbol{\vartheta})}{\partial \boldsymbol{\vartheta}}|_{\boldsymbol{\vartheta}=\boldsymbol{\vartheta}_0} \ . \tag{14.384}$$

As the estimate $\hat{\boldsymbol{\vartheta}}_N$ will be close to the true parameter vector $\boldsymbol{\vartheta}_0$ for large N, one can now use the linearized expressions

$$\mathbf{f}_N(\hat{\boldsymbol{\vartheta}}_N) \approx \mathbf{f}_N(\boldsymbol{\vartheta}_0) + \mathbf{F}\left(\hat{\boldsymbol{\vartheta}}_N - \boldsymbol{\vartheta}_0\right) \ , \tag{14.385}$$

$$\mathbf{g}_N(\hat{\boldsymbol{\vartheta}}_N) \approx \mathbf{g}_N(\boldsymbol{\vartheta}_0) + \mathbf{G}\left(\hat{\boldsymbol{\vartheta}}_N - \boldsymbol{\vartheta}_0\right) \ . \tag{14.386}$$

Making use of (14.385) and (14.386) in (14.381) and (14.382) now leads to

$$\mathbf{0} \approx \left[\mathbf{f}_N^T(\boldsymbol{\vartheta}_0) + \left(\hat{\boldsymbol{\vartheta}}_N - \boldsymbol{\vartheta}_0\right)^T \mathbf{F}^T\right]\mathbf{F} + \boldsymbol{\lambda}^T\mathbf{G} \ ,$$

$$\mathbf{0} \approx \mathbf{g}_N(\boldsymbol{\vartheta}_0) + \mathbf{G}\left(\hat{\boldsymbol{\vartheta}}_N - \boldsymbol{\vartheta}_0\right) \ ,$$

which is easily rewritten as

$$\mathbf{0} = \mathbf{F}^T\mathbf{f}_N + \left(\mathbf{F}^T\mathbf{F}\right)\left(\hat{\boldsymbol{\vartheta}}_N - \boldsymbol{\vartheta}_0\right) + \mathbf{G}^T\boldsymbol{\lambda} \ , \tag{14.387}$$

$$\mathbf{0} \approx \mathbf{g}_N + \mathbf{G}\left(\hat{\boldsymbol{\vartheta}}_N - \boldsymbol{\vartheta}_0\right) \ , \tag{14.388}$$

where, for convenience only, the arguments of the functions are skipped. Note that (14.387), (14.388) is a linear system of equations with $\hat{\boldsymbol{\vartheta}}_N - \boldsymbol{\vartheta}_0$ and $\boldsymbol{\lambda}$ as unknowns.

The solution can be written in the form

$$\begin{pmatrix} \hat{\boldsymbol{\vartheta}}_N - \boldsymbol{\vartheta}_0 \\ \boldsymbol{\lambda} \end{pmatrix} = -\begin{pmatrix} \mathbf{F}^T\mathbf{F} & \mathbf{G}^T \\ \mathbf{G} & \mathbf{0} \end{pmatrix}^{-1} \begin{pmatrix} \mathbf{F}^T\mathbf{f}_N \\ \mathbf{g}_N \end{pmatrix}$$
$$= -\left[\begin{pmatrix} \left(\mathbf{F}^T\mathbf{F}\right)^{-1} & \mathbf{0} \\ \mathbf{0} & \mathbf{0} \end{pmatrix} - \begin{pmatrix} -\left(\mathbf{F}^T\mathbf{F}\right)^{-1}\mathbf{G}^T \\ \mathbf{I} \end{pmatrix} \right.$$
$$\left. \times \left[\mathbf{G}\left(\mathbf{F}^T\mathbf{F}\right)^{-1}\mathbf{G}^T\right]^{-1} \left(-\mathbf{G}\left(\mathbf{F}^T\mathbf{F}\right)^{-1} \ \mathbf{I}\right)\right] \begin{pmatrix} \mathbf{F}^T\mathbf{f}_N \\ \mathbf{g}_N \end{pmatrix} ,$$

leading to

$$\hat{\boldsymbol{\vartheta}}_N - \boldsymbol{\vartheta}_0 = -\left(\mathbf{F}^T\mathbf{F}\right)^{-1}\mathbf{F}^T\mathbf{f}_N + \left(\mathbf{F}^T\mathbf{F}\right)^{-1}\mathbf{G}^T$$
$$\times \left[\mathbf{G}\left(\mathbf{F}^T\mathbf{F}\right)^{-1}\mathbf{G}^T\right]^{-1} \left[-\mathbf{G}\left(\mathbf{F}^T\mathbf{F}\right)^{-1}\mathbf{F}^T\mathbf{f}_N + \mathbf{g}_N\right] . \quad (14.389)$$

14.E Asymptotic Distribution for the Covariance Matching Method

14.E.1 Covariance Matrix of the Extended Parameter Vector

In this section the results of Theorem 14.1 are extended to give the covariance matrix not only of the estimate $\hat{\boldsymbol{\theta}}$ but also of $\hat{\mathbf{r}}_z$ and of $\hat{\boldsymbol{\rho}} = \left(\hat{\lambda}_y \ \hat{\lambda}_u\right)^T$. How $\hat{\boldsymbol{\rho}}$ can be found is described in Remark 8.6.

In fact, the augmented model then used has precisely the same algebraic form as in (8.21):

$$\hat{\bar{\mathbf{r}}} = \begin{pmatrix} \hat{\mathbf{r}} \\ \hat{\mathbf{r}}_2 \end{pmatrix} , \quad (14.390)$$

$$\bar{\mathbf{F}}(\boldsymbol{\theta}) = \begin{pmatrix} \mathbf{F}(\boldsymbol{\theta}) & \mathbf{0} \\ \mathbf{F}_2(\boldsymbol{\theta}) & \mathbf{I} \end{pmatrix} \quad \overline{\mathbf{W}} = \begin{pmatrix} \mathbf{W}_{11} & \mathbf{W}_{12} \\ \mathbf{W}_{21} & \mathbf{W}_{22} \end{pmatrix} , \quad (14.391)$$

$$\bar{\mathbf{r}}_\mathbf{z} = \begin{pmatrix} \mathbf{r}_\mathbf{z} \\ \boldsymbol{\rho} \end{pmatrix} \quad \mathbf{r}_2 = \begin{pmatrix} r_y(0) \\ r_u(0) \end{pmatrix} \quad \boldsymbol{\rho} = \begin{pmatrix} \lambda_y \\ \lambda_u \end{pmatrix} . \quad (14.392)$$

In particular, assume there is no cross-weighting, so $\mathbf{W}_{12} = \mathbf{0}$ in (14.391). Then the estimate of the noise variances can be treated separately and is indeed given by

$$\hat{\boldsymbol{\rho}} = \begin{pmatrix} \hat{r}_y(0) \\ \hat{r}_u(0) \end{pmatrix} - \mathbf{F}_2(\hat{\boldsymbol{\theta}})\hat{\mathbf{r}}_\mathbf{z} . \quad (14.393)$$

Concerning the accuracy of the estimate (14.393) one needs to relate it to the covariance matrix of $\hat{\overline{\mathbf{r}}}$ in (14.390). To this aim introduce

$$\overline{\mathbf{R}} = \lim_{N\to\infty} N\mathsf{E}\left\{\left[\hat{\overline{\mathbf{r}}} - \overline{\mathbf{r}}\right]\left[\hat{\overline{\mathbf{r}}} - \overline{\mathbf{r}}\right]^T\right\} = \begin{pmatrix} \mathbf{R} & \mathbf{R}_{12} \\ \mathbf{R}_{21} & \mathbf{R}_2 \end{pmatrix} \tag{14.394}$$

and note that it can be computed using the algorithm for $\mathbf{R}$, see Sect. 14.5.2. (The modification is only that a few running indices, such as μ, ν, have to start at 0 instead of 1.)

The following result applies.

Lemma 14.15 *Consider the joint estimate $\hat{\boldsymbol{\theta}}$, $\hat{\mathbf{r}}_{\mathbf{z}}$, and $\hat{\boldsymbol{\rho}}$. The normalized asymptotic joint covariance matrix can be written as*

$$\begin{aligned}
&\lim_{N\to\infty} N\mathsf{E}\left\{\begin{pmatrix} \hat{\theta} - \theta \\ \hat{\mathbf{r}}_{\mathbf{z}} - \mathbf{r} \\ \hat{\rho} - \rho \end{pmatrix}\left(\left(\hat{\theta} - \theta\right)^T \; (\hat{\mathbf{r}}_{\mathbf{z}} - \mathbf{r})^T \; (\hat{\rho} - \rho)^T\right)\right\} \\
&= \begin{pmatrix} \mathbf{P}_\theta & \mathbf{P}_{\theta\mathbf{z}} & \mathbf{P}_{\theta\rho} \\ \mathbf{P}_{\mathbf{z}\theta} & \mathbf{P}_{\mathbf{z}} & \mathbf{P}_{\mathbf{z}\rho} \\ \mathbf{P}_{\rho\theta} & \mathbf{P}_{\rho\mathbf{z}} & \mathbf{P}_\rho \end{pmatrix} ,
\end{aligned} \tag{14.395}$$

where

$$\mathbf{P}_\theta = \mathbf{L}_1\mathbf{R}\mathbf{L}_1^T , \tag{14.396}$$

$$\mathbf{P}_{\theta\mathbf{z}} = \mathbf{L}_1\mathbf{R}\mathbf{L}_2^T , \tag{14.397}$$

$$\mathbf{P}_{\theta\rho} = \mathbf{L}_1\left(\mathbf{R}_{12} - \mathbf{R}\mathbf{L}_1^T\mathbf{S}_2^T - \mathbf{R}\mathbf{L}_2^T\mathbf{F}_2^T\right) , \tag{14.398}$$

$$\mathbf{P}_{\mathbf{z}} = \mathbf{L}_2\mathbf{R}\mathbf{L}_2^T , \tag{14.399}$$

$$\mathbf{P}_{\mathbf{z}\rho} = \mathbf{L}_2\left(\mathbf{R}_{12} - \mathbf{R}\mathbf{L}_1^T\mathbf{S}_2^T - \mathbf{R}\mathbf{L}_2^T\mathbf{F}_2^T\right) , \tag{14.400}$$

$$\begin{aligned}
\mathbf{P}_\rho &= (\mathbf{S}_2\mathbf{L}_1 + \mathbf{F}_2\mathbf{L}_2)\,\mathbf{R}\,(\mathbf{S}_2\mathbf{L}_1 + \mathbf{F}_2\mathbf{L}_2)^T + \mathbf{R}_2 \\
&\quad - (\mathbf{S}_2\mathbf{L}_1 + \mathbf{F}_2\mathbf{L}_2)\,\mathbf{R}_{12} - \mathbf{R}_{21}\,(\mathbf{S}_2\mathbf{L}_1 + \mathbf{F}_2\mathbf{L}_2)^T ,
\end{aligned} \tag{14.401}$$

$$\mathbf{L}_1 = \left(\mathbf{S}^T\mathbf{M}\mathbf{S}\right)^{-1}\mathbf{S}^T\mathbf{M} , \tag{14.402}$$

$$\mathbf{L}_2 = \left(\mathbf{F}^T\mathbf{M}_{\mathbf{z}}\mathbf{F}\right)^{-1}\mathbf{F}^T\mathbf{M}_{\mathbf{z}} , \tag{14.403}$$

$$\mathbf{S}_2 = \left(\frac{\partial \mathbf{F}_2}{\partial \theta_1}\mathbf{r}_{\mathbf{z}} \; \dots \; \frac{\partial \mathbf{F}_2}{\partial \theta_{na+nb}}\mathbf{r}_{\mathbf{z}}\right) , \tag{14.404}$$

$$\mathbf{M}_{\mathbf{z}} = \mathbf{W} - \mathbf{W}\mathbf{S}\left(\mathbf{S}^T\mathbf{W}\mathbf{S}\right)^{-1}\mathbf{S}^T\mathbf{W} . \tag{14.405}$$

Further, $\mathbf{M}$ *is given by (14.142),* $\mathbf{S}$ *by (14.128), and* $\mathbf{F}_2$ *by (8.29).*

Proof Note from (14.8) and (14.132), (14.133) that asymptotically (for large N)

$$\tilde{\theta} = \left(\mathbf{S}^T\mathbf{M}\mathbf{S}\right)^{-1}\mathbf{S}^T\mathbf{M}\tilde{\mathbf{r}} = \mathbf{L}_1\tilde{\mathbf{r}} . \tag{14.406}$$

Similarly to (14.8), cf. also (14.126),

$$\begin{aligned}
\tilde{\mathbf{r}}_z &= (\mathbf{0}\ \mathbf{I})\left(\mathbf{L}^T\mathbf{W}\mathbf{L}\right)^{-1}\mathbf{L}^T\mathbf{W}\tilde{\mathbf{r}} \\
&= (\mathbf{0}\ \mathbf{I})\begin{pmatrix}\mathbf{S}^T\mathbf{W}\mathbf{S} & \mathbf{S}^T\mathbf{W}\mathbf{F} \\ \mathbf{F}^T\mathbf{W}\mathbf{S} & \mathbf{F}^T\mathbf{W}\mathbf{F}\end{pmatrix}^{-1}\begin{pmatrix}\mathbf{S}^T\mathbf{W} \\ \mathbf{F}^T\mathbf{W}\end{pmatrix}\tilde{\mathbf{r}} \\
&= (\mathbf{0}\ \mathbf{I})\begin{pmatrix}-\left(\mathbf{S}^T\mathbf{W}\mathbf{S}\right)^{-1}\mathbf{S}^T\mathbf{W}\mathbf{F} \\ \mathbf{I}\end{pmatrix} \\
&\quad\times\left(\mathbf{F}^T\mathbf{W}\mathbf{F}-\mathbf{F}^T\mathbf{W}\mathbf{S}\left(\mathbf{S}^T\mathbf{W}\mathbf{S}\right)^{-1}\mathbf{S}^T\mathbf{W}\mathbf{F}\right)^{-1} \\
&\quad\times\left(-\mathbf{F}^T\mathbf{W}\mathbf{S}\left(\mathbf{S}^T\mathbf{W}\mathbf{S}\right)^{-1}\ \ \mathbf{I}\right)\begin{pmatrix}\mathbf{S}^T\mathbf{W} \\ \mathbf{F}^T\mathbf{W}\end{pmatrix}\tilde{\mathbf{r}} \\
&= \left(\mathbf{F}^T\mathbf{M}_z\mathbf{F}\right)^{-1}\left(\mathbf{F}^T\mathbf{W}-\mathbf{F}^T\mathbf{W}\mathbf{S}\left(\mathbf{S}^T\mathbf{W}\mathbf{S}\right)^{-1}\mathbf{S}^T\mathbf{W}\right)\tilde{\mathbf{r}} \\
&= \left(\mathbf{F}^T\mathbf{M}_z\mathbf{F}\right)^{-1}\mathbf{F}^T\mathbf{M}_z\tilde{\mathbf{r}} = \mathbf{L}_2\tilde{\mathbf{r}}\ .
\end{aligned} \tag{14.407}$$

Linearizing (14.393) around the true values leads to

$$\tilde{\rho} = \tilde{\mathbf{r}}_2 - \mathbf{S}_2\tilde{\theta} - \mathbf{F}_2\tilde{\mathbf{r}}_z\ . \tag{14.408}$$

Combining (14.406)–(14.408) leads to

$$\begin{aligned}
\begin{pmatrix}\tilde{\theta} \\ \tilde{\mathbf{r}}_z \\ \tilde{\rho}\end{pmatrix} &= \begin{pmatrix}\mathbf{L}_1\tilde{\mathbf{r}} \\ \mathbf{L}_2\tilde{\mathbf{r}} \\ -\left(\mathbf{S}_2\mathbf{L}_1+\mathbf{F}_2\mathbf{L}_2\right)\tilde{\mathbf{r}}+\tilde{\mathbf{r}}_2\end{pmatrix} \\
&= \begin{pmatrix}\mathbf{L}_1 & \mathbf{0} \\ \mathbf{L}_2 & \mathbf{0} \\ -\left(\mathbf{S}_2\mathbf{L}_1+\mathbf{F}_2\mathbf{L}_2\right) & \mathbf{I}\end{pmatrix}\begin{pmatrix}\tilde{\mathbf{r}} \\ \tilde{\mathbf{r}}_2\end{pmatrix}\ ,
\end{aligned} \tag{14.409}$$

and the statement follows directly by evaluating the covariance matrix of the vector in (14.409). ■

14.E.2 Proof of Theorem 14.2

Consider first the block matrix $\mathbf{R}_{11}$. One can write

$$\hat{\mathbf{r}}_y = \frac{1}{N}\sum_{t=1}^{N}\varphi(t)y(t)\ , \tag{14.410}$$

where, in this particular case,

$$\boldsymbol{\varphi}(t) = \begin{pmatrix} y(t) \\ \vdots \\ y(t-p_y) \end{pmatrix} . \tag{14.411}$$

Furthermore, let $\mathbf{r}_y$ denote the true (and expected) value of $\hat{\mathbf{r}}_y$:

$$\mathbf{r}_y = \mathsf{E}\left\{\boldsymbol{\varphi}(t)y(t)\right\} = \begin{pmatrix} r_y(0) \\ \vdots \\ r_y(p_y) \end{pmatrix} . \tag{14.412}$$

Consider the asymptotic normalized covariance matrix of $\hat{\mathbf{r}}$:

$$\mathbf{R}_{11} = \lim_{N\to\infty} N\mathsf{E}\left\{[\hat{\mathbf{r}}_y - \mathbf{r}_y][\hat{\mathbf{r}}_y - \mathbf{r}_y]^T\right\} . \tag{14.413}$$

Using (14.412)

$$\begin{aligned} \mathbf{R}_{11} &= \lim_{N\to\infty} N\mathsf{E}\left\{\hat{\mathbf{r}}_y\hat{\mathbf{r}}_y^T - \mathbf{r}_y\mathbf{r}_y^T\right\} \\ &= \lim_{N\to\infty} N\left[\frac{1}{N^2}\sum_{t=1}^{N}\sum_{s=1}^{N}\mathsf{E}\left\{\boldsymbol{\varphi}(t)y(t)\boldsymbol{\varphi}^T(s)y(s)\right\} - \mathbf{r}_y\mathbf{r}_y^T\right] . \end{aligned} \tag{14.414}$$

Use Assumptions **AN2** and **AI3** and apply the general rule for product of Gaussian variables, as given in Lemma A.9. Using the result (A.68) in (14.414) leads to

$$\begin{aligned} \mathbf{R}_{11} = \lim_{N\to\infty}\frac{1}{N}\sum_{t=1}^{N}\sum_{s=1}^{N}\Big(&\left[\mathsf{E}\left\{\boldsymbol{\varphi}(t)\boldsymbol{\varphi}^T(s)\right\}\right]\left[\mathsf{E}\left\{y(t)y(s)\right\}\right] \\ &+ \left[\mathsf{E}\left\{\boldsymbol{\varphi}(t)y(s)\right\}\right]\left[\mathsf{E}\left\{\boldsymbol{\varphi}^T(s)y(t)\right\}\right]\Big) . \end{aligned} \tag{14.415}$$

Due to Assumption **AS1** the covariance function $r_y(\tau)$ decays exponentially with τ. Therefore one can write

$$\| \lim_{N\to\infty}\frac{1}{N}\sum_{\tau=-N}^{N}|\tau|\mathbf{R}_{\varphi}(\tau)r_y(\tau) \| \leq \lim_{N\to\infty}\frac{1}{N}\sum_{\tau=0}^{N}2\tau C\alpha^{\tau} = 0 \tag{14.416}$$

for some $|\alpha| < 1$. Using this result leads to

$$\begin{aligned} \mathbf{R}_{11} &= \lim_{N\to\infty}\frac{1}{N}\sum_{\tau=-N}^{N}(N-|\tau|)\left[\mathbf{R}_{\varphi}(\tau)r_y(\tau) + \mathbf{r}_{\varphi y}(\tau)\mathbf{r}_{\varphi y}^T(-\tau)\right] \\ &= \sum_{\tau=-\infty}^{\infty}\left[\mathbf{R}_{\varphi}(\tau)r_y(\tau) + \mathbf{r}_{\varphi y}(\tau)\mathbf{r}_{\varphi y}^T(-\tau)\right] . \end{aligned} \tag{14.417}$$

In order to proceed one will need a technique for computing sums of the form

$$\sum_{\tau=-\infty}^{\infty} r_y(\tau) r_y(\tau+k) , \tag{14.418}$$

where k is an arbitrary integer. The measured output signal can be written as

$$y(t) = y_0(t) + \tilde{y}(t) = \frac{B(q^{-1})}{A(q^{-1})} \frac{C(q^{-1})}{D(q^{-1})} e(t) + \tilde{y}(t) \tag{14.419}$$

and Lemma A.10 can be applied: Write the generic element of the upper left partition $\mathbf{R}_{11}$ as follows, using (14.419):

$$\begin{aligned}
(\mathbf{R}_{11})_{\mu\nu} &= \sum_{\tau=-\infty}^{\infty} \left[r_y(\tau+\mu-\nu) r_y(\tau) + r_y(\tau-\mu) r_y(\tau+\nu) \right] \\
&= \sum_{\tau=-\infty}^{\infty} \Big[\left\{ r_{y_0}(\tau+\mu-\nu) + \lambda_y \delta_{\tau+\mu-\nu,0} \right\} \left\{ r_{y_0}(\tau) + \lambda_y \delta_{\tau,0} \right\} \\
&\quad + \left\{ r_{y_0}(\tau-\mu) + \lambda_y \delta_{\tau-\mu,0} \right\} \left\{ r_{y_0}(\tau+\nu) + \lambda_y \delta_{\tau+\nu,0} \right\} \Big] \\
&= \lambda_y^2 \left[\delta_{\mu,\nu} + \delta_{\mu,0}\delta_{\nu,0} \right] + \lambda_y \left[r_{y_0}(\mu-\nu) + r_{y_0}(\nu-\mu) \right. \\
&\quad \left. + r_{y_0}(\mu+\nu) + r_{y_0}(-\nu-\mu) \right] \\
&\quad + \sum_{\tau=-\infty}^{\infty} \left[r_{y_0}(\tau+\mu-\nu) r_{y_0}(\tau) + r_{y_0}(\tau-\mu) r_{y_0}(\tau+\nu) \right] .
\end{aligned} \tag{14.420}$$

Then it follows from Lemma A.10 that the sought matrix element can be computed as

$$(\mathbf{R}_{11})_{\mu\nu} = \lambda_y^2 \left[\delta_{\mu,\nu} + \delta_{\mu,0}\delta_{\nu,0} \right] + 2\lambda_y \left[r_{y_0}(\mu-\nu) + r_{y_0}(\mu+\nu) \right] + \beta_{\mu-\nu} + \beta_{\mu+\nu} , \tag{14.421}$$

which is (14.145).

Next consider the remaining matrix blocks of $\mathbf{R}$. As $\mathbf{R}$ is symmetric, it remains to find the elements of the block matrices $\mathbf{R}_{12}$, $\mathbf{R}_{13}$, $\mathbf{R}_{22}$, $\mathbf{R}_{23}$, and $\mathbf{R}_{33}$.

The block $\mathbf{R}_{12}$ can be written as

$$\begin{aligned}
\mathbf{R}_{12} &= \lim_{N\to\infty} N \mathsf{E} \left\{ [\hat{\mathbf{r}}_y - \mathbf{r}_y][\hat{\mathbf{r}}_u - \mathbf{r}_u]^T \right\} \\
&= \lim_{N\to\infty} N \left[\frac{1}{N^2} \sum_{t=1}^{N} \sum_{s=1}^{N} \mathsf{E} \left\{ \boldsymbol{\varphi}_y(t) y(t) \boldsymbol{\varphi}_u^T(s) u(s) \right\} - \mathbf{r}_y \mathbf{r}_u^T \right] ,
\end{aligned} \tag{14.422}$$

where $\boldsymbol{\varphi}_y(t)$ is as given by (14.411) and

$$\boldsymbol{\varphi}_u(t) = \left(u(t) \ldots u(t-p_u)\right)^T . \tag{14.423}$$

Proceeding as before one obtains

$$\mathbf{R}_{12} = \sum_{\tau=-\infty}^{\infty} \left[\mathbf{R}_{\varphi_y\varphi_u}(\tau) r_{yu}(\tau) + \mathbf{r}_{\varphi_y u}(\tau)\mathbf{r}^T_{\varphi_u y}(-\tau)\right] , \tag{14.424}$$

compare (14.417). In contrast to the previous developments there will in this case be no contribution involving the measurement noise variances. A generic element $(0 \le \mu \le p_y,\ 0 \le \nu \le p_u)$ of the matrix $\mathbf{R}_{12}$ can be written as

$$(\mathbf{R}_{12})_{\mu\nu} = \sum_{\tau=-\infty}^{\infty} \left[r_{yu}(\tau-\mu+\nu) r_{yu}(\tau) + r_{yu}(\tau-\mu) r_{yu}(\tau+\nu)\right] . \tag{14.425}$$

Invoking Lemma A.10 and (14.148) one can write

$$(\mathbf{R}_{12})_{\mu\nu} = \beta^{(2)}_{-\mu+\nu} + \beta^{(2)}_{\mu+\nu} , \tag{14.426}$$

which is (14.147).

In a similar fashion one obtains for the block $\mathbf{R}_{13}$

$$\begin{aligned}
\mathbf{R}_{13} &= \lim_{N\to\infty} N \mathsf{E}\left\{[\hat{\mathbf{r}}_y - \mathbf{r}_y][\hat{\mathbf{r}}_{yu} - \mathbf{r}_{yu}]^T\right\} \\
&= \lim_{N\to\infty} N \left[\frac{1}{N^2}\sum_{t=1}^{N}\sum_{s=1}^{N} \mathsf{E}\left\{\boldsymbol{\varphi}_y(t) y(t) \boldsymbol{\varphi}^T_{yu}(s) y(s)\right\} - \mathbf{r}_y \mathbf{r}^T_{yu}\right] ,
\end{aligned} \tag{14.427}$$

where this time $\varphi_y(t)$ is given by (14.411) and it holds

$$\boldsymbol{\varphi}_{yu}(t) = \left(u(t-p_1) \ldots u(t-p_2)\right)^T . \tag{14.428}$$

This leads, as above, to

$$\mathbf{R}_{13} = \sum_{\tau=-\infty}^{\infty} \left[\mathbf{R}_{\varphi_y\varphi_{yu}}(\tau) r_y(\tau) + \mathbf{r}_{\varphi_y y}(\tau)\mathbf{r}^T_{\varphi_{yu} y}(-\tau)\right] , \tag{14.429}$$

compare (14.417). The generic element of $\mathbf{R}_{13}$ $(0 \le \mu \le p_y,\ p_1 \le \nu \le p_2)$ can be written as

$$
\begin{aligned}
(\mathbf{R}_{13})_{\mu\nu} &= \lambda_y r_{yu}(-\mu+\nu) + \lambda_y r_{yu}(\mu+\nu) \\
&\quad + \sum_{\tau=-\infty}^{\infty} \left[r_{yu}(\tau-\mu+\nu) r_{y_0}(\tau) + r_{yu}(\tau+\nu) r_{y_0}(\tau-\mu) \right] \\
&= \lambda_y [r_{yu}(-\mu+\nu) + r_{yu}(\mu+\nu)] + \beta^{(3)}_{\nu-\mu} + \beta^{(3)}_{\mu+\nu} ,
\end{aligned}
\tag{14.430}
$$

which is (14.149).

For the block $\mathbf{R}_{22}$ it holds

$$
\begin{aligned}
\mathbf{R}_{22} &= \lim_{N\to\infty} N \mathsf{E}\left\{ [\hat{\mathbf{r}}_u - \mathbf{r}_u][\hat{\mathbf{r}}_u - \mathbf{r}_u]^T \right\} \\
&= \lim_{N\to\infty} N \left[\frac{1}{N^2} \sum_{t=1}^{N} \sum_{s=1}^{N} \mathsf{E}\left\{ \boldsymbol{\varphi}_u(t) u(t) \boldsymbol{\varphi}_u^T(s) u(s) \right\} - \mathbf{r}_u \mathbf{r}_u^T \right] ,
\end{aligned}
\tag{14.431}
$$

where $\boldsymbol{\varphi}_u(t)$ is as given by (14.423). Therefore

$$
\mathbf{R}_{22} = \sum_{\tau=-\infty}^{\infty} \left[\mathbf{R}_{\varphi_u}(\tau) r_u(\tau) + \mathbf{r}_{\varphi_u u}(\tau) \mathbf{r}^T_{\varphi_u u}(-\tau) \right] . \tag{14.432}
$$

The generic element ($0 \le \mu \le p_u$, $0 \le \nu \le p_u$) of the matrix $\mathbf{R}_{22}$ becomes

$$
\begin{aligned}
(\mathbf{R}_{22})_{\mu\nu} &= \sum_{\tau=-\infty}^{\infty} \Big[\left(\lambda_u \delta_{\mu,\tau+\nu} + r_{u_0}(\tau+\mu-\nu) \right) \left(\lambda_u \delta_{\tau,0} + r_{u_0}(\tau) \right) \\
&\quad + \left(\lambda_u \delta_{\tau-\mu,0} + r_{u_0}(\tau-\mu) \right) \left(\lambda_u \delta_{-\tau-\nu,0} + r_{u_0}(-\tau-\nu) \right) \Big] \\
&= \lambda_u^2 \left[\delta_{\mu,\nu} + \delta_{\mu,0}\delta_{\nu,0} \right] + 2\lambda_u \left[r_{u_0}(\mu-\nu) + r_{u_0}(\mu+\nu) \right] \\
&\quad + \sum_{\tau=-\infty}^{\infty} \left[r_{u_0}(\tau+\mu-\nu) r_{u_0}(\tau) + r_{u_0}(\tau-\mu) r_{u_0}(-\tau-\nu) \right] \\
&= \lambda_u^2 \left[\delta_{\mu,\nu} + \delta_{\mu,0}\delta_{\nu,0} \right] + 2\lambda_u \left[r_{u_0}(\mu-\nu) + r_{u_0}(\mu+\nu) \right] + \beta^{(4)}_{\mu-\nu} + \beta^{(4)}_{\mu+\nu} ,
\end{aligned}
\tag{14.433}
$$

which is (14.151) and where $\beta_k^{(4)}$ is given by (14.152).

The block $\mathbf{R}_{23}$ can be written as

$$
\begin{aligned}
\mathbf{R}_{23} &= \lim_{N\to\infty} N \mathsf{E}\left\{ [\hat{\mathbf{r}}_u - \mathbf{r}_u][\hat{\mathbf{r}}_{yu} - \mathbf{r}_{yu}]^T \right\} \\
&= \lim_{N\to\infty} N \left[\frac{1}{N^2} \sum_{t=1}^{N} \sum_{s=1}^{N} \mathsf{E}\left\{ \boldsymbol{\varphi}_u(t) u(t) \boldsymbol{\varphi}_{yu}^T(s) y(s) \right\} - \mathbf{r}_u \mathbf{r}_{yu}^T \right] ,
\end{aligned}
\tag{14.434}
$$

where $\boldsymbol{\varphi}_u(t)$ is as given by (14.423) and $\boldsymbol{\varphi}_{yu}(t)$ is defined in (14.428).

Proceeding as before one obtains

$$\mathbf{R}_{23} = \sum_{\tau=-\infty}^{\infty} \left[\mathbf{R}_{\varphi_u \varphi_{yu}}(\tau) r_{uy}(\tau) + \mathbf{r}_{\varphi_u y}(\tau) \mathbf{r}^T_{\varphi_{yu} u}(-\tau) \right] . \tag{14.435}$$

A generic element $(0 \le \mu \le p_u,\ p_1 \le \nu \le p_2)$ of the matrix $\mathbf{R}_{23}$ can be written as

$$(\mathbf{R}_{23})_{\mu\nu} = \sum_{\tau=-\infty}^{\infty} \left[r_u(\tau - \mu + \nu) r_{yu}(-\tau) + r_{yu}(-\tau + \mu) r_u(\tau + \nu) \right] . \tag{14.436}$$

Invoking Lemma A.10 and (14.436) one can write

$$\begin{aligned}(\mathbf{R}_{23})_{\mu\nu} &= \lambda_u \left[r_{yu}(\nu - \mu) + r_{yu}(\nu + \mu) \right] \\ &\quad + \sum_{\tau=-\infty}^{\infty} \left[r_{u_0}(\tau - \mu + \nu) r_{yu}(-\tau) + r_{yu}(-\tau + \mu) r_{u_0}(\tau + \nu) \right] \\ &= \lambda_u \left[r_{yu}(\nu - \mu) + r_{yu}(\nu + \mu) \right] + \beta^{(5)}_{-\mu+\nu} + \beta^{(5)}_{\mu+\nu} ,\end{aligned} \tag{14.437}$$

which is (14.153).

Finally, the block $\mathbf{R}_{33}$ can be written as

$$\begin{aligned}\mathbf{R}_{33} &= \lim_{N\to\infty} N \mathsf{E} \left\{ [\hat{\mathbf{r}}_{yu} - \mathbf{r}_{yu}][\hat{\mathbf{r}}_{yu} - \mathbf{r}_{yu}]^T \right\} \\ &= \lim_{N\to\infty} N \left[\frac{1}{N^2} \sum_{t=1}^{N} \sum_{s=1}^{N} \mathsf{E} \left\{ \boldsymbol{\varphi}_{yu}(t) y(t) \boldsymbol{\varphi}^T_{yu}(s) y(s) \right\} - \mathbf{r}_{yu} \mathbf{r}^T_{yu} \right] ,\end{aligned} \tag{14.438}$$

where $\boldsymbol{\varphi}_{yu}(t)$ is as given by (14.428). Proceeding again as before one obtains

$$\mathbf{R}_{33} = \sum_{\tau=-\infty}^{\infty} \left[\mathbf{R}_{\varphi_{yu}}(\tau) r_y(\tau) + \mathbf{r}_{\varphi_{yu} y}(\tau) \mathbf{r}^T_{\varphi_{yu} y}(-\tau) \right] , \tag{14.439}$$

compare (14.417). A generic element $(p_1 \le \mu \le p_2,\ p_1 \le \nu \le p_2)$ of the matrix $\mathbf{R}_{33}$ can be written as

$$\begin{aligned}(\mathbf{R}_{33})_{\mu\nu} &= \sum_{\tau=-\infty}^{\infty} \left[r_u(\tau - \mu + \nu) r_y(\tau) + r_{uy}(\tau - \mu) r_{yu}(\tau + \nu) \right] \\ &= \lambda_y \lambda_u \delta_{\mu,\nu} + \lambda_u r_{y_0}(\mu - \nu) + \lambda_y r_{u_0}(\nu - \mu) \\ &\quad + \sum_{\tau=-\infty}^{\infty} \left[r_{u_0}(\tau - \mu + \nu) r_{y_0}(\tau) + r_{uy}(\tau - \mu) r_{yu}(\tau + \nu) \right] .\end{aligned} \tag{14.440}$$

Invoking Lemma A.10, (14.156) and (14.157) one can write

$$(\mathbf{R}_{33})_{\mu\nu} = \lambda_y \lambda_u \delta_{\mu,\nu} + \lambda_u r_{y_0}(\mu - \nu) + \lambda_y r_{u_0}(\nu - \mu) + \beta^{(6)}_{-\mu+\nu} + \gamma^{(6)}_{\mu+\nu} \,, \quad (14.441)$$

which is (14.155). This observation completes the proof.

14.E.3 Proof of Lemma 14.9

The key tool to use in the proof is Lemma A.12. As that lemma applies directly for any specific element of the matrix **R**, the result follows directly from equation (A.80).

14.F Asymptotic Distribution for PEM and ML Estimates

14.F.1 Asymptotic Covariance Matrix of the Parameter Estimates

14.F.1.1 Proof of Lemma 14.11

It holds

$$V'_\infty(\boldsymbol{\vartheta}) = \mathsf{E}\{-\ell_{\boldsymbol{\varepsilon}}(\boldsymbol{\varepsilon}, \boldsymbol{\vartheta}, t)\boldsymbol{\psi}^T(t, \boldsymbol{\vartheta}) + \ell_{\boldsymbol{\vartheta}}(\boldsymbol{\varepsilon}, \boldsymbol{\vartheta}, t)\} \,.$$

Noting that $\boldsymbol{\psi}(t, \boldsymbol{\vartheta})$ depends on old data, has zero mean, and is independent of $\boldsymbol{\varepsilon}(t, \boldsymbol{\vartheta})$, and using (9.27) one finds that

$$V''_\infty(\boldsymbol{\vartheta}_0) = \mathsf{E}\{\boldsymbol{\psi}(t, \boldsymbol{\vartheta}_0)\ell_{\boldsymbol{\varepsilon\varepsilon}}\boldsymbol{\psi}^T(t, \boldsymbol{\vartheta}_0) + \ell_{\boldsymbol{\vartheta\vartheta}}\} \,, \quad (14.442)$$

which is (14.168). Similarly,

$$\mathbf{P}_0 = \lim_{N\to\infty} \frac{1}{N} \sum_{t=1}^{N} \sum_{s=1}^{N} \mathsf{E}\left\{[-\boldsymbol{\psi}(t, \boldsymbol{\vartheta}_0)\ell^T_{\boldsymbol{\varepsilon}}(\boldsymbol{\varepsilon}(t, \boldsymbol{\vartheta}_0), \boldsymbol{\vartheta}_0, t) + \ell^T_{\boldsymbol{\vartheta}}(\boldsymbol{\varepsilon}(t, \boldsymbol{\vartheta}_0), \boldsymbol{\vartheta}_0, t)]\right.$$
$$\left. \times[-\ell_{\boldsymbol{\varepsilon}}(\boldsymbol{\varepsilon}(s, \boldsymbol{\vartheta}_0), \boldsymbol{\vartheta}_0, s)\boldsymbol{\psi}^T(s, \boldsymbol{\vartheta}_0) + \ell_{\boldsymbol{\vartheta}}(\boldsymbol{\varepsilon}(s, \boldsymbol{\vartheta}_0), \boldsymbol{\vartheta}_0, s)]\right\} \,. \quad (14.443)$$

Note that the factors $\ell_{\boldsymbol{\vartheta}}(t)$ and $\ell_{\boldsymbol{\varepsilon}}(t)$ are uncorrelated in time and independent of $\boldsymbol{\psi}(s)$ for $s \leq t$. This gives that in the double sum (14.443) only terms with $t = s$ contribute. Cf also (9.27). Therefore

$$\mathbf{P}_0 = \mathsf{E}\left\{\boldsymbol{\psi}(t, \boldsymbol{\vartheta}_0)\{\mathsf{E}\ell^T_{\boldsymbol{\varepsilon}}\ell_{\boldsymbol{\varepsilon}}\}\boldsymbol{\psi}^T(t, \boldsymbol{\vartheta}_0)\right\} + \mathsf{E}\{\ell^T_{\boldsymbol{\vartheta}}\ell_{\boldsymbol{\vartheta}}\} \,, \quad (14.444)$$

which is (14.169).

14.F.1.2 Proof of Corollary 14.2

Consider the criterion (9.28):

$$\ell(\boldsymbol{\varepsilon}, \boldsymbol{\vartheta}, t) = \frac{1}{2} \log \det \mathbf{Q}(\boldsymbol{\vartheta}) + \frac{1}{2} \boldsymbol{\varepsilon}^T(t, \boldsymbol{\vartheta}) \mathbf{Q}^{-1}(\boldsymbol{\vartheta}) \boldsymbol{\varepsilon}(t, \boldsymbol{\vartheta}) . \tag{14.445}$$

One needs to evaluate the expectations

$$\mathsf{E}\{\ell_{\boldsymbol{\varepsilon}\boldsymbol{\varepsilon}}\}, \quad \mathsf{E}\{\ell_{\boldsymbol{\vartheta}\boldsymbol{\vartheta}}\}, \quad \mathsf{E}\left\{\ell_{\boldsymbol{\varepsilon}}^T \ell_{\boldsymbol{\varepsilon}}\right\}, \quad \mathsf{E}\left\{\ell_{\boldsymbol{\vartheta}}^T \ell_{\boldsymbol{\vartheta}}\right\} .$$

Set

$$\mathbf{Q}_{jk} = \frac{\partial^2}{\partial \boldsymbol{\vartheta}_j \partial \boldsymbol{\vartheta}_k} \mathbf{Q} . \tag{14.446}$$

In this case

$$\begin{aligned}
\ell_{\boldsymbol{\varepsilon}} &= \boldsymbol{\varepsilon}^T \mathbf{Q}^{-1} , \\
\ell_{\boldsymbol{\varepsilon}\boldsymbol{\varepsilon}} &= \mathbf{Q}^{-1} , \\
\ell_{\boldsymbol{\vartheta}_j} &= \frac{1}{2} \mathrm{tr}[\mathbf{Q}^{-1}\mathbf{Q}_j] - \frac{1}{2} \boldsymbol{\varepsilon}^T \mathbf{Q}^{-1} \mathbf{Q}_j \mathbf{Q}^{-1} \boldsymbol{\varepsilon} , \\
\ell_{\boldsymbol{\vartheta}_j \boldsymbol{\vartheta}_k} &= \frac{1}{2} \mathrm{tr}[-\mathbf{Q}^{-1}\mathbf{Q}_k \mathbf{Q}^{-1}\mathbf{Q}_j + \mathbf{Q}^{-1}\mathbf{Q}_{jk}] \\
&\quad - \frac{1}{2} \boldsymbol{\varepsilon}^T [-\mathbf{Q}^{-1}\mathbf{Q}_k \mathbf{Q}^{-1}\mathbf{Q}_j \mathbf{Q}^{-1} + \mathbf{Q}^{-1}\mathbf{Q}_{jk}\mathbf{Q}^{-1} - \mathbf{Q}^{-1}\mathbf{Q}_j \mathbf{Q}^{-1}\mathbf{Q}_k \mathbf{Q}^{-1}] \boldsymbol{\varepsilon} .
\end{aligned}$$

Apparently,

$$\mathsf{E}\{\ell_{\boldsymbol{\varepsilon}\boldsymbol{\varepsilon}}\} = \mathbf{Q}^{-1} , \tag{14.447}$$

$$\mathsf{E}\left\{\ell_{\boldsymbol{\varepsilon}}^T \ell_{\boldsymbol{\varepsilon}}\right\} = \mathsf{E}\left\{\mathbf{Q}^{-1} \boldsymbol{\varepsilon}\boldsymbol{\varepsilon}^T \mathbf{Q}^{-1}\right\} = \mathbf{Q}^{-1} . \tag{14.448}$$

Recall that

$$\mathsf{E}\left\{\boldsymbol{\varepsilon}^T \mathbf{Q}^{-1}\mathbf{Q}_j \mathbf{Q}^{-1} \boldsymbol{\varepsilon}\right\} = \mathsf{E}\left\{ \mathrm{tr}(\mathbf{Q}^{-1}\mathbf{Q}_j \mathbf{Q}^{-1} \boldsymbol{\varepsilon}\boldsymbol{\varepsilon}^T)\right\} = \mathrm{tr}(\mathbf{Q}^{-1}\mathbf{Q}_j) ,$$

so that $\mathsf{E}\left\{\ell_{\boldsymbol{\vartheta}_j}\right\} = \mathbf{0}$ as it should. Next one finds

$$\begin{aligned}
\mathsf{E}\left\{\ell_{\boldsymbol{\vartheta}_j} \ell_{\boldsymbol{\vartheta}_k}\right\} &= -\frac{1}{4} \left(\mathrm{tr}[\mathbf{Q}^{-1}\mathbf{Q}_j]\right) \left(\mathrm{tr}[\mathbf{Q}^{-1}\mathbf{Q}_k]\right) \\
&\quad + \frac{1}{4} \mathsf{E}\left\{[\boldsymbol{\varepsilon}^T \mathbf{Q}^{-1}\mathbf{Q}_j \mathbf{Q}^{-1} \boldsymbol{\varepsilon}][\boldsymbol{\varepsilon}^T \mathbf{Q}^{-1}\mathbf{Q}_k \mathbf{Q}^{-1} \boldsymbol{\varepsilon}]\right\} \\
&= \frac{1}{2} \mathrm{tr}[\mathbf{Q}^{-1}\mathbf{Q}_j \mathbf{Q}^{-1}\mathbf{Q}_k] ,
\end{aligned} \tag{14.449}$$

where Lemma A.9 is used.

Further,

$$\begin{aligned} \mathsf{E}\left\{\ell_{\vartheta_j \vartheta_k}\right\} &= -\frac{1}{2}\mathrm{tr}(\mathbf{Q}^{-1}\mathbf{Q}_j\mathbf{Q}^{-1}\mathbf{Q}_k) + \frac{1}{2}\mathrm{tr}(\mathbf{Q}^{-1}\mathbf{Q}_{jk}) \\ &\quad + \mathrm{tr}(\mathbf{Q}^{-1}\mathbf{Q}_j\mathbf{Q}^{-1}\mathbf{Q}_k) - \frac{1}{2}\mathrm{tr}(\mathbf{Q}^{-1}\mathbf{Q}_{jk}) \\ &= \frac{1}{2}\mathrm{tr}(\mathbf{Q}^{-1}\mathbf{Q}_j\mathbf{Q}^{-1}\mathbf{Q}_k) \, . \end{aligned} \tag{14.450}$$

Finally, inserting (14.447)–(14.450) into (14.168), (14.169) gives (14.177).

14.F.2 Asymptotic Distribution for Frequency Domain ML Estimates

To fully describe the matrix $\mathbf{P}_\theta$, it is sufficient to have explicit expressions for $\mathbf{P}_1$ and $\mathbf{P}_2$, see (14.198), (14.199).

Introduce the notations

$$\begin{aligned} &A^\mu = \frac{\partial A}{\partial \theta_u}, \quad B^\mu = \frac{\partial G}{\partial \theta_u}, \quad F^\mu = \frac{\partial F}{\partial \theta_u} \, , \\ &\mu = 1, \ldots, n_a + n_b \, . \end{aligned} \tag{14.451}$$

To proceed, differentiate (14.185) to get

$$rA^\mu A^* + rAA^{*\mu} + B^\mu B^* + BB^{*\mu} = F^\mu F^* + FF^{*\mu} \, . \tag{14.452}$$

For the second-order derivatives, introduce the notation

$$F^{\mu\nu} = \frac{\partial}{\partial \boldsymbol{\theta}_\nu} F^\mu = \frac{\partial^2}{\partial \boldsymbol{\theta}_\nu \partial \boldsymbol{\theta}_\mu} F \, , \tag{14.453}$$

and note that $A^{\mu\nu} = 0$, $B^{\mu\nu} = 0$, as A^μ and B^μ by construction are constant polynomials and do not depend on $\boldsymbol{\theta}$. Differentiation of (14.452) with respect to $\boldsymbol{\theta}_\nu$ gives

$$rA^\mu A^{*\nu} + rA^\nu A^{*\mu} + B^\mu B^{*\nu} + B^\mu B^{*\nu} = F^{\mu\nu}F^* + F^\nu F^{*\mu} + F^\mu F^{*\nu} + FF^{*\mu\nu} \, . \tag{14.454}$$

Note that (14.452) and (14.454) can be seen as linear Diophantine equations for the unknown polynomials F^μ and $F^{\mu\nu}$, respectively. Equating the different powers of z leads to linear systems of equations in the unknown polynomial coefficients of F^μ and $F^{\mu\nu}$.

To compute $\mathbf{P}_1$ and $\mathbf{P}_2$, proceed as follows. First find F by the spectral factorization (14.185). Next find F^{μ} by solving the Diophantine equation (14.452) for $\mu = 1, \ldots, n_a + n_b$ and $F^{\mu\nu}$ by solving the Diophantine equation (14.454) for $\mu, \nu = 1, \ldots, n_a + n_b$.

Next examine the correlation between $\varepsilon(t)$ and $\boldsymbol{\psi}(s)$, which appears in the third terms of (14.193). Differentiating equation (14.187) gives

$$\begin{aligned}
\frac{\partial}{\partial \boldsymbol{\theta}_{\mu}} \varepsilon(t, \boldsymbol{\theta}) &= \frac{A^{\mu}}{F} y(t) - \frac{B^{\mu}}{F} u(t) - \frac{F^{\mu}}{F^2} \left(A y(t) - B u(t) \right), \\
&= \frac{A^{\mu} F - A F^{\mu}}{F^2} \left(\frac{B}{A} u_0(t) + \tilde{y}(t) \right) - \frac{B^{\mu} F - B F^{\mu}}{F^2} \left(u_0(t) + \tilde{u}(t) \right) \\
&= \frac{A^{\mu} B - A B^{\mu}}{A F} u_0(t) + \frac{A^{\mu} F - A F^{\mu}}{F^2} \tilde{y}(t) - \frac{B^{\mu} F - B F^{\mu}}{F^2} \tilde{u}(t) \\
&\stackrel{\Delta}{=} \frac{A^{\mu} B - A B^{\mu}}{A F} u_0(t) + \frac{\alpha^{\mu}}{F^2} \tilde{y}(t) - \frac{\beta^{\mu}}{F^2} \tilde{u}(t) \, .
\end{aligned} \tag{14.455}$$

By construction, the first term is always uncorrelated with $\varepsilon(t + \tau, \boldsymbol{\theta}_0)$ for any value of τ.

Now, let specifically $\tau \geq 0$. Equations (14.195), (14.455) lead to

$$\begin{aligned}
\mathsf{E} \left\{ \varepsilon(t + \tau, \boldsymbol{\theta}_0) \frac{\partial}{\partial \boldsymbol{\theta}_{\mu}} \varepsilon(t, \boldsymbol{\theta}_0) \right\} &= r \lambda_u \frac{1}{2\pi i} \oint z^{-\tau} \frac{A^*}{F^*} \frac{A^{\mu} F - A F^{\mu}}{F^2} \frac{dz}{z} \\
&\quad + \lambda_u \frac{1}{2\pi i} \oint z^{-\tau} \frac{B^*}{F^*} \frac{B^{\mu} F - B F^{\mu}}{F^2} \frac{dz}{z} \, .
\end{aligned} \tag{14.456}$$

Using (14.452) in (14.456) now leads to

$$\begin{aligned}
&\mathsf{E} \left\{ \varepsilon(t + \tau, \boldsymbol{\theta}_0) \frac{\partial}{\partial \boldsymbol{\theta}_{\mu}} \varepsilon(t, \boldsymbol{\theta}_0) \right\} \\
&= \lambda_u \frac{1}{2\pi i} \oint z^{-\tau} \frac{1}{F^* F} \left(r A^* A^{\mu} + B^* B^{\mu} \right) \frac{dz}{z} \\
&\quad + \lambda_u \frac{1}{2\pi i} \oint z^{-\tau} \frac{F^{\mu}}{F^* F^2} \underbrace{\left(-r A^* A - B^* B \right)}_{-F F^*} \frac{dz}{z} \\
&= \lambda_u \frac{1}{2\pi i} \oint z^{-\tau} \frac{1}{F^* F} \left(-r A A^{*\mu} - B B^{*\mu} + F F^{*\mu} \right) \frac{dz}{z} \\
&= -\lambda_u \frac{1}{2\pi i} \oint z^{\tau} \frac{1}{F^* F} \left(r A^* A^{\mu} + B^* B^{\mu} - F^* F^{\mu} \right) \frac{dz}{z} \\
&= -\mathsf{E} \left\{ \varepsilon(t - \tau, \boldsymbol{\theta}_0) \frac{\partial}{\partial \boldsymbol{\theta}_{\mu}} \varepsilon(t, \boldsymbol{\theta}_0) \right\} \, .
\end{aligned} \tag{14.457}$$

When considering (14.457) specifically for $\tau = 0$ one finds that

$$\mathsf{E}\left\{\varepsilon(t, \boldsymbol{\theta}_0)\frac{\partial}{\partial \boldsymbol{\theta}_\mu}\varepsilon(t, \boldsymbol{\theta}_0)\right\} = 0 \,, \tag{14.458}$$

which is though something already known (as $\boldsymbol{\theta}_0$ is the minimizing point of $V_{31}(\boldsymbol{\theta})$), see (14.189).

Introduce also the square Sylvester matrix of dimension $n_a + n_b$, cf. Sect. A.1.5

$$\mathscr{S}(B, -A) = \begin{pmatrix} 0 & b_1 & \dots & b_{n_b} & & \mathbf{0} \\ & & \ddots & & \ddots & \\ & \mathbf{0} & & b_1 & \dots & b_{n_b} \\ -1 & -a_1 & \dots & -a_{n_a} & & \mathbf{0} \\ & \ddots & & & \ddots & \\ & \mathbf{0} & -1 & & \dots & -a_{n_a} \end{pmatrix} \,, \tag{14.459}$$

and the matrices

$$\mathbf{A}^\mu = \begin{pmatrix} \alpha^1 \\ \vdots \\ \alpha^{n_a+n_b} \end{pmatrix}, \quad \mathbf{B}^\mu = \begin{pmatrix} \beta^1 \\ \vdots \\ \beta^{n_a+n_b} \end{pmatrix} \,, \tag{14.460}$$

where, with some abuse of notations, the row vectors consist of the individual polynomial coefficients. Then it holds

$$\begin{aligned} \mathsf{E}\left\{\boldsymbol{\psi}(t)\boldsymbol{\psi}^T(t)\right\} &= \mathscr{S}(B, -A) \ \mathrm{cov}\begin{pmatrix} \frac{1}{AF}u_0(t-1) \\ \vdots \\ \frac{1}{AF}u_0(t-n_a-n_b) \end{pmatrix} \mathscr{S}^T(B, -A) \\ &+ r\lambda_u \mathbf{A}^\mu \mathrm{cov}\begin{pmatrix} \frac{1}{F^2}v(t-1) \\ \vdots \\ \frac{1}{F^2}v(t-n_a-n_b) \end{pmatrix} \mathbf{A}^{\mu T} \\ &+ \lambda_u \mathbf{B}^\mu \mathrm{cov}\begin{pmatrix} \frac{1}{F^2}v(t-1) \\ \vdots \\ \frac{1}{F^2}v(t-n_a-n_b) \end{pmatrix} \mathbf{B}^{\mu T} \,, \end{aligned} \tag{14.461}$$

where $v(t)$ denotes a white noise process of unit variance.

To find a way to compute the second term in (14.198), differentiate (14.455), evaluate it for $\boldsymbol{\theta} = \boldsymbol{\theta}_0$, and neglect the part consisting of filtered $u_0(t)$ (recall that $\varepsilon(t, \boldsymbol{\theta}_0)$ which is uncorrelated with any filtered version of $u_0(t)$). This procedure gives

$$\frac{\partial^2}{\partial \boldsymbol{\theta}_\nu \partial \boldsymbol{\theta}_\mu}\varepsilon(t, \boldsymbol{\theta}) = -\frac{A^\mu F^\nu}{F^2}y(t) + \frac{B^\mu F^\nu}{F^2}u(t) - \frac{F^{\mu\nu}}{F^2}\left(Ay(t) - Bu(t)\right)$$

$$
\begin{aligned}
&+\frac{2F^{\mu}F^{\nu}}{F^3}(Ay(t)-Bu(t))-\frac{F^{\mu}}{F^2}(A^{\nu}y(t)-B^{\nu}u(t)) \\
&=\left(-\frac{A^{\mu}F^{\nu}}{F^2}-\frac{A^{\nu}F^{\mu}}{F^2}-\frac{AF^{\mu\nu}}{F^2}+2\frac{AF^{\mu}F^{\nu}}{F^3}\right)\tilde{y}(t) \\
&+\left(\frac{B^{\mu}F^{\nu}}{F^2}+\frac{B^{\nu}F^{\mu}}{F^2}+\frac{BF^{\mu\nu}}{F^2}-2\frac{BF^{\mu}F^{\nu}}{F^3}\right)\tilde{u}(t) \\
&\triangleq \frac{\alpha^{\mu\nu}}{F^3}\tilde{y}(t)-\frac{\beta^{\mu\mu}}{F^3}\tilde{u}(t)\,. \qquad (14.462)
\end{aligned}
$$

It is clear that the expression in (14.462) can be interpreted as the sum of two independent ARMA processes, driven by the white noise processes $\tilde{y}(t)$ and $\tilde{u}(t)$, respectively. To compute the second term in (14.198) is then a standard problem for computing cross-covariances.

The term $\mathsf{E}\left\{\varepsilon(t)\varepsilon''(t)\right\}$ can thus be evaluated componentwise as

$$
\mathsf{E}\left\{\varepsilon(t)\varepsilon''(t)_{\mu\nu}\right\}=r\lambda_u\mathsf{E}\left\{\frac{A}{F}v(t)\frac{\alpha^{\mu\nu}}{F^3}v(t)\right\}+\lambda_u\mathsf{E}\left\{\frac{B}{F}v(t)\frac{\beta^{\mu\nu}}{F^3}v(t)\right\}, \tag{14.463}
$$

where again $v(t)$ is a white noise process of unit variance.

Next consider the sum in (14.199). As $\varepsilon(t)$ is uncorrelated of $u_0(s)$ for all t and s, it holds

$$
\mathsf{E}\left\{\varepsilon(t+\tau)\psi_{\mu}(t)\right\}=\mathsf{E}\left\{\frac{A}{F}\tilde{y}(t+\tau)\frac{\alpha^{\mu}}{F^2}\tilde{y}(t)\right\}+\mathsf{E}\left\{\frac{B}{F}\tilde{u}(t+\tau)\frac{\beta^{\mu}}{F^2}\tilde{u}(t)\right\}. \tag{14.464}
$$

Lemma A.10 can be conveniently applied for the evaluation of an infinite sum of cross-covariances. Using that lemma leads to

$$
\begin{aligned}
&\left[\sum_{\tau=-\infty}^{\infty}\mathsf{E}\left\{\varepsilon(t+\tau)\psi(t)\right\}\mathsf{E}\left\{\varepsilon(t-\tau)\psi^T(t)\right\}\right]_{\mu\nu} \\
&=\sum_{\tau=-\infty}^{\infty}r_{\varepsilon\psi_{\mu}}(\tau)r_{\varepsilon\psi_{\nu}^T}(-\tau) \\
&=r^2\lambda_u^2\mathsf{E}\left\{\frac{\alpha^{\mu}\alpha^{\nu}}{F^4}v(t)\frac{A^2}{F^2}v(t)\right\}+\lambda_u^2\mathsf{E}\left\{\frac{\beta^{\mu}\beta^{\nu}}{F^4}v(t)\frac{B^2}{F^2}v(t)\right\} \\
&+r\lambda_u^2\mathsf{E}\left\{\frac{\alpha^{\mu}\beta^{\nu}}{F^4}v(t)\frac{AB}{F^2}v(t)\right\}+r\lambda_u^2\mathsf{E}\left\{\frac{\alpha^{\nu}\beta^{\mu}}{F^4}v(t)\frac{AB}{F^2}v(t)\right\}, \qquad (14.465)
\end{aligned}
$$

with $v(t)$ being a white noise of unit variance.

14.F.3 Asymptotic Distribution for the Extended ML Approach

The asymptotic distribution is characterized by the matrices **F** and **H** introduced in (14.206) and (14.208), respectively. To evaluate them, introduce the notations

$$W_k \stackrel{\Delta}{=} Y_k - G_k U_k = \tilde{Y}_k - G_k \tilde{U}_k \ , \tag{14.466}$$

$$N_k \stackrel{\Delta}{=} r + |G_k|^2 \ , \tag{14.467}$$

$$G_k^{\mu} \stackrel{\Delta}{=} \frac{\partial G_k}{\partial \boldsymbol{\theta}_{\mu}} \ . \tag{14.468}$$

It then holds, cf. also Example A.3

$$V_{31} = \frac{1}{N} \sum_k \frac{|W_k|^2}{N_k} \ , \tag{14.469}$$

$$(\mathbf{f}_1)_{\mu} = \frac{\partial V_{31}}{\partial \boldsymbol{\theta}_{\mu}} = \frac{1}{N} \sum_k \frac{1}{N_k} \left(-G_k^{\mu *} U_k^* W_k - W_k^* G_k^{\mu} U_k \right)$$

$$- \frac{1}{N} \sum_k \frac{|W_k|^2}{N_k^2} \left(G_k^{\mu *} G_k + G_k^* G_k^{\mu} \right) \ , \tag{14.470}$$

$$\mathbf{f}_2 = \frac{1}{N} \sum_k \frac{1}{N_k} - \frac{1}{N \lambda_u} \sum_k \frac{|W_k|^2}{N_k^2} \ , \tag{14.471}$$

$$\mathbf{f}_3 = \lambda_u - V_{31} = \lambda_u - \frac{1}{N} \sum_k \frac{|W_k|^2}{N_k} \ , \tag{14.472}$$

$$\mathsf{E}\left\{W_k\right\} = 0 \ , \tag{14.473}$$

$$\mathsf{E}\left\{|W_k|^2\right\} = \lambda_u N_k \ , \tag{14.474}$$

$$\mathsf{E}\left\{|W_j|^2 |W_k|^2\right\} = \begin{cases} \lambda_u^2 N_j N_k & j \neq k \\ 2\lambda_u^2 N_k^2 & j = k \end{cases} \ , \tag{14.475}$$

$$\mathsf{E}\left\{W_k^2\right\} = 0 \ . \tag{14.476}$$

Here the extension of Lemma A.9 to circular complex Gaussian variables was used. Another useful result is Lemma A.13 which implies

$$\mathsf{E}\left\{U_j^* U_k\right\} = (\lambda_u + \phi_k)\, \delta_{j,k}, \quad \phi_k = \phi(\mathrm{e}^{\mathrm{i}\omega_k}) \ , \tag{14.477}$$

where ϕ is the spectral density of the noise-free input $u_0(t)$. It thus also holds

$$\mathsf{E}\left\{U_k W_k\right\} = 0 \ , \tag{14.478}$$

$$\mathsf{E}\left\{U_k^* W_k\right\} = -G_k \lambda_u \ . \tag{14.479}$$

The two matrices **F** and **H** can be expressed in partitioned form as

$$\mathbf{F} = \lim_{N\to\infty} \begin{pmatrix} \frac{\partial \mathbf{f}_1}{\partial \boldsymbol{\theta}} & \frac{\partial 2\mathbf{f}_1}{\partial r} & \frac{\partial \mathbf{f}_1}{\partial \lambda_u} \\ \frac{\partial \mathbf{f}_2}{\partial \boldsymbol{\theta}} & \frac{\partial \mathbf{f}_2}{\partial r} & \frac{\partial \mathbf{f}_2}{\partial \lambda_u} \\ \frac{\partial \mathbf{f}_3}{\partial \boldsymbol{\theta}} & \frac{\partial \mathbf{f}_3}{\partial r} & \frac{\partial \mathbf{f}_3}{\partial \lambda_u} \end{pmatrix} \stackrel{\Delta}{=} \begin{pmatrix} \mathbf{F}_{11} & \mathbf{F}_{12} & \mathbf{F}_{13} \\ \mathbf{F}_{21} & \mathbf{F}_{22} & \mathbf{F}_{23} \\ \mathbf{F}_{31} & \mathbf{F}_{32} & \mathbf{F}_{33} \end{pmatrix}, \tag{14.480}$$

$$\mathbf{H} = \lim_{N\to\infty} N\mathsf{E}\left\{ \begin{pmatrix} \mathbf{f}_1 \\ \mathbf{f}_2 \\ \mathbf{f}_3 \end{pmatrix} \begin{pmatrix} \mathbf{f}_1^T & \mathbf{f}_2 & \mathbf{f}_3 \end{pmatrix} \right\} \stackrel{\Delta}{=} \begin{pmatrix} \mathbf{H}_{11} & \mathbf{H}_{12} & \mathbf{H}_{13} \\ \mathbf{H}_{21} & \mathbf{H}_{22} & \mathbf{H}_{23} \\ \mathbf{H}_{31} & \mathbf{H}_{32} & \mathbf{H}_{33} \end{pmatrix}. \tag{14.481}$$

Set

$$L_k^\mu = G_k^{\mu *} G_k + G_k^* G_k^\mu . \tag{14.482}$$

The blocks of the matrix **F** can now be evaluated as follows.

$$\begin{aligned}
(\mathbf{F}_{11})_{\mu\nu} &= \lim_{N\to\infty} \frac{\partial}{\partial \boldsymbol{\theta}_\nu} (\mathbf{f}_1)_\mu \\
&= \lim_{N\to\infty} \frac{1}{N} \sum_k \frac{1}{N_k} \left(-G_k^{\mu\nu *} U_k^* W_k + G_k^{\mu *} U_k^* G_k^\nu U_k + G_k^{\nu *} U_k^* G_k^\mu U_k - W_k^* G_k^{\mu\nu} U_k\right) \\
&\quad - \lim_{N\to\infty} \frac{1}{N} \sum_k \frac{L_k^\nu}{N_k^2} \left(-G_k^{\mu *} U_k^* W_k - W_k^* G_k^\mu U_k\right) \\
&\quad + \lim_{N\to\infty} \frac{2}{N} \sum_k \frac{|W_k|^2}{N_k^3} L_k^\nu L_k^\mu - \lim_{N\to\infty} \frac{\lambda_u}{N} \sum_k \frac{L_k^\mu L_k^\nu}{N_k^2} - \lim_{N\to\infty} \frac{1}{N} \sum_k \frac{|W_k|^2}{N_k^2} \frac{\partial}{\partial \boldsymbol{\theta}_\nu} L_k^\mu \\
&= \lim_{N\to\infty} \frac{1}{N} \sum_k \left[\frac{1}{N_k} \left(G_k^{\mu\nu *} G_k \lambda_u + G_k^{\mu *} G_k^\nu (\lambda_u + \phi_k) + G_k^{\nu *} G_k^\mu (\lambda_u + \phi_k) \right.\right. \\
&\quad \left.\left. + G_k^* G_k^{\mu\nu} \lambda_u \right) - \frac{L_k^\nu}{N_k^2} \left(G_k^{\mu *} G_k \lambda_u + G_k^* G_k^\mu \lambda_u\right) + L_k^\mu L_k^\nu \frac{\lambda_u}{N_k^2} - \frac{\lambda_u}{N_k} \frac{\partial}{\partial \boldsymbol{\theta}_\nu} L_k^\mu \right] \\
&= \lim_{N\to\infty} \frac{1}{N} \sum_k \frac{\lambda_u}{N_k} \left[\left(G_k^{\mu\nu *} G_k + G_k^{\mu *} G_k^\nu + G_k^{\nu *} G_k^\mu + G_k^* G_k^{\mu\nu}\right) - \frac{\lambda_u}{N_k} \frac{\partial}{\partial \boldsymbol{\theta}_\nu} L_k^\mu \right] \\
&\quad + \lim_{N\to\infty} \frac{1}{N} \sum_k \frac{\phi_k}{N_k} \left(G_k^{\mu *} G_k^\nu + G_k^{\nu *} G_k^\mu\right) \\
&= \lim_{N\to\infty} \frac{1}{N} \sum_k \frac{1}{N_k} \left[\left(G_k^{\mu\nu *} G_k \lambda_u + G_k^{\mu *} G_k^\nu \lambda_u + G_k^{\nu *} G_k^\mu \lambda_u + G_k^* G_k^{\mu\nu} \lambda_u\right) \right. \\
&\quad \left. - \lambda_u \left(G_k^{\mu\nu *} G_k + G_k^{\mu *} G_k^\nu + G_k^{\nu *} G_k^\mu + G_k^* G_k^{\mu\nu}\right) \right] \\
&\quad + \lim_{N\to\infty} \frac{1}{N} \sum_k \frac{\phi_k}{N_k} \left(G_k^{\mu *} G_k^\nu + G_k^{\nu *} G_k^\mu\right)
\end{aligned}$$

$$= \lim_{N\to\infty} \frac{1}{N} \sum_k \frac{\phi_k}{N_k} \left(G_k^{\mu *} G_k^{\nu} + G_k^{\nu *} G_k^{\mu} \right) , \tag{14.483}$$

$$\begin{aligned}
(\mathbf{F}_{12})_\mu &= \lim_{N\to\infty} \frac{\partial}{\partial r} (\mathbf{f}_1)_\mu \\
&= - \lim_{N\to\infty} \frac{1}{N} \sum_k \frac{1}{N_k^2} \left(-G_k^{\mu *} U_k^* W_k - W_k^* G_k^{\mu} U_k \right) + \lim_{N\to\infty} \frac{2}{N} \sum_k \frac{|W_k|^2}{N_k^3} L_k^{\mu} \\
&= \lim_{N\to\infty} \frac{1}{N} \sum_k \left[-\frac{1}{N_k^2} \left(G_k^{\mu *} G_k \lambda_u + G_k^* G_k^{\mu} \lambda_u \right) + \frac{2}{N_k^3} \lambda_u N_k L_k^{\mu} \right] \\
&= \lim_{N\to\infty} \frac{\lambda_u}{N} \sum_k \frac{L_k^{\mu}}{N_k^2} ,
\end{aligned} \tag{14.484}$$

$$(\mathbf{F}_{13})_\mu = \lim_{N\to\infty} \frac{\partial}{\partial \lambda_u} (\mathbf{f}_1)_\mu = 0 , \tag{14.485}$$

$$\begin{aligned}
(\mathbf{F}_{21})_\nu &= \lim_{N\to\infty} \frac{\partial}{\partial \boldsymbol{\theta}_\nu} \mathbf{f}_2 \\
&= - \lim_{N\to\infty} \frac{1}{N} \sum_k \frac{1}{N_k^2} L_k^{\nu} + \lim_{N\to\infty} \frac{2}{N\lambda_u} \sum_k \frac{|W_k|^2}{N_k^3} L_k^{\nu} \\
&\quad + \lim_{N\to\infty} \frac{1}{N\lambda_u} \sum_k \frac{1}{N_k^2} \left(U_k^* G_k^{\nu *} W_k + W_k^* G_k^{\nu} U_k \right) \\
&= \lim_{N\to\infty} \frac{1}{N} \sum_k \left[-\frac{L_k^{\nu}}{N_k^2} + \frac{2}{\lambda_u} \frac{\lambda_u N_k}{N_k^3} L_k^{\nu} + \frac{\lambda_u}{\lambda_u N_k^2} \left(-G_k^{\nu *} G_k - G_k^* G_k^{\nu} \right) \right] \\
&= 0 ,
\end{aligned} \tag{14.486}$$

$$\begin{aligned}
\mathbf{F}_{22} &= \lim_{N\to\infty} \frac{\partial}{\partial r} \mathbf{f}_2 \\
&= \lim_{N\to\infty} \left(-\frac{1}{N} \sum_k \frac{1}{N_k^2} + \frac{2}{N\lambda_u} \sum_k \frac{|W_k|^2}{N_k^3} \right) \\
&= \lim_{N\to\infty} \frac{1}{N} \sum_k -\frac{1}{N_k^2} + \frac{2}{\lambda_u} \frac{\lambda_u N_k}{N_k^3} \\
&= \lim_{N\to\infty} \frac{1}{N} \sum_k \frac{1}{N_k^2} ,
\end{aligned} \tag{14.487}$$

$$\begin{aligned}
\mathbf{F}_{23} &= \lim_{N\to\infty} \frac{\partial}{\partial \lambda_u} \mathbf{f}_2 \\
&= \lim_{N\to\infty} \frac{1}{N\lambda_u^2} \sum_k \frac{|W_k|^2}{N_k^2} = \lim_{N\to\infty} \frac{1}{N} \sum_k \frac{1}{\lambda_u^2} \frac{\lambda_u N_k}{N_k^2} \\
&= \lim_{N\to\infty} \frac{1}{N} \sum_k \frac{1}{\lambda_u N_k} ,
\end{aligned} \tag{14.488}$$

$$(\mathbf{F}_{31})_\nu = \lim_{N\to\infty} \frac{\partial}{\partial \boldsymbol{\theta}_\nu}\mathbf{f}_3 = 0\,, \tag{14.489}$$

$$\begin{aligned}\mathbf{F}_{32} &= \lim_{N\to\infty} \frac{\partial}{\partial r}\mathbf{f}_3 \\ &= \lim_{N\to\infty} \frac{1}{N}\sum_k \frac{|W_k|^2}{N_k^2} \\ &= \lim_{N\to\infty} \frac{1}{N}\sum_k \frac{\lambda_u}{N_k}\,,\end{aligned} \tag{14.490}$$

$$\mathbf{F}_{33} = \lim_{N\to\infty} \frac{\partial}{\partial \lambda_u}\mathbf{f}_3 = 1\,. \tag{14.491}$$

Hence the matrix $\mathbf{F}$ gets the structure

$$\mathbf{F} = \begin{pmatrix} \mathbf{F}_{11} & \mathbf{F}_{12} & \mathbf{0} \\ \mathbf{0} & \mathbf{F}_{22} & \mathbf{F}_{23} \\ \mathbf{0} & \mathbf{F}_{32} & 1 \end{pmatrix}. \tag{14.492}$$

To proceed the analysis of the matrix $\mathbf{F}$, introduce the unique polynomial F as the stable, but not monic, factor through the spectral factorization

$$FF^* = rAA^* + BB^*\,. \tag{14.493}$$

Then obviously

$$N_k = F_k F_k^*/(A_k A_k^*)\,. \tag{14.494}$$

Further, as $G_k = B_k/A_k$ it holds

$$G_k^\mu = \frac{B_k^\mu A_k - A_k^\mu B_k}{A_k^2}\,. \tag{14.495}$$

Using (14.494) and (14.495) in (14.483) one gets

$$\begin{aligned}(\mathbf{F}_{11})_{\mu\nu} = \lim_{N\to\infty} \frac{1}{N}\sum_k \frac{\phi_k}{|F_k|^2}|A_k|^2 \Bigg[&\left(\frac{B_k^\mu A_k - A_k^\mu B_k}{A_k^2}\right)^* \left(\frac{B_k^\nu A_k - A_k^\nu B_k}{A_k^2}\right) \\ &+ \left(\frac{B_k^\nu A_k - A_k^\nu B_k}{A_k^2}\right)^* \left(\frac{B_k^\mu A_k - A_k^\mu B_k}{A_k^2}\right)\Bigg]\,.\end{aligned} \tag{14.496}$$

Invoking Parseval's relation and using the Sylvester matrix

$$\mathscr{S}(-B, A) = \begin{pmatrix} 0 & -b_1 & \dots & -b_{n_b} & & \mathbf{0} \\ & \ddots & & & & \\ \mathbf{0} & & 0 & -b_1 & \dots & -b_{n_b} \\ 1 & a_1 & \dots & a_{n_a} & & \mathbf{0} \\ & \ddots & & & & \\ \mathbf{0} & & 1 & a_1 & \dots & a_{n_a} \end{pmatrix} \tag{14.497}$$

it follows that

$$\mathbf{F}_{11} = 2\mathscr{S}(-B, A) \operatorname{cov} \left(\frac{1}{A(q^{-1})F(q^{-1})} \begin{pmatrix} u_0(t-1) \\ \vdots \\ u_0(t-n_a-n_b) \end{pmatrix} \right) \mathscr{S}^T(-B, A) . \tag{14.498}$$

The block matrix $\mathbf{F}_{11}$ is positive definite and hence non-singular. To verify that $\mathbf{F}$ itself is non-singular, it is therefore enough to ensure that

$$\mathbf{F}_{22} - \mathbf{F}_{23}\mathbf{F}_{32} \neq 0 ,$$

see (14.492). However,

$$\mathbf{F}_{22}\mathbf{F}_{33} - \mathbf{F}_{23}\mathbf{F}_{32} = \lim_{N\to\infty} \left[\left(\frac{1}{N} \sum_k \frac{1}{N_k^2} \right) \left(\frac{1}{N} \sum_k 1 \right) - \left(\frac{1}{N} \sum_k \frac{1}{N_k} \right)^2 \right] , \tag{14.499}$$

which is positive due to Cauchy–Schwarz inequality. Thus the matrix $\mathbf{F}$ has been proved to be non-singular.

By calculations similar to going from (14.496)–(14.498) one can also show

$$\mathbf{F}_{12} = 2\lambda_u \mathscr{S}(-B, A) \mathsf{E} \left\{ \frac{1}{F^2(q^{-1})} \begin{pmatrix} v(t-1) \\ \vdots \\ v(t-n_a-n_b) \end{pmatrix} \frac{A(q^{-1})B(q^{-1})}{F^2(q^{-1})} v(t) \right\} , \tag{14.500}$$

$$\mathbf{F}_{22} = \mathsf{E} \left\{ \left[\frac{A^2(q^{-1})}{F^2(q^{-1})} v(t) \right]^2 \right\} , \tag{14.501}$$

$$\mathbf{F}_{23} = \frac{1}{\lambda_u} \mathsf{E} \left\{ \left[\frac{A(q^{-1})}{F(q^{-1})} v(t) \right]^2 \right\} , \tag{14.502}$$

$$\mathbf{F}_{32} = \lambda_u \mathsf{E} \left\{ \left[\frac{A(q^{-1})}{F(q^{-1})} v(t) \right]^2 \right\} , \tag{14.503}$$

where $v(t)$ is white noise of zero mean and unit variance.

When evaluating the symmetric matrix $\mathbf{H}$, note that all the factors $\mathbf{f}_1$, $\mathbf{f}_2$, $\mathbf{f}_3$ have zero mean (for finite N), and this holds of course also when they are multiplied with a deterministic factor.

In order to evaluate the different blocks of $\mathbf{H}$, a number of auxiliary results are needed. It holds

$$\mathsf{E}\left\{U_k^* W_k U_j^* W_j\right\} = G_k G_j \lambda_u^2 (1 + \delta_{j,k}) \,, \tag{14.504}$$

$$\mathsf{E}\left\{U_k^* W_k U_j W_j^*\right\} = G_k G_j^* \lambda_u^2 + \delta_{j,k} N_k \lambda_u (\phi_k + \lambda_u) \,, \tag{14.505}$$

$$\mathsf{E}\left\{U_k^* W_k W_j^* W_j\right\} = -G_k \lambda_u^2 N_j - \delta_{j,k} N_j \lambda_u^2 G_j \,. \tag{14.506}$$

Using (14.504)–(14.506) repeatedly, one gets

$$\begin{aligned}
(\mathbf{H}_{11})_{\mu\nu} &= \lim_{N\to\infty} \frac{1}{N} \sum_k \sum_j \mathsf{E}\left\{\left[\frac{1}{N_k}\left(-G_k^{\mu *} U_k^* W_k - W_k^* G_k^{\mu} U_k\right) - \frac{|W_k|^2}{N_k^2} L_k^{\mu}\right]\right. \\
&\quad \left. \times \left[\frac{1}{N_j}\left(-G_j^{\nu *} U_j^* W_j - W_j^* G_j^{\nu} U_j\right) - \frac{|W_j|^2}{N_j^2} L_j^{\nu}\right]\right\} \\
&= \lim_{N\to\infty} \frac{1}{N} \sum_k \sum_j \Bigg[\\
&\quad \frac{1}{N_k N_j}\Big[G_k^{\mu *} G_j^{\nu *} G_k G_j \lambda_u^2 (1 + \delta_{j,k}) + G_k^{\mu} G_j^{\nu} G_k^* G_j^* \lambda_u^2 (1 + \delta_{j,k}) \\
&\quad + G_k^{\mu *} G_j^{\nu} \left(G_k G_j^* \lambda_u^2 + \delta_{j,k} N_k \lambda_u (\phi_k + \lambda_u)\right) \\
&\quad + G_k^{\mu} G_j^{\nu *} \left(G_j G_k^* \lambda_u^2 + \delta_{j,k} N_j \lambda_u (\phi_j + \lambda_u)\right)\Big] \\
&\quad - \frac{L_j^{\nu}}{N_k N_j^2}\Big[G_k^{\mu *} G_k \lambda_u^2 N_j \left(1 + \delta_{j,k}\right) + G_k^{\mu}\left(G_k^* \lambda_u^2 N_j + \delta_{j,k} N_j \lambda_u^2 G_j^*\right)\Big] \\
&\quad - \frac{L_k^{\mu}}{N_k^2 N_j}\left[G_j^{\nu *} G_j \lambda_u^2 N_k \left(1 + \delta_{j,k}\right) + G_j^{\nu}\left(G_j^* \lambda_u^2 N_k + \delta_{j,k} N_k \lambda_u^2 G_k^*\right)\right] \\
&\quad + \frac{L_k^{\mu} L_j^{\nu}}{N_k^2 N_j^2} \lambda_u^2 N_j N_k \left(1 + \delta_{j,k}\right)\Bigg] \\
&= \lim_{N\to\infty} \frac{1}{N} \sum_k \sum_j \Bigg[\frac{1}{N_k N_j}\left[L_k^{\mu} L_j^{\nu} \lambda_u^2 \left(1 + \delta_{j,k}\right)\right. \\
&\quad \left. + \delta_{j,k} \phi_k \lambda_u N_k \left(G_k^{\mu *} G_k^{\nu} + G_k^{\mu} G_k^{\nu *}\right)\right] \\
&\quad - \frac{L_j^{\nu}}{N_k N_j^2} \lambda_u^2 N_j L_k^{\mu} (1 + \delta_{j,k}) - \frac{L_k^{\mu}}{N_k^2 N_j} \lambda_u^2 N_k L_j^{\nu} (1 + \delta_{j,k})
\end{aligned}$$

$$+ \frac{L_k^{\mu} L_j^{\nu}}{N_k^2 N_j^2} \lambda_u^2 N_j N_k \left(1 + \delta_{j,k}\right) \Bigg]$$

$$= \lim_{N\to\infty} \frac{1}{N} \sum_k \frac{\lambda_u \phi_k}{N_k} \left(G_k^{\mu *} G_k^{\nu} + G_k^{\nu *} G_k^{\mu} \right) . \tag{14.507}$$

By the procedure leading to (14.498) one then finds

$$\mathbf{H}_{11} = \mathscr{S}(-B, A) \mathbf{P} \mathscr{S}^T(-B, A) \tag{14.508}$$

$$\mathbf{P} = 2\lambda_u \text{cov} \left(\frac{1}{A(q^{-1})F(q^{-1})} \begin{pmatrix} u_0(t-1) \\ \vdots \\ u_0(t-n_a-n_b) \end{pmatrix} \right) . \tag{14.509}$$

Next examine the block $\mathbf{H}_{12}$:

$$(\mathbf{H}_{12})_{\mu} = \lim_{N\to\infty} N\mathsf{E} \left\{ \left[\frac{1}{N} \sum_k \frac{1}{N_k} \left(-G_k^{\mu *} U_k^* W_k - W_k^* G_k^{\mu} U_k \right) - \frac{1}{N} \sum_k \frac{|W_k|^2}{N_k^2} L_k^{\mu} \right] \right.$$
$$\left. \times \left[\frac{1}{N} \sum_j \frac{1}{N_j} - \frac{1}{N\lambda_u} \sum_j \frac{|W_j|^2}{N_j^2} \right] \right\}$$
$$= \lim_{N\to\infty} \frac{1}{N\lambda_u} \sum_k \sum_j \left[\frac{1}{N_k N_j^2} G_k^{\mu *} \left(-G_k \lambda_u^2 N_j (1 + \delta_{j,k}) \right) \right.$$
$$\left. + \frac{1}{N_k N_j^2} G_k^{\mu} \left(-G_k^* \lambda_u^2 N_j (1 + \delta_{j,k}) \right) + \frac{1}{N_k^2 N_j^2} L_k^{\mu} \lambda_u^2 N_k N_j (1 + \delta_{j,k}) \right]$$
$$= \lim_{N\to\infty} \frac{1}{N\lambda_u} \sum_k \sum_j \frac{1}{N_k N_j} \lambda_u^2 (1 + \delta_{j,k}) \left[-G_k^{\mu *} G_k - G_k^{\mu} G_k^* + L_k^{\mu} \right]$$
$$= 0 . \tag{14.510}$$

This leads to

$$\mathbf{H}_{12} = \mathbf{0} . \tag{14.511}$$

The block $\mathbf{H}_{13}$ can be evaluated using very similar calculations as follows:

$$(\mathbf{H}_{13})_{\mu} = \lim_{N\to\infty} N\mathsf{E} \left\{ \left[\frac{1}{N} \sum_k \frac{1}{N_k} \left(-G_k^{\mu *} U_k^* W_k - W_k^* G_k^{\mu} U_k \right) - \frac{1}{N} \sum_k \frac{|W_k|^2}{N_k^2} L_k^{\mu} \right] \right.$$
$$\left. \times \left[\lambda_u - \frac{1}{N} \sum_j \frac{|W_j|^2}{N_j} \right] \right\}$$

$$
\begin{aligned}
&= \lim_{N\to\infty} \frac{1}{N} \sum_k \sum_j \left[\frac{1}{N_k N_j} G_k^{\mu *} \left(-G_k \lambda_u^2 N_j (1+\delta_{j,k})\right) \right. \\
&\quad \left. + \frac{1}{N_k N_j} G_k^{\mu} \left(-G_k^* \lambda_u^2 N_j (1+\delta_{j,k})\right) + \frac{1}{N_k^2 N_j} L_k^{\mu} \lambda_u^2 N_k N_j (1+\delta_{j,k}) \right] \\
&= \lim_{N\to\infty} \frac{1}{N} \sum_k \sum_j \frac{1}{N_k} \lambda_u^2 (1+\delta_{j,k}) \left[-G_k^{\mu *} G_k - G_k^{\mu} G_k^* + L_k^{\mu} \right] \\
&= 0 ,
\end{aligned}
\tag{14.512}
$$

leading to

$$
\mathbf{H}_{13} = \mathbf{0} . \tag{14.513}
$$

Further calculations lead to

$$
\begin{aligned}
\mathbf{H}_{22} &= \lim_{N\to\infty} \frac{1}{N} \left[\frac{1}{N_k^2} - 2\frac{\lambda_u N_k}{\lambda_u N_k^3} + 2\frac{\lambda_u^2 N_k^2}{\lambda_u^2 N_k^4} \right] \\
&= \lim_{N\to\infty} \frac{1}{N} \sum_k \frac{1}{N_k^2} \;=\; \mathsf{E}\left\{ \left[\frac{A^2(q^{-1})}{F^2(q^{-1})} v(t) \right]^2 \right\} ,
\end{aligned}
\tag{14.514}
$$

$$
\begin{aligned}
\mathbf{H}_{23} &= \lim_{N\to\infty} \frac{1}{N} \mathsf{E}\left\{ \left[\frac{\lambda_u}{N_k} - \frac{|W_k|^2}{N_k^2} - \frac{\lambda_u |W_k|^2}{\lambda_u N_k^2} + \frac{|W_k|^4}{\lambda_u N_k^3} \right] \right\} \\
&= \lim_{N\to\infty} \frac{1}{N} \sum_k \left[\frac{\lambda_u}{N_k} - \frac{\lambda_u N_k}{N_k^2} - \frac{\lambda_u^2 N_k}{\lambda_u N_k^2} + 2\frac{\lambda_u^2 N_k^2}{\lambda_u N_k^3} \right] \\
&= \lim_{N\to\infty} \frac{\lambda_u}{N} \sum_k \frac{1}{N_k} \;=\; \mathsf{E}\left\{ \left[\frac{A(q^{-1})}{F(q^{-1})} v(t) \right]^2 \right\} ,
\end{aligned}
\tag{14.515}
$$

$$
\begin{aligned}
\mathbf{H}_{33} &= \lim_{N\to\infty} \frac{1}{N} \sum_k \mathsf{E}\left\{ \lambda_u^2 - \frac{2\lambda_u |W_k|^2}{N_k} + \frac{|W_k|^4}{N_k^2} \right\} \\
&= \lim_{N\to\infty} \frac{1}{N} \sum_k \left(\lambda_u^2 - \frac{2\lambda_u^2 N_k}{N_k} + \frac{2\lambda_u^2 N_k^2}{N_k^2} \right) \;=\; \lambda_u^2 .
\end{aligned}
\tag{14.516}
$$

14.G Asymptotic Distribution Results for Periodic Data

14.G.1 Proof of Lemma 14.12

It follows from the general theory of instrumental variable estimation for multivariable system, see Lemma 14.3, Söderström and Stoica (1983), Söderström and Stoica (1989), that the estimation error is asymptotically Gaussian distributed as

$$\sqrt{MN}(\hat{\boldsymbol{\theta}}_{\text{EIV}} - \boldsymbol{\theta}_0) \xrightarrow{\text{dist}} \mathcal{N}(\mathbf{0}, \mathbf{P}_{\text{EIV}}) , \tag{14.517}$$

where

$$\mathbf{P}_{\text{EIV}} = \mathbf{P}(\mathbf{W}) = (\mathbf{R}^T\mathbf{W}\mathbf{R})^{-1}\mathbf{R}^T\mathbf{W}\mathbf{C}_{\text{EIV}}\mathbf{W}\mathbf{R}(\mathbf{R}^T\mathbf{W}\mathbf{R})^{-1} , \tag{14.518}$$

and

$$\mathbf{C}_{\text{EIV}} = \mathsf{E}\left\{\left[\sum_{j=0}^{\infty}\mathbf{Z}(t+j)\mathbf{H}_j\right]\boldsymbol{\Lambda}\left[\sum_{k=0}^{\infty}\mathbf{H}_k^T\mathbf{Z}^T(t+k)\right]\right\} . \tag{14.519}$$

In (14.519) $\{\mathbf{H}_j\}_{j=0}^{\infty}$ and $\boldsymbol{\Lambda}$ are defined by a spectral factorization:

$$\boldsymbol{\Phi}_{\mathbf{V}}(\omega) = \mathbf{H}(\mathrm{e}^{\mathrm{i}\omega})\boldsymbol{\Lambda}\mathbf{H}^*(\mathrm{e}^{-\mathrm{i}\omega}) \tag{14.520}$$

along with the condition that $\mathbf{H}_0 = \mathbf{I}$, $\mathbf{H}(q^{-1}) = \sum_{j=0}^{\infty}\mathbf{H}_jq^{-j}$ and $\mathbf{H}^{-1}(q^{-1})$ being asymptotically stable. In (14.520), $\boldsymbol{\Phi}_V(\omega)$ denotes the spectral density matrix of the vector $\mathbf{V}(t)$, see (12.20). Note that all quantities in (14.520) as well as $\{\mathbf{H}_j\}$ are $M \times M$ matrices.

Due to Assumption **AI4**, the measurement noise sequences in different periods are uncorrelated. Hence the different components $v_j(t)$ of $\mathbf{V}(t)$ are uncorrelated. Therefore the spectral density matrix $\boldsymbol{\Phi}_{\mathbf{V}}(\omega)$ is diagonal, and, in fact, its diagonal elements are equal. Now write this as

$$\boldsymbol{\Phi}_{\mathbf{V}}(\omega) = \phi_v(\omega)\mathbf{I}_M , \tag{14.521}$$

and it follows that the spectral factorization of (14.520) can be substituted by a scalar spectral factorization:

$$\phi_v(\omega) = H(\mathrm{e}^{\mathrm{i}\omega})\lambda H(\mathrm{e}^{-\mathrm{i}\omega}) , \tag{14.522}$$

$$H(q^{-1}) = \sum_{k=0}^{\infty} h_k q^{-k}, \qquad h_0 = 1 , \tag{14.523}$$

and it follows that

$$\mathbf{H}_k = h_k\mathbf{I}_M ,$$
$$\boldsymbol{\Lambda} = \lambda\mathbf{I}_M .$$

Therefore, the matrix $\mathbf{C}_{\text{EIV}}$ in (14.519) can be simplified:

$$\mathbf{C}_{\text{EIV}} = \mathsf{E}\left\{\left[\sum_{j=0}^{\infty}h_j\mathbf{Z}(t+j)\right]\lambda\mathbf{I}_M\left[\sum_{k=0}^{\infty}h_k\mathbf{Z}^T(t+k)\right]\right\}$$

$$
\begin{aligned}
&= \mathsf{E}\left\{\lambda \sum_{j=0}^{\infty}\sum_{k=0}^{\infty}\left[h_k \mathbf{Z}(t'-k)h_j \mathbf{Z}^T(t'-j)\right]\right\} \\
&= \lambda \mathsf{E}\left\{\left[H(q^{-1})\mathbf{Z}(t)\right]\left[H(q^{-1})\mathbf{Z}(t)\right]^T\right\} ,
\end{aligned} \tag{14.524}
$$

which is (14.224).

14.G.2 Proof of Lemma 14.13

First recall that

$$
\mathbf{C}_{\mathrm{EIV}} = \lambda \mathsf{E}\left\{\left[H(q^{-1})\left[\mathbf{z}_1(t)\ldots\mathbf{z}_M(t)\right]\right]\left[H(q^{-1})\left[\mathbf{z}_1(t)\ldots\mathbf{z}_M(t)\right]\right]^T\right\} . \tag{14.525}
$$

Split the instrumental vector $\mathbf{z}_j(t)$ into a noise-free part and a noise contribution as, see (12.8),

$$
\begin{aligned}
\mathbf{z}_j(t) &= \mathbf{z}_j^0(t) + \tilde{\mathbf{z}}_j(t) \\
&= \mathbf{g}_L \otimes \boldsymbol{\varphi}_0(t) + \begin{pmatrix} \tilde{\boldsymbol{\varphi}}_{j+1}(t) \\ \vdots \\ \tilde{\boldsymbol{\varphi}}_{j+L}(t) \end{pmatrix} .
\end{aligned} \tag{14.526}
$$

Using Assumption **AI4** leads to

$$
\begin{aligned}
\mathbf{C}_{\mathrm{EIV}} &= \lambda \mathsf{E}\left\{\left[H(q^{-1})\left(\mathbf{g}_L \otimes \boldsymbol{\varphi}_0(t)\right)\right]\left[H(q^{-1})\left(\mathbf{g}_L^T \otimes \boldsymbol{\varphi}_0^T(t)\right)\right]\right\} \\
&\quad + \lambda \mathsf{E}\left\{\frac{1}{M}\sum_{j=1}^{M}\left[H(q^{-1})\begin{pmatrix} \tilde{\boldsymbol{\varphi}}_{j+1}(t) \\ \vdots \\ \tilde{\boldsymbol{\varphi}}_{j+L}(t) \end{pmatrix}\right]\left[H(q^{-1})\left(\tilde{\boldsymbol{\varphi}}_{j+1}^T(t)\ldots\tilde{\boldsymbol{\varphi}}_{j+L}^T(t)\right)\right]\right\} \\
&= \lambda \mathbf{g}_L \mathbf{g}_L^T \otimes \mathbf{C}_1 + \lambda \mathbf{I}_L \otimes \mathbf{C}_2 ,
\end{aligned} \tag{14.527}
$$

which proves (14.228).

By using properties of the Kronecker product and the matrix inversion lemma

$$
\begin{aligned}
\mathbf{C}_{\mathrm{EIV}}^{-1} &= \left[\lambda \mathbf{g}_L \mathbf{g}_L^T \otimes \mathbf{C}_1 + \lambda \mathbf{I}_L \otimes \mathbf{C}_2\right]^{-1} \\
&= \frac{1}{\lambda}[\mathbf{I}_L \otimes \mathbf{C}_2 + (\mathbf{g}_L \otimes \mathbf{I})(1 \otimes \mathbf{C}_1)(\mathbf{g}_L^T \otimes \mathbf{I})]^{-1} \\
&= \frac{1}{\lambda}\Big[(\mathbf{I}_L \otimes \mathbf{C}_2^{-1}) - (\mathbf{I}_L \otimes \mathbf{C}_2^{-1})(\mathbf{g}_L \otimes \mathbf{I})
\end{aligned}
$$

$$\times[(1\otimes\mathbf{C}_1)^{-1}+(\mathbf{g}_L^T\otimes\mathbf{I})(\mathbf{I}_L\otimes\mathbf{C}_2^{-1})(\mathbf{g}_L\otimes\mathbf{I})]^{-1}\,(\mathbf{g}_L^T\otimes\mathbf{I})(\mathbf{I}_L\otimes\mathbf{C}_2^{-1})\Big]$$
$$=\frac{1}{\lambda}\Big[(\mathbf{I}_L\otimes\mathbf{C}_2^{-1})-(\mathbf{g}_L\otimes\mathbf{C}_2^{-1})[\mathbf{C}_1^{-1}+L\mathbf{C}_2^{-1}]^{-1}(\mathbf{g}_L^T\otimes\mathbf{C}_2^{-1})\Big]$$
$$=\frac{1}{\lambda}\Big[\mathbf{I}_L\otimes\mathbf{C}_2^{-1}-\mathbf{g}_L\mathbf{g}_L^T\otimes\mathbf{C}_2^{-1}[\mathbf{C}_1^{-1}+L\mathbf{C}_2^{-1}]^{-1}\mathbf{C}_2^{-1}\Big]$$
$$=\frac{1}{\lambda}\Big[\mathbf{I}_L\otimes\mathbf{C}_2^{-1}-\mathbf{g}_L\mathbf{g}_L^T\otimes[\mathbf{C}_2\mathbf{C}_1^{-1}\mathbf{C}_2+L\mathbf{C}_2]^{-1}\Big]\,. \tag{14.528}$$

Using (14.528) in (14.226) gives

$$\begin{aligned}(\mathbf{P}_{\text{EIV}}^{\text{opt}})^{-1}&=(\mathbf{g}_L^T\otimes\mathbf{R}_0)\mathbf{C}_{\text{EIV}}^{-1}(\mathbf{g}_L\otimes\mathbf{R}_0)\\
&=\frac{1}{\lambda}\Big[\mathbf{g}_L^T\mathbf{g}_L\otimes(\mathbf{R}_0\mathbf{C}_2^{-1}\mathbf{R}_0)-\mathbf{g}_L^T\mathbf{g}_L\mathbf{g}_L^T\mathbf{g}_L\\
&\qquad\otimes\mathbf{R}_0[\mathbf{C}_2\mathbf{C}_1^{-1}\mathbf{C}_2+L\mathbf{C}_2]^{-1}\mathbf{R}_0\Big]\\
&=\frac{1}{\lambda}\Big[L(\mathbf{R}_0\mathbf{C}_2^{-1}\mathbf{R}_0)-L^2\mathbf{R}_0[\mathbf{C}_2\mathbf{C}_1^{-1}\mathbf{C}_2+L\mathbf{C}_2]^{-1}\mathbf{R}_0\Big]\\
&=\frac{L}{\lambda}\mathbf{R}_0\Big[\mathbf{C}_2^{-1}-L[\mathbf{C}_2\mathbf{C}_1^{-1}\mathbf{C}_2+L\mathbf{C}_2]^{-1}\Big]\mathbf{R}_0\\
&=\frac{1}{\lambda}\mathbf{R}_0\left[\mathbf{C}_1+\frac{\mathbf{C}_2}{L}\right]^{-1}\mathbf{R}_0\,.\end{aligned} \tag{14.529}$$

By inverting this (14.230) is finally obtained.

14.G.3 Proof of Corollary 14.3

For the IV variant of Example 12.3, when $\mathbf{W}=\mathbf{I}$, using (12.30) and (14.228), equation (14.223) shows

$$\begin{aligned}\mathbf{P}_{\text{EIV}}(\mathbf{I})&=\left(\mathbf{R}^T\mathbf{R}\right)^{-1}\mathbf{R}^T\mathbf{C}\mathbf{R}\left(\mathbf{R}^T\mathbf{R}\right)^{-1}\\
&=\left[(\mathbf{g}_L\otimes\mathbf{R}_0)^T(\mathbf{g}_L\otimes\mathbf{R}_0)\right]^{-1}(\mathbf{g}_L\otimes\mathbf{R}_0)^T\\
&\qquad\times\lambda(\mathbf{g}_L\mathbf{g}_L^T\otimes\mathbf{C}_1+\mathbf{I}_L\otimes\mathbf{C}_2)(\mathbf{g}_L\otimes\mathbf{R}_0)\left[(\mathbf{g}_L\otimes\mathbf{R}_0)^T(\mathbf{g}_L\otimes\mathbf{R}_0)\right]^{-1}\\
&=\lambda\left[(\mathbf{g}_L^T\mathbf{g}_L)^{-1}\mathbf{g}_L^T\otimes(\mathbf{R}_0^T\mathbf{R}_0)^{-1}\mathbf{R}_0^T\right]\\
&\qquad\times(\mathbf{g}_L\mathbf{g}_L^T\otimes\mathbf{C}_1+\mathbf{I}_L\otimes\mathbf{C}_2)\left[\mathbf{g}_L(\mathbf{g}_L^T\mathbf{g}_L)^{-1}\otimes\mathbf{R}_0(\mathbf{R}_0^T\mathbf{R}_0)^{-1}\right]\\
&=\frac{\lambda}{L^2}[\mathbf{g}_L^T\mathbf{g}_L\mathbf{g}_L^T\mathbf{g}_L\otimes\mathbf{R}_0^{-1}\mathbf{C}_1\mathbf{R}_0^{-1}+\mathbf{g}_L^T\mathbf{I}_L\mathbf{g}_L\otimes\mathbf{R}_0^{-1}\mathbf{C}_2\mathbf{R}_0^{-1}]\\
&=\frac{\lambda}{L^2}[L^2\mathbf{R}_0^{-1}\mathbf{C}_1\mathbf{R}_0^{-1}+L\mathbf{R}_0^{-1}\mathbf{C}_2\mathbf{R}_0^{-1}]\end{aligned}$$

$$= \lambda \mathbf{R}_0^{-1} \left(\mathbf{C}_1 + \frac{\mathbf{C}_2}{L} \right) \mathbf{R}_0^{-1} . \tag{14.530}$$

Hence, (14.231) is proved.

14.G.4 Proof of Lemma 14.14

In order to compute the covariance matrix $\mathbf{P}_{\text{FML}}$ one needs to find the first- and second -order derivatives of the loss function. Note that in (14.234) the parameter vector $\boldsymbol{\theta}$ appears only in the factors G_j. A direct differentiation gives

$$\begin{aligned}
\frac{\partial V_N}{\partial \boldsymbol{\theta}_k} &= \frac{4\pi}{N} \sum_{j=0}^{N-1} \frac{1}{\left(\lambda_y + G_j^* G_j \lambda_u\right)^2} \left[(\lambda_y + G_j^* G_j \lambda_u) \left(-U_j^* \frac{\partial G_j^*}{\partial \boldsymbol{\theta}_k} (Y_j - G_j U_j) \right) \right. \\
&\quad \left. - \frac{\partial G_j^*}{\partial \boldsymbol{\theta}_k} G_j \lambda_u \left(Y_j^* - U_j^* G_j^*\right) (Y_j - G_j U_j) \right] \\
&= \frac{4\pi}{N} \sum_{j=0}^{N-1} \frac{1}{\left(\lambda_y + G_j^* G_j \lambda_u\right)^2} \frac{\partial G_j^*}{\partial \boldsymbol{\theta}_k} (Y_j - G_j U_j) \\
&\quad \times \left[- (\lambda_y + G_j^* G_j \lambda_u) U_j^* - G_j \lambda_u \left(Y_j^* - U_j^* G_j^*\right) \right] \\
&= \frac{4\pi}{N} \sum_{j=0}^{N-1} \beta_j(k) w_j ,
\end{aligned} \tag{14.531}$$

where

$$\beta_j(k) = \frac{1}{\left(\lambda_y + G_j^* G_j \lambda_u\right)^2} \frac{\partial G_j^*}{\partial \boldsymbol{\theta}_k} , \tag{14.532}$$

$$w_j = (Y_j - G_j U_j) \left(-\lambda_y U_j^* - \lambda_u G_j Y_j^*\right) . \tag{14.533}$$

Note that $\beta_j(k)$ is a deterministic quantity and that w_j is random and has zero mean.
It holds that

$$\begin{aligned}
\mathsf{E}\left\{w_j w_j^*\right\} &= \mathsf{E}\left\{ \left(\tilde{Y}_j - G_j \tilde{U}_j\right) \left(\tilde{Y}_j^* - \tilde{U}_j^* G_j^*\right) \right. \\
&\quad \left. \times \left(-\lambda_y U_j^* - \lambda_u G_j Y_j^*\right) \left(-\lambda_y U_j - \lambda_u Y_j G_j^*\right) \right\} \\
&= \mathsf{E}\left\{ \left(\tilde{Y}_j - G_j \tilde{U}_j\right) \left(\tilde{Y}_j^* - \tilde{U}_j^* G_j^*\right) \right\} \\
&\quad \times \mathsf{E}\left\{ \left(-\lambda_y U_j^* - \lambda_u G_j Y_j^*\right) \left(-\lambda_y U_j - \lambda_u Y_j G_j^*\right) \right\}
\end{aligned}$$

$$= \left(\lambda_y + G_j G_j^* \lambda_u\right) \left(\lambda_y^2 \phi_{0j} + \lambda_y^2 \lambda_u + \lambda_u^2 G_j G_j^* \left(G_j G_j^* \phi_{0j} + \lambda_y\right) + \lambda_u \lambda_y G_j^* G_j \phi_{0j} + \lambda_u \lambda_y G_j G_j^* \phi_{0j}\right)$$
$$= \left(\lambda_y + G_j G_j^* \lambda_u\right) \left[\lambda_y \lambda_u \left(\lambda_y + G_j G_j^* \lambda_u\right) + \phi_{0j} \left(\lambda_y + G_j G_j^* \lambda_u\right)^2\right]$$
$$= \left(\lambda_y + G_j G_j^* \lambda_u\right)^2 \left[\lambda_y \lambda_u + \phi_{0j} \left(\lambda_y + G_j G_j^* \lambda_u\right)\right] . \quad (14.534)$$

Therefore

$$\mathsf{E}\left\{\frac{\partial V_N}{\partial \boldsymbol{\theta}_k} \frac{\partial V_N}{\partial \boldsymbol{\theta}_\ell}\right\} = \left(\frac{4\pi}{N}\right)^2 \sum_i \sum_j \mathsf{E}\left\{\beta_i(k) w_i w_j^* \beta_j^*(\ell)\right\}$$
$$= \left(\frac{4\pi}{N}\right)^2 \sum_j \beta_j(k) \beta_j^*(\ell) \mathsf{E}\left\{w_j w_j^*\right\}$$
$$= \left(\frac{4\pi}{N}\right)^2 \sum_j \frac{1}{\left(\lambda_y + G_j G_j^* \lambda_u\right)^2} \frac{\partial G_j}{\partial \boldsymbol{\theta}_k} \frac{\partial G_j^*}{\partial \boldsymbol{\theta}_\ell} \times \left[\lambda_y \lambda_u + \phi_{0j} \left(\lambda_y + G_j G_j^* \lambda_u\right)\right] . \quad (14.535)$$

Taking further derivatives in (14.531),

$$\frac{\partial^2 V_N}{\partial \boldsymbol{\theta}_k \partial \boldsymbol{\theta}_\ell} = \frac{4\pi}{N} \sum_j \left(\frac{\partial \beta_j(k)}{\partial \boldsymbol{\theta}_\ell} w_j + \beta_j(k) \frac{\partial w_j}{\partial \boldsymbol{\theta}_\ell}\right) \quad (14.536)$$

leads to

$$\lim_{N\to\infty} \mathsf{E}\left\{\frac{\partial^2 V_N}{\partial \boldsymbol{\theta}_k \partial \boldsymbol{\theta}_\ell}\right\} = \lim_{N\to\infty} \frac{4\pi}{N} \sum_j \beta_j(k) \mathsf{E}\left\{\frac{\partial w_j}{\partial \boldsymbol{\theta}_\ell}\right\} . \quad (14.537)$$

The expectation in (14.537) can be evaluated as follows:

$$\mathsf{E}\left\{\frac{\partial w_j}{\partial \boldsymbol{\theta}_\ell}\right\} = \mathsf{E}\left\{-\frac{\partial G_j}{\partial \boldsymbol{\theta}_\ell} U_j \left(-\lambda_y U_j^* - \lambda_u G_j Y_j^*\right) + \left(Y_j - G_j U_j\right) \left(-\lambda_u \frac{\partial G_j}{\partial \boldsymbol{\theta}_\ell} Y_j^*\right)\right\}$$
$$= \frac{\partial G_j}{\partial \boldsymbol{\theta}_\ell} \mathsf{E}\left\{-U_j \left(-\lambda_y U_j^* - \lambda_u G_j Y_j^*\right) + \left(\tilde{Y}_j - G_j \tilde{U}_j\right) \left(-\lambda_u Y_j^*\right)\right\}$$
$$= \frac{\partial G_j}{\partial \boldsymbol{\theta}_\ell} \left[\lambda_y \left(\phi_{0j} + \lambda_u\right) + \lambda_u G_j G_j^* \phi_{0j} - \lambda_u \lambda_y\right]$$
$$= \phi_{0j} \left(\lambda_y + G_j G_j^* \lambda_u\right) \frac{\partial G_j}{\partial \boldsymbol{\theta}_\ell} . \quad (14.538)$$

This gives

$$\lim_{N\to\infty} \mathsf{E}\left\{\frac{\partial^2 V_N}{\partial \boldsymbol{\theta}_k \partial \boldsymbol{\theta}_\ell}\right\} = 2\int_{-\pi}^{\pi} \frac{1}{\left(\lambda_y + GG^*\lambda_u\right)^2} \frac{\partial G^*}{\partial \boldsymbol{\theta}_k} \frac{\partial G}{\partial \boldsymbol{\theta}_\ell} \phi_{u_0} \left(\lambda_y + GG^*\lambda_u\right) \mathrm{d}\omega$$

$$= 2\int_{-\pi}^{\pi} \frac{\phi_{u_0}}{\left(\lambda_y + GG^*\lambda_u\right)} \frac{\partial G^*}{\partial \boldsymbol{\theta}_k} \frac{\partial G}{\partial \boldsymbol{\theta}_\ell} \mathrm{d}\omega \,. \tag{14.539}$$

Combining the results (14.235), (14.535), and (14.539) gives the results (14.236)–(14.238).

14.H The Cramér–Rao Lower Bound for the Frequency Domain ML Problem

Applying the results (14.274)–(14.276) gives

$$\mathbf{J}_{1,1} = \mathsf{E}\left\{\left[\frac{2}{r\lambda_u}\text{Real}\left(\mathbf{G}^*\tilde{\mathbf{Y}}\right) + \frac{2}{\lambda_u}\text{Real}\left(\tilde{\mathbf{U}}\right)\right] \times \left[\frac{2}{r\lambda_u}\text{Real}\left(\mathbf{G}^*\tilde{\mathbf{Y}}\right) + \frac{2}{\lambda_u}\text{Real}\left(\tilde{\mathbf{U}}\right)\right]^T\right\}$$

$$= \frac{4}{r^2\lambda_u^2}\mathsf{E}\left\{\frac{1}{2}\text{Real}\left(\mathbf{G}^*\tilde{\mathbf{Y}}\tilde{\mathbf{Y}}^*\mathbf{G}\right)\right\} + \frac{4}{\lambda_u^2}\mathsf{E}\left\{\frac{1}{2}\text{Real}\left(\tilde{\mathbf{U}}\tilde{\mathbf{U}}^*\right)\right\}$$

$$= \frac{2}{r\lambda_u}\text{Real}\left(\mathbf{G}^*\mathbf{G}\right) + \frac{2}{\lambda_u}\mathbf{I}\,, \tag{14.540}$$

$$\mathbf{J}_{1,2} = \mathsf{E}\left\{\left[\frac{2}{r\lambda_u}\text{Real}\left(\mathbf{G}^*\tilde{\mathbf{Y}}\right) + \frac{2}{\lambda_u}\text{Real}\left(\tilde{\mathbf{U}}\right)\right] \times \left[\frac{2}{r\lambda_u}\text{Imag}\left(\mathbf{G}^*\tilde{\mathbf{Y}}\right) + \frac{2}{\lambda_u}\text{Imag}\left(\tilde{\mathbf{U}}\right)\right]^T\right\}$$

$$= \frac{4}{r^2\lambda_u^2}\mathsf{E}\left\{-\frac{1}{2}\text{Imag}\left(\mathbf{G}^*\tilde{\mathbf{Y}}\tilde{\mathbf{Y}}^*\mathbf{G}\right)\right\} + \frac{4}{\lambda_u^2}\mathsf{E}\left\{-\frac{1}{2}\text{Imag}\left(\tilde{\mathbf{U}}\tilde{\mathbf{U}}^*\right)\right\}$$

$$= -\frac{2}{r\lambda_u}\text{Imag}\left(\mathbf{G}^*\mathbf{G}\right) - \frac{2}{\lambda_u}\text{Imag}(\mathbf{I}) = \mathbf{0}\,, \tag{14.541}$$

$$\left(\mathbf{J}_{1,3}\right)_{:,\mu} = -\mathsf{E}\left\{\left[\frac{2}{r\lambda_u}\text{Real}\left(\mathbf{G}^*\tilde{\mathbf{Y}}\right) + \frac{2}{\lambda_u}\text{Real}\left(\tilde{\mathbf{U}}\right)\right]\left[\frac{2}{r\lambda_u}\text{Real}\left(\mathbf{U}_0^*\mathbf{G}_\mu^*\tilde{\mathbf{Y}}\right)\right]\right\}$$

$$= -\frac{4}{r^2\lambda_u^2}\mathsf{E}\left\{\frac{1}{2}\text{Real}\left(\mathbf{G}^*\tilde{\mathbf{Y}}\tilde{\mathbf{Y}}^*\mathbf{G}_\mu\mathbf{U}_0\right)\right\} = -\frac{2}{r\lambda_u}\text{Real}\left(\mathbf{G}^*\mathbf{G}_\mu\mathbf{U}_0\right)\,, \tag{14.542}$$

$$
\begin{aligned}
\mathbf{J}_{2,2} &= \mathsf{E}\left\{\left[\frac{2}{r\lambda_u}\text{Imag}\left(\mathbf{G}^*\tilde{\mathbf{Y}}\right)+\frac{2}{\lambda_u}\text{Imag}\left(\tilde{\mathbf{U}}\right)\right]\right. \\
&\quad \times \left.\left[\frac{2}{r\lambda_u}\text{Imag}\left(\mathbf{G}^*\tilde{\mathbf{Y}}\right)+\frac{2}{\lambda_u}\text{Imag}\left(\tilde{\mathbf{U}}\right)\right]^T\right\} \\
&= \frac{4}{r^2\lambda_u^2}\mathsf{E}\left\{\frac{1}{2}\text{Real}\left(\mathbf{G}^*\tilde{\mathbf{Y}}\tilde{\mathbf{Y}}^*\mathbf{G}\right)\right\}+\frac{4}{\lambda_u^2}\mathsf{E}\left\{\frac{1}{2}\text{Real}\left(\tilde{\mathbf{U}}\tilde{\mathbf{U}}^*\right)\right\} \\
&= \frac{2}{r\lambda_u}\text{Real}\left(\mathbf{G}^*\mathbf{G}\right)+\frac{2}{\lambda_u}\mathbf{I}\,, \qquad (14.543)
\end{aligned}
$$

$$
\begin{aligned}
\left(\mathbf{J}_{2,3}\right)_{:,\mu} &= -\mathsf{E}\left\{\left[\frac{2}{r\lambda_u}\text{Imag}\left(\mathbf{G}^*\tilde{\mathbf{Y}}\right)+\frac{2}{\lambda_u}\text{Imag}\left(\tilde{\mathbf{U}}\right)\right]\left[\frac{2}{r\lambda_u}\text{Real}\left(\mathbf{U}_0^*\mathbf{G}_\mu^*\tilde{\mathbf{Y}}\right)\right]\right\} \\
&= -\frac{4}{r^2\lambda_u^2}\mathsf{E}\left\{-\frac{1}{2}\text{Imag}\left(\mathbf{G}^*\tilde{\mathbf{Y}}\tilde{\mathbf{Y}}^*\mathbf{G}_\mu\mathbf{U}_0\right)\right\} = \frac{2}{r\lambda_u}\text{Imag}\left(\mathbf{G}^*\mathbf{G}_\mu\mathbf{U}_0\right)\,, \qquad (14.544)
\end{aligned}
$$

$$
\begin{aligned}
\left(\mathbf{J}_{3,3}\right)_{\mu,\nu} &= \mathsf{E}\left\{\left[\frac{2}{r\lambda_u}\text{Real}\left(\mathbf{U}_0^*\mathbf{G}_\mu^*\tilde{\mathbf{Y}}\right)\right]\left[\frac{2}{r\lambda_u}\text{Real}\left(\mathbf{U}_0^*\mathbf{G}_\nu^*\tilde{\mathbf{Y}}\right)\right]\right\} \\
&= \frac{4}{r^2\lambda_u^2}\mathsf{E}\left\{\frac{1}{2}\text{Real}\left(\mathbf{U}_0^*\mathbf{G}_\mu^*\tilde{\mathbf{Y}}\tilde{\mathbf{Y}}^*\mathbf{G}_\nu\mathbf{U}_0\right)\right\} \\
&= \frac{2}{r\lambda_u}\text{Real}\left(\mathbf{U}_0^*\mathbf{G}_\mu^*\mathbf{G}_\nu\mathbf{U}_0\right)\,. \qquad (14.545)
\end{aligned}
$$

To evaluate the remaining blocks, one needs also expressions for the fourth-order moments. It holds

$$
\begin{aligned}
\mathsf{E}\left\{\left(\tilde{\mathbf{U}}^*\tilde{\mathbf{U}}\right)^2\right\} &= \mathsf{E}\left\{\sum_{k=1}^{N}\sum_{j=1}^{N}\tilde{U}_k^*\tilde{U}_k\tilde{U}_j^*\tilde{U}_j\right\} \\
&= \sum_{k=1}^{N}\sum_{j=1}^{N}\left(\mathsf{E}\left\{\tilde{U}_k^*\tilde{U}_k\right\}\mathsf{E}\left\{\tilde{U}_j^*\tilde{U}_j\right\}+\mathsf{E}\left\{\tilde{U}_k^*\tilde{U}_j\right\}\mathsf{E}\left\{\tilde{U}_j^*\tilde{U}_k\right\}\right) \\
&= \sum_{k=1}^{N}\sum_{j=1}^{N}\left(\lambda_u^2+\lambda_u^2\delta_{j,k}\right) \\
&= \left(N^2+N\right)\lambda_u^2\,. \qquad (14.546)
\end{aligned}
$$

Using (14.546) leads quickly to

$$
\begin{aligned}
\mathbf{J}_{4,4} &= \mathsf{E}\left\{\left(-\frac{2N}{\lambda_u}+\frac{1}{r\lambda_u^2}\tilde{\mathbf{Y}}^*\tilde{\mathbf{Y}}+\frac{1}{\lambda_u^2}\tilde{\mathbf{U}}^*\tilde{\mathbf{U}}\right)^2\right\} \\
&= \frac{4N^2}{\lambda_u^2}+\frac{1}{r^2\lambda_u^4}(N^2+N)r^2\lambda_u^2+\frac{1}{\lambda_u^4}(N^2+N)\lambda_u^2
\end{aligned}
$$

$$-\frac{4N}{r\lambda_u^3}Nr\lambda_u - \frac{4N}{\lambda_u^3}N\lambda_u + \frac{2}{r\lambda_u^4}N^2r\lambda_u^2$$
$$= \frac{2N}{\lambda_u^2}, \tag{14.547}$$

$$\mathbf{J}_{4,5} = \mathsf{E}\left\{\left(-\frac{2N}{\lambda_u} + \frac{1}{r\lambda_u^2}\tilde{\mathbf{Y}}^*\tilde{\mathbf{Y}} + \frac{1}{\lambda_u^2}\tilde{\mathbf{U}}^*\tilde{\mathbf{U}}\right)\left(-\frac{N}{r} + \frac{1}{r^2\lambda_u}\tilde{\mathbf{Y}}^*\tilde{\mathbf{Y}}\right)\right\}$$
$$= \frac{2N^2}{r\lambda_u} - \frac{N}{r^2\lambda_u^2}Nr\lambda_u - \frac{N}{r\lambda_u^2}N\lambda_u - \frac{2N}{r^2\lambda_u^2}Nr\lambda_u$$
$$+\frac{1}{r^3\lambda_u^3}(N^2+N)r^2\lambda_u^2 + \frac{1}{r^2\lambda_u^3}N^2r\lambda_u^2$$
$$= \frac{N}{r\lambda_u}, \tag{14.548}$$

$$\mathbf{J}_{5,5} = \mathsf{E}\left\{\left(-\frac{N}{r} + \frac{1}{r^2\lambda_u}\tilde{\mathbf{Y}}^*\tilde{\mathbf{Y}}\right)^2\right\}$$
$$= \frac{N^2}{r^2} + \frac{1}{r^4\lambda_u^2}(N^2+N)r^2\lambda_u^2 - \frac{2N}{r^3\lambda_u}Nr\lambda_u$$
$$= \frac{N}{r^2}. \tag{14.549}$$

Noting that $\mathbf{J}_{1,1} = \mathbf{J}_{2,2}$ is diagonal and $\mathbf{J}_{1,2} = \mathbf{0}$, it is straightforward to evaluate the matrix $\mathbf{X}$ as

$$\mathbf{X}_{\mu,\nu} = \left(\mathbf{J}_{3,1}\right)_{\mu,:}\mathbf{J}_{1,1}^{-1}\left(\mathbf{J}_{1,3}\right)_{:,\nu} + \left(\mathbf{J}_{3,2}\right)_{\mu,:}\mathbf{J}_{2,2}^{-1}\left(\mathbf{J}_{2,3}\right)_{:,\nu}$$
$$= \sum_k \frac{2}{r\lambda_u}\text{Real}\left(G_k^*G_k^\mu U_{0,k}\right)\frac{1}{\frac{2}{r\lambda_u}|G_k|^2 + \frac{2}{\lambda_u}}\frac{2}{r\lambda_u}\text{Real}\left(G_k^*G_k^\nu U_{0,k}\right)$$
$$+\sum_k \frac{2}{r\lambda_u}\text{Imag}\left(G_k^*G_k^\mu U_{0,k}\right)\frac{1}{\frac{2}{r\lambda_u}|G_k|^2 + \frac{2}{\lambda_u}}\frac{2}{r\lambda_u}\text{Imag}\left(G_k^*G_k^\nu U_{0,k}\right)$$
$$= \frac{2}{r\lambda_u}\sum_k \frac{1}{|G_k|^2+r}\Big[\text{Real}\left(G_k^*G_k^\mu U_{0,k}\right)\text{Real}\left(G_k^*G_k^\nu U_{0,k}\right)$$
$$+\ \text{Imag}\left(G_k^*G_k^\mu U_{0,k}\right)\text{Imag}\left(G_k^*G_k^\nu U_{0,k}\right)\Big]$$
$$= \frac{2}{r\lambda_u}\sum_k \frac{1}{|G_k|^2+r}\text{Real}\left(G_k^*G_k^\mu U_{0,k}\left(G_k^*G_k^\nu U_{0,k}\right)^*\right). \tag{14.550}$$

Using (14.293), (14.294) leads to

$$\mathbf{X}_{\mu,\nu} = \frac{2}{r\lambda_u} \sum_k \frac{|A_k|^2}{|B_k|^2 + r|A_k|^2} \text{Real} \left(\frac{B_k^*}{A_k^*} \frac{A_k B_k^{\mu} - A_k^{\mu} B_k}{A_k^2} U_{0,k} \frac{B_k}{A_k} \frac{A_k^* B_k^{\nu *} - A_k^{\nu *} B_k^*}{(A_k^2)^*} U_{0,k}^* \right) . \tag{14.551}$$

Introduce the polynomial $F(q^{-1})$ of degree $\max(n_a, n_b)$, with all zeros inside the unit circle, from the spectral factorization, see also (14.185),

$$FF^* = rAA^* + BB^* . \tag{14.552}$$

Note that the polynomial F is not monic. Paralleling the derivation of (14.297), (14.298) gives

$$\lim_{N\to\infty} \frac{1}{N}\mathbf{X} = \frac{2}{r\lambda_u} \mathscr{S}(-B, A) \mathsf{E}\left\{\boldsymbol{\psi}(t)\boldsymbol{\psi}^T(t)\right\} \mathscr{S}^T(-B, A) , \tag{14.553}$$

$$\boldsymbol{\psi}(t) = \frac{B(q^{-1})}{A^2(q^{-1})F(q^{-1})} \begin{pmatrix} u_0(t-1) \\ \vdots \\ u_0(t - n_a - n_b) \end{pmatrix} . \tag{14.554}$$

Note that due to (14.552) it holds

$$\left| \frac{B(\mathrm{e}^{\mathrm{i}\omega})}{F(\mathrm{e}^{\mathrm{i}\omega})} \right| < 1 \ \forall \omega , \tag{14.555}$$

and

$$\boldsymbol{\psi}(t) = \frac{B(q^{-1})}{F(q^{-1})} \boldsymbol{\varphi}(t) , \tag{14.556}$$

with $\boldsymbol{\varphi}(t)$ introduced in (14.298). Therefore it holds

$$\mathsf{E}\left\{\boldsymbol{\psi}(t)\boldsymbol{\psi}^T(t)\right\} < \mathsf{E}\left\{\boldsymbol{\varphi}(t)\boldsymbol{\varphi}^T(t)\right\} , \tag{14.557}$$

which leads to

$$\mathbf{J}_{3,3} > \mathbf{X} \tag{14.558}$$

as desired.

An explicit calculation shows also, where $\boldsymbol{\Phi}_{\varphi}(\omega)$ denotes the spectrum of the vector $\boldsymbol{\varphi}(t)$, (14.298),

$$\lim_{N\to\infty} \frac{1}{N} \left(\mathbf{J}_{3,3} - \mathbf{X}\right) = \frac{2}{r\lambda_u} \mathscr{S}(-B, A) \mathbf{P}_0 \mathscr{S}^T(-B, A) ,$$
$$\mathbf{P}_0 = \mathsf{E}\left\{\boldsymbol{\varphi}(t)\boldsymbol{\varphi}^T(t)\right\} - \mathsf{E}\left\{\boldsymbol{\psi}(t)\boldsymbol{\psi}^T(t)\right\}$$

$$
\begin{aligned}
&= \int_{-\pi}^{\pi} \boldsymbol{\Phi}_{\varphi}(\omega)\mathrm{d}\omega - \int_{-\pi}^{\pi} \left| \frac{B(\mathrm{e}^{-\mathrm{i}\omega})}{F(\mathrm{e}^{-\mathrm{i}\omega})} \right|^2 \boldsymbol{\Phi}_{\varphi}(\omega)\mathrm{d}\omega \\
&= \int_{-\pi}^{\pi} r \left| \frac{A(\mathrm{e}^{-\mathrm{i}\omega})}{F(\mathrm{e}^{-\mathrm{i}\omega})} \right|^2 \boldsymbol{\Phi}_{\varphi}(\omega)\mathrm{d}\omega \\
&= r \operatorname{cov} \left(\frac{1}{A(q^{-1})F(q^{-1})} \begin{pmatrix} u_0(t-1) \\ \vdots \\ u_0(t-n_a-n_b) \end{pmatrix} \right) \\
&= r\mathsf{E} \left\{ \frac{A(q^{-1})}{F(q^{-1})} \boldsymbol{\varphi}(t) \frac{A(q^{-1})}{F(q^{-1})} \boldsymbol{\varphi}^T(t) \right\} ,
\end{aligned}
\tag{14.559}
$$

which proves (14.305).

Chapter 15
Errors-in-Variables Problems in Practice

So far in the book, the general errors-in-variables problem has been defined and analyzed. A large number of identification methods have been presented and described, and the properties of the estimates have been analyzed. In this chapter the perspective is different; it is more user-oriented. With the results described so far, how should the user proceed in a practical situation when facing an errors-in-variables problem? One cannot expect to have *a general recipe* to follow in an automatic fashion. Rather, the purpose of the present chapter is to use the results of the previous part of the book to present some *guidelines for the user*.

The chapter starts in Sect. 15.1 where quite a number of the methods studied so far are compared in terms of performance. Section 15.2 describes the user choices in some of the more commonly used identification methods for errors-in-variables problems. Then follows a discussion in Sect. 15.3 of how the methods differ in what basic assumptions they rely on. This consideration then leads to a number of guidelines and questions for the user to think of when proceeding with a practical application. These guidelines are given in framed boxes in the following. Some general considerations are provided in Sect. 15.4, while aspects on the assumptions on the experiment are provided in Sect. 15.5, on the assumptions on the measurement noise in Sect. 15.6, and on the assumptions on the noise-free input in Sect. 15.7.

15.1 Comparing Performance of Some Estimators

Earlier in the book several estimators applicable to solve the EIV problem for dynamic systems have been presented. Their behaviors differ considerably. One extreme is the IV method, which is very simple computationally and therefore very fast. On the other hand, the results often have poor quality. For example, the standard deviations

T. Söderström, *Errors-in-Variables Methods in System Identification*, Communications and Control Engineering,
https://doi.org/10.1007/978-3-319-75001-9_15

of the parameter estimates are high. The other extreme can be represented by PEM and ML, as described in Chap. 9. Then the standard deviations of the parameter estimates are some magnitudes smaller, but the computation time is much greater than for the IV case.

In the following example, several different methods are compared in terms of standard deviations and computation time.

Example 15.1 In this example several of the estimators previously introduced in the book are compared, using the standard data set. The computation time is measured using the implementations made. Needless to say, this is a crude measure of the computational complexity, as implementations can always be done more or less sophisticated. No specific aspects for making the computations extra efficient were added. The figures should thus be regarded as approximate values and primarily be used for qualitatively comparing different methods.

The performance was examined using the standard deviations of the parameter estimates. In particular, the standard deviations of the estimates $\hat{a}_1$ and $\hat{b}_1$ were computed.

The methods listed in Table 15.1 were considered.

The numerical results are displayed in Figs. 15.1 and 15.2. Note the logarithmic scales.

From the figures one can make the following observations. As already mentioned one should consider the results qualitatively and not as exact quantitative measures. The precise results may change (to an unknown amount) for other systems, other implementations of the estimators, etc.

- There is certainly a trade-off between two extremes. One extreme is represented by instrumental variable methods. The IV estimates require very modest computation, but are not very accurate at all. The other extreme is represented by PEM and ML. Then the computational load is (very) large, which is combined with high accuracy. This finding thus illustrates that 'there is no free lunch,' or 'you get what you pay for.'

Table 15.1 Acronyms used for comparing different estimators

IV	Standard IV estimator, cf. Example 6.2
IVm	Standard IV estimator, using statistics from 97 out of 100 realizations, see Table 6.2
IV4	Overdetermined IV estimator, using $p = 4$, see Example 6.3
GIVE	Standard GIVE estimator, cf. Example 7.10
GIVE2	GIVE estimator, using an overdetermined system of equations see Example 7.11
CM	Standard CM estimator, cf. Example 8.3
CM2	CM estimator, using an overdetermined system of equations see Example 8.4
SPEC	Spectral factor estimator, see Example 10.1
PEM	Prediction error estimator, modeling $u_0(t)$ as an ARMA process, see Example 9.5
ML	Maximum likelihood estimator, modeling $u_0(t)$ as an ARMA process, see Example 9.6

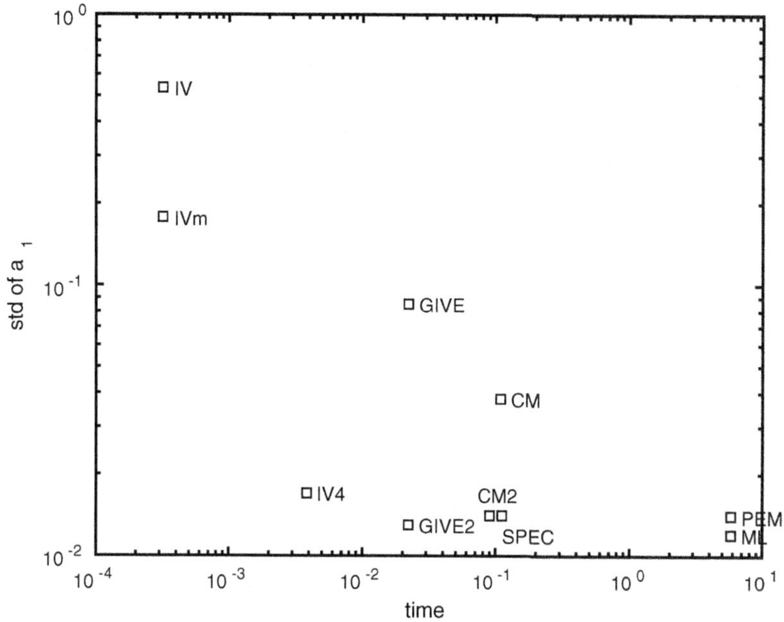

Fig. 15.1 Accuracy measured as standard deviation of $\hat{a}_1$ versus computational complexity measured as computational time. Acronyms as in Table 15.1

- GIVE, which in contrast to IV, is based on optimization, requires much more computations than IV, but gives also much more accurate results. GIVE with additional equations improves the accuracy significantly (GIVE2 has better performance than GIVE).
- CM, which is also based on optimization but over a higher dimension than GIVE, improves the accuracy considerably compared to GIVE. The estimates are basically based on the same set of covariance elements, but the information in these covariances is exploited in a more efficient way in CM. To use some additional covariance elements improves the estimates (CM2 gives better performance than CM).
- SPEC requires about the same amount of computations as CM, but produces less accurate parameter estimates. ■

Remark 15.1 One should not exaggerate the aspect of the computational load, as the computational power nowadays is very large anyway. An exception may occur when recursive processing is to be carried out in real time and the sampling interval is very small. ■

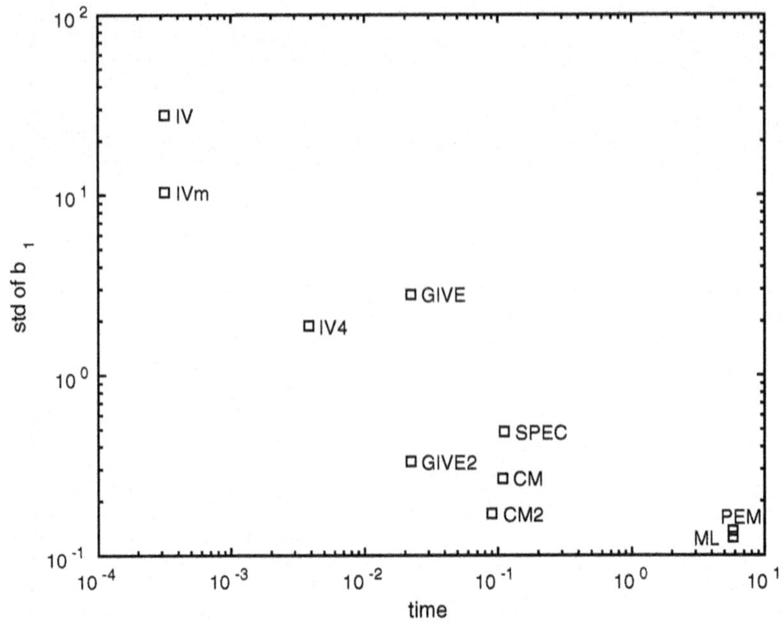

Fig. 15.2 Accuracy measured as standard deviation of $\hat{b}_1$ versus computational complexity measured as computational time. Acronyms as in Table 15.1

15.2 User Choices in the Algorithms

Most of the identification methods described in the book have some user choices. This means that some parameters have to be selected by the user when applying the method. This section is devoted to a summarized description of the user choices, including some general guidelines for the choices for some of the more common identification methods.

To begin with, all the methods concerned here (which are IV, GIVE, CM, and PEM/ML) are all based on some common assumptions. In particular:

- The true system (the dynamics between the unperturbed data) is linear, causal, and of finite order. Hence it admits a parametric model of the form

$$A(q^{-1})y_0(t) = B(q^{-1})u_0(t) \; . \tag{15.1}$$

  (This was labeled as Assumptions **AS1**, **AS2**, **AS3** in the past.)
- There is additive measurement noise, so the available data are

$$\begin{aligned} u(t) &= u_0(t) + \tilde{u}(t) \; , \\ y(t) &= y_0(t) + \tilde{y}(t) \; , \end{aligned} \tag{15.2}$$

with $\tilde{u}(t)$, $\tilde{y}(t)$, and $u_0(t)$ being stationary processes and mutually uncorrelated. (Assumptions **AN1**, **AI1**, **AI2**, **AI3**).

In (15.1), the polynomials are

$$A(q^{-1}) = 1 + a_1 q^{-1} + \cdots + a_{n_a} q^{-n_a} , \tag{15.3}$$
$$B(q^{-1}) = b_1 q^{-1} + \cdots + b_{n_b} q^{-n_b} , \tag{15.4}$$

and the unknown parameter vector is

$$\boldsymbol{\theta} = \begin{pmatrix} a_1 & \ldots & a_{n_a} & b_1 & \ldots & b_{n_b} \end{pmatrix}^T . \tag{15.5}$$

Thus the system can be written as a linear regression

$$y(t) = \boldsymbol{\varphi}^T(t)\boldsymbol{\theta} + v(t) , \tag{15.6}$$

where the regressor vector is

$$\boldsymbol{\varphi}(t) = \begin{pmatrix} -y(t-1) & \ldots & -y(t-n_a) & u(t-1) & \ldots & u(t-n_b) \end{pmatrix}^T , \tag{15.7}$$

and the noise term

$$v(t) = A(q^{-1})\tilde{y}(t) - B(q^{-1})\tilde{u}(t) \tag{15.8}$$

describes effects of both output noise and input noise.

Next some of the methods in Table 15.1 are described with a specific emphasis on the user choices and how these can be made.

1. **Instrumental variable, IV**. This method is described in Chap. 6. The estimated parameter vector $\hat{\boldsymbol{\theta}}$ is determined as the solution to the system of equations

$$\hat{\mathbf{R}}_{\mathbf{z}\varphi}\hat{\boldsymbol{\theta}} = \hat{\mathbf{r}}_{y\varphi} , \tag{15.9}$$

which may be overdetermined. The instrumental variable (the elements of the vector $\mathbf{z}(t)$) can be chosen in a number of ways (which indeed is a user choice). A common option is to take

$$\mathbf{z}(t) = \begin{pmatrix} y(t-n_a-1) \\ \vdots \\ y(t-n_a-p_y) \\ u(t-n_b-1) \\ \vdots \\ u(t-n_b-p_u) \end{pmatrix} . \tag{15.10}$$

Then the user choice is the two integers p_y and p_u. To get at least as many equations as unknowns in (15.9) one must take

$$p_y + p_u \geq n_a + n_b \,. \tag{15.11}$$

In case the output noise $\tilde{y}(t)$ is assumed to be autocorrelated, no delayed outputs should appear in $\mathbf{z}(t)$, and then one should substitute (15.11) by

$$p_y = 0, \quad p_u \geq n_a + n_b \,. \tag{15.12}$$

Often it pays to let $p_y + p_u - n_a - n_b$ to be greater than zero, say a small positive integer.
Aspects on the accuracy of IV estimates, including their asymptotic distribution, are given in Sect. 14.3.

2. **Generalized instrumental variable, GIVE**. This approach is similar to IV, but there is now no constraint that noise parts of $\mathbf{z}(t)$ and $\boldsymbol{\varphi}(t)$ must be uncorrelated. GIVE is presented and analyzed in Chap. 7. The estimate $\hat{\boldsymbol{\theta}}$ is given as the solution to

$$\left(\hat{\mathbf{R}}_{\mathbf{z}\varphi} - \mathbf{R}_{\tilde{\mathbf{z}}\tilde{\varphi}}(\boldsymbol{\rho})\right)\hat{\boldsymbol{\theta}} = \left(\hat{\mathbf{r}}_{\mathbf{z}y} - \mathbf{r}_{\tilde{\mathbf{z}}\tilde{y}}(\boldsymbol{\rho})\right) \,, \tag{15.13}$$

where

$$\boldsymbol{\rho} = \begin{pmatrix} \lambda_y & \lambda_u \end{pmatrix}^T \,. \tag{15.14}$$

The equation may be overdetermined and is nonlinear in the unknowns $\hat{\boldsymbol{\theta}}$ and $\boldsymbol{\rho}$. It has though a lot of internal structure and can be solved using a variable projection algorithm, see Sect. 7.4. This means that the estimation is converted into a nonlinear optimization problem over the two-dimensional vector $\boldsymbol{\rho}$. The vector $\mathbf{z}(t)$ is a user choice, and a common option is to take

$$\mathbf{z}(t) = \begin{pmatrix} y(t) \\ \vdots \\ y(t - n_a - p_y) \\ u(t) \\ \vdots \\ u(t - n_b - p_u) \end{pmatrix} \,. \tag{15.15}$$

To get at least as many equations in (15.13) as unknowns require

$$\left(n_a + p_y + 1\right) + (n_b + p_u + 1) \geq n_a + n_b + 2 \,, \tag{15.16}$$

that is,

$$p_y + p_u \geq 0 \,. \tag{15.17}$$

If the output noise $\tilde{y}$ is autocorrelated, all elements of delayed output in (15.15) must be abandoned and p_u increased accordingly. The system of equation (15.13) may be overdetermined. It often pays to let $p_y + p_u$ be a small positive integer. Aspects on the accuracy and asymptotic distribution of GIVE are provided in Sect. 14.4.

3. **Covariance matching, CM**. The covariance matching method is developed and analyzed in Chap. 8. It adjusts the model parameters to match the covariance elements

$$r_y(0), \ldots, r_y(p_y),\ r_u(0), \ldots, r_u(p_u),\ r_{yu}(p_1), \ldots, r_{yu}(p_2)\ , \tag{15.18}$$

and the user choice of the method is the four integers

$$p_y \geq 0,\ \ p_u \geq 0,\ \ p_1 \leq 0,\ \ p_2 \geq 0\ . \tag{15.19}$$

These integers must be chosen so that the corresponding equations (number of elements in (15.18)) are at least as large as the number of unknowns. The $n_a + n_b$ elements of $\boldsymbol{\theta}$ are included in the unknowns but so are also some additional quantities, as explained in some detail in Chap. 8. A specific discussion on the user choices (15.19) is provided in Sect. 8.4. A common alternative is to let $p_y - n_a$, $p_u - n_b$, $-p_1$, p_2 all be some small positive integers.

In case the output noise is autocorrelated, no extra terms $r_y(\cdot)$ should appear, and then $p_y = 0$.

The underlying system of equations, see (8.22), can be overdetermined. In any case, the specific nonlinear system of equations can be converted using a variable projection algorithm to a nonlinear optimization problem over $n_a + n_b$ variables. The computational complexity of CM is therefore larger than for GIVE.

Results on the asymptotic distribution of CM estimates are given in Sect. 14.5.

4. **Prediction error method and maximum likelihood, PEM and ML**. These two methods require a parametric model of the noise-free input $u_0(t)$. See Chap. 9, where it is assumed that $u_0(t)$ can be accurately modeled as an ARMA model. Phrased differently, it is assumed that the noise-free input is a stationary stochastic process and its spectral density is a rational function of $e^{i\omega}$. The user choice is the model order of the input model, that is, the two degrees of the ARMA polynomials. The coefficients of the ARMA polynomials describing the noise-free input are estimated along with $\boldsymbol{\theta}$. This leads to an optimization problem of higher dimension than that of CM. The two-dimensional measurement vector $\zeta(t) = \begin{pmatrix} y(t) & u(t) \end{pmatrix}^T$ is regarded as a multivariable ARMA process with specific internal structure. Its unknown parameters (i.e., $\boldsymbol{\theta}$, the variances of the input and output noises, and the ARMA model parameters of $u_0(t)$) are then estimated

using the maximum likelihood or possible some minor modification thereof.

If the order of the ARMA model is chosen too large, one may loose identifiability (and see difficulties in determining the ARMA model parameters). If the order of the ARMA model is chosen too small, consistency does not take place, and there will be some systematic errors. When it is appropriately chosen, the accuracy of all parameters is usually very high.

Results on the asymptotic distribution of PEM and ML estimates are given in Sect. 14.6. See also Example 15.6 later in this chapter for an illustration of the choice of model order for the ARMA model.

15.3 The Role of Assumptions

The comparisons made in Sect. 15.1 show that the methods treated differ to a large extent in their performance. This concerns both the asymptotic accuracy (what is the variance of the parameter estimates) and the computational load. An essential factor for determining the computational load is how complex optimization problem that is set up to be solved. In general terms, the more free variables to be optimized over, the larger is the computational load.

However, with the current development of computational power, the computational load is most likely not a decisive factor anymore for choosing an estimation method. Another, and more important aspect, that distinguishes the different methods is what basic assumptions that are needed to be fulfilled for the methods to work as designed. This issue was discussed mainly in Chaps. 3 and 4.

Quite a number of assumptions are generally applied; that is, they are utilized for almost all methods. See Sect. 15.2. In most cases it is also assumed that there is one single experiment and that the input noise $\tilde{u}(t)$ is white (Assumption **AN3b**).

In the methods used in Example 15.1 the additional assumptions differ as follows:

- Instrumental variable, IV: No particular additional assumption.
- Generalized instrumental variable, GIVE, and covariance matching, CM: Normally it is assumed that the output noise $\tilde{y}(t)$ is white (Assumption **AN3c**).
- Spectral factor estimate, SPEC: The output noise $\tilde{y}(t)$ is white.
- Prediction error method, PEM, and maximum likelihood method, ML: Normally it is assumed that the output noise $\tilde{y}(t)$ is white (Assumption **AN3c**). It is also assumed that the noise-free input $u_0(t)$ is an ARMA process (Assumption **AI4**).
- Some other methods, see Chap. 12, are designed to deal with periodic data or to indeed exploit multiple experiments (Assumption **AE2**). Then the assumption of white input noise (Assumption **AE3b**) can be relaxed. It is instead assumed that $u_0(t)$ is correlated between experiments (or periods), while the noises are assumed to be uncorrelated between experiments (or periods).

Since the methods are based on somewhat different sets of assumptions it makes sense in practice when solving EIV problems to first consider what assumptions that can be assumed to be valid. This will also be the basis for the remaining sections of this chapter.

One can note though that GIVE, CM, and SPEC rely on the same set of assumptions. As illustrated in Example 15.1 one can see that the information in the data is best utilized when the CM method is applied.

> The methods GIVE, CM, and SPEC are based on the same set of assumptions. Generally, CM exploits the information in the data more efficiently than GIVE and SPEC.

One can also note from examples in Chaps. 6, 7, and 8 that IV, GIVE, and CM are based on solving a set of equations that in most cases is overdetermined. (When the number of equations is increased, the number of unknowns remains the same for IV and GIVE, while it is increasing slowly for CM). Does it pay to use additional equations?

> The methods IV, GIVE, and CM are based on solving an overdetermined set of equations. In most cases it pays (in terms of quality of the estimates) to use a few more than the minimal number of equations. There is an option to use an optimal weighting of the equations, but this weighting depends on the unknown parameters and can at best be approximated. It is reasonable to try equal weighting of the equations. Using far too many equations can lead to bad quality of the parameter estimates.

15.4 Some General Guidelines

As explained in Sect. 15.3 the different methods are based on different sets of the general assumptions.

This section contains some general considerations.

> A first consideration is to find out if the problem at hand is indeed an EIV problem. How large is the noise/uncertainty on the input measurements? Will its impact on the final result be significant enough to bother about?

> Assuming that the problem is indeed an EIV problem, one should consider carefully what a priori information is available, and how it can be exploited.

For all the estimation methods described, some of the basic assumptions have to be fulfilled, but different methods rely on different types of assumptions.

An illustration of what can happen if the presence of input noise is neglected (i.e., the problem is not treated as an EIV one) is given in Example 3.2. If there is a significant amount of input noise (the signal-to-noise ratio is low on the input side) the parameter estimates will be biased for most standard methods. The bias is small for low noise levels.

In general terms, when basic assumptions are fulfilled in practice, the performance of the estimators is improved when these assumptions are exploited. On the other hand, when assumptions are used and not really fulfilled in practice, the performance deteriorates and systematic errors (bias) are introduced in the estimates. Examples of both these phenomena are given in the coming sections.

One should also be aware that general aspects for how to deal with and solve identification problems apply also in the EIV case.

Combine estimation and model validation. Try simple things first. If they work, be happy with that and do not make things more complicated and 'vulnerable' by applying more complex techniques. Sometimes this experience is labeled more brutally as the KISS principle (KISS = 'keep it simple, stupid').

The guidelines in the coming sections are organized around different types of general assumptions:

- Assumptions on the experiment in Sect. 15.5.
- Assumptions on the noise in Sect. 15.6.
- Assumptions on the noise-free input in Sect. 15.7.

15.5 Guidelines Related to the Experimental Setup

Some general questions to consider about the experiment include the following:

Which signals are measured? Are there more signals than the input $u(t)$ and the output $y(t)$ that are known? For example, is there any recorded reference signal $r(t)$ available?

Is any additional experiment available? This would include cases with periodic excitation. In the case of multiple experiments, what can be assumed about the signals? Is the noise-free input correlated between the experiments? (It would be identical if the experiments indeed correspond to different periods and there is periodic excitation.)

Is any feedback present in the experiment?

When additional signals are available, they should be exploited in the identification. For example, if the reference signal (or external excitation signal) $r(t)$ is known, the system transfer function can be estimated directly using

$$G(\mathrm{e}^{\mathrm{i}\omega}) = \phi_{yr}(\omega)/\phi_{ur}(\omega) \ . \tag{15.20}$$

Below is another example showing that using the knowledge that the noise-free input is periodic will indeed improve the performance. When it is known that the excitation is periodic, the sample maximum likelihood (SML) method would be a natural choice, see Sect. 12.3. See also Chap. 12 for further options on using periodic data. Recall that EIV problems with data from feedback experiments are discussed in Sect. 4.4. The conclusion there is that the precise formulation and details of the assumptions are quite important for how to proceed.

Example 15.2 In this example the standard system is used, see Sect. 3.2. The second-order system is still given by the equation

$$y_0(t) - 1.5 y_0(t-1) + y_0(t-2) = 2.0 u_0(t-1) + 1.0 u_0(t-2) \ . \tag{15.21}$$

The noise-free input $u_0(t)$ is taken as a second-order ARMA process:

$$u_0(t) - 1.0 u_0(t-1) + 0.2 u_0(t-2) = v(t) + 0.7 v(t-1) \ , \tag{15.22}$$

where $v(t)$ is Gaussian white noise of variance $\lambda_v = 1$. The specifics here is that the noise-free input is an ARMA process, as in the general case, but that it is repeated (concatenated) so that two periods, each of length $N = 1000$, are used. Both input and output measurements are contaminated with white noise as in the standard case, see Sect. 3.3.

In the example a standard instrumental variable (IV) technique, where the information about periodic excitation is not exploited, is considered. The instrumental variable vector $\mathbf{z}(t)$ is chosen as in Example 6.3,

$$\mathbf{z}(t) = \begin{pmatrix} u(t-3) & \ldots & u(t-6-p) \end{pmatrix}^T \ , \tag{15.23}$$

where the user parameter p varies from $p = 0$ to $p = 10$.

A further alternative is another IV estimator where the a priori information about periodic excitation is taken into account. The approach described in Sect. 12.1 is considered. Let the regressor vector in experiment/period i be denoted $\boldsymbol{\varphi}^{(i)}(t)$. For period i ($i = 1, 2$ in this case) it holds

$$\left(\frac{1}{N}\sum_t \varphi^{(2)}(t)\varphi^{(1)T}(t)\right)\hat{\theta} = \left(\frac{1}{N}\sum_t \varphi^{(2)}(t)y^{(1)}(t)\right), \tag{15.24}$$

$$\left(\frac{1}{N}\sum_t \varphi^{(1)}(t)\varphi^{(2)T}(t)\right)\hat{\theta} = \left(\frac{1}{N}\sum_t \varphi^{(1)}(t)y^{(2)}(t)\right). \tag{15.25}$$

As the matrices in the left-hand sides of both (15.24) and (15.25) are expected to converge to e $\left\{\varphi^{(1)}(t)\varphi^{(1)T}\right\}$ (recall that $u_0(t)$ is periodic), it makes sense to ensure that the left-hand sides become symmetric. This is done by finally choosing the estimator as the solution to

$$\left[\frac{1}{N}\sum_t \left(\varphi^{(1)}(t)\varphi^{(2)T}(t) + \varphi^{(2)}(t)\varphi^{(1)T}(t)\right)\right]\hat{\theta}$$
$$= \left[\frac{1}{N}\sum_t \left(\varphi^{(1)}(t)y^{(2)}(t) + \varphi^{(2)}(t)y^{(1)}(t)\right)\right]. \tag{15.26}$$

The numerical results are presented in Table 15.2. They show that the performance is indeed improved when exploiting the information that the noise-free input is periodic. ■

The following consideration can be drawn from this section and from Example 12.4.

> If the noise-free input is periodic, it pays indeed in performance to take this fact into account in the estimation. Further, SML is in such cases a natural choice of estimation method.

15.6 Guidelines Related to the Measurement Noise

The general questions related to these aspects include

> What is known about the measurement noise on the input and the output sides? Is it possible to record data without external excitations and thereby have data sets of the measurement noise only?

> Is the input noise white?

Is the output noise white?

Is the noise variance ratio $r = \lambda_y/\lambda_u$ known?

Again, any knowledge of the noise statistics should be exploited in the estimator with the aim to improve the performance. On the other hand, if assumptions are applied but not fulfilled, systematic errors can be expected. These aspects are illustrated in the examples to follow.

Table 15.2 Instrumental variable estimates using two periods of the data set. The first approach uses overdetermined IV with equal weighting. The periodic nature of the data is not exploited. The second approach is a simple basic IV estimator which exploits that the noise-free data consists of two periods

Parameter	a_1	a_2	b_1	b_2
True value	−1.5	0.7	2.0	1.0
First approach p				
0	−1.486 (0.52)	0.684 (0.36)	3.46 (19.4)	−0.08 (18.1)
1	−1.502 (0.057)	0.702 (0.036)	1.94 (4.59)	1.02 (3.91)
2	−1.498 (0.026)	0.700 (0.016)	1.61 (2.51)	1.33 (2.12)
3	−1.499 (0.019)	0.700 (0.013)	1.78 (1.75)	1.18 (1.45)
4	−1.500 (0.013)	0.700 (0.010)	1.79 (1.02)	1.17 (0.85)
5	−1.500 (0.013)	0.700 (0.010)	1.89 (0.94)	1.08 (0.78)
6	−1.501 (0.013)	0.701 (0.010)	1.87 (0.91)	1.09 (0.76)
7	−1.500 (0.014)	0.701 (0.010)	1.86 (0.94)	1.10 (0.78)
8	−1.501 (0.013)	0.701 (0.010)	1.80 (0.89)	1.15 (0.74)
9	−1.501 (0.013)	0.701 (0.010)	1.86 (0.88)	1.10 (0.73)
10	−1.501 (0.013)	0.701 (0.009)	1.83 (0.92)	1.12 (0.75)
Second approach	−1.502 (0.018)	0.701 (0.013)	2.05 (0.22)	0.93 (0.30)

Example 15.3 In this example the standard system is treated. Both the input and the output noises are white. It will be shown that performance is improved when the noise variance ratio is known and exploited.

In this case the GIVE estimator (see Chap. 7) simplifies as the optimization problem for the concentrated loss function, see (7.104), has only one unknown. For certain simple case of instruments GIVE then becomes equal to the classical Koopmans–Levin estimator.

The instrumental vector was chosen as

$$\mathbf{z}(t) = \left(y(t) \dots y(t-2-p)\ u(t) \dots u(t-2-p) \right)^T , \tag{15.27}$$

Table 15.3 GIVE using no knowledge of the noise variance ratio versus GIVE with knowledge of the noise variance ratio, TLS, and frequency domain ML

Parameter	a_1	a_2	b_1	b_2
True value	−1.5	0.7	2.0	1.0
r unknown				
p				
0	−1.528 (0.085)	0.709 (0.040)	3.57 (2.82)	−0.69 (3.30)
1	−1.500 (0.018)	0.696 (0.011)	2.36 (1.01)	0.65 (1.08)
2	−1.498 (0.013)	0.698 (0.010)	2.21 (0.33)	0.81 (0.37)
3	−1.499 (0.014)	0.698 (0.010)	2.22 (0.40)	0.79 (0.44)
10	−1.500 (0.019)	0.699 (0.012)	2.15 (1.25)	0.84 (1.23)
r known				
p				
0	−1.445 (0.025)	0.660 (0.017)	2.66 (0.47)	0.85 (0.57)
1	−1.495 (0.017)	0.691 (0.012)	3.11 (0.45)	0.05 (0.50)
2	−1.501 (0.014)	0.695 (0.011)	3.22 (0.40)	−0.10 (0.42)
3	−1.500 (0.015)	0.694 (0.011)	3.17 (0.38)	−0.05 (0.41)
10	−1.498 (0.019)	0.696 (0.012)	2.65 (0.99)	0.43 (1.01)
TLS	−1.499 (0.029)	0.699 (0.019)	2.03 (0.41)	0.98 (0.58)
ML	−1.496 (0.014)	0.696 (0.010)	2.09 (0.22)	0.94 (0.29)

and the integer p is considered for the values $p = 0, 1, 2, 3$ and 10. The GIVE method is applied both for the case when λ_y and λ_u are considered as two unknowns and for the case when the noise variance ratio $r = \lambda_y/\lambda_u$ is known.

There are also other estimation method designed (or known) to work for the case when the noise variance ratio is available. Examples of such methods include

- The total least squares estimator, see Chap. 11 where one first scales the signals to have equal amount of noise.
- The frequency domain maximum likelihood method developed in Sect. 9.4.

The numerical results are presented in Table 15.3. The results for r being unknown appear also in Example 7.11. The results show that the performance is indeed improved when exploiting the information that the noise-free input is periodic. It is further noted that when r is known, the best accuracy is obtained for the frequency domain ML method. ■

Example 15.4 In this example the standard system is treated. Both the input and the output noises are white. The covariance matching approach is applied. However, in the estimation phase, two alternatives are compared. In the first alternative white output noise is assumed. In the second alternative arbitrarily autocorrelated output noise is assumed.

The covariance matching method is applied, see Chap. 8 for a detailed description. The user parameters in (15.18) are chosen as

$$\begin{aligned} p_y &= 3 + p\ , \\ p_u &= 4 + p\ , \\ p_1 &= -2 - p\ , \\ p_2 &= 2 + p\ , \end{aligned} \qquad p = 2, \ldots, 5\ . \tag{15.28}$$

When the output noise is considered as generally correlated, the choices in (15.28) are modified to $p_y = 0$.

The results are displayed in Table 15.4. They show that best results are clearly obtained when using the a priori information that in the data the output noise is indeed white. ■

Example 15.5 This example is in a sense the reverse to Example 15.4. The two alternative CM variants of Example 15.4 are applied. In this case, though, the true output noise is correlated. It is a very low frequent noise, but the SNR value on the output remains the same as in the standard data sets.

Hence it is examined what happens with the parameter estimates if the output noise is modeled as generally correlated (consistency should still apply) or if it is modeled as white noise (a simplification that should lead to some systematic errors).

The user parameters were chosen as in (15.18), (15.28).

The results, which are displayed in Table 15.5, show that best results are obtained when using the a priori information that in the data the output noise is correlated. ■

Based on the examples one can conclude that exploiting any a priori information about the measurement noise leads to improved estimates.

15.7 Guidelines Related to the Noise-Free Input Signal

For several methods (say instrumental variable, generalized instrumental variable, and covariance matching) no specific input model is used. Concerning the noise-free input signal $u_0(t)$, 'only' the assumption of $u_0(t)$ being uncorrelated with the input noise is applied.

One of the possible assumptions on the noise-free input signal $u_0(t)$ is that it is a realization of a stationary stochastic process. Such a parametric model of $u_0(t)$ is used in the maximum likelihood method, see Chap. 9. In case the input is a stationary process, it is a rather reasonable assumption. Then the signal has a well-defined spectrum, and the assumption means that it can be accurately approximated by a rational function.

Table 15.4 Covariance matching method, exploiting output noise to be white, compared to allowing an arbitrary correlated output noise. The true output noise is white

Parameter	a_1	a_2	b_1	b_2
True value	−1.5	0.7	2.0	1.0
White $\tilde{y}$ assumed				
p				
2	−1.488 (0.018)	0.692 (0.013)	1.976 (0.158)	1.115 (0.250)
3	−1.493 (0.013)	0.695 (0.010)	1.991 (0.170)	1.063 (0.229)
4	−1.495 (0.013)	0.697 (0.010)	2.001 (0.186)	1.033 (0.243)
5	−1.497 (0.014)	0.698 (0.010)	2.011 (0.208)	1.011 (0.271)
Correlated $\tilde{y}$ assumed				
p				
2	−1.537 (0.238)	0.737 (0.204)	1.966 (0.430)	0.931 (1.387)
3	−1.499 (0.125)	0.702 (0.098)	1.965 (0.375)	1.074 (0.987)
4	−1.499 (0.071)	0.700 (0.052)	1.992 (0.307)	1.032 (0.689)
5	−1.501 (0.033)	0.701 (0.024)	2.026 (0.177)	0.979 (0.352)

Table 15.5 Covariance matching method, exploiting output noise to be white, compared to allowing an arbitrary correlated output noise. The true output noise is correlated and of low-pass character

Parameter	a_1	a_2	b_1	b_2
True value	−1.5	0.7	2.0	1.0
White $\tilde{y}$ assumed p				
2	−1.444 (0.067)	0.674 (0.027)	1.799 (1.52)	1.673 (2.22)
3	−1.468 (0.045)	0.688 (0.016)	1.845 (1.01)	1.464 (1.56)
4	−1.482 (0.028)	0.696 (0.011)	1.910 (0.47)	1.297 (0.83)
5	−1.490 (0.020)	0.698 (0.011)	1.993 (0.35)	1.136 (0.57)
Correlated $\tilde{y}$ assumed p				
2	−1.574 (0.152)	0.757 (0.114)	2.205 (0.348)	0.506 (0.868)
3	−1.525 (0.083)	0.719 (0.055)	2.153 (0.349)	0.746 (0.757)
4	−1.508 (0.049)	0.706 (0.031)	2.144 (0.294)	0.827 (0.547)
5	−1.500 (0.033)	0.701 (0.020)	2.139 (0.300)	0.867 (0.477)

What happens when the assumption of an ARMA model for $u_0(t)$ is applied for other types of input? One should then expect some degradation of the performance. This is the topic of the next example.

Example 15.6 The system and the general setup for the standard data set are used. The maximum likelihood method of Chap. 9 is applied, using an ARMA(2,1) model for $u_0(t)$. Two different situations are applied.

One case is the standard case as already treated in Chap. 9.

The other case is when the input $u_0(t)$ instead is a multi-sine signal, composed as a weighted sum of three sine waves in the low frequency region. The SNR value on the input side remains unchanged. The noise-free input signal is in this case equal to

$$u_0(t) = \sum_{k=1} A_k \sin(\omega_k t + \phi_k) \,, \tag{15.29}$$

where the parameters are

$$\begin{aligned} &A_1 = A_2 = A_3 = 2.46 \,, \\ &\omega_1 = 0.1, \omega_2 = 0.3, \omega_3 = 0.5 \,, \\ &\phi_1 = 0, \ \phi_2 = 2\pi/3, \ \phi_3 = 0 \,. \end{aligned} \tag{15.30}$$

Table 15.6 Maximum likelihood estimates for the standard data sets. Left: true $u_0(t)$ is an ARMA process. Right: true $u_0(t)$ is a multi-sine function

Parameter	True value	$u_0(t)$ ARMA		$u_0(t)$ multi-sine	
		Mean	Standard dev.	Mean	Standard dev.
a_1	−1.5	−1.497	0.012	−1.524	0.011
a_2	0.7	0.698	0.010	0.718	0.009
b_1	2.0	1.990	0.125	2.108	0.166
b_2	1.0	1.028	0.165	0.746	0.208

The numerical results are presented in Table 15.6. The case with the true noise-free input being an ARMA process was also presented earlier in Example 9.6. The results when the true noise-free input is a multi-sine function, show some bias in the estimates. ■

Appendix A
General Background Results

In this appendix, some general background results are collected. It is assumed that the reader has a working knowledge of system identification, dynamic systems, and linear algebra. For general treatments of system identification, see Söderström and Stoica (1989) and Ljung (1999).

The topics treated in this appendix concern matrix results on the one hand and some statistical issues on the other.

A.1 Some Matrix Results

For a large collection of many results on matrices, the reader may consult Bernstein (2005) or Bernstein (2009).

A.1.1 Nonnegative Definite Matrices

Convention A.1 For a symmetric matrix $\mathbf{A}$, use the notation

$$\mathbf{A} \geq \mathbf{0} \tag{A.1}$$

to denote a nonnegative definite matrix. This means that $\mathbf{A}$ has eigenvalues larger or equal to zero, and hence

$$\mathbf{x}^T \mathbf{A} \mathbf{x} \geq 0, \quad \forall \mathbf{x} \, . \tag{A.2}$$

Further,

$$\mathbf{A} > \mathbf{0} \tag{A.3}$$

will denote a positive definite matrix (all eigenvalues are strictly positive). ■

T. Söderström, *Errors-in-Variables Methods in System Identification*, Communications and Control Engineering,
https://doi.org/10.1007/978-3-319-75001-9

Convention A.2 For two symmetric matrices **A** and **B** the notation

$$\mathbf{A} \geq \mathbf{B} \tag{A.4}$$

means

$$\mathbf{A} - \mathbf{B} \geq \mathbf{0} \,. \tag{A.5}$$

■

Lemma A.1 *Let* **Q** *and* **R** *be symmetric, positive definite matrices, and* **S** *a matrix of full column rank. Consider*

$$\mathbf{P}(\mathbf{Q}) = (\mathbf{S}^T\mathbf{Q}\mathbf{S})^{-1}(\mathbf{S}^T\mathbf{Q}\mathbf{R}\mathbf{Q}\mathbf{S})(\mathbf{S}^T\mathbf{Q}\mathbf{S})^{-1} \tag{A.6}$$

as a function of **Q***. Then* **P**(**Q**) *can be minimized with respect to* **Q***, in the sense*

$$\mathbf{P}(\mathbf{Q}) \geq \mathbf{P}(\mathbf{R}^{-1}) = \left(\mathbf{S}^T\mathbf{R}^{-1}\mathbf{S}\right)^{-1} \,, \tag{A.7}$$

where equality holds if

$$\mathbf{Q} = \mathbf{R}^{-1} \,. \tag{A.8}$$

Proof It holds that

$$\begin{pmatrix} \mathbf{S}^T\mathbf{R}^{-1}\mathbf{S} & \mathbf{S}^T\mathbf{Q}\mathbf{S} \\ \mathbf{S}^T\mathbf{Q}\mathbf{S} & \mathbf{S}^T\mathbf{Q}\mathbf{R}\mathbf{Q}\mathbf{S} \end{pmatrix} = \begin{pmatrix} \mathbf{S}^T\mathbf{R}^{-1} \\ \mathbf{S}^T\mathbf{Q} \end{pmatrix} \mathbf{R} \left(\mathbf{R}^{-1}\mathbf{S} \;\; \mathbf{Q}\mathbf{S}\right) \geq \mathbf{0} \tag{A.9}$$

apparently is nonnegative definite. It then follows from properties of partitioned matrices; see, for example, Lemma A.3 of Söderström and Stoica (1989) that

$$\left(\mathbf{S}^T\mathbf{R}^{-1}\mathbf{S}\right) \geq \left(\mathbf{S}^T\mathbf{Q}\mathbf{S}\right)\left(\mathbf{S}^T\mathbf{Q}\mathbf{R}\mathbf{Q}\mathbf{S}\right)^{-1}\left(\mathbf{S}^T\mathbf{Q}\mathbf{S}\right) \tag{A.10}$$

and (A.7) follows. ■

A.1.2 Vectorization and Kronecker Product

Convention A.3 If a matrix is written as

$$\mathbf{A} = \left(\mathbf{a}_1 \; \ldots \; \mathbf{a}_n\right) \tag{A.11}$$

with columns $\{\mathbf{a}_i\}, i = 1, \ldots, n$, then vec(**A**) denote the tall vector obtained by stacking the columns on top of each other,

$$\text{vec}(\mathbf{A}) = \begin{pmatrix} \mathbf{a}_1 \\ \vdots \\ \mathbf{a}_n \end{pmatrix} . \tag{A.12}$$

■

Convention A.4 Let $\mathbf{A} \in \mathscr{R}^{n\times m}$ and $\mathbf{B} \in \mathscr{R}^{k\times \ell}$ be two matrices. The Kronecker product $\mathbf{A} \otimes \mathbf{B} \in \mathscr{R}^{nk\times m\ell}$ is the partitioned matrix

$$\mathbf{A} \otimes \mathbf{B} = \begin{pmatrix} \mathbf{A}_{11}\mathbf{B} \dots \mathbf{A}_{1m}\mathbf{B} \\ \vdots \\ \mathbf{A}_{n1}\mathbf{B} \dots \mathbf{A}_{nm}\mathbf{B} \end{pmatrix} . \tag{A.13}$$

■

Lemma A.2 *It holds that*

$$\text{vec}(\mathbf{ABC}) = (\mathbf{C}^T \otimes \mathbf{A})\text{vec}(\mathbf{B}) . \tag{A.14}$$

Proof See, for example, Lemma A.34 of Söderström and Stoica (1989), or Proposition 7.1.9 of Bernstein (2005). ■

Lemma A.3 *Let* **A**, **B**, **C**, **D** *be square matrices of the same size. Then it holds*

$$\text{tr}(\mathbf{ABCD}) = \left(\text{vec}(\mathbf{A}^T)\right)^T \left(\mathbf{D}^T \otimes \mathbf{B}\right) \text{vec}(\mathbf{C}) . \tag{A.15}$$

Proof The proof of the lemma is based on Lemma A.2 and the result:

$$\text{tr}(\mathbf{AB}) = \left(\text{vec}(\mathbf{A}^T)\right)^T \text{vec}(\mathbf{B}) . \tag{A.16}$$

When these are applied to the left hand side of (A.15) one gets

$$\begin{aligned} \text{tr}(\mathbf{ABCD}) &= \left(\text{vec}(\mathbf{A}^T)\right)^T \text{vec}(\mathbf{BCD}) \\ &= \left(\text{vec}(\mathbf{A}^T)\right)^T \left(\mathbf{D}^T \otimes \mathbf{B}\right) \text{vec}(\mathbf{C}) , \end{aligned} \tag{A.17}$$

which proves (A.15). ■

Convention A.5 Introduce the so-called Kronecker permutation matrix

$$\mathbf{J}^{(n,m)} = \left(\mathbf{I}_n \otimes \mathbf{e}_1 \ \mathbf{I}_n \otimes \mathbf{e}_2 \dots \mathbf{I}_n \otimes \mathbf{e}_m\right) , \tag{A.18}$$

where $\mathbf{e}_i,\ i = 1, \dots, m$, are the unit vectors in the Euclidean space $\mathscr{R}^m$. ■

Next some properties of the matrix $\mathbf{J}^{(n,m)}$ introduced in (A.18) are established.

Lemma A.4 *The following results hold.*

1. *The matrix* $\mathbf{J}^{(n,m)}$ *is a permutation matrix.*
2. *It can alternatively be expressed as*

$$\mathbf{J}^{(n,m)} = \sum_{k=1}^{n} \sum_{j=1}^{m} \left(\mathbf{e}_k \mathbf{f}_j^T\right) \otimes \left(\mathbf{f}_j \mathbf{e}_k^T\right) , \tag{A.19}$$

where $\{\mathbf{e}_k\}_{k=1}^{n}$ *are the unit vectors in* $\mathscr{R}^n$, *and* $\left\{\mathbf{f}_j\right\}_{j=1}^{m}$ *are the unit vectors in* $\mathscr{R}^m$.

3. *It holds*

$$\mathbf{J}^{(n,m)} \left(\mathbf{J}^{(n,m)}\right)^T = \mathbf{I}_{nm}, \quad \left(\mathbf{J}^{(n,m)}\right)^T = \mathbf{J}^{(m,n)} . \tag{A.20}$$

4. *If* $\mathbf{A}$ *is an* $n \times m$ *matrix, then*

$$\text{vec}\left(\mathbf{A}^T\right) = \mathbf{J}^{(n,m)} \text{vec}\left(\mathbf{A}\right) . \tag{A.21}$$

Proof See Sect. 7.4 of Bernstein (2005). ■

A.1.3 Matrix Inverses

Lemma A.5 *Let* $\mathbf{F}$ *and* $\mathbf{G}$ *be two full column rank matrices, such that*

$$\begin{aligned} &\left(\mathbf{F}\ \mathbf{G}\right) \text{ non-singular} , \\ &\mathbf{F}^T \mathbf{G} = \mathbf{0} , \end{aligned} \tag{A.22}$$

and let $\mathbf{R}$ *be a symmetric, positive definite matrix. Then it holds*

$$\mathbf{R}^{-1} = \mathbf{R}^{-1}\mathbf{F}\left(\mathbf{F}^T\mathbf{R}^{-1}\mathbf{F}\right)^{-1}\mathbf{F}^T\mathbf{R}^{-1} + \mathbf{G}\left(\mathbf{G}^T\mathbf{R}\mathbf{G}\right)^{-1}\mathbf{G}^T . \tag{A.23}$$

Proof Let $\mathbf{R}^{1/2}$ denote a symmetric square root of $\mathbf{R}$, and set

$$\begin{aligned} \overline{\mathbf{F}} &= \mathbf{R}^{-1/2}\mathbf{F} , \\ \overline{\mathbf{G}} &= \mathbf{R}^{1/2}\mathbf{G} . \end{aligned} \tag{A.24}$$

Then the right hand side of (A.23) can be written as

$$\mathbf{R}^{-1/2}\left[\overline{\mathbf{F}}\left(\overline{\mathbf{F}}^T\overline{\mathbf{F}}\right)^{-1}\overline{\mathbf{F}}^T + \overline{\mathbf{G}}\left(\overline{\mathbf{G}}^T\overline{\mathbf{G}}\right)^{-1}\overline{\mathbf{G}}^T\right]\mathbf{R}^{-1/2} .$$

Therefore, (A.23) is equivalent to

$$\mathbf{I} = \overline{\mathbf{F}}\left(\overline{\mathbf{F}}^T\overline{\mathbf{F}}\right)^{-1}\overline{\mathbf{F}}^T + \overline{\mathbf{G}}\left(\overline{\mathbf{G}}^T\overline{\mathbf{G}}\right)^{-1}\overline{\mathbf{G}}^T . \tag{A.25}$$

Next examine the difference between the left and right hand sides of (A.25) by computing

$$\left[\mathbf{I}-\overline{\mathbf{F}}\left(\overline{\mathbf{F}}^T\overline{\mathbf{F}}\right)^{-1}\overline{\mathbf{F}}^T-\overline{\mathbf{G}}\left(\overline{\mathbf{G}}^T\overline{\mathbf{G}}\right)^{-1}\overline{\mathbf{G}}^T\right]\left(\overline{\mathbf{F}}\;\overline{\mathbf{G}}\right)$$
$$=\left(-\overline{\mathbf{G}}\left(\overline{\mathbf{G}}^T\overline{\mathbf{G}}\right)^{-1}\overline{\mathbf{G}}^T\overline{\mathbf{F}}\;-\overline{\mathbf{F}}\left(\overline{\mathbf{F}}^T\overline{\mathbf{F}}\right)^{-1}\overline{\mathbf{F}}^T\overline{\mathbf{G}}\right). \tag{A.26}$$

As $\overline{\mathbf{F}}$ and $\overline{\mathbf{G}}$ have full column rank, (A.22) implies

$$\overline{\mathbf{F}}^T\overline{\mathbf{G}}=\mathbf{0}, \tag{A.27}$$

and $\left(\overline{\mathbf{F}}\;\overline{\mathbf{G}}\right)$ is non-singular. Thus the expression in (A.26) is zero, and it can be concluded that (A.25), and thus also (A.23), holds. ■

Lemma A.6 *Consider the matrix*

$$\mathbf{B}=\begin{pmatrix}\mathbf{K}^T & \mathbf{Q}\\ \mathbf{0} & \mathbf{K}\end{pmatrix}, \tag{A.28}$$

where $\mathbf{K}$ *is a* $m\times n$ *matrix of rank* m, *and* $\mathbf{Q}$ *is an* $n\times n$ *symmetric and nonnegative definite matrix. Set*

$$\mathbf{P}=\mathbf{K}^T\left(\mathbf{K}\mathbf{K}^T\right)^{-1}\mathbf{K}=\mathbf{K}^T\mathbf{K}^{T^\dagger}=\mathbf{K}^\dagger\mathbf{K}, \tag{A.29}$$
$$\mathbf{P}_\perp=\mathbf{I}-\mathbf{P}, \tag{A.30}$$
$$\mathbf{D}_0=\mathbf{P}+\mathbf{P}_\perp\mathbf{Q}\mathbf{P}_\perp, \tag{A.31}$$

where $\mathbf{K}^\dagger$ *denotes the pseudo-inverse of* $\mathbf{K}$. *Assume that* $\mathbf{D}_0$ *is invertible. Then* $\mathbf{B}$ *is invertible and its inverse is given by*

$$\mathbf{B}^{-1}=\begin{pmatrix}\mathbf{H}_{11} & \mathbf{H}_{12}\\ \mathbf{H}_{21} & \mathbf{H}_{22}\end{pmatrix}, \tag{A.32}$$

where

$$\mathbf{H}_{11}=\left(\mathbf{K}\mathbf{K}^T\right)^{-1}\mathbf{K}(\mathbf{I}-\mathbf{Q}\mathbf{D}), \tag{A.33}$$
$$\mathbf{H}_{12}=-\left(\mathbf{K}\mathbf{K}^T\right)^{-1}\mathbf{K}\mathbf{Q}(\mathbf{I}-\mathbf{D}\mathbf{Q})\mathbf{K}^T\left(\mathbf{K}\mathbf{K}^T\right)^{-1}, \tag{A.34}$$
$$\mathbf{H}_{21}=\mathbf{D}, \tag{A.35}$$
$$\mathbf{H}_{22}=(\mathbf{I}-\mathbf{D}\mathbf{Q})\mathbf{K}^T\left(\mathbf{K}\mathbf{K}^T\right)^{-1}, \tag{A.36}$$
$$\mathbf{D}=\mathbf{D}_0^{-1}\mathbf{P}_\perp. \tag{A.37}$$

Proof The proof is based on direct multiplication. Equation (A.32) holds precisely when

$$\begin{pmatrix} \mathbf{K}^{T^\dagger}(\mathbf{I}-\mathbf{QD}) & -\mathbf{K}^{T^\dagger}\mathbf{Q}(\mathbf{I}-\mathbf{DQ})\mathbf{K}^\dagger \\ \mathbf{D} & (\mathbf{I}-\mathbf{DQ})\mathbf{K}^\dagger \end{pmatrix} \begin{pmatrix} \mathbf{K}^T & \mathbf{Q} \\ \mathbf{0} & \mathbf{K} \end{pmatrix} = \begin{pmatrix} \mathbf{I} & \mathbf{0} \\ \mathbf{0} & \mathbf{I} \end{pmatrix},$$

or in block form

$$\mathbf{K}^{T^\dagger}(\mathbf{I}-\mathbf{QD})\mathbf{K}^T = \mathbf{I}\,, \tag{A.38}$$

$$\mathbf{DK}^T = \mathbf{0}\,, \tag{A.39}$$

$$\mathbf{K}^{T^\dagger}(\mathbf{I}-\mathbf{QD})\mathbf{Q} - \mathbf{K}^{T^\dagger}\mathbf{Q}(\mathbf{I}-\mathbf{DQ})\mathbf{K}^\dagger\mathbf{K} = \mathbf{0}\,, \tag{A.40}$$

$$\mathbf{DQ} + (\mathbf{I}-\mathbf{DQ})\mathbf{K}^\dagger\mathbf{K} = \mathbf{I}\,. \tag{A.41}$$

The left hand side of (A.39) is easily evaluated:

$$\mathbf{DK}^T = \mathbf{D}_0^{-1}\mathbf{P}_\perp\mathbf{K}^T = \mathbf{D}_0^{-1}\left[\mathbf{I} - \mathbf{K}^T\mathbf{K}^{T^\dagger}\right]\mathbf{K}^T = \mathbf{0}\,,$$

and hence (A.39) is proved. Using (A.39) the left hand side of (A.38) becomes

$$\mathbf{K}^{T^\dagger}(\mathbf{I}-\mathbf{QD})\mathbf{K}^T = \mathbf{K}^{T^\dagger}\mathbf{K}^T = \mathbf{I}\,,$$

and (A.38) is proved. Next consider the left hand side of (A.41):

$$\begin{aligned} &\mathbf{DQ} + (\mathbf{I}-\mathbf{DQ})\mathbf{K}^\dagger\mathbf{K} \\ &= \mathbf{D}_0^{-1}\left[\mathbf{P}_\perp\mathbf{Q} + (\mathbf{D}_0 - \mathbf{P}_\perp\mathbf{Q})\mathbf{K}^\dagger\mathbf{K}\right] \\ &= \mathbf{D}_0^{-1}\left[\mathbf{P}_\perp\mathbf{Q} + (\mathbf{P} + \mathbf{P}_\perp\mathbf{Q}\mathbf{P}_\perp - \mathbf{P}_\perp\mathbf{Q})\,\mathbf{P}\right] \\ &= \mathbf{D}_0^{-1}\left[\mathbf{P}_\perp\mathbf{Q} + \mathbf{P} - \mathbf{P}_\perp\mathbf{Q}\mathbf{P}\right] \\ &= \mathbf{D}_0^{-1}\left[\mathbf{P} + \mathbf{P}_\perp\mathbf{Q}\,(\mathbf{I}-\mathbf{P})\right] = \mathbf{D}_0^{-1}\mathbf{D}_0 = \mathbf{I}\,, \end{aligned}$$

and (A.41) is proved. Finally, using (A.41), consider the left hand side of (A.40):

$$\begin{aligned} &\mathbf{K}^{T^\dagger}(\mathbf{I}-\mathbf{QD})\mathbf{Q} - \mathbf{K}^{T^\dagger}\mathbf{Q}(\mathbf{I}-\mathbf{DQ})\mathbf{K}^\dagger\mathbf{K} \\ &= \mathbf{K}^{T^\dagger}\,(\mathbf{Q} - \mathbf{QDQ})\,\left(\mathbf{I} - \mathbf{K}^\dagger\mathbf{K}\right) \\ &= \mathbf{K}^{T^\dagger}\mathbf{Q}\,(\mathbf{I}-\mathbf{DQ})\,\left(\mathbf{I} - \mathbf{K}^\dagger\mathbf{K}\right) \\ &= \mathbf{K}^{T^\dagger}\mathbf{Q}\,(\mathbf{I}-\mathbf{DQ}) - \mathbf{K}^{T^\dagger}\mathbf{Q} \times (\mathbf{I}-\mathbf{DQ}) = \mathbf{0}\,, \end{aligned}$$

and the lemma is proven. ■

A.1.4 Variable Projection Algorithms

In several applications where a system of nonlinear equations is to be solved, it can be formulated as a separable nonlinear least squares problem. This means that the vectors $\boldsymbol{\theta}$ and $\mathbf{x}$ are to be found such that

$$\mathbf{A}(\boldsymbol{\theta})\mathbf{x} \approx \mathbf{b} , \tag{A.42}$$

where the matrix $\mathbf{A}$ is $m \times n$, $\mathbf{x}$ is a vector of dimension n, and $\boldsymbol{\theta}$ is a vector of dimension $\leq m - n$. The solution is formulated as finding the minimizing arguments of the loss function

$$V(\boldsymbol{\theta}, \mathbf{x}) = \| \mathbf{b} - \mathbf{A}(\boldsymbol{\theta})\mathbf{x} \|_{\mathbf{W}}^2 , \tag{A.43}$$

where $\mathbf{W}$ is an $m \times m$ positive definite weighting matrix. The solution sought to (A.42) is

$$(\hat{\boldsymbol{\theta}}, \hat{\mathbf{x}}) = \arg \min_{\boldsymbol{\theta}, \mathbf{x}} V(\boldsymbol{\theta}, \mathbf{x}) . \tag{A.44}$$

The system is called *separable* because V can be minimized in two steps. Indeed, define

$$\begin{aligned} \hat{\mathbf{x}} \stackrel{\Delta}{=} \hat{\mathbf{x}}(\boldsymbol{\theta}) &= \arg \min_{\mathbf{x}} V(\boldsymbol{\theta}, \mathbf{x}) \\ &= \left(\mathbf{A}^T(\boldsymbol{\theta})\mathbf{W}\mathbf{A}(\boldsymbol{\theta})\right)^{-1} \mathbf{A}^T(\boldsymbol{\theta})\mathbf{W}\mathbf{b} . \end{aligned} \tag{A.45}$$

Inserting this expression into (A.43) gives

$$\begin{aligned} V_1(\boldsymbol{\theta}) &\stackrel{\Delta}{=} V(\boldsymbol{\theta}, \hat{\mathbf{x}}(\boldsymbol{\theta})) \\ &= \min_{\mathbf{x}} V(\boldsymbol{\theta}, \mathbf{x}) \\ &= \mathbf{b}^T\mathbf{b} - \mathbf{b}^T\mathbf{W}\mathbf{A}(\boldsymbol{\theta}) \left(\mathbf{A}^T(\boldsymbol{\theta})\mathbf{W}\mathbf{A}(\boldsymbol{\theta})\right)^{-1} \mathbf{A}^T(\boldsymbol{\theta})\mathbf{W}\mathbf{b} . \end{aligned} \tag{A.46}$$

The function $V_1(\boldsymbol{\theta})$ is called *the concentrated loss function*. Minimizing it leads to

$$\hat{\boldsymbol{\theta}} = \arg \min_{\boldsymbol{\theta}} V_1(\boldsymbol{\theta}) , \tag{A.47}$$

and then one has

$$\hat{\mathbf{x}} = \hat{\mathbf{x}}(\hat{\boldsymbol{\theta}}) \tag{A.48}$$

using (A.45). Notice that the minimization in the original loss function (A.43) is over higher dimension than the minimization in (A.47). This reduction in dimension most often implies a reduction in the computational complexity.

More details about variable projection algorithms can be found in Golub and Pereyra (1973) and Golub and Pereyra (2003). See also Kaufman (1975) and Kaufman et al. (1994).

A.1.5 Sylvester Matrices

This section is patterned after Söderström and Stoica (1989). Sylvester matrices are defined, and a result on their rank properties is given.

Definition A.1 Consider the two polynomials

$$\begin{aligned} A(z) &= a_0 z^{n_a} + a_1 z^{n_a-1} + \cdots a_{n_a} , \\ B(z) &= b_0 z^{n_b} + b_1 z^{n_b-1} + \cdots b_{n_b} . \end{aligned} \tag{A.49}$$

The Sylvester matrix $\mathscr{S}(A, B)$ associated with A and B has dimension $(n_a + n_b) \times (n_a + n_b)$ and is defined as

$$\mathscr{S}(A, B) = \begin{pmatrix} a_0 & a_1 & \dots & a_{n_a} & & \mathbf{0} \\ & \ddots & & & \ddots & \\ \mathbf{0} & & a_0 & a_1 & \dots & a_{n_a} \\ b_0 & b_1 & \dots & b_{n_b} & & \mathbf{0} \\ & \ddots & & & \ddots & \\ \mathbf{0} & & b_0 & b_1 & \dots & b_{n_b} \end{pmatrix} \begin{matrix} \left.\vphantom{\begin{matrix} a \\ a \\ a \end{matrix}}\right\} n_b \text{ rows} \\ \left.\vphantom{\begin{matrix} a \\ a \\ a \end{matrix}}\right\} n_a \text{ rows} \end{matrix} . \tag{A.50}$$

■

The following lemma describes the rank properties of the Sylvester matrix.

Lemma A.7 *The rank of the Sylvester matrix $\mathscr{S}(A, B)$ is given by*

$$\operatorname{rank} \mathscr{S}(A, B) = n_a + n_b - k , \tag{A.51}$$

where k is the number of common zeros of the polynomials $A(z)$ and $B(z)$.

Proof Consider the equation (whose solution characterizes the null space of $\mathscr{S}^T(A, B)$)

$$\mathbf{x}^T \mathscr{S}(A, B) = \mathbf{0} , \tag{A.52}$$

where

$$\mathbf{x}^T = \left(\tilde{b}_1 \dots \tilde{b}_{n_b} \; \tilde{a}_1 \dots \tilde{a}_{n_a} \right) , \tag{A.53}$$

and introduce the associated polynomials

$$\begin{aligned} \tilde{A}(z) &= \tilde{a}_1 z^{n_a-1} + \cdots + \tilde{a}_{n_a} , \\ \tilde{B}(z) &= \tilde{b}_1 z^{n_b-1} + \cdots + \tilde{b}_{n_b} . \end{aligned} \tag{A.54}$$

Some simple manipulations show that (A.52) is equivalent to

$$\tilde{B}(z)A(z) + \tilde{A}(z)B(z) = 0\,. \tag{A.55}$$

As $A(z)$ and $B(z)$ have exactly k common zeros, one can write

$$\begin{aligned} A(z) &= A_0(z)L(z)\,, \\ B(z) &= B_0(z)L(z)\,, \qquad \text{(A.56)} \\ L(z) &= \ell_0 z^k + \ell_1 z^{k-1} + \cdots + \ell_k\,, \qquad \text{(A.57)} \end{aligned}$$

with the polynomials $A_0(z)$ and $B_0(z)$ being coprime (i.e., they do not have any common zero). Thus (A.55) is equivalent to

$$\tilde{B}(z)A_0(z) \equiv -\tilde{A}(z)B_0(z)\,. \tag{A.58}$$

As both sides of (A.58) must have the same set of zeros, it follows that

$$\begin{aligned} \tilde{A}(z) &= A_0(z)M(z)\,, \\ \tilde{B}(z) &= -B_0(z)M(z)\,, \end{aligned} \tag{A.59}$$

where the polynomial

$$M(z) = m_1 z^{k-1} + \cdots + m_k$$

has arbitrary coefficients. This means precisely that the solution $\mathbf{x}$ to (A.52) lies in a subspace of dimension k. However, this subspace is the null-space $\mathcal{N}(\mathcal{S}^T(A, B))$, and therefore its dimension is also equal to $n_a + n_b - \text{rank}\, \mathcal{S}^T(A, B)$. This observation completes the proof of (A.51). ■

Corollary A.1 *If $A(z)$ and $B(z)$ are coprime, then the Sylvester matrix $\mathcal{S}^T(A, B)$ is non-singular.* ■

Remark A.1 If, for example, $a_0 = 0$, then it is interpreted as $A(z)$ having a zero at infinity. With this interpretation the above results hold also if the leading coefficients happen to be zero. ■

A.2 Discrete Fourier Transform

This section presents some basic results for discrete Fourier transforms (DFTs).

Definition A.2 Given a time domain sequence $y(t), t = 0, \ldots, N-1$, its DFT is defined as

$$Y_k = Y(e^{i\omega_k}) = \frac{1}{\sqrt{N}} \sum_{t=0}^{N-1} y(t)e^{-i\omega_k t}\,, \quad k = 0, \ldots, N-1\,, \tag{A.60}$$

where

$$\omega_k = k\frac{2\pi}{N} . \tag{A.61}$$

■

Next some standard properties of DFT's are formulated as a lemma.

Lemma A.8 *(a) All DFTs satisfy the antisymmetry relation*

$$Y_{N-k} = Y_k^*, \quad k = 0, \ldots, N-1 . \tag{A.62}$$

(b) The underlying time domain signal can be computed from its DFT according to

$$y(t) = \frac{1}{\sqrt{N}} \sum_{k=0}^{N-1} Y_k e^{i\omega_k t} \quad t = 0, \ldots, N-1 . \tag{A.63}$$

Proof To prove part (a), apply the definition (A.60) to derive

$$\begin{aligned} Y_{N-k} &= \frac{1}{\sqrt{N}} \sum_{t=0}^{N-1} y(t) e^{-i\frac{2\pi}{N}t} \\ &= \frac{1}{\sqrt{N}} \sum_{t=0}^{N-1} y(t) e^{-i2\pi t} e^{i\frac{2\pi}{N}kt} \\ &= \frac{1}{\sqrt{N}} \sum_{t=0}^{N-1} y(t) e^{i\omega_k t} = Y_k^* . \end{aligned}$$

To prove part (b), apply the definition (A.60) to the right hand side of (A.63) which leads to

$$\begin{aligned} \frac{1}{\sqrt{N}} \sum_{k=0}^{N-1} Y_k e^{i\omega_k t} &= \frac{1}{\sqrt{N}} \sum_{k=0}^{N-1} \frac{1}{\sqrt{N}} \sum_{s=0}^{N-1} y(s) e^{-i\omega_k s} e^{i\omega_k t} \\ &= \frac{1}{N} \sum_{s=0}^{N-1} y(s) \sum_{k=0}^{N-1} e^{i(t-s)\frac{2\pi}{N}k} \\ &= \frac{1}{N} \sum_{s=0}^{N-1} y(s) N \delta_{t,s} = y(t) . \end{aligned}$$

■

Remark A.2 Part (b) of Lemma A.8 implies that the time sequence $y(0), \ldots, y(N-1)$ and its DFT $Y_0, \ldots, Y_{N-1}$ carry the same information. Further, part (a) implies that this information is indeed also carried by the first half of the DFT sequence, $Y_0, \ldots, Y_{N/2-1}$. (Some minor adjustments apply when N is an odd number.) ■

Example A.1 Let $u(t)$ be a time domain signal with DFT $\{U_k\}_{k=0}^{N-1}$. Consider a delayed signal

$$y(t) = u(t-1) . \tag{A.64}$$

Then its DFT is given by

$$Y_k = \mathrm{e}^{-\mathrm{i}\omega_k} U_k + \frac{1}{\sqrt{N}} [u(-1) - u(N-1)] , \quad k = 0, \ldots, N-1 . \tag{A.65}$$

To verify (A.65), the DFT sequence Y_k is computed:

$$\begin{aligned}
Y_k &= \frac{1}{\sqrt{N}} \sum_{t=0}^{N-1} y(t) \mathrm{e}^{-\mathrm{i}\omega_k t} \\
&= \frac{1}{\sqrt{N}} \sum_{t=0}^{N-1} u(t-1) \mathrm{e}^{-\mathrm{i}\omega_k (t-1)} \mathrm{e}^{-\mathrm{i}\omega_k} \\
&= \mathrm{e}^{-\mathrm{i}\omega_k} \frac{1}{\sqrt{N}} \left[\sum_{t'=0}^{N-1} u(t') \mathrm{e}^{-\mathrm{i}\omega_k t'} + u(-1) \mathrm{e}^{-\mathrm{i}\omega_k (-1)} - U(N-1) \mathrm{e}^{-\mathrm{i}\omega_k (N-1)} \right] \\
&= \mathrm{e}^{-\mathrm{i}\omega_k} U_k + \mathrm{e}^{-\mathrm{i}\omega_k} \frac{1}{\sqrt{N}} \mathrm{e}^{\mathrm{i}\omega_k} [u(-1) - u(N-1)] .
\end{aligned}$$

■

This example can be generalized to an arbitrary (finite) delay.

Example A.2 Let $u(t)$ be a time domain signal with DFT $\{U_k\}_{k=0}^{N-1}$. Consider a delayed signal

$$y(t) = u(t-d), \quad d \geq 1 , \tag{A.66}$$

where d is an integer. The DFT of the signal $y(t)$ can be computed as

$$\begin{aligned}
Y_k &= \frac{1}{\sqrt{N}} \sum_{t=0}^{N-1} y(t) \mathrm{e}^{-\mathrm{i}\omega_k t} \\
&= \frac{1}{\sqrt{N}} \sum_{t=0}^{N-1} u(t-d) \mathrm{e}^{-\mathrm{i}\omega_k (t-d)} \mathrm{e}^{-\mathrm{i}\omega_k d} \\
&= \mathrm{e}^{-\mathrm{i}\omega_k d} \frac{1}{\sqrt{N}} \left[\sum_{s=0}^{N-1} u(s) \mathrm{e}^{-\mathrm{i}\omega_k s} + \sum_{s=-d}^{-1} u(s) \mathrm{e}^{-\mathrm{i}\omega_k s} - \sum_{s=N-d}^{N-1} u(s) \mathrm{e}^{-\mathrm{i}\omega_k s} \right] \\
&= \mathrm{e}^{-\mathrm{i}\omega_k d} U_k + \mathrm{e}^{-\mathrm{i}\omega_k d} \sum_{s=-d}^{-1} u(s) \mathrm{e}^{-\mathrm{i}\omega_k s} - \mathrm{e}^{-\mathrm{i}\omega_k d} \sum_{s=N-d}^{N-1} u(s) \mathrm{e}^{-\mathrm{i}\omega_k (N-s)}
\end{aligned}$$

$$= \mathrm{e}^{-\mathrm{i}\omega_k d} U_k + \frac{1}{\sqrt{N}} \begin{pmatrix} 1 & \mathrm{e}^{-\mathrm{i}\omega_k} & \dots & \mathrm{e}^{-\mathrm{i}\omega_k (d-1)} \end{pmatrix} \begin{pmatrix} u(-d) - u(N-d) \\ \vdots \\ u(-1) - u(N-1) \end{pmatrix} . \tag{A.67}$$

■

Remark A.3 One can note that a delay q^{-d} gives raise to multiplication of the DFT by a factor $\mathrm{e}^{-\mathrm{i}\omega_k d}$. Further, an additional term appears. It is a polynomial of degree $d-1$ and comprises the effects of both initial and final values of the time series. ■

A.3 Results from Probability and Statistics

The concepts and results in this section can be found in many textbooks.

For books in the mathematical statistics area, see, for example, Bickel and Doksum (2001), Hogg and Craig (1970), and Larsen and Marx (2012).

For texts on the topic from the systems and control area, some examples include Ljung (1999), Söderström (2002), and Stoica and Moses (2005).

A.3.1 Results on Expectation

Lemma A.9 *Let x_1, x_2, x_3, x_4 be jointly Gaussian distributed. Then*

$$\mathsf{E}\{x_1 x_2 x_3 x_4\} = \mathsf{E}\{x_1 x_2\}\mathsf{E}\{x_3 x_4\} + \mathsf{E}\{x_1 x_3\}\mathsf{E}\{x_2 x_4\} + \mathsf{E}\{x_1 x_4\}\mathsf{E}\{x_2 x_3\} . \tag{A.68}$$

Proof See, for example, Lemma B.9 of Söderström and Stoica (1989). ■

Lemma A.10 *Consider the ARMA processes*

$$\begin{aligned} x_1(t) &= \frac{B_1(q^{-1})}{A_1(q^{-1})} e(t) , \quad x_2(t) = \frac{B_2(q^{-1})}{A_2(q^{-1})} e(t) , \\ x_3(t) &= \frac{C_1(q^{-1})}{D_1(q^{-1})} v(t) , \quad x_4(t) = \frac{C_2(q^{-1})}{D_2(q^{-1})} v(t) , \end{aligned} \tag{A.69}$$

where $e(t)$ and $v(t)$ are zero mean white noise sequences, possibly correlated, and of unit variances. Then it holds that

$$\sum_{\tau=-\infty}^{\infty} r_{x_1x_2}(\tau) r_{x_3x_4}(\tau) = \mathsf{E}\left\{\left[\frac{B_2(q^{-1})}{A_2(q^{-1})}\frac{C_1(q^{-1})}{D_1(q^{-1})}e_0(t)\right]\right\} \times \left[\frac{B_1(q^{-1})}{A_1(q^{-1})}\frac{C_2(q^{-1})}{D_2(q^{-1})}e_0(t)\right], \tag{A.70}$$

where $e_0(t)$ is white noise with unit variance.

Proof The trick, which is presented in Söderström and Stoica (1989), is to embed the problem into something more general. For this purpose, set

$$\beta_k = \sum_{\tau=-\infty}^{\infty} r_{x_1x_2}(\tau) r_{x_3x_4}(\tau + k) \tag{A.71}$$

for an arbitrary integer k. (It is of interest to determine β_0, but the analysis shows how to find β_k for any value of k.) Use now the following definition of a spectrum,

$$\phi(z) = \sum_{\tau=-\infty}^{\infty} r(\tau) z^{-\tau} . \tag{A.72}$$

Then it holds

$$r(0) = \frac{1}{2\pi i}\oint \phi(z)\frac{dz}{z} , \tag{A.73}$$

where the integration is counterclockwise around the unit circle; see, for example, Söderström (2002).

Next form

$$\begin{aligned}
\sum_{k=-\infty}^{\infty} \beta_k z^{-k} &= \sum_{k=-\infty}^{\infty}\sum_{\tau=-\infty}^{\infty} r_{x_1x_2}(\tau) r_{x_3x_4}(\tau + k) z^{-k} \\
&= \sum_{\tau=-\infty}^{\infty} r_{x_1x_2}(\tau) z^{\tau} \sum_{k=-\infty}^{\infty} r_{x_3x_4}(\tau + k) z^{-(k+\tau)} \\
&= \sum_{\tau=-\infty}^{\infty} r_{x_1x_2}(\tau) z^{\tau} \sum_{p=-\infty}^{\infty} r_{x_3x_4}(p) z^{-p} \\
&= \phi_{x_1x_2}(z^{-1}) \phi_{x_3x_4}(z) \\
&= \phi_{x_2x_1}(z) \phi_{x_3x_4}(z) \\
&= \frac{B_2(z)}{A_2(z)}\frac{B_1(z^{-1})}{A_1(z^{-1})}\frac{C_1(z)}{D_1(z)}\frac{C_2(z^{-1})}{D_2(z^{-1})} \\
&= \frac{B_2(z)}{A_2(z)}\frac{C_1(z)}{D_1(z)} \times \frac{B_1(z^{-1})}{A_1(z^{-1})}\frac{C_2(z^{-1})}{D_2(z^{-1})} ,
\end{aligned}$$

from which the result (A.70) follows after invoking (A.72). ■

Lemma A.11 *Let the vector* $\mathbf{z}(t)$ *and the scalar* $\varepsilon(t)$ *be jointly Gaussian distributed. Assume their covariance functions* $\mathbf{R}_{\mathbf{z}}(\tau)$, $r_{\varepsilon}(\tau)$, $\mathbf{r}_{\mathbf{z}\varepsilon}(\tau)$ *all decay exponentially when* $|\tau| \to \infty$. *Consider*

$$\mathbf{Q} = \lim_{N\to\infty} N\mathsf{E}\left\{\left[\frac{1}{N}\sum_{t=1}^{N}\mathbf{z}(t,\boldsymbol{\theta}_0)\varepsilon(t,\boldsymbol{\theta}_0) - \mathbf{r}_{\mathbf{z}\varepsilon}(0)\right]\right.$$
$$\left.\times\left[\frac{1}{N}\sum_{s=1}^{N}\mathbf{z}^T(s,\boldsymbol{\theta}_0)\varepsilon(s,\boldsymbol{\theta}_0) - \mathbf{r}_{\mathbf{z}\varepsilon}^T(0)\right]\right\} . \tag{A.74}$$

Then it holds

$$\mathbf{Q} = \sum_{\tau=-\infty}^{\infty}\left[\mathbf{R}_{\mathbf{z}}(\tau)r_{\varepsilon}(\tau) + \mathbf{r}_{\mathbf{z}\varepsilon}(\tau)\mathbf{r}_{\mathbf{z}\varepsilon}^T(-\tau)\right] , \tag{A.75}$$

Proof Use Lemma A.9, to find

$$\mathbf{Q} = \lim_{N\to\infty}\left[\frac{1}{N}\sum_{s,t=1}^{N}\left\{\mathbf{r}_{\mathbf{z}\varepsilon}(0)\mathbf{r}_{\mathbf{z}\varepsilon}^T(0) + \mathbf{R}_{\mathbf{z}}(t-s)r_{\varepsilon}(t-s)\right.\right.$$
$$\left. + \mathbf{r}_{\mathbf{z}\varepsilon}(t-s)\mathbf{r}_{\varepsilon\mathbf{z}}(t-s)\right\}$$
$$\left. - N\mathbf{r}_{\mathbf{z}\varepsilon}(0)\mathbf{r}_{\mathbf{z}\varepsilon}^T(0) - N\mathbf{r}_{\mathbf{z}\varepsilon}(0)\mathbf{r}_{\mathbf{z}\varepsilon}^T(0) + N\mathbf{r}_{\mathbf{z}\varepsilon}(0)\mathbf{r}_{\mathbf{z}\varepsilon}^T(0)\right]$$
$$= \lim_{N\to\infty}\left[\frac{1}{N}\sum_{s,t=1}^{N}\mathbf{R}_{\mathbf{z}}(t-s)r_{\varepsilon}(t-s) + \mathbf{r}_{\mathbf{z}\varepsilon}(t-s)\mathbf{r}_{\varepsilon\mathbf{z}}(t-s)\right]$$
$$= \lim_{N\to\infty}\frac{1}{N}\sum_{\tau=-N}^{N}(N-|\tau|)\left[\mathbf{R}_{\mathbf{z}}(\tau)r_{\varepsilon}(\tau) + \mathbf{r}_{\mathbf{z}\varepsilon}(\tau)\mathbf{r}_{\varepsilon\mathbf{z}}(\tau)\right]$$
$$= \sum_{\tau=-\infty}^{\infty}\left[\mathbf{R}_{\mathbf{z}}(\tau)r_{\varepsilon}(\tau) + \mathbf{r}_{\mathbf{z}\varepsilon}(\tau)\mathbf{r}_{\mathbf{z}\varepsilon}^T(-\tau)\right] , \tag{A.76}$$

which is (A.75).

Note that due to the assumption of exponentially decaying covariance functions, it holds for some $C > 0$, and α, with $|\alpha| < 1$,

$$\| \frac{1}{N}\sum_{\tau=-N}^{N}|\tau|\left[\mathbf{R}_{\mathbf{z}}(\tau)r_{\varepsilon}(\tau) + \mathbf{r}_{\mathbf{z}\varepsilon}(\tau)\mathbf{r}_{\varepsilon\mathbf{z}}(\tau)\right] \|$$
$$\leq \frac{1}{N}\sum_{\tau=-N}^{N}|\tau|C\alpha^{|\tau|} = \frac{2C}{N}\sum_{\tau=0}^{N}\tau\alpha^{\tau} \leq \frac{2C}{N}\sum_{\tau=0}^{\infty}\tau\alpha^{\tau}$$

$$= \frac{2C}{N} \frac{\alpha}{(1-\alpha)^2} \to 0, \quad N \to \infty . \tag{A.77}$$

This proves the last equality in (A.76). ■

Lemma A.12 *Consider the processes*

$$x_k(t) = \sum_{i=0}^{\infty} h_i^{(k)} e(t-i), \quad k = 1, \ldots, 4 . \tag{A.78}$$

Let $e(t)$ be white noise satisfying

$$\mathsf{E}\{e(t)\} = 0, \ \mathsf{E}\left\{e^2(t)\right\} = \lambda, \ \mathsf{E}\left\{e^4(t)\right\} = \mu . \tag{A.79}$$

The filters in (A.78) are assumed to be exponentially stable, so $|h_i^{(k)}| \le C\alpha^i$ for some $C > 0$ and some α, $|\alpha| < 1$. Then the following result holds

$$\begin{aligned} Q \triangleq \lim_{N\to\infty} N\mathsf{E}\Bigg\{ & \left(\frac{1}{N} \sum_{t=1}^{N} x_1(t)x_2(t) - \mathsf{E}\{x_1(t)x_2(t)\} \right) \\ & \times \left(\frac{1}{N} \sum_{s=1}^{N} x_3(s)x_4(s) - \mathsf{E}\{x_3(s)x_4(s)\} \right) \Bigg\} \\ = \sum_{\tau=-\infty}^{\infty} & \Big[\mathsf{E}\{x_1(t+\tau)x_3(t)\}\mathsf{E}\{x_2(t+\tau)x_4(t)\} \\ & + \mathsf{E}\{x_1(t+\tau)x_4(t)\}\mathsf{E}\{x_2(t+\tau)x_3(t)\} \Big] \\ & + \frac{\mu - 3\lambda^2}{\lambda^2} \mathsf{E}\{x_1(t)x_2(t)\}\mathsf{E}\{x_3(t)x_4(t)\} . \end{aligned} \tag{A.80}$$

Proof Direct calculations give

$$\begin{aligned} Q = \lim_{N\to\infty} N & \left[\mathsf{E}\left\{ \left(\frac{1}{N} \sum_{t=1}^{N} x_1(t)x_2(t) \right) \left(\frac{1}{N} \sum_{s=1}^{N} x_3(s)x_4(s) \right) \right\} \right. \\ & \left. - \mathsf{E}\{x_1(t)x_2(t)\}\mathsf{E}\{x_3(t)x_4(t)\} \right] \\ = \lim_{N\to\infty} N & \left[\frac{1}{N^2} \sum_{t=1}^{N} \sum_{s=1}^{N} \sum_{i_1=0}^{\infty} \sum_{i_2=0}^{\infty} \sum_{i_3=0}^{\infty} \sum_{i_4=0}^{\infty} h_{i_1}^{(1)} h_{i_2}^{(2)} \right. \\ & \times h_{i_3}^{(3)} h_{i_4}^{(4)} \mathsf{E}\left\{e(t-i_1)e(t-i_2)e(s-i_3)e(s-i_4)\right\} \\ & \left. - \mathsf{E}\{x_1(t)x_2(t)\}\mathsf{E}\{x_3(t)x_4(t)\} \right] . \end{aligned} \tag{A.81}$$

However, as $e(t)$ is *white* noise, the expectation of the product of four noise factors in (A.81) is nonzero precisely when the time arguments are either pairwise equal or all equal. This principle gives that

$$\begin{aligned}
&\mathsf{E}\{e(t-i_1)e(t-i_2)e(s-i_3)e(s-i_4)\} \\
&= \lambda^2\left[\delta_{i_1,i_2}\delta_{i_3,i_4} + \delta_{t-s,i_1-i_3}\delta_{t-s,i_2-i_4} + \delta_{t-s,i_1-i_4}\delta_{t-s,i_2-i_3}\right] \\
&\quad + (\mu - 3\lambda^2)\delta_{i_1,i_2}\delta_{i_3,i_4}\delta_{t-s,i_1-i_3}\,. \qquad \text{(A.82)}
\end{aligned}$$

Therefore

$$\begin{aligned}
Q &= \lambda^2 \lim_{N\to\infty} N \Bigg[\frac{1}{N^2}\sum_{t=1}^{N}\sum_{s=1}^{N}\sum_{i_1=0}^{\infty}\sum_{i_2=0}^{\infty}\sum_{i_3=0}^{\infty}\sum_{i_4=0}^{\infty} \\
&\times h_{i_1}^{(1)}h_{i_2}^{(2)}h_{i_3}^{(3)}h_{i_4}^{(4)}\lambda^2\Bigg(\delta_{i_1,i_2}\delta_{i_3,i_4} + \delta_{t-s,i_1-i_3}\delta_{t-s,i_2-i_4} + \delta_{t-s,i_1-i_4}\delta_{t-s,i_2-i_3} \\
&+ \frac{\mu-3\lambda^2}{\lambda^2}\delta_{i_1,i_2}\delta_{i_3,i_4}\delta_{t-s,i_1-i_3}\Bigg) \\
&- \mathsf{E}\{x_1(t)x_2(t)\}\,\mathsf{E}\{x_3(t)x_4(t)\}\Bigg] \\
&= \lim_{N\to\infty}\frac{1}{N}\sum_{\tau=-N}^{N}(N-|\tau|)\,\lambda^2\Bigg[\sum_{i_1=0}^{\infty}h_{i_1}^{(1)}h_{i_1-\tau}^{(3)}\sum_{i_2=0}^{\infty}h_{i_2}^{(2)}h_{i_2-\tau}^{(4)} \\
&+\sum_{i_1=0}^{\infty}h_{i_1}^{(1)}h_{i_1-\tau}^{(4)}\sum_{i_2=0}^{\infty}h_{i_2}^{(2)}h_{i_2-\tau}^{(3)} + (\mu-3\lambda^2)\sum_{i_1=0}^{\infty}h_{i_1}^{(1)}h_{i_1}^{(2)}h_{i_1-\tau}^{(3)}h_{i_1-\tau}^{(4)}\Bigg]\,. \qquad \text{(A.83)}
\end{aligned}$$

Due to the assumptions on stability, one gets as in the proof of Lemma A.11 that terms including the factor $|\tau|$ vanish, and

$$\begin{aligned}
Q &= \sum_{\tau=-\infty}^{\infty}\Bigg[\mathsf{E}\{x_1(t+\tau)x_3(t)\}\,\mathsf{E}\{x_2(t+\tau)x_4(t)\} \\
&+\mathsf{E}\{x_1(t+\tau)x_4(t)\}\,\mathsf{E}\{x_2(t+\tau)x_3(t)\} + (\mu-3\lambda^2)\sum_{i_1=0}^{\infty}h_{i_1}^{(1)}h_{i_1}^{(2)}h_{i_1-\tau}^{(3)}h_{i_1-\tau}^{(4)}\Bigg] \\
&= \sum_{\tau=-\infty}^{\infty}\Bigg[\mathsf{E}\{x_1(t+\tau)x_3(t)\}\,\mathsf{E}\{x_2(t+\tau)x_4(t)\} \\
&+\mathsf{E}\{x_1(t+\tau)x_4(t)\}\,\mathsf{E}\{x_2(t+\tau)x_3(t)\}\Bigg] + (\mu-3\lambda^2)\sum_{i=0}^{\infty}h_i^{(1)}h_i^{(2)}\sum_{\tau'=0}^{\infty}h_{\tau'}^{(3)}h_{\tau'}^{(4)} \\
&= \sum_{\tau=-\infty}^{\infty}\Bigg[\mathsf{E}\{x_1(t+\tau)x_3(t)\}\,\mathsf{E}\{x_2(t+\tau)x_4(t)\}
\end{aligned}$$

$$+\mathsf{E}\left\{x_1(t+\tau)x_4(t)\right\}\mathsf{E}\left\{x_2(t+\tau)x_3(t)\right\}\Big] + \frac{\mu-3\lambda^2}{\lambda^2}\mathsf{E}\{x_1(t)x_2(t)\}\mathsf{E}\{x_3(t)x_4(t)\}\,, \tag{A.84}$$

which proves (A.80). ■

A.3.2 *Circular Symmetric Complex Gaussian Distribution*

Complex-valued Gaussian variables are treated in this section. The scalar case, which is relatively straightforward, is analyzed first. The multivariable case, which contains some technicalities, is treated next. The analysis is restricted to the subclass that is circular symmetric.

The Scalar Case

Let x and y be zero mean Gaussian distributed random variables and set $w = x + \mathrm{i}y$. Then define the variance of w as

$$\mathsf{E}\{w\overline{w}\} = \mathsf{E}\left\{|w|^2\right\} = \mathsf{E}\{(x+\mathrm{i}y)(x-\mathrm{i}y)\} = \mathsf{E}\left\{x^2+y^2\right\}\,. \tag{A.85}$$

Assume further that x and y *are independent* and *have equal variance*, say $\mathsf{E}\left\{x^2\right\} = \mathsf{E}\left\{y^2\right\} = \sigma^2/2$. Then $\mathsf{E}\left\{|w|^2\right\} = \sigma^2$ and

$$w \sim \mathscr{C}N(0, \sigma^2) \tag{A.86}$$

is said to be circular symmetric complex Gaussian distributed with zero mean and variance σ^2. Note that in such a case

$$\mathsf{E}\left\{w^2\right\} = \mathsf{E}\left\{(x+\mathrm{i}y)^2\right\} = \mathsf{E}\left\{x^2-y^2\right\} + 2\mathrm{i}\mathsf{E}\{xy\} = 0\,. \tag{A.87}$$

The pdf for w can be derived from

$$p_w(w) = p\left(\begin{pmatrix} x \\ y \end{pmatrix}\right)\,.$$

Note that

$$\begin{pmatrix} x \\ y \end{pmatrix} \sim \mathscr{N}\left(\begin{pmatrix} 0 \\ 0 \end{pmatrix}, \frac{\sigma^2}{2}\begin{pmatrix} 1 & 0 \\ 0 & 1 \end{pmatrix}\right)\,,$$

and therefore

$$p_w(w) = \frac{1}{(2\pi)^{2/2}\sqrt{\det(\sigma^2 I/2)}}\exp\left(-\frac{1}{2}\begin{pmatrix} x \\ y \end{pmatrix}^T\left(\frac{\sigma^2}{2}\mathbf{I}_2\right)^{-1}\begin{pmatrix} x \\ y \end{pmatrix}\right)$$

$$= \frac{1}{2\pi \frac{1}{2}\sigma^2} \exp\left(-(x^2+y^2)/\sigma^2\right)$$
$$= \frac{1}{\pi\sigma^2} e^{-(w^* w)/\sigma^2} . \tag{A.88}$$

The Multivariable Case

The multivariable case is similar to the scalar case, but technically a bit more tricky. Now x and y will be substituted by vectors. The circular symmetry does not imply in the vector case that these two vectors are independent.

Set in this case

$$\mathbf{z} = \mathbf{x} + \mathrm{i}\mathbf{y} , \tag{A.89}$$

where $\mathbf{x}$ and $\mathbf{y}$ are zero mean, real-valued, n-dimensional and jointly Gaussian distributed. For circular symmetry it is required, cf. (A.85), (A.87)

$$\mathsf{E}\left\{\mathbf{z}\mathbf{z}^T\right\} = \mathbf{0}_{n\times n} , \tag{A.90}$$
$$\mathsf{E}\left\{\mathbf{z}\mathbf{z}^*\right\} = \boldsymbol{\Gamma} , \tag{A.91}$$

where $\boldsymbol{\Gamma}$ is a Hermitian matrix. In what follows the pdf of the random vector $\mathbf{z}$ is derived. Inserting (A.89) into (A.90) and (A.91) leads to

$$\mathsf{E}\left\{(\mathbf{x}+\mathrm{i}\mathbf{y})\left(\mathbf{x}^T+\mathrm{i}\mathbf{y}^T\right)\right\} = \mathsf{E}\left\{\left(\mathbf{x}\mathbf{x}^T-\mathbf{y}\mathbf{y}^T\right)\right\} + \mathrm{i}\mathsf{E}\left\{\left(\mathbf{y}\mathbf{x}^T+\mathbf{x}\mathbf{y}^T\right)\right\} , \tag{A.92}$$
$$\mathsf{E}\left\{(\mathbf{x}+\mathrm{i}\mathbf{y})\left(\mathbf{x}^T-\mathrm{i}\mathbf{y}^T\right)\right\} = \mathsf{E}\left\{\left(\mathbf{x}\mathbf{x}^T+\mathbf{y}\mathbf{y}^T\right)\right\} + \mathrm{i}\mathsf{E}\left\{\left(\mathbf{y}\mathbf{x}^T-\mathbf{x}\mathbf{y}^T\right)\right\} . \tag{A.93}$$

These relations imply that the covariance matrix of the $(\mathbf{x}^T\ \mathbf{y}^T)^T$ has the structure

$$\mathbf{R} = \mathsf{E}\left\{\begin{pmatrix}\mathbf{x}\\ \mathbf{y}\end{pmatrix}\left(\mathbf{x}^T\ \mathbf{y}^T\right)\right\} = \begin{pmatrix}\mathbf{R}_1 & \mathbf{R}_{12}\\ \mathbf{R}_{21} & \mathbf{R}_1\end{pmatrix} , \tag{A.94}$$

and that

$$\boldsymbol{\Gamma} = 2\left(\mathbf{R}_1 - \mathrm{i}\mathbf{R}_{12}\right), \quad \mathbf{R}_1^T = \mathbf{R}_1, \quad \mathbf{R}_{21}^T = -\mathbf{R}_{12}^T . \tag{A.95}$$

The pdf of $(\mathbf{x}^T\ \mathbf{y}^T)^T$ is

$$p\left(\begin{pmatrix}\mathbf{x}\\ \mathbf{y}\end{pmatrix}\right) = \frac{1}{(2\pi)^n\sqrt{\det(\mathbf{R})}} \exp\left(-\frac{1}{2}\left(\mathbf{x}^T\ \mathbf{y}^T\right)\mathbf{R}^{-1}\begin{pmatrix}\mathbf{x}\\ \mathbf{y}\end{pmatrix}\right) . \tag{A.96}$$

To proceed one needs to evaluate both $\mathbf{R}^{-1}$ and $\det(\mathbf{R})$ and relate them to $\boldsymbol{\Gamma}$. To this aim introduce the matrix

$$\mathbf{J} = \begin{pmatrix}\mathbf{I}_n & \mathrm{i}\mathbf{I}_n\\ \mathrm{i}\mathbf{I}_n & \mathbf{I}_n\end{pmatrix} . \tag{A.97}$$

It is easy to verify that

$$\mathbf{J}^{-1} = \frac{1}{2}\begin{pmatrix} \mathbf{I}_n & -\mathrm{i}\mathbf{I}_n \\ -\mathrm{i}\mathbf{I}_n & \mathbf{I}_n \end{pmatrix} = \frac{1}{2}\bar{\mathbf{J}}\,, \tag{A.98}$$

and

$$\det(\mathbf{J}) = \det(\mathbf{I}_n)\det\left(\mathbf{I}_n - (-\mathrm{i}\mathbf{I}_n)\mathbf{I}_n^{-1}(-\mathrm{i}\mathbf{I}_n)\right) = \det(\mathbf{I}_n)\det(2\mathbf{I}_n) = 2^n\,. \tag{A.99}$$

Next evaluate the product

$$\begin{aligned}\bar{\mathbf{J}}\mathbf{R}\mathbf{J} &= \begin{pmatrix} \mathbf{I}_n & -\mathrm{i}\mathbf{I}_n \\ -\mathrm{i}\mathbf{I}_n & \mathbf{I}_n \end{pmatrix}\begin{pmatrix} \mathbf{R}_1 & \mathbf{R}_{12} \\ -\mathbf{R}_{12} & \mathbf{R}_1 \end{pmatrix}\begin{pmatrix} \mathbf{I}_n & \mathrm{i}\mathbf{I}_n \\ \mathrm{i}\mathbf{I}_n & \mathbf{I}_n \end{pmatrix} \\ &= \begin{pmatrix} \mathbf{I}_n & -\mathrm{i}\mathbf{I}_n \\ -\mathrm{i}\mathbf{I}_n & \mathbf{I}_n \end{pmatrix}\begin{pmatrix} \mathbf{R}_1 + \mathrm{i}\mathbf{R}_{12} & \mathbf{R}_{12} + \mathrm{i}\mathbf{R}_1 \\ -\mathbf{R}_{12} + \mathrm{i}\mathbf{R}_1 & \mathbf{R}_1 - \mathrm{i}\mathbf{R}_{12} \end{pmatrix} \\ &= 2\begin{pmatrix} \mathbf{R}_1 + \mathrm{i}\mathbf{R}_{12} & \mathbf{0} \\ \mathbf{0} & \mathbf{R}_1 - \mathrm{i}\mathbf{R}_{12} \end{pmatrix} = \begin{pmatrix} \bar{\boldsymbol{\Gamma}} & \mathbf{0} \\ \mathbf{0} & \boldsymbol{\Gamma} \end{pmatrix}.\end{aligned} \tag{A.100}$$

It follows from (A.100)

$$\begin{aligned}\det(\bar{\mathbf{J}}\mathbf{R}\mathbf{J}) &= \det\bar{\mathbf{J}}\det\mathbf{R}\det\mathbf{J} = (\det\mathbf{J})^2\det\mathbf{R} \\ &= 2^{2n}\det(\mathbf{R}) \\ &= \det\bar{\boldsymbol{\Gamma}}\det\boldsymbol{\Gamma} = (\det\boldsymbol{\Gamma})^2\,,\end{aligned} \tag{A.101}$$

and hence

$$\det(\mathbf{R}) = \frac{(\det\boldsymbol{\Gamma})^2}{2^{2n}}\,, \tag{A.102}$$

$$(2\pi)^n\sqrt{\det\mathbf{R}} = (2\pi)^n\frac{\det\boldsymbol{\Gamma}}{2^n} = \pi^n\det(\boldsymbol{\Gamma})\,. \tag{A.103}$$

Equation (A.100) leads also to

$$\begin{aligned}&\frac{1}{2}\begin{pmatrix}\mathbf{x}\\ \mathbf{y}\end{pmatrix}^T\mathbf{R}^{-1}\begin{pmatrix}\mathbf{x}\\ \mathbf{y}\end{pmatrix} \\ &= \frac{1}{2}\begin{pmatrix}\mathbf{x}\\ \mathbf{y}\end{pmatrix}^T\mathbf{J}\left(\bar{\mathbf{J}}\mathbf{R}\mathbf{J}\right)^{-1}\bar{\mathbf{J}}\begin{pmatrix}\mathbf{x}\\ \mathbf{y}\end{pmatrix} \\ &= \frac{1}{2}\left(\mathbf{x}^T\ \mathbf{y}^T\right)\begin{pmatrix} \mathbf{I}_n & \mathrm{i}\mathbf{I}_n \\ \mathrm{i}\mathbf{I}_n & \mathbf{I}_n \end{pmatrix}\begin{pmatrix} \bar{\boldsymbol{\Gamma}}^{-1} & \mathbf{0} \\ \mathbf{0} & \boldsymbol{\Gamma}^{-1} \end{pmatrix}\begin{pmatrix} \mathbf{I}_n & -\mathrm{i}\mathbf{I}_n \\ -\mathrm{i}\mathbf{I}_n & \mathbf{I}_n \end{pmatrix}\begin{pmatrix}\mathbf{x}\\ \mathbf{y}\end{pmatrix} \\ &= \frac{1}{2}\left(\mathbf{z}^T\ \mathrm{i}\bar{\mathbf{z}}^T\right)\begin{pmatrix} \bar{\boldsymbol{\Gamma}}^{-1} & \mathbf{0} \\ \mathbf{0} & \boldsymbol{\Gamma}^{-1} \end{pmatrix}\begin{pmatrix}\bar{\mathbf{z}}\\ -\mathrm{i}\mathbf{z}\end{pmatrix} \\ &= \frac{1}{2}\left(\mathbf{z}^T\bar{\boldsymbol{\Gamma}}^{-1}\bar{\mathbf{z}} + \bar{\mathbf{z}}^T\boldsymbol{\Gamma}^{-1}\mathbf{z}\right) = \bar{\mathbf{z}}^T\boldsymbol{\Gamma}^{-1}\mathbf{z}\,.\end{aligned} \tag{A.104}$$

Inserting the expressions (A.103) and (A.104) into (A.96) finally gives

$$p(z) = \frac{1}{\pi^n \det(\boldsymbol{\Gamma})} \exp\left(-\bar{\mathbf{z}}^T \boldsymbol{\Gamma}^{-1} \mathbf{z}\right) . \tag{A.105}$$

One can add the comment that linear and affine transformations of random variables that are circular symmetric Gaussian remain in that class of variables. More specifically, if

$$\mathbf{x} \sim \mathscr{C}N(\mathbf{0}, \boldsymbol{\Gamma}) , \tag{A.106}$$

$$\mathbf{y} = \mathbf{A}\mathbf{x} + \mathbf{b} , \tag{A.107}$$

where $\mathbf{A}$ is a matrix of full row rank, and $\mathbf{b}$ is a vector, then

$$\mathbf{y} \sim \mathscr{C}N(\mathbf{b}, \mathbf{A}\boldsymbol{\Gamma}\mathbf{A}^*) . \tag{A.108}$$

Example A.3 For a scalar $\mathscr{C}N(0, \sigma^2)$ variable $\omega = x + \mathrm{i}y$, it holds

$$\mathsf{E}\left\{w^2\right\} = 0 , \tag{A.109}$$

$$\mathsf{E}\left\{|w|^2\right\} = \sigma^2 , \tag{A.110}$$

$$\mathsf{E}\left\{|w|^4\right\} = 2\sigma^4 . \tag{A.111}$$

The relation (A.109) follows directly from (A.87). As

$$\begin{pmatrix} x \\ y \end{pmatrix} \sim \mathscr{N}\left(\mathbf{0}, \frac{\sigma^2}{2}\mathbf{I}\right) ,$$

it also holds

$$\mathsf{E}\left\{|w|^2\right\} = \mathsf{E}\left\{x^2 + y^2\right\} = 2\frac{\sigma^2}{2} = \sigma^2 ,$$

and

$$\begin{aligned} \mathsf{E}\left\{|w|^4\right\} &= \mathsf{E}\left\{(x^2 + y^2)^2\right\} = \mathsf{E}\left\{x^4 + 2x^2y^2 + y^4\right\} \\ &= 3\left(\frac{\sigma^2}{2}\right)^2 + 2\frac{\sigma^2}{2}\frac{\sigma^2}{2} + 3\left(\frac{\sigma^2}{2}\right)^2 = 8\left(\frac{\sigma^2}{2}\right)^2 = 2\sigma^4 . \end{aligned}$$

■

For the multivariable case it follows from (A.91), (A.94), (A.95) that

$$\begin{aligned} &\mathbf{z} \sim \mathscr{C}N(\mathbf{0}, \boldsymbol{\Gamma}) \Rightarrow \boldsymbol{\Gamma} \text{ Hermitian} , \\ &\boldsymbol{\Gamma} = 2\mathbf{R}_1 - 2\mathrm{i}\mathbf{R}_{12} , \\ &\mathbf{R}_1 \text{ symmetric} , \\ &\mathbf{R}_{12} \text{ antisymmetric} . \end{aligned}$$

Further, if $\boldsymbol{\Gamma}$ is real-valued, this is equivalent to $\mathbf{R}_{12} = \mathbf{0}$ and also equivalent to $\mathbf{x}$ and $\mathbf{y}$ being independent.

An important case of scalar $\mathscr{C}N$ variables are treated in the following example.

Example A.4 Let $e(t), t = 0, \ldots, N-1$ be (time domain) white noise, with zero mean and variance σ^2. Let $e(t)$ have a DFT $E_k, k = 0, \ldots, N-1$. Then

$$E_k = \frac{1}{\sqrt{N}} \sum_{t=0}^{N-1} e(t) \mathrm{e}^{-\mathrm{i}\omega_k t} , \tag{A.112}$$

$$\begin{aligned}
\mathsf{E}\left\{E_j E_k\right\} &= \frac{1}{N} \mathsf{E} \left\{ \sum_{t=0}^{N-1} e(t) \mathrm{e}^{-\mathrm{i}\omega_k t} \sum_{s=0}^{N-1} e(s) \mathrm{e}^{-\mathrm{i}\omega_k s} \right\} \\
&= \frac{1}{N} \sum_t \sum_s \sigma^2 \delta_{t,s} \mathrm{e}^{-\mathrm{i}\omega_k t} \mathrm{e}^{-\mathrm{i}\omega_k s} \\
&= \frac{\sigma^2}{N} \sum_{t=0}^{N-1} \mathrm{e}^{-\mathrm{i}(\omega_j + \omega_k)t} \\
&= \frac{\sigma^2}{N} N \delta_{j+k,N} = \begin{cases} 0 & j+k \neq N , \\ \sigma^2 & j+k = N . \end{cases}
\end{aligned} \tag{A.113}$$

$$\begin{aligned}
\mathsf{E}\left\{E_j E_k^*\right\} &= \frac{1}{N} \sum_t \sum_s \sigma^2 \delta_{t,s} \mathrm{e}^{-\mathrm{i}\omega_k t} \mathrm{e}^{\mathrm{i}\omega_k s} \\
&= \frac{\sigma^2}{N} \sum_{t=0}^{N-1} \mathrm{e}^{\mathrm{i}(\omega_k - \omega_j)t} = \frac{\sigma^2}{N} N \delta_{j,k} = \begin{cases} \sigma^2 & j = k , \\ 0 & j \neq k . \end{cases}
\end{aligned} \tag{A.114}$$

In particular, if the first half of the DFT sequence is considered, that is, $E_k, k = 0, \ldots, N/2-1$, cf. Sect. A.2, it is found that E_k is white circular symmetric complex Gaussian white noise. ■

Remark A.4 Lemma A.9 on the product of four jointly Gaussian variables can be extended to circular complex Gaussian variables; see Janssen and Stoica (1988). ■

Lemma A.13 *Let $y(t)$, $t = 1, \ldots, N$ be a realization of a stationary stochastic process, and let its DFT be denoted $Y_0, \ldots, Y_{N-1}$. Then it holds*

$$\mathsf{E}\left\{Y_j Y_k\right\} = o(1) , \tag{A.115}$$

$$\mathsf{E}\left\{|Y_j|^2\right\} = \phi\left(j \frac{2\pi}{N}\right) + o(1) , \tag{A.116}$$

$$\mathsf{E}\left\{Y_j Y_k^*\right\} = o(1), \quad j \neq k , \tag{A.117}$$

where the terms $o(1) \to 0$ as $N \to \infty$, and $\phi(\omega)$ denotes the spectral density of $y(t)$.

Furthermore, the sequence $Y_k, k = 0, \ldots, N-1$, is asymptotically complex circular symmetric Gaussian distributed.

Proof See Brillinger (1981), or its reprinted form Brillinger (2001). ■

Remark A.5 More properties of circular symmetric complex Gaussian random variables can be found, for example, in Picinbono (1996), Picinbono (1993). ■

A.3.3 General Estimation Concepts

Consider a parametric model with a parameter vector $\boldsymbol{\theta}$ to be estimated. Let its true value be denoted $\boldsymbol{\theta}_0$. Assume further that N data points are available, so the data are

$$\mathbf{y} = \begin{pmatrix} y_1 & \ldots & y_N \end{pmatrix}^T . \tag{A.118}$$

An estimate of $\boldsymbol{\theta}$ is then a function of the data,

$$\hat{\boldsymbol{\theta}} = \mathbf{f}(\mathbf{y}) . \tag{A.119}$$

Convention A.6 The estimate $\hat{\boldsymbol{\theta}}$ is unbiased, if

$$\mathsf{E}\{\hat{\boldsymbol{\theta}}\} = \boldsymbol{\theta}_0 . \tag{A.120}$$

The estimate $\hat{\boldsymbol{\theta}}$ is consistent, if

$$\hat{\boldsymbol{\theta}} \to \boldsymbol{\theta}_0, \quad \text{as } N \to \infty . \tag{A.121}$$

The limit in (A.121) is to be interpreted in an appropriate sense, which may be convergence in mean square, in probability, or with probability one. ■

Convention A.7 The maximum likelihood (ML) estimate of $\boldsymbol{\theta}$ is the parameter argument that maximizes the a posteriori probability density function of the observed variables, that is,

$$\hat{\boldsymbol{\theta}}_{\text{ML}} \stackrel{\Delta}{=} \arg\min_{\boldsymbol{\theta}} p(\mathbf{y}|\boldsymbol{\theta}) . \tag{A.122}$$

ML estimates can be said to select the parameters so that the observed values become as probable (likely) as possible. In order to compute the ML estimate, obviously the distribution of the observations must be specified, as a function of $\boldsymbol{\theta}$. ML estimates enjoy many useful properties. They often provide accurate estimates, in the sense that the covariance matrix of the estimates is small. The function

$$L(\mathbf{y}, \boldsymbol{\theta}) = p(\mathbf{y}|\boldsymbol{\theta}) \tag{A.123}$$

is called the likelihood function. ■

Convention A.8 A statistic $\mathbf{t} = \mathbf{T}(\mathbf{X})$ of data $\mathbf{X}$ is said to be a *sufficient statistic* for the underlying model parameter vector $\boldsymbol{\theta}$ precisely if the conditional probability distribution of the data $\mathbf{X}$ given the statistic $\mathbf{t} = \mathbf{T}(\mathbf{X})$ does not depend on $\boldsymbol{\theta}$, i.e.,

$$\Pr(\mathbf{x}|\mathbf{t}, \boldsymbol{\theta}) = \Pr(\mathbf{x}|\mathbf{t}) . \tag{A.124}$$

■

Example A.5 Assume that X_i, $i = 1, \ldots, N$, are independent and jointly Gaussian distributed, with mean m and variance σ. Then

$$\mathbf{X} = \begin{pmatrix} X_1 \ldots X_N \end{pmatrix}^T$$

has a pdf

$$p_{\mathbf{X}}(\mathbf{x}) = \frac{1}{(2\pi\sigma)^{N/2}} \exp\left(-\frac{1}{2\sigma} (\mathbf{x} - m\mathbf{e})^T (\mathbf{x} - m\mathbf{e}) \right)$$

with $\mathbf{e} = (1 \ldots 1)^T$. Hence

$$\log p_{\mathbf{X}}(\mathbf{x}) = -\frac{N}{2} \log(2\pi\sigma) - \frac{1}{2\sigma} \left(\sum_{i_1}^{N} x_i^2 - 2m \sum_{i=1}^{N} x_i + Nm^2 \right) . \tag{A.125}$$

Therefore the two-dimensional vector

$$\mathbf{T}(\mathbf{x}) = \begin{pmatrix} \sum_i x_i \ \sum_i x_i^2 \end{pmatrix}^T \tag{A.126}$$

is a sufficient statistic. The dependence of the pdf $p_{\mathbf{X}}(\mathbf{x})$ on $\mathbf{x}$ (an N-dimensional vector) can be expressed fully as a function of the statistic $\mathbf{T}(\mathbf{x})$ (a two-dimensional vector). ■

A.3.4 Central Limit Theorem

The central limit theorem is useful when deriving results on asymptotic distributions for various estimates, when the number of observations becomes large. In its simple form it states that a large number of independent and equally distributed random variable is with good approximation Gaussian distributed. A more advanced form is useful for analyzing system identification methods.

Lemma A.14 *Consider*

$$\mathbf{x}_N = \frac{1}{\sqrt{N}} \sum_{t=1}^{N} \mathbf{z}(t) , \tag{A.127}$$

where the vector $\mathbf{z}(t)$ *is a zero mean stationary stochastic process given by*

$$\mathbf{z}(t) = \boldsymbol{\Psi}(t)\mathbf{v}(t) , \tag{A.128}$$

where the matrix $\boldsymbol{\Psi}(t)$ *and the vector* $\mathbf{v}(t)$ *have entries that are stationary, possibly correlated ARMA processes with zero means, and underlying innovation sequences with finite fourth-order moments. The elements of* $\boldsymbol{\Psi}(t)$ *may also contain a bounded deterministic term.*

Assume that the limit

$$\mathbf{P} = \lim_{N\to\infty} \mathsf{E}\{\mathbf{x}_N\mathbf{x}_N^T\} \tag{A.129}$$

exists and is nonsingular. Then $\mathbf{x}_N$ *is asymptotically Gaussian distributed,*

$$\mathbf{x}_N \overset{\text{dist}}{\longrightarrow} \mathcal{N}(\mathbf{0}, \mathbf{P}), \quad N \to \infty . \tag{A.130}$$

Proof See Ljung (1977), Ljung (1999). ■

The following result is often called Slutsky's lemma.

Lemma A.15 *Let* $\{\mathbf{x}_n\}$ *be a sequence of random vectors that converge in probability to a constant* $\mathbf{x}$. *Let* $\mathbf{f}(\mathbf{x})$ *be a rational function, and suppose that* $\mathbf{f}(\mathbf{x})$ *is finite. Then* $\mathbf{f}(\mathbf{x}_n)$ *converges in probability to* $\mathbf{f}(\mathbf{x})$.

Proof See Cramér (1946b). ■

A.3.5 The Cramér–Rao Lower Bound

There is a famous lower bound on the covariance matrix of unbiased parameter estimates.

Lemma A.16 *Let* $\mathbf{y}$ *be a stochastic vector-valued variable, the distribution of which depends on the unknown parameter vector* $\boldsymbol{\theta}$. *Let* $\hat{\boldsymbol{\theta}}(\mathbf{y})$ *denote an arbitrary unbiased estimate of* $\boldsymbol{\theta}$ *determined from* $\mathbf{y}$. *Then*

$$\text{cov}(\hat{\boldsymbol{\theta}}) \geq \left[\mathsf{E}\left\{\left(\frac{\partial \log L}{\partial \boldsymbol{\theta}}\right)^T \left(\frac{\partial \log L}{\partial \boldsymbol{\theta}}\right)\right\}\right]^{-1} = -\left[\mathsf{E}\left\{\frac{\partial^2 \log L}{\partial \boldsymbol{\theta}^2}\right\}\right]^{-1} \triangleq \mathbf{J}^{-1} . \tag{A.131}$$

Proof An early reference for the theorem is Cramér (1946b), which proves the result for the case when $\mathbf{y}$ has one component (i.e., a scalar) or two. The general result is described, but the detailed proof for that case appears in Cramér (1946a). The book Cramér (1946b) describes also the extension of the result to biased estimators. Proofs of the theorem appear also in many books published later, see, for example, Lemma B.15 of Söderström and Stoica (1989). ■

Convention A.9 The matrix

$$\mathbf{J} = \mathsf{E}\left\{\left(\frac{\partial \log L}{\partial \boldsymbol{\theta}}\right)^T \left(\frac{\partial \log L}{\partial \boldsymbol{\theta}}\right)\right\} \tag{A.132}$$

is called the Fisher information matrix.

The inverse of $\mathbf{J}$ is apparently a lower bound on the covariance matrix of any unbiased estimate. This bound is known as the Cramér–Rao lower bound (CRB). ■

Remark A.6 An unbiased estimate is said to be *statistically efficient* if equality applies in (A.131). Similarly, it is said to be asymptotically statistically efficient, if equality holds on (A.131) when the number of data points, N, grows to infinity. Both sides of (A.131) typically go to zero as $N \to \infty$, but for asymptotic statistical efficiency it is required that their difference is converging at a faster rate than any of the two sides. ■

Remark A.7 Under weak assumptions, the ML estimate, see (A.122), attains the Cramér–Rao lower bound, at least asymptotically as $N \to \infty$. ■

In order to compute the CRB, one needs to know the distribution of the observations. There are some convenient ways to compute the CRB for Gaussian distributed observations.

Lemma A.17 *Let* $\mathbf{y}(t), t = 1, \ldots, N$ *be observations of a possibly vector-valued stationary stochastic Gaussian process with spectral density* $\boldsymbol{\Phi}$*. Then, asymptotically as* $N \to \infty$*, the information matrix is given in component form by*

$$\mathbf{J}_{jk} = \frac{N}{4\pi i}\mathrm{tr}\left(\oint \boldsymbol{\Phi}^{-1}(z)\boldsymbol{\Phi}'_j(z)\boldsymbol{\Phi}^{-1}(z)\boldsymbol{\Phi}'_k(z)\frac{dz}{z}\right), \quad j, k = 1, 2, \ldots, \tag{A.133}$$

where $\boldsymbol{\Phi}'_j = (\partial/\partial_{\theta_k})\boldsymbol{\Phi}$*, and* $\oint$ *denotes integration counterclockwise around the unit circle.*

Proof See Whittle (1953), or, for example, Kay (1993). ■

Remark A.8 The relation (A.133) is known as Whittle's formula. ■

Remark A.9 An extension of (A.133) so that $\mathbf{y}(t)$ also contains a deterministic component that depends on the parameter vector $\boldsymbol{\theta}$ is given in Karlsson et al. (2000). ■

Lemma A.18 *Let* $\mathbf{y}(t), t = 1, \ldots, N$ *be observations of a possibly vector-valued stationary stochastic Gaussian process with zero mean. Set*

$$\mathbf{R} = \mathsf{E}\left\{\begin{pmatrix}\mathbf{y}(1)\\ \vdots \\ \mathbf{y}(N)\end{pmatrix}\begin{pmatrix}\mathbf{y}^T(1) \ldots \mathbf{y}^T(N)\end{pmatrix}\right\}. \tag{A.134}$$

Then the information matrix can be expressed in component form as

$$\mathbf{J}_{jk} = \frac{1}{2}\mathrm{tr}\left(\mathbf{R}^{-1}\mathbf{R}_j'\mathbf{R}^{-1}\mathbf{R}_k'\right) , \tag{A.135}$$

where $\mathbf{R}_j' = (\partial/\partial\boldsymbol{\theta}_j)\mathbf{R}$.

Proof See Slepian (1954), Bangs (1971), or Stoica and Moses (2005). ■

Remark A.10 The relation (A.135) is known as Slepian–Bangs' formula. ■

A third result on computing the Cramér–Rao lower bound is for an arbitrary linear stochastic state-space model. The details of this approach are given in Sect. 14.8.2.

References

Abatzoglou TJ (1989) Analysis of the constrained total least squares method and its relation to harmonic superresolution. SPIE 1152:158–169

Abatzoglou TJ, Mendel JM (1991) The constrained total least squares technique and its applications to harmonic superresolution. IEEE Trans Signal Process 39(5):1070–1087

Adcock RJ (1877) Note on the methods of least squares. Anal 4(6):183–184

Adcock RJ (1878) A problem in least squares. Anal 5(2):53–54

Agüero JC (2005) System identification methodologies incorporating constraints. Ph.D. thesis, The University of Newcastle, Callaghan, NSW, Australia

Agüero JC, Goodwin GC (2006) Identifiability of errors in variables dynamic systems. In: Proceedings of the 14th IFAC symposium on system identification, Newcastle, Australia

Agüero JC, Goodwin GC (2008) Identifiability of errors in variables dynamic systems. Automatica 44:371–382

Agüero JC, Goodwin GC, Salgado ME (2005) On the optimal estimation of errors in variables models for robust control. In: Proceedings of the 16th IFAC world congress, Prague, Czech Republic

Agüero JC, Yuz JI, Goodwin GC (2008) Identifiability of EIV dynamic systems with non-stationary data. In: Proceedings of the 17th IFAC world congress, Seoul, Korea

Agüero JC, Goodwin GC, Söderström T, Yuz JI (2009) Sampled data errors in variables systems. In: Proceedings of the 15th IFAC symposium on system identification, Saint-Malo, France

Agüero JC, Yuz JI, Goodwin GC, Delgado RA (2010) On the equivalence of time and frequency domain maximum likelihood estimation. Automatica 46(2):260–270

Agüero JC, Rojas C, Hjalmarsson H, Goodwin GC (2012) Accuracy of linear multiple-input multiple-output (MIMO) models obtained by maximum likelihood estimation. Automatica 48:632–637

Agüero JC, Conzález K, Carvajal G (2017) EM-based identification of ARX systems having quantized output data. In: Proceedings of the 20th IFAC world congress, Toulouse, France

Akaike H (1977) On entropy maximization principle. In: Krishnaiah PR (ed) Applications of statistics. North-Holland, Amsterdam

ALMutawa J (2008) Identification of errors-in-variables models using the EM algorithm. In: Proceedings of the 17th IFAC world congress, Seoul, Korea

ALMutawa J (2009) Identification of errors-in-variables state space models with observation outliers based on minimum covariance determinant. J Process Control 19:879–887

Amemiya Y, Fuller WA (1988) Estimation for the nonlinear functional relationship. Ann Stat 16(1):147–160

T. Söderström, *Errors-in-Variables Methods in System Identification*, Communications and Control Engineering,
https://doi.org/10.1007/978-3-319-75001-9

Anderson BDO (1985) Identification of scalar errors-in-variables models with dynamics. Automatica 21:709–716
Anderson BDO (1999) From Wiener to hidden Markov models. Control Syst Mag 19:41–51
Anderson BDO, Deistler M (1984) Identifiability of dynamic errors-in-variables models. J Time Ser Anal 5:1–13
Anderson BDO, Deistler M (1987) Dynamic errors-in-variables systems in three variables. Automatica 23:611–616
Anderson BDO, Deistler M (2008a) Generalized linear dynamic factor models - a structure theory. In: Proceedings of the 47th IEEE conference on decision and control, Cancun, Mexico
Anderson BDO, Deistler M (2008b) Properties of zero-free transfer function matrices. SICE J Control Meas Syst Integr 1:284–292
Anderson BDO, Green M, Deistler M (1987) Errors-in-variables identification: linear-Gaussian stationary results. In: Proceedings of the 10th IFAC world congress, Munich, Germany
Anderson BDO, Deistler M, Scherrer W (1996) Solution set properties for static errors-in-variables problems. Automatica 32(7):1031–1035
Anderson BDO, Deistler M, Chen W, Filler A (2009) AR models of singular spectral matrices. In: Proceedings of the 48th IEEE conference on decision and control, Shanghai, P. R. China
Anderson TW (1984) Estimating linear statistical relationships. Ann Math Stat 12(1):1–45
Anderson TW (1989) Linear latent variable models and covariance structures. J Econom 41:91–119
Anderson TW, Hsiao C (1981) Estimation of dynamic models with error components. J Am Stat Assoc 76(375):598–606
Anderson TW, Rubin H (1956) Statistical inference in factor analysis. In: Proceedings of the third Berkeley symposium on mathematical statistics and probability, vol 5. University of California Press, Berkeley, pp 111–150
Andersson BDO, Deistler M (2009) Properties of zero-free spectral matrices. IEEE Trans Autom Control 54(10):2365–2375
Andersson BDO, Deistler M, Chen W, Filler A (2012) Autoregressive models of singular spectral matrices. Automatica 48:2843–2849
Andersson BDO, Deistler M, Felsenstein E, Koelbl L (2017) The structure of multivariate AR and ARMA systems: regular and singular systems; the single and mixed frequency case. J Econom
Aoki M, Yue PC (1970) On a priori error estimate of some identification methods. IEEE Trans Autom Control AC-15:541–548
Arablouei R, Doğançay K, Adali T (2014) Unbiased recursive least-squares estimation utilizing dichotomous coordinate-descent iterations. IEEE Trans Signal Process 62(11):2973–2983
Arablouei R, Doğançay K, Werner S (2015) Recursive total least-squares algorithm based on inverse power method and dichotomous coordinate-descent iterations. IEEE Trans Signal Process 63(8):1941–1949
Åström KJ, Wittenmark B (1997) Computer-controlled systems: theory and design. Prentice Hall, Upper Saddle River
Bagozzi R, Yi Y (2012) Specification, evaluation, and interpretation of structural equation models. J Acad Mark Sci 40(1):8–34
Bangs WJ (1971) Array processing with generalized beamformers. Ph.D. thesis, Yale University, New Haven, CT, USA
Barbé K, Schoukens J, Pintelon R (2007) Frequency domain errors-in-variables estimation of linear dynamic systems using data from overlapping subrecords. In: Proceedings of the IEEE instrumentation and measurement technology conference, Warsaw, Poland
Barbé K, Schoukens J, Pintelon R (2008) Frequency domain errors-in-variables estimation of linear dynamic systems using data from overlapping subrecords. IEEE Trans Instrum Meas 57:1529–1536
Barbé K, Schoukens J, Pintelon R (2012) The use of non-parametric noise models extracted from overlapping subrecords for system identification. IEEE Trans Signal Process 59(10):4635–4647
Barhoumi K, Damé O, Ferrara L (2014) Dynamic factor models: a review of the literature. OECD J Bus Cycle Meas Anal 2013/2:73–107

Bartholomew D, Knott M, Moustaki I (2011) Latent variable models and factor analysis, 3rd edn. Wiley, Chichester
Basseville M, Benveniste A, Moustakides GV, Rougée A (1987) Optimal sensor location for detecting changes in dynamical behavior. IEEE Trans Autom Control 32(12):1067–1075
Batselier K, Dreesen P, De Moor B (2012) Maximum likelihood estimation and polynomial system solving. In: Proceedings of the ESANN 2012 European symposium on artificial neural networks, computational intelligence and machine learning, Bruges, Belgium
Batselier K, Dreesen P, De Moor B (2014) A fast recursive orthogonalization scheme for the Macaulay matrix. J Comput Appl Math 267:20–32
Beck A (2007) The matrix-restricted total least-squares problem. Signal Process 87:2303–2312
Beghelli S, Soverini U (1992) Identification of linear relations from noisy data: geometrical aspects. Syst Control Lett 18(5):339–346
Beghelli S, Guidorzi R, Soverini U (1990) The Frisch scheme in dynamic system identification. Automatica 26:171–176
Beghelli S, Castaldi P, Guidorzi R, Soverini U (1993) A robust criterion for model selection in identification from noisy data. In: Proceedings of the 9th international conference on systems engineering, Las Vegas, Nevada, USA, pp 480–484
Beghelli S, Castaldi P, Soverini U (1997) A frequential approach for errors-in-variables models. In: Proceedings of the 4th European control conference, ECC'97, Brussels, Belgium
Ben Rosen J, Park H, Glick J (1996) Total least norm formulation and solution for structured problems. SIAM J Matrix Anal Appl 17(1):110–126
Bengtsson M, Ottersten B (2001) A generalization of weighted subspace fitting to full-rank models. IEEE Trans Signal Process 49(5):1002–1112
Bernstein DS (2005) Matrix mathematics. Princeton University Press, Princeton
Bernstein DS (2009) Matrix mathematics, 2nd edn. Princeton University Press, Princeton
Bickel PJ, Doksum KA (2001) Mathematical statistics, vol 1, 2nd edn. Prentice Hall, Upper Saddle River
Bloch AM (1989) Identification and estimation of dynamic errors-in-variables models. J Econom 41:145–158
Bollen KA (1989) Structural equations with latent variables. Wiley, New York
Bottegal G, Picci G (2011) A note on generalized factor analysis models. In: Proceedings of the 50th IEEE conference on decision and control, Orlando, Florida, USA, pp 1485–1490
Bottegal G, Picci G (2015) Modeling complex systems by generalized factor analyisis. IEEE Trans Autom Control 60(3):759–774
Bottegal G, Picci G, Pinzoni S (2011) On the identifiability of errors-in-variables models with white measurement errors. Automatica 47(3):545–551
Bottegal G, Risuleo RS, Zamani M, Ninness B, Hjalmarsson H (2017) On maximum likelihood identification of errors-in-variables models. In: Proceedings of the 20th IFAC world congress, Toulouse, France
Brillinger DR (1981) Time series: data analysis and theory. Holden-Day Inc, San Francisco
Brillinger DR (2001) Time series: data analysis and theory. SIAM (Classics in applied mathematics), Philadelphia
Browne MW (1974) Generalized least squares estimators in the analysis of covariance structures. S Afr Stat J 8:1–24
Browne MW (1984) Asymptotically distribution-free methods for the analysis of covariance structures. Br J Math Stat Psychol 37:62–83
Cadzow JA, Solomon OM (1986) Algebraic approach to system identification. IEEE Trans Acoust Speech Signal Process ASSP-34(3):462–469
Camba-Mendez G, Kapeanios G (2004) Bootstrap statistical tests of rank determination for system identification. IEEE Trans Autom Control 49(2):238–243
Cappé O, Moulines E, Rydén T (2005) Inference in hidden Markov models. Springer, San Francisco

Carvajal R, Delgado R, Agüero JC, Goodwin GC (2012) An identification method for errors-in-variables system using incomplete data. In: Proceedings of the 16th IFAC symposium on system identification, Belgium, Brussels, pp 1359–1364

Casini M, Garulli A, Vicino A (2014) Feasible parameter set approximation for linear models with bounded uncertain regressors. IEEE Trans Autom Control 59(11):2910–2920

Castaldi P, Soverini U (1996) Identification of dynamic errors-in-variables models. Automatica 32(4):631–636

Castaldi P, Soverini U (1998) Identification of errors-in-variables models and optimal output reconstruction. In: Proceedings of the 14th international symposium on mathematical theory of networks and systems (MTNS), Padova, Italy

Castaldi P, Soverini U, Beghelli S (1995) Identification of ARX models in presence of input noise. In: Proceedings of the international conference on systems science, Wroclaw, Poland

Castaldi P, Soverini U, Beghelli S (1996) Rank reducibility of a covariance matrix in the Frisch scheme. In: Proceedings of the 13th IFAC world congress, San Francisco, USA

Castaldi P, Diversi R, Guidorzi R, Soverini U (1999) Identification of multivariable errors-in-variables models. In: Proceedings of the 5th European control conference, Karlsruhe, Germany

Castaldi P, Montanari M, Tilli A (2002) Induction motor model identification via frequency-domain Frisch scheme. In: Proceedings of the 15th IFAC world congress, Barcelona, Spain

Cedervall M, Stoica P (1996) System identification from noisy measurements by using instrumental variables and subspace fitting. Circuits Syst Signal Process 15(2):275–290

Cerone V, Piga D, Regruto D (2012) Set-membership error-in-variables identification through convex relaxation techniques. IEEE Trans Autom Control 57(2):517–522

Cerone V, Fadda E, Regruto D (2017a) A robust optimization approach to kernel-based nonparametric errors-in-variables identification in the presence of bounded noise. In: Proceedings of the American control conference (ACC), Seattle, WA, USA

Cerone V, Regruto D, Abuabiah M (2017b) Direct data-driven control design through set-membership errors-in-variables identification techniques. In: Proceedings of the American control conference (ACC), Seattle, WA, USA

Cerone V, Razza V, Regruto D (2018) Set-membership errors-in-variables identification of MIMO linear systems. Automatica 90:25–37

Chan LK, Mak TK (1984) Maximum likelihood estimation in multivariate structural relationships. Scand J Stat 11(1):45–50

Chen CC, Lin HW (2011) The advantages of dynamic factor models as techniques for forecasting: evidence from taiwanese macroeconomic data. Int J Econ Financ 3(5):208–21,542

Chen HF (2007) Recursive identification for multivariate errors-in-variables systems. Automatica 43:1234–1242

Chen HF (2008) Recursive identification of EIV ARMA processes. In: Proceedings of the 17th IFAC world congress, Seoul, Korea, pp 1366–1371

Chen HF, Yang JM (2005) Strongly consistent coefficient estimates for errors-in-variables models. Automatica 41(6):1025–1033

Chen HY, Lee AC, Lee CF (2015) Alternative errors-in-variables models and their applications in finance research. Q Rev Econ Financ 58:213–227

Chen XM, Gao C (2015) Recursive nonparametric regression with errors in variables. In: Proceedings of the 34th Chinese control conference (CCC), Hangzhou, China, pp 2088–2092

Cheng CL, Van Ness JW (1999) Kendall's library of statistics: statistical regression with measurement error, vol 6. Edward Arnold, London

Chetoui M, Malti R, Thomassin M, Aoun M, Najar S, Oustaloup A, Abdelkrim MN (2012) EIV methods for system identification with fractional models. In: Proceedings of the 16th IFAC symposium on system identification, Belgium, Brussels, pp 1641–1646

Chetoui M, Malti R, Thomassin M, Najar S, Aoun M, Abdelkrim MN, Oustaloup A (2013a) Fourth-order cumulants based method for continuous-time EIV fractional model identification. In: Proceedings of the 10th international multi-conference on systems, signals and devices (SSD), Hammarmet, Tunisia

Chetoui M, Thomassin M, Malti R, Aoun M, Najar S, Abdelkrim MN, Oustaloup A (2013b) New consistent methods for order and coefficient estimation of continuous-time errors-in-variables fractional models. Comput Math Appl 66(5):860–872

Child D (2006) The essential of factor analysis, 3rd edn. Bloomsbury Academic Press, London

Chou CT, Verhaegen M (1997) Subspace algorithms for the identification of multivariable dynamic errors-in-variables models. Automatica 33(10):1857–1869

Cramér H (1946a) A contribution to the theory of statistical estimation. Skand Aktuarietidskr 29(1–2):85–94

Cramér H (1946b) Mathematical methods of statistics. Princeton University Press, Princeton

Criens CHA, Willems FPT, van Keulen TAC, Steinbuch M (2015) Disturbance rejection in diesel engines for low emissions and high fuel efficiency. IEEE Trans Control Syst Technol 23(2):662–669

Croux C, Fekri M, Ruiz-Gazen A (2010) Fast and robust estimation of the multivariate errors in variables model. TEST 19:286–303

Dankers A, Van den Hof PMJ, Bombois X, Heuberger PSC (2014) Errors-in-variables identification in dynamic networks by an instrumental variable approach. In: Proceedings of the 19th IFAC world congress, Cape Town, South Africa, pp 2335–2340

Dankers A, Van den Hof PMJ, Bombois X, Heuberger PSC (2015) Errors-in-variables identification in dynamic networks - consistency results for an instrumental variable approach. Automatica 62:39–50

Dankers A, Van den Hof PMJ, Bombois X, Heuberger PSC (2016) Identification of dynamic models in networks with prediction error methods: predictor input selection. IEEE Trans Autom Control 61(4):937–952

Dankers A, Van den Hof PMJ, Materassi D, Weerts HHM (2017) Conditions for handling confounding variables in dynamic networks. In: Proceedings of the 20th IFAC world congress, Toulouse, France

Davila CE (1998) A subspace approach to estimation of autoregressive parameters from noisy measurements. IEEE Trans Signal Process 46(2):531–534

De Moor B (1993) The singular value decompostion and long and short spaces of noisy matrices. IEEE Trans Signal Process 41:2826–2838

De Moor B (1994) Total least squares for affinely structured matrices and the noisy realization problem. IEEE Trans Signal Process 42(11):3104–3113

De Moor B, Roorda B (1994) l_2-optimal linear system identification structured total least squares for SISO systems. In: Proceedings of the 33th IEEE conference on decision and control, Lake Buena Vista, FL, USA

De Moor B, Vandewalle J (1986) A geometrical approach to the maximal corank problem in the analysis of linear relation. In: Proceedings of the 25th IEEE conference on decision and control, Athens, Greece

De Moor B, Vandewalle J (1990) A unifying theorem for linear and total least squares. IEEE Trans Autom Control 35(5):563–566

De Nadai B, Lewbei A (2016) Nonparametric errors in variables models with measurement errors on both sides of the equation. J Econom 191(1):19–32

Deistler M (1986) Linear dynamic errors-in-variables models. J Appl Probab 23A:23–39

Deistler M, Anderson BDO (1989) Linear dynamic errors-in-variables models. Some structure theory. J Econom 41:39–63

Deistler M, Zinner C (2007) Modelling high-dimensional time series by generalized linear dynamic factor models: an introductory survey. Commun Inf Syst 7(2):153–166

Deistler M, Anderson BDO, Filler A, Zinner C, Chen W (2010a) Generalized dynamic factor models - an approach via singular autoregressions. Eur J Control 16(3):211–224

Deistler M, Filler A, Anderson BDO, Chen W, Felsenstein E (2010b) Singular autoregressions for generalized dynamic factor models. In: Proceedings of the 49th IEEE conference on decision and control, Atlanta, GA, USA

Delaigle A, Hall P (2008) Using SIMEX for smoothing-parameter choice in errors-in-variables problems. J Am Stat Assoc 103(481):280–287–205
Delgado RA, Wigren T, Lau K, Middleton RH (2017) Data flow delay equalization for feedback control applications using 5G wireless dual connectivity. In: Proceedings of the IEEE VTC 2017 spring, Sydney, Australia
Dempster AP, Laird NM, Rubin DB (1977) Maximum likelihood from incomplete data via the EM algorithm. J R Stat Soc Ser B Stat Methodol 39(1):1–38
Devereux PJ (2007) Improved errors-in-variables estimators for grouped data. J Bus Econ Stat 25(3):278–287
D'Haultfoeuille X, Fevier P (2015) Identification of mixture models using support variations. J Econom 189(1):70–82
Ding F, Chen T (2005a) Hierarchical gradient-based identification of multivariable discrete-time systems. Automatica 41(2):315–325
Ding F, Chen T (2005b) Hierarchical least squares identification method for multivariable systems. IEEE Trans Autom Control 50(3):397–402
Ding F, Shi Y, Chen T (2006) Performance analysis of estimation algorithms of nonstationary ARMA processes. IEEE Trans Signal Process 54(3):1041–1053
Diversi R (2009) Noisy FIR identification as a quadratic eigenvalue problem. IEEE Trans Signal Process 57(11):4563–4568
Diversi R (2012) A fast algorithm for errors-in-variables filtering. IEEE Trans Autom Control 57(5):1303–1309
Diversi R (2013) Bias-eliminating least-squares identification of errors-in-variables models with mutually correlated noises. Int J Adapt Control Signal Process 27:915–924
Diversi R, Guidorzi R (2009) Frisch scheme-based identification of multivariable errors-in-variables models. In: Proceedings of the 15th IFAC symposium on system identification, Saint-Malo, France, pp 1563–1567
Diversi R, Guidorzi R (2010) Combining the Frisch scheme and Yule-Walker equations for identifying multivariable errors-in-variables models. In: Proceedings of the 19th international symposium on mathematical theory of networks and systems (MTNS), Budapest, Hungary, pp 2407–2413
Diversi R, Guidorzi R (2012) A covariance matching criterion in the Frisch scheme identification of MIMO EIV models. In: Proceedings of the 16th IFAC symposium on system identification, Belgium, Brussels, pp 1647–1652
Diversi R, Guidorzi R (2013) Identification of errors-in-variables models as a quadratic eigenvalue problem. In: Proceedings of the 12th European control conference (ECC), Zürich, Switzerland, pp 1896–1901
Diversi R, Guidorzi R (2017) The Frisch scheme in multivariable errors-in-variables identification. Eur J Control 37:43–53
Diversi R, Soverini U (2015) Identification of errors-in-variables models with colored output noise. In: Proceedings of the 14th European control conference (ECC), Linz, Austria, pp 1784–1789
Diversi R, Guidorzi R, Soverini U (2003a) Algorithms for optimal errors-in-variables filtering. Syst Control Lett 48:1–13
Diversi R, Guidorzi R, Soverini U (2003b) A new criterion in EIV identification and filtering applications. In: Proceedings of the 13th IFAC symposium on system identification, Rotterdam, The Netherlands, pp 1993–1998
Diversi R, Guidorzi R, Soverini U (2004) Frisch scheme–based algorithms for EIV identification. In: Proceedings of the 12th IEEE mediterranean conference on control and automation, Kusadasi, Turkey
Diversi R, Guidorzi R, Soverini U (2005a) Kalman filtering in extended noise environments. IEEE Trans Autom Control 50:1396–1402
Diversi R, Guidorzi R, Soverini U (2005b) Optimal errors-in-variables filtering in the MIMO case. In: Proceedings of the 16th IFAC world congress, Prague, Czech Republic

Diversi R, Guidorzi R, Soverini U (2006) Yule-Walker equations in the Frisch scheme solution of errors-in-variables identification problems. In: Proceedings of the 17th international symposium on mathematical theory of networks and systems (MTNS), Kyoto, Japan, pp 391–395

Diversi R, Guidorzi R, Soverini U (2007) Maximum likelihood identification of noisy input-output models. Automatica 43:464–472

Diversi R, Guidorzi R, Soverini U (2009) Identification of ARMAX models with additive output noise. In: Proceedings of the 15th IFAC symposium on system identification, Saint-Malo, France, pp 1574–1579

Diversi R, Guidorzi R, Soverini U (2010) Identification of ARX and ARARX models in the presence of input and output noises. Eur J Control 16(3):242–255

Diversi R, Guidorzi R, Soverini U (2011) Identification of ARMAX models with noisy input and output. In: Proceedings of the 18th IFAC world congress, Milan, Italy, pp 4272–4277

Diversi R, Guidorzi R, Soverini U (2012) Identification of errors-in-variables models with mutually correlated input and output noises. In: Proceedings of the 16th IFAC symposium on system identification, Belgium, Brussels, pp 1353–1358

Diversi R, Tilli A, Bartolini A, Benini L (2013a) Errors-in-variables identification of thermal models for many-core computing systems. In: Proceedings of the 12th European control conference (ECC), Zürich, Switzerland, pp 1908–1915

Diversi R, Tilli A, Bartolini A, Benini L (2013b) Identification of many-core systems-on-chip with input and output noise. In: Proceedings of the 52nd IEEE conference on decision and control (CDC), Florence, Italy, pp 6481–6488

Diversi R, Tilli A, Bartolini A, Beneventi F, Benini L (2014) Bias-compensated least squares identification of distributed thermal models for many-core systems-on-chip. IEEE Trans Circuits Syst - I 61(9):2663–2676

Diversi R, Grivel E, Merchan F (2017) ARMA model identification from noisy observations based on a two-step errors-in-variables approach. In: Proceedings of the 20th IFAC world congress, Toulouse, France

Dreesen P, De Moor B (2009) Polynomial optimization problems are eigenvalue problems. In: Van den Hof PMJ, Scherer C, Heuberger P (eds) Model-based control - bridging rigorous theory and advanced technology. Springer, Berlin

Dreesen P, Batselier K, De Moor B (2012a) Back to the roots: polynomial system solving, linear algebra, systems theory. In: Proceedings of the 16th IFAC symposium on system identification, Brussels, Belgium

Dreesen P, Batselier K, De Moor B (2012b) Weighted/structured total least squares problems and polynomial system solving. In: Proceedings of the ESANN 2012 European symposium on artificial neural networks, computational intelligence and machine learning, Bruges, Belgium

Durbin J (1959) Efficient estimation of parameters in moving average models. Biometrika 46:306–316

Eckart G, Young G (1936) The approximation of one matrix by another of lower rank. Psychometrika 1:211–218

Eising R, Linssen N, Rietbergen H (1983) System identification from noisy measurements of inputs and outputs. Syst Control Lett 2:348–353

Ekman M (2005) Identification of linear systems with errors in variables using separable nonlinear least squares. In: Proceedings of the 16th IFAC world congress, Prague, Czech Republic

Ekman M, Hong M, Söderström T (2006) A separable nonlinear least squares approach for identification of linear systems with errors in variables. In: Proceedings of the 14th IFAC symposium on system identification, Newcastle, Australia

Elliott RJ, Aggoun L, Moore JB (1995) Hidden Markov models. Springer, New York

Erickson T, Jiang CH, Whited TM (2014) Minimum distance estimation of the errors-in-variables model using linear cumulant equations. J Econom 183(2):211–221

Fan D, Lo K (2013) Recursive identification for dynamic linear systems from noisy input-output measurements. J Appl Math 2013:article ID 318,786

Fan D, Luo G (2010) Frisch scheme identification for errors-in-variables systems. In: Proceedings of the 9th IEEE conference on cognitive informatics, ICCI'10, Beijing, P. R. China, pp 794–799

Fan J, Truong YK (1993) Nonparametric regression with errors in variables. Ann Stat 21(4):1900–1925

Fedorov VV (1974) Regression problems with controllable variables subject to error. Biometrika 61(1):49–56

Feng CB, Zheng WX (1991) Robust identification of stochastic linear systems with correlated noise. IEE Proc Part D 138(5):484–492

Fernando KV, Nicholson H (1985) Identification of linear systems with input and output noise: the Koopmans–Levin method. IEE Proc Part D 132(1):30–36

Feuer A, Goodwin GC (1996) Sampling in digital signal processing and control. Birkhäuser, Boston

Finigan BM, Rowe IH (1974) Strongly consistent parameter estimation by the introduction of strong instrumental varibles. IEEE Trans Autom Control AC-19:825–830

Forni M, Lippi M (2001) The generalized dynamic factor model: representation theory. Econom Theory 17:1113–1141

Forni M, Hallin M, Lippi M, Reichlin L (2000) The generalized dynamic-factor model: identification and estimation. Rev Econ Stat 82(4):540–544

Forni M, Hallin M, Lippi M, Reichlin L (2004) The generalized dynamic factor model consistency and rates. J Econom 119:231–255

Forssell U, Gustafsson F, McKelvey T (1999) Time-domain identification of dynamic errors-in-variables systems using periodic excitation signals. In: Proceedings of the 14th IFAC world congress, Beijing, P. R. China

Friedlander B (1984) The overdetermined recursive instrumental variable method. IEEE Trans Autom Control AC-29:353–356

Frisch R (1934) Statistical confluence analysis by means of complete regression systems. Technical report 5, University of Oslo, Economics Institute, Oslo, Norway

Fuchs JJ, Maria S (2007) A new approach to variable selection using the TLS approach. IEEE Trans Signal Process 55(1):410–419

Fujikoshi Y (1974) The likelihood ratio tests for the dimensionality of regression coefficients. J Multivar Anal 4(3):327–340

Fuller WA (1980) Properties of some estimators for the errors-in-variables model. Ann Stat 8(2):407–422

Fuller WA (1987) Measurement error models. Wiley, New York

Galrinho M, Rojas C, Hjalmarsson H (2015) On estimating initial conditions in unstructured models. In: Proceedings of the 54th IEEE conference on decision and control, Osaka, Japan

Galrinho M, Everitt N, Hjalmarsson H (2017) Incorporating noise modeling in dynamic networks using non-parametric models. In: Proceedings of the 20th IFAC world congress, Toulouse, France

Garnier H, Gilson M, Zheng WX (2000) A bias-eliminated least-squares method for continuous-time model identification of closed-loop systems. Int J Control 71:38–48

Gauss KF (1809) Teoria Motus Corporum Coelestium in Sectionibus Conicus Solem Ambientieum. Reprinted translation: Theory of the motion of heavenly bodies moving about the sun in conic sections. Dover, New York

Geng LH, Xiao DY, Zhang T, Song JY (2011) Worst-case identification of errors-in-variables models in closed loop. IEEE Trans Autom Control 56(4):762–771

Geng LH, Geng LY, Lu SL, Cui SG (2012) l_2-optimal identification of MIMO errors-in-variables models from the v-gap geometrical interpretation. Int J Control 85:898–905

Gevers M, Wertz V (1984) Uniquely identifiable state-space and ARMA parameterizations for multivariable linear systems. Automatica 20(3):333–347

Gevers M, Bazanella A, Parraga A (2017) On the identifiability of dynamical networks. In: Proceedings of the 20th IFAC world congress, Toulouse, France

Gevers MR (1986) ARMA models, their Kronecker indices and their McMillan degree. Int J Control 43(6):1745–1761

Gewecke J (1977) The dynamic factor analysis of economic time series. Latent variables in socio-economic models. North-Holland, New York, pp 365–383
Ghosh D (1989) Maximum likelihood estimation of the dynamic shock-error model. J Econom 41:121–143
Gibson S, Ninness BM (2005) Robust maximum likelihood estimation of multivariable dynamic systems. Automatica 41(10):1667–1682
Gillard J (2010) An overview of linear structural models in errors in variables regression. REVSTAT - Stat J Am Stat Assoc 8(1):57–80
Gilson M, Van den Hof P (2001) On the relation between a bias-eliminated least-squares (BELS) and an IV estimator in closed-loop identification. Automatica 37:1593–1600
Gleser LJ (1981) Estimation in a multivariable "errors in variables" regression model: large sample results. Ann Stat 9(1):24–44
Godfrey KR, Tan AH, Barker HA, Chong B (2005) A survey of readily accessible perturbation signals for system identification in the frequency domain. Control Eng Pract 13(11):1391–1402
Golub GH, Pereyra V (1973) The differentiation of pseudoinverses and nonlinear least squares problems whose variables separate. SIAM J Numer Anal 10(2):413–432
Golub GH, Pereyra V (2003) Separable nonlinear least squares: the variable projection method and its applications. Inverse Probl 19(2):R1–R26
Golub GH, Van Loan CF (1980) An analysis of the total least squares problem. SIAM J Numer Anal 17:883–893
Golub GH, Hoffman A, Stewart GW (1987) A generalization of the Eckart-Young-Mirsky theorem. Linear Algebra Appl 88(89):317–327
Goodwin GC, Sin KS (1984) Adaptive filtering prediction and control. Prentice Hall, Englewood Cliffs
Goodwin GC, Agüero JC, Feuer A (2002) State estimation for systems having random measurement delays using errors in variables. In: Proceedings of the 15th IFAC world congress, Barcelona, Spain
Goos J, Lataire J, Lourroudi E, Pintelon R (2017) Frequency domain weighted nonlinear least squares estimation of parameter-varying differential equations. Automatica 75:191–199
Green M, Anderson BDO (1986) Identification of multivariable errors in variables models with dynamics. IEEE Trans Autom Control 31(5):467–471
Guidorzi R (1975) Canonical structures in the identification of multivariable systems. Automatica 11(4):361–374
Guidorzi R (1981) Invariants and canonical forms for systems structural and parametric identification. Automatica 17(1):117–133
Guidorzi R (1991) Certain models from uncertain data: the algebraic case. Syst Control Lett 17(6):415–424
Guidorzi R (1995) Identification of the maximal number of linear relations from noisy data. Syst Control Lett 24(3):159–165
Guidorzi R (1996) Identification of multivariable processes in the Frisch scheme context. In: Proceedings of the 12th international symposium on mathematical theory of networks and systems (MTNS), St Louis, USA
Guidorzi R, Diversi R (2009) A geometric approach to multivariable errors-in-variables identification. In: Proceedings of the 15th IFAC symposium on system identification, Saint-Malo, France, pp 1615–1619
Guidorzi R, Diversi R (2013) Structural health monitoring application of errors-in-variables identification. In: Proceedings of the 21st IEEE mediterranean conference on control and automation (MED), Platanias, Greece
Guidorzi R, Pierantoni M (1995) A new parameterization of Frisch scheme solutions. In: Proceedings of the XII international conference on systems science, Wroclaw, Poland
Guidorzi R, Diversi R, Soverini U (2003) Optimal errors-in-variables filtering. Automatica 39:281–289

Guidorzi R, Diversi R, Soverini U (2008) The Frisch scheme in algebraic and dynamic identification problems. Kybernetika 44(5):585–616

Guidorzi R, Diversi R, Vincenzi L, Simioli V (2015) AR + noise versus AR and ARMA models in SHM-oriented identification. In: Proceedings of the 23rd IEEE mediterranean conference on control and automation (MED), Torremolinos, Spain

Guillaume P, Pintelon R, Schoukens J (1995) Robust parametric transfer function estimation using complex logarithmic frequency response data. IEEE Trans Autom Control 40:1180–1190

Guo F, Hariprasad K, Huang B, Ding YS (2017a) Robust identification for nonlinear errors-in-variables systems using the EM algorithm. J Process Control 54:129–137

Guo F, Wu Q, Ding Y, Huang B (2017b) A data-based augmented model identification method for linear errors-in-variables systems based on EM algorithm. IEEE Trans Ind Electron 64(11):8657–8665

Hannan EJ (1963) Regression for time-series with errors of measurement. Biometrika 50(3 and 4):293–302

Haykin S (ed) (1985) Array signal processing. Prentice Hall, Englewood Cliffs

Haykin S (1991a) Adaptive filter theory, 2nd edn. Prentice Hall, Englewood Cliffs

Haykin S (1991b) Advances in spectrum analysis and array processing, vol 1 and 2. Prentice Hall, Englewood Cliffs

Heath WP (2005) The variance of nonparametric errors-in-variables estimates. IEEE Trans Instrum Meas 54(1):228–236

Heij C, Scherrer W (1999) Consistency of system identification by global total least squares. Automatica 35(6):993–1008

Heij C, Scherrer W, Deistler M (1997) System identification by dynamic factor models. SIAM J Control Optim 35(6):1924–1951

Hjalmarsson H, Mårtensson J, Rojas C, Söderström T (2011) On the accuracy in errors-in-variables identification compared to prediction-error identification. Automatica 47(12):2704–2712

Hogg RV, Craig AT (1970) Introduction to mathematical statistics, 3rd edn. MacMillan, New York

Hong M, Söderström T (2009) Relations between bias-eliminating least squares, the Frisch scheme and extended compensated least squares methods for identifying errors-in-variables systems. Automatica 45(1):277–282

Hong M, Söderström T, Zheng WX (2006) Accuracy analysis of bias-eliminating least squares estimates for errors-in-variables identification. In: Proceedings of the 14th IFAC symposium on system identification, Newcastle, Australia

Hong M, Söderström T, Zheng WX (2007a) Accuray analysis of bias-eliminating least squares estimates for errors-in-variables systems. Automatica 43(9):1590–1596

Hong M, Söderström T, Zheng WX (2007b) A simplified form of the bias-eliminating least squares method for errors-in-variables identification. IEEE Trans Autom Control 52(9):1754–1756

Hong M, Söderström T, Schoukens J, Pintelon R (2008) Accuracy analysis of time domain maximum likelihood method and sample maximum likelihood method in errors-in-variables identification. In: Proceedings of the 17th IFAC world congress, Seoul, Korea

Hostettler R, Nordenvaad Lundberg M, Birk W (2014) The pavement as a waveguide: modeling, system identification and parameter estimation. IEEE Trans Instrum Meas 63(8):2052–2063

Hostettler R, Birk W, Nordenvaad Lundberg M (2016) Maximum likelihood estimation of the non-parametric FRF for pulse-like excitations. IEEE Trans Autom Control 61(8):2276–2281

Hsiao C (1977) Identification for a linear dynamic simultaneous error-shock model. Int Econ Rev 18(1):181–194

Hsiao C (1989) Consistent estimation for some nonlinear errors-in-variables models. J Econom 41:159–185

Hsiao C, Robinson PM (1978) Efficient estimation of dynamic error-shock model. Int Econ Rev 19(2):467–479

Hsiao C, Wang L, Wang Q (1997) Estimation of nonlinear errors-in-variables models: an approximate solution. Stat Pap 38:1–25

Hu Y (2008) Identification and estimation of nonlinear models with misclassification error using instrumental variables: a general solution. J Econom 144:27–61

Hu Y, Schennach SM (2008) Instrumental variable treatment of nonclassical measurement error models. Econometrica 76(1):195–216

Huwang L, Huang YHS, Wang YHT (2009) Uniformly robust tests in errors-in-variables models. Ann Inst Stat Math 61:789–810

IEEE (2017) 5G the new wireless frontier. Inst 41(1):

Ikenoue M, Wada K (2015) Generalized eigenvector method for errors-in-variables model identification. In: Proceedings of the 54th IEEE conference on decision and control. Osaka, Japan, pp 777–782

Ikenoue M, Kanae S, Yang ZJ, Wada K (2005) Identification of noisy input-output system using bias-compensated least-squares method. In: Proceedings of the 16th IFAC world congress, Prague, Czech Republic

Ikenoue M, Kanae S, Yang ZJ, Wada K (2008) Bias-compensated based method for errors-in-variables model identification. In: Proceedings of the 17th IFAC world congress, Seoul, Korea

Ivanov DV (2013) Identification discrete fractional order linear dynamic systems with errors-in-variables. In: Proceedings of the IEEE East-West design and test symposium (EWDTS), Rostov on Don, Russia

Jamoos A, Grivel E, Bobillet W, Guidorzi R (2007) Errors-in-variables-based approach for the identification of AR time-varying fading channels. IEEE Signal Process Lett 11:793–796

Janssen PHM, Stoica P (1988) On the expectation of the product of four matrix-valued Gaussian random variables. IEEE Trans Autom Control AC-33(9):867–879

Jansson M, Swindlehurst AL, Ottersten B (1998) Weighted subspace fitting for general array error models. IEEE Trans Signal Process 46(9):2484–2498

Jia LJ, Ikenoue M, Jin CZ, Wada K (2001) On bias compensated least squares method for noisy input-output system identification. In: Proceedings of the 40th IEEE conference on decision and control, Orlando, Florida, USA, pp 3332–3337

Jia LJ, Hanada T, Jin CZ, Yang ZJ, Wada K (2002) On the relationship between BELS and IV methods. In: Proceedings of the 41th IEEE conference on decision and control, Las Vegas, Nevada, USA, pp 3440–3445

Jia LJ, Kanae S, Yang ZJ, Wada K (2003) On bias compensation estimation for noisy AR process. In: Proceedings of the 42th IEEE conference on decision and control, Maui, Hawaii, USA, pp 405–410

Jia LJ, Tao R, Kanae S, Yang ZJ, Wada K (2011) A unified framework for bias compensation based methods in correlated noise case. IEEE Trans Autom Control 56(3):625–629

Jin CZ, Jia LJ, Yang ZJ, Wada K (2001) On convergence of a BCLS algorithm for noisy autoregressive process estimation. In: Proceedings of the 40th IEEE conference on decision and control, Orlando, Florida, USA, pp 4252–4257

Jöreskog KG (1967) Some contributions to maximum likelihood factor analysis. Psychometrika 43(4):443–482

Jöreskog KG (1970) A general method for analysis of covariance structure. Biometrika 57:239–251

Jöreskog KG (1972) Factor analysis by generalized least squares. Psychometrika 37(3):243–260

Jöreskog KG, Goldberger AS (1975) Estimation of a model with multiple indicators and multiple causes of a single latent variable. J Am Stat Assoc 70(351):631–639

Jöreskog KG, Andersen EB, Laake P, Cox DR, Schweder T (1981) Analysis of covariance structures. Scand J Stat 8(2):65–92

Jukić D, Marković D (2010) On nonlinear weighted errors-in-variables parameter estimation problem in the three-parameter Weibull model. Appl Math Comput 215:3599–3609

Jun BE, Bernstein DS (2006) Least-correlation estimates for errors-in-variables models. Int J Adapt Control Signal Process 20:337–351

Jun BE, Bernstein DS (2007) Extended least-correlation estimates for errors-in-variables non-linear models. Int J Control 80(2):256–267

Kalman RE (1982a) Identification from real data. In: Hazewinkel M, Kan AHGR (eds) Current developments in the interface: economics, econometrics, mathematics, D. Reidel, Dordrecht

Kalman RE (1982b) System identification from noisy data. In: Bednarek AR, Cesari L (eds) Dynamical systems II. Academic Press, New York

Karimi A, van Heusden K, Bonvin D (2007) Non-iterative data-driven controller tuning using the correlation approach. In: Proceedings of the European control conference, Kos, Greece

Karlsson E, Söderström T, Stoica P (2000) The Cramér-Rao lower bound for noisy input-output systems. Signal Process 80(11):2421–2447

Katayama T (2005) Subspace methods for system identification. Springer, London

Kaufman L (1975) A variable projection method for solving separable nonlinear least squares problems. BIT 15:49–57

Kaufman L, Sylvester G, Wright MH (1994) Structured linear least-squares problems in system identification and separable nonlinear data fitting. SIAM J Optim 4:847–871

Kay SM (1979) The effects of noise on the autoregressive spectral estimator. IEEE Trans Acoust Speech Signal Process ASSP-27(5):478–485

Kay SM (1980) Noise compensation for autoregressive spectral estimates. IEEE Trans Acous Speech Signal Process ASSP-28(3):292–303

Kay SM (1988) Modern spectral estimation. Prentice-Hall, Englewood Cliffs

Kay SM (1993) Fundamentals of statistical signal processing: estimation theory. Prentice-Hall, Upper Saddle River

Kennedy RA, Chung SH (1992) Identification of coupled Markov chain model with application. In: Proceedings of the 31st IEEE conference on decision and control, Tuscon, Arizona, USA, pp 3429–3534

Kennedy RA, Chung SH (1996) Modelling and identification of coupled Markov chain model with application. Int J Adapt Control Signal Process 10:623–634

Klepper S, Leamer EE (1984) Consistent sets of estimates for regressions with errors in all variables. Econometrics 52(1):163–184

Koelbl L, Braumann A, Felsenstein E, Deistler M (2016) Estimation of VAR systems from mixed-frequency data: the stock and the flow case. Dyn Factor Models Adv Econom 35:43–73

Koopmans TJ (1937) Linear regression analysis of economic time series. N. V. Haarlem, The Netherlands

Kreiberg D, Söderström T, Wallentin F (2013) Errors-in-variables identification using covariance matching and structural equation modelling. In: Proceedings of the 52th IEEE conference on decision and control, Florence, Italy

Kreiberg D, Söderström T, Fan-Wallentin F (2016) Errors-in-variables identification using structural equation modeling. Automatica 66:218–230

Krim H, Viberg M (1996) Two decades of array signal processing research. The parametric approach. IEEE Signal Process Mag 13(4):67–94

Krishnamurthy V (1994) On-line estimation of dynamic shock-error models based on the Kullback Leibler information measure. IEEE Trans Autom Control 39(5):1129–1135

Krishnamurthy V (1995) Estimation of quantized linear errors-in-variables models. Automatica 31(10):1459–1464

Krishnamurthy V, Logothetis A (1996) Iterative and recursive estimators for hidden Markov errors-in-variables models. IEEE Trans Signal Process 44(3):629–639

Krusevac ZB, Kennedy RA, Rapajic PB (2006) Optimal implicit channel estimation for finite state Markov communication channels. In: Proceedings of the 2006 IEEE international symposium on information theory, Seattle, USA, pp 2657–2662

Kukush A, Tsaregorodtsev Y (2016) Goodness-of-fit test in a multivariate errors-in-variables model $AX = B$. Mod Stat: Theory Appl 3:287–302

Kukush A, Van Huffel S (2004) Consistency of element-wise weighted total least squares estimator in a multivariate errors-in-variables model $AX = B$. Metrika 59(1):75–97

Kukush A, Markovsky I, Van Huffel S (2004) Consistent estimation in an implicit quadratic measurement model. Comput Stat Data Anal 47:123–147

Kukush A, Markovsky I, Van Huffel S (2005) Consistency of the structured total least squares estimator in a multivariate errors-in-variables model. J Stat Plan Inference 133:315–358

Kukush A, Markovsky I, Van Huffel S (2007) Estimation in a linear multivariate measurement error model with a change point in the data. Comput Stat Data Anal 52:1167–1182

Kučera V (1979) Discrete linear control. Wiley, Chichester

Labarre D, Grivel E, Berthoumieu Y, Todini E, Najim M (2006) Consistent estimation of autogressive parameters from noisy observations based on two interacting Kalman filters. Signal Process 86:2863–2876

Larkowski T, Linden JG, Vinsonneau B, Burnham KJ (2008a) Identification of dynamic errors-in-variables models for bilinear systems. In: Proceedings of the 7th international conference on technical informatics, Timisoara, Romania

Larkowski T, Linden JG, Vinsonneau B, Burnham KJ (2008b) Identification of errors-in-variables systems via extended compensated least squares for the case of coloured output noise. In: Proceedings of the 19th international conference on systems engineering, Las Vegas, NV, USA

Larkowski T, Linden JG, Vinsonneau B, Burnham KJ (2008c) Novel algorithms based on conjunction of the Frisch scheme and extended least squares. In: Proceedings of the UKACC international conference on control 2008, Manchester, UK

Larkowski T, Linden JG, Burnham KJ (2009a) Frisch scheme based identification of dynamic errors-in-variables bilinear systems. In: Proceedings of the 15th IFAC symposium on system identification, Saint-Malo, France

Larkowski T, Linden JG, Burnham KJ (2009b) Identification of dynamic nonlinear polynomial models in the errors-in-variables framework. In: Proceedings of the 15th IFAC symposium on system identification, Saint-Malo, France

Larkowski T, Linden JG, Vinsonneau B, Burnham KJ (2009c) Frisch scheme identification for dynamic diagonal bilinear models. Int J Control 82(9):1591–1604

Larsen RJ, Marx ML (2012) An introduction to mathematical statistics and its applications, 5th edn. Prentice Hall, Upper Saddle River

Lataire J, Pintelon R (2009) Estimating a nonparametric colored-noise model for linear, slowly time-varying systems. IEEE Trans Instrum Meas 58:1535–1545

Lataire J, Pintelon R (2011) Frequency domain errors-in-variables identification of a time-varying, discrete-time system. In: Proceedings of the 18th IFAC world congress, Milan, Italy, pp 4272–4277

Lau K, Braslavsky JH, Goodwin GC (2007) Errors-in-variables problems in transient electromagnetic mineral exploration. In: Proceedings of the 46th IEEE conference on decision and control, New Orleans, LA, USA

Lau K, Braslavsky JH, Agüero JC, Goodwin GC (2008) Application of non-stationary EIV methods to transient electromagnetic mineral exploration. In: Proceedings of the 17th IFAC world congress, Seoul, Korea

Lau K, Braslavsky JH, Agüero JC, Goodwin GC (2009) An errors-in-variables method for non-stationary data with application to mineral exploration. Automatica 45:2971–2976

Lawrence RR (1989) A tutorial on hidden Markov models and selected applications in speech recognition. Proc IEEE 77(2):257–286

Lee SY, Jennrich RI (1979) A study of algorithms for covariance structure analysis with specific comparisons using factor analysis. Psychometrika 44:99–113

Lemmerling P, De Moor B (2001) Misfit versus latency. Automatica 37:2057–2067

Lemmerling P, Van Huffel S (2002) Structured total least squares. In: Van Huffel S, Lemmerling P (eds) Total least squares and errors-in-variables modeling. Kluwer Academic Publishers, Dordrecht

Lemmerling P, De Moor B, Van Huffel S (1996) On the equivalence of constrained total least squares and structured total least squares. IEEE Trans Signal Process 44(11):2908–2911

Lemmerling P, Vanhamme L, Van Huffel S, De Moor B (2001) IQML-like algorithms for solving structured total least squares problems: a unified view. Signal Process 81:1935–1945

Levin MJ (1964) Estimation of a system pulse transfer function in the presence of noise. IEEE Trans Autom Control AC–9:229–235

Li G, Xue L (2008) Empirical likelihood confidence region for the parameter in a partially linear errors-in-variables model. Commun Stat Theory Methods 37:1552–1564

Li J, Xu L, Stoica P, Forsythe K, Buss DW (2008) Range compression and waveform optimization for MIMO radar: a Cramér-Rao bound based study. IEEE Trans Signal Process 56(1):218–232

Li W, Qin SJ (2001) Consistent dynamic PCA based on errors-in-variables subspace identification. J Process Control 11:661–678

Li Z, Kruger U, Wang X, Xie L (2014) An error-in-variable projection to latent structure framework for monitoring technical systems with orthogonal signal components. Chemom Intell Lab Syst 133:70–83

Lim JS (2017) Fast ℓ_1 recursive total least squares algorithm for sparse system identification. Digital Signal Process 70:24–29

Lim JS, Pang HS (2016a) l_1 regularized recursive total least squares based sparse system identification for the errors-in-variables. Springerplus 5(1460):

Lim JS, Pang HS (2016b) Mixed norm regularized total least squares for group sparse system identification. Int J Adapt Control Signal Process 30:664–673

Linden J (2009) Algorithms for recursive Frisch scheme identification and errors-in-variables filtering. Ph.D. thesis, Coventry University, UK

Linden JG, Burnham KJ (2009) A recursive Frisch scheme algorithm for coloured output noise. In: ICINCO, Proceedings of the 5th international conference on informatics in control, automation and robotics, Funchal, Portugal

Linden JG, Vinsonneau B, Burnham KJ (2007) Fast algorithms for recursive Frisch scheme system identification. In: Proceedings of the 22nd IAR and ACD workshop, Grenoble, France

Linden JG, Larkowski T, Burnham KJ (2008a) Recursive Frisch scheme identification via variable projection. In: Proceedings of the mechatronics 2008, University of Limerick, Ireland

Linden JG, Vinsonneau B, Burnham KJ (2008b) Gradient based approaches for recursive Frisch scheme identification. In: Proceedings of the 17th IFAC world congress, Seoul, Korea

Linden JG, Larkowski T, Burnham KJ (2009) An improved recursive Frisch scheme identification algorithm. In: ICSENG 2008, international conference on systems engineering, Las Vegas, Nevada

Linden JG, Larkowski T, Burnham KJ (2012) Algorithms for resursive/semi-recursive bias-compensating least squares system identification within the errors-in-variables framework. Int J Control 85(11):1625–1643

Linder J (2017) Indirect system identification for unknown input problems with applications to ships. Ph.D. thesis, Linköping University, Sweden

Linder J, Enqvist M (2017a) Identification and prediction in dynamic networks with unobservable nodes. In: Proceedings of the 20th IFAC world congress, Toulouse, France

Linder J, Enqvist M (2017b) Identification of systems with unknown inputs using indirect input measurements. Int J Control 90(4):729–745

Lindley DV (1947) Regression lines and the linear functional relationship. Suppl J R Stat Soc 9(2):218–244

Liu X, Zhu Y (2015) A method of ARX model estimation of errors-in-variables systems. In: Proceedings of the 17th IFAC symposium on system identification, Beijing, China

Liu X, Zhu Y (2016) The asymptotic method for the identification of errors-in-variables systems. In: Proceedings of the 35th Chinese control conference, Chengdu, China

Liu X, Zhu Y (2017) Identification of errors-in-variables systems: an asymptotic approach. Int J Adapt Control Signal Process 31(8):1126–1138

Liu X, Zhu Y, Xu Z (2017) Model order selection in errors-in-variables system identification. In: Proceedings of the 20th IFAC world congress, Toulouse, France

Ljung L (1977) Some limit results for functionals of stochastic processes. Technical report LiTH-ISY-I-0167, Department of Electrical Engineering, Linköping University, Sweden

Ljung L (1999) System identification - theory for the user, 2nd edn. Prentice Hall, Upper Saddle River

Ljung L, Söderström T (1983) Theory and practice of recursive identification. MIT Press, Cambridge

Louarroudi E, Lataire J, Pintelon R, Janssens P, Swevers J (2014) Frequency domain, parametric estimation of the evolution of the time-varying dynamics of periodically time-varying systems from noisy input-output observations. Mech Syst Signal Process 47(1–2):151–174

Madansky A (1959) The fitting of straight lines when both variables are subject to error. J Am Stat Assoc 54(285):173–205

Mahata K (2007) An improved bias-compensation approach for errors-in-variables model identification. Automatica 43:1339–1354

Mahata K, Garnier H (2005) Direct identification of continuous-time errors-in-variables models. In: Proceedings of the 14th IFAC world congress, Prague, Czech Republic

Mahata K, Garnier H (2006) Identification of continuous-time errors-in-variables models. Automatica 42:1477–1490

Mahata K, Söderström T (2002) Identification of dynamic errors-in-variables model using prefiltered data. In: Proceedings of the 15th IFAC world congress, Barcelona, Spain

Mahata K, Söderström T (2004) Large sample properties of separable nonlinear least squares estimators. IEEE Trans Signal Process 52(6):1650–1658

Mahata K, Mousavi S, Söderström T, Mossberg M, Valdek U, Hillström L (2003) On the use of flexural wave propagation experiments for identification of complex modulus. IEEE Trans Control Syst Technol 11(6):863–874

Mahmoudi A, Karimi M (2010) Parameter estimation of autoregressive signals from observations corrupted with colored noise. Signal Process 90:157–164

Mansoori M, Dankers A, Van den Hof PMJ (2014) Errors-in-variables identification in bilaterally coupled systems with application to oil well testing. In: Proceedings of the 19th IFAC world congress, Cape Town, South Africa, pp 4656–4661

Maravall A (1979) Identification in dynamic shock error models. Springer, Berlin

Markovsky I (2008) Structured low-rank approximation and its application. Automatica 44:891–909

Markovsky I (2009) Applications of structured low-rank approximation. In: Proceedings of the 15th IFAC symposium on system identification, Saint-Malo, France

Markovsky I (2014) Recent progress on variable projection methods for structured low-rank approximation. Signal Process 96:406–419

Markovsky I (2015) System identification in the behavioral setting: a structured low-rank approximation approach. In: Proceedings of the 12th international conference on latent variable analysis and signal separation (LVA/ICA), Liberec, Czech Republic, vol 9237. Lecture notes in computer science, pp 235–242

Markovsky I, De Moor B (2005) Linear dynamic filtering with noisy input and output. Automatica 41(1):167–171

Markovsky I, Pintelon R (2015) Identification of linear time-invariant systems from multiple experiments. IEEE Trans Signal Process 63(13):3549–3554

Markovsky I, Van Huffel S (2007) Overview of total least-squares methods. Signal Process 87:2283–2302

Markovsky I, Willems JC, De Moor B (2002) Continuous-time errors-in-variables filtering. In: Proceedings of the 41st IEEE conference on decision and control (CDC 2002), Las Vegas, Nevada, pp 2576–2581

Markovsky I, Kukush A, Van Huffel S (2004) On the computation of the structured total least squares estimator. Numer Linear Algebra Appl 11:591–608

Markovsky I, Van Huffel S, Pintelon R (2005a) Block-Toeplitz/Hankel structured total least squares. SIAM J Matrix Anal Appl 26(4):1083–1099

Markovsky I, Willems JC, Van Huffel S, De Moor B, Pintelon R (2005b) Application of structured total least squares for system identification and model reduction. IEEE Trans Autom Control 50(10):1490–1500, special issue on system identification: linear vs. nonlinear

Markovsky I, Kukush A, Van Huffel S (2006a) On errors-in-variables estimation with unknown noise variance ratio. In: Proceedings of the 14th IFAC symposium on system identification, Newcastle, Australia
Markovsky I, Rastello ML, Premoli A, Kukush A, Van Huffel S (2006b) The element-wise weighted total least-squares problem. Comput Stat Data Anal 50:181–209
Markovsky I, Willems JC, Van Huffel S, De Moor B (2006c) Exact and approximate modeling of linear systems. A behavioral approach. SIAM, Philadelphia
Matei B, Meer P (2006) Estimation of nonlinear errors-in-variables models for computer vision applications. IEEE Trans Pattern Anal Mach Intell 28(10):1537–1552
Maurya D, Tangirala AK, Narasimhan S (2016) Identification of linear dynamic systems using dynamic iterative principle component analysis. In: Proceedings of the 11th IFAC symposium on dynamics and control of process systems including biosystems, Trondheim, Norway
Mayne DQ (1967) A method for estimating discrete time transfer functions. In: Advances in computer control, second UKAC control convention, Bristol, UK
Mayne DQ, Firoozan F (1982) Linear identification of ARMA processes. Automatica 18:461–466
McKelvey T (1995) Identification of state-space models from time and frequency data. Ph.D. thesis, Linköping University
Meirovitch L (1997) Principles and techniques of vibrations. Prentice Hall, Upper Saddle River
Mendel J (1991) Tutorial on high-order statistics (spectra) in signal processing and system theory: theoretical results and some applications. Proc IEEE 79(3):278–305
Milanese M, Vicino A (1991) Optimal estimation theory for dynamic systems with set membership uncertainty: an overview. Automatica 27(6):997–1009
Mirsky L (1960) Symmetric gauge functions and unitarily invariant norms. Quart J Math Oxf 11:50–59
Moore TJ, Kozick RJ, Sadler BM (2007) The constrained Cramér-Rao bound from the perspective of fitting a model. IEEE Signal Process Lett 14(8):564–567
Moran PAP (1971) Estimating structural and functional relationships. J Multivar Anal 1:232–255
Mossberg M (2007a) Analysis of a covariance matching method for continuous-time errors-in-variables identification. In: Proceedings of the 46th IEEE conference on decision and control, LA, USA, New Orleans, pp 5511–5515
Mossberg M (2007b) Analysis of a covariance matching method for discrete-time errors-in-variables identification. In: Proceedings of the 14th IEEE statistical signal processing workshop, Madison, WI, USA, pp 759–763
Mossberg M (2008) Errors-in-variables identification through covariance matching: analysis of a colored measurement noise case. In: Proceedings of the American control conference, Seattle, WA, USA, pp 1310–1315
Mossberg M (2015) On the use of two sampling stragegies for solving an errors-in-variables problem. In: Proceedings of the 14th European control conference (ECC 15), Linz, Austria, pp 1778–1783
Mossberg M (2016) Performance analysis and filter choice for an errors-in-variables method. In: Proceedings of the 15th European control conference (ECC 16), Aalborg, Denmark, pp 1371–1376
Mossberg M, Söderström T (2012) On covariance matching for multiple input multiple output errors-in-variables systems. In: Proceedings of the 16th IFAC symposium on system identification, Brussels, Belgium
Mossberg M, Hillström L, Söderström T (2001) Non-parametric identification of viscoelastic materials from wave propagation experiments. Automatica 37(4):511–521
Mu BQ, Chen HF (2013) Recursive identification of errors-in-variables Wiener systems. Automatica 49:2744–2753
Mu BQ, Chen HF (2014) Recursive identification of errors-in-variables Wiener-Hammerstein systems. Eur J Control 20:14–23
Mu BQ, Chen HF (2015) Recursive identification of multi-input multi-output errors-in-variables Hammerstein systems. IEEE Trans Autom Control 60:843–849

Musekiwa A (2005) Estimating the slope in the simple linear errors-in-variables model. Master thesis, Mathematical Statistics, Faculty of Science, University of Johannesburg

Narasimhan S, Shah SL (2008) Model identification and error covariance estimation from noisy data using PCA. Control Eng Pract 16:146–155

Nazari S, Zhao Q, Huang B (2015) Detecting and isolating abrupt changes in linear switching systems. Int J Control 88(4):801–814

Nehorai A, Stoica P (1988) Adaptive algorithms for constrained ARMA signals in the presence of noise. IEEE Trans Acoust Speech Signal Process ASSP-36:1282–1291

Nikias CL, Mendel JM (1993) Signal processing with higher-order spectra. IEEE Signal Process Mag 10(3):10–37

Ning L, Georgiou TT, Tannenbaum A, Boyd SP (2015) Linear models based on noisy data and the Frisch scheme. SIAM Rev 57(2):167–197

Ninness BM, Wills A (2010) Discussion on: "Generalized linear dynamic factor models: an approach via singular autoregressions". Eur J Control 16(3):225–227

Ninness BM, Wills A, Schön T (2010) Estimation of general nonlinear state-space systems. In: Proceedings of the 49th IEEE conference on decision and control, Atlanta, GA, USA

Nowak E (1985) Global identification of the dynamic shock-error model. J Econom 27:211–219

Nowak E (1992) Identifiability in multivariate dynamic linear errors-in-variables models. J Am Stat Assoc 87(419):713–723

Nowak E (1993) The identification of multivariate linear dynamic errors-in-variables models. J Econom 59:213–227

Nyquist H (1987) Robust estimation of the structural errors-in-variables model. Metrika 34:177–183

Oomen T, van Herpen R, Quist S, van de Wal M, Bosgra O, Steinbuch M (2014) Connecting system identification and robust control for next generation motion control of a wafer stage. IEEE Trans Control Syst Technol 22(1):102–118

Otter PW (1986) Dynamic structural systems under indirect observation: identifiability and estimation aspects from a system theoretic perspective. Psychometrika 51(3):415–428

Ottersten B, Viberg M, Kailath T (1992) Analysis of subspace fitting and ML techniques for parameter estimation from sensor array data. IEEE Trans Signal Process 40(3):590–599

Paige CC, Strakoš Z (2002) Scaled total least squares fundamentals. Numer Math 91:117–146

Paige CC, Strakoš Z (2006) Core problems in linear algebraic systems. SIAM J Matrix Anal Appl 27(3):861–875

Pan G, Zhou Y, Sun H, Gao W (2015) Linear observation based total least squares. Surv Rev 47(340):18–27

Patwardhan SC, Narasimhan S, Jagadeesan P, Gopaluni B, Shah SL (2012) Nonlinear Bayesian state estimation: a review of recent developments. Control Eng Pract 20:933–953

Pearson AE, Lee FC (1985) On the identification of polynomial input-output differential systems. IEEE Trans Autom Control AC-30:778–782

Pearson AE, Shen Y (1993) Weighted least squares/MFT algorithms for linear differential system identification. In: Proceedings of the 32nd IEEE confererence on decision and control, San Antonio, Texas, pp 2032–2037

Picci G (1989) Parameterization of factor analysis models. J Econom 41:17–38

Picci G, Pinzoni S (1986a) Dynamic factor analysis models for stationary processes. IMA J Math Control Inf 3:185–210

Picci G, Pinzoni S (1986b) Factor analysis models for stationary stochastic processes. In: Bensoussan A et al (eds) Analysis and optimization of systems. Springer, New York, pp 411–423

Picci G, Gei F, Pinzoni S (1993) Identification of errors-in-variables models with white measurement errors. In: Proceedings of the 2nd European control conference, Groningen, The Netherlands

Picinbono B (1993) Random signals and systems. Prentice-Hall, Englewood Cliffs

Picinbono B (1996) Second-order complex random vectors and normal distributions. IEEE Trans Signal Process 44(10):2637–2640

Pierre JW, Trudnowski D, Donnelly M, Zhou N, Tuffner FK, Dosiek L (2012) Overview of system identification for power systems from measured responses. In: Proceedings of the 16st IFAC symposium on system identification, Brussels, Belgium

Pillonetto G, Dinuzzo F, Chen T, De Nicolao G, Ljung L (2014) Kernel methods in system identification, machine learning and function estimation: a survey. Automatica 50:657–682

Pintelon R, Hong M (2007) Asymptotic uncertainty of transfer function estimates using nonparametric models. IEEE Trans Instrum Meas 56(6):2599–2605

Pintelon R, Schoukens J (1997) Identification of continuous-time systems using arbitrary signals. Automatica 33(5):991–994

Pintelon R, Schoukens J (2001) System identification. A frequency domain approach. IEEE Press, New York

Pintelon R, Schoukens J (2007) Frequency domain maximum likelihood estimation of linear dynamic errors-in-variables models. Automatica 43(4):621–630

Pintelon R, Schoukens J (2012a) Errors-in-variables identification of linear dynamic systems using periodic excitation. In: Proceedings of the 16th IFAC symposium on system identification, Brussels, Belgium

Pintelon R, Schoukens J (2012b) System identification. A frequency domain approach, 2nd edn. Wiley, Hoboken

Pintelon R, Guillaume P, Rolain Y, Verbeyst F (1992) Identification of linear systems captured in a feedback loop. IEEE Trans Instrum Meas 41(6):747–754

Pintelon R, Guillaume P, Rolain Y, Schoukens J, Van Hamme H (1994) Parametric identification of transfer functions in the frequency domain - a survey. IEEE Trans Autom Control 39(11):2245–2260

Pintelon R, Schoukens J, Vandersteen G (1997) Frequency domain system identification using arbitrary signals. IEEE Trans Autom Control 42(12):1717–1720

Pintelon R, Louarroudi E, Lataire J (2013a) Detecting and quantifying the nonlinear and time-variant effects in the FRF measurements using periodic excitations. IEEE Trans Instrum Meas 62(12):3361–3373

Pintelon R, Louarroudi E, Lataire J (2013b) Detection and quantification of the influence of time variation in closed-loop frequency-response-function measurements. IEEE Trans Instrum Meas 62(4):853–863

Pintelon R, Louarroudi E, Lataire J (2017) Time-variant frequency response function measurement of multivariate time-variant systems operating in feedback. IEEE Trans Instrum Meas 66(1):177–190

Polderman J, Willems JC (1998) Introduction to mathematical systems theory. Springer, New York

Porat B, Friedlander B (1986) Bounds on the accuracy of Gaussian ARMA parameter estimation method based on sample covariances. IEEE Trans Autom Control AC-31(6):579–582

Pulford G, Kennedy RA, Chung SH (1995) Identification of individual channel kinetics from recordings containing many identical channels. Signal Process 43:207–221

Rasmussen CE, Williams CKI (2006) Gaussian processes for machine learning. MIT Press, Cambridge

Reiersøl O (1941) Confluence analysis by means of lag moments and other methods of confluence analysis. Econometrica 9:1–24

Reiersøl O (1950) Identifiability of a linear relation between variables which are subject to error. Econometrica 18(4):375–389

Rensfelt A, Söderström T (2010) Structure testing of wave propagation models used in identification of viscoelastic materials. Automatica 46(4):728–734

Rhode S, Bleimund F, Gauterin F (2014) Recursive generalized total least squares with noise covariance estimation. In: Proceedings of the 19th IFAC world congress, Cape Town, South Africa, pp 4637–4643

Risuleo RS (2016) System identification with input uncertainties: an EM kernel-based approach, Licentiate thesis, Royal Institute of Technology

Risuleo RS, Bottegal G, Hjalmarsson H (2015) Kernel-based system identification from noisy and incomplete input-output data. In: Proceedings of the 55th IEEE conference on decision and control, Las Vegas, NV, USA, pp 2061–2066

Risuleo RS, Bottegal G, Hjalmarsson H (2017a) Approximate inference of nonparametric Hammerstein models. In: Proceedings of the 20th IFAC world congress, Toulouse, France

Risuleo RS, Bottegal G, Hjalmarsson H (2017b) Modeling and identification of uncertain-input systems. Technical report, School of Electrical Engineering, Royal Institute of Technology, Stockholm, Sweden. arXiv:1709.03421

Risuleo RS, Bottegal G, Hjalmarsson H (2017c) A nonparametric kernel-based approach to Hammerstein system identification. Automatica 85:234–247

Risuleo RS, Bottegal G, Hjalmarsson H (2017d) Variational Bayes identification of acyclic dynamic networks. In: Proceedings of the 20th IFAC world congress, Toulouse, France

Roorda B (1995) Algorithms for global total least squares modelling of finite multivariable time series. Automatica 31(3):391–404

Roorda B, Heij C (1995) Global total least squares modeling of multivariate time series. IEEE Trans Autom Control 40(1):50–63

Rougée A, Basseville M, Benveniste A, Moustakides GV (1987) Optimum robust detection of changes in the AR part of a multivariable ARMA process. IEEE Trans Autom Control 32:1116–1120

Rowe IH (1970) A bootstrap method for the statistical estimation of model parameters. Int J Control 12:721–738

Sadeghi P, Kennedy RA, Rapajic PB, Shams R (2008) Finite-state Markov modeling of fading channels. IEEE Signal Process Mag 25(5):57–80

Sagara S, Wada K (1977) On-line modified least-squares parameter estimation of linear discrete dynamics systems. Int J Control 25:329–343

Sakai H, Arase M (1979) Recursive parameter estimation of an autoregressive process disturbed by white noise. Int J Control 30:949–966

Samad T (2016) Control systems and the internet of things. IEEE Control Syst Mag 36(1):13–16

Sandell Jr NR, Yared KI (1978) Maximum likelihood identification of state space models for linear systems. Technical report ESL-R-814, Electronic Systems Laboratory, Department of Electrical Engineering and Computer Science, Massachusetts Institute of Technology, Cambridge

Sayed AH (2003) Fundamentals of adaptive filtering. Wiley Interscience, Hoboken

Schaffrin B, Felus YA (2009) An algorithm approach to the total least-squares problem with linear and quadratic constraints. Studia Geophysica et Geodaetica 53(1):1–16

Schennach SM (2004) Estimation of nonlinear models with measurement error. Econometrica 72(1):33–75

Schennach SM (2007) Instrumental variable estimation of nonlinear errors-in-variables models. Econometrica 75(1):201–239

Schennach SM (2013) Regression with Berkson errors in covariate - a nonparametric approach. Ann Stat 41(3):1642–1668

Schennach SM (2014) Entropic latent variable integration via simulation. Econometrica 82(1):345–385

Schennach SM, Hu Y (2013) Nonparametric identification and semiparametric estimation of classical measurement error models without side information. J Am Stat Assoc 108(501):177–186

Scherrer W, Deistler M (1998) A structure theory for linear dynamic errors-in-variables models. SIAM J Control Optim 36(6):2148–2175

Schneeweiss H (1976) Consistent estimation of a regression with errors in the variables. Metrika 23:101–115

Schön T, Wills A, Ninness B (2011) System identification of nonlinear state-space models. Automatica 47(1):39–49

Schoukens J, Pintelon R, Vanhamme H (1994) Identification of linear dynamic systems using piecewise-constant excitation - use, misuse and alternatives. Automatica 30(7):1153–1169

Schoukens J, Pintelon R, Vandersteen G, Guillaume P (1997) Frequency domain system identification using non-parametric noise models estimated from a small number of data sets. Automatica 33:1073–1086
Schoukens J, Pintelon R, Rolain Y (1999a) Study of conditional ML estimators in time and frequency-domain system identification. Automatica 35:91–100
Schoukens J, Vandersteen G, Pintelon R, Guillaume P (1999b) Frequency domain identification of linear systems using arbitrary excitations and a non-parametric noise model. IEEE Trans Autom Control 44(2):343–347
Schoukens J, Vandersten G, Barbé K, Pintelon R (2009) Nonparametric preprocessing in system identification: a powerful tool. Eur J Control 3–4:260–274
Schwarz G (1978) Estimating the dimension of a model. Ann Stat 6(2):461–464
Shapiro A (1985) Identifiability of factor analysis: some results and open problems. Linear Algebra Appl 70:1–7
Shi Y, Xu P, Liu J, Shi C (2015) Alternative formulae for parameter estimation in partial errors-in-variables models. J Geod 89:13–16
Shumway RH, Stoffer DS (1982) An approach to time series smoothing and forecasting using the EM algorithm. J Time Ser Anal 3(4):253–264
Skrondal A, Rabe-Hesketh S (2004) Generalized latent variable modeling: multilevel, longitudinal, and structural equation models. Chapman and Hall/CRC, Boca Raton
Slepian D (1954) Estimation of signal parameters in the presence of noise. Trans IRE Prof Group Inf Theory 3:68–69
Söderström T (1974) Convergence properties of the generalized least squares identification method. Automatica 10:617–626
Söderström T (1979) Some methods for identification of linear systems with noisy input/output data. In: Proceedings of the 5th IFAC/IFORS symposium on identification and system parameter estimation, Darmstadt, Germany
Söderström T (1980) Spectral decomposition with application to identification. In: Archetti F, Cugiani M (eds) Numerical techniques for stochastic systems. NorthHolland, Amsterdam
Söderström T (1981) Identification of stochastic linear systems in presence of input noise. Automatica 17:713–725
Söderström T (2002) Discrete-time stochastic systems - estimation and control, 2nd edn. Springer, London
Söderström T (2003) Why are errors-in-variables problems often tricky? In: Proceedings of the 7th European control conference, Cambridge, UK
Söderström T (2005) Accuracy analysis of the Frisch estimates for identifying errors-in-variables systems. In: Proceedings of the 44th IEEE conference on decision and control and 8th European control conference, Seville, Spain
Söderström T (2006) On computing the Cramér-Rao bound and covariance matrices for PEM estimates in linear state space models. In: Proceedings of the 14th IFAC symposium on system identification, Newcastle, Australia
Söderström T (2007a) Errors-in-variables methods in system identification. Automatica 43(6):939–958, survey paper
Söderström T (2007b) A Frisch scheme for correlated output noise errors-in-variables identification. In: Proceedings of the 9th European control conference, Kos, Greece
Söderström T (2008) Extending the Frisch scheme for errors-in-variables identification to correlated noise. Int J Adapt Control Signal Process 22(1):55–73
Söderström T (2009) Sampling approximations for continuous-time identification. In: Proceedings of the 15th IFAC symposium on system identification, Saint-Malo, France
Söderström T (2011) A generalized instrumental variable estimation method for errors-in-variables identification problems. Automatica 47(8):1656–1666
Söderström T (2012) A generalized instrumental variable estimator for multivariable errors-in-variables problems. Int J Control 85(3):287–303

Söderström T (2012a) Model order determination based on rank properties of almost singular covariance matrices. In: Proceedings of the 16th IFAC symposium on system identification, Brussels, Belgium

Söderström T (2012b) System identification for the errors-in-variables problem. Trans Inst Meas Control 34(7):780–792

Söderström T (2013) Comparing some classes of bias-compensating least squares methods. Automatica 49(3):840–845

Söderström T, Hong M (2004) Identification of dynamic errors-in-variables systems with periodic data. Technical report 2004-037, Department of Information Technology, Uppsala University, Uppsala, Sweden

Söderström T, Hong M (2005) Identification of dynamic errors-in-variables systems with periodic data. In: Proceedings of the 16th IFAC world congress, Prague, Czech Republic

Söderström T, Mahata K (2002) On instrumental variable and total least squares approaches for identification of error-in-variables models. Int J Control 75:709–718

Söderström T, Mossberg M (2011) Accuracy analysis of a covariance matching approach for identifying errors-in-variables systems. Automatica 47(2):272–282

Söderström T, Soverini U (2016) Errors-in-variables identification using maximum likelihood in the frequency domain. Technical report 2016-016, Department of Information Technology, Uppsala University, Uppsala, Sweden. http://www.it.uu.se/research/publications/reports/2016-016

Söderström T, Soverini U (2017) Errors-in-variables identification using maximum likelihood in the frequency domain. Automatica 79:131–143

Söderström T, Stoica P (1983) Instrumental variable methods for system identification. Springer, Berlin

Söderström T, Stoica P (1989) System identification. Prentice Hall International, Hemel Hempstead

Söderström T, Wang L (2012) On model order determination for errors-in-variables estimation. In: Proceedings of the 16th IFAC symposium on system identification, Brussels, Belgium

Söderström T, Yuz J (2013) Model validation methods for errors-in-variables estimation. In: Proceedings of the 52th IEEE conference on decision and control, Florence, Italy

Söderström T, Zheng WX, Stoica P (1999) Comments on 'On a least-squares-based algorithm for identification of stochastic linear systems'. IEEE Trans Signal Process 47:1395–1396

Söderström T, Mahata K, Soverini U (2002) Identification of dynamic errors-in-variables model using a frequency domain Frisch scheme. In: Proceedings of the 15th IFAC world congress, Barcelona, Spain

Söderström T, Soverini U, Mahata K (2002) Perspectives on errors-in-variables estimation for dynamic systems. Signal Process 82(8):1139–1154

Söderström T, Mahata K, Soverini U (2003) Identification of dynamic errors-in-variables model: approaches based on two-dimensional ARMA modelling of the data. Automatica 39(5):929–935

Söderström T, Hong M, Zheng WX (2005) Convergence properties of bias-eliminating algorithms for errors-in-variables identification. Int J Adapt Control Signal Process 19(9):703–722

Söderström T, Larsson EK, Mahata K, Mossberg M (2006) Using continuous-time modeling for errors-in-variables identification. In: Proceedings of the 14th IFAC symposium on system identification, Newcastle, Australia

Söderström T, Mossberg M, Hong M (2009) A covariance matching approach for identifying errors-in-variables systems. Automatica 45(9):2018–2031

Söderström T, Hong M, Schoukens J, Pintelon R (2010) Accuracy analysis of time-domain maximum likelihood method and sample maximum likelihood method for errors-in-variables and output error identification. Automatica 46(4):721–727

Söderström T, Irshad Y, Mossberg M, Zheng WX (2012) Accuracy analysis of a covariance matching method for continuous-time errors-in-variables system identification. In: Proceedings of the 16th IFAC symposium on system identification, Brussels, Belgium

Söderström T, Irshad Y, Mossberg M, Zheng WX (2013) Accuracy analysis of a covariance matching method for continuous-time errors-in-variables system identification. Automatica 49(10):2982–2993

Söderström T, Wang L, Pintelon R, Schoukens J (2013) Can errors-in-variables systems be identified from closed-loop experiments? Automatica 49(2):681–684

Söderström T, Diversi R, Soverini U (2014a) A unified framework for EIV identification methods in the presence of mutually correlated noises. In: Proceedings of the 19th IFAC world congress, Cape Town, South Africa

Söderström T, Diversi R, Soverini U (2014b) A unified framework for EIV identification methods when the measurement noises are correlated. Automatica 50(12):3216–3223

Söderström T, Kreiberg D, Mossberg M (2014c) Extended accuracy analysis of a covariance matching approach for identifying errors-in-variables systems. Automatica 50(10):2597–2605

Solari ME (1969) The "maximum likelihood solution" of the problem of estimating a linear functional relationship. J R Stat Soc Ser 31(2):372–375

Solo V (1986) Identifiability of time series models with errors in variables. J Appl Probab 23A:63–71

Song QJ (2013) Identification of errors-in-variables systems with nonlinear output observations. Automatica 49:987–992

Song QJ, Chen HF (2008) Identification of errors-in-variables systems with ARMA observation noises. Syst Control Lett 57:420–424

Song S, Schennach SM, White H (2015) Estimating nonseparable models with mismeasured endogenous variables. Quant Econ 6(3):749–794

Song W (2007) Model checking in errors-in-variables regression. J Multivar Anal 99:2406–2443

Song W (2009) Lack of fit in errors-in-variables regression models with validation data. Stat Probab Lett 79:765–773

Sorelius J, Stoica P (1998) Noisy input-output system identification as a Frisch problem. In: Proceedings of the Swedish national meeting in control (Reglermöte), Lund, Sweden

Soverini U, Beghelli S (2001) Identification of static errors-in-variables models: the rank reducibility problem. Automatica 37:1079–1084

Soverini U, Söderström T (2000) Identification methods of dynamic systems in presence of input noise. In: Proceedings of the 12th IFAC symposium on system identification, Santa Barbara, California

Soverini U, Söderström T (2014a) Frequency domain EIV identification: a Frisch scheme approach. In: Proceedings of the 19th IFAC world congress, Cape Town, South Africa

Soverini U, Söderström T (2014b) Frequency domain maximum likelihood identification of noisy input-output models. In: Proceedings of the 19th IFAC world congress, Cape Town, South Africa

Soverini U, Söderström T (2015a) Frequency domain EIV identification combining the Frisch scheme and Yule-Walker equations. In: Proceedings of the 14th European control conference, Linz, Austria, pp 2038–2043

Soverini U, Söderström T (2015b) Frequency domain identification of autoregressive models in the presence of additive noise. Technical report 2015-027, Department of Information Technology, Uppsala University, Uppsala, Sweden. http://www.it.uu.se/research/publications/reports/2015-027

Stanković MS, Stanković SS, Stipanović DM (2015) Consensus-based decentralized real-time identification of large-scale systems. Automatica 60:219–226

Stock JH, Watson MW (2002) Macroeconomic forecasting using diffusion indexes. J Bus Econ Stat 20(2):147–162

Stoica P, Babu P (2012) On the proper forms of BIC for model order selection. IEEE Trans Signal Process 60(5):4956–4961

Stoica P, Jansson M (2009) On maximum likelihood estimation in factor analysis. An algebraic derivation. Signal Process 89:1260–1262

Stoica P, Larsson EG (2001) Comments on 'Linearization method for finding Cramér-Rao bounds in signal processing'. IEEE Trans Signal Process 49(12):3168–3169

Stoica P, Li J (1995) Study of the Cramér-Rao bound as the numbers of observations and unknown parameters increase. IEEE Signal Process Lett 3(11):299–300

Stoica P, Li J (2005) On nonexistence of the maximum likelihood estimate in blind multichannel identification. IEEE Signal Process Mag 22(4):99–101

Stoica P, Moses RL (2005) Spectral analysis of signals. Prentice-Hall, Upper Saddle River

Stoica P, Nehorai A (1986) An asymptotically efficient ARMA estimator based on sample covariances. IEEE Trans Autom Control AC-31(11):1068–1071

Stoica P, Nehorai A (1987) On the uniqueness of prediction error models for systems with noisy input-output data. Automatica 23(4):541–543

Stoica P, Nehorai A (1991) Statistical efficiency study of direction estimation methods, part I: analysis of MUSIC and preliminary study of MLM. In: Haykin S (ed) Advances in spectrum analysis and array processing, vol II. Prentice-Hall, Englewood Cliffs, pp 263–306

Stoica P, Ng BC (1998) On the Cramér-Rao bound under parametric constraints. IEEE Signal Process Lett 5(7):177–179

Stoica P, Söderström T (1982) Bias correction in least-squares identification. Int J Control 35(3):449–457

Stoica P, Söderström T (1985) Optimization with respect to covariance sequence parameters. Automatica 21:671–675

Stoica P, Söderström T, Ahlén A, Solbrand G (1985) On the convergence of pseudo-linear regression algorithms. Int J Control 41:1429–1444

Stoica P, Cedervall M, Eriksson A (1995a) Combined instrumental variable and subspace fitting approach to parameter estimation of noisy input-output systems. IEEE Trans Signal Process 43:2386–2397

Stoica P, Söderström T, Šimonytė V (1995b) Study of a bias-free least squares parameter estimator. IEE Proc Control Theory Appl 142:1–6

Stoica P, Sorelius J, Cedervall M, Söderström T (1996) Errors-in-variables modeling: an instrumental variable approach. In: Proceedings of second international workshop on total least squares and errors-in-variables modeling, K. U. Leuven, Belgium

Stoica P, Li J, Söderström T (1997) On the inconsistency of IQML. Signal Process 56:185–190

Stoica P, Besson O, Gershman AB (2001) Direction-of-arrival estimation of an amplitude-distorted wavefront. IEEE Trans Signal Process 49(2):269–276

Sumisławska M, Larkowski TM, Burnham KJ (2011) Design of unknown input reconstruction filter based on parity equations for errors-in-variables case. In: Proceedings of the 18th IFAC world congress, Italy, Milan, pp 4272–4277

Sumisławska M, Larkowski TM, Burnham KJ (2012) Unknown input reconstruction observer for Hammerstein-Wiener systems in the errors-in-variables framework. In: Proceedings of the 16th IFAC symposium on system identification, Belgium, Brussels, pp 1377–1382

Sun J (1998) Stability and accuracy. Perturbation analysis of algebraic eigenproblems (book manuscript). Technical report UMINF 98-07, Umeå University, Department of Computing Science

Svensson A, Schön T (2017) A flexible state-space model for learning nonlinear dynamical systems. Automatica 80:189–199

Tanaka H, Katayama T (2006) Identification of dynamic errors-in-variables models from discrete time frequency domain power spectra. In: Proceedings of the 14th IFAC symposium on system identification, Newcastle, Australia, pp 202–207

Thil S (2007) Contributions à l'identification de modèles avec des erreurs en les variables. Ph.D. thesis, Nancy Université, France

Thil S, Garnier H, Gilson M, Mahata K (2007) Continuous-time model identification from noisy input/output measurements using fourth-order cumulants. In: Proceedings of the 46th IEEE conference on decision and control, New Orleans, LA, USA, pp 4257–4262

Thil S, Garnier H, Gilson M (2008a) Third-order cumulants based methods for continuous-time errors-in-variables model identification. Automatica 44:647–658

Thil S, Gilson M, Garnier H (2008b) On instrumental variable-based methods for errors-in-variables model identification. In: Proceedings of the 17th IFAC world congress, Seoul, Korea, pp 426–431

Thil S, Hong M, Söderström T, Gilson M, Garnier H (2008c) Statistical analysis of a third-order cumulants based algorithm for discrete errors-in-variables identification. In: Proceedings of the 17th IFAC world congress, Seoul, Korea

Thil S, Zheng WX, Gilson M, Garnier H (2009) Unifying some higher-order statistics methods for errors-in-variables model identification. Automatica 45:1937–1942

Tugnait JK (1992) Stochastic system identification with noisy input using cumulant statistics. IEEE Trans Autom Control AC-37:476–485

Tugnait JK, Ye Y (1995) Stochastic system identification with noisy input-output measurement using polyspectra. IEEE Trans Autom Control AC-40:670–683

Ugryomova D, Pintelon R, Vandersteen G (2015) Frequency response matrix estimation from missing input-output data. IEEE Trans Instrum Meas 64(11):3124–3136

Vajk I (2005) Identification methods in a unified framework. Automatica 41:1385–1393

Vajk I (2008) Parameter estimation from noisy measurements. Int J Syst Sci 39(4):437–447

Vajk I, Hetthéssy J (2003) Identification of nonlinear errors-in-variables models. Automatica 39:2099–2107

Vajk I, Hetthéssy J (2005) On the generalization of the Koopmans-Levin estimation methods. In: Proceedings of the 44th IEEE conference on decision and control, Seville, Spain, pp 4134–4139

Vajk I, Hetthéssy J (2006) Covariance matching criteria in EIV identification. In: Proceedings of the 14th IFAC symposium on system identification, Newcastle, Australia, pp 184–189

Vajk I, Hetthéssy J (2008) Efficient estimation of errors-in-variables models. In: Proceedings of the 17th IFAC world congress, Seoul, Korea, pp 1384–1389

van Berkel M, Vandersteen G, Geerardyn E, Pintelon R, Zwart H, de Baar M (2014) Frequency domain sample maximum likelihood estimation for spatially dependent parameter estimation in PDEs. Automatica 50(8):2113–2119

van den Bos A (1992) Identification of continuous-time systems using multiharmonic test signals. In: Sinha NK, Rao GP (eds) Identification of continuous-time systems. Kluwer Academic, Dordrecht

van Heusden K, Karimi A, Bonvin D (2011a) Data-driven model reference control with asymptotically guaranteed stability. J Adapt Control Signal Process 25(4):331–351

van Heusden K, Karimi A, Söderström T (2011b) On identification methods for direct data-driven controller tuning. Int J Adapt Control Signal Process 25(5):448–465

Van Huffel S (ed) (1997) Recent advances in total least squares techniques and errors-in-variables modelling. SIAM, Philadelphia

Van Huffel S, Lemmerling P (eds) (2002) Total least squares and errors-in-variables modelling. Analysis, algorithms and applications. Kluwer, Dordrecht

Van Huffel S, Vandewalle J (1989) Comparison of total least squares and instrumental variable methods for parameter estimation of transfer function models. Int J Control 50:1039–1056

Van Huffel S, Vandewalle J (1991) The total least squares problem: computation aspects and analysis. SIAM, Philadelphia

Van Overschee P, De Moor B (1996) Subspace identification for linear systems: theory, implementation, applications. Kluwer Academic Publishers, Boston

Van Overschee P, De Moor B, Dehandschutter W, Swevers J (1997) A subspace algorithm for the identification of discrete time frequency domain power spectra. Automatica 33:2147–2157

van Schuppen JH (1989) Stochastic realization problems. In: Nijmeijer N, Schumacher JM (eds) Three decades of mathematical system theory. Springer, Berlin

Vandersteen G, Barbé K, Pintelon R, Schoukens J (2009) Bootstrapped total least squares estimator using (circular) overlap for errors-in-variables identification. In: Proceedings of the 15th IFAC symposium on system identification, Saint-Malo, France

Viberg M, Ottersten B (1991) Sensor array processing based on subspace fitting. IEEE Trans Signal Process 39(5):1110–1121

Viberg M, Ottersten B, Kailath T (1991) Detection and estimation in sensor arrays using weighted subspace fitting. IEEE Trans Signal Process 39(11):2436–2449

Victor S, Diudichi A, Melchior P (2017) Closed-loop continuous-time model identification with noisy input-output. In: Proceedings of the 20th IFAC world congress, Toulouse, France

Wada K, Eguchi M, Sagara S (1990) Estimation of pulse transfer function via bias-compensated least-squares method in the presence of input and output noise. Syst Sci 16(3):57–70

Wahlberg B, Hjalmarsson H, Mårtensson J (2008) On identification of cascade systems. In: Proceedings of the 17th IFAC world congress, Seoul, Korea, pp 1384–1389

Wahlberg B, Hjalmarsson H, Mårtensson J (2009a) Variance results for identification of cascade systems. Automatica 45:1443–1448

Wahlberg B, Stoica P, Babu P (2009b) On estimation of cascade systems with common dynamics. In: Proceedings of the 15th IFAC symposium on system identification, Saint-Malo, France

Wahlberg B, Welsh J, Ljung L (2014) Identification of Wiener systems with process noise is a nonlinear errors-in-variables problem. In: Proceedings of the 53th IEEE conference on decision and control, Los Angeles, CA, USA

Wald A (1940) The fitting of straight lines if both variables are subject to error. Ann Math Stat Ser B 11(3):284–300

Wang J, Chen T, Huang B (2007) FIR modelling for errors-in-variables/closed-loop systems by exploiting cyclo-stationarity. Int J Adapt Control Signal Process 21:603–622

Wang Z, Wang Y, Ji Z (2017) A novel two-stage estimation algorithm for nonlinear Hammerstein-Wiener systems from noisy input and output data. J Frankl Inst 354:1937–1944

Watson GA (2007) Robust counterparts of errors-in-variables problems. Comput Stat Data Anal 52:1080–1089

Weerts HHM, Van den Hof PMJ, Dankers AG (2017) Identification of dynamic networks with rank-reduced process noise. In: Proceedings of the 20th IFAC world congress, Toulouse, France

Weerts HHM, Van den Hof PMJ, Dankers A (2018) Identifiability of linear dynamic networks. Automatica 89:247–258

Westland JC (2015) Structural equation modeling: from paths to networks. Springer, New York

Whittle P (1953) The analysis of multiple stationary time series. J R Stat Soc 15:125–139

Whittle P (1963) On the fitting of multivariate autoregressions and the approximate canonical factorization of a spectral density matrix. Biometrika 50:129–134

Wiesel A, Eldar Y, Yeredor A (2008) Linear regression with Gaussian model uncertainty: algorithms and bounds. IEEE Trans Signal Process 56(6):2194–2205

Wiggins RA, Robinson EA (1966) Recursive solution to the multichannel filtering problem. J Geophys Res 70:1885–1891

Wigren T (1998) Adaptive filtering using quantized output measurements. IEEE Trans Signal Process 46(12):3423–3426

Wigren T (2017) Networked and delayed recursive identification of nonlinear systems. In: Proceedings of 56th IEEE conference on decision and control, Melbourne, VA, Australia

Willems J (1986) From time series to linear systems. Automatica 37:561–580

Wills A, Schön TB, Ninness B (2010) Estimating state-space models in innovations form using the expectation maximization algorithm. In: Proceedings of the 49th IEEE conference on decision and control, Atlanta, GA, USA, pp 5524–5529

Wolter KM, Fuller WA (1982) Estimation of nonlinear errors-in-variables models. Ann Stat 10(2):539–548

Wong KY, Polak E (1967) Identification of linear discrete time systems using the instrumental variable approach. IEEE Trans Autom Control 12:707–718

Woodgate KG (1995) An upper bound on the number of linear relations identified from noisy data by the Frisch scheme. Syst Control Lett 24(3):153–158

Wu Q, Hariprasad K, Huang B, Forbes JF (2016) Identification of linear dynamic errors-in-variables systems with a dynamic uncertain input using the EM algorithm. In: Proceedings of the 55th IEEE conference on decision and control, Las Vegas, NV, USA, pp 1229–1234

Xia Y, Zheng WX (2015) Novel parameter estimation of autoregressive signals in the presence of noise. Automatica 62:98–105

Xin J, Aoki Y, Ohmori H, Sano A (1996) Order-determination of transfer function model from noisy input-output signals. In: Proceedings of the 13th IFAC world congress, San Francisco, USA

Xu P, Liu J, Shi C (2012) Total least squares adjustment in partial errors-in-variables models: algorithm and statistical analysis. J Geod 86:661–675

Xu X, Zhang Y, Williams J, Antoniou E, McCombie WR, Wu S, Zhu W, Davidson NO, Denoya P, Li W (2013) Parallel comparison of illumna RNA-seq and affymetrix microarray platforms on transciptomic profiles generated from 5-asa-deoxy-cytidine treated HT-29 colon cancer cells and simulated datasets. BMC Bioinform 14, article S1
Yan Z, Wu D, Zhou J, Hao L (2014) Recursive subspace identification of AUV dynamic model under general noise assumption. Math Probl Eng Article 547539
Yasini S, Wigren T (2018) Counterexamples to parametric convergence in networked identification. In: Proceedings of American control conference. Milwaukee, USA. (to appear)
Yeredor A (2001) The extended least squares criterion: minimization algorithms and applications. IEEE Trans Signal Process 49(1):74–86
Youcef A, Diversi R, Grivel E (2015) Errors-in-variables identification of noisy moving average models. In: Proceedings of the 23rd European signal processing conference (EUSIPCO), Nice, France
Young PC (1965) On a weighted steepest descent method of process parameter estimation. Technical report, Engineering Laboratory, Cambridge University, United Kingdom
Young PC (1970) An instrumental variable method for real-time identification of a noisy process. Automatica 6:271–287
Yu C, Zhang C, Xie L (2013) Blind system identification using precise and quantized observations. Automatica 49(9):2822–2830
Yuan KH, Bentler PM (2010) Two simple approximations to the distributions of quadratic form. Br J Math Stat Psychol 63:273–191
Yue Z, Thunberg J, Pan W, Ljung L, Gonçalves H (2017) Linear dynamic network reconstruction from heterogeneous datasets. In: Proceedings of the 20th IFAC world congress, Toulouse, France
Yuz JI, Goodwin GC (2014) Sampled-data models for linear and nonlinear systems. Springer (Communications and control engineering series), London
Zavala AAZ, Bolfarine H, de Castro M (2007) Consistent estimation and testing in heteroscedastic polynomial errors-in-variables method. Ann Inst Stat Math 59(3):515–530
Zhang E, Pintelon R (2017) Identification of multivariable dynamic errors-in-variables system with arbitrary inputs. Automatica 82:69–78
Zhang E, Pintelon R (2018) Nonparametric identification of linear dynamic errors-in-variables systems. Automatica to appear
Zhang E, Pintelon R, Schoukens J (2013) Errors-in-variables identification of dynamic systems excited by arbitrary non-white input. Automatica 49(10):3032–3041
Zhang Y, Zhao Z, Cui G (2015) Auxiliary model method for transfer function estimation from noisy input and output data. Appl Math Model 39(15):4257–4265
Zheng WX (1998a) Transfer function estimation form noisy input and output data. Int J Adapt Control Signal Process 12:365–380
Zheng WX (1998b) Unbiased identification of autoregressive signals observed in colored noise. In: Proceedings of the 23rd IEEE conference on acoustics, speech and signal processing, Seattle, pp 2329–2332
Zheng WX (1999a) An efficient algorithm for stochastic system identification with noisy input. In: Proceedings of the 38th IEEE conference on decision and control, Phoenix, AZ, pp 3657–3662
Zheng WX (1999b) A least-squares based method for autoregressive signals in the presence of noise. IEEE Trans Circuits Syst II: Analog Digit Signal Process 46:81–85
Zheng WX (1999c) On least-squares identification of stochastic linear systems with noisy input-output data. Int J Adapt Control Signal Process 13:131–143
Zheng WX (1999d) Parametric identification of noisy closed-loop linear systems. In: Proceedings of the 14th IFAC world congress, Beijing, P. R. China
Zheng WX (2000a) Autoregressive parameter estimation from noisy data. IEEE Trans Circuits Syst II: Analog Digital Signal Process 47(1):71–75
Zheng WX (2000b) Unbiased identification of stochastic linear systems from noisy input and output measurements. In: Proceedings of the 39th IEEE conference on decision and control, Australia, Sydney, pp 2710–2715

Zheng WX (2002a) A bias correction method for identification of linear dynamic errors-in-variables models. IEEE Trans Autom Control 47(7):1142–1147

Zheng WX (2002b) On least-squares identification of ARMAX models. In: Proceedings of the 15th IFAC world congress, Barcelona, Spain

Zheng WX (2005) Fast identification of autoregressive signals from noisy observations. IEEE Trans Circuits Syst II: Express Briefs 52(a):43–48

Zheng WX (2006a) A new look at parameter estimation of autoregressive signals from noisy observations. In: Proceedings of the IEEE international symposium on circuits and systems, Kos, Greece, pp 3778–3781

Zheng WX (2006b) On estimation of autoregressive signals in the presence of noise. IEEE Trans Circuits Syst II: Express Briefs 523(12):1471–1475

Zheng WX, Feng CB (1989) Unbiased parameter estimation of linear systems in presence of input and output noise. Int J Adapt Control Signal Process 3:231–251

Zheng WX, Feng CB (1990) Identification of stochastic time lag systems in the presence of colored noise. Automatica 26(4):769–779

Zheng WX, Feng CB (1992) Identification of a class of dynamic errors-in-variables models. Int J Adapt Control Signal Process 6:431–440

Zheng WX, Feng CB (1995) A bias-correction method for indirect identification of closed-loop systems. Automatica 31(7):1019–1024

Zinde-Walsh V (2014) Measurement error and deconvolution in spaces of generalized functions. Econom Theory 30(6):1207–1246

Zorzi M, Chiuso A (2017) Sparse plus low rank network identification: a nonparametric approach. Automatica 76:355–366

Zorzi M, Ferrante A (2012) On the estimation of structured covariance matrices. Automatica 48:2145–2151

Index of Cited Authors

A

Abatzoglou, T. J., 248, 251
Abdelkrim, N. M., 111
Abuabiah, M., 4
Adali, T., 168, 284
Adcock, R. J., 30, 133, 237, 248
Aggoun, L., 34
Agüero, J. C., 8, 13, 75, 82, 87, 88, 110, 205, 219, 221, 236, 354
Ahlén, A., 284
Akaike, H., 107
ALMutawa, J., 133
Amemiya, Y., 31
Andersen, E. B., 26, 34
Anderson, B. D. O., 31, 33, 34, 72, 87
Anderson, T. W., 31, 65, 87
Antoniou, E., 8
Aoki, M., 12, 139, 166
Aoki, Y., 164
Aoun, M., 111
Arablouei, R., 168, 249, 284
Arase, M., 69
Åström, K. J., 113

B

Babu, P., 88, 111
Bagozzi, R., 34
Bangs, W. J., 446
Barbé, K., 266, 267
Barhoumi, K., 33
Barker, A. H., 267
Bartholomew, D., 21, 28, 33
Bartolini, A., 8
Basseville, M., 8
Batselier, K., 251
Bazanella, A., 88
Beck, A., 248
Beghelli, S., 30, 32, 140, 145, 147, 157, 158, 166, 167, 177
Beneventi, F., 8
Bengtsson, M., 33
Benini, L., 8
Ben Rosen, J., 248
Bentler, P. M., 108
Benveniste, A., 8
Bernstein, D. S., 111, 168, 421, 423, 424
Berthoumieu, Y., 69
Besson, O., 354
Bickel, P. J., 432
Birk, W., 8, 266
Bleimund, F., 249
Bloch, A. M., 87
Bobillet, W., 8
Bolfarine, H., 34
Bollen, K. A., 26, 33
Bombois, X., 3, 88
Bonvin, D., 4
Bosgra, O., 267
Bottegal, G., 22, 32, 33, 88, 198, 199, 203, 218, 219
Boyd, S. P., 32
Braslavsky, J. H., 8
Braumann, A., 33
Brillinger, D. R., 442
Browne, M. W., 353
Burnham, K. J., 13, 111, 167, 284
Buss, D. W., 354

C

Cadzow, J. A., 220
Camba-Mendez, G., 108
Cappé, O., 34
Carvajal, G., 219
Carvajal, R., 236
Casini, M., 88

T. Söderström, *Errors-in-Variables Methods in System Identification*, Communications and Control Engineering,
https://doi.org/10.1007/978-3-319-75001-9

Castaldi, P., 32, 67, 87, 88, 145, 157, 158, 166, 167, 226
Cedervall, M., 31, 131, 132, 311
Cerone, V., 4, 55, 75
Chan, L. K., 304
Chen, C. C., 3
Chen, H. F., 69, 111, 218, 284
Chen, H. Y., 31
Chen, T., 88, 198, 284
Chen, W., 33
Chen, X. M., 32
Cheng, C. L., 31
Chetoui, M., 111
Child, D., 32
Chiuso, A., 33
Chong, B., 267
Chou, C. T., 133
Chung, S. H., 188
Conzález, K., 219
Cox, D. R., 26, 34
Craig, A. T., 432
Cramér, H., 304, 444
Criens, C. H. A., 267
Croux, C., 31
Cui, G., 168
Cui, S. G., 88

D
Damé, O., 33
Dankers, A., 3, 8, 88
Davidson, N. O., 8
Davila, C. E., 69
de Baar, M., 267
de Castro, M., 34
Dehandschutter, W., 228
Deistler, M., 2, 30, 33, 72, 87, 88, 128
Delaigle, A., 34
Delgado, R. A., 6, 110, 205, 219, 236
De Moor, B., 2, 13, 32, 110, 228, 239, 248, 250, 251
Dempster, A. P., 219
De Nadai, B., 31
De Nicolao, G., 198
Denoya, P., 8
Devereux, P. J., 31
D'Haultfoeuille, X., 32
Ding, F., 284
Ding, Y., 219
Dinuzzo, F., 198
Diudichi, A., 88
Diversi, R., 8, 13, 67–69, 110, 111, 145, 147, 148, 157–159, 165–168, 177, 187, 219
Doğançay, K., 168, 249, 284
Doksum, K. A., 432
Donnelly, M., 267
Dosiek, L., 267
Dreesen, P., 251
Durbin, J., 224

E
Eckart, G., 251, 252
Eguchi, M., 165, 284
Eising, R., 65, 139
Ekman, M., 159, 167
Eldar, Y., 34
Elliott, R. J., 34
Enqvist, M., 88, 218
Erickson, T., 133
Eriksson, A., 131, 132
Everitt, N., 88

F
Fadda, E., 55
Fan, D., 284, 301
Fan, J., 31
Federov, V. V., 31
Fekri, M., 30
Felsenstein, E., 33
Felus, Y. A., 248
Feng, C. B., 138, 139, 164–166, 284, 285
Fernando, K. V., 139, 166
Ferrante, A., 33
Ferrara, L., 33
Feuer, A., 13, 110
Fevier, P., 32
Filler, A., 33
Finigan, B. M., 133
Firoozan, F., 224
Forbes, J. F., 219
Forni, M., 32, 33
Forssell, U., 266
Forsythe, K., 354
Friedlander, B., 188, 280
Frisch, R., 2, 30–32, 145, 166
Fuchs, J. J., 251
Fujikoshi, Y., 108
Fuller, W. A., 1, 31, 34, 353

G
Galrinho, M., 88, 97
Gao, C., 32
Gao, W., 249

Garnier, H., 88, 93, 110, 130, 133, 139, 166, 167
Garulli, A., 88
Gauss, K. F., 133
Gauterin, F., 249
Geerardyn, E., 267
Gei, F., 88
Geng, L. H., 88
Geng, L. Y., 88
Georgiou, T. T., 32
Gershman, A. B., 354
Gevers, M., 88, 101, 229
Gewecke, J., 33
Ghosh, D., 218
Gibson, S., 283
Gillard, J., 31
Gilson, M., 130, 133, 139, 166, 167
Gleser, L. J., 31, 242
Glick, J., 248
Godfrey, K. R., 267
Goldberger, A. S., 25, 34, 304
Golub, G. H., 248, 251, 283, 427
Gonçalves, H., 88
Goodwin, G. C., 8, 13, 75, 82, 87, 88, 110, 205, 219, 221, 236, 283, 354
Goos, J., 236
Gopaluni, B., 219
Green, M., 87
Grivel, E., 8, 69
Guidorzi, R., 8, 13, 30, 32, 67, 68, 101, 110, 111, 139, 140, 145, 147, 148, 157, 158, 166, 167, 177, 187, 219
Guillaume, P., 84, 88, 213, 221, 261, 264, 266, 267, 342
Guo, F., 219
Gustafsson, F., 266

H
Hallin, M., 33
Hall, P., 34
Hanada, T., 166
Hannan, E. J., 31
Hao, L., 8
Hariprasad, K., 219
Haykin, S., 33, 187, 276
Heath, W. P., 236
Heij, C., 2, 250
Hetthéssy, J., 34, 164, 166, 219
Heuberger, P. S. C., 3, 88
Hillström, L., 187
Hjalmarsson, H., 88, 97, 198, 199, 203, 218, 219, 332, 354
Hoffman, A., 251
Hogg, R. V., 432
Hong, M., 87, 130, 155, 159, 165, 167, 177, 195, 267, 274, 284–286, 311, 341, 353, 354
Hostettler, R., 8, 266
Hsiao, C., 34, 111
Hsio, C., 87
Huang, B., 8, 88, 219
Huang, Y. H. S., 353
Huwang, L., 353
Hu, Y., 32, 34

I
Ikenoue, M., 165, 168
Irshad, Y., 353
Ivanov, D. V., 111

J
Jagadeesan, P., 219
Jamoos, A., 8
Janssen, P. H. M., 441
Janssens, P., 266
Jansson, M., 33, 34
Jennrich, R. I., 32
Jia, L. J., 165, 166
Jiang, C. H., 133
Jin, C. Z., 165, 166
Ji, Z., 111
Jöreskog, K. G., 21, 25, 26, 34, 304
Jukić, D., 34
Jun, B. E., 111, 168

K
Kailath, T., 33
Kalman, R. E., 2, 32
Kanae, S., 165, 166, 168
Kapeanios, G., 108
Karimi, A., 4
Karimi, M., 69
Karlsson, E., 354, 445
Katayama, T., 110, 228
Kaufman, L., 427
Kay, S. M., 69, 223, 445
Kennedy, R. A., 188
Klepper, S., 31
Knott, M., 22, 28, 33
Koelbl, L., 33
Koopmans, T. J., 31, 69, 139, 166
Kozick, R. J., 354
Kreiberg, D., 34, 188, 324

Krim, H., 33
Krishnamurthy, V., 7, 111, 284
Kruger, U., 31
Krusevac, Z. B., 188
Kučera, V., 97, 201
Kukush, A., 31, 34, 83, 242, 244, 248, 251, 266

L

Laake, P., 26, 34
Labarre, D., 69
Laird, N. M., 219
Larkowski, T., 13, 111, 167, 284
Larsen, R. J., 432
Larsson, E. G., 354
Larsson, E. K., 175, 203
Lataire, J., 88, 236, 266, 267
Lau, K., 6, 8
Lawrence, R. R., 34
Leamer, E. E., 31
Lee, A. C., 31
Lee, C. F., 31
Lee, F. C., 93
Lee, S. Y., 32
Lemmerling, P., 1, 248, 251
Levin, M. J., 12, 69, 139, 166
Lewbei, A., 32
Li, G., 31
Li, J., 34, 65, 68, 248, 354
Lim, J. S., 250
Linden, J. G., 111, 167, 284
Linder, J., 8, 88, 218
Lindley, D. V., 20, 31, 45, 214
Lin, H. W., 3
Linssen, N., 65, 139
Lippi, M., 33
Liu, J., 249
Liu, X., 111, 165, 236
Li, W., 8, 133
Li, Z., 31
Ljung, L., 1, 10, 12, 66, 78, 88, 97, 104, 111, 122, 133, 198–200, 203, 218, 272, 276–278, 282, 283, 302, 309, 331, 335, 421, 432, 444
Logothetis, A., 111
Lo, K., 284
Louarroudi, E., 88, 236, 266
Luo, G., 284, 301
Lu, S. L., 88

M

Madansky, A., 31
Mahata, K., 13, 65, 68, 88, 93, 107, 110, 131, 133, 145, 167, 175, 187, 203, 218, 225, 249
Mahmoudi, A., 69
Mak, T. K., 304
Malti, R., 111
Mansoori, M., 8, 88
Maravall, A., 111
Maria, S., 251
Marković, D., 34
Markovsky, I., 2, 13, 34, 83, 239, 242, 244, 248—251, 266
Mårtensson, J., 88, 332
Marx, M. L., 432
Matei, B., 34
Materassi, D., 88
Maurya, D., 251
Mayne, D. Q., 133, 224
McCombie, W. R., 8
McKelvey, T., 218, 266
Meer, P., 34
Meirovitch, L., 2
Melchior, P., 88
Mendel, J. M., 133, 248
Merchan, F., 69
Middleton, R. H., 6
Milanese, M., 75
Mirsky, L., 251
Montanari, M., 226
Moore, J. B., 34
Moore, T. J., 354
Moran, P. A. P., 31
Moses, R. L., 223, 432, 446
Mossberg, M., 110, 175, 176, 187, 188, 195, 203, 218, 274, 324, 353
Moulines, E., 34
Mousavi, S., 187
Moustakides, G. V., 8
Moustaki, I., 21, 28, 33
Mu, B. Q., 111
Musekiwa, A., 31

N

Najar, S., 111
Najim, M., 69
Narasimhan, S., 219, 251
Nazari, S., 8
Nehorai, A., 69, 88, 188, 347
Ng, B. C., 354
Nicholson, H., 139, 166
Nikias, C. L., 133
Ning, L., 32

Ninness, B., 33, 203, 219, 283
Nordenvaad Lundberg, M., 8, 266
Nowak, E., 88
Nyquist, H., 31

O
Ohmori, H., 164
Oomen, T., 267
Otter, P. W., 188
Ottersten, B., 33, 187
Oustaloup, A., 111

P
Paige, C. C., 248
Pan, G., 249
Pang, H. S., 250
Pan, W., 88
Park, H., 248
Parraga, A., 88
Patwardhan, S. C., 219
Pearson, A. E., 93
Pereyra, V., 283, 427
Picci, G., 32, 33, 88
Picinbono, B., 442
Pierantoni, M., 32
Pierre, J. W., 267
Piga, D., 75
Pillonetto, G., 198
Pintelon, R., 2, 13, 84, 85, 87, 88, 96, 110, 114, 199, 205–207, 213, 214, 217, 219, 221, 236, 239, 242, 248, 261, 264, 266, 267, 342, 353, 354
Pinzoni, S., 33, 88
Polak, E., 133
Polderman, J., 2, 187
Porat, B., 188
Premoli, A., 248
Pulford, G., 188

Q
Qin, S. J., 133
Quist, S., 267

R
Rabe-Hesketh, S., 32
Rapajic, P. B., 188
Rasmussen, C. E., 198
Rastello, M. L., 248
Razza, V., 75
Regruto, D., 4, 55, 75
Reichlin, L., 33
Reiersøl, O., 31, 133
Rensfelt, A., 108, 118
Rhode, S., 249
Rietbergen, H., 65, 139
Risuleo, R. S., 88, 198, 199, 203, 218, 219
Robinson, E. A., 101
Robinson, P. M., 111
Rojas, C., 97, 332, 354
Rolain, Y., 88, 219, 221, 266
Roorda, B., 250
Rougée, A., 8
Rowe, I. H., 133
Rubin, D. B., 219
Rubin, H., 65
Ruiz-Gazen, A., 31
Rydén, T., 34

S
Sadeghi, P., 188
Sadler, B. M., 354
Sagara, S., 138, 165, 166, 284
Sakai, H., 69
Salgado, M., 75, 87, 221
Samad, T., 6
Sandell Jr, N. R., 218
Sano, A., 164
Sayed, A. H., 276
Schaffrin, B., 248
Schennach, S. M., 31, 34
Scherrer, W., 2, 31, 87, 250
Schneeweiss, H., 353
Schön, T. B., 219, 283
Schoukens, J., 2, 84, 85, 87, 88, 96, 110, 114, 199, 205–207, 213, 214, 219, 221, 236, 261, 264, 266, 267, 342, 353, 354
Schwarz, G., 107
Schweder, T., 26, 34
Shah, S. L., 219, 251
Shams, R., 188
Shapiro, A., 32
Shen, Y., 93
Shi, C., 248, 249
Shi, Y., 248, 249, 284
Shumway, R. H., 283
Simioli, V., 8
Šimonytė, V., 69, 139, 166
Sin, K. S., 283
Skrondal, A., 32
Slepian, D., 446
Söderström, T., 1, 4, 7, 10, 12, 13, 31, 34, 60, 65, 66, 68, 69, 80, 85, 87, 88,

93, 97–99, 101, 104, 107, 108, 110, 111, 116–119, 122, 125, 130, 131, 133, 137–139, 145, 147, 150, 153, 155–157, 159, 165–168, 175, 176, 187, 189, 190, 195, 199–201, 203, 209, 218, 219, 225, 248, 249, 259, 266, 267, 272–274, 276–278, 280, 282–286, 302, 309–312, 315, 319, 323, 324, 331, 332, 335, 341, 346, 353, 354, 360, 391, 421–423, 428, 432, 433, 444, 445
Solari, M. E., 34, 65
Solbrand, G., 284
Solomon, O. M., 250
Solo, V., 88
Song, J. Y., 88
Song, Q. J., 111, 284
Song, S., 31
Song, W., 34
Sorelius, J., 31, 150, 151
Soverini, U., 13, 30, 32, 65, 67–69, 87, 88, 107, 110, 111, 131, 133, 140, 145, 147, 148, 157, 158, 166–168, 177, 187, 218, 219, 225, 354
Stanković, M. S., 133
Stanković, S. S., 133
Steinbuch, M., 267
Stewart, G. W., 251
Stipanović, D. M., 133
Stock, J. H., 3
Stoffer, D. S., 283
Stoica, P., 1, 10, 12, 31, 34, 64, 66, 68, 88, 97, 99, 101, 104, 107, 111, 122, 125, 131—133, 137–139, 150, 151, 166, 168, 188, 190, 199, 200, 203, 218, 223, 248, 259, 273, 280, 284, 302, 309–312, 315, 323, 331, 335, 347, 354, 360, 391, 421–423, 428, 432, 433, 441, 444–446
Strakoš, Z., 248
Sumisławska, M., 13, 111
Sun, H., 249
Sun, J-g., 118
Svensson, A., 219
Swevers, J., 228, 266
Swindlehurst, A. L., 33
Sylvester, G., 427

T
Tan, A. H., 267
Tanaka, H., 228
Tangirala, A. K., 251
Tannenbaum, A., 32
Tao, R., 166
Thil, S., 110, 130, 133, 167
Thomassin, M., 111
Thunberg, J., 88
Tilli, A., 8, 226
Todini, E., 69
Trudnowski, D., 267
Truong, Y. K., 31
Tsaregorodtsev, Y., 242
Tuffner, F. K., 267
Tugnait, J. K., 76, 128, 133

U
Ugryomova, D., 236

V
Vajk, I., 34, 164, 166, 219
Valdek, U., 187
van Berkel, M., 267
van den Bos, A., 266
Van den Hof, P. M. J., 3, 8, 88, 139, 166
Vandersteen, G., 84, 96, 213, 236, 261, 264, 266, 267, 342
Vandewalle, J., 133, 238, 248, 249
van de Wal, M., 267
Van hamme, H., 114, 221, 266, 267
Vanhamme, L., 248
van Herpen, R., 267
van Heusden, K., 4
Van Huffel, S., 1, 2, 13, 31, 34, 83, 133, 238, 239, 242, 244, 248–251, 266
van Keulen, T. A. C., 267
Van Loan, C. F., 248
Van Ness, J. W., 31
Van Overschee, P., 110, 228
van Schuppen, J. H., 1
Verbeyst, F., 88
Verhaegen, M., 133
Viberg, M., 33, 187
Vicino, A., 75, 88
Victor, S., 88
Vincenzi, L., 8
Vinsonneau, B., 111, 167, 284

W
Wada, K., 138, 165, 166, 168, 284
Wahlberg, B., 88, 111
Wald, A., 19, 31, 266
Wallentin, F., 34, 188
Wang, J., 88

Wang, L., 34, 85, 88, 111
Wang, Q., 34
Wang, X., 31
Wang, Y., 111
Wang, Y. H. T., 353
Wang, Z., 111
Watson, G. A., 31
Watson, M. W., 3
Weerts, H. H. M., 88
Welsh, J., 111
Werner, S., 249
Wertz, V., 229
Westland, J. C., 34
White, H., 31
Whited, T. M., 133
Whittle, P., 101, 445
Wiesel, A., 34
Wiggins, R. A., 101
Wigren, T., 6, 7
Willems, F. P. T., 267
Willems, J. C., 1, 2, 13, 187, 239, 248, 250
Williams, C. K. I., 198
Williams, J., 8
Wills, A., 33, 219, 283
Wittenmark, B., 113
Wolter, K. M., 34
Wong, K. Y., 133
Woodgate, K. G., 32
Wright, M. H., 427
Wu, D., 8
Wu, Q., 219
Wu, S., 8

X
Xiao, D. Y., 88
Xia, Y., 69
Xie, L., 8, 31
Xin, J., 164
Xue, L., 31
Xu, L., 354
Xu, P., 248, 249
Xu, X., 8
Xu, Z., 111

Y
Yang, J. M., 218
Yang, Z. J., 165, 166, 168
Yan, Z., 8
Yared, K. I., 218
Yasini, S., 7
Yeredor, A., 34, 249
Ye, Y., 76, 128, 133
Yi, Y., 33
Youcef, A., 69
Young, G., 252
Young, P. C., 133
Yuan, K. H., 108
Yu, C., 8
Yue, P. C., 12, 139, 166
Yue, Z., 88
Yuz, J. I., 87, 111, 205, 219

Z
Zamani, M., 203, 219
Zavala, A. A. Z., 34
Zhang, C., 8
Zhang, E., 2, 199, 214, 217, 219
Zhang, T., 88
Zhang, Y., 8, 168
Zhao, Q., 8
Zhao, Z., 168
Zheng, W. X., 69, 133, 138, 139, 142, 155, 157, 164–166, 177, 284–286, 353
Zhou, J., 8
Zhou, N., 267
Zhou, Y., 249
Zhu, W., 8
Zhu, Y., 111, 165, 236
Zinde-Walsh, V., 32
Zinner, C., 33
Zorzi, M., 33
Zwart, H., 267

Subject Index

A
alternating projection, 150
antisymmetry relation, 430
application, 1, 2, 8
approximate dynamic factor model, 3
ARMA process, 78, 103, 223
assumption, xxv, 50, 410
asymptotic distribution, 299
autoregressive process observed with noise, 69, 133

B
basic instrumental variable, 309
behavioral modeling, 1, 250, 266
bias-compensation, 135
bias-eliminating least squares, 138, 157, 164
bilinear, 141

C
canonical form, 164, 168
canonical parameterization, 101
central limit theorem, 443
circular symmetric Gaussian distribution, 437
closed-loop operation, 84, 88
comparison, 306, 403
complex Gaussian distribution, 437
concentrated loss function, 155, 427
confirmatory factor analysis (CFA), 21, 22, 46, 185, 188, 296, 302, 353, 354
consistency, 9, 45, 124, 214, 243, 258, 272, 442
continuous-time model, 88, 92, 110, 112
contraction mapping, 270
corank, 30, 32
correlated output noise, 142, 150, 363
covariance matching, 171, 280, 320, 353, 370
Cramér–Rao bound, 336, 342, 343, 352, 354, 397, 444

D
data least squares, 16
data-driven controller design, 4
decomposition, 30
degrees of freedom, 75, 87, 166
Diophantine equation, 380
discrete Fourier transform (DFT), 205, 429
dynamic networks, 3, 88

E
Eckart-Young-Mirsky lemma, 251
empirical transfer function estimator (ETFE), 236
ergodic process, 10, 12, 46
errors-in-variables filtering, 13
errors-in-variables problem, 49
estimating the noise-free input, 62
exogenous variable, 23
expectation, 432
expectation-maximization (EM) algorithm, 133, 199, 203, 219, 283
experimental setup, 412
extended compensated least squares, 159
extended instrumental variable, 123, 310
extended maximum likelihood in frequency domain, 214, 335, 384

F
factor analysis, 25, 32
factor model, 1, 32
filtered pulse-train, 114

T. Söderström, *Errors-in-Variables Methods in System Identification*, Communications and Control Engineering,
https://doi.org/10.1007/978-3-319-75001-9

first-order-hold, 114
Fisher information matrix, 445
fit function, 25
forecasting, 3
frequency domain, 96, 110, 204, 221, 261, 333, 341, 346, 380, 395, 397
frequency domain model, 96, 110
Frisch scheme, 29, 32, 145, 157–159, 166, 225, 228, 284
Frobenius norm, 29, 151
full polynomial form, 99, 101

G
Gaussian, 437
general estimation concepts, 442
generalized dynamic factor models, 32
generalized factor analysis, 31
generalized instrumental variable (GIVE), 153, 167, 231, 279, 284, 314, 353, 361
generalized least squares, 26
generic, 238
generic consistency, 125
global total least squares, 250
guidelines, 403

H
hidden Markov model, 111
higher-order statistics, 31, 76, 128, 133
hyperparameters, 115, 198

I
identifiability, 10, 71, 182, 244, 272
initial condition, 63, 96, 206, 234
innovations, 224
innovations form, 98, 199, 201, 343
instrumental variable, 31, 123, 133, 257, 266, 309, 339, 358, 391

K
Kalman filter, 201, 343
kernel methods, 198
known noise variance ratio, 20, 42, 57, 63, 69, 76, 139, 166, 207, 243, 244, 349
Kronecker indices, 101
Kronecker permutation matrix, 423
Kronecker product, 100, 422

L
latent variable, 1, 21, 34, 199
least squares, 16, 61, 121, 135
line fitting, 15, 35
linear regression, 90
loss function, 25
lower bound, 308, 312, 332

M
Markov chain model, 188
matrix results, 421
maximum likelihood, 34, 62, 197, 203, 204, 207, 282, 303, 330, 332, 333, 346, 354, 378, 380, 395, 397, 442, 445
model order, 51
model order determination, 103, 111
modeling aspects, 89
modeling noise properties, 95
models with linear constraints, 368
Monte Carlo simulations, 56
multiple experiments, 82, 255
multivariable model, 99, 110, 161, 289, 318
multivariate statistics, 21

N
networked feedback control, 6
noise model, 110
noise-free input signal, 49, 62
non-Gaussian distribution, 128, 317, 325, 326, 366
non-parametric methods, 222
nonlinear dynamics, 199
nonlinear model, 31, 34, 111
nonnegative definite matrices, 421
notations, xix, xxiii, 11, 89
nuisance parameters, 101

O
optimal instrumental variable, 311
orthogonal regression, 17, 237
over-determined systems of equations, 275, 291
overlapping experiments, 256

P
parameter vector, 50, 90
parameterization, 192
parametric estimator, 223
parametric model, 76, 78
Parseval's relation, 205, 233, 350, 388
periodic data, 76, 255, 336, 354, 391
permutation matrix, 424
persistent excitation, 10, 261

practical aspects, 403
prediction error method, 68, 197, 204, 330, 354, 378
prejudices, 2, 75

Q
quantization, 6

R
rank test, 107
realization theory, 106
recursive algorithm, 276, 284
regressor vector, 90
Riccati equation, 344
root-mean-square (RMS), 56, 128, 160, 236

S
sample maximum likelihood, 213, 261, 266, 341
sampling, 110, 112
sensitivity matrix, 344, 361
sensor array processing, 22, 28, 33
separable least squares, 154, 176, 187, 426
set membership estimation, 75
shock error model, 111, 218, 284
Slepian–Bang's formula, 446
small singular values, 105, 107, 118
spectral density, 33, 72, 222
spectral factor estimator, 228
spectral factorization, 11, 333, 353, 381, 387, 392, 400
spectrum estimation, 223
state space model, 98, 200, 343
static system, 15, 31, 214
statistically efficient, 445
structural equation modeling, 33, 353
structured total least squares, 240, 248
subspace system identification, 133
sufficient statistic, 102, 443
Sylvester matrix, 152, 195, 351, 353, 382, 388, 428
system leakage term, 97, 207

T
time delay, 92
time domain, 199
Toeplitz, 25, 26
total least squares, 18, 31, 237, 252
transfer function operator, 49
transient, 63, 96, 206, 234

U
unbiased estimate, 442
unweighted least squares, 26
user choices, 26, 125, 174, 181, 271, 311, 406

V
variable projection, 187, 274, 283, 286, 289, 426
vectorization, 422

W
weighted least squares, 26, 155
weighted subspace fitting, 28, 33, 131
Whittle's formula, 445

Y
Yule–Walker equations, 147, 158

Z
zero-order-hold, 113

The manufacturer's authorised representative in the EU is Springer Nature Customer Service Centre GmbH, Europaplatz 3, 69115 Heidelberg, Germany. If you have any concerns regarding our products, please contact ProductSafety@springernature.com

Printed and bound by CPI Group (UK) Ltd, Croydon, CR0 4YY
01/10/2026
02225860-0017